ANNUAL REVIEW OF ECOLOGY, EVOLUTION, AND SYSTEMATICS

ANNUAL REVIEW OF ECOLOGY, EVOLUTION, AND SYSTEMATICS

VOLUME 34, 2003

DOUGLAS J. FUTUYMA, *Editor*
University of Michigan

H. BRADLEY SHAFFER, *Associate Editor*
University of California, Davis

DANIEL SIMBERLOFF, *Associate Editor*
University of Tennessee

www.annualreviews.org science@annualreviews.org 650-493-4400

ANNUAL REVIEWS
4139 El Camino Way • P.O. Box 10139 • Palo Alto, California 94303-0139

AR
ANNUAL REVIEWS
Palo Alto, California, USA

International Standard Serial Number: 1543-592X
International Standard Book Number: 0-8243-1434-4
Library of Congress Catalog Card Number: 71-135616

∞ The paper used in this publication meets the minimum requirements of American National Standards for Information Sciences—Permanence of Paper for Printed Library Materials. ANSI Z39.48-1992.

TYPESET BY TECHBOOKS, FAIRFAX, VA
PRINTED AND BOUND BY MALLOY INCORPORATED, ANN ARBOR, MI

Annual Review of Ecology, Evolution, and Systematics
Volume 34, 2003

CONTENTS

ERRATA

An online log of corrections to *Annual Review of Ecology, Evolution, and Systematics* chapters may be found at http://ecolsys.annualreviews.org/errata.shtml

RELATED ARTICLES

From the ***Annual Review of Earth and Planetary Sciences***, Volume 31 (2003)

Phanerozoic Atmospheric Oxygen, Robert A. Berner, David J. Beerling, Robert Dudley, Jennifer M. Robinson, and Richard A. Wildman, Jr.

The Role of Decay and Mineralization in the Preservation of Soft-Bodied Fossils, Derek E.G. Briggs

Phylogenetic Approaches Toward Crocodylian History, Christopher A. Brochu

From the ***Annual Review of Environment and Resources***, Volume 28 (2003)

Climate Change, Climate Modes, and Climate Impacts, Guiling Wang and David Schimel

The Cleansing Capacity of the Atmosphere, Ronald G. Prinn

Evaluating Uncertainties in Regional Photochemical Air Quality Modeling, James Fine, Laurent Vuilleumier, Steve Reynolds, Philip Roth, and Nancy Brown

Transport of Energy, Information, and Material Through the Biosphere, William A. Reiners and Kenneth L. Driese

Global State of Biodiversity and Loss, Rodolfo Dirzo and Peter H. Raven

Patterns and Mechanisms of the Forest Carbon Cycle, Stith T. Gower

From the ***Annual Review of Entomology***, Volume 48 (2003)

Tomato, Pests, Parasitoids, and Predators: Tritrophic Interactions Involving the Genus Lycopersicon, George G. Kennedy

Molecular Systematics of Anopheles: From Subgenera to Subpopulations, Jaroslaw Krzywinski and Nora J. Besansky

Feather Mites (Acari: Astigmata): Ecology, Behavior, and Evolution, Heather C. Proctor

Comparative Social Biology of Basal Taxa of Ants and Termites, Barbara L. Thorne and James F.A. Traniello

The Evolution of Alternative Genetic Systems in Insects, Benjamin B. Normark

Belowground Herbivory by Insects: Influence on Plants and Aboveground Herbivores, Bernd Blossey and Tamaru R. Hunt-Joshi

From the ***Annual Review of Microbiology***, Volume 57 (2003)

The Uncultured Microbial Majority, Michael S. Rappé and Stephen J. Giovannoni

Gene Organization: Selection, Selfishness, and Serendipity, Jeffrey G. Lawrence

Natural Selection and the Emergence of a Mutation Phenotype: An Update of the Evolutionary Synthesis Considering Mechanisms that Affect Genome Variation, Lynn Helena Caporale

From the ***Annual Review of Phytopathology***, Volume 41 (2003)

Parasitic Nematode Interactions with Mammals and Plants, Douglas P. Jasmer, Aska Goverse, and Geert Smant

Ecology of Mycorrhizae: A Conceptual Framework for Complex Interactions Among Plants and Fungi, M.F. Allen, W. Swenson, J.I. Querejeta, L.M. Egerton-Warburton, and K.K. Treseder

The Ecological Significance of Biofilm Formation by Plant-Associated Bacteria, Cindy E. Morris and Jean-Michel Monier

ANNUAL REVIEWS is a nonprofit scientific publisher established to promote the advancement of the sciences. Beginning in 1932 with the *Annual Review of Biochemistry*, the Company has pursued as its principal function the publication of high-quality, reasonably priced *Annual Review* volumes. The volumes are organized by Editors and Editorial Committees who invite qualified authors to contribute critical articles reviewing significant developments within each major discipline. The Editor-in-Chief invites those interested in serving as future Editorial Committee members to communicate directly with him. Annual Reviews is administered by a Board of Directors, whose members serve without compensation.

Annual Reviews of

Anthropology
Astronomy and Astrophysics
Biochemistry
Biomedical Engineering
Biophysics and Biomolecular Structure
Cell and Developmental Biology
Earth and Planetary Sciences
Ecology, Evolution, and Systematics
Entomology
Environment and Resources
Fluid Mechanics
Genetics
Genomics and Human Genetics
Immunology
Materials Research
Medicine
Microbiology
Neuroscience
Nuclear and Particle Science
Nutrition
Pharmacology and Toxicology
Physical Chemistry
Physiology
Phytopathology
Plant Biology
Political Science
Psychology
Public Health
Sociology

SPECIAL PUBLICATIONS
Excitement and Fascination of Science, Vols. 1, 2, 3, and 4

Annu. Rev. Ecol. Evol. Syst. 2003. 34:1–26
doi: 10.1146/annurev.ecolsys.34.011802.132355

First published online as a Review in Advance on September 29, 2003

EFFECTS OF INTRODUCED BEES ON NATIVE ECOSYSTEMS

Dave Goulson
Division of Biodiversity and Ecology, School of Biological Sciences, University of Southampton, Bassett Crescent East, Southampton, SO16 7PX, United Kingdom; email: dg3@soton.ac.uk

Key Words competition, pollination, weeds, *Apis*, *Bombus*

■ **Abstract** Bees are generally regarded as beneficial insects for their role in pollination, and in the case of the honeybee *Apis mellifera*, for production of honey. As a result several bee species have been introduced to countries far beyond their home range, including *A. mellifera*, bumblebees (*Bombus* sp.), the alfalfa leafcutter bee *Megachile rotundata*, and various other solitary species. Possible negative consequences of these introductions include: competition with native pollinators for floral resources; competition for nest sites; co-introduction of natural enemies, particularly pathogens that may infect native organisms; pollination of exotic weeds; and disruption of pollination of native plants. For most exotic bee species little or nothing is known of these possible effects. Research to date has focused mainly on *A. mellifera*, and has largely been concerned with detecting competition with native flower visitors. Considerable circumstantial evidence has accrued that competition does occur, but no experiment has clearly demonstrated long-term reductions in populations of native organisms. Most researchers agree that this probably reflects the difficulty of carrying out convincing studies of competition between such mobile organisms, rather than a genuine absence of competitive effects. Effects on seed set of exotic weeds are easier to demonstrate. Exotic bees often exhibit marked preferences for visiting flowers of exotic plants. For example, in Australia and New Zealand many weeds from Europe are now visited by European honeybees and bumblebees. Introduced bees are primary pollinators of a number of serious weeds. Negative impacts of exotic bees need to be carefully assessed before further introductions are carried out.

INTRODUCTION

The devastating impacts that some exotic organisms have wreaked on native ecosystems surely ought to have taught us a lesson as to the perils of allowing release of alien species. The introduction of Nile perch to Lake Victoria, and of cane toads, prickly pear, rabbits, foxes, and cats among numerous others to Australia, are perhaps some of the best known examples, but they constitute only the tip of the iceberg. Australia alone had 24 introduced mammal species, 26 birds, 6 reptiles,

1543-592X/03/1215-0001$14.00

1 amphibian, 31 fish, more than 200 known invertebrates, and no less than 2,700 non-native plants at the last count (Alexander 1996; reviewed in Low 1999). A strong case can be made that exotic species represent the biggest threat to global biodiversity after habitat loss (Pimm et al. 1995, Low 1999).

Whereas the threat posed by exotic species is now widely appreciated, exotic bees appear to have received disproportionately little attention. Bees are widely perceived to be beneficial, for their role in the pollination of crops and wildflowers and, in the case of the honeybee *Apis mellifera* (L.) (Apidae), for the production of honey. Because of these economic benefits there is reluctance to regard bees as potentially damaging to the environment. As long ago as 1872, Darwin stated that honeybees in Australia were "rapidly exterminating the small, stingless native bee." In fact the bee he refers to, presumably *Trigona carbonaria* Sm., is still abundant. However, almost no research was carried out upon the impact of honeybees until the 1980s, by which time they had long since become established on every continent except Antarctica.

Here I review the scale on which bees have been artificially distributed around the globe. Three bee species, the honeybee *A. mellifera*, the bumblebee *Bombus terrestris* (L.) (Apidae), and the alfalfa leaf-cutter bee *Megachile rotundata* (Fabr.), are of particular concern because their range has been considerably expanded owing to both deliberate and accidental releases. I examine the potential consequences of this range expansion.

The possible undesirable effects of exotic bees include:

1. Competition with native flower visitors for floral resources;
2. Competition with native organisms for nest sites;
3. Transmission of parasites or pathogens to native organisms;
4. Changes in seed set of native plants (either increases or decreases);
5. Pollination of exotic weeds.

I examine evidence for each of these processes in turn. Reviewing studies to date serves to highlight the substantial gaps in our knowledge. I suggest further experimental approaches that may provide less equivocal answers as to the threat posed by these exotic organisms.

DISTRIBUTION AND ABUNDANCE OF INTRODUCED BEES

The honeybee is thought to be native to Africa, western Asia, and southeast Europe (Michener 1974), although its association with man is so ancient that it is hard to be certain of its origins. It has certainly been domesticated for at least 4000 years (Crane 1990a), and has been introduced to almost every country in the world. It is now among the most widespread and abundant insects on earth. The European strain of the honeybee appears to be adapted to temperate and Mediterranean climates, and flourishing feral populations occur throughout much of Asia, North America, the southern half of South America, and Australia. Major events in this range expansion include its introduction to North America in about 1620 (Buchmann & Nabhan 1996), to Australia in 1826 (Doull 1973), and to New Zealand in 1839 (Hopkins 1911). The African race, *A. mellifera scutellata* Lepeletier, is associated with tropical forests and savannas, and has spread throughout the neotropics and into North America following its introduction to Brazil in 1957.

More recently, bumblebees (*Bombus* spp.), a group whose natural range is largely confined to the temperate northern hemisphere, have been introduced to various countries to enhance crop pollination. New Zealand has four established *Bombus* species native to the U.K., *B. hortorum* (L.), *B. terrestris, B. subterraneus* (L.), and *B. ruderatus* (F.), following introductions in 1885 and 1906 intended to improve pollination of red clover, *Trifolium repens* (Hopkins 1914). *B. hortorum* and *B. subterraneus* have restricted distributions within New Zealand, whereas *B. terrestris* and *B. ruderatus* have become ubiquitous (Macfarlane & Gurr 1995). *B. terrestris* spread into Israel in the 1960s (Dafni & Shmida 1996), perhaps as a result of the presence of introduced weeds. This species has also become established in the wild in Japan following escapes from commercial colonies used for pollination in glass houses (Dafni 1998). Most recently, *B. terrestris* arrived in Hobart, Tasmania, in 1992, perhaps accidentally transported in cargo, and has since spread out to occupy a substantial portion of the island (Buttermore 1997, Stout & Goulson 2000, Hingston et al. 2002). *B. ruderatus* was introduced to Chile in 1982 and 1983 for pollination of red clover (Arretz & Macfarlane 1986), and by 1994 had spread to Argentina (Abrahamovich et al. 2001).

The only other group of bees to have been deliberately redistributed around the globe in substantial numbers are the Megachilidae. Perhaps because of the importance of alfalfa as a crop in the United States, a plant which is not adequately pollinated by honeybees, this country has shown particular enthusiasm for introducing exotic pollinators. The most widespread is *M. rotundata*, a native of Eurasia that appeared in North America in the 1930s, and which is now widely used commercially for pollination of alfalfa (Bohart 1972). A range of other species have been imported to pollinate various crops, including *Osmia cornuta* Latr. from Spain for pollination of almonds (Torchio 1987), *Osmia cornifrons* (Radoszkowski) from Japan for pollination of fruit trees (Batra 1979), *Osmia coerulescens* (L.) from Europe for pollination of red clover (Parker 1981), and *Megachile apicalis* Spinola from Europe for pollination of alfalfa (Cooper 1984, Stephen 1987). Furthermore, species have been moved within the United States and established in regions far from their home ranges; *Osmia ribifloris biedermannii* Michener from the west coast has been released in Maine to pollinate blueberries (Stubbs et al. 1994). At least three exotic Megachilidae are now established in California, *M. apicalis*, *M. rotundata*, and *M. concinna* (Smith) (Frankie et al. 1998), and *M. rotundata* even occurs in the Everglades National Park, Florida (Pascarella et al. 1999). New introductions continue to occur; for example *M. sculpturalis* Smith, a native of China and Japan was recently recorded in North Carolina (Mangum & Brooks 1997).

The fate of some deliberate introductions is not known. For example *Chalicodoma nigripes* from Egypt and *Pithitis smaragulda* F. from India were introduced to the United States in the 1970s for pollination of alfalfa, but to my knowledge it is not known if these species became established (Daly et al. 1971, Parker et al. 1976). One megachilid, *M. rotunda* was introduced to New Zealand in 1971 for pollination of alfalfa and flourished (Donavan 1975). Recently this species also became established in southern Australia (Woodward 1996).

One final bee that has expanded its range with the deliberate help of man is the alkali bee, *Nomia melanderi* (Cockerell) (Halictidae). This native of North America was introduced to New Zealand in 1971 for pollination of alfalfa and has become established at restricted sites (Donovan 1975, 1979). A summary of the distribution of exotic bee species is given in Table 1.

Bees are a large group (about 20,000 species are known), but little is known about basic aspects of the ecology of most species. For the majority we have only a rudimentary knowledge of their natural distribution. It is almost certain that other species have been transported by man to new locations, but that these events have gone unrecorded.

Both *B. terrestris* and *A. mellifera* are social species, with colonies attaining sizes of up to 500 and 50,000 individuals, respectively. In their natural range, nest density estimates for *A. mellifera* vary from 0.5 to >70 nests/km^2 in Europe (Visscher & Seeley 1982, Oldroyd et al. 1995) and 4.2 nests/km^2 in Botswana (McNally & Schneider 1996). Where honeybees have been introduced, estimates include 50–150 nests/km^2 in southern Australia (Oldroyd et al. 1997) and 6–100 nests/km^2 for Africanized bees in the neotropics (Roubik 1983, 1988; Otis 1991). Densities are no doubt greatly influenced by variation in habitat quality and availability of nest sites. Given the large numbers of workers per nest, even the lowest estimates indicate substantial densities of foragers. No information is available on densities of nests of *B. terrestris*, either within their natural range or where they are introduced, because they are notoriously hard to locate.

In general both honeybees and *B. terrestris* appear to maintain higher population densities than semisocial and solitary species across a broad range of habitats and geographic regions (South Australia, Pyke & Balzer 1985; California, Dobson 1993; Brazil, Wilms et al. 1997; New Zealand, Donovan 1980; Israel, Dafni 1998). It is often impossible to determine how large the equilibrium feral population of honeybees would be because wild populations are supplemented by swarms from commercial hives, and foragers observed in the field are likely to originate from both managed and wild colonies. Little information is available on populations of the introduced Megachilidae, but one study suggests that these solitary species do not attain high densities in Australia (Woodward 1996).

Because introduced bees are widespread, any deleterious effects of their presence are now occurring on a large scale. The abundance of honeybees and bumblebees makes such effects more probable. Some researchers have concluded that competition with native organisms is inevitable (Roubik 1978, Roubik & Buchmann 1984, Sugden et al. 1996).

TABLE 1 The distribution and origins of known exotic bee species

Species	Family	Introduced range	Origin	References
Apis mellifera	Apidae	North and South America, Eastern Asia, Australia, New Zealand	Eastern Europe, Western Asia, Africa	Hopkins 1911, Doull 1973, Buchmann & Nabhan 1996
Bombus terrestris	Apidae	Israel, Japan, New Zealand, Tasmania	Europe	Hopkins 1914, Dafni & Shmida 1996, Buttermore 1997, Dafni 1998, Stout & Goulson 2000
Bombus ruderatus	Apidae	New Zealand, Chile, Argentina	Europe	Hopkins 1914, Arretz & Macfarlane 1986
Bombus hortorum	Apidae	New Zealand, Iceland	Europe	Hopkins 1914, Prys-Jones et al. 1981
Bombus lucorum	Apidae	Iceland	Europe	Prys-Jones et al. 1981
Bombus subterraneus	Apidae	New Zealand	Europe	Hopkins 1914
Megachile rotundata	Megachilidae	North America, Australia, New Zealand	Eurasia	Bohart 1972, Donovan 1975, Woodward 1996, Frankie et al. 1998, Pascarella et al. 1999
Megachile apicalis	Megachilidae	United States	Europe	Cooper 1984; Stephen 1987
Megachile concinna	Megachilidae	California	Europe	Frankie et al. 1998
Megachile sculpturalis	Megachilidae	North Carolina	China, Japan	Mangum & Brooks 1997
Osmia coerulescens	Megachilidae	United States	Europe	Parker 1981
Osmia cornifrons	Megachilidae	United States	Japan	Batra 1979
Osmia cornuta	Megachilidae	United States	Europe	Torchio 1987
Osmia ribifloris biedermannii	Megachilidae	Maine	Southwestern United States	Stubbs et al. 1994
Pithitis smaragulda	Megachilidae	United States	India	Daly et al. 1971
Chalicodoma nigripes	Megachilidae	United States (establishment unknown)	Egypt	Parker et al. 1976
Nomia melanderi	Halictidae	New Zealand	North America	Donovan 1975, 1979

DIET BREADTH OF INTRODUCED BEES AND NICHE OVERLAP WITH NATIVE POLLINATORS

The diet of all bee species consists more or less exclusively of pollen and nectar collected from flowers (occasionally supplemented by honeydew, plant sap, waxes and resins, and water) (Michener 1974). The two bee species that have proved to be most adaptable in colonizing new habitats, *A. mellifera* and *B. terrestris*, have done so because they are generalists. The colonies of both species are relatively long lived and so must be able to adapt to a succession of different flower sources as they become available. *A. mellifera* usually visits a hundred or more different species of plant within any one geographic region (Pellet 1976, O'Neal & Waller 1984, Wills et al. 1990, Roubik 1991, Butz Huryn 1997, Coffey & Breen 1997), and in total has been recorded visiting nearly 40,000 different species (Crane 1990b). *B. terrestris* is similarly polylectic. It has been recorded visiting 66 native plants of 21 families in Tasmania (Hingston & McQuillan 1998) and 419 introduced and native plants in New Zealand (Macfarlane 1976). It has been argued that such statistics are misleading, because of the many flower species visited, most visits are targeted at a few favored species (Butz Huryn 1997). For example, Menezes Pedro & Camargo (1991) found that of 47 species of flower visited by honeybees in Brazil, 65% of visits were to only 9 plant species. However, even minor sources of forage for honeybees can receive substantial numbers of visits, simply because honeybees are often very abundant.

Rather less is known of the diet breadth of introduced Megachilidae. Thorp (1996) suggests that the exotic species found in California each show strong preferences for one plant family; *M. rotundata* and *M. concinna* primarily visit members of the Fabaceae, while *M. apicalis* visit Asteraceae. However, Donovan (1980) described *M. rotundata* as being polylectic in New Zealand, visiting a broad range of Fabaceae and Asteraceae.

A diverse range of different organisms collect pollen and/or nectar from flowers, including birds, bats, mammals, and insects. Of the insects, the main groups are the bees and wasps (Hymenoptera), butterflies and moths (Lepidoptera), beetles (Coleoptera), and flies (Diptera). The wide distribution and polylectic diet of most introduced bees means that potentially they might compete with many thousands of different native species. Even introduced Megachilidae no doubt overlap with many native species, because their preferred plant families are large and favored by many other flower visitors. It seems reasonable to predict that introduced bees are most likely to compete with native bee species, because these are likely to be most similar in terms of their ecological niche. Studies of niche overlap in terms of flowers visited have all concluded that both honeybees and bumblebees overlap substantially with native bees and with other flower visitors such as nectivorous birds (Donovan 1980, Roubik 1982a, Roubik et al. 1986, Menezes Pedro & Camargo 1991, Thorp et al. 1994, Wilms & Wiechers 1997, Hingston & McQuillan 1998).

Honeybees and bumblebees differ from many other flower visitors in having a prolonged flight season; honeybees remain active for all of the year in warmer climates, while bumblebees commonly forage throughout the spring and summer in the temperate climates where they naturally occur. Thus, in terms of the time of year at which they are active, they overlap with almost all other flower visitors with which they co-occur.

COMPETITION WITH NATIVE ORGANISMS FOR FLORAL RESOURCES

Demonstration of niche overlap does not prove that competition is occurring. In fact it is notoriously difficult to provide unambiguous evidence of competition, particularly in mobile organisms. Because of this there is no clear agreement as to whether non-native bees have had a significant negative impact upon native pollinator populations (compare Robertson et al. 1989, Buchmann & Nabhan 1996, Sugden et al. 1996 with Butz Huryn 1997).

The majority of studies to date have been carried out in the neotropics, stimulated by the arrival of Africanized honeybees, and in Australia, where awareness of the possible impacts of introduced species is unusually high. Australia also has a large native bee fauna of over 1500 species (Cardale 1993) that is arguably the most distinctive in the world (Michener 1965). Most work has focused on the effects of honeybees.

Effects on Foraging Behavior

Each honeybee nest harvests 10–60 kg/yr of pollen and also requires 20–150 kg/yr of honey (Stanley & Liskens 1974, Roubik et al. 1984, Buchmann 1996). Crude extrapolation from the range of nest densities that have been recorded suggests that honeybees may gather 5–9000 kg pollen and 10–22,500 kg honey km^2/yr. In New Zealand, 8000 tons of honey is harvested from about 227,000 commercial hives every year; (this does not take into account honey used by the bees themselves, or that gathered by feral colonies) (Donovan 1980). I am unaware of any estimates of the total amounts of pollen or nectar available in natural habitats over a year, and it no doubt varies enormously, but common sense suggests that honeybees must use a substantial proportion of the available floral resources.

Honeybees commonly deter other bee species from foraging on the richest sources of forage (Wratt 1968; Eickwort & Ginsberg 1980; Roubik 1978, 1980, 1996a; Wilms & Wiechers 1997; Gross 2001) (although in one instance the converse had been reported, Menke 1954). Native organisms are often displaced to less profitable forage (Holmes 1964; Schaffer et al. 1979, 1983; Ginsberg 1983). In Panama, the presence of Africanized honeybees effectively eliminated foraging peaks of Meliponine bees because these native species were prevented from visiting their preferred sources of forage; as a result, the rate at which pollen was accrued in the nest was lower (Roubik et al. 1986). Displacement of native organisms has been attributed to the larger size of honeybee when compared to the majority of bee species (Roubik 1980), but is not necessarily size related. For example, the presence of honeybees has been found to deter foraging by hummingbirds

(Schaffer et al. 1983). Similarly, in a year when honeybees were naturally scarce, native bumblebees in Colorado were found to expand their diet breadth to include flowers usually visited mainly by honeybees (Pleasants 1981).

Hingston & McQuillan (1999) examined interactions between bumblebees and native bees in Tasmania and concluded that native bees were deterred from foraging by the presence of bumblebees, perhaps because bumblebees depressed availability of floral resources. Honeybees have been shown to depress availability of nectar and pollen (Paton 1990, 1996; Wills et al. 1990; Horskins & Turner 1999), which may explain why other flower visitors then choose to forage elsewhere.

Most authors concur that honeybees are not particularly aggressive to other insects while foraging, so that impacts on other species occur primarily through exploitative competition (Schaffer et al. 1979, 1983; Thorp 1987; Roubik 1991). However, honeybees have been found to displace smaller species from flowers by physical disturbance (Gross & Mackay 1998). Honeybees do attack nests of other honey-storing species to steal the honey, a behavior that may have contributed to the decline of *Apis cerana* in Japan (Sakagami 1959).

Both honeybees and bumblebees begin foraging earlier in the morning than many native bee species (Corbet et al. 1993, Dafni & Shmida 1996, Horskins & Turner 1999). Honeybees are able to achieve this owing to their large size (compared to most bees) and also owing to heat retention within their large nests (Roubik 1989). Bumblebees are able to begin foraging earlier still because of their great size and densely hairy body. It has been argued that depletion of nectar before native bees begin to forage may result in a significant asymmetry in competition in favor of these introduced species (Matthews 1984, Hopper 1987, Anderson 1989, Dafni & Shmida 1996, Schwarz & Hurst 1997).

Asymmetries in competition may also occur because of the ability of honeybees and bumblebees to communicate the availability and/or location of valuable food sources with nest mates, so improving foraging efficiency (von Frisch 1967, Dornhaus & Chittka 1999) (the majority of bee species are solitary, and each individual must discover the best places to forage by trial and error). Thus social species are collectively able to locate new resources more quickly, which again may enable them to gather the bulk of the resources before solitary species arrive (Roubik 1980, 1981; Schwarz & Hurst 1997).

Honeybees and bumblebees appear to be unusual in the distances over which they are capable of foraging. Honeybees are known to forage over 10 km from their nest, on occasion up to 20 km (Seeley 1985, Schwarz & Hurst 1997), and *B. terrestris* up to at least 4 km (Goulson & Stout 2001). Little is known of the foraging range of most other bee species, but those estimates that are available suggest that they are generally lower. For example *Melipona fasciata* travels up to 2.4 km (Roubik & Aluja 1983) and Trigonini over 1 km (Roubik et al. 1986). Solitary bee species are generally thought to travel only a few hundred meters at most (Schwarz & Hurst 1997).

Managed honeybee hives have further advantages over wild bee species; they are often given supplementary feeds when floral resources are scarce, and they are moved to track changing patterns of floral abundance. In this way populations of honeybees may be elevated far above those that could naturally persist in particular habitats.

Asymmetries in competition may not be stable, because the relative competitive abilities of bee species are likely to vary during the day according to temperature and resource availability, and are likely to vary spatially according to the types of flowers available (Corbet et al. 1995). The main exotic bees are large compared to most of the native species with which they might compete; *B. terrestris* weighs 109–315 mg (Prys-Jones 1982), and *A. mellifera* workers 98 $\pm$ 2.8 mg (Corbet et al. 1995). They also have longer tongues than many native species, particularly in Australia where most native species are short tongued (Armstrong 1979). Large bees are at a competitive advantage in cool conditions because of their ability to maintain a body temperature considerably higher than the ambient air temperature. They can thus forage earlier and later in the day than most smaller bees, and during cooler weather. Bees with longer tongues can also extract nectar from deeper flowers. However, large bees are not always at an advantage. The energetic cost of foraging is approximately proportional to weight (Heinrich 1979). Thus large bees burn energy faster. As nectar resources decline, the marginal rate of return will be reached more quickly by large bees. Also, long tongues are inefficient at handling shallow flowers. Thus large bees are likely to be at a competitive advantage early in the day and during cool weather, and they will be favored by the presence of deep flowers that provide them with a resource that other bees cannot access. But small bees with short tongues can forage profitably on shallow flowers even when rewards per flower are below the minimum threshold for large bees.

Although in general honeybees and bumblebees are able to forage at cooler temperatures than native bees, there may be exceptions. The Australian native *Exoneura xanthoclypeata* is adapted for foraging in cool conditions (Tierney 1994). It has been argued that this species is specialized for foraging on (naturally) uncontested resources early in the day, and that this species may be particularly susceptible to competition with exotic bees that forage at the same time (Schwarz & Hurst 1997).

The outcome of interactions between exotic and native flower visitors depends upon whether floral resources are limiting. Resource availability is likely to vary greatly during the year as different plant species come into flower (Carpenter 1978). When an abundant or large plant flowers, it may provide a nectar flush. Competition is unlikely to occur during such periods (Tepedino & Stanton 1981).

Overall, it seems probable that depression of resources by introduced bees is likely to have negative effects on native bee species, at least at some times of the year. To determine whether these effects are largely trivial (such as forcing native bees to modify their foraging preferences) or profound (resulting in competitive exclusion), population-level studies are necessary.

Evidence for Population-Level Changes

The only way to test unequivocally whether floral resources are limiting is to conduct experiments in which the abundance of the introduced bee species is artificially manipulated, and the population size of native species is then monitored. If populations are significantly higher in the absence of the introduced bee, then competition is occurring. Such experiments have proved to be exceedingly hard to accomplish. Excluding bees from an area is difficult. Within- and between-season variation is likely to be large, so such experiments need to be well replicated, with replicates situated many kilometers apart, and conducted over several years. No such study has been carried out.

An alternative approach, which is far easier but provides more equivocal data, is to correlate patterns of diversity of native bees with abundance of exotic bees, without manipulating their distribution. Aizen & Feinsinger (1994) found that fragmentation of forests in Argentina resulted in a decline in native flower visitors and an increase in honeybee populations. Similarly, Kato et al. (1999) studied oceanic islands in the northwest Pacific, and found that indigenous bees were rare or absent on islands where honeybees were numerous. On Mt Carmel in Israel, Dafni & Shmida (1996) reported declines in abundance of medium- and large-sized native bees (and also of honeybees) following arrival of *B. terrestris* in 1978. Conversely, Goulson et al. (2002) found no evidence for reduced abundance or diversity of native Tasmanian bees in areas colonized by *B. terrestris*, but did find that native bee abundance was considerably higher in the few sites where honeybees were absent. However, such studies can be criticized on the grounds that the relationship between exotic bee abundance and declining native bee populations (if found) need not be causative (Butz Huryn 1997). Increasing honeybee populations are often associated with increased environmental disturbance by man, which may explain declines in native bees.

Some researchers have attempted to manipulate numbers of introduced bees, either enhancing populations in experimental plots by placing hives within them, or conversely by remove hives from experimental plots in areas where hives have traditionally been placed. Areas without hives usually still have some honeybees, since there are likely to be some feral nests, and also because honeybees can forage over great distances. Replicates of the treatment without hives need to be sited many kilometers from replicates with hives to ensure that bees do not travel between the two, so many studies have been carried out without replication (e.g., Sugden & Pyke 1991). Despite these limitations, some interesting results have been obtained. Wenner & Thorp (1994) found that removal of feral nests and hives from part of Santa Cruz Island in California resulted in marked increases in numbers of native bees and other flower-visiting insects. Addition of honeybee hives caused the Australian nectivorous bird *Phylidonyris novaehollandiae* to expand its home foraging range and to avoid parts of inflorescences favored by honeybees (Paton 1993), but a comparison of areas with and without hives found no difference in the density of this bird species (Paton 1995). Roubik (1978) found a decrease in abundance of native insects when he placed hives of the Africanized honeybee in forests in French Guiana. However, Roubik (1982a, 1983) found no consistent detrimental effects on brood size, honey, and pollen stores in nests of two Meliponine bee species in Panama when Africanized honeybee hives were placed nearby for 30 days.

Monitoring of numbers of native bee species using light traps over many years since the arrival of Africanized bee has not revealed any clear declines in abundance (Wolda & Roubik 1986, Roubik 1991, Roubik & Wolda 2001). Roubik (1996a) describes the introduction of Africanized honeybees to the neotropics as a vast experiment, but it is an experiment without replicates or controls, so interpreting the results is difficult. Sugden & Pyke (1991) and Schwarz et al. (1991, 1992a,b) failed to find clear evidence for a link between abundance of honeybees and reproductive success of anthophorid bees belonging to the genus *Exoneura* in Australia in experiments in which they greatly enhanced honeybee numbers at experimental sites. However, the native species that they studied are themselves polylectic (Schwarz & Hurst 1997). As such they are the species least likely to be affected by competition. The majority of bee species are more specialized; in a review of data for 960 solitary bee species, Schemske (1983) found that 64% gathered pollen from only one plant family, often only one genus. For example, some Australian halictine bees have only been recorded on flowers of *Wahlenbergia* sp. (Michener 1965). Very little is known about such species, and no studies have been carried out to determine whether they are adversely affected by exotic bees (Schwarz & Hurst 1997). Also, the Australian studies of Sugden & Pyke (1991) and Schwarz et al. (1991, 1992a,b) were carried out in flower-rich heathlands; floral resources are more likely to be limiting in arid regions of Australia (Schwarz & Hurst 1997), and these areas often contain the highest native bee diversity (Michener 1979, O'Toole & Raw 1991). The *Exoneura* species studied in Australia had coexisted with honeybees for 180 years, so it is not surprising that they are not greatly affected by competition with this species. If there are species that are excluded by competition with exotic bees, honeybees in particular, there is no point looking for them in places where these bees are abundant. Unfortunately this leaves rather few places where they might occur.

Overall, there is no indisputable evidence that introduced bees have had a substantial impact via competition with native species. Given the difficulties involved in carrying out rigorous manipulative experiments, this should not be interpreted as the absence of competition. The abundance of exotic bees, the high levels of niche overlap, and evidence of resource depression and displacement of native pollinators, all point to the likelihood that competition is occurring. But we do not know whether such competition results (or resulted) in competitive exclusion. The best way to test for such competition is to carry out replicated experiments in which exotic bee numbers are manipulated and native pollinator numbers and reproductive success monitored over long periods. Ideally, such studies should target native species that are not generalists, and areas where floral resources are not abundant.

COMPETITION FOR NEST SITES

Honeybees nest in cavities, usually in old trees, and there is clear potential for competition. Many other organisms, including bees, mammals, and birds use such cavities for shelter or for nesting. In managed woodland, old trees with cavities are often in short supply. Hence it seems likely that honeybees may compete with native organisms for these sites, but rigorous studies are scarce. Both Oldroyd et al. (1994) and Moller & Tilley (1989) found that nesting holes were not in limiting supply in particular forests in Victoria and New Zealand, respectively. However, both studies were confined to small geographic areas, and it is hard to draw any general conclusions without further work.

B. terrestris generally nests in existing cavities below ground, often using abandoned rodent holes (Donovan & Weir 1978), and spaces beneath man-made structures such as garden sheds (personal observation). To my knowledge there have been no studies to determine whether such sites are used by native organisms in any of the countries to which this species has been introduced, although Donovan (1980) considered it unlikely that bumblebees compete with native bee species for nest sites in New Zealand.

Megachilidae nest in small cavities in wood. Donovan (1980) reported that nests sites used by *M. rotundata* overlap with those used by native bees belonging to the Hylaeinae, and also with mason wasps and spiders in New Zealand. However, it is not known whether availability of sites is limiting. Barthell & Thorp (1995) found that introduced *M. apicalis* in California aggressively usurp native species from nests sites, and concluded that competition was likely. However, subsequent work suggested that differences in habitat preferences between native and introduced species, and an abundance of nest sites may mean that competition is weak or absent (Barthell et al. 1998). Nothing is known of niche overlap in nesting requirements between introduced Megachilidae and native species elsewhere in North America or in Australia.

TRANSMISSION OF PARASITES OR PATHOGENS TO NATIVE ORGANISMS

A great deal is known about the pathogens and parasites of honeybees, and to a lesser extent bumblebees and leafcutter bees, since these species are of economic importance. Bees and their nests support a diverse microflora including pathogenic, commensal and mutualistic organisms (Gilliam & Taber 1991, Goerzen 1991, Gilliam 1997). Many pathogens are likely to have been transported to new regions with their hosts, particularly where introductions were made many years ago when awareness of bee natural enemies was low. Thus for example the honeybee diseases chalkbrood, caused by the fungus *Ascosphaera apis*, foulbrood, caused by the bacteria *Paenibacillus larvae*, the microsporidian *Nosema apis*, and the mite *Varroa destructor* now occur throughout much of the world. Hive beetles, *Aethina tumida*, were recently transported from Africa to North America, where they are proving to be serious pests of commercial honeybee colonies (Evans et al. 2000). Similarly, bumblebees in New Zealand are host to a parasitic nematode and three mite species, all of which are thought to have come from the U.K. with the original introduction of bees (Donovan 1980). During more recent deliberate introductions of exotic bees, such as that of *N. melanderi* to New Zealand, care has been taken to eliminate pathogens or parasites before bees were released (Donovan 1979). However, parasites are easily overlooked. Queens of *Bombus ignitus* are currently sent from their native Japan to the Netherlands, where they are induced to found colonies. The colonies are then returned to Japan for commercial purposes. Goka et al. (2001) recently discovered that the returned colonies are infested with a European race of the tracheal mite *Locustacarus buchneri*.

It is hard to exaggerate our ignorance of the natural enemies of most bee species, particularly their pathogens. We do not know what species infect them, or what the host ranges of these pathogens are. Very little is known of the susceptibility of native organisms to the parasites and pathogens that have been introduced with exotic bees. In a survey of natural enemies of native and introduced bees in New Zealand, Donovan (1980) concluded that no enemies of introduced bees were attacking native bees, but that the converse was true. A chalcidoid parasite of native bees was found to attack *M. rotundata* and, rarely, *B. terrestris*. One fungus, *Bettsia alvei*, which is a pathogen of honeybee hives elsewhere in the world was recorded infecting a native bee in New Zealand, but it is not known whether the fungus is also native to New Zealand. Indeed the natural geographic range of bee pathogens is almost wholly unknown. Some bee pathogens have a broad host range; for example, chalkbrood (*A. apis*), is also known to infect *A. cerana* (Gilliam et al. 1993) and the distantly related *Xylocopa californica* (Gilliam et al. 1994). The related chalkbrood fungus *Ascosphaera aggregata* is commonly found infecting *M. rotundata*; in Canada, where *M. rotundata* is an exotic species, this fungus infects the native bees *Megachile pugnata* Say (Goerzen et al. 1992) and *M. relativa* Cresson (Goerzen et al. 1990).

It seems likely that these few recorded instances of exotic bee pathogens infecting native species are just the tip of the iceberg, since so few studies have been carried out. As to whether these pathogens have had, or are having, a significant impact on native species, we do not know; if the introduction of a new pathogen were to lead to an epizootic in native insects, it would almost certainly go unnoticed. In other better known organisms, exotic pathogens have had disastrous impacts; for example the introduction of several crayfish species from North America has led to elimination of the native species *Astacus astacus* and *Austropotamobius pallipes* from large portions of Europe. The native species have little resistance to the exotic fungal pathogen *Aphanomyces astaci* that is carried by the introduced crayfish (Butler & Stein 1985). Studies of the incidence and identity of pathogen and parasite infestations of wild populations of native bees are urgently needed.

EFFECTS ON POLLINATION OF NATIVE FLORA

Recently, concerns have been expressed that exotic bees may reduce pollination of native plants, or alter the population structure of these plants by mediating different patterns of pollen transfer to native pollinators (Butz Huryn 1997, Gross & Mackay 1998). Efficient pollination requires a match between the morphology of the flower and that of the pollinator (reviews in Ramsey 1988, Burd 1994). If there is a mismatch, then floral rewards may be gathered without efficient transfer of pollen, a process known as floral parasitism (McDade & Kinsman 1980). Specialized obligate relationship between plants and pollinators do exist (reviewed in Goulson 1999) but are the exception (Waser et al. 1996). Most flowers are visited by a range of pollinator species, each of which will provide a different quality of pollinator service.

The efficiency of honeybees as pollinators of native plants in Australia and North America was reviewed by Butz Huryn (1997). She concluded that honeybees provide an effective pollination service to the majority of the flower species that they visit, although they do act as floral parasites when visiting a small number of plant species such as *Grevillea X gaudichaudii* in Australia (Taylor & Whelan 1988) and *Impatiens capensis* and *Vaccinium ashei* in North America (Wilson & Thomson 1991, Cane & Payne 1988). Similar results have been found for honeybees visiting Jamaican flora (Percival 1974). That honeybees are effective pollinators of many plants, even ones with which they did not coevolve is not surprising. After all, they have been used for centuries to pollinate a broad range of crops. Thus pollination of the native Australian *Banksia ornata* was increased by the presence of honeybee hives (Paton 1995), and honeybees have proved to be as effective as native bees in pollinating wild cashews, *Anacardium occidentale* in South America (Freitas & Paxton 1998). However, their presence may result in reduced seed set of some native plants. Roubik (1996b) reported declining seed set in the neotropical plant *Mimosa pudica* when honeybees were the dominant visitors, compared to sites where native bees were the more abundant, while Aizen & Feinsinger (1994) found reduced pollination of a range of Argentinian plant species in areas where forests were fragmented and honeybees more abundant. Gross & Mackay (1998) demonstrated that honeybees were poor pollinators of the Australian native *Melastoma affine*, so that when honeybees were the last visitors to a flower, seed set was reduced. As Roubik (1996b) points out, if native pollinators are lost (be it through competition with exotic bees, habitat loss, or use of pesticides) then we cannot expect honeybees to provide an adequate replacement pollination service for all wild plants and crops.

No studies have yet been reported of the effects of exotic bumblebees on the seed set of native plants. *B. terrestris* has the potential to disrupt pollinator services in a different way. This bee species is known to rob flowers. When the structure of the flower renders the nectaries inaccessible, *B. terrestris* (and some other bee species) may use their powerful mandibles to bite through the base of the corolla (Inouye 1983). In this way they act as floral parasites, removing nectar without effecting pollination. In Tasmania they rob some bird-pollinated plants in this way (personal observation). The effects of this behavior are hard to predict. Robbers have been found to reduce the amount of reward available, resulting in decreased visitation rates by pollinators (McDade & Kinsman 1980) and a reduction in seed set (Roubik 1982b, Roubik et al. 1985, Irwin & Brody 1999). Robbing can damage floral tissues preventing seed production (Galen 1983). However, nectar robbing may have little influence on plant fecundity if nectar robbers also collect pollen and in doing so effect pollination, or if other pollinators are present (Newton & Hill 1983, Arizmendi et al. 1995, Morris 1996, Stout et al. 2000). Some plants may actually benefit from the activity of nectar robbers by forcing legitimate foragers to make more long-distance flights hence increasing genetic variability through outcrossing (Zimmerman & Cook 1985).

A second possible detrimental effect of exotic bees is that they may alter the population structure by effecting a different pattern of pollen transport to native pollinators. In South Australia, Paton (1990, 1993) found that honeybees extracted more nectar and pollen from a range of flower species than did birds, the primary native pollinators. However, honeybees moved between plants far less than did birds, and so were less effective in cross-pollinating, resulting in decreased seed set. Several other studies have reported that interplant movement by honeybees is lower than that of other visitors (McGregor et al. 1959, Heinrich & Raven 1972, Silander & Primack 1978). Of course other pollinators often also move small distances, and it has been argued that honeybees are not unusual in this respect (Butz Huryn 1997). However, this is not true. Workers of social bees are unusual in that they are not constrained in their foraging behavior by the need to find mates, locate oviposition sites or guard a territory. In contrast, for example, butterflies intersperse visits to flowers with long patrolling flights in which they search for mates or oviposition sites (Goulson et al. 1997). Thus honeybees, bumblebees, and other social bees do tend to engage in fewer long flights than other species (Schmitt 1980, Waser 1982). The most obvious possible effect of exotic social bees in this respect is increased self-pollination, which could result in reduced seed set if the plant is self-infertile. Reduced interpatch pollen movement could result in reproductive fragmentation of plant populations. There are at present no data available on the impact of exotic bees on the genetic structure of plant populations.

Clearly it is not possible to generalize as to the effects that exotic bees will have on seed set of native flowers. For some species they will provide effective pollination, for others they will not. Where native pollinators have declined for other reasons, for example as a result of habitat loss and fragmentation, exotic bees may provide a valuable replacement pollinator service for native flowers. Where exotic bees are floral parasites, the effect will depend on whether rates of parasitism are sufficient to deter native pollinators. Any change in seed set (including increases) of plant species within a community could lead to long-term ecological change, but such effects would be exceedingly hard to detect among the much larger environmental changes that are currently taking place.

POLLINATION OF EXOTIC WEEDS

As we have seen, both honeybees and bumblebees visit a broad range of flowers. They also appear to prefer to visit exotic flowers (Telleria 1993, Thorp et al. 1994). For example, in Ontario, 75% of pollen collected by honeybees was from introduced plants (Stimec et al. 1997). In New Zealand, *B. terrestris* has been recorded visiting 400 exotic plant species but only 19 native species (Macfarlane 1976). The three other introduced *Bombus* species also visit mainly introduced plants (Donovan 1980). In the highlands of New Zealand, honeybees rely almost exclusively on introduced plants for pollen during most of the season (Pearson & Braiden 1990). Introduced *Megachile rotunda* appear to feed exclusively on introduced plants in Australia (Woodward 1996).

Do visits by exotic bees improve seed set of weeds? In general, rather little is known of the pollination biology of non-native plants, and it is unclear whether inadequate pollination is commonly a limiting factor (Richardson et al. 2000). By virtue of their abundance and foraging preferences, exotic bees often make up a very large proportion of insect visits to weeds. For example in a site dominated by European weeds in Tasmania, honeybees and bumblebees were the major flower visitors and comprised 98% of all insect visits to creeping thistle, *Cirsium arvense* (D. Goulson, unpublished data). In North America, honeybees increase seed set of the yellow star thistle, *Centaurea solstitialis* (Barthell et al. 2001) and are the main pollinators of two important weeds, purple loosestrife, *Lythrum salicaria* (Mal et al. 1992) and *Raphanus sativus* (Stanton 1987). Donovan (1980) reports that bumblebees are major pollinators of introduced weeds in New Zealand. It thus seems obvious and inevitable that exotic bees will prove to be important pollinators of various weeds (Sugden et al. 1996).

Remarkably, this view has been challenged. It is hard to agree with the conclusions of Butz Huryn & Moller (1995) that "Although honey bees may be important pollinators of some weeds, they probably do not contribute substantially to weed problems." Butz Huryn (1997) argues that most weeds do not rely on insect pollination, either because they are anemophilous, self-pollinating, apomictic, or primarily reproduce vegetatively. This is undoubtedly true of some weed species. For example of the 33 worst environmental weeds in New Zealand (Williams & Timmins 1990), nine fall into one of these categories (Butz Huryn & Moller 1995). However, 16 require pollination and are visited by honeybees, and one is pollinated more or less exclusively by them (the barberry shrub, *Berberis darwinii*). Eight more are listed as having unknown pollination mechanisms (Butz Huryn & Moller 1995). This group includes the tree lupin, *Lupinus arboreus*, and broom, *Cytisus scoparius*, which are self-incompatible and rely on pollination by bumblebees (Stout et al. 2002, Stout 2000). It also includes gorse, *Ulex europeaus*, which is thought to depend on honeybee pollination, and in which seed set is greatly reduced by a lack of pollinators in the Chatham Islands where honeybees and bumblebees are absent (McFarlane et al. 1992). Thus at least four major weeds in New Zealand are pollinated primarily by exotic bees.

L. arboreus is currently a minor weed in Tasmania. However, seed set in areas recently colonized by *B. terrestris* has increased dramatically, and it is likely that *L. arboreus* may become as problematic in Tasmania as it is in New Zealand now that it has an effective pollinator (Stout et al. 2002). Its zygomorphic flowers have to be forced apart to expose the stamens and stigma; only a large, powerful bee is able to do this, and no such bees are native to Tasmania. *L. arboreus* is only one of many weeds in Tasmania, New Zealand, and southern Australia that originated in the temperate northern hemisphere and are coadapted for pollination by bumblebees.

Demonstrating that exotic bees increase seed set of weeds is not sufficient in itself to conclusively show that the action of the bees will increase the weed population (Butz Huryn 1997). No long-term studies of weed population dynamics in relation to the presence or absence of exotic bees have been carried out. Because most weed species are short-lived and dependent on high reproductive rates, it seems probable that seed production is a crucial factor in determining their abundance. Key factor analysis of the life history could reveal whether seed set is directly related to population size.

At present, Australia alone has 2700 exotic weed species, and the costs of control and loss of yields due to these weeds costs an estimated AU$3 billion per year (Commonwealth of Australia 1997). The environmental costs are less easy to quantify but are certainly large. Most of these weed species are at present of trivial importance. The recent arrival of the bumblebee may awake some of these "sleeper" weeds, particularly if they are adapted for bumblebee pollination. Positive feedback between abundance of weeds and abundance of bumblebees is probable, since an increase in weed populations will encourage more bumblebees, and vice versa. If even one new major weed occurs in Australia due to the presence of bumblebees, the economic and environmental costs could be substantial.

CONCLUSIONS

Both *A. mellifera* and *B. terrestris* are now abundant over large areas where they naturally did not occur. They are both polylectic, and thus use resources utilized by a broad range of native species. Various Megachilidae have been introduced to North America and one species to Australia and New Zealand, but very little is known about their impacts.

It seems almost certain that abundant and widespread exotic organisms that single-handedly utilize a large proportion of the available floral resources do impact on local flower-visiting fauna. Consider, for example, the Tasmania native bee community. One hundred and eighty years ago this presumably consisted of a large number of small, solitary and subsocial species. Over 100 species are known to be present today, and many more probably exist. Nowadays, by far the most abundant flower-visiting insects at almost every site is the honeybee, often outnumbering all other flower-visiting insects by a factor of 10 or more (D. Goulson, unpublished data). In the southeast, the second most abundant flower visitor is usually the bumblebee, *B. terrestris*. The majority of floral resources are gathered by these bees, often during the morning before native bees have become active. It is hard to conceive how the introduction of these exotic species and their associated pathogens could not have substantially altered the diversity and abundance of native bees. Unfortunately we will never know what the abundance and diversity of the Tasmanian bee fauna were like before the introduction of the honeybee. Of course the same applies to most other regions such as North America where the honeybee has now been established for nearly 400 years. It is quite possible that some, perhaps many, native bee species were driven to extinction by the introduction of this numerically dominant species or by exotic pathogens that arrived with it. Even were it practical or considered desirable to eradicate honeybees from certain areas, it would be too late for such species.

Similarly, the introduction of exotic bees must increase seed set and hence weediness of some exotic plants, particularly when, as in the case of the bumblebee in Australia, many of the weeds were introduced from the same geographic region and are co-adapted with the introduced bee.

It must be remembered that introduced bees provide substantial benefits to man in terms of pollination of crops, and in the case of the honeybee, in providing honey. These quantifiable benefits should be weighed against the likely costs. In areas where weeds pollinated by exotic bees are a serious threat, and/or where native communities of flora and fauna are particularly valued, it may be that the benefits provided by these species are outweighed by the costs. Clearly further research, particularly rigorous manipulative experiments, are needed to determine how much introduced bees contribute to weed problems and whether they do substantially impact upon native pollinator communities. The cautionary principle argues that in the meantime we should at the very least prevent further deliberate release of exotic bee species (such as of bumblebees in mainland Australia, and speculative introductions of various solitary bee species in the United States). Unlike many of the other impacts that man has on the environment, introduction of exotic species is usually irreversible. It would also seem sensible to avoid placing honeybee hives within environmentally sensitive areas where possible, particularly areas where the native flora is threatened by invasion with weed species.

ACKNOWLEDGMENTS

I would like to thank David Roubik and Caroline Gross for discussion and comments.

The *Annual Review of Ecology, Evolution, and Systematics* is online at http://ecolsys.annualreviews.org

LITERATURE CITED

Abrahamovich AH, Telleria MC, Díaz NB. 2001. *Bombus* species and their associated flora in Argentina. *Bee World* 82:76–87

Aizen MA, Feinsinger P. 1994. Forest fragmentation, pollination, and plant reproduction in a Chaco dry forest, Argentina. *Ecology* 75:330–51

Alexander N, ed. 1996. *Australia: State of the Environment 1996.* Melbourne: CSIRO. 78 pp.

Anderson JME. 1989. Honeybees in Natural Ecosystems. In *Mediterranean Landscapes in Australia: Mallee Ecosystems and their Management*, ed. JC Noble,

RA Bradstock, pp. 300–4. East Melbourne: CSIRO

Arizmendi MC, Dominguez CA, Dirzo R. 1995. The role of an avian nectar robber and of hummingbird pollinators in the reproduction of two plant species. *Funct. Ecol.* 10:119–27

Armstrong JA. 1979. Biotic pollination mechanisms in the Australian flora—a review. *NZ J. Bot.* 17:467–508

Arretz PV, Macfarlane RP. 1986. The introduction of *Bombus ruderatus* to Chile for red clover pollination. *Bee World* 67:15–22

Barthell JF, Frankie GW, Thorp RW. 1998. Invader effects in a community of cavity nesting megachilid bees (Hymenoptera: Megachilidae). *Environ. Entomol.* 27:240–47

Barthell JF, Randall JM, Thorp RW, Wenner AM. 2001. Promotion of seed set in yellow star-thistle by honey bees: Evidence of an invasive mutualism. *Ecol. Appl.* 11:1870–83

Barthell JF, Thorp RW. 1995. Nest usurpation among females of an introduced leaf-cutter bee, *Megachile apicalis*. *Southwest. Entomol.* 20:117–24

Batra SWT. 1979. *Osmia cornifrons* and *Pithitis smaragulda*, two Asian bees introduced into the United States for crop pollination. In *Proc. IV Int. Symp. Pollination*, ed. MC Dewey, pp. 79–83. College Park: Univ. Md. Agric. Exp. Stn. Spec. Misc. Publ. I.

Bohart GE. 1972. Management of wild bees for the pollination of crops. *Annu. Rev. Entomol.* 17:287–312

Buchmann SL. 1996. Competition between honey bees and native bees in the Sonoran Desert and global bee conservation issues. See Matheson et al. 1996, pp. 125–42

Buchmann SL, Nabhan GP. 1996. *The Forgotten Pollinators*. Washington, DC: Island

Burd M. 1994. Bateman's principle and plant reproduction: the role of pollen limitation in fruit and seed set. *Bot. Rev.* 60:83–39

Butler MJ, Stein RA. 1985. An analysis of the mechanisms governing species replacements in crayfish. *Oecologia* 66:168–77

Buttermore RE. 1997. Observations of successful *Bombus terrestris* (L.) (Hymenoptera: Apidae) colonies in southern Tasmania. *Aust. J. Entomol.* 36:251–54

Butz Huryn VM. 1997. Ecological impacts of introduced honey bees. *Q. Rev. Biol.* 72:275–97

Butz Huryn VM, Moller H. 1995. An assessment of the contribution of honeybees (*Apis mellifera*) to weed reproduction in New Zealand protected natural areas. *NZ J. Ecol.* 19:111–22

Cane JH, Payne JA. 1988. Foraging ecology of the bee *Habropoda laboriosa* (Hymenoptera: Anthophoridae), an oligolege of blueberries (Ericaceae: *Vaccimium*) in the southeastern United States. *Ann. Entomol. Soc. Am.* 81: 419–27

Cardale JC. 1993. Hymenoptera: Apoidea. In *Zoological Catalogue of Australia*, ed. WWK Houston, GV Maynard, Vol. 10. Canberra: Aust. GPS

Carpenter FL. 1978. A spectrum of nectar-eater communities. *Am. Zool.* 18:809–19

Coffey MF, Breen J. 1997. Seasonal variation in pollen and nectar sources of honey bees in Ireland. *J. Apic. Res.* 36:63–76

Commonw. Aust. 1997. *The National Weeds Strategy*. Canberra: Commonw. Aust.

Cooper KW. 1984. Discovery of the first resident population of the European bee, *Megachile apicalis*, in the United States (Hymenoptera: Megachilidae). *Entomol. News* 95:225–26

Corbet SA, Fussell M, Ake R, Fraser A, Gunson C, et al. 1993. Temperature and the pollinating activity of social bees. *Ecol. Entomol.* 18:17–30

Corbet SA, Saville NM, Fussell M, Prys-Jones OE, Unwin DM. 1995. The competition box: a graphical aid to forecasting pollinator performance. *J. Appl. Ecol.* 32:707–19

Crane E. 1990a. *Bees and Beekeeping: Science, Practice, and World Resources*. Ithaca, NY: Cornell Univ. Press/Cornstock

Crane E. 1990b. *Bees and Beekeeping*. Oxford: Heinemann Newnes

Dafni A. 1998. The threat of *Bombus terrestris* spread. *Bee World* 79:113–14

Dafni A, Shmida A. 1996. The possible

ecological implications of the invasion of *Bombus terrestris* (L.) (Apidae) at Mt Carmel, Israel. See Matheson et al. 1996, pp. 183–200

Daly HV, Bohart GE, Thorp RW. 1971. Introduction of small carpenter bees in California for pollination. I. Release of *Pithitis smaragulda*. *J. Econ. Entomol.* 64:1145–50

Darwin C. 1872. *The Origin of Species by Means of Natural Selection: Or the Preservation of Favored Races in the Struggle for Life*. New York: Appleton

Dobson HEM. 1993. Bee fauna associated with shrubs in 2 California chaparral communities. *Pan-Pac. Entomol.* 69:77–94

Donovan BJ. 1975. Introduction of new bee species for pollinating lucerne. *Proc. NZ Grasslands Assoc.* 36:123–28

Donovan BJ. 1979. Importation, establishment and propagation of the alkali bee *Nomia melanderi* Cockerell (Hymenoptera: Halictidae) in New Zealand. *Proc. Int. Symp. Pollinat., 4th,* Maryland Agric. Exp. Stn. Spec. Misc. Publ. 1:257–68

Donovan BJ. 1980. Interactions between native and introduced bees in New Zealand. *NZ J. Ecol.* 3:104–16

Donovan BJ, Weir SS. 1978. Development of hives for field population increase, and studies on the life cycle of the four species of introduced bumble bees in New Zealand. *NZ J. Agric. Res.* 21:733–56

Dornhaus A, Chittka L. 1999. Insect behaviour—Evolutionary origins of bee dances. *Nature* 401:38

Doull K. 1973. Bees and their role in pollination. *Aust. Plants* 7:223–36

Eickwort GC, Ginsberg HS. 1980. Foraging and mating behaviour in Apoidea. *Annu. Rev. Entomol.* 25:421–26

Evans JD, Pettis JS, Shimanuki H. 2000. Mitochondrial DNA relationships in an emergent pest of honey bees: *Aethina tumida* (Coleoptera : Nitidulidae) from the United States and Africa. *Ann. Entomol. Soc. Am.* 93:415–20

Frankie GW, Thorp RW, Newstrom-Lloyd LE, Rizzardi MA, Barthell JF, et al. 1998. Monitoring solitary bees in modified wildland habitats: implications for bee ecology and conservation. *Environ. Entomol.* 27:1137–48

Freitas BM, Paxton RJ. 1998. A comparison of two pollinators: the introduced honey bee *Apis mellifera* and an indigenous bee *Centris tarsata* on cashew *Anacardium occidentale* in its native range of NE Brazil. *J. Appl. Ecol.* 35:109–21

Galen C. 1983. The effects of nectar thieving ants on seedset in floral scent morphs of *Polemonium viscosum*. *Oikos* 41:245–49

Gilliam M. 1997. Identification and roles of non-pathogenic microflora associated with honey bees. *FEMS Microbiol. Lett.* 155:1–10

Gilliam M, Lorenz BJ, Buchmann SL. 1994. *Ascosphaera apis*, the chalkbrood pathogen of the honeybee, *Apis mellifera*, from larvae of a carpenter-bee, *Xylocopa californica arizonensis*. *J. Invertebr. Pathol.* 63:307–9

Gilliam M, Lorenz BJ, Prest DB, Camazine S. 1993. *Ascosphaera apis* from *Apis cerana* from South Korea. *J. Invertebr. Pathol.* 61:111–12

Gilliam M, Taber S. 1991. Diseases, pests, and normal microflora of honeybees, *Apis mellifera*, from feral colonies. *J. Invertebr. Pathol.* 58:286–89

Ginsberg HS. 1983. Foraging ecology of bees in an old field. *Ecology* 64:165–75

Goerzen DW. 1991. Microflora associated with the alfalfa leafcutting bee, *Megachile rotundata* (Fab) (Hymenoptera, Megachilidae) in Saskatchewan, Canada. *Apidologie* 22:553–61

Goerzen DW, Dumouchel L, Bissett J. 1992. Occurrence of chalkbrood caused by *Ascosphaera aggregata* Skou in a native leafcutting bee, *Megachile pugnata* Say (Hymenoptera, Megachilidae), in Saskatchewan. *Can. Entomol.* 124:557–58

Goerzen DW, Erlandson MA, Bissett J. 1990. Occurrence of chalkbrood caused by *Ascosphaera aggregata* Skou in a native leafcutting bee, *Megachile relativa* Cresson (Hymenoptera, Megachilidae), in Saskatchewan. *Can. Entomol.* 122:1269–70

Goka K, Okabe K, Yoneda M, Niwa S. 2001. Bumblebee commercialization will cause worldwide migration of parasitic mites. *Mol. Ecol.* 10:2095–99

Goulson D. 1999. Foraging strategies for gathering nectar and pollen in insects. *Perspect. Plant Ecol. Evol. Syst.* 2:185–209

Goulson D, Ollerton J, Sluman C. 1997. Foraging strategies in the small skipper butterfly, *Thymelicus flavus*; when to switch? *Anim. Behav.* 53:1009–16

Goulson D, Stout JC. 2001. Homing ability of bumblebees; evidence for a large foraging range? *Apidologie* 32:105–12

Goulson D, Stout JC, Kells AR. 2002. Do alien bumblebees compete with native flower-visiting insects in Tasmania? *J. Insect Conserv.* 6:179–89

Gross CL. 2001. The effect of introduced honeybees on native bee visitation and fruit-set in *Dillwynia juniperina* (Fabaceae) in a fragmented ecosystem. *Biol. Conserv.* 102:89–95

Gross CL, Mackay D. 1998. Honeybees reduce fitness in the pioneer shrub *Melastoma affine* (Melastomataceae). *Biol. Conserv.* 86:169–78

Halvorson WL, Maender GJ, eds. 1994. *The Fourth Californian Islands Symposium: Update on the Status of Resources.* Santa Barbara, CA: Santa Barbara Mus. Nat. Hist. 628 pp.

Heinrich B. 1979. *Bumblebee Economics.* Cambridge, MA: Harvard Univ. Press

Heinrich B, Raven PH. 1972. Energetics and pollination ecology. *Science* 176:597–602

Hingston AB, Marsden-Smedley J, Driscoll DA, Corbett S, Fenton J, et al. 2002. Extent of invasion of Tasmanian native vegetation by the exotic bumblebee *Bombus terrestris* (Apoidea: Apidae). *Aust. Ecol.* 27:162–72

Hingston AB, McQuillan PB. 1999. Displacement of Tasmanian native megachilid bees by the recently introduced bumblebee *Bombus terrestris* (Linnaeus, 1758) (Hymenoptera: Apidae). *Aust. J. Zool.* 47:59–65

Hingston AB, McQuillan PB. 1998. Does the recently introduced bumblebee *Bombus terrestris* (Apidae) threaten Australian ecosystems? *Aust. J. Ecol.* 23:539–49

Holmes FO. 1964. The distribution of honey bees and bumblebees on nectar secreting plants. *Am. Bee J.* January: pp. 12–13

Hopkins I. 1911. *Australasian Bee Manual.* Wellington, NZ: Gordon & Gotch. 173 pp.

Hopkins I. 1914. History of the bumblebee in New Zealand: its introduction and results. *New Zealand Dept. Agric., Ind. Commer.* 46:1–29

Hopper SD. 1987. Impact of honeybees on Western Australia's nectarivorous fauna. In *Beekeeping and Land Management,* ed. J Blyth, pp. 59–71. Albany: West. Aust. CALM

Horskins K, Turner VB. 1999. Resource use and foraging patterns of honeybees, *Apis mellifera,* and native insects on flowers of *Eucalyptus costata. Aust. J. Ecol.* 24:221–27

Inouye DW. 1983. The ecology of nectar robbing. In *The Biology of Nectarines,* ed. TS Elias, B Bentley, pp. 152–73. New York: Columbia Univ. Press

Irwin RE, Brody AK. 1999. Nectar-robbing bumble bees reduce the fitness of *Ipomopsis aggregata* (Polemoniaceae). *Ecology* 80:1703–12

Kato M, Shibata A, Yasui T, Nagamasu H. 1999. Impact of introduced honeybees, Apis mellifera, upon native bee communities in the Bonin (Ogasawara) Islands. *Res. Popul. Ecol.* 2:217–28

Low T. 1999. *Feral Future.* Ringwood, Aust: Penguin Books

Macfarlane RP. 1976. Bees and pollination. In *New Zealand Insect Pests,* ed. DN Ferro, pp. 221–29. NZ: Lincoln Univ. Coll. Agric.

Macfarlane RP, Gurr L. 1995. Distribution of bumble bees in New Zealand. *NZ Entomol.* 18:29–36

Mal TK, Lovett-Doust J, Lovett-Doust L, Mulligan GA. 1992. The biology of Canadian weeds. 100. *Lythrum salicaria. Can. J. Plant Sci.* 72:1305–30

Mangum WA, Brooks RW. 1997. First records of *Megachile (Callomegachile) sculpturalis* Smith (Hymenoptera: Megachilidae) in the continental United States. *J. Kans. Entomol. Soc.* 70:140–42

Matheson A, Buchmann SL, O'Toole C, Westrich P, Williams IH, eds. 1996. *The Conservation of Bees.* London: Academic. 628 pp.

Matthews E. 1984. *To Bee or Not? Bees in National Parks—The Introduced Honeybee in Conservation Parks in South Australia.* Adelaide: Mag. South Aust. Natl. Parks Assoc., pp. 9–14

McDade LA, Kinsman S. 1980. The impact of floral parasitism in two neotropical hummingbird-pollinated plant species. *Evolution* 34:944–58

McFarlane RP, Grundell JM, Dugdale JS. 1992. Gorse on the Chatham Islands: seed formation, arthropod associates and control. *Proc. NZ Plant Prot. Conf., 45th,* pp. 251–55

McGregor SE, Alcorn EB, Kuitz EB Jr, Butler GD Jr. 1959. Bee visitors to Saguaro flowers. *J. Econ. Entomol.* 52:1002–4

McNally LC, Schneider SS. 1996. Spatial distribution and nesting biology of colonies of the African honey bee *Apis mellifera scutellata* (Hymenoptera: Apidae) in Botswana, Africa. *Environ. Entomol.* 25:643–52

Menezes Pedro SR, Camargo JMF. 1991. Interactions on floral resources between the Africanized honey bee *Apis mellifera* L and the native bee community (Hymenoptera: Apoidea) in a natural "cerrado" ecosystem in southeast Brazil. *Apidologie* 22:397–415

Menke HF. 1954. Insect pollination in relation to alfalfa seed production in Washington. *Wash. Agric. Exp. Stn. Bull.* 555:1–24

Michener CD. 1965. A classification of the bees of the Australian and South Pacific regions. *Bull. Am. Mus. Nat. Hist.* 130:1–324

Michener CD. 1974. *The Social Behavior of the Bees: A Comparative Study.* Cambridge, MA: Harvard Univ. Press. 404 pp. 2nd ed.

Michener CD. 1979. Biogeography of bees. *Ann. Mo. Bot. Gard.* 66:277–347

Moller H, Tilley JAV. 1989. Beech honeydew: seasonal variation and use by wasps, honey bees and other insects. *NZ J. Zool.* 16:289–302

Morris WF. 1996. Mutualism denied—nectar-robbing bumble bees do not reduce female or male success of bluebells. *Ecology* 77:1451–62

Newton SD, Hill GD. 1983. Robbing of field bean flowers by the short-tongued bumble bee *Bombus terrestris* L. *J. Apicult. Res.* 22:124–29

Oldroyd BP, Lawler SH, Crozier RH. 1994. Do feral honey-bees (*Apis mellifera*) and regent parrots (*Polytelis anthopeplus*) compete for nest sites. *Aust. J. Ecol.* 19:444–50

Oldroyd BP, Smolenski A, Lawler S, Estoup A, Crozier R. 1995. Colony aggregations in *Apis mellifera* L. *Apidologie* 26:119–30

Oldroyd BP, Thexton EG, Lawler SH, Crozier RH. 1997. Population demography of Australian feral bees (*Apis mellifera*). *Oecologia* 111:381–87

O'Neal RJ, Waller GD. 1984. On the pollen harvest by the honey bee (*Apis mellifera* L.) near Tucson, Arizona (1976–1981). *Desert Plants* 6:81–94

Otis GW. 1991. Population biology of the Africanized honey bee. See Spivak et al. 1991, pp. 213–34

O'Toole C, Raw A. 1991. *Bees of the World.* London: Blandford

Parker FD. 1981. A candidate for red clover, *Osmia coerulescens* L. *J. Apicult. Res.* 20:62–65

Parker FD, Torchio PF, Nye WP, Pedersen M. 1976. Utilization of additional species and populations of leafcutter bees for alfalfa pollination. *J. Apicult. Res.* 15:89–92

Pascarella JB, Waddington KD, Neal PR. 1999. The bee fauna (Hymenoptera: Apoidea) of Everglades National Park, Florida and adjacent areas: Distribution, phenology, and biogeography. *J. Kans. Entomol. Soc.* 72:32–45

Paton DC. 1990. Budgets for the use of floral resources in mallee heath. In *The Mallee Lands: A Conservation Perspective*, ed. JC Noble, PJ Joss, GK Jones, pp. 189–93. Melbourne: CSIRO

Paton DC. 1993. Honeybees in the Australian Environment—does *Apis mellifera* disrupt or benefit the native biota? *BioScience* 43:95–103

Paton DC. 1995. Impact of honeybees on the flora and fauna of Banksia heathlands in Ngarkat Conservation Park. *SASTA J.* 95:3–11

Paton DC. 1996. *Overview of feral and managed honeybees in Australia: distribution, abundance, extent of interactions with native biota, evidence of impacts and future research.* Canberra: Aust. Nat. Conserv. Agency

Pearson WD, Braiden V. 1990. Seasonal pollen collection by honeybees from grass shrub highlands in Canterbury, New Zealand. *J. Apicult. Res.* 29:206–13

Pellet FC. 1976. *American Honey Plants.* Hamilton, IL: Dadant & Sons. 5th ed.

Percival M. 1974. Floral ecology of coastal scrub in Southeast Jamaica. *Biotropica* 6: 104–29

Pimm SL, Russell GJ, Gittleman JL, Brookes TM. 1995. The future of biodiversity. *Science* 269:347–50

Pleasants JM. 1981. Bumblebee response to variation in nectar availability. *Ecology* 62:1648–61

Prys-Jones OE. 1982. *Ecological studies of foraging and life history in bumblebees.* PhD thesis. Univ. Cambridge, UK

Prys-Jones OE, Ólafsson E, Kristjánsson K. 1981. The Icelandic bumble bee fauna (Bombus Latr., Apidae) and its distributional ecology. *J. Apicult. Res.* 20:189–97

Pyke GH, Balzer L. 1985. The effects of the introduced honey-bee on Australian native bees. *NSW Natl. Parks Wildl. Serv. Occas. Pap. No. 7*

Ramsey MW. 1988. Differences in pollinator effectiveness of birds and insects visiting *Banksia menziesii* (Protaceae). *Oecologia* 76:119–24

Richardson DM, Allsop N, D'Antonio CM, Milton SJ, Rejmanek M, 2000. Plant invasions—the role of mutualisms. *Biol. Rev. Camb. Philos. Soc.* 75:65–93

Robertson P, Bennett AF, Lumsden LF, Silveira CE, Johnson PG, et al. 1989. Fauna of the Mallee study area north-western Victoria. Natl. Parks Wildl. Div. Tech. Rep. Ser., No. 87. Victoria, Australia: Dept. Conserv. Forests, Lands, pp. 41–42

Roubik DW. 1978. Competitive interactions between neotropical pollinators and Africanized honey bees. *Science* 201:1030–32

Roubik DW. 1980. Foraging behavior of commercial Africanized honeybees and stingless bees. *Ecology* 61:8336–45

Roubik DW. 1981. Comparative foraging behaviour of *Apis mellifera* and *Trigona corvina* (Hymenoptera: Apidae) on *Baltimora recta* (Compositae). *Rev. Biol. Trop.* 29:177–84

Roubik DW. 1982a. Ecological impact of Africanized honeybees on native neotropical pollinators. In *Social Insects of the Tropics*, ed. P Jaisson, pp. 233–47. Paris: Université Paris-Nord

Roubik DW. 1982b. The ecological impact of nectar robbing bees and pollinating hummingbirds on a tropical shrub. *Ecology* 63: 354–60

Roubik DW. 1983. Experimental community studies: time series tests of competition between African and neotropical bees. *Ecology* 64:971–78

Roubik DW. 1988. An overview of Africanized honey-bee populations: reproduction, diet and competition. In *Africanized Honey Bees and Bee Mites*, ed. GR Needham, RE Page Jr, M Delfinado-Baker, C Bowman, pp. 45–54. Boulder, CO: Westview

Roubik DW. 1989. *Ecology and Natural History of Tropical Bees.* Cambridge, UK: Cambridge Univ. Press

Roubik DW. 1991. Aspects of Africanized honey bee ecology in tropical America. See Spivak et al. 1991, pp.259–81

Roubik DW. 1996a. Measuring the meaning of honeybees. See Matheson et al. 1996, pp. 163–72

Roubik DW. 1996b. African honey bees as exotic pollinators in French Guiana. See Matheson et al. 1996, pp. 173–82

Roubik DW, Aluja M. 1983. Flight ranges of

Melipona and *Trigona* in tropical forests. *J. Kans. Entomol. Soc.* 56: 217–22

Roubik DW, Buchmann SL. 1984. Nectar selection by *Melipona* and *Apis mellifera* (Hymenoptera: Apidae) and the ecology of nectar intake by bee colonies in a tropical forest. *Oecologia* 61:1–10

Roubik DW, Holbrook NM, Parrav G. 1985. Roles of nectar robbers in reproduction of the tropical treelet *Quassia amara* (Simaroubaceae). *Oecologia* 66:161–67

Roubik DW, Moreno JE, Vergara C, Wittman D. 1986. Sporadic food competition with the African honey bee: projected impact on neotropical social species. *J. Trop. Ecol.* 2:97–111

Roubik DW, Schmalzel RJ, Moreno JE. 1984. Estudio apibotanico de Panamá: cosecha y fuentes de polen y nectar usados por *Apis mellifera* y sus patrones estacionales y anuales. *Bol. Téc. No. 24.* Org. Int. Reg. Sanidad Agropecuaria Mex., Centro Am. Panama. 73 pp.

Roubik DW, Wolda H. 2001. Do competing honey bees matter? Dynamics and abundance of native bees before and after honey bee invasion. *Popul. Ecol.* 43:53–62

Sakagami SF. 1959. Some interspecific relations between Japanese and European honeybees. *J. Anim. Ecol.* 28: 51–68

Schaffer WM, Jensen DB, Hobbs DE, Gurevitch J, Todd JR, Valentine Schaffer M. 1979. Competition, foraging energetics, and the cost of sociality in three species of bees. *Ecology* 60:976–87

Schaffer WM, Zeh DW, Buchmann SL, Kleinhans S, Valentine Schaffer M, Antrim J. 1983. Competition for nectar between introduced honey bees and native North American bees and ants. *Ecology* 64:564–77

Schemske DW. 1983. Limits to specialization and coevolution in plant-animal mutualisms. In *Coevolution,* ed. MH Nitecki, pp. 67–109. Chicago: Univ. Chicago Press

Schmitt D. 1980. Pollinator foraging behaviour and gene dispersal in *Senecio* (Compositae). *Evolution* 34:934–43

Schwarz MP, Gross CL, Kukuk PF. 1991. Assessment of competition between honeybees and native bees. July 1991. Prog. Rep. World Wildl. Fund, Aust. Proj. P158

Schwarz MP, Gross CL, Kukuk PF. 1992a. Assessment of competition between honeybees and native bees. January 1992. Prog. Rep. World Wildl. Fund, Aust. Proj. P158

Schwarz MP, Gross CL, Kukuk PF. 1992b. Assessment of competition between honeybees and native bees. July 1992. Prog. Rep. World Wildl. Fund, Aus. Proj. P158

Schwarz MP, Hurst PS. 1997. Effects of introduced honey bees on Australia's native bee fauna. *Vic. Nat.* 114:7–12

Seeley TD. 1985. The information-center strategy of honey bee foraging. In *Experimental Behavioral Ecology and Sociobiology,* ed. B Hölldobler, M Lindauer, pp. 75–90. Sunderland, MA: Sinauer

Silander JA, Primack RB. 1978. Pollination intensity and seed set in the Evening Primrose (*Oenothera fruticosa*). *Am. Midl. Nat.* 100:213–16

Spivak M, Fletcher DJC, Breed MD, eds. 1991. *The "African" Honey Bee.* Boulder, CO: Westview

Stanley RG, Liskens HF. 1974. *Pollen: Biology, Biochemistry, Management.* Berlin: Springer-Verlag

Stanton ML. 1987. Reproductive biology of petal color variants in wild populations of *Raphanus sativus* II: Factors limiting seed production. *Am. J. Bot.* 74:188–96

Stephen WP. 1987. *Megachile (Eutricharea) apicalis,* an introduced bee with potential as a domesticable alfalfa pollinator. *J. Kans. Entomol. Soc.* 60:583–84

Stimec J, ScottDupree CD, McAndrews JH. 1997. Honey bee, *Apis mellifera,* pollen foraging in southern Ontario. *Can. Field-Nat.* 111:454–56

Stout JC. 2000. Does size matter? Bumblebee behaviour and the pollination of *Cytisus scoparius* L. (Fabaceae). *Apidologie* 31:129–39

Stout JC, Allen JA, Goulson D. 2000. Nectar robbing, forager efficiency and seed set: bumblebees foraging on the self incompatible plant *Linaria vulgaris* Mill. (Scrophulariaceae). *Acta Oecol.* 21:277–83

Stout JC, Goulson D. 2000. Bumblebees in Tasmania: their distribution and potential impact on Australian flora and fauna. *Bee World* 81:80–86

Stout JC, Kells AR, Goulson D. 2002. Pollination of a sleeper weed, *Lupinus arboreaus*, by introduced bumblebees in Tasmania. *Biol. Conserv.* 106: 425–34

Stubbs CS, Drummond FA, Osgood EA. 1994. *Osmia ribifloris biedermannii* and *Megachile rotundata* (Hymenoptera: Megachilidae) introduced into the lowbush blueberry agroecosystem in Maine. *J. Kans. Entomol. Soc.* 67:173–85

Sugden EA, Pyke GH. 1991. Effects of honey bees on colonies of *Exoneura asimillima*, an Australian native bee. *Aust. J. Ecol.* 16:171–81

Sugden EA, Thorp RW, Buchmann SL. 1996. Honey bee native beecompetition: focal point for environmental change and apicultural response in Australia. *Bee World* 77:26–44

Taylor G, Whelan RJ. 1988. Can honeybees pollinate *Grevillea*? *Aust. Zool.* 24:193–96

Telleria MC. 1993. Flowering and pollen collection by the honeybee (*Apis mellifera* L. var *ligustica*) in the Pampas region of Argentina. *Apidologie* 24:109–20

Tepedino VJ, Stanton NL. 1981. Diversity and competition in bee-plant communities on short-grass prairie. *Oikos* 36:35–44

Thorp RW. 1987. World overview of the interactions between honeybees and other flora and fauna. In *Beekeeping and Land Management*, ed. JD Blyth, pp. 40–47. Como, Aust: Dept. Conserv. Land Manag.

Thorp RW. 1996. Resource overlap among native and introduced bees in California. See Matheson et al. 1996, pp. 143–51

Thorp RW, Wenner AM, Barthell JF. 1994. Flowers visited by honeybees and native bees on Santa Cruz Island. See Halvorson & Maender 1994, pp. 351–65

Tierney SM. 1994. Life cycle and social organisation of two native bees in the subgenus *Brevineura*. Unpublished BSc (Hons) thesis, Flinders Univ. South Aust.

Torchio PF. 1987. Use of non-honey bee species as pollinators of crops. *Proc. Entomol. Soc. Ont.* 118:111–24

Visscher PK, Seeley TD. 1982. Foraging strategy of honeybee colonies in a temperate deciduous forest. *Ecology* 63:1790–801

von Frisch K. 1967. *The Dance Language and Orientation of Bees*. Cambridge, MA: Harvard Univ. Press

Waser NM. 1982. A comparison of distances flown by different visitors to flowers of the same species. *Oecologia* 55:251–57

Waser NM, Chittka L, Price MV, Williams NM, Ollerton J. 1996. Generalization in pollination systems, and why it matters. *Ecology* 77:1043–60

Wenner AM, Thorp RW. 1994. Removal of feral honey bee (*Apis mellifera*) colonies from Santa Cruz Island. See Halvorson & Maender 1994, pp. 513–22

Williams PA, Timmins SM. 1990. *Weeds in New Zealand Protected Natural Areas: a Review for the Department of Conservation.* Sci. Res. Ser. No. 14, Wellington, NZ: Dept. Conserv. 114 pp.

Wills RT, Lyons MN, Bell DT. 1990. The European honey bee in Western Australian kwongan: foraging preferences and some implications for management. *Proc. Ecol. Soc. Aust.* 16:167–76

Wilms W, Wendel L, Zillikens A, Blochtein B, Engels W. 1997. Bees and other insects recorded on flowering trees in a subtropical Araucaria forest in southern Brazil. *Stud. Neotrop. Fauna Environ.* 32:220–26

Wilms W, Wiechers B. 1997. Floral resource partitioning between native *Melipona* bees and the introduced Africanized honey bee in the Brazilian Atlantic rain forest. *Apidologie* 28:339–55

Wilson P, Thomson JD. 1991. Heterogeneity among floral visitors leads to discordance between removal and deposition of pollen. *Ecology* 72:1503–7

Wolda H, Roubik DW. 1986. Nocturnal bee

abundance and seasonal bee activity in a Panamanian forest. *Ecology* 67:426–33
Woodward DR. 1996. Monitoring for impact of the introduced leafcutting bee, *Megachile rotundata* (F) (Hymenoptera: Megachilidae), near release sites in South Australia. *Aust. J. Entomol.* 35:187–91
Wratt EC. 1968. The pollinating activities of bumble bees and honey bees in relation to temperature, competing forage plants, and competition from other foragers. *J. Apicult. Res.* 7:61–66
Zimmerman M, Cook S. 1985. Pollinator foraging, experimental nectar-robbing and plant fitness in *Impatiens capensis*. *Am. Midl. Nat.* 113:84–91

Annu. Rev. Ecol. Evol. Syst. 2003. 34:27–49
doi: 10.1146/annurev.ecolsys.34.011802.132441

First published online as a Review in Advance on July 8, 2003

Avian Sexual Dichromatism in Relation to Phylogeny and Ecology

Alexander V. Badyaev[1]* and Geoffrey E. Hill[2]

[1]*Department of Ecology and Evolutionary Biology, University of Arizona, Tucson, Arizona 85721; email: abadyaev@email.arizona.edu*

[2]*Department of Biological Sciences, Auburn University, Auburn, Alabama 36849; email: ghill@acesag.auburn.edu*

Key Words plumage coloration, sexual selection, natural selection, sexual ornaments

■ **Abstract** The extent and diversity of sexual dichromatism in birds is thought to be due to the intensity of current sexual selection on the plumage ornamentation of males and females. This view leads to an expectation of concordance between ecological conditions and sexual dichromatism. Yet, because expression of dichromatism is the result of not only current selection, but also historical patterns of development, function, and selection, the concordance between ecology and current sexual dichromatism is not straightforward. Recent studies have revealed a number of trends in the evolution of avian sexual ornamentation that seem contrary to what is expected if current sexual selection is the primary force shaping dichromatism. For example, change in sexual dichromatism is often the result of evolutionary changes in female rather than male ornamentation. Moreover, sexual dichromatism is often an ancestral rather than a derived state; current expression of dichromatism is frequently the result of selection for lesser ornamentation in one sex and not for ornament elaboration. Loss and gain of sexual ornamentation sometimes precedes changes in preference for sexual ornamentation, and sexual ornaments can have high evolutionary lability despite their developmental and functional complexity. These findings emphasize that phylogenetic reconstructions must play a central role in attempts to understand the function and evolution of sexual dichromatism. With a historical perspective, one can test the relative importance of direct selection, indirect selection, and drift in relation to changes of sexual dichromatism. If sexual selection is invoked, the mechanisms of sexual selection can be explored by examining the concordance between the elaboration of ornamentation and the preferences for ornamentation across species and by tracing phylogenetic trajectories of sexual ornaments. Finally, placing physiological, genetic, and developmental mechanisms of sexual ornamentation into such a phylogenetic framework will enable greater inference about the past evolution and current function of sexual dichromatism in birds.

*Corresponding author

1543-592X/03/1215-0027$14.00

INTRODUCTION

Sexual dichromatism, defined as differences in the coloration of males and females of the same species, is thought to have evolved in response to selection pressures that differ between the sexes. In turn, the selection pressures on the sexes are influenced by the environment in which breeding occurs. Changes in predation, parasitism, or the distribution and abundance of resources can shift the balance between the benefits of exaggerated ornamental plumage and the cost of maintaining and developing such traits, and these environmental conditions often influence male and female plumage differently. Thus, diversity of ecological conditions commonly leads to substantial intra- and interspecific variability in sexual dichromatism.

Whereas a general relationship between sexual dichromatism and ecological factors has been addressed in many studies (Andersson 1994, Bennett & Owens 2002), major questions remain. First, it is unclear why some families and orders of birds show extensive variation in sexual dichromatism while other taxa, often apparently experiencing a similar range of environments, are remarkably conservative in their sexual ornamentation and degree of dichromatism. Second, it remains poorly understood to what degree the high genetic correlations typically observed between the sexes for plumage traits affect the evolution and diversification of sexual ornamentation. Because most of the physiological and developmental processes that produce sexual ornamentation are shared between the sexes (e.g., Kimball & Ligon 1999), high between-sex genetic correlations themselves might be a product of long-term selection. For example, when selection for sex-biased expression of a trait is not consistent, it may be advantageous for the developmental program of each sex not to respond rapidly to environmental change (Badyaev 2002). This would limit the speed of change in ornamentation in each sex and explain the lack of concordance between current ecology and sexual dichromatism within a species. Third, it is unclear whether sexual dichromatism is generally a derived state, as has been traditionally assumed, or if it can also be an ancestral state. Similarly, in most cases it remains uncertain whether sexual dichromatism is due to selection for greater ornamentation in males, as is commonly assumed, or due to selection for reduced ornamentation in females. Furthermore, it remains poorly understood to what degree ancestral dimorphic traits, such as pigment type and developmental patterns of plumage coloration, bias the evolution of derived ornamental traits and whether such constraints differ between taxa. Finally, the role of sexual selection versus other selective forces in the evolution of dichromatism and, when sexual selection is implicated, the roles of various mechanisms of sexual selection in the production of sexual dichromatism are highly debated issues. As we emphasize in this review, the most fruitful approaches to addressing these questions are comparative analyses of sexual dichromatism in relation to ecological pressures accompanied by reconstruction of phylogenetic pathways of change in dichromatism.

ECOLOGICAL CORRELATES OF SEXUAL DICHROMATISM

Latitudinal Distribution and Migratory Tendencies

A strong association with latitude of breeding and migratory tendencies is one of the most frequently documented ecological patterns of sexual dichromatism. Bird species that are migratory that have a wider geographic distribution, and that breed at higher latitudes are more sexually dimorphic than species that are resident, have restricted geographic ranges, and that breed at lower latitudes (Bailey 1978, Fitzpatrick 1994, Grant 1965, Hamilton 1961, Mayr 1942, Peterson 1996, Price 1998, Scott & Clutton-Brock 1989). Understanding the mechanisms behind these patterns, however, remains elusive.

Three major explanations have been proposed: (*a*) the patterns are driven by geographical variation in the strength of sexual and natural selection (e.g., geographical variation in mate sampling or the importance of mate recognition). (*b*) The patterns are the result of nonselective factors, such as genetic drift. For example, small, resident, and isolated populations might be more prevalent at lower latitudes and more susceptible to the effects of drift. Or, (*c*) the patterns are due to a combination of (*a*) and (*b*). For example, if the intensity of sexual selection (e.g., competition for extra-pair mates) is influenced by the amount of genetic variation in populations, then low genetic diversity in small populations could decrease the intensity of sexual selection and lead, ultimately, to lesser sexual ornamentation (Burke et al. 1998, Petrie & Kempenaers 1998).

To derive testable explanations for latitudinal and migratory patterns of sexual dichromatism, it is essential to know the ancestral state of sexual dichromatism, the sex bias in evolutionary transitions in plumage elaboration, and the relative frequency of sexual dichromatism transformations across lineages. Furthermore, intraspecific studies that examine the development and function of sexual ornamentation in relation to population size, migratory tendencies, and latitude might be informative for understanding the mechanisms behind the interspecific patterns.

Hamilton (1961) documented that, in warblers (Parulidae) and orioles (Icteridae), species at lower latitudes were less sexually dichromatic than their relatives at higher latitudes, and he attributed the pattern to a decrease in female coloration at higher latitudes. Noting that species at low latitudes are often resident and maintain longer pair bonds than species at high latitudes, Hamilton suggested that the duller coloration of females may reduce intrasexual aggression at the time of pair formation and that increased sexual dichromatism could facilitate reliable species and mate recognition. In turn, this would speed up reestablishment of territories and pair bonds favored by the short northern breeding season (Hamilton 1961). Similarly, Bailey (1978) investigated latitudinal variation in coloration across 787 passerine species in North and Central America and found that sexual dichromatism is more pronounced in high-latitude species. Contrary to Hamilton's explanation, however, in species that were dichromatic, females as well as males were more ornamented at higher latitudes.

Several studies have corroborated Hamilton's (1961) idea that greater dichromatism may be associated with a shorter mate-sampling period. For example, resident species and species that mate while in winter flocks may have more opportunities and a longer time to evaluate and compare potential mates based on their actual performance rather than on sexual ornamentation (Slagsvold & Lifjeld 1997). The differences in the time that females have to make mate choices and the costs of mate sampling affect selection on sexual ornamentation (Badyaev & Qvarnström 2002) and might account for the greater dichromatism of migratory birds across geographical regions (Badyaev 1997a, Fitzpatrick 1994).

Alternatively, latitudinal variation in sexual dichromatism could be due to geographical variation in the patterns of natural selection, such as latitudinal differences in predation (Martin 1996; see also below). Bailey (1978) suggested that background matching is a predation-avoidance strategy that favors brighter colors for both sexes at lower latitudes. Alternatively, the reduced coloration of females at higher latitudes could be due to higher predation pressure. Important to understanding the role of predation in shaping latitudinal gradients in sexual dichromatism is the phylogenetic information on whether latitudinal transitions in plumage brightness and predation risk are sex biased.

Several studies documented that sexually dichromatic taxa tend to have wider geographic distributions than monochromatic taxa (e.g., Badyaev & Ghalambor 1998, Price 1998). This pattern is puzzling because other studies found that sexually dimorphic species have higher extinction rates and are less able to colonize novel environments than monomorphic species (McLain 1993, McLain et al. 1995, Sorci et al. 1998), presumably because resources allocated toward elaboration of sexual ornamentation might compromise an organism's ability to track environmental changes (McLain 1993, McLain et al. 1995, Sorci et al. 1998). However, recent study of nonpasserine European birds documented no difference in risk of extinction or in population declines between monomorphic and dimorphic species (Prinzing et al. 2002).

Fitzpatrick (1994) proposed that sexual ornamentation might indicate the ability of individuals to withstand the energetic demands of long migration and to select good quality wintering habitats, and this might be responsible for the interspecific association between migratory tendency and sexual dichromatism. If evolution of sexual dichromatism is related to migratory abilities, then a shift from migratory to resident status, such as in island populations should be followed by a transition from sexual dichromatism to monomorphism. In this case, loss of dichromatism is the result of weaker selection on male sexual ornamentation and thus lesser ornamentation of males. Under this scenario, a gain in sexual dichromatism following the transition from resident to migratory status is as likely as the loss of sexual dichromatism in a resident population (Fitzpatrick 1994).

Crucial to understanding the mechanisms behind latitudinal variation in sexual dichromatism is knowledge of the ancestral state of sexual dimorphism in a lineage. For example, sexual dichromatism in dabbling ducks (Anatidae) is most common in species that have a wide geographic distribution, that breed at higher latitudes, and that occur on continents, whereas monochromatism prevails among nonmigratory, southern species that have restricted, isolated ranges (Figuerola & Green 2000, Omland 1997, Scott & Clutton-Brock 1989). For these reasons, monochromatism is particularly common among island taxa. Using phylogenetic reconstruction of sexual dichromatism in these birds, Omland (1997) showed that sexual dichromatism is an ancestral stage. Therefore, widely distributed and migratory species, when settling on islands and becoming isolated, might form monochromatic populations because of the effects of genetic drift and inbreeding (Burke et al. 1998, Omland 1997, Peterson 1996). Yet, whereas both genetic drift and natural selection can produce equal gains and losses in sexual dichromatism following shifts in migratory tendencies, given the complexity and sex-bias of many color patterns, genetic drift alone would be more likely to lead to loss of rather than gain in sexual dichromatism (e.g., Omland 1997).

Another proposed reason that island-dwelling species could be less dichromatic than their mainland-dwelling counterparts is that the risk of hybridization is commonly lower and species recognition is less important on islands than on the mainland. Support for this explanation of reduced sexual dichromatism on islands is mixed with some studies finding no association (e.g., Owens & Clegg 1999) while others are finding significant trends (Figuerola & Green 2000).

Mating Systems and Parental Care

Sexual selection arising from difference in the reproductive success and parental investment of males exerts strong selection on sexual ornamentation (Andersson 1994, Kirkpatrick & Ryan 1991, Payne 1984, Owens & Bennett 1997). Thus, the ecological conditions that affect paternal investment should affect sexual dichromatism (Andersson 1994).

Because variance in male reproductive success is expected to be higher in polygynous than in monogamous species, it is commonly assumed that sexual dichromatism should be greater in polygynous mating systems. To the contrary, however, while a close correlation between mating system and ecological conditions is well established in birds (Bennett & Owens 2002), only a few studies have documented a direct association between mating system and sexual dichromatism (Cuervo & Møller 1999, Dunn et al. 2001, Figuerola & Green 2000). The examples in this section address this apparent paradox and illustrate three points. First, the expected association between mating system and sexual dichromatism is often documented only when mating systems are defined at a very detailed scale (Dunn et al. 2001; Møller & Birkhead 1994; Owens & Bennett 1994, 1997; Møller & Cuervo 1998; Scott & Clutton-Brock 1989) and sexual dichromatism is partitioned into developmentally distinct components such as carotenoid-, melanin-, or structurally based coloration (Owens & Hartley 1998). Second, phylogenetic information on sex-biased transitions in ornament elaboration helps to identify what exactly needs to be explained—change in male coloration or change in female coloration—and thus facilitates an understanding of the association between mating systems and sexual dichromatism (Irwin 1994, Burns 1998, Cuervo & Møller 1999, Figuerola & Green 2000). Finally, a hierarchical approach afforded by phylogenetic studies of mating systems allows examination of temporal concordance in changes of sexual dichromatism and mating systems (Dunn et al. 2001; Owens & Bennett 1995, 1997).

In one of the first studies of the association between mating system and sexual dichromatism, Crook (1964) showed that the monogamous weavers (Ploceidae) were monomorphic, whereas polygynous species were dichromatic. He attributed this pattern to the distribution of food and nesting habitat. Most recent studies, however, have found that the association between mating system and dichromatism is not straightforward. Indeed, most passerines are sexually dimorphic regardless of their social mating system; polygynous European passerines are not more often sexually dimorphic in plumage than monogamous species (Møller 1986). Similarly, despite their polygynous mating system, many species of hummingbirds (Trochilidae) are monomorphic with female coloration showing the most variation (Bleiweiss 1992).

In one of the few studies that documented the association between mating systems and sexual dichromatism, Scott & Clutton-Brock (1989) examined plumage variation in 146 species of Anatidae. They thoroughly delineated mating systems based on frequency of pairing, duration of pair bond, and partitioning of parental care and found that sexual dichromatism was greater in species with shorter pair bonds and with distinct parental roles. Male plumage brightness was most strongly correlated with pair bond duration, paternal care, and nest dispersion (e.g., with mating opportunities), whereas female brightness varied the most with nest placement and features of nesting habitat (e.g., with predation risk) (Scott & Clutton-Brock 1989). These results corroborated Kear's (1970) findings that in the majority of monochromatic species of waterfowl both sexes shared parental duties, while in most dimorphic species females raised the young alone. Figuerola & Green (2000) examined evolutionary changes in sexual dichromatism and concluded that changes in mating system are significantly correlated with changes in dichromatism. Similarly, in passerines, males of monochromatic species were more likely to participate in nest building (Soler et al. 1998) and to share incubation with females (Verner & Willson 1969) than males of dichromatic species.

Extensive paternal care may both reduce mating opportunities for males and increase risk of predation. To distinguish between these two factors it is necessary to know whether variation in sexual dichromatism is due to change in male or female coloration. Owens & Bennett (1994) documented that adult mortality closely covaried with parental care, but not with sexual dichromatism across 37 Palearctic bird species. The association between sexual dichromatism and parental care was mostly due to variation in mating opportunities among species with different paternal care. Similarly, among socially monogamous passerines, male plumage brightness was associated with the frequency of extra-pair paternity; species with higher levels of extra-pair paternity had more ornamented males and greater sexual dichromatism (Møller & Birkhead 1994).

Owens & Hartley (1998) surveyed sexual dimorphism across 73 bird species and found that different types of dimorphism were affected by distinct selection pressures. Sexual dimorphism in size was strongly associated with social mating system and parental roles (Björklund 1990, 1991; Webster 1992), whereas sexual dichromatism was most closely associated with levels of extra-pair paternity and only weakly with parental roles (e.g., Verner & Willson 1969). Given the apparently distinct patterns of selection on different components of dimorphism, it is interesting to examine the evolutionary lability of these components, which may be greater in environment-dependent traits, such as diet-derived coloration (Badyaev & Hill 2000, Gray 1996, Hill 1996), compared to sexual dimorphism in body size and in complex patterns of coloration that may be more developmentally integrated and thus less labile phylogenetically (Badyaev 2002, Price 2002, Price & Pavelka 1996, Omland & Lanyon 2000).

In a series of comparative studies, Owens and colleagues (Owens & Bennett 1995, 1997; Owens & Hartley 1998) suggested that patterns of diversification in mating systems and life history strategies are historically nested. They argued that phylogenetically distant taxa may have converged on similar mating systems despite different evolutionary histories. Thus, ancestral evolutionary events, such as changes in the partitioning of parental care, nesting, and feeding habits, may determine the response of a lineage to current ecological conditions (Owens & Bennett 1997). Such phylogenetic constraints that limit a taxon to a specific range of mating behaviors could also limit variation in sexual dichromatism and contribute to the lack of contemporary associations between sexual dichromatism and mating systems.

Dichromatism can also vary with ecological factors such as climate or the distribution of nest sites because male parental care, and hence the intensity of sexual selection, changes with such factors. Male parental investment varies with ecological factors such as climate or the distribution of food or nest sites (Badyaev & Ghalambor 2001). For example, colder nest microclimate and spatial separation of nesting and feeding resources, such as is found at high elevations, is associated with greater male care (Badyaev 1997a, Badyaev & Ghalambor 2001). Thus, in monogamous species, the intensity of sexual selection should vary with the breeding elevation. This association was documented across 126 species of Cardueline finches; species occupying lower elevations were more sexually dichromatic than species at higher elevations, and the altitudinal variation was largely due to increased ornamentation of males at lower elevations (Badyaev 1997a).

Irwin (1994) showed that sexual dichromatism varied with mating system (polygynous species were more dimorphic) across family Icteridae and that the association was owing largely to changes in female plumage. She suggested that variation in sexual dichromatism in this group resulted from social selection on females rather than sexual selection on males. More generally, sexual selection on females to display brighter plumage should be greater in monogamous systems (Irwin 1994, Moreau 1960). In turn, mutual mate choice and female-female interactions associated with monogamous breeding may contribute to the association between female plumage brightness, sexual dichromatism, and mating system (Bleiweiss 1992, Irwin 1994, Johnson 1988, Trail 1990, West-Eberhard 1983). In one of the most comprehensive studies to date, Dunn et al. (2001) found strong and consistent associations between sexual dichromatism and social mating systems across 1,031 species of birds; sexual dichromatism was greater in polygynous and lekking species than in monogamous species.

These studies emphasize the importance of distinguishing between monomorphism when both sexes are ornamented and monomorphism where both sexes have reduced sexual ornamentation. "Dull" monomorphism may arise from monogamous mating systems in which selection pressures associated with breeding are similar between sexes and in which mates have an extended opportunity to evaluate each other based on performance and direct comparisons (references in Badyaev & Qvarnström 2002). "Bright" monomorphism might be more prevalent in monogamous mating systems with short mate-sampling periods (West-Eberhard 1983, Fitzpatrick 1994).

It is commonly expected that sexual dichromatism should be associated with lek breeding, because variance in male reproductive success and hence sexual selection is assumed to be very strong in this mating system (Darwin 1871, Kirkpatrick 1987, Payne 1984). Interestingly, however, lekking species are not more likely to be sexually dichromatic than nonlekking species (Höglund 1989, Payne 1984, Trail 1990; but see Dunn et al. 2001). Studies of the association between lekking and sexual dichromatism illustrate that in addition to examination of the current selection on both males and females it is important to know the historical sequence of transitions such as whether a shift to or from lekking behavior preceded or followed the change in sexual dichromatism. Moreover, one needs phylogenetic information about the ancestral state of the sexual ornamentation of both sexes to generate hypotheses about the patterns of sexual ornamentation in relation to lekking. For example, prior to evolution of sex-biased expression, a transition from monomorphic dull to monomorphic bright states is expected under correlated response of females to selection on male ornamentation (Lande 1980). Increased risk of predation associated with evolutionary transition to lekking may explain changes in plumage coloration from sexually dimorphic or monomorphic bright to monomorphic dull. For example, Bleiweiss (1997) examined covariation of sexual dichromatism and plumage brightness with occurrence of lekking behavior across 415 bird species by analyzing evolutionary transitions of plumage brightness in both sexes. He found that in addition to sexual selection, predation risks and foraging behaviors associated with lekking are likely to constrain ornament elaboration (Bleiweiss 1997). In a recent analysis of phylogenetic transitions of sexual ornamentation, however, Cuervo & Møller (1999) found that acquisition of elaborate plumage ornaments was more closely associated with transition from monogamy to lekking than with change in male parental care, diet, or predation risk.

Ecological Factors Affecting Mortality and Parasitism

One explanation for sexual dichromatism is that it evolved through selection for crypsis in females because of their greater vulnerability to predators around the nest (Baker & Parker 1979, Butcher & Rohwer 1993, Götmark 1999, Wallace 1889). Sexual dichromatism in birds is generally thought to arise from sexual selection favoring conspicuous coloration in males, although natural selection (e.g., predation) is thought to ultimately limit conspicuousness (Darwin 1871, Fisher 1930). Alternatively, bright coloration may be favored by predation because it advertises that a prey is unprofitable, and the degree of sexual dichromatism of a species may be owing to the difference between the sexes in their profitability to a predator (Baker & Parker 1979, Butcher & Rohwer 1993, Cott 1947, Götmark 1994). Promislow et al. (1992, 1994) examined variation in sex-specific mortality due to sexual ornamentation in passerines and waterfowl. They suggested that female mortality may constrain the upper limit of sexual dichromatism in a population by limiting the maximum mortality rate of males. In turn, the ornamentation of males could be further constrained by mortality due to elaborated plumage and more intensive sexual competition (Promislow et al. 1992, 1994).

Götmark et al. (1997) showed that predation on adult chaffinches (*Fringilla coelebs*) exerts greater pressure on female than on male coloration, ultimately leading to variation in sexual dichromatism. Similarly, Burns (1998) attributed more frequent evolutionary changes in female versus male sexual ornamentation in tanagers (Thraupidae) to greater predation risk of females associated with nest predation. In cardueline finches, sexual dichromatism and plumage ornamentation in both sexes closely covaried with life history traits, but in opposite directions: fecundity covaried negatively with male sexual ornamentation but positively with female ornamentation (Badyaev 1997b).

Badyaev (1997c) examined variation in the sex-specific costs of plumage elaboration along an elevational gradient in finches and found that the association between fecundity and sexual ornamentation was more similar between the sexes in high-elevation species than in low-elevation species (Badyaev 1997a). Furthermore, Badyaev & Ghalambor (Badyaev 1997b, Badyaev & Ghalambor 2001) suggested that elevational variation in sexual dichromatism was associated with lower juvenile mortality at higher elevations. Such an association between elevation and sexual dichromatism would result when low-elevation environments favor increased and more elaborated sexual ornamentation, but when the development of such traits commonly results in reduced juvenile survival (Owens & Bennett 1994).

While a relationship between sexual dichromatism and mortality is well established, two problems persist: (*a*) identifying the specific factors behind this relationship, and (*b*) determining what is cause and what is effect in the relationship. If nest predation limits ornament elaboration (Wallace 1889, Baker & Parker 1979, Shutler & Weatherhead 1990, Johnson 1991), then male and female ornamentation should vary with the time that each sex spends at the nest vicinity. In particular, because females typically incubate eggs and brood nestlings, we expect reduced female ornamentation when the nest environment exposes females to predators. In contrast, male birds typically do not incubate or brood young, so male ornamentation might not vary as strongly with predation at nests. By separately examining male and female plumage across Parulidae and Carduelinae, Martin & Badyaev (1996) found that female plumage brightness varied with nest placement and was negatively correlated with nest predation. These results suggested that nest predation may place greater constraints on female than male plumage brightness, at least in taxa where only females incubate eggs and brood young. Moreover,

Martin & Badyaev (1996) found that female sexual ornamentation varied at least partly independently of male ornamentation, emphasizing the need to consider variation in both sexes in tests of plumage dimorphism. In warblers and finches, sexual dichromatism differed between ground- and off-ground-nesting species, but the relationship between sexual dichromatism and nest predation was positive rather than negative (Johnson 1991, Shutler & Weatherhead 1990). Specifically, differences in sexual dichromatism between ground- and off-ground-nesting birds resulted only partially from decreased male brightness (Dunn et al. 2001, Johnson 1991, Shutler & Weatherhead 1990). Most of the patterns were the result of an increase in female brightness in ground-nesting birds, which was related to their reduced risk of nest predation compared to off-ground nesters (Martin & Badyaev 1996). Effects of nest predation on sexual dichromatism are most evident when one separately examines sexual dichromatism in different body parts. For example, dichromatism of upper body parts but not lower body parts strongly covaried with nest placement across cardueline finches (Badyaev 1997a). In addition, variation in parasite prevalence across nesting and foraging strata contributed to vertical stratification of sexual dichromatism in birds (Gavrin & Remsen 1997).

The importance of current variation in nesting biology in shaping sexual dichromatism was questioned by Owens & Bennett (1995). Based on comparative analysis of current sexual dichromatism and phylogenetic history of avian groups, the authors concluded that current variation in nesting and feeding habits have little effect on current avian life history strategies, which are almost entirely due to ancient evolutionary events. If ancient and hierarchically nested evolutionary diversifications, such as changes in nest placement, were responsible for changes in sexual dichromatism, we would expect to see concordant and similarly historically nested patterns of divergence in sexual dichromatism. Other studies, however, showed that large-scale diversification in life histories are produced by more recent ecological changes (e.g., Martin & Clobert 1996). These examples illustrate the need to examine historical transitions in sexual dichromatism and plumage ornamentation in relation to changes in nesting strata or parental behavior in order to properly test the association between nesting and foraging habits and sexual dichromatism (e.g., Owens & Bennett 1994, 1997).

Sensory Characteristics, Physical Features of Habitat, and Diet

Exploitation of new habitats by birds is often accompanied by changes in plumage ornamentation. The evolution of novel sexual ornamentation may be favored by both preexisting sensory biases within lineages and characteristics of new environments that make some ornaments more easily perceived (Endler 1992, Endler & Théry 1996, Schluter & Price 1993). Physical characteristics, such as substrate abrasiveness, ultraviolet (UV) radiation, and temperature might affect sexual dichromatism by favoring specific patterns of pigmentation (Burtt 1986). The examples in this section emphasize that comparative studies need to show that color patterns are indeed preceded by habitat shifts (e.g., Marchetti 1993) and that divergence into different habitats promotes divergence in sexually selected traits

(Badyaev & Snell-Rood 2003, Barraclough et al. 1995, Møller & Cuervo 1998, Price 1998, Schluter & Price 1993).

Physical features of habitats may favor certain plumage pigmentation and thereby constrain distribution of other types of pigments or structural colors. For example, melanin pigmentation makes feathers more resistant to mechanical damage and birds living in environments with more abrasive substrates have more melanin-based colors in their plumage (Burtt 1986). Moreover, within the plumage of an individual bird, feathers that are subjected to more wear and abrasion have a higher proportion of melanin pigmentation (Fitzpatrick 1998). The presence of melanin, in turn, might bias the distribution of structural- (reviewed in Prum 1999) and carotenoid-based coloration (references in Savalli 1995). It was also suggested that high absorption qualities of some pigments might protect birds from UV radiation; Brush (1970) attributed more intense pigmentation and sexual dichromatism in tanagers breeding at higher elevations to the greater need for protection from UV.

Price (1996) examined variation in sexual dichromatism across finch species and found that drier and more open habitats had a lower proportion of dichromatic species than did moister, denser habitats (see also Badyaev 1997a). Habitat influences, however, may be confounded by the effects of nest dispersion because greater plumage dichromatism in finches is associated with solitary nesting and most open-habitat species are semicolonial. Price (1996) considered habitat density as a correlate rather than a cause of sexual dichromatism. He suggested that finches in closed habitats may breed at higher densities and thus have increased potential for extra-pair paternity (Møller & Birkhead 1993; but see Westneat & Sherman 1997).

Endler and colleagues (Endler & Théry 1996, Endler & Wescott 1998) reported a high degree of ambient-light specificity in display behaviors of several tropical species. It is unclear, however, whether such behaviors followed existing coloration patterns to maximize their function, or if patterns of coloration evolved as a result of the light environment or display behaviors of a species. McNaught & Owens (2002) found that differentiation in sexual ornamentation among 40 avian species is strongly affected by features of habitat that influence signal transmission. Similarly, differences among habitats and geographical locations in food composition may affect diet-dependent components of sexual dichromatism. For example, geographical variation in intensity of red coloration among populations of the house finch (*Carpodacus mexicanus*) was influenced by local access to carotenoids (Hill et al. 2002).

PHYLOGENETIC STUDIES OF SEXUAL DICHROMATISM

Historical Patterns of Complexity of Sexual Ornamentation

A key starting point in studies of sexual dichromatism is an understanding of the signal content of plumage displays. This requires an examination of the source of plumage coloration (melanin, carotenoid, or structural) and the factors that might influence their displays. Such studies help researchers understand not just the proximate control of color ornaments but also the roles that developmental and phylogenetic constraints play in evolution of sexual dichromatism.

The examples in this section illustrate three points. First, different types of color display—carotenoid pigmentation, melanin pigmentation, and structural coloration—have different evolutionary lability. Second, knowledge of the phylogeny is essential for an understanding of the sequence of transitions in color patterns and ornament structure. Finally, color traits may differ in their detectability and in the information that they provide in a given environment, and these differences may bias the evolution of sexual dichromatism in such traits.

Hill & Badyaev (Badyaev & Hill 2000, Hill 1996) suggested that because carotenoid-based plumage coloration is more dependent on environment and less constrained developmentally than is melanin-based coloration, variation in sexual dichromatism should be driven more by changes in carotenoid coloration than by changes in melanin coloration. They found that across all cardueline finch species: (*a*) carotenoid-derived coloration has changed more frequently than melanin-based coloration; (*b*) in both sexes an increase in carotenoid-based coloration, but not in melanin-based coloration, was strongly associated with increase in sexual dichromatism; and, (*c*) sexual dichromatism in carotenoid-based coloration contributed more to overall dichromatism than sexual dichromatism in melanin-based plumage (Badyaev & Hill 2000, Hill 1996).

These findings supported the results of Gray's (1996) analyses of male plumage across North American passerines that the extent of carotenoid pigmentation in male plumage was positively associated with overall dichromatism, whereas the extent of melanin and structural coloration in male plumage was not related to overall dichromatism. Gray (1996) noted that carotenoids appear to be used as ornamental signals by granivorous and insectivorous taxa (for which carotenoids are present in the diet but not overly abundant) but rarely used by frugivorous (for which carotenoids are overly abundant in the diet) or carnivorous taxa (for which carotenoids are rare). Subsequently, Johnson & Lanyon (2000) showed that carotenoid-based ornaments are evolutionary labile in New World Icteridae such that transitions to greater carotenoid ornamentation closely followed historical shifts into different environments.

The similarity of coloration patterns and pigment distribution across a wide range of species within taxa suggests common developmental mechanisms and constraints. In their comprehensive study of the evolution of color patterns in *Phylloscopus* warblers, Price & Pavelka (1996) showed that components of melanin-based coloration were repeatedly gained and lost during evolution. They suggested that once a pattern of coloration evolved in a lineage it could persist even if it was not expressed phenotypically and in this way complex patterns of ornamentation could reappear (e.g., under hormonal control) when favored by selection (see below). Moreover, the colors and patterns that are currently expressed necessarily affect the development and evolution of components that are derived from the ornaments such as symmetry (Price & Pavelka 1996; see also Badyaev et al. 2001). In an analysis of hybrids of domesticated birds, Price (2002) documented a strong historical hierarchy of divergence in sexual ornamentation and high evolutionary lability of plumage color ornamentation. Omland & Lanyon (2000) also reported high evolutionary lability of plumage characteristics within the oriole genus *Icterus*. Among oriole genera, however, the patterns of plumage change provided evidence for developmental constraints. Both examples emphasize that identification of the evolutionary sequences of coloration patterns is essential to the study of sexual dichromatism.

Schluter & Price (1993) noted that sexual selection will favor ornamental traits with more sex-biased genetic and phenotypic variance, greater condition-dependence, and easier detection in a local environment. Under certain conditions, traits like song or behavioral displays will be more likely to invade a sexually dichromatic population and in this way bias the evolution of other sexually dimorphic traits. For example, predation may limit variation in sexual dichromatism in Parulinae warblers, and song complexity may replace plumage characteristics as the target of sexual selection (Shutler & Weatherhead 1990). Badyaev et al. (2002) examined the relationship between song and plumage elaborations in cardueline finches and found that across species song complexity was strongly negatively related to elaboration of plumage ornamentation. Moreover, when plumage coloration was partitioned into carotenoid-based and melanin-based components, song complexity was negatively related to elaboration of male carotenoid-based coloration but unrelated to elaboration of melanin-based coloration. The trade-off between carotenoid plumage and song complexity might be due to their high costs and environmental dependency (Badyaev & Leaf 1997; Snell-Rood & Badyaev, in review). Similarly, Bailey (1978) suggested that structural colors are favored by selection in the tropics because structural colors are easily changed by behavioral displays depending on variable light conditions in dark habitats.

Phylogenetic Inferences About the Origin of Sexual Dichromatism

Sexual dichromatism arises from sex-biased genetic expression or from selection acting on traits with sex-limited or sex-biased genetic and phenotypic variation (Lande 1980). Once sex-specific expression of ornamentation is established, variation in sexual dimorphism can be affected by various forces (Badyaev 2002) that can be revealed by phylogenetic methods. The sources of sex-biased expression of plumage ornamentation could range from mutations on sex chromosomes to sex-limited expression of genes. A majority of expression of sexually dimorphic ornaments, however, is owing to sex-biased expression of developmental programs that are shared between the sexes. Consequently, comparative studies often find little evidence of long-term constraints on the evolution of sexual ornamentation imposed by high between-sex genetic correlations. Phylogenetic studies of the developmental processes that enable sex-biased expression of shared organismal processes hold great potential to further our understanding of the evolution and current function of sexual displays (Badyaev 2003, Kimball & Ligon 1999, Reinhold 1999).

Expression of sex-biased plumage ornamentation often depends on sex-specific hormonal profiles (reviewed in Owens & Short 1995). In some avian groups, dull coloration develops in the presence of estrogen, whereas bright coloration develops in the absence of estrogen. In other groups, expression of sexual ornamentation is regulated by testosterone—bright coloration develops under the influence of testosterone, whereas dull coloration develops when circulating testosterone is absent (Kimball & Ligon 1999). Kimball & Ligon (1999) studied the hormonal control of plumage dimorphism in a phylogenetic context and concluded that estrogen-dependent dichromatism is ancestral and testosterone-dependent dichromatism is a derived state. They suggested that the most parsimonious evolutionary sequence for the evolution of sexual dichromatism was bright monomorphism followed by selection for duller coloration in one sex (Kimball & Ligon 1999). General implication of epigenetic control of sexual ornamentation is that we would predict (*a*) easier and faster loss than gain of male sexual ornamentation, and (*b*) more frequent phylogenetic transition from dichromatism to monochromatism than from monochromatism to dichromatism (e.g., Omland 1997, Price & Birch 1996). If sexual dichromatism results from mutations on sex chromosomes that are magnified by selection favoring dichromatism, no directional bias between loss and gain is expected.

Once sex-limitation is established, genetic drift, selection, and gene interactions could influence the evolution of sexual dichromatism. On a macroevolutionary scale, genetic drift is not expected to produce consistent associations across lineages between sexual dichromatism and factors such as ecological conditions (Leroi et al. 1994, Sheldon & Whittingham 1997). On the contrary, if sexual dichromatism evolved in response to selection, change in sexual dichromatism should follow specific sequences in which shifts to new environments or changes in behaviors are followed by transitions in plumage coloration (Sheldon & Whittingham 1997).

Phylogenetic Reconstructions of Plumage Dichromatism

PHYLOGENETIC RECONSTRUCTION OF TRANSFORMATIONS IN SEXUAL DICHROMATISM One way to explore biases in the evolution of sexual dichromatism is to examine the relative frequency of changes between monochromatism and dichromatism as well as differences in the rates of evolution of male and female ornamentation (e.g., Price & Birch 1996). Recent phylogenetic studies of avian coloration have revealed that losses of sexual ornaments are more common than gains, but that most of the transitions from dichromatism to monomorphism involve females gaining male-like ornamentation. This is surprising because sexual ornamentation is often assumed to be maintained by current sexual selection on male ornamentation (Wiens 2001).

Price & Birch (1996) estimated the frequency of evolutionary transitions in dichromatism across 5,298 species of passerines and found that sexual dichromatism evolved numerous times independently and that transitions from dimorphism to monomorphism were more likely than transition in the opposite direction. Omland (1997) reached similar conclusions in his study of ducks (Anatidae). He showed that sexual dichromatism is an ancestral trait and that the evolution of sexual dichromatism was biased toward loss of dichromatism. Similarly, in a phylogenetic study of 47 genera of tanagers (Thraupidae), Burns (1998) found that in males a transition from bright to dull coloration is five times more likely than a transition from dull to bright, and that tanagers descended from an ancestor that was dichromatic with colorful males and dull females. These findings are corroborated in a study by Peterson (1996) in which he examined geographical variation in sexual dichromatism in 158 species of birds representing 43 families. He concluded that sexual monomorphism with bright males and dull females is a likely ancestral stage in birds. Similarly, Kimball et al. (2001), on the basis of thorough molecular analyses, concluded that the two least-ornamented species of pheasants (Phasianidae) are the most derived, implying that the sexual dimorphism and elaborated ornamentation in this clade is an ancestral state.

Whereas several studies suggested that loss of male ornamentation can be favored by adaptive female preference (Badyaev & Qvarnström 2002, Qvarnström et al. 2000, Saetre et al. 1997), comparative studies reveal that loss and gain of sexual ornamentation often precedes changes in preference for sexual ornamentation (reviewed in Wiens 2001) emphasizing the role of genetic drift and developmental processes in the evolution of dichromatism (Lande 1981, see below).

PHYLOGENETIC RECONSTRUCTIONS OF TRANSFORMATIONS IN MALE AND FEMALE PLUMAGE Several phylogenetic studies have addressed relative changes in male and female ornamentation in relation to evolutionary lability of dichromatism (reviewed in Amundsen 2000). Peterson (1996) examined the relative frequency of "bright" and "dull" monomorphism and concluded that the evolution of female plumage contributed to the evolution of sexual dichromatism as frequently as did the evolution of male plumage. Males were much more likely to lose than to gain bright plumage, whereas in females the trend was the opposite. The fact that loss of sexual dichromatism occurs in both directions (to "dull" and to "bright" monomorphism) makes it unlikely that selection can explain the majority of cases, leading Peterson (1996) to propose genetic drift as the evolutionary force behind variation in sexual dichromatism. Similarly, Björklund (1991) documented that in two lineages of blackbirds, sexual dichromatism resulted from a loss of female coloration rather than a gain in male coloration (see also Burns 1998, Irwin 1994). These recent studies corroborate original observations that an association between plumage brightness and mating systems is mostly due to variation in female plumage (Moreau 1960). A recent study of waterfowl (Anseriformes), however, concluded that evolutionary changes in plumage ornamentation were more frequent in males than females, presumably due to greater sexual selection on male ornamentation (Figuerola & Green 2000).

Sexual Dichromatism in Relation to Mechanism of Sexual Selection

Phylogenetic analyses provide a powerful means of distinguishing between different mechanisms of sexual selection. First, hypotheses of the mechanisms of sexual selection can be tested by experimentally examining the congruence between current expression of sexual ornamentation and current preference for this ornamentation. Second, different models of selection make distinct predictions of diversification patterns, hierarchical complexity, and convergence among lineages, thus allowing insight into sexual selection mechanisms.

In runaway models of sexual selection, a genetic correlation develops between male ornamentation and female preference and greater expression leads to greater preference until either ornamentation or preference are limited by natural selection (Andersson 1994). Hill (1994a) proposed that the runaway models of sexual selection cannot account for reduction in sexual ornamentation in the absence of changes in female preferences or in viability costs. By examining these predictions in relation to geographic variation in male appearance and female preference across subspecies of the house finch, Hill concluded that the models of runaway mate choice can be rejected in this species. Later modifications of the runaway model showed that a cyclic gain and loss of female preferences could occur without changes in natural selection on female preference (Iwasa & Pomiankowski 1995).

Studies of bowerbirds (Ptilonorhynchidae) by Kusmierski et al. (1997) and manakins (Pipridae) by Prum (1997) showed that patterns of ornament distribution and differential evolutionary lability of ornaments could be used to uncover mechanisms of selection operating within a lineage. In the runaway model, drift along equilibria lines between the male ornamentation and the female preference produces periods of rapid evolution resulting in large-scale diversifications and elaboration of male sexual ornamentation (Lande 1980). Thus, the runaway model predicts rapid differentiation in sexual ornamentation and the evolution of multiple sexual traits among lineages. By this model there should be little convergence between lineages, but one should see a historically nested distribution of traits that are shared among lineages within a clade (Prum 1997). In bowerbirds, sexually dimorphic plumage characters were extremely labile and sexual dichromatism appeared to be largely unconstrained (Kusmierski et al. 1997). Similarly, Prum (1997) found that diversity of manakin displays was explosive, indicating that evolution of these traits is largely unconstrained. Patterns of diversification and hierarchical structure of displays within these lineages is most consistent with the predictions of runaway and sensory bias mechanisms (Endler 1992; see also Irwin 1996) and also may be consistent with phylogenetic predictions of the "chase-away" model of sexual selection (Holland & Rice 1998).

In contrast, evolution of multiple quality indicator traits is constrained because evolution of a new indicator would favor elimination of previous indicators (Hill 1994b, Iwasa & Pomiankowski 1994). Consequently, indicator models predict sequential evolution of increasingly informative and increasingly constrained sets of ornaments within lineages (Badyaev et al. 2002, Hill 1994b, Prum 1997). Johnson (1999) found support for this model in the transitions from more costly to less costly ornamental displays in dabbling ducks, as well as in the gains and loses in some sexual ornaments in relation to presence of other ornaments.

The "chase-away" process of sexual selection also predicts sequential evolution of more exaggerated traits. It also predicts that evolution should be accompanied by selection for retention of existing sexual ornaments. Sensory bias models predict frequent convergence in ornaments across lineages that have similar preexisting biases (Andersson 1994, Endler 1992, Ryan 1990). In addition, the sensory drive hypothesis predicts convergence of preferences and ornaments across lineages with similar ecological conditions (Hill 1994b, Prum 1997, Boughman 2002). Similarly, if sexual traits evolve to minimize the costs associated with mate sampling and selection, strong convergences in sexual traits among lineages that share similar ecological conditions are expected (Price 1998, Prum 1997, Schluter & Price 1993). Finally, direct selection for species recognition should favor displays that are unique and should select against similar ornaments among lineages, thus resulting in decreased ornament diversity and their reduced hierarchical structure within a lineage (Grant & Grant 1997, Prum 1997).

Virtually all of the studies that we have reviewed used field guides or study skins and assessments by human observers to rank or score the plumage coloration of individuals and to determine the similarities and differences between the sexes. One criticism that could be leveled at these studies of sexual dichromatism is that assessments of dichromatism that are made with human visual systems ignore the UV component of coloration (Bennett et al. 1994). All diurnal birds tested to date perceive UV light and many plumage color displays, particularly color displays that appear violet or blue to human observers, have a substantial UV component (Cuthill et al. 2000). Failure to consider UV coloration could lead to the misclassification of some bird species as monochromatic when in fact they are dichromatic. For example, recent studies in which plumage coloration was measured with a reflectance spectrometer showed that there are substantial differences between the sexes in UV coloration in some species (Andersson et al. 1998, Hunt et al. 1998). The fact that so many interesting patterns related to sexual dichromatism have been revealed in comparative studies that ignore the UV portion of the spectrum suggests that the visible portion of the spectrum must, in many cases, be a reasonable approximation of the overall coloration and dimorphism of a species. At the same time, the revelation that substantial variation in plumage coloration and dimorphism might be missed in studies that ignore the UV component of color displays raises the intriguing possibility that patterns in comparative studies may become clearer and indeed new patterns and explanations might be uncovered if the UV component of color displays is considered in future studies.

We have emphasized in this review that to advance our understanding of sexual dichromatism in birds there is a pressing need to take a historical approach when considering the proximate mechanisms behind sexual dichromatism, life history variation in relation to sexual selection, and patterns of mate choice. Studies that have attempted such syntheses (e.g., Kimball & Ligon 1999, Price 2002, Price & Pavelka 1996) have produced powerful insights into the evolution of sexual ornamentation and sexual dichromatism in birds. Moreover we need more experimental studies of mate choice to be carried in the phylogenetic context with special focus on temporal correspondence between male ornamentation and female preference for such ornamentation (Wiens 2001, Hill 1994a). Phylogenetic reconstructions will continue to play a central role in uncovering the function and evolution of sexual dichromatism.

ACKNOWLEDGMENTS

We thank the National Science Foundation (DEB-0075388, DEB-0077804, IBN-9722171) for funding this work.

LITERATURE CITED

Amundsen T. 2000. Why are female birds ornamented? *TREE* 15:149–55

Andersson M. 1994. *Sexual Selection*. Princeton: Princeton Univ. Press

Andersson S, Ornborg J, Andersson, M. 1998. Ultraviolet sexual dimorphism and assortative mating in blue tits. *Proc. R. Soc. London Ser. B* 265:445–50

Badyaev AV. 1997a. Altitudinal variation in sexual dimorphism: A new pattern and alternative hypotheses. *Behav. Ecol.* 8:675–90

Badyaev AV. 1997b. Avian life history variation along altitudinal gradients: an example with cardueline finches. *Oecologia* 111:365–74

Badyaev AV. 1997c. Covariation between life history and sexually selected traits: an example with cardueline finches. *Oikos* 80:128–38

Badyaev AV. 2002. Growing apart: an ontogenetic perspective on the evolution of sexual size dimorphism. *TREE* 17:369–78

Badyaev AV. 2003. Integration and modularity in the evolution of sexual ornaments: an overlooked perspective. In *The Evolutionary Biology of Complex Phenotypes*, ed. M Pigliucci, K Preston. Oxford: Oxford Univ. Press

Badyaev AV, Ghalambor CK. 1998. Does a trade-off exist between sexual ornamentation and ecological plasticity? Sexual dichromatism and occupied elevational range in finches. *Oikos* 82:319–24

Badyaev AV, Ghalambor CK. 2001. Evolution of life histories along elevational gradients: Evidence for a trade-off between parental care and fecundity in birds. *Ecology* 82:2948–60

Badyaev AV, Hill GE. 2000. Evolution of sexual dichromatism: Contribution of carotenoid- versus melanin-based coloration. *Biol. J. Linn. Soc.* 69:153–72

Badyaev AV, Hill GE, Dunn PO, Glen JC. 2001. Plumage color as a composite trait: Developmental and functional integration of sexual ornamentation. *Am. Nat.* 158:221–35

Badyaev AV, Hill GE, Weckworth BV. 2002. Species divergence in sexually selected traits: Increase in song elaboration is related to decrease in plumage ornamentation in finches. *Evolution* 56:412–19

Badyaev AV, Leaf ES. 1997. Habitat associations of song characteristics in *Phylloscopus* and *Hippolais* warblers. *Auk* 114:40–46

Badyaev AV, Qvarnström A. 2002. Putting sexual traits into the context of an organism: a life-history perspective in studies of sexual selection. *Auk* 119:301–10

Badyaev AV, Snell-Rood EC. 2003. Rapid evolutionary divergence of environment-dependent sexual traits in speciation: A paradox? In *Proc. XXIII Int. Ornith. Cong. Acta Zool. Sin., Beijing*, ed. WJ Bock, R Schodde, pp. 1–19

Bailey, SF. 1978. Latitudinal gradients in colours and patterns of passerine birds. *Condor* 80:372–81

Baker RR, Parker GA. 1979. The evolution of bird coloration. *Philos. Trans. R. Soc. London Ser. B* 287:65–130

Barraclough TG, Harvey PH, Lee S. 1995. Sexual selection and taxonomic diversity in passerine birds. *Proc. R. Soc. London Ser. B* 259:211–15

Bennett ATD, Cuthill IC, Norris KJ. 1994. Sexual selection and the mismeasure of color. *Am. Nat.* 144:848–60

Bennett PM, Owens IPF. 2002. *Evolutionary Ecology of Birds: Life Histories, Mating Systems and Extinction.* Oxford: Oxford Univ. Press

Björklund M. 1990. A phylogenetic interpretation of sexual dimorphism in body size and ornament in relation to mating system in birds. *J. Evol. Biol.* 3:171–83

Björklund M. 1991. Evolution, phylogeny, sexual dimorphism, mating system in the grackles *Quiscalus*-spp Icterinae. *Evolution* 45:608–21

Bleiweiss R. 1992. Widespread polychromatism in female sunangel hummingbirds (Heliangelus: Trochilidae). *Biol. J. Linn. Soc.* 45:291–314

Bleiweiss R. 1997. Covariation of sexual dichromatism and plumage colours in lekking and non-lekking birds: A comparative analysis. *Evol. Ecol.* 11:217–35

Boughman JW. 2002. How sexual drive can promote speciation. *TREE* 17:571–77

Brush AH. 1970. Pigments in hybrid, variant and melanistic tanagers (birds). *Comp. Biochem. Physiol.* 36:785–93

Burke T, Griffith SC, Smith P. 1998. Identifying the factors driving extra-pair paternity in passerine birds: island populations are more monogamous, but is the reason genetic? *Proc. 7th Int. Behav. Ecol. Congr.,* Pacific Grove, CA

Burns KJ. 1998. A phylogenetic perspective on the evolution of sexual dichromatism in tanagers (Thraupidae): The role of female versus male plumage. *Evolution* 52:1219–24

Burtt EH Jr. 1986. An analysis of physical, physiological, and optical aspects of avian coloration with emphasis on wood-warblers. *Ornithol. Monogr.* 38:1–126

Butcher GS, Rohwer S. 1993. The evolution of conspicuous and distinctive coloration for communication in birds. *Curr. Ornithol.* 6:61–108

Cott JB. 1947. The edibility of birds: illustrated by five years' experiments and observations (1941–1946) on the food preferences of the hornet, cat and man; and considered with special reference to the theories of adaptive colouration. *Proc. Zool. Soc. London* 116:371–524

Crook JH. 1964. The evolution of social organization and visual communication in the weaver birds (Ploceinae). *Behaviour* Suppl. 10

Cuervo JJ, Møller AP. 1999. Ecology and evolution of extravagant feather ornaments. *J. Evol. Biol.* 12:986–98

Cuthill IC, Partridge JC, Bennett ATD. 2000. Avian UV vision and sexual selection. In *Animal Signals: Signaling and Signal Design in Animal Communication,* ed. Y Epsmark, T Amundsen, G Rosenqvist, pp. 61–82. Trondheim, Nor: Tapir Acad.

Darwin C. 1871. *The Descent of Man and Selection in Relation to Sex.* London: John Murray

Dunn PO, Whittingham LA, Pitcher TE. 2001. Mating systems, sperm competition, and the evolution of sexual dimorphism in birds. *Evolution* 55:161–75

Endler JA. 1992. Signals, signal conditions, and the direction of evolution. *Am. Nat.* 139:S125–53

Endler JA, Théry M. 1996. Interacting effects of lek placement, display behavior, ambient light, and colour patterns in three neotropical forest-dwelling birds. *Am. Nat.* 148:421–52

Endler JA, Wescott D. 1998. Ornaments, microhabitat choice, and microhabitat construction. *Proc. 7th Int. Behav. Ecol. Congr., Pacific Grove, CA*

Figuerola J, Green AJ. 2000. The evolution of sexual dimorphism in relation to mating patterns, cavity nesting, insularity and sympatry in the Anseriformes. *Funct. Ecol.* 14:701–10

Fisher RA. 1930. *The Genetical Theory of Natural Selection.* Oxford: Clarendon

Fitzpatrick S. 1994. Colourful migratory birds: Evidence for a mechanism other than parasite resistance for the maintenance of 'good genes' sexual selection. *Proc. R. Soc. London Ser. B* 257:155–60

Fitzpatrick S. 1998. Birds' tails as signaling devices: Markings, shape, length, and feather quality. *Am. Nat.* 151:157–73

Gavrin MC, Remsen JV. 1997. An alternative hypothesis for heavier parasite load of brightly colored birds: exposure at the nest. *Auk* 114:179–91

Götmark F. 1994. Are bright birds distasteful? A reanalysis of H.B. Cott's data on the edibility of birds. *J. Avian Biol.* 25:184–97

Götmark F. 1999. The importance of non-reproductive functions of bird colouration, especially anti-predator adaptations: a review. *Proc. XXII Int. Ornithol. Congr.*, pp. 1706–18. Durban, So. Afr.: Univ. Natal

Götmark F, Post P, Olsson AJ, Himmelmann D. 1997. Natural selection and sexual dimorphism: sex-biased sparrowhawk predation favours crypsis in female chaffinches. *Oikos* 80:540–48

Grant PR. 1965. Plumage and evolution of birds on islands. *Syst. Zool.* 14:47–52

Grant PR, Grant BR. 1997. Genetics and the origin of bird species. *Proc. Natl. Acad. Sci. USA* 94:7768–75

Gray DA. 1996. Carotenoids and sexual dichromatism in North American passerine birds. *Am. Nat.* 148:453–80

Hamilton TH. 1961. On the functions and causes of sexual dimorphism in breeding plumage characteristics of North American species of warblers and orioles. *Am. Nat.* 45:121–23

Hill GE. 1994a. Geographic variation in male ornamentation and female preferences in the house finch: a comparative test of models of sexual selection. *Behav. Ecol.* 5:64–73

Hill GE. 1994b. Trait elaboration via adaptive mate choice: sexual conflict in the evolution of signals of male quality. *Ethol. Ecol. Evol.* 6:351–70

Hill GE. 1996. Redness as a measure of the production cost of ornamental coloration. *Ethol. Ecol. Evol.* 8:157–75

Hill GE, Inouye CY, Montgomerie R. 2002. Dietary carotenoids predict plumage coloration in wild house finches. *Proc. R. Soc. London Ser. B* 269:1119–24

Höglund J. 1989. Size and plumage dimorphism in lek-breeding birds: A comparative analysis. *Am. Nat.* 134:72–87

Holland B, Rice WR. 1998. Chase-away sexual selection: Antagonistic seduction versus resistance. *Evolution* 52:1–7

Hunt S, Bennett ATD, Cuthill IC, Griffiths R. 1998. Blue tits are ultraviolet tits. *Proc. R. Soc. London Ser. B* 265:451–55

Irwin RE. 1994. The evolution of plumage dichromatism in the New World blackbirds: social selection on female brightness? *Am. Nat.* 144:890–907

Irwin RE. 1996. The phylogenetic content of avian courtship display and song evolution. In *Phylogenies and the Comparative Method in Animal Behavior*, ed. EP Martins, pp. 234–52 New York: Oxford Univ. Press

Iwasa Y, Pomiankowski A. 1994. The evolution of mate preferences for multiple sexual ornaments. *Evolution* 48:853–67

Iwasa Y, Pomiankowski A. 1995. Continual change in mate preferences. *Nature* 377:420–422

Johnson K. 1988. Sexual selection in pinyon jays II: Male choice and female-female competition. *Anim. Behav.* 36:1048–53

Johnson KP. 1999. The evolution of bill coloration and plumage dimorphism supports the tranference hypothesis in dabbling ducks. *Behav. Ecol.* 10:63–67

Johnson KP, Lanyon SM. 2000. Evolutionary changes in color patches of blackbirds are associated with marsh nesting. *Behav. Ecol.* 11:515–19

Johnson SG. 1991. Effects of predation, parasites, and phylogeny on the evolution of bright coloration in North American male passerines. *Evol. Ecol.* 5:52–62

Kear J. 1970. Adaptive radiation of parental care in waterfowl. In *Social Behavior in Birds and Mammals: Essays on the Social Ethology of Animals and Man*, ed. JH Crook. pp. 357–92. New York: Academic

Kimball RT, Braun EL, Ligon JD, Lucchini V, Randi E. 2001. A molecular phylogeny of the peacock-pheasants (Galliformes: Polyplectron spp.) indicates loss and reduction of ornamental traits and display behaviors. *Biol. J. Linn. Soc.* 73:187–198

Kimball RT, Ligon JD. 1999. Evolution of avian plumage dichromatism from a proximate perspecitve. *Am. Nat.* 154:182–193

Kirkpatrick M. 1987. Sexual selection by female choice in polygynous animals. *Annu. Rev. Ecol. Syst.* 18:43–70

Kirkpatrick M, Ryan MJ. 1991. The evolution of mating preferences and the paradox of lek. *Nature* 350:33–38

Kusmierski R, Borgia G, Uy A, Crozier RH. 1997. Labile evolution of display traits in bowerbirds indicates reduced effects of phylogenetic constraints. *Proc. R. Soc. London Ser. B* 264:307–11

Lande R. 1980. Sexual dimorphism, sexual selection and adaptation in polygenic characters. *Evolution* 34:292–305

Lande R. 1981. Models of speciation by sexual selection on polygenic traits. *Proc. Natl. Acad. Sci. USA* 78:3721–25

Leroi AM, Rose MR, Lauder GV. 1994. What does the comparative method reveal about adaptation? *Am. Nat.* 143:381–402

Marchetti K. 1993. Dark habitats and bright birds illustrate the role of the environment in species divergence. *Nature* 362:149–52

Martin TE. 1996. Life history evolution in tropical and south temperate birds: What do we really know? *J. Avian Biol.* 27:263–72

Martin TE, Badyaev AV. 1996. Sexual dichromatism in birds: Importance of nest predation and nest location for females versus males. *Evolution* 50:2454–60

Martin TE, Clobert J. 1996. Nest predation and avian life history evolution in Europe versus North America: a possible role of humans. *Am. Nat.* 147:1028–46

Mayr E. 1942. *Systematics and the Origin of Species.* New York: Dover

McLain DK. 1993. Cope's rules sexual selection and the loss of ecological plasticity. *Oikos* 68:490–500

McLain DK, Moulton MP, Redfearn TP. 1995. Sexual selection and the risk of extinction of introduced birds on oceanic islands. *Oikos* 74:27–34

McNaught MK, Owens IPF. 2002. Interspecific variation in plumage colour among birds: species recognition or light environment? *J. Evol. Biol.* 15:505–14

Møller AP. 1986. Mating systems among European passerines: A review. *Ibis* 128:234–50

Møller AP, Birkhead TR. 1993. Cuckoldry and sociality: A comparative study of birds. *Am. Nat.* 142:118–40

Møller AP, Birkhead TR. 1994. The evolution of plumage brightness in birds is related to extrapair paternity. *Evolution* 48:1089–1100

Møller AP, Cuervo JJ. 1998. Speciation and feather ornamentation in birds. *Evolution* 52:859–69

Moreau RE. 1960. Conspectus and classification of the Ploceinae weaverbirds. *Ibis* 102:298–321

Omland KE. 1997. Examining two standard assumptions of ancestral reconstructions: Repeated loss of dichromatism in dabbling ducks (Anatini). *Evolution* 51:1636–46

Omland KE, Lanyon SM. 2000. Reconstructing plumage evolution in orioles (*Icterus*): Repeated convergence and reversal in patterns. *Evolution* 54:2119–33

Owens IPF, Bennett PM. 1994. Mortality costs of parental care and sexual dimorphism in birds. *Proc. R. Soc. London Ser. B* 257:1–8

Owens IPF, Bennett PM. 1995. Ancient ecological diversification explains life-history variation among living birds. *Proc. R. Soc. London Ser. B* 261:227–32

Owens IPF, Bennett PM. 1997. Variation in mating systems among birds: ecological basis revealed by hierarchical comparative analysis of mate desertion. *Proc. R. Soc. London Ser. B* 264:1103–10

Owens IPF, Clegg SM. 1999. Species-specific sexual plumage: species-isolating mechanisms or sexually selected ornaments? *Proc.22th Int. Congr.*, pp. 1141–53

Owens IPF, Hartley IR. 1998. Sexual dimorphism in birds: Why are there so many different forms of dimorphism? *Proc. R. Soc. London Ser. B* 265:397–407

Owens IPF, Short RV. 1995. Hormonal basis of sexual dimorphism in birds: implications for new theories of sexual selection. *TREE* 10:44–47

Payne RB. 1984. Sexual selection, lek and arena behavior, and sexual dimorphism in birds. *Ornithol. Monogr.* 33:1–52

Peterson AT. 1996. Geographic variation in sexual dichromatism in birds. *Bull. Br. Ornithol. Club* 116:156–72

Petrie M, Kempenaers B. 1998. Extra-pair paternity in birds: explaining variation between species and populations. *TREE* 13:52–58

Price T. 1996. An association of habitat with color dimorphism in finches. *Auk* 113:256–57

Price T. 1998. Sexual selection and natural selection in bird speciation. *Philos. Trans. R. Soc. London Ser. B* 353:251–60

Price T. 2002. Domesticated birds as a model for the genetics of speciation by sexual selection. *Genetica* 116:311–27

Price T, Birch GL. 1996. Repeated evolution of sexual color dimorphism in passerine birds. *Auk* 113:842–48

Price T, Pavelka M. 1996. Evolution of a colour pattern: History, development, and selection. *J. Evol. Biol.* 9:451–70

Prinzing A, Brandle M, Pfeifer R, Brandl R. 2002. Does sexual selection influence population trends in European birds? *Evol. Ecol. Res.* 4:49–60

Promislow D, Montgomerie R, Martin TE. 1994. Sexual selection and survival in North-American waterfowl. *Evolution* 48:2045–50

Promislow D, Montgomerie R, Martin TE. 1992. Mortality costs of sexual dimorphism in birds. *Proc. R. Soc. London Ser. B* 250:143–50

Prum R. 1997. Phylogenetic tests of alternative intersexual selection mechanisms: trait macroevolution in a polygynous clade (Aves: Pipridae). *Am. Nat.* 149:668–92

Prum R. 1999. Structural colours of birds: Anatomy, production mechanisms, and evolution. *Proc. 22nd Int. Ornithol. Congr.*, Durban, So. Afr.: Univ. Natal

Qvarnström A, Part T, Sheldon BC. 2000. Adaptive plasticity in mate preference linked to differences in reproductive effort. *Nature* 405:344–47

Reinhold K. 1999. Evolutionary genetics of sex-limited traits under fluctuating selection. *J. Evol. Biol.* 12:897–902

Saetre G-P, Moum T, Bures S, Kral M, Adamjan M, Moreno J. 1997. A sexually selected character displacement in flycatchers reinforces premating isolation. *Nature* 387:589–92

Savalli UM. 1995. The evolution of bird coloration and plumage elaboration: A review of hypothesis. *Curr. Ornithol.* 12:141–90

Schluter D, Price T. 1993. Honesty, perception and population divergence in sexually selected traits. *Proc. R. Soc. London Ser. B* 253:117–22

Scott DK, Clutton-Brock TH. 1989. Mating systems, parasites and plumage dimorphism in waterfowl. *Behav. Ecol. Sociobiol.* 26:261–73

Sheldon FH, Whittingham LA. 1997. Phylogeny in studies of bird ecology, behavior, and morphology. In *Avian Molecular Evolution and Systematics*, ed. DP Mindell, pp. 279–99. New York: Academic

Shutler D, Weatherhead PJ. 1990. Targets of sexual selection: song and plumage of wood warblers. *Evolution* 44:1967–77

Slagsvold T, Lifjeld J. 1997. Incomplete knowledge of male quality may explain variation in extra-pair paternity in birds. *Behaviour* 134:353–71

Snell-Rood EC, Badyaev AV. 2003. Ecological gradient of sexual selection: elevation and song complexity in finches. *Behav. Ecol.* In press.

Soler JJ, Møller AP, Soler M. 1998. Nest building, sexual selection and parental investment. *Evol. Ecol.* 12:427–41

Sorci G, Møller AP, Clobert J. 1998. Plumage dichromatism of birds predicts introduction success in New Zealand. *J. Anim. Ecol.* 67:263–69

Trail PW. 1990. Why should lek-breeders be monomorphic? *Evolution* 44:1837–52

Verner J, Willson MF. 1969. Mating systems,

sexual dimorphism and the role of male North American passerine birds in the nesting cycle. *Ornithol. Monogr.* 9:1–76
Wallace AR. 1889. *Darwinism*. London: Macmillan
Webster MS. 1992. Sexual dimorphism, mating system and body size in New World blackbirds (Icterinae). *Evolution* 46:1627–41
West-Eberhard MJ. 1983. Sexual selection, social competition, and speciation. *Q. Rev. Biol.* 58:155–83
Westneat DF, Sherman PW. 1997. Density and extra-pair fertilizations in birds: a comparative analysis. *Behav. Ecol. Sociobiol.* 41:205–15
Wiens J. 2001. Widespread loss of sexually selected traits: how the peacock lost its spots. *TREE* 16:517–23

Annu. Rev. Ecol. Evol. Syst. 2003. 34:51–69
doi: 10.1146/annurev.ecolsys.34.121101.153549

First published online as a Review in Advance on July 8, 2003

PALEOBIOGEOGRAPHY: The Relevance of Fossils to Biogeography

Bruce S. Lieberman
Department of Geology and Department of Ecology and Evolutionary Biology, University of Kansas, Lawrence, Kansas 06045; email: blieber@ku.edu

Key Words biogeography, geo-dispersal, vicariance, phylogenetics, extinction

■ **Abstract** Paleobiogeography has advanced as a discipline owing to the increasing utilization of a phylogenetic approach to the study of biogeographic patterns. Coupled with this, there has been an increasing interdigitation of paleontology with molecular systematics because of the development of techniques to analyze ancient DNA and because of the use of sophisticated methods to utilize molecules to date evolutionary divergence events. One pervasive pattern emerging from several paleontological and molecular analyses of paleobiogeographic patterns is the recognition that repeated episodes of range expansion or geo-dispersal occur congruently in several different lineages, just as congruent patterns of vicariance also occur in independent lineages. The development of new analytical methods based on a modified version of Brooks Parsimony Analysis makes it possible to analyze both geo-dispersal and vicariance in a phylogenetic context, suggesting that biogeography as a discipline should focus on the analysis of a variety of congruent phenomena, not just vicariance. The important role that extinction plays in influencing apparent biogeographic patterns among modern and fossil groups suggests that this is another area ripe for new methodological developments.

INTRODUCTION

Biogeography is a dynamic discipline that has led to expansive growth in our understanding of the evolution and distribution of the Earth's biota. Not restricted to one research school or mode of thought, it is organized into two subdisciplines aimed at probing patterns in the genealogical and economic hierarchies: phylogenetic and ecological biogeography, respectively (Brooks & McLennan 1991, 2002; Lieberman 2000, 2003a). The analysis of biogeographic patterns in fossil taxa, paleobiogeography, is also an important area of research (Hallam 1994, Lieberman 2000, Brooks & McLennan 2002). There have been recent important gains and developments in paleobiogeography. These have occurred in three broad thematic areas. The first involves the appearance of new studies of fossil groups that incorporate a greater degree of analytical rigor; joined to this theme has been a growing unification of molecular evolution with paleontology, representing a rapprochement between these two fields. The second thematic area involves methodological developments generated specifically for the analysis of fossil data. Concomitant with these methodological developments, paleobiogeographers recognized that range expansion can occur congruently in several different groups and can be studied in a phylogenetic context. The final thematic area is associated with the recognition that extinction has left an indelible stamp on the modern biota, potentially challenging our ability to retrieve biogeographic patterns among organisms. The developments in each of these thematic areas suggests the need for a deeper consideration of the fossil record and its relevance to biogeography.

IMPORTANT PALEOBIOGEOGRAPHIC STUDIES

Most work in paleobiogeography has emphasized phylogenetic rather than ecological patterns. However, because ecological biogeography is such a rich discipline (e.g., Brown & Maurer 1989, Rosenzweig 1995, Brown & Lomolino 1998, Maurer 1999), it is anticipated that ecological paleobiogeographic studies are likely to represent an area of future growth for the field (Lieberman 2000). Still, the contributions of phylogenetic paleobiogeographic studies have been substantial.

Early Contributions

Fossils have long been an important factor in our understanding of the distribution of ancient continents and to the recognition that the Earth's plates have been in continuous motion (Hallam 1967, 1981). One of the real visionaries in this area is Tony Hallam (e.g., Hallam 1977, 1983, 1994). Paleobiogeographic studies have also demonstrated that some modern continents are built up of a melange of different terranes (Ross & Ross 1985, Hallam 1986). One example is North America, which appears stable and unitary but in fact has a mixture of different terranes along its western margin. Many of these different terranes may have originally contained different biotas that traveled great distances (Hallam 1986, Ward et al. 1997). Thus, it is conceivable that some elements of the modern terrestrial biota of North America may represent forms that arrived rafted on terranes. However, thus far, biogeographic patterns from such groups as modern land snails (e.g., Scott 1997) suggest this may not be the case.

Phylogenetic Approaches

Building on these early contributions, paleobiogeography has in the past decade entered a phase where the underlying data and framework are explicitly phylogenetic, and the analytical methods used rely on phylogenetic approaches to biogeography. Diverse studies indicate the existence of one fundamental pattern: an oscillation between episodes of vicariance and episodes of range expansion throughout the history of life. Some of the most important paleobiogeographic studies have focused on dinosaurs. For instance, Sereno et al. (1996) documented biogeographic patterns in the largest, most awesome terrestrial predators to have stalked the earth:

The patterns showed oscillating cycles of vicariance (in the Early Jurassic), followed by range expansion (in the Early Cretaceous), followed by vicariance (in the Late Cretaceous). Other well-known dinosaurs, such as the ceratopsians, have similar patterns (Sereno 1997). The early episodes of vicariance are associated with the breakup of Pangea during the Jurassic, and some of the strongest evidence for this comes from the vicariant links between dinosaurian (and also mammalian) faunas of Africa, South America, and Madagascar (Krause et al. 1997, Sampson et al. 1998).

Although the patterns of vicariance are important, they are not as prominent as the congruent range expansions (Sereno 1999). Although some (e.g., Upchurch et al. 2002) have argued for the dominant role of vicariance in dinosaur paleobiogeography, there are significant concerns about the quality and quantity of the data they utilized (Sereno 1997). The recognition of congruent, temporally correlated range expansion in independent clades matches the phenomenon of geo-dispersal identified by Lieberman & Eldredge (1996) and Lieberman (1997, 2000). Lieberman & Eldredge (1996) coined the term geo-dispersal to distinguish it from traditional dispersal [sensu (Humphries & Parenti 1986)] that involves individual species moving over geographic barriers. Such traditional dispersal is by its very nature incongruent and not replicated in independent clades. The episodes of congruent range expansion or geo-dispersal appear to have been associated with large-scale plate tectonic events, such as continent-continent collisions, which facilitate concurrent range expansion by species in many independent clades because they eliminate geographic barriers. Wholesale climatic changes can also effectively do the same thing.

Analyses of such groups of fossil vertebrates as the mammals reiterate the pattern recovered from dinosaurs: Again, the mammals show evidence of congruent episodes of vicariance, followed by geo-dispersal, followed by vicariance. One of the pioneers in this area was McKenna (1975, 1983), who recognized that lineages in several independent clades of mammals moved between Europe and North America and between North America and Asia throughout the Cenozoic. Beard (1998) identified three consecutive waves in the Late Paleocene and Early Eocene from Asia into North America, involving the movement of primates, rodents, perissodactyls, artiodactyls, and other groups. It appears that climate changes, especially warming events, rather than tectonic events per se (which may have been the predominant factors for the earlier dinosaurs) caused these geo-dispersal events (Beard 1998, 2002; Bowen et al. 2002). Analysis of modern groups provides more evidence for repeated episodes of vicariance and geo-dispersal involving the biotas of North America, Asia, and Europe (Sanmartin et al. 2001).

Murray (2001), working with cichlid fishes, provides another intriguing study of vertebrate paleobiogeography that used a phylogenetic approach. By augmenting modern distributions of cichlids with those of fossil taxa, and mapping these distributions parsimoniously onto a phylogeny, she found evidence of numerous episodes of Cenozoic range expansion (Figure 1). This result is supported by the molecular clock studies of Vences et al. (2001).

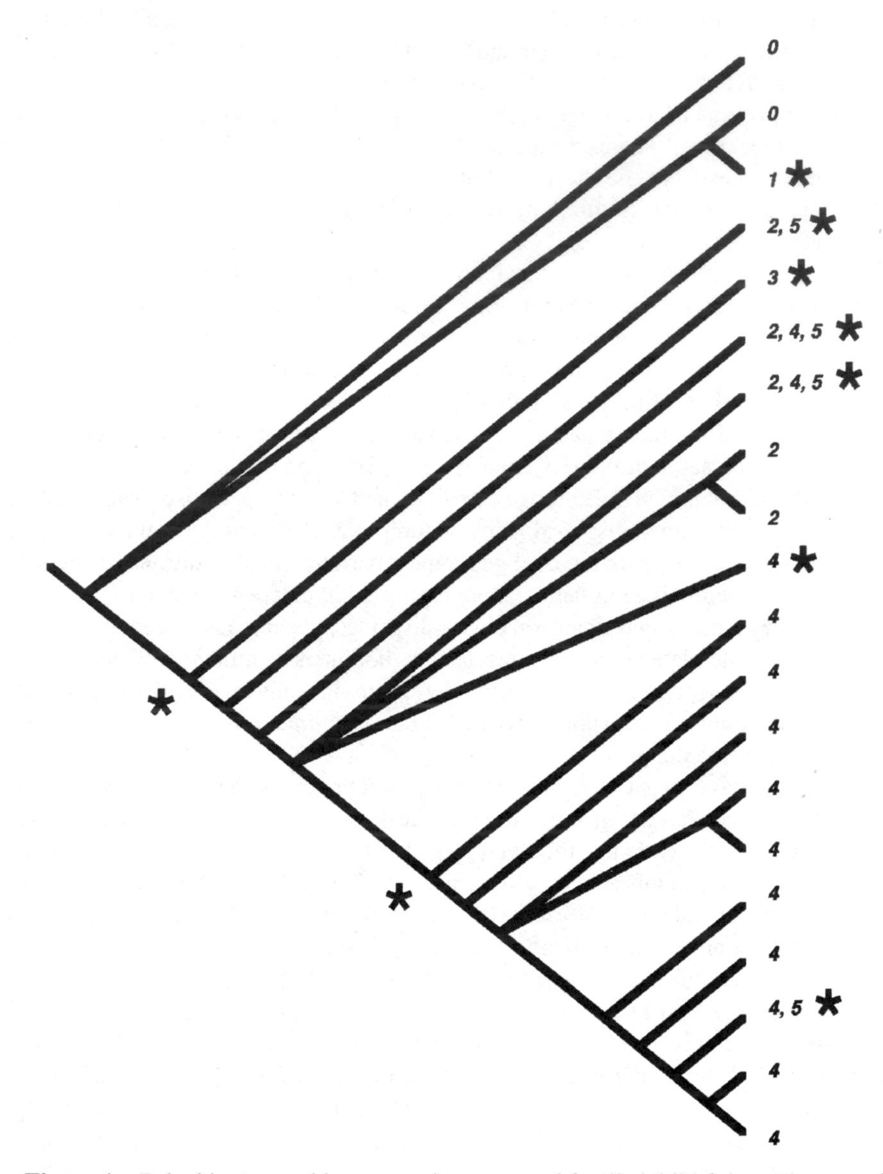

Figure 1 Paleobiogeographic patterns in extant and fossil cichlid fishes. The area of occurrence is substituted for the taxon name, with the area cladogram modified from Murray (2001), where 0 = Madagascar; 1 = India; 2 = West Africa; 3 = North America and South America; 4 = East Africa; and 5 = North Africa, Europe, and the Middle East. Mapping the geographic states to the ancestral nodes using phases one and two of Fitch (1971) optimization suggests that there were several range expansion events (each denoted by *) during the evolutionary history of the group.

Phylogenetic approaches to paleobiogeography have not only focused on the charismatic vertebrates. Several important studies have examined paleobiogeographic patterns of fossil arthropods. Grimaldi (1992) used insect fossils to test hypotheses of vicariance. Hasiotis (1999) considered how the breakup of Pangea influenced paleobiogeographic patterns of fossil freshwater crayfish. Rode & Lieberman (2002) evaluated the paleobiogeography of Devonian crustaceans and found evidence for vicariance and range expansion related to tectonic collision and sea-level rise and fall.

Phylogenetic paleobiogeographic analyses of the diverse marine invertebrate trilobites have also revealed important insights. An early study of these was conducted by Fortey & Cocks (1992). Although they lacked detailed trilobite phylogenies to consider paleobiogeographic patterns, their study used a parsimony analysis of endemism (PAE). (See Bisconti et al. 2001 for an innovative application of PAE to modern organisms.) Ebach & Edgecombe (2001) used trilobites to evaluate paleobiogeographic patterns while considering different analytical methods. Lieberman & Eldredge (1996) applied a modified version of Brooks Parsimony Analysis (BPA), a method that is described in greater detail below, to five clades of Devonian trilobites. They recognized significant evidence of congruent episodes of vicariance and geo-dispersal. Patterns of vicariance and patterns of geo-dispersal can be expressed as trees that show, respectively, the relative times that geographic barriers formed, separating regions and encouraging vicariance, and fell, joining regions and encouraging geo-dispersal (Figure 2). The two trees are actually fairly similar for Devonian trilobites, suggesting that the processes controlling vicariance also governed geo-dispersal. This implicates repeated cycles of sea-level rise and fall as the arbiter of trilobite paleobiogeographic patterns. These cycles occurred extensively during the Devonian. Moreover, the Middle Devonian was a time of extensive continental collision, with the world's cratons approaching a Pangea-like configuration (Scotese 1997).

Using a similar approach, Lieberman (2002a; 2003b,c) conducted a phylogenetic paleobiogeographic analysis of Early Cambrian trilobites. Again, as with the Middle Devonian, the vicariance tree was well resolved, suggesting an important earth history control of evolution, with the formation of geographic barriers owing to tectonic rifting and sea-level fall leading to speciation. However, unlike in the Middle Devonian, the Early Cambrian geo-dispersal tree is relatively poorly resolved (Figure 3). This makes sense, as the Early Cambrian was a time when the Earth's continents were splitting apart (Torsvik et al. 1996, Dalziel 1997). The excessive vicariance in the Early Cambrian may have elevated speciation rates during the Cambrian radiation (Lieberman 1997), a time of rapid evolution when abundant skeletonized animals first appeared in the fossil record. By contrast, in the Devonian, rates of speciation of trilobites (and other groups) were more muted, possibly owing to the pervasive opportunities for range expansion at that time. The fact that clades of trilobites from both time periods reveal abundant evidence of paleobiogeographic congruence suggests that such Earth history events as tectonic and climatic changes exercise a fundamental control on evolution. Furthermore,

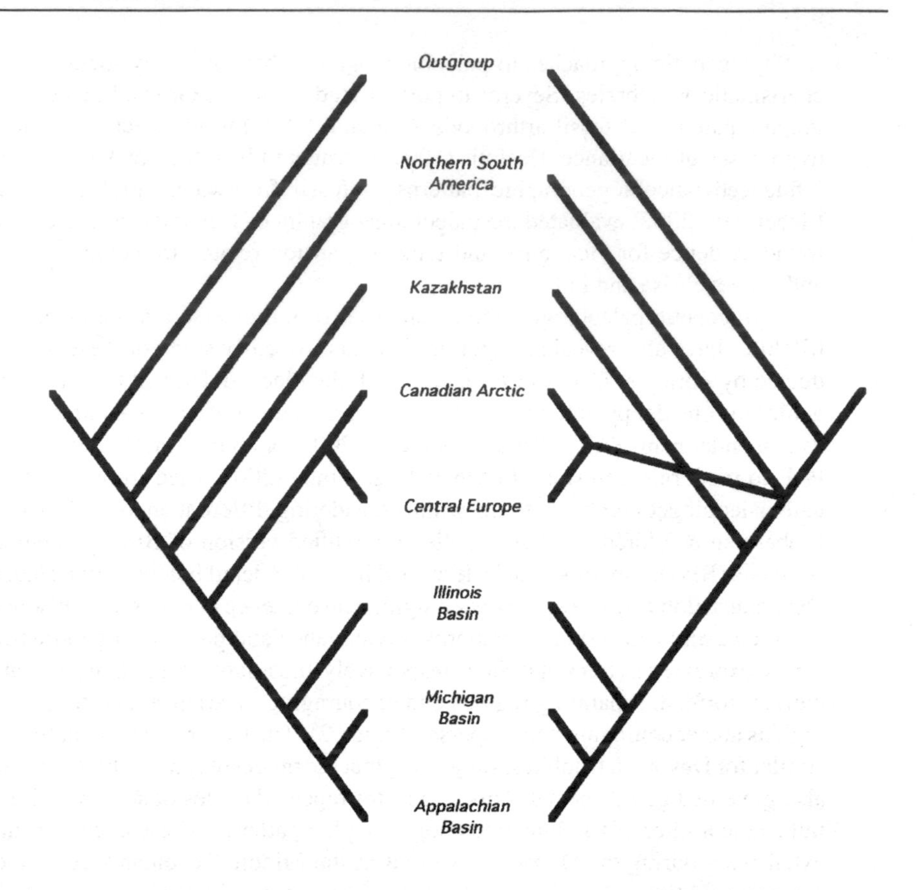

Figure 2 Paleobiogeographic patterns in Devonian trilobites based on Lieberman (2000), derived using the modified version of BPA described in the text. The most parsimonious pattern of vicariance and the strict consensus of the most parsimonious patterns of geo-dispersal are labeled. Also see discussion in Brooks & McLennan (2002).

during times when the geological or climatic signatures differ, the corresponding evolutionary and paleobiogeographic signatures also differ.

Another important paleobiogeographic study is Waggoner's (1999) analysis of ediacarans. These are an enigmatic set of organisms that are most abundant in the Late Neoproterozoic, immediately prior to the Cambrian radiation. It is a matter of debate whether they represent a new kingdom of organisms, mostly modern phyla, or a combination of extinct and modern forms. Thus far recalcitrant to phylogenetic analysis because of their unusual anatomies, they cannot be subjected to phylogenetic paleobiogeographic analysis, so Waggoner (1999) analyzed paleobiogeographic patterns using PAE. The analysis revealed important

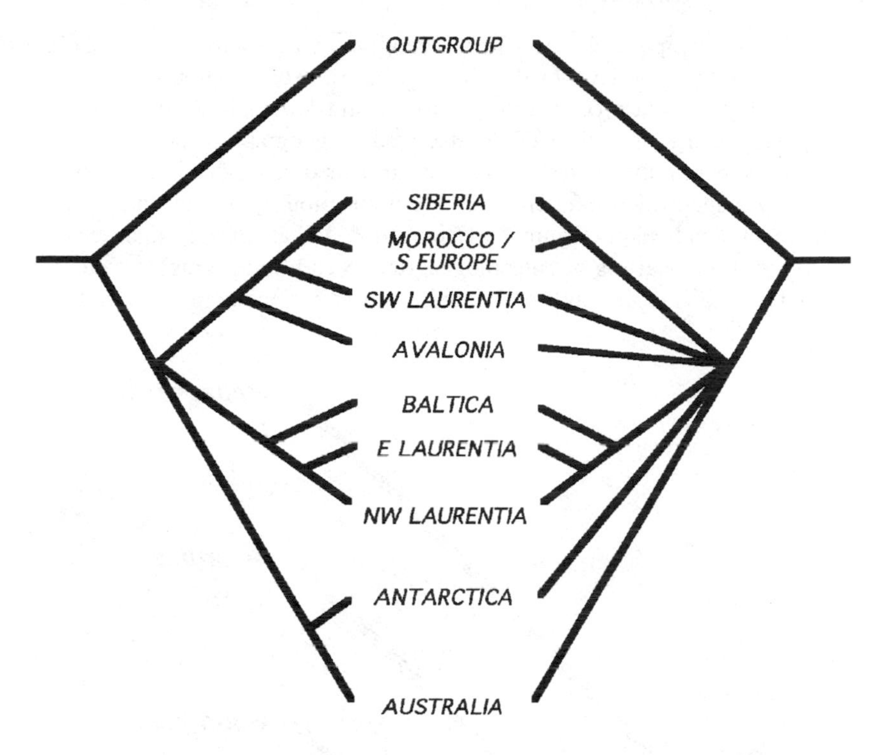

Figure 3 Paleobiogeographic patterns in Cambrian trilobites based on Lieberman (2003b,c), derived using the modified version of BPA described in the text. The most parsimonious pattern of vicariance and the strict consensus of the most parsimonious patterns of geo-dispersal are labeled. Laurentia refers to North America and Greenland, which were once conjoined; Baltica was a separate continent in the Cambrian, comprising Scandinavia and the eastern European platform; Avalonia refers to what was, in the Cambrian, a series of closely associated microcontinents including principally the eastern part of Newfoundland and much of Great Britain.

information about the breakup sequence of the supercontinent Rodinia, an early analog of the supercontinent Pangea but with the continents in different positions, which began to split apart shortly before the Cambrian radiation.

Paleomolecular Analyses

Understanding biogeographic patterns connecting extinct and modern lineages remains a major challenge (Lieberman 2000, 2002b; Murphy et al. 2001), yet important progress has been made. In particular, the analysis of ancient DNA and the refinement of approaches for dating divergence events using molecules enabled several new approaches to paleobiogeography. In a sense, such techniques have expanded the purview of paleobiogeography and biogeography and allowed more interdigitation between the two fields. Austin & Arnold (2001), using molecular phylogenetics of ancient DNA, analyzed biogeographic patterns in extinct land tortoises from the Mascarene Islands, an island chain that lies east of Malagasy. They found evidence for episodes of colonization from Mauritius to other islands in the island chain (Figure 4). Haddrath & Baker (2001), in an exciting study, merged the analysis of paleobiogeographic and biogeographic patterns through analysis of complete mtDNA genomes of two extinct large flightless birds (moas)

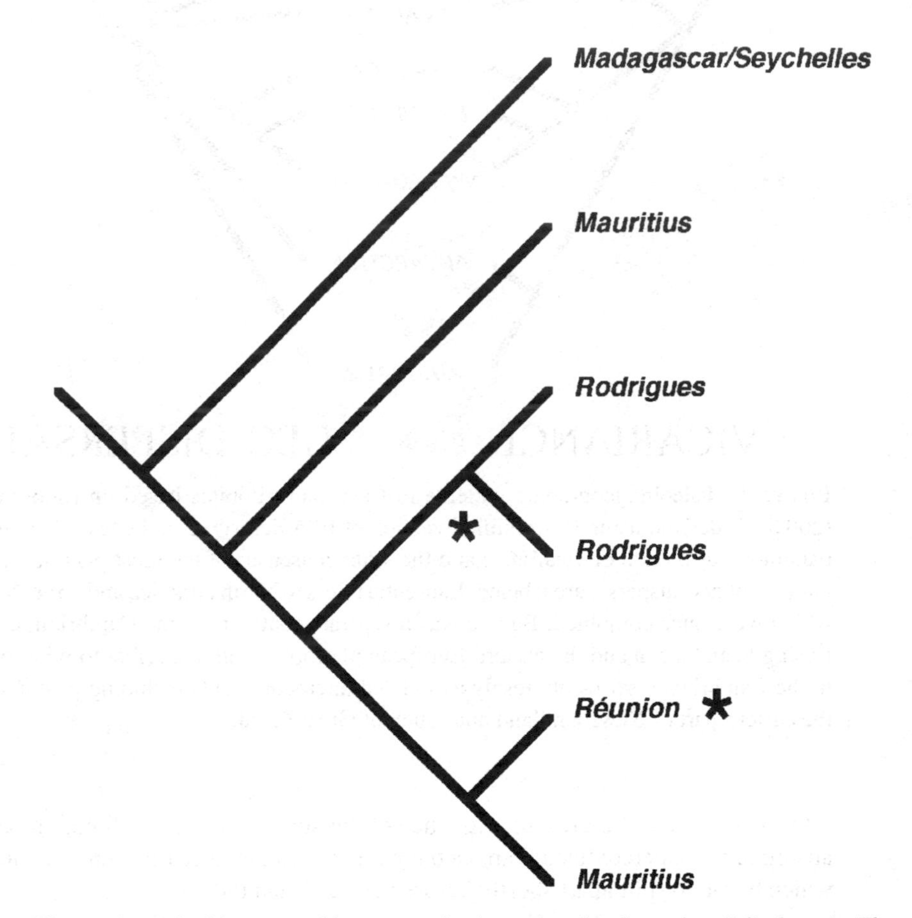

Figure 4 Paleobiogeographic patterns in the extinct land tortoise *Cylindraspis*. The area of occurrence is substituted for the species name, and the area cladogram is modified from Austin & Arnold (2001). Mapping the geographic states to the ancestral nodes using phases one and two of Fitch (1971) optimization suggests that there were two range expansion events (each denoted by *) during the evolutionary history of the group.

from New Zealand. Their molecular systematic study of large flightless ratite birds included five extant taxa along with the two extinct species. They used molecular sequence data to reconstruct evolutionary relationships and time divergence events. Haddrath & Baker (2001) found that some of the divergence events in the ratites were likely episodes of vicariance driven by the fragmentation of Gondwana, as Cracraft (2001) has posited; however, other episodes of secondary range expansion into New Zealand were also indicated. A similar pattern was recovered in the flying passerines by Ericson et al. (2002), providing support for the idea that different groups that occur together evolve congruently across geographic space in spite of differences in dispersal abilities.

Of course ancient DNA can be isolated only rarely. Molecular studies, however, need not be constrained by their inability to incorporate extinct taxa, and such studies can still make important contributions to our understanding of paleobiogeography. One such instance is revealed in the work of Murphy et al. (2001). They used Bayesian and maximum likelihood methods to look at the interordinal relationships of mammals. They found evidence for a basal split in the group that appears to be related to the split of South America and Africa 100 Mya, thereby correlating this key evolutionary event with a major geological event.

In another interesting study, Conti et al. (2002) used molecular systematics and molecular dating techniques to show how the Indian plate, during its independent history in the Cretaceous and Cenozoic, picked up a plant lineage in Africa and deposited it in Asia, an excellent example of geo-dispersal.

Even the lowly cockroach, a lineage well adapted to modern urban environments, shows the stamp of major episodes of geological and climatic change in its evolutionary history (Clark et al. 2001). Because of the relatively limited fossil record of this group, insight into the timing of biogeographic and paleobiogeographic patterns can be afforded only by analysis of degrees of molecular divergence.

These molecular analyses are in many ways complementary to paleobiogeographic studies that rely on information from organismal phylogenies and distribution gleaned from fossils. One area, however, where paleobiogeographic analysis of fossils is handicapped is in the analysis of patterns of intraspecific differentiation, which is difficult to analyze using a phylogenetic approach in fossil organisms. Such patterns are, however, particularly important if we seek to understand the biogeographic and evolutionary effects of the more recent major climatic changes that have occurred in the past few million years. One important area in which the field of biogeography has expanded in recent years, as documented by Brooks & McLennan (2002), is in the development of molecular phylogeography. This is a vibrant research area pioneered by Avise (1992), Zink (1996), and others. Even though the focus of phylogeography is on the differentiation of populations within species rather than species within clades, as is typically the case with paleobiogeography, it is remarkable how on the whole phylogeographic patterns conform well with paleobiogeographic patterns in one particular respect: Both show evidence for repeated episodes of vicariance and range expansion (geo-dispersal) occurring in an oscillatory fashion, albeit operating at different hierarchical levels and different timescales. For example, Lieberman (2001) found molecular phylogeographic evidence for repeated episodes of vicariance and geo-dispersal in a freshwater bivalve distributed in the central United States, a region buffeted by major climatic changes in the past two million years. This mollusc appears to have responded to the profound climatic changes by tracking habitat, rather than by speciating. The same result is reiterated in disparate groups including fish (Wiley & Mayden 1985) and birds (Klicka & Zink 1997), and it also emerged upon analysis of a broad swath of the flora and fauna in Europe that responded to a similar scope of climatic changes (Taberlet et al. 1998).

METHODOLOGICAL DEVELOPMENTS IN PALEOBIOGEOGRAPHY

Evidence seems to be accumulating from paleobiogeographic studies (including the studies already emphasized) that the history of clades and biotas involves cycles of vicariance and geo-dispersal. Indeed, the recognition that cycles of congruent range expansion and vicariance have oscillated repeatedly has a long historical heritage extending back to Lyell (1832), the Darwinian notebooks of 1837–1838 (in Barrett et al. 1987), Huxley (1870), and Wallace (1876) (see review in Lieberman 2000). More recently, several authors that operate within an explicitly phylogenetic perspective have recognized the existence of a similar pattern (e.g., Brundin 1988; Cracraft 1988; Noonan 1988; Wiley 1988a; Bremer 1992; Ronquist 1994, 1998; Lieberman & Eldredge 1996; Riddle 1996; Hovenkamp 1997; Lieberman 1997, 2000; Voelker 1999; Waggoner 1999; Bisconti et al. 2001; Brooks & McLennan 2002).

The existence of congruent patterns of geo-dispersal indicates that it would be beneficial to have a biogeographic method designed to retrieve both vicariance and geo-dispersal. Fortunately, such a method has been developed recently expressly for the analysis of paleobiogeographic data, although it can just as readily be used for analysis of biogeographic patterns in extant taxa (Lieberman & Eldredge 1996; Lieberman 1997, 2000). In an important sense, this method is based on the recognition that phylogenetic biogeography and paleobiogeography must be about the study of evolutionary and geographic congruence, not just vicariance. Instead of limiting phylogenetic biogeography simply to the analysis of one type of congruent biogeographic pattern, vicariance, scientists should welcome the opportunity to identify and study other types of Earth history events also related to plate tectonics or climate change that can congruently influence evolution and geographic distribution. Further, if geo-dispersal is ignored and biogeographic data analyzed using only a vicariance approach, results may be difficult to interpret, appear unresolved, or be simply inaccurate because of the overlapping signature of two different types of biogeographic processes.

The method for using phylogenetic information in conjunction with distributional data to identify geo-dispersal side by side with vicariance is based on a modified version of BPA. BPA was described in detail by Brooks (1981, 1985, 1990), Wiley (1988a,b), Funk & Brooks (1990), Wiley et al. (1991), and Brooks & McLennan (1991, 2002). The modified version of BPA is methodologically and philosophically similar to standard BPA but differs in a few key details. The method starts with an area cladogram: a phylogeny with the geographic distributions substituted for the taxon names. Then, the geographic states are optimized onto the nodes of the tree using phases one and two of the Fitch (1971) parsimony algorithm, which allows unordered transformation between character states (see discussion in Lieberman 2000). The area cladogram is subsequently converted into two separate data matrices: one designed to retrieve congruent episodes of vicariance and one designed to retrieve congruent episodes of geo-dispersal. Each data matrix has the areas of interest and an all-zero (absent) outgroup to polarize biogeographic character change.

The characters of the matrices code the nodes and the terminals of the tree along with the biogeographic transitions between ancestral and descendant nodes and ancestral nodes and descendant terminals. The manner in which the transitions are coded differs in the two data matrices. For instance, transitions between ancestral nodes and descendant nodes or terminals that involve range contraction are treated as putative episodes of vicariance and coded as an ordered multistate character in the vicariance matrix but not in the geo-dispersal matrix. By contrast, transitions between ancestral nodes and descendant nodes or terminals that involve range shifts or expansions are treated as putative episodes of geo-dispersal and coded as an ordered multistate character in the geo-dispersal matrix but not in the vicariance matrix (Lieberman & Eldredge 1996; Lieberman 1997, 2000) (see Figure 5 and Tables 1 and 2). After all area cladograms are coded into the two data matrices, each matrix is analyzed separately using the parsimony algorithm PAUP (Swofford 1998).

Repeated patterns of vicariance involving the same areas are treated as support for a particular area history in the vicariance tree. Patterns of range expansion do not figure into the construction of this tree and do not create homoplasy. By contrast, repeated patterns of geo-dispersal involving the same areas are treated as support for a particular area history in the geo-dispersal tree, and patterns of vicariance do not figure into the construction of this tree and do not create homoplasy. If there is limited evidence for vicariance or geo-dispersal in either matrix, the resultant tree(s) will be poorly resolved or only weakly supported.

The results are expressed as two most parsimonious trees, the vicariance and geo-dispersal trees, depicting patterns of the common history of areas and biotas. The vicariance tree gives information about the relative times that geographic barriers formed separating regions: The closer two areas are on the tree the more recently they shared a common history and the more recently barriers formed between them. The geo-dispersal tree gives information about the relative time that geographic barriers fell, joining regions: The closer two areas are on the tree the more recently barriers fell, joining regions.

In some instances the geo-dispersal trees and the vicariance trees are both well resolved, such as the example from Lieberman & Eldredge (1996) discussed above. By contrast, in examples involving different time periods, such as

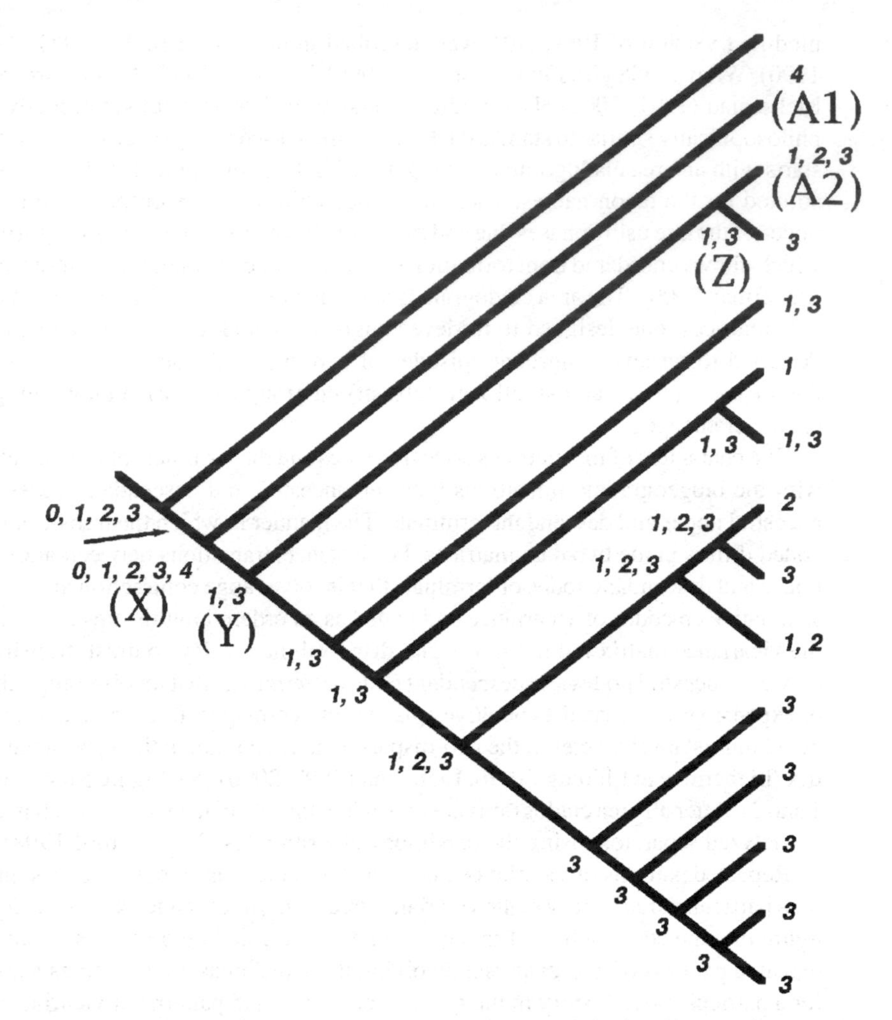

Figure 5 Area cladogram, with biogeographic states depicted as numbers, to illustrate the mechanics of the modified version of BPA described herein. These states were mapped to the ancestral nodes using phases one and two of Fitch (1971) optimization. The ancestral biogeographic state of the clade is indicated by the arrow. The letters next to the nodes and terminals are used in Tables 1 and 2. Figure based on Lieberman (2000).

Lieberman's (1997, 2003b,c) studies of the Early Cambrian described above, only the vicariance tree is well resolved. Although this method has thus far only been applied to the analysis of paleobiogeographic data, it can be applied to extant taxa without modification.

Ronquist (1997) developed dispersal-vicariance analysis, which is another attempt to consider episodes of range expansion within a phylogenetic framework. The method uses a model to consider the prevalence of different modes of

TABLE 1 Modified BPA coding of five characters of the vicariance matrix representing, in order, the basal node, node X, terminal taxon A1, nodes Y and Z, and terminal taxon A2, all from the area cladogram in Figure 5

	Characters					
Taxon	**1**	**2**	**3**	**4**	**5**	**6**
Outgroup	0	0	0	0	0	0
Area 0	1	1	1	1	0	0
Area 1	1	1	1	2	1	1
Area 2	1	1	1	1	0	1
Area 3	1	1	1	2	1	1
Area 4	0	1	2	1	0	0

0 indicates absent from a region; 1 and 2 indicate present in a region; outgroup refers to the ancestral biogeographic condition for the clade; multistate characters are ordered.

speciation in evolution and was recently implemented in a study by Zink et al. (2000). Although the method cannot be used to synthesize biogeographic patterns from different groups and does not distinguish traditional dispersal from geo-dispersal, unlike Lieberman & Eldredge's (1996) and Lieberman's (1997, 2000) method, it demonstrates the growing interest that phylogenetic biogeographers have in looking beyond vicariance as the sole process producing significant biogeographic patterns.

TABLE 2 Modified BPA coding of five characters of the geo-dispersal matrix representing, in order, the basal node, node X, terminal taxon A1, nodes Y and Z, and terminal taxon A2, all from the area cladogram in Figure 5

	Characters					
Taxon	**1**	**2**	**3**	**4**	**5**	**6**
Outgroup	0	0	0	0	0	0
Area 0	1	1	0	0	0	0
Area 1	1	1	0	1	1	1
Area 2	1	1	0	0	0	2
Area 3	1	1	0	1	1	1
Area 4	0	2	1	0	0	0

0 indicates absent from a region; 1 and 2 indicate present in a region; outgroup refers to the ancestral biogeographic condition for the clade; multistate characters are ordered.

THE EFFECTS OF EXTINCTION ON THE ANALYSIS OF PALEOBIOGEOGRAPHIC AND BIOGEOGRAPHIC PATTERNS

Extinction has the potential to obscure or bias paleobiogeographic patterns because through time the biotas of regions may appear different simply because one region has experienced differential extinction. For example, Sereno (1999) argued that this phenomenon underlies the apparent increase in endemism seen in some dinosaur faunas through time. Also, Van Oosterzee (1997) argued that extinction of island biotas precipitated by volcanic cataclysms underlies some modern biotic disparities in the Malay Archipelago. She suggested that eruption amplified apparent degrees of endemism by wholesale elimination of various taxa that were biotic intermediates.

Extinction can also complicate biogeographic studies of the extant biota (Lieberman 2000, 2002b). This is due to a situation analogous to one recognized by Gauthier et al. (1988) and Donoghue et al. (1989) for phylogenetic analyses of extant organisms. These may produce incorrect topologies because critical extinct taxa are excluded. Lieberman (2002b) suggested, based on simulation studies, that extant clades may appear biogeographically incongruent artificially, simply owing to accumulated extinction. Such artificial incongruence is especially prevalent in long-lived clades or in clades with high extinction probabilities; this may mean that at times it will be difficult to reconstruct the stamp of temporally distant tectonic events on the modern biota.

In essence, Wiley & Mayden (1985) and Brooks & McLennan (1991) identified this same phenomenon when they recognized that only geologically young clades that had not accumulated too much extinction should be the subjects of phylogenetic biogeographic analyses. Paleobiogeographers need not be as concerned with this second type of effect of extinction, as generally all of their subjects are extinct. However, they must be concerned with the quality of the fossil record. The incompleteness of the fossil record biases the preservation of fossil taxa and thus operates in a manner analogous to the effects extinction has on the extant biota (Lieberman 2002b). Fossil groups or time periods with poor fossil records also make the analysis of paleobiogeographic patterns problematic because a poor fossil record can also create artificial incongruence (Lieberman 2002b).

Because both biogeographic and paleobiogeographic studies may have a predilection toward artificial incongruence, potentially ideal biogeographic studies would be those that incorporate both fossil and extant taxa. This is not possible in some groups of organisms. Still, even in these groups, consideration of the fossil record may be helpful, for this is the place to turn to identify the ages of clades or other parameters such as the extinction probabilities of lineages within these clades. This information could in turn be used in simulation studies to evolve hypothetical clades comparable to a clade of interest to estimate to what extent the apparent biogeographic patterns in the modern clade may have been distorted by artificial biogeographic incongruence. These and other considerations represent a compelling argument for a theoretical unification of biogeography with paleobiogeography (also see Brooks & McLennan 1991, 2002). It is clear that the analysis of how extinction affects our ability to retrieve biogeographic patterns is still in its early stages, but this is likely to be an area of future important insights (Lieberman 2000, 2002b).

CONCLUSIONS

Paleobiogeography is a discipline that has recently shown renewed growth. This partly reflects an increasing emphasis on taking phylogenetic approaches to the study of paleobiogeography. It also results from an expansion of the scope of the field, which now includes studies of ancient DNA and studies that incorporate molecular methods to date biogeographic events. Paleobiogeographic analyses have spurred the recognition that vicariance is not the only process capable of producing biogeographic congruence. Geo-dispersal of organisms related to the fall of geographic barriers, driven by tectonism or climate change, can also structure geographic distributions in independent clades simultaneously. This is a very different process from traditional dispersal, which involves movement over a barrier that is unlikely to occur congruently in different lineages. Concomitant with the recognition of the potential significance of geo-dispersal, methodological advances in paleobiogeographic analysis have also occurred, including modifications of such well-known techniques as BPA to allow geo-dispersal to be treated within an analytical framework. Finally, insights from paleobiogeography suggest that extinction is a process that may influence our ability to retrieve biogeographic patterns in extant organisms, and thus extinct fossil taxa should be considered in biogeographic analysis whenever possible. All of these suggest that some excellent opportunities for conceptual and methodological advances may occur on the interface between paleobiogeography and biogeography.

ACKNOWLEDGMENTS

Thanks to Dan Brooks, Tony Hallam, Roger Kaesler, Alycia Rode, Paul Sereno, the scientific editor of this series, and Ed Wiley for helpful comments on earlier versions of this manuscript. This research was supported by NSF OPP-9909302; NSF EAR-0106885; and a Self Faculty Fellowship.

LITERATURE CITED

Austin JJ, Arnold EN. 2001. Ancient mitochondrial DNA and morphology elucidate an extinct island radiation of Indian Ocean giant tortoises (Cylindraspis). *Proc. R. Soc. London Ser. B* 268:2515–23

Avise JC. 1992. Molecular population structure and the biogeographic history of a regional fauna: a case history with lessons for conservation biology. *Oikos* 63:62–76

Barrett PH, Gautrey PJ, Herbert S, Kohn D,

Smith S. 1987. *Charles Darwin's Notebooks, 1836–1844*. Ithaca, NY: Cornell Univ. Press

Beard KC. 1998. East of Eden: Asia as an important center of taxonomic origination in mammalian evolution. *Bull. Carnegie Mus. Nat. Hist.* 34:5–39

Beard KC. 2002. East of Eden at the Paleocene/Eocene boundary. *Science* 295:2028–29

Bisconti M, Landini W, Bianucci G, Cantalamessa G, Carnevale G, et al. 2001. Biogeographic relationships of the Galapagos terrestrial biota: parsimony analyses of endemicity based on reptiles, land bards and *Scalesia* land plants. *J. Biogeogr.* 28:495–510

Bowen GJ, Clyde WC, Koch PL, Ting S, Alroy J, et al. 2002. Mammalian dispersal at the Paleocene/Eocene boundary. *Science* 295:2062–65

Bremer K. 1992. Ancestral areas: a cladistic reinterpretation of the center of origin concept. *Syst. Biol.* 41:436–45

Brooks DR. 1981. Hennig's parasitological method: a proposed solution. *Syst. Zool.* 30:229–49

Brooks DR. 1985. Historical ecology: a new approach to studying the evolution of ecological associations. *Ann. Mo. Bot. Gard.* 72:660–80

Brooks DR. 1990. Parsimony analysis in historical biogeography and coevolution: methodological and theoretical update. *Syst. Zool.* 39:14–30

Brooks DR, McLennan DA. 1991. *Phylogeny, Ecology, and Behavior*. Chicago: Univ. Chicago Press

Brooks DR, McLennan DA. 2002. *The Nature of Diversity*. Chicago: Univ. Chicago Press

Brown JH, Lomolino MV. 1998. *Biogeography*. Sunderland, MA: Sinauer. 2nd ed.

Brown JH, Maurer BA. 1989. Macroecology: the division of food and space among species on continents. *Science* 243:1143–50

Brundin LZ. 1988. Phylogenetic biogeography. In *Analytical Biogeography*, ed. AA Myers, PS Giller, pp. 343–69. New York: Chapman & Hall

Clark JW, Hossain S, Burnside CA, Kambhampati S. 2001. Coevolution between a cockroach and its bacterial endosymbiont: a biogeographical perspective. *Proc. R. Soc. London Ser. B* 268:393–98

Conti E, Eriksson T, Schonenberger J, Systsma KJ, Baum DA. 2002. Early Tertiary out of India dispersal of Crypteroniaceaea: evidence from phylogeny and molecular dating. *Evolution* 56:1931–42

Cracraft J. 1988. Deep-history biogeography: retrieving the historical pattern of evolving continental biotas. *Syst. Zool.* 37:221–36

Cracraft J. 2001. Avian evolution, Gondwana biogeography and the Cretaceous-Tertiary mass extinction event. *Proc. R. Soc. London Ser. B* 268:459–69

Dalziel IWD. 1997. Neoproterozoic-Paleozoic geography and tectonics: review, hypothesis, and environmental speculations. *Geol. Soc. Am. Bull.* 109:16–42

Donoghue MJ, Doyle J, Gauthier J, Kluge A, Rowe T. 1989. The importance of fossils in phylogeny reconstruction. *Annu. Rev. Ecol. Syst.* 20:431–60

Ebach MC, Edgecombe GD. 2001. Cladistic biogeography: component-based methods and paleontological application. In *Fossils, Phylogeny, and Form: An Analytical Approach*, ed. JM Adrain, GD Edgecombe, BS Lieberman, pp. 235–89. New York: Kluwer Academic/Plenum

Ericson PGP, Christidis L, Cooper A, Irestedt M, Jackson J, et al. 2002. A gondwanan origin of passerine birds supported by DNA sequences of the endemic New Zealand wrens. *Proc. R. Soc. London Ser. B* 269:235–41

Fitch WM. 1971. Toward defining the course of evolution: minimum change for a specific tree topology. *Syst. Zool.* 20:406–16

Fortey RA, Cocks LRM. 1992. The early Paleozoic of the North Atlantic region as a test case for the use of fossils in continental reconstruction. *Tectonophysics* 206:147–58

Funk VA, Brooks DR. 1990. Phylogenetic systematics as the basis of comparative biology. *Smithson. Contrib. Bot.* 73:1–45

Gauthier J, Kluge AG, Rowe T. 1988. Amniote phylogeny and the importance of fossils. *Cladistics* 4:105–209

Grimaldi DA. 1992. Vicariance biogeography, geographic extinctions, and the North American Oligocene Tsetse flies. In *Extinction and Phylogeny*, ed. MJ Novacek, QD Wheeler, pp. 184–204. New York: Columbia Univ. Press

Haddrath O, Baker AJ. 2001. Complete mitochondrial DNA genome sequences of extinct birds: ratite phylogenetics and the vicariance biogeography hypothesis. *Proc. R. Soc. London Ser. B* 268:939–45

Hallam A. 1967. The bearing of certain paleogeographic data on continental drift. *Palaeogeogr. Palaeoclimatol. Palaeoecol.* 3:201–24

Hallam A. 1977. Jurassic bivalve biogeography. *Paleobiology* 3:58–73

Hallam A. 1981. *Great Geological Controversies.* New York: Oxford Univ. Press

Hallam A. 1983. Early and mid-Jurassic molluscan biogeography and the establishment of the central Atlantic seaway. *Palaeogeogr. Palaeoclimatol. Palaeoecol.* 43:181–93

Hallam A. 1986. Evidence of displaced terranes from Permian to Jurassic faunas around the Pacific margins. *J. Geol. Soc. London* 143:209–16

Hallam A. 1994. *An Outline Of Phanerozoic Biogeography*, Vol. 10. Oxford: Oxford Univ. Press

Hasiotis ST. 1999. The origin and evolution of freshwater crayfish based on crayfish body and trace fossils. *Freshw. Crayfish* 12:49–70

Hovenkamp P. 1997. Vicariance events, not areas, should be used in biogeographical analysis. *Cladistics* 13:67–79

Humphries CJ, Parenti L. 1986. Cladistic biogeography. *Oxford Mon. Biogeogr.* 2:1–98

Huxley TH. 1870. Anniversary address. In *The Scientific Memoirs of Thomas Henry Huxley*, ed. M Foster, ER Lankester, pp. 510–50. London: Macmillan

Klicka J, Zink RM. 1997. The importance of recent Ice Ages in speciation: a failed paradigm. *Science* 277:1666–69

Krause DW, Prasad GVR, von Koenigswald W, Sahni A, Grine FE. 1997. Cosmopolitanism among gondwanan Late Cretaceous mammals. *Nature* 390:504–7

Lieberman BS. 1997. Early Cambrian paleogeography and tectonic history: a biogeographic approach. *Geology* 25:1039–42

Lieberman BS. 2000. *Paleobiogeography*. New York: Plenum/Kluwer Academic

Lieberman BS. 2001. Applying molecular phylogeography to test paleoecological hypotheses: a case study involving *Amblema plicata* (Mollusca: Unionidae). In *Evolutionary Paleoecology*, ed. WD Allmon, DJ Bottjer, pp. 83–103. New York: Columbia Univ. Press

Lieberman BS. 2002a. Phylogenetic analysis of some basal Early Cambrian trilobites, the biogeographic origins of the Eutrilobita, and the timing of the Cambrian radiation. *J. Paleontol.* 76:692–708

Lieberman BS. 2002b. Phylogenetic biogeography with and without the fossil record: gauging the effects of extinction and paleontological incompleteness. *Palaeogeogr. Palaeoclimatol. Palaeoecol.* 178:39–52

Lieberman BS. 2003a. Unifying theory and methodology in biogeography. *Evol. Biol.* 33:1–25

Lieberman BS. 2003b. Biogeography of the Cambrian radiation: deducing geological processes from trilobite evolution. *Spec. Pap. Palaeontol.* In press

Lieberman BS. 2003c. Taking the pulse of the Cambrian radiation. *J. Int. Comp. Biol.* 43(1):In press

Lieberman BS, Eldredge N. 1996. Trilobite biogeography in the Middle Devonian: geological processes and analytical methods. *Paleobiology* 22:66–79

Lyell C. 1832. *Principles of Geology*, Vol. 2. Chicago: Univ. Chicago Press. 2nd ed.

Maurer BA. 1999. *Untangling Ecological Complexity*. Chicago: Univ. Chicago Press

McKenna MC. 1975. Fossil mammals and Early Eocene North Atlantic land continuity. *Ann. Mo. Bot. Gard.* 62:335–53

McKenna MC. 1983. Holarctic landmass rearrangement, cosmic events, and Cenozoic terrestrial organisms. *Ann. Mo. Bot. Gard.* 70:459–89

Murphy WJ, Eizirik E, O'Brien SJ, Madsen

O, Scally M, et al. 2001. Resolution of the early placental mammal radiation using Bayesian phylogenetics. *Science* 294:2348–51

Murray AM. 2001. The fossil record and biogeography of the Cichlidae (Actinopterygii: Labroidei). *Biol. J. Linn. Soc.* 74:517–32

Noonan GR. 1988. Biogeography of North American and Mexican insects, and a critique of vicariance biogeography. *Syst. Zool.* 37:366–84

Riddle BR. 1996. The molecular phylogeographic bridge between deep and shallow history in continental biotas. *Trends Ecol. Evol.* 11:207–11

Rode A, Lieberman BS. 2002. Phylogenetic and biogeographic analysis of Devonian phyllocarid crustaceans. *J. Paleontol.* 76:271–86

Ronquist F. 1994. Ancestral areas and parsimony. *Syst. Biol.* 43:267–74

Ronquist F. 1997. Dispersal-vicariance analysis: a new approach to the quantification of historical biogeography. *Syst. Biol.* 46:195–203

Ronquist F. 1998. Phylogenetic approaches in coevolution and biogeography. *Zool. Scr.* 26:313–22

Rosenzweig ML. 1995. *Species Diversity in Space and Time.* New York: Cambridge Univ. Press

Ross A, Ross JRP. 1985. Carboniferous and early Permian biogeography. *Geology* 13:27–30

Sampson SD, Witmer LM, Forster CA, Krasue DW, O'Connor PM, et al. 1998. Predatory dinosaur remains from Madagascar: implications for the Cretaceous biogeography of Gondwana. *Science* 280:1048–51

Sanmartin I, Enghoff H, Ronquist F. 2001. Patterns of animal dispersal, vicariance and diversification in the Holarctic. *Biol. J. Linn. Soc.* 73:345–90

Scotese CR. 1997. *Paleogeographic Atlas*. Arlington, TX: Paleomap Proj.

Scott B. 1997. Biogeography of the Helicoidea (Mollusca: Gastropoda: Pulmonata): land snails with a Pangean distribution. *J. Biogeogr.* 24:399–407

Sereno PC. 1997. The origin and evolution of dinosaurs. *Annu. Rev. Earth Planet. Sci* 25:435–89

Sereno PC. 1999. The evolution of dinosaurs. *Science* 284:2137–47

Sereno PC, Dutheil DB, Iarochene M, Larsson HCE, Lyon GH, et al. 1996. Predatory dinosaurs from the Sahara and Late Cretaceous faunal differentiation. *Science* 272:986–91

Swofford DL. 1998. *PAUP (Phylogenetic Analysis Using Parsimony)*, Version 4.0. Sunderland, MA: Sinauer

Taberlet P, Fumagalli L, Wust-Saucy A-G, Cosson J-F. 1998. Comparative phylogeography and postglacial colonization routes in Europe. *Mol. Ecol.* 7:453–64

Torsvik TH, Smethurst MA, Meert JG, Van der Voo R, McKerrow WS, et al. 1996. Continental breakup and collision in the Neoproterozoic and Paleozoic—a tale of Baltica and Laurentia. *Earth Sci. Rev.* 40:229–58

Upchurch P, Hunn CA, Norman DB. 2002. An analysis of dinosaurian biogeography: evidence for the existence of vicariance and dispersal patterns caused by geological patterns. *Proc. R. Soc. London Ser. B* 269:613–21

Van Oosterzee P. 1997. *Where Worlds Collide.* Ithaca, New York: Cornell Univ. Press

Vences M, Freyhoff J, Sonnenberg R, Kosuch J, Veith M. 2001. Reconciling fossils and molecules: Cenozoic divergence of cichlid fishes and the biogeography of Madagascar. *J. Biogeogr.* 28:1091–99

Voelker G. 1999. Dispersal, vicariance, and clocks: historical biogeography and speciation in a cosmopolitan passerine genus (*Anthus*: Motacillidae). *Evolution* 53:1536–52

Waggoner B. 1999. Biogeographic analyses of the Ediacara biota: a conflict with paleotectonic reconstructions. *Paleobiology* 25:440–58

Wallace AR. 1876. *The Geographical Distribution of Animals*. New York: Harpers

Ward PD, Hurtado JM, Kirschvink JL, Verosub KL. 1997. Measurements of the Cretaceous paleolatitude of Vancouver Island: consistent with the Baja-British Columbia hypothesis. *Science* 277:1642–45

Wiley EO. 1988a. Parsimony analysis and vicariance biogeography. *Syst. Zool.* 37:271–90

Wiley EO. 1988b. Vicariance biogeography. *Annu. Rev. Ecol. Syst.* 19:513–42

Wiley EO, Mayden RL. 1985. Species and speciation in phylogenetic systematics, with examples from the North American fish fauna. *Ann. Mo. Bot. Gard.* 72:596–635

Wiley EO, Siegel-Causey D, Brooks DR, Funk VA. 1991. *The Compleat Cladist*. Lawrence, KS: Univ. Kansas Press

Zink RM. 1996. Comparative phylogeography in North American birds. *Evolution* 50:308–17

Zink RM, Blackwell-Rago RC, Ronquist F. 2000. The shifting roles of dispersal and vicariance in biogeography. *Proc. R. Soc. London Ser. B* 267:497–503

Annu. Rev. Ecol. Evol. Syst. 2003. 34:71–98
doi: 10.1146/annurev.ecolsys.34.011802.132353

First published online as a Review in Advance on July 8, 2003

THE ECOLOGY OF BIRD INTRODUCTIONS

Richard P. Duncan,[1] Tim M. Blackburn,[2] and Daniel Sol[3]
[1]*Ecology and Entomology Group, Soil, Plant and Ecological Sciences Division, P.O. Box 84, Lincoln University, Canterbury, New Zealand; email: duncanr@lincoln.ac.nz*
[2]*School of Biosciences, University of Birmingham, Edgbaston, Birmingham B15 2TT, United Kingdom; email: t.blackburn@bham.ac.uk*
[3]*Department of Biology, McGill University, 1205 avenue Docteur Penfield, Montréal, Québec, H3A 1B1 Canada; email: dsolru@po-box.mcgill.ca*

Key Words establishment success, exotic species, introduced birds, biological invasion, biotic resistance, invasive species

■ **Abstract** A growing number of species have been transported and introduced by humans to new locations and have established self-sustaining wild populations beyond their natural range limits. Many of these species go on to have significant environmental or economic impacts. However, not all species transported and introduced to new locations succeed in establishing wild populations, and of the established species only some become widespread and abundant. What factors underlie this variation in invasion success? Here, we review progress that has been made in identifying factors underpinning invasion success from studies of bird introductions. We review what is known about the introduction, establishment, and spread of introduced bird species, focusing on comparative studies that use historical records to test hypotheses about what factors determine success at different stages in the invasion process. We close with suggestions for future research.

INTRODUCTION

As humans have spread around the globe, they have deliberately or accidentally transported a huge variety of plant and animal species to locations beyond their natural ranges (Elton 1958; Lodge 1993; Williamson 1996, 1999). By design or accident many of these introduced species have established self-sustaining wild populations, which have subsequently increased and spread to varying degrees. Many of these species have profoundly affected the ecosystems they have invaded, and some have imposed substantial economic and health costs on human societies (Elton 1958, Ebenhard 1988, Lever 1994, Simberloff 1995, Williamson 1996, Vitousek et al. 1997, Parker et al. 1999, Dalmazzone 2000, Mack et al. 2000, Perrings et al. 2000, McNeely 2001). Thus, the establishment and spread of introduced species is currently recognized as a major risk worldwide. This risk is likely to increase as greater volumes of transport and trade increase the rate at which novel species are introduced to new locations (Ricciardi et al. 2000).

Nevertheless, only a small proportion of the species introduced to a new location establish wild populations, and of these only a small proportion become abundant or widespread and have significant impact (Lodge 1993, Williamson 1996, Williamson & Fitter 1996). Given that once established the eradication or control of introduced species is costly, the most effective way to minimize their impact is to prevent establishment or spread in the first place (Ricciardi & Rassmusen 1998, Mack et al. 2000). This approach requires that we understand the factors underlying success at different stages in the invasion process (Figure 1) so that we can identify situations where invasion risk is high. Specifically, what are the factors that allow certain species to establish and spread when introduced to locations outside their natural range?

Despite concerted effort, the field of invasion ecology has been criticized for its lack of success in answering this question (Ehrlich 1989, Vermeij 1996, Mack et al. 2000). In this review we hope to show that considerable progress has been made in identifying the factors underpinning invasion success using historical data on bird introductions. We discuss what is known about the introduction, establishment, and spread of introduced birds, focusing on comparative studies that use the historical record of bird introductions to test hypotheses about the factors determining invasion success.

Why Study Bird Introductions?

Data on historical bird introductions provide a rare opportunity to test hypotheses about invasion success for at least three reasons. First, there is an excellent record of the bird species introduced to locations around the world (Long 1981, Lever 1987). Most bird introductions occurred in the eighteenth and nineteenth centuries during the major period of European expansion and settlement. Introductions were often associated with the formation of acclimatization societies that aimed to establish beneficial or desirable species in the new settlements (see especially Thomson 1922) or with private individuals who sometimes attempted to naturalize a variety of species (e.g., Eastham Guild on Tahiti; Guild 1938, 1940). Birds were prominent among the species introduced by settlers, especially for hunting, biocontrol, and aesthetic reasons. Many societies or individuals kept records of the birds they introduced, sometimes including details such as the numbers of individuals released, the exact location of release, and the origin of the birds introduced (Thomson 1922, Long 1981, Lever 1987). Because most of these introductions occurred decades to centuries ago, and because birds are conspicuous and well studied, introduction outcomes can be determined with reasonable certainty. We can therefore compile comprehensive lists of the bird species introduced to new locations, whether or not those introduced species established wild populations, and the extent to which the established species have spread.

Second, because many attempts were made to introduce birds, there are data on a large and taxonomically diverse set of species introduced to a wide range of locations with which to test hypotheses about invasion success. Worldwide, there have been recorded more than 1400 attempts to introduce about 400 species

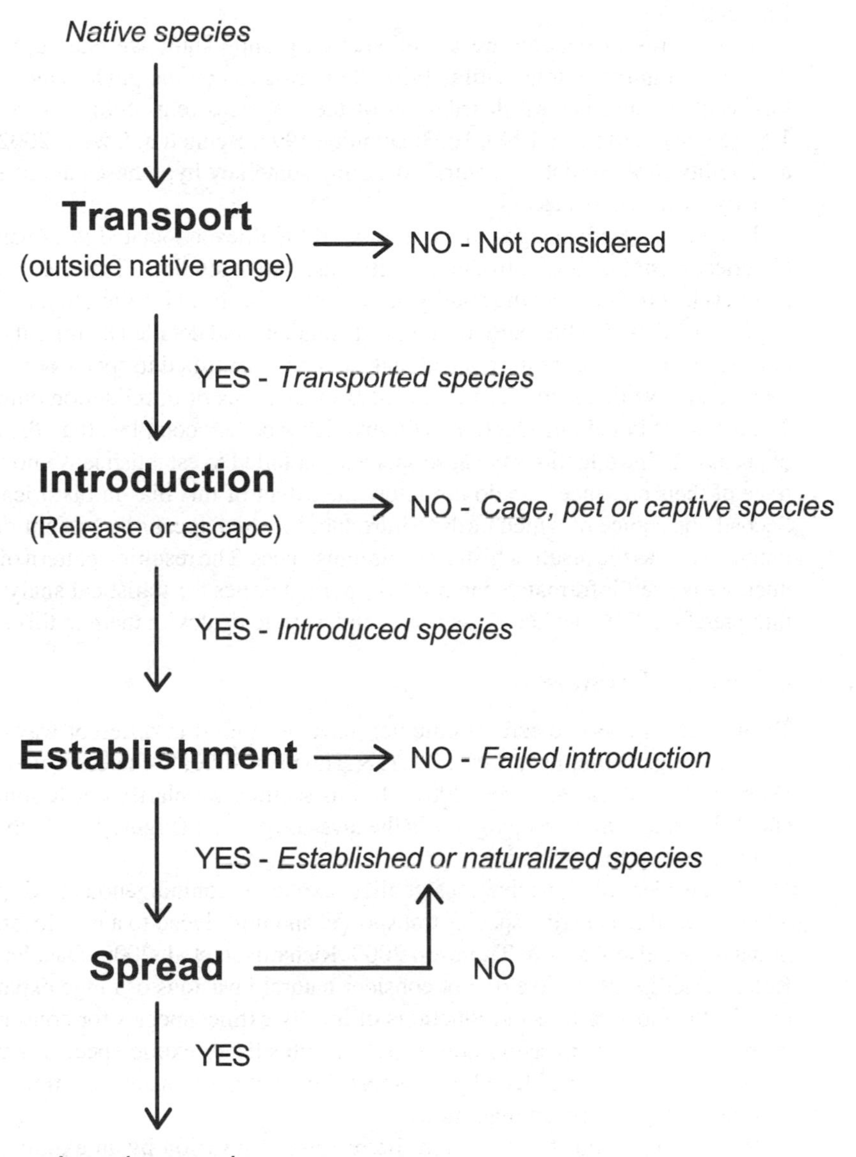

Figure 1 Schematic representation of stages in the process of a human-caused invasion that we recognize in this review. The stages through which a species must pass are shown in bold. The names of species that do or do not pass through each stage are shown in italics.

of birds (Long 1981, Lever 1987, Lockwood 1999, Blackburn & Duncan 2001b, Cassey 2002).

Third, birds in general are a well-studied group. Thus, we can supplement data on bird introductions with published information on the phylogeny, life history, ecology, and natural distribution of these species (e.g., Sibley & Ahlquist 1990; Sibley & Monroe 1990, 1993; Dunning 1993; Bennett & Owens 2002). The availability of these data is central to testing some key hypotheses about factors underlying invasion success.

Despite these advantages, there are two difficulties associated with using the historical record of bird introductions in quantitative studies. First, the historical record is incomplete, and the quality of the data varies from location to location. In New Zealand, which has perhaps the most complete and detailed historical record, at least seven birds were introduced that cannot be identified to species (Thomson 1922) and have therefore been excluded from analyses of introduction outcomes. The record of failed introductions will invariably be less complete than the record of successful introductions because species that failed to establish leave no further trace of their presence. We do not know the extent of this bias in historical data. Second, the choice of which birds to introduce to which locations was not made at random but was the result of historical circumstances. The resulting pattern of introductions is itself informative but raises important issues for statistical analysis and interpretation. We consider these issues and ways to deal with them in this review.

Definitions and Framework

Terms such as invasive and introduction have been used in different ways in the literature (e.g., Williamson 1996, Davis & Thomson 2000, Richardson et al. 2000, Daehler 2001, Kolar & Lodge 2001). In this section, we clarify our terminology and define stages that we recognize in the invasion process (Figure 1), which serves as a framework for our review.

We consider only introductions of alien, exotic, or nonindigenous bird species, all terms used to describe species transported and introduced to a new location by humans (see also Davis & Thomson 2000, Richardson et al. 2000, Daehler 2001, Kolar & Lodge 2001). We do not consider natural invasions or range expansions nor, for the most part, re-introductions of locally extinct species for conservation purposes. The term invasion here refers specifically to exotic species that have been deliberately or accidentally transported by humans to a new location beyond their normal geographic range limits.

We recognize four transitions in the process of invasion by an exotic species (Figure 1) (see Williamson 1996, Richardson et al. 2000, Davis & Thomson 2000, Daehler 2001, Kolar & Lodge 2001). First, the exotic species must be transported from its native geographic range to a new location. Second, the species must be released or escape into the environment. Third, the species must succeed in establishing a self-sustaining wild population following release. Finally, species that establish successfully may increase in abundance and spread beyond the release point; the extent of this spread defines their geographic range in the new environment. We term these stages transport, introduction, establishment, and spread (see also Lockwood 1999, Kolar & Lodge 2001, Sakai et al. 2001). For a species to have reached a given stage in the invasion process, it must have passed through all prior stages.

An introduced species is one that passes through the first two stages and is released (or escapes) into a new environment. An established species is one that establishes a self-sustaining wild population following introduction (also termed naturalized). An established species that succeeds in spreading beyond the site of introduction is termed invasive. The introduction of a species to a location is termed an introduction event. When we use the term introduction, we refer to a species that has been transported to, and released or escaped into, an alien environment, regardless of whether it establishes or spreads.

Not all stages in the invasion process have been equally well studied, a fact reflected in the attention we give them in this review. Most effort has gone into identifying factors that influence the successful transition from introduction to establishment. No studies have attempted to quantify the first stage in the process: which species of birds have been selected for transport to new locations, regardless of whether they have been introduced or not (but see Guix et al. 1997 for an analysis of the pet trade of Neotropical parrots in Barcelona). We start, therefore, by addressing two questions associated with the introduction stage: (*a*) Which species were selected for transport followed by introduction to new locations; and (*b*) to which locations were birds introduced? We then examine factors associated with success in establishment and spread.

TRANSPORT WITH INTRODUCTION

Which Species Were Selected for Introduction?

Fewer than 5% of all bird species have been transported and introduced to a new location (Blackburn & Duncan 2001b). It is important to understand why some species and not others were chosen for introduction because biases in the type of species chosen, termed selectivity, may limit the generality of conclusions we draw regarding the factors affecting success in later stages of the invasion process. If only certain types of birds were chosen for introduction, then the factors that determine invasion success in this group may not necessarily apply to other groups of birds. Selectivity also has important implications for the statistical analysis of success at later stages in the invasion process (see below). Selectivity in the birds chosen for introduction has been examined with regard to the taxonomic affiliations of species, the geographic origin of species, and specific characteristics of the taxa chosen.

TAXONOMIC SELECTIVITY Taxonomic selectivity in global bird introductions has been examined for all introduced species (Blackburn & Duncan 2001b) and for the subset of species that were both introduced and established (Lockwood 1999, Lockwood et al. 2000). The results of both analyses are similar in that families over-represented at the introduction stage are also the families over-represented among established species. This fact implies that a major cause of taxonomic selectivity among established species is taxonomic selectivity in the species chosen for introduction. Two thirds of all species chosen for introduction belong to just 6 of the 145 bird families: Anatidae, Columbidae, Fringillidae, Passeridae, Phasianidae, and Psittacidae (Blackburn & Duncan 2001b). The over-representation of species in these families almost certainly reflects two of the major motivations behind introducing birds. Species in the families Anatidae and Phasianidae were primarily introduced for hunting, while species in the families Fringillidae, Passeridae, and Psittacidae are frequently kept as cage birds, and many were introduced for aesthetic reasons (see Long 1981, Lever 1987). Nevertheless, between 69% and 94% of the species in these families have never been the subject of an introduction event (Blackburn & Duncan 2001b).

GEOGRAPHICAL SELECTIVITY As for many other taxa, bird species richness is higher nearer the Equator. Nevertheless, most birds chosen for introduction have their native geographic ranges centered in temperate regions, between 30° and 50° in both hemispheres (Blackburn & Duncan 2001b). Only around 8% of introduction events involve species from the Neotropics, although the Neotropics are home to around 30% of all bird species (Rahbek 1997). Conversely, about 16% of introduction events involve species from the Palaearctic, despite the fact that only about 10% of bird species have geographic ranges that include this region (Evans 1997; Sibley & Monroe 1990, 1993). Although there has been no formal test, species from tropical regions appear to be greatly under-represented in lists of introduced species.

Much of this geographical selectivity can be attributed to introductions carried out by European settlers, who mostly colonized other temperate parts of the world. Thus, opportunities for the transport and trade of birds during the eighteenth and nineteenth centuries would have been greatest between Western Europe and the predominantly temperate locations where Europeans settled, including North America, South Africa, Australia, and New Zealand. Around 60% of the bird species introduced to New Zealand originate from the Palaearctic and Australasian regions (Duncan et al. 2003), the two places with which trade and transport were most frequent during early European settlement of New Zealand.

CHARACTER SELECTIVITY These taxonomic and geographical patterns of selectivity imply that the birds chosen for introduction will be nonrandom with respect to their life history and ecological traits, because these traits are not randomly distributed across taxa or regions (see, e.g., Bergmann 1847; Lack 1947, 1948; Owens & Bennett 1995; Gaston & Blackburn 2000; Cardillo 2002). Thus, we would expect introduced birds predominantly to possess the characteristics of temperate game birds (Anatidae and Phasianidae) or cage birds (Psittacidae, Fringillidae, and Passeridae). There have been few tests of character selectivity for introduced birds, although for those introduced to Australia, a high proportion of species are ground nesters that use grassland, cultivated, or suburban habitats and have largely vegetarian diets (seeds, fruit, vegetation) (Newsome & Noble 1986). Introduced species are also larger-bodied, on average, than would be expected if they were a random sample of bird species (116.6 g versus 50.5 g for introduced and all bird species, respectively; Blackburn & Gaston 1994, Cassey 2001a).

Character selectivity also occurs independently of taxonomic and geographical selectivity. For both a geographic (species in the British avifauna) and a taxonomic (order Anseriformes; wildfowl) subset of the world's birds, species with larger population sizes were significantly more likely to have been chosen for introduction (Blackburn & Duncan 2001b). In addition, larger-bodied wildfowl and resident British birds that were widely distributed had a higher probability of selection. Species with large population size, wide distribution, and resident status would have been the most readily available for capture and transport to new locations. While most species were released at very few locations, a few species, typically widely distributed and abundant in their native ranges, were widely introduced (Long 1981, Lever 1987, Cassey 2002).

Together, these results suggest a general hierarchy of causes that have contributed to the selective introduction of certain birds. First, the species chosen for introduction were concentrated in geographic regions that in large part reflect the origin of European settlers and their subsequent patterns of settlement and trade. Second, from within geographic regions, people chose certain kinds of bird for introduction for a variety of reasons. An emphasis on birds for hunting and aesthetic purposes has resulted in birds from five families being significantly over-represented among those chosen. Finally, given that birds were chosen for particular purposes and from certain regions, the species that were finally caught, transported, and introduced tend to be those that were common in the source locations. People preferentially selected abundant, widely distributed species, either because they were most readily available for capture or because they were species with which they were most familiar, and therefore most desired to introduce.

To Which Locations Were Birds Introduced?

People mostly introduced birds to islands. Although islands make up about 3% of ice-free land area (Mielke 1989), around 70% of introduction events have been to islands (Blackburn & Duncan 2001b). Slightly over one half of all introductions were to Pacific islands (especially the Hawaiian islands) and Australasia. The greatest number of mainland introductions was to the Nearctic, followed by the Palaearctic, and continental Australia. Birds have been introduced to all major regions of the world and to most latitudes with ice-free land. However, relatively few introductions have been to equatorial regions; most were to latitudes between 10° and 40° on both sides of the equator (Blackburn & Duncan 2001b).

ESTABLISHMENT

A species introduced to a new location will either succeed or fail to establish a self-sustaining population. There are numerous reasons why introductions could fail, making it unlikely that we will ever be able to predict with certainty the outcome of any one introduction event (Ehrlich 1989, Gilpin 1990). Nevertheless, a major goal of invasion ecology is to identify factors that have some ability to explain the outcome of introduction events, to rank factors in terms of their importance, and to determine whether different factors are important under different circumstances, and why (Williamson 1999).

Factors hypothesized to influence establishment can be grouped into three categories: (*a*) characteristics of the species introduced, such as their population growth rate or migratory strategy; (*b*) features of the introduction location, such as its environment or insularity; and (*c*) factors associated with, and often unique to, each introduction event, such as the number of individuals released. We term these species-, location-, and event-level factors, respectively (Blackburn & Duncan 2001a; see Williamson 1999, 2001).

Two crucial consequences of the selectivity in historical introductions (see above) are that factors at different levels are likely to be confounded, and that introduction events are unlikely to represent statistically independent data points when analyses include introduction data from more than one location. We first consider these issues because they are central to how we analyze and interpret comparative data on establishment success in any taxonomic group.

Statistical Analysis of Historical Introduction Outcomes

Species-, location-, and event-level factors are likely to be confounded because selectivity means that introductions are nonrandom with regard to the taxonomy and characteristics of the species introduced and their locations of origin and introduction. For example, species that are more common in Britain were introduced in greater numbers to New Zealand (Blackburn & Duncan 2001b). Species that are common in Britain also possess traits that distinguish them from less common species (e.g., smaller body mass and typical niche position; Nee et al. 1991, Gregory & Gaston 2000). Hence, for introductions of British birds to New Zealand, factors at the species-level (certain life-history traits) and event-level (the number of individuals introduced) are confounded. The solution in this case is to identify confounding factors and to control for them statistically using techniques such as multiple regression.

The second issue is that, when we consider data from several introduction locations, each introduction event is unlikely to represent a statistically independent data point (Blackburn & Duncan 2001a,b). This is because species were often introduced to multiple locations, and multiple species were introduced to each location. If some species are better at establishing than others, or if it is easier to establish at some locations than others, then the outcome of introductions of the same species, or of introductions to the same location, will be correlated. In these circumstances, the introduction of 5 species to each of 10 different locations does not provide 50 independent pieces of evidence to assess how species- or location-level factors affect establishment (see also Hurlbert 1984, McArdle 1996).

Introduction events are unlikely to be independent if they can be clustered by both species and location. Furthermore, species and locations may themselves cluster into higher-level units. In particular, closely related species are more likely to share life-history, morphological, and behavioral traits than are more distantly related species. If these shared traits affect the likelihood of establishment, then introduction outcomes will be clustered by phylogenetic or taxonomic relatedness.

Clustering, leading to correlated responses in groups of observations, typically violates a core assumption of standard statistical models: that the error terms in the model are independent (i.e., uncorrelated). This leads to standard error estimates that are smaller than the true values, resulting in overestimates of the significance of factors included in the model (a greater frequency of Type I errors). In plain terms, we are more likely to obtain models containing spuriously significant factors if we fail correctly to account for the clustering of observations in our data.

There are at least two approaches to dealing with this problem when analyzing historical introduction data. First, it is less of an issue if we consider introductions of multiple species to a single location (for example, the outcome of bird introductions to Tahiti), because these data are not clustered by species and location—we have only one event per species at a single location (likewise for introductions of a single species to multiple locations). While this tactic largely overcomes the problem, it limits the questions we can address, and we still have to consider nonindependence owing to phylogenetic relatedness. Alternatively, there has been renewed interest in developing statistical models that deal directly with the problem of nonindependence in clustered data. These approaches include the use of generalized least squares, linear mixed models, generalized linear mixed models and generalized estimating equations (Diggle et al. 1994, Goldstein 1995). These approaches are appropriate for analyzing spatially, temporally, and phylogenetically clustered data that are regularly encountered in ecological studies. Generalized linear mixed models have been applied to a global analysis of establishment success in birds, to account for the clustering of introduction events by species, taxonomic groups, and geographic locations (Blackburn & Duncan 2001a). Clearly, problems of clustering and nonindependence in introduction data are not specific to birds, and comparative studies of invasion success in any taxonomic group should carefully consider these issues.

Event-Level Influences on Establishment Success

INTRODUCTION EFFORT A feature of introduced species is that they are typically released in low numbers and so start with small founding populations. Smaller populations are at greater risk of extinction from stochastic fluctuations owing to demographic, environmental, or genetic stochasticity (Pimm 1991). In addition, small populations may suffer from Allee effects (for example, an inability to find mates at low population densities) and are more prone to extinction from natural catastrophes (e.g., several populations of introduced birds appear to have died out as a result of hurricanes; see Long 1981). Given that they start out with small founding populations, it is not surprising that most introductions fail. For birds, 744 of 1466 (51%) introduction events in the global data set analyzed by Blackburn

& Duncan (2001a) did not result in establishment, and this success rate is high relative to other taxa (Williamson 1996, Williamson & Fitter 1996).

A straightforward explanation for the success of some introductions is that they involved the release of a greater number of individuals and so were more likely to escape the threats facing small founding populations. That is, establishment success increases with greater introduction effort, or propagule pressure. Because information on release sizes has been recorded in many cases, it is possible to use historical bird introductions to test this hypothesis. All studies that have addressed this question so far have shown that species introduced with greater effort (typically measured as the total number of individuals released) have a higher probability of establishment [Dawson 1984 (cited in Williamson 1996); Newsome & Noble 1986; Griffith et al. 1989; Pimm 1991; Veltman et al. 1996; Duncan 1997; Green 1997; Sol & Lefebvre 2000; Duncan et al. 2001, 2003; Cassey 2001b; Duncan & Blackburn 2002], although all but two of these studies use data from just two locations, New Zealand and Australia. Nevertheless, observational and experimental evidence in other animals has also shown that the size of the founder population increases the chance of establishment (e.g., Hopper & Roush 1993, Berggren 2001, Forsyth & Duncan 2001). These results are consistent with the hypothesis that small populations are less likely to establish owing to stochastic events and that these play a major role in invasions (Green 1997). Nevertheless, there are exceptions. Despite the introduction of over half a million common quail, *Coturnix coturnix*, to more than 30 states in North America between 1875 and 1958 (Bump 1970), no introduction succeeded (Lever 1987). Clearly, factors other than propagule pressure can be critical in determining the fate of introductions (see below).

The role of demographic stochasticity in avian introductions has been explicitly modeled by Legendre et al. (1999). Because demographic stochasticity can generate random fluctuations in sex ratio leading to difficulty in finding a mate, they constructed a two-sex model with an explicit mating system and calculated resulting extinction probabilities via Monte Carlo simulations. The results showed that demographic uncertainty imposes high extinction risk on monogamous as compared to polygynous birds, and on short-lived as compared to long-lived birds. The model fit reasonably well to the extinction probabilities calculated for passerine birds introduced to New Zealand.

Nevertheless, stochasticity is not the only mechanism that could explain the link between introduction effort and establishment success. A greater number of release attempts may improve establishment by increasing the chance that at least one release population encounters favorable conditions for population growth (Crawley 1986). Griffith et al. (1989) found that introductions involving releases over a greater number of years had a greater probability of establishing, independently of the number of individuals released, although Green (1997) failed to find a similar relationship for New Zealand bird introductions. In a more direct test, Veltman et al. (1996) reported that species successfully introduced to New Zealand were released at more localities than unsuccessful species, although the number of localities and the total number of individuals released are correlated in these data.

Green (1997) also suggested that the link between introduction effort and establishment success could arise indirectly if, for example, people directed more effort toward releasing species that seemed pre-adapted to the conditions they would encounter at their location of introduction. Similarly, introduction effort could be related to establishment success indirectly through relationships with abundance in the source location and associated species traits. Nevertheless, studies that have attempted to control for these indirect relationships by including species traits, environmental tolerances, and overseas range size in multiple regression models have consistently found that introduction effort remains by far the strongest independent explanation for establishment success in birds (e.g., Duncan et al. 2001).

CLIMATE/ENVIRONMENTAL MATCHING One of the most frequently stated hypotheses in the biological invasion literature is that species should have a better chance of establishing if the climate and physical environment at the location of introduction and in the species' natural range are closely matched (Brown 1989, Williamson 1996). Data from bird introductions provide some of the very few quantitative tests of the hypothesis that a close climate/environmental match between a target and source location enhances establishment success. We include climate/environmental matching under event-level effects because this is a feature unique to each introduction event.

Avian establishment success is significantly greater when the difference between a species' latitude of origin and its latitude of introduction is small (Blackburn & Duncan 2001a, Cassey 2001b), when climatic conditions in the locations of origin and introduction are more similar (Duncan et al. 2001), and when species are introduced to locations within their native biogeographical region (Blackburn & Duncan 2001a, Cassey 2003). Likewise, for re-introductions, Wolf et al. (1998) found that species re-introduced in the core of their original range had greater establishment success than species released in the periphery. Regions at similar latitudes or within the same biogeographic region are likely to be similar in climatic and habitat conditions. Hence, these results support the hypothesis that introduction success is enhanced if species are matched with suitable environments (Brown 1989).

Also related to the idea that climate/environmental matching is important is the finding in some studies that bird species with larger geographic ranges are more likely to establish following introduction (Moulton & Pimm 1986, Blackburn & Duncan 2001a, Duncan et al. 2001). Species may have large geographic ranges because they can exploit a broad range of conditions (they have large niche breadth), or use conditions that are themselves widespread (they have a typical niche position; Gaston 1994). In a similar vein, it has been suggested that being native to a relatively variable abiotic environment could also enhance establishment success (Ehrlich 1989), because species that survive in such environments can generally cope with a wide range of environmental conditions (see also Stevens 1989). Any of these situations would increase the chance that a species introduced to a new location would encounter conditions favorable to its survival, although only the range size effect has received empirical support.

Location-Level Influences on Establishment Success

Three characteristics of a location are likely to influence establishment: enemies, resources, and the physical environment (Shea & Chesson 2002). Establishment should be favored at locations where there are fewer enemies (predators, parasites, or diseases), more available resources (owing to higher resource availability or an absence of competitors), or a more benign environment. It is difficult to identify precisely which enemies, resources, or aspects of the physical environment influence establishment, so studies of location-level effects often consider features of the environment that ought to be correlated with these characteristics. For example, species richness is commonly used as a surrogate for the number of enemies, competitors or vacant niches likely to be present at a location (Shea & Chesson 2002). Species richness itself tends to co-vary with latitude and whether or not the location is an island (Elton 1958).

In fact, there is little evidence from bird introductions to support the hypothesis (Elton 1958) that species-poor locations are easier to invade than species-rich locations. Case (1996) showed that the number of established bird species does not vary significantly with the richness of the native avifauna nor the variety of potential mammalian predators. Whereas New Zealand has fewer native but a greater number of exotic bird species than Australia, there is no difference in establishment probability for birds introduced to these locations (Sol 2000a). Likewise, a global analysis of bird introductions failed to find a relationship between establishment success and latitude of introduction, implying no significant effect of resident species richness on establishment (Blackburn & Duncan 2001a).

Islands are often viewed as more vulnerable to invasion because of lower species richness, but also owing to factors such as an absence of certain functional groups, reduced competitive ability of island species, intensive human exploitation, or reduced habitat diversity (e.g., Vitousek 1988, Dalmazzone 2000). In support of this notion, Newsome & Noble (1986) reported a higher failure rate for bird species introduced to mainland Australia compared with bird species introduced to Australia's offshore islands. However, this result could be an artifact of differences in invasive ability between the species introduced to mainlands and islands. A species-by-species examination of introduced birds in two independent island-mainland comparisons (New Zealand versus Australia, and Hawaiian Islands versus U.S. mainland) found no evidence that islands were easier to invade (Sol 2000a). This result has been generalized in global analyses of bird introductions that failed to find a relationship between establishment success and whether the introduction was to a mainland or island location (Blackburn & Duncan 2001a, Cassey 2003). The high proportions of exotic bird species found on islands appears primarily to be a consequence of the many attempts to introduce birds to islands (see above To Which Locations Were Birds Introduced?) rather than any inherent feature of islands that make them easy to invade (Sol 2000a, Blackburn & Duncan 2001a).

Correlative studies testing for a relationship between species richness and invasion success have been criticized because variation in species richness may be confounded with variation in other factors that affect establishment, making it difficult to isolate species richness effects (Shea & Chesson 2002). For birds, the studies described above suggest that factors other than resident species richness are overwhelmingly important in determining the outcome of introductions, even after statistically controlling for confounding factors. This lack of an effect is probably real. Competition between native and introduced birds does not appear to regulate establishment because most exotic birds establish in habitats highly modified by humans, such as farmland and urban areas, which are little used by native species (Diamond & Veitch 1981, Simberloff 1992, Smallwood 1994, Case 1996). Such modified habitats may favor pre-adapted invaders through release of resources and enemy reduction (Mack et al. 2000, Shea & Chesson 2002). Thus, many introduced birds may succeed because they exploit either vacant niches created by human activities or existing niches that humans have expanded.

Some authors have suggested that establishment is more likely for species that can exploit niche opportunities left by extinct or declining native species (Herbold & Moyle 1986, Case 1996, Mack et al. 2000). This is less likely if native and exotic species show strong habitat segregation. Nevertheless, Cassey (2001b) showed that, whereas extinct bird species in New Zealand tend to be larger-bodied than extant species, they do not differ in size from established exotic species. He suggested that some established invaders could be occupying the niches of extinct species. Case (1996) likewise reported a positive relationship between the numbers of established and recently extinct bird species at locations around the world. This relationship does not arise because introduced birds cause the extinction of native species: Most extinctions occurred prior to bird introductions. More likely, high extinction rates are associated with high levels of human disturbance, which in turn creates habitat favorable for the establishment of introduced birds (Case 1996).

Habitat segregation suggests that competition between native and exotic species has little role in determining whether or not introduced birds establish. However, it leaves open the question as to whether interactions with native species can prevent introduced species from spreading into more pristine habitats. Diamond & Veitch (1981) reported that, in New Zealand, exotic birds that are abundant on the main islands are virtually absent from unmodified forests on offshore islands, despite having ample opportunities to colonize these. There is no evidence that this segregation results from competitive interactions. More likely, introduced species establish in habitats to which they are pre-adapted but to which many native species are not, and vice versa (Case 1996, Sax & Brown 2000).

Whereas competition between exotic and native bird species does not appear to influence establishment, it is possible that competition between the introduced species may be important at locations with many introductions. In a series of studies, Moulton and coworkers (e.g., Moulton & Pimm 1983; Moulton 1985, 1993; Lockwood et al. 1993; Lockwood & Moulton 1994; Brooke et al. 1995; Moulton & Sanderson 1999; Moulton et al. 2001b) have used two lines of evidence to argue that this is the case. First, for passerine introductions to Hawaii and Saint Helena, the probability that a species will establish declines through time as the number of established species accumulates. Moulton and coworkers argue that it is harder for progressively later introductions to establish because they face increasing competition from a greater number of already established species. Second, at several locations successfully established species overlap less in their morphological characteristics than would be expected if a random selection of the introduced species had established. Moulton and coworkers argue that this pattern arises because species are more likely to fail due to competition with other introduced species of similar morphology, leading to a pattern of morphological overdispersion among the established species.

However, these conclusions have been questioned by studies showing that the same patterns could arise through various artifacts (Simberloff & Boecklen 1991, Duncan 1997, Duncan & Blackburn 2002). Whereas later introductions of passerine birds to New Zealand faced a greater number of already established species and were less likely to succeed, this relationship was confounded by introduction effort (Duncan 1997). Later introductions tended to be of fewer individuals and so were less likely to succeed for that reason. Similarly, Duncan & Blackburn (2002) showed that significant morphological overdispersion among gamebirds established in New Zealand (Moulton et al. 2001b) could not have resulted from competition, but could be explained by greater effort being applied to the introduction of species that were morphologically different from one another. In addition to facing a greater diversity of already established birds, later introductions to New Zealand, and probably elsewhere, would also have faced a greater diversity and abundance of introduced predators, making it impossible to isolate the effect of competition (Duncan et al. 2003). These results for New Zealand do not rule out the importance of inter-specific competition in other introduced bird assemblages, but they do question the evidence presented in support of this mechanism. Overall, there is little unequivocal support for the hypothesis that competition affects the outcome of bird introductions.

The idea that establishment is affected by competitive and predatory interactions has suggested at least three further hypotheses. First, social foraging species should be better invaders than solitary species (Mayr 1965, Ehrlich 1989), because social foraging can increase the probability of detecting predators, locating food, and learning about new food sources. This hypothesis has not received empirical support (Sol 2000b, Duncan et al. 2001). Second, ground-nesting birds should have lower probabilities of establishing than canopy-, shrub-, or hole-nesters, because nest predation is generally higher in ground-nesters (see Reed 1999). There is some support for this hypothesis (Newsome & Noble 1986; McLain et al. 1999; but see Sol et al. 2002). Third, herbivores have been predicted to invade new habitats more easily than carnivores (Hairston et al. 1960, Crawley 1986), because competition is thought to be less intense among herbivores than among carnivores. The hypothesis has not received empirical support for birds (e.g., Veltman et al. 1996), and its assumptions are questionable (Crawley 1986).

A further location-level factor proposed to influence establishment is the area of suitable habitat (Smallwood 1994, Case 1996). Larger areas can support more individuals and are thus more likely to maintain self-sustaining populations in the long-term. However, this fact may be of little relevance to establishment when populations generally start out small. Indeed larger areas may result in more dispersed founding populations and a lower probability of success owing, for example, to greater difficulty finding mates. Case (1996) found a weak, but significant, positive correlation between island area and the number of species introduced, but no correlation with establishment success. Site invasibility was not found to decrease with area in California's nature reserves (Smallwood 1994).

Finally, in a global analysis of bird introductions, Blackburn & Duncan (2001a) found significant variation in establishment rate among biogeographic regions after having controlled for other confounding factors such as latitude. It is unclear what underlies such large-scale geographic variation in establishment probability.

Species-Level Influences on Establishment Success

The widespread success of certain introduced birds, such as the starling (*Sturnus vulgaris*), house sparrow (*Passer domesticus*), and rock dove (*Columba livia*; Long 1981), suggests that some species may simply be good at establishing in novel environments. Whether species differ inherently in their probability of establishing, and if so, what characteristics regulate this difference, are two questions that have long interested ecologists (Mayr 1965, Ehrlich 1989).

DO SPECIES DIFFER IN THEIR PROBABILITY OF ESTABLISHING? Simberloff & Boecklen (1991) noted what they termed an all-or-none (AON) pattern for bird introductions to the Hawaiian islands. Bird species tended to establish successfully or to fail repeatedly on all islands to which they were introduced, with mixed outcomes, where species established on some but not other islands, being rare (but see Moulton 1993; Brooke et al. 1995; Moulton & Sanderson 1996, 1999; Duncan 1997; Duncan & Young 1999). Such a pattern would suggest that species-level attributes are a key determinant of establishment. Some species are particularly good invaders, and so succeed everywhere they are introduced, while other species are poor invaders and fail regardless of the location.

A brief perusal of any compendium of avian introductions (e.g., Long 1981, Lever 1987) is enough to show that birds do not follow a strict AON pattern. This is expected even if species-level attributes strongly influence establishment success (Duncan & Young 1999). Species are unlikely to be either absolutely good or poor invaders because event- and location-level factors will also affect establishment. Hence, even an exceptionally good invader might occasionally fail because of unfavorable circumstances, and vice versa. The key question is whether there is significant variation in establishment probability among introduced species, the alternative being that all species establish with equal probability. In testing this hypothesis, we need to control for confounding factors—an inherently poor invader might appear better than it really is if introduced to locations that are easy to invade. Fitting a mixed model to data on global bird introductions, and having controlled for location- and event-level factors likely to affect establishment, Blackburn & Duncan (2001a) showed that there are highly significant differences among species in establishment probability. Given this variation, do certain attributes of the species themselves distinguish the good from the poor invaders?

WHICH TRAITS INFLUENCE ESTABLISHMENT SUCCESS? The idea that certain species attributes influence establishment success is supported by comparative studies linking the outcome of historical bird introductions to behavioral, life history, and morphological traits (e.g., Newsome & Noble 1986, McLain et al. 1995, Veltman et al. 1996, Green 1997, Cassey 2001a, Sol 2003). However, two problems make it difficult to draw firm general conclusions from the studies to date. First, some studies present results from univariate statistical models while other studies present results from multivariate models. Multivariate models are more powerful in correlative studies because they reduce the likelihood that a significant relationship between two variables results from their correlation with a third variable, assuming that important confounding variables have been included in the analysis. Second, some studies suffer from the problems of non-independence we discussed above. In particular, the importance of traits in explaining establishment is often assessed by fitting a generalized linear model with establishment included as either a binary outcome (the result of each introduction event coded as success or failure) or a binomial outcome (the number of successful events / the total number of introduction events for each species). In either model, a species is represented in the analysis as many times as it has been introduced. However, when the data include multiple introductions of the same species to different locations, then each event will not represent an independent data point because the outcomes of introductions of the same species to different locations are almost certainly correlated. In these circumstances, it is important to consider extensions of generalized linear models (such as generalized linear mixed models or generalized estimating equations) that take into account the clustering of events and minimize the problem of inflated type I error rates.

The traits identified as influencing establishment success are of three types: (*a*) traits that preadapt species to the new environment, (*b*) traits that favor population increase from a low level, and (*c*) traits that constrain establishment success.

Preadaptations to establishment One reason for failing to establish is that a species fails to find a suitable niche in the new environment, either because the niche does not exist or because it is occupied by other species. The chance of finding a favorable niche will depend on the presence of preadaptations to exploit the resources and escape enemies in the new environment.

A generalist species with a wide niche breadth should have, on average, a better chance of finding appropriate resources and environmental conditions wherever it is introduced than a more specialized species (Ehrlich 1989, Forys & Allen 1999). A problem with testing this hypothesis is the difficulty in quantifying niche breadth (Gaston 1994). Using a coarse measure of dietary breadth, McLain et al.

(1999) found that, for 132 passerine species introduced to nine islands, those with a broader dietary range were more likely to establish (see also Wolf et al. 1998, Cassey 2001b). Similarly, Brooks (2001) found that introduced birds categorized as habitat specialists were less likely to establish successfully. However, other studies have failed to find such relationships (Veltman et al. 1996, Sol et al. 2002), and it is difficult to determine if these discrepancies result from a weak influence of niche breadth or reflect the noise associated with using coarse measures to quantify breadth.

A species released into a new environment will face a variety of novel challenges for which it may not be well adapted (Sol 2003). Animals may partly compensate for this lack of adaptive fit by inventing new behaviors or adjusting established behaviors to the novel conditions (Klopfer 1962, Plotkin & Odling-Smee 1979, Morse 1980, Arcese et al. 1997, Brooker et al. 1998, Berger et al. 2001, Lefebvre et al. 2003, Sol 2003). This behavioral flexibility may aid establishment through, for example, the ready adoption of unexploited new food resources, the adjustment of breeding to the prevailing breeding conditions, or rapid behavioral changes to avoid novel enemies (Sol 2003). Because behaviorally flexible species are believed to be more exploratory (Greenberg & Mettke-Hofmann 2001) and ecologically generalist (Sol 2003), they may also have higher chances of discovering and adopting new habitats or new resources that may enhance survival and reproduction in the new environment. The hypothesis that behavioral flexibility enhances establishment (Mayr 1965) is supported by the finding that established birds tend to have a larger brain size per unit body mass and to show more innovative behaviors in their region of origin than failed species (Sol & Lefebvre 2000, Sol et al. 2002).

Certain traits associated with generalist behavior may enhance establishment for some species, whereas other traits may preadapt species to specific habitats. The latter may explain why birds that inhabit human-modified habitats in their native range and those with a history of close association with humans tend to be successful invaders (Sol et al. 2002; see also Mayr 1965, Brown 1989, Sax & Brown 2000).

Securing the bridgehead Most introduced populations start off small and face the threat of stochastic extinction. One argument is that species with higher rates of population growth should have a higher probability of establishing because they can more quickly escape the risks associated with remaining at small population size (Moulton & Pimm 1986, Pimm et al. 1988, Pimm 1991). Because population growth rates are difficult to measure, most studies that test this hypothesis use life-history traits known to be correlated with population growth rate as surrogates. Thus, species with small body mass, short development times, multiple broods per season, and large clutch sizes are expected to have higher establishment. However, small-bodied species with high rates of population growth also tend to have more variable population sizes, which could increase their risk of extinction because population densities will drop to low levels more often (Pimm 1991). In these circumstances, larger-bodied, longer-lived, species with slower rates of population growth that are relatively unaffected by environmental fluctuations are expected to have higher establishment (Legendre et al. 1999, Forsyth & Duncan 2001).

These contradictory theoretical predictions are mirrored in empirical findings. Whereas some studies report a positive relationship between clutch size and establishment success (Green 1997, Cassey 2001b), others have reported negative relationships or no relationship at all (Veltman et al. 1996, Duncan et al. 2001). Cassey (2001b) found that birds with longer generation times were more likely to establish following introduction to New Zealand, but Griffith et al. (1989) reported higher establishment success for re-introduced species classified as "early breeders" than for those considered "late breeders." Results for body mass are similarly equivocal (e.g., Veltman et al. 1996, Green 1997, Sol & Lefebvre 2000, Blackburn & Duncan 2001b, Sol et al. 2002). Moreover, the relationship between body mass and establishment appears to depend on the taxonomic level considered. Cassey (2001a) found that global introduction success is significantly negatively related to body size across species, families, and higher family nodes. However, within taxa, larger-bodied species are more likely to establish.

Population growth rate has also been invoked in the suggestion that colonial species should establish less readily than noncolonial. Colonial species may have, at low population densities, a reduced per capita growth rate owing to the Allee effect, which would increase the probability of stochastic extinction (Reed 1999). The only study that has examined this prediction failed to find support for it (Sol 2000b).

Constraints on success Some traits may lower establishment probability regardless of other traits the species possess. The two most commonly cited such traits are migratory behavior and sexual color dimorphism. In New Zealand, migratory species are less likely to establish than nonmigratory species (Veltman et al. 1996, Duncan et al. 2003; but see Duncan et al. 2001, Sol et al. 2002). The reasons for this result are unclear, but they could be related to physiological costs associated with preparing for migration, or because suitable habitats to migrate to are not available or cannot be found from the new location (Veltman et al. 1996). In birds, sexual dimorphism in plumage color appears to be the result of sexual selection, and in some studies sexually dimorphic species are less likely to establish following introduction than sexually monomorphic species (McLain et al. 1995, 1999; Sorci et al. 1998; Sol et al. 2002; but see Sol & Lefebvre 2000; Duncan et al. 2001, 2003). The reasons for this may be associated with the cost to males of producing and maintaining secondary sexual characters (A.P. Møller, M.C. Gontard-Danek, unpublished manuscript).

TAXONOMIC VARIATION IN ESTABLISHMENT PROBABILITY Given that species differ significantly in establishment probability, how is this variation partitioned among different levels in the taxonomic hierarchy? If significant variation resides at high levels in the hierarchy, say among families, then it should be possible to predict establishment from knowledge of a species' taxonomic status alone (Daehler & Strong 1993). In addition, knowing which families, or other higher taxa, have a high establishment rate may suggest traits shared by members of those groups that contribute to their success.

Applying methods such as nested ANOVA and mixed models to bird introduction data reveals that most variation in establishment probability resides at low rather than high taxonomic levels (Blackburn & Duncan 2001a, Sol et al. 2002). That is, closely related species are likely to differ substantially in their probability of establishment. Nevertheless, the extent to which at least some variation in establishment resides at higher taxonomic levels remains unclear. Sol et al. (2002) reported that 21% of variation in establishment rate was found at order level. Having controlled for variation in location-level effects, Blackburn & Duncan (2001a) found no similar significant effect in their global analysis of bird introductions. In Australia, introduced gamebirds (order Galliformes) were significantly less likely to establish than other species (Duncan et al. 2001), with similar low rates of establishment observed in New Zealand and the United States. This observation could, in part, reflect hunting pressure, although in the United States long moratorium periods were implemented before the hunting of gamebird species commenced (Long 1981). Moulton et al. (2001a) also reported lower establishment success in Galliformes and Columbiformes relative to Passeriformes.

The general pattern of variance partitioning suggests that interspecific variation in establishment probability is driven primarily by traits that vary among closely related species. It follows that traits shared by closely related species, including phylogenetically conserved life history traits such as body mass and clutch size, should poorly explain establishment. The equivocal results reported for many such traits, along with their failure significantly to explain establishment for global bird introductions (Blackburn & Duncan 2001a), bear this out. The general failure to identify life history traits that consistently explain establishment may reflect the overwhelming importance of factors such as introduction effort or environmental matching. Equally, however, there may be no simple set of traits that favor establishment. Instead, different traits may enhance establishment under different conditions, and at any one location there may be establishment opportunities for species possessing a variety of trait combinations.

SPREAD

The European starling (*Sturnus vulgaris*) established following introduction to New York state in the late nineteenth century and has subsequently spread to become among the most widely distributed and abundant birds in North America. At about the same time, the crested myna (*Acridotheres cristatellus*) was introduced and became established in Vancouver. This close relative of the starling became reasonably numerous but spread little, remaining largely confined to Vancouver and its environs, with the population now having dwindled to low numbers. Relative to other taxa, birds are generally good dispersers. Why then, following establishment, do bird species differ so markedly in the rate at which they spread and the final extent of their distribution?

Because birds are well studied, sufficient records exist quantitatively to model range expansion in several introduced species (Hengeveld 1989, Van den Bosch et al. 1992, Veit & Lewis 1996, Williamson 1996, Shigesada & Kawasaki 1997, Lensink 1998, Gammon & Maurer 2002, Silva et al. 2002). Whereas these models allow us to examine factors affecting the rate of expansion in a few typically widespread species, no study has yet quantified or examined the reasons underlying interspecific variation in rate of spread.

Three studies have nevertheless examined the determinants of current geographic range size in two introduced avifaunas: New Zealand (Duncan et al. 1999) and Australia (Duncan et al. 2001, Williamson 2001). The extent of habitat or environmental conditions suitable for an introduced species should ultimately constrain its distribution in a new location. This hypothesis is supported by data from both avifaunas. In New Zealand, the species with the largest geographic ranges were those whose preferred habitat was most widespread, specifically species that use extensive human-modified habitats such as farmland. On the larger and more climatically varied continent of Australia, species with larger geographic ranges were those with a greater area of more climatically suitable habitat available, with this variable explaining 69% of the variation in range sizes.

All three studies show that species with a larger native geographic range size achieved a larger range size in the location of introduction (Duncan et al. 1999, 2001; Williamson 2001). At least two explanations could underlie this relationship. First, as we have discussed, species with larger native ranges may have broader environmental tolerances or use more widespread resources, enhancing success at both establishment and spread. We suspect this is the primary cause of this relationship. The second explanation follows from the finding that species with larger native range sizes tend to be introduced more often and in greater numbers (Blackburn & Duncan 2001b). In New Zealand, the species introduced with greater effort were not only more likely to establish but were also more likely to spread and achieve a larger geographic range size (Duncan et al. 1999). One explanation for this finding is that species with large founding populations might be able to capture a greater proportion of shared resources from species with smaller founding populations. This initial advantage could have compounded itself—those species initially able to capture a greater share of resources would have had faster population growth and rate of spread, allowing them further to pre-empt resources at newly colonized sites as their ranges expanded. As we would expect, this effect is most pronounced among closely related species that are more likely to compete for similar resources (Duncan et al. 1999). This explanation implies that competition may play a role in limiting the range sizes of established species but not in influencing establishment success (see above). Nevertheless, it requires further testing even to establish if the relationship between introduction effort and range size holds more generally (it does not for Australia; Duncan et al. 2001).

Finally, several life history traits consistently relate to the geographic range size of established species in New Zealand and Australia (Duncan et al. 1999, 2001; Cassey 2001b; but see Williamson 2001). More widespread species tend to be those with life history traits associated with high population growth rates, characteristically small-bodied, rapidly developing species with high fecundity. It has been suggested that species with fast population growth rates have larger ranges because they may be less vulnerable to local extinction when colonizing unoccupied sites (Gaston 1988).

FUTURE DIRECTIONS

Throughout this review, we have emphasized how the quantity and quality of data on historical bird introductions have provided opportunities to test key hypotheses about the factors underpinning invasion success. We finish by highlighting five promising research directions.

First, attempts to analyze data on bird introductions have highlighted issues that are likely to be a feature of all historical introduction data: confounding of explanatory variables and non-independence of observations. Given that historical introduction data provide us with opportunities to test hypotheses that we cannot test experimentally (because of cost or ethical considerations), a priority is to identify and apply appropriate statistical methods to analyzing such data for all taxonomic groups. Undoubtedly, the degree and type of non-independence will depend on the circumstances surrounding the transport and introduction of different taxa. Birds and plants selected for transport and introduction, for example, probably show different patterns of taxonomic clustering. Analyses that do not consider issues of non-independence should be interpreted with caution.

Second, while comparative studies of bird introductions have provided useful tests of several invasion hypotheses, others remain for which bird introductions appear ideally suited. For example, the "enemy release" hypothesis posits that some invaders do better in their location of introduction than in their native range owing to a lack of natural enemies (e.g., competitors, predators, and pathogens). Because birds are well studied, a considerable amount of detailed information exists on the factors regulating the populations of many bird species in their native ranges. Similarly detailed studies identifying the factors regulating populations of the same species that have established at new locations could test, for example, whether release from predation or competition is the explanation underlying enhanced success in the new location.

Third, measuring and explaining the impact of invaders remains a major unresolved issue in invasion biology (Ricciardi et al. 2000). The impacts of bird invaders have generally received less attention than those associated with other taxa (see, e.g., Ebenhard 1988). One possible reason is the perception that, because most avian invaders occur in human-modified habitats rather than in pristine habitats (Case 1996), their ecological impact should be relatively less important (e.g., Diamond & Veitch 1981). However, birds have also been reported to generate serious ecological impacts on recipient communities, including hybridization and introgression with native species (Rhymer & Simberloff 1996), transmission of diseases (van Riper et al. 1986), competition and predation (Penny 1974; but see Koenig 2003), and habitat alteration (Ebenhard 1988). In addition, introduced birds have major economic impacts in many locations (Lever 1994, Bomford & Sinclair 2002). Defining and measuring precisely what we mean by impact remains a challenging task, but one that is critical to setting priorities for managing invasive species (Parker et al. 1999; Smith et al. 1999; Williamson 1999, 2001; Ricciardi et al. 2000). Assuming that we can quantify impacts, we can, in principle, apply comparative methods to identify why some species have greater impact than others, as has been done for other invasion transitions, and use this information in explanatory models. Here again it will be important to consider how attributes of the invader and characteristics of the community interact to determine impact. For example, islands are hypothesized to be more vulnerable to the impacts of exotic invaders because island species have not been exposed to mainland selective pressures (Loope et al. 1988, Simberloff 1995). Data on introduced birds may provide opportunities to test these and related hypotheses.

Fourth, the field of invasion ecology has been criticized for drawing a distinction between invasions resulting from human-caused introductions and natural invasions (as we do in this study) when the underlying processes may be very similar and mutually informative (Davis et al. 2001). Because many natural bird invasions have been well documented (e.g., Hengeveld 1989, Clegg et al. 2002), birds provide opportunities to compare these invasion pathways and to assess whether an understanding of invasion dynamics gleaned from human-caused introductions can reliably inform us of natural invasion processes, and vice versa. We caution that this need not be the case. In particular, introduced birds are a distinctly nonrandom subset of the world's birds so that conclusions about how these species behave may not transfer to other bird groups.

Finally, species introduced to novel habitats provide unique opportunities to investigate the evolutionary process (Mooney & Cleland 2001). Established populations of many introduced species are isolated from source populations and may diverge rapidly from their ancestors through a combination of divergent natural selection, genetic drift and divergence under uniform selection (e.g., Selander & Johnston 1967, Baker & Moeed 1987, Baker 1992, Reznick & Ghalambor 2001, Badyaev et al. 2002). In their classic studies on geographic variation in house sparrows introduced to North America, Selander & Johnston (1967) found major geographic divergence in coloration and morphometric characters within less than 100 years following establishment. More recent studies in other birds have also demonstrated rapid genetic and life history differentiation in introduced populations (e.g., Baker & Mooed 1987, Baker 1992, Badyaev et al. 2002). Given the well-documented history of many bird introductions, often detailing dates and numbers of individuals released, and the range of locations at which some species have established, populations of introduced birds provide a remarkable and largely untapped resource for addressing questions about evolutionary change.

ACKNOWLEDGMENTS

We thank Julie Lockwood and Mark Williamson for comments on the manuscript. TMB especially thanks M. Bergman. DS was supported by a Québec Ministry of Education Postdoctoral Fellowship and a NSERC (Canada) grant to Louis Lefebvre.

The *Annual Review of Ecology, Evolution, and Systematics* is online at http://ecolsys.annualreviews.org

LITERATURE CITED

Arcese P, Keller LF, Cary JR. 1997. Why hire a behaviorist into a conservation or management team? In *Behavioral Approaches to Conservation in the Wild*, ed. JR Clemmons, R Buchholz, pp. 48–71. Cambridge, UK: Cambridge Univ. Press

Badyaev AV, Hill GE, Beck ML, Dervan AE, Duckworth RA, et al. 2002. Sex-biased hatching order and adaptive population divergence in a passerine bird. *Science* 295:316–18

Baker AJ. 1992. Genetic and morphometric divergence in ancestral European and descendent New Zealand populations of chaffinches (*Fringilla coelebs*). *Evolution* 46:1784–800

Baker AJ, Moeed A. 1987. Rapid genetic differentiation and founder effects in colonizing populations of common mynas (*Acridotheres tristis*). *Evolution* 41:525–38

Bennett PM, Owens IPF. 2002. *The Evolutionary Ecology of Birds*. Oxford: Oxford Univ. Press

Berger J, Swenson JE, Persson I-L. 2001. Recolonizing carnivores and native prey: conservation lessons from Pleistocene extinctions. *Science* 291:1036–39

Berggren A. 2001. Colonization success in Roesel's bush-cricket *Metrioptera roeseli*: the effect of propagule size. *Ecology* 82:274–80

Bergmann C. 1847. Ueber die Verhältnisse der Wärmeökonomie der Thiere zu ihrer Grösse. *Göttinger studien* 3:595–708

Blackburn TM, Duncan RP. 2001a. Determinants of establishment success in introduced birds. *Nature* 414:195–97

Blackburn TM, Duncan RP. 2001b. Establishment patterns of exotic birds are constrained by non-random patterns in introduction. *J. Biogeogr.* 28:927–39

Blackburn TM, Gaston KJ. 1994. The distribution of body sizes of the worlds bird species. *Oikos* 70:127–30

Bomford M, Sinclair R. 2002. Australian research on bird pests: impact, management and future directions. *Emu* 102:29–45

Brooke RK, Lockwood JL, Moulton MP. 1995. Patterns of success in passeriform bird introductions on Saint Helena. *Oecologia* 103:337–42

Brooker ML, Davies NB, Noble DG. 1998. Rapid decline of host defences in response to reduced cuckoo parasitism: behavioural flexibility of reed warblers in a changing world. *Proc. R. Soc. London Ser. B* 265:1277–82

Brooks T. 2001. Are unsuccessful avian invaders rarer in their native range than successful invaders? In *Biotic Homogenization*, ed. JL Lockwood, ML McKinney, pp. 125–55. New York: Kluwer

Brown JH. 1989. Patterns, modes and extents of invasions by vertebrates. See Drake et al. 1989, pp. 85–109

Bump G. 1970. *The Coturnix or Old World Quails*. US Dep. Int. Fish, Wildl. Serv., Bur. Sport, Fish. Wildl, FGL-10. Washington, DC: GPO

Cardillo M. 2002. The life-history basis of latitudinal diversity gradients: how do species traits vary from the poles to the equator? *J. Anim. Ecol.* 71:79–87

Case TJ. 1996. Global patterns in the establishment and distribution of exotic birds. *Biol. Conserv.* 78:69–96

Cassey P. 2001a. Are there body size implications for the success of globally introduced land birds? *Ecography* 24:413–20

Cassey P. 2001b. Determining variation in the success of New Zealand land birds. *Global Ecol. Biogeogr.* 10:161–72

Cassey P. 2002. *Comparative analysis of successful establishment among introduced landbirds*. PhD thesis. Griffith Univ., Aust. 172 pp.

Cassey P. 2003. A comparative analysis of the relative success of introduced landbirds on islands. *Evol. Ecol. Res.* In press

Clegg SM, Degnan SM, Kikkawa J, Moritz C, Estoup A, Owens IPF. 2002. Genetic consequences of sequential founder events by an island-colonizing bird. *Proc. Natl. Acad. Sci. USA* 99:8127–32

Crawley MJ. 1986. The population biology of invaders. *Philos. Trans. R. Soc. London B* 314:711–31

Daehler CC. 2001. Two ways to be an invader, but one is more suitable for ecology. *Bull. Ecol. Soc. Amer.* 82:101–2

Daehler CC, Strong DR. 1993. Prediction and biological invasions. *Trends Ecol. Evol.* 8:380–81

Dalmazzone S. 2000. Economic factors affecting vulnerability to biological invasions. See Perrings et al. 2000, pp. 17–30

Davis MA, Thompson K. 2000. Eight ways to be a colonizer; two ways to be an invader: A proposed nomenclature scheme for invasion ecology. *Bull. Ecol. Soc. Am.* 81:226–30

Davis MA, Thompson K, Grime JP. 2001. Charles S. Elton and the dissociation of invasion ecology from the rest of ecology. *Divers. Distrib.* 7:97–102

Dawson JC. 1984. *A statistical analysis of species characteristics affecting the success of bird introductions*. BSc. thesis. Univ. York, UK. 47 pp.

Diamond JM, Veitch CR. 1981. Extinctions and introductions in the New Zealand avifauna: cause and effect? *Science* 211:499–501

Diggle P, Liang K, Zeger SL. 1994. *Analysis of Longitudinal Data*. Oxford: Clarendon

Drake JA, Mooney HA, di Castri F, Groves RH, Kruger FG, eds. 1989. *Biological Invasions: a Global Perspective*. Chichester: Wiley. 525 pp.

Duncan RP. 1997. The role of competition and introduction effort in the success of passeriform birds introduced to New Zealand. *Am. Nat.* 149:903–15

Duncan RP, Blackburn TM. 2002. Morphological over-dispersion in game birds (Aves: Galliformes) successfully introduced to New Zealand was not caused by interspecific competition. *Evol. Ecol. Res.* 4:551–61

Duncan RP, Blackburn TM, Cassey P. 2003. Factors affecting the release, establishment and spread of introduced birds in New Zealand. In *Biological Invasions in New Zealand*, ed. RB Allen, WG Lee. Berlin: Springer-Verlag. In press

Duncan RP, Blackburn TM, Veltman CJ. 1999. Determinants of geographical range sizes: a test using introduced New Zealand birds. *J. Anim. Ecol.* 68:963–75

Duncan RP, Bomford M, Forsyth DM, Conibear L. 2001. High predictability in introduction outcomes and the geograpical range size of introduced Australian birds: a role for climate. *J. Anim. Ecol.* 70:621–32

Duncan RP, Young JR. 1999. The fate of Passeriform introductions on oceanic islands. *Conserv. Biol.* 13:934–36

Dunning JB. 1993. *CRC Handbook of Avian Body Masses*. Boca Raton, FL: CRC

Ebenhard T. 1988. Introduced birds and mammals and their ecological effects. *Swed. Wildl. Res.* 13:1–107

Ehrlich PH. 1989. Attributes of invaders and the

invading processes: vertebrates. See Drake et al. 1989, pp. 315–28
Elton CS. 1958. *The Ecology of Invasions by Animals and Plants*. London: Methuen
Evans LGR. 1997. *The Definitive Checklist of the Birds of the Western Palaearctic*. Amersham: LGRE
Forsyth DM, Duncan RP. 2001. Propagule size and the relative success of exotic ungulate and bird introductions to New Zealand. *Am. Nat.* 157:583–95
Forys JM, Allen CR. 1999. Biological invasions and deletions: community change in south Florida. *Biol. Conserv.* 87:341–47
Gammon DE, Maurer BA. 2002. Evidence for non-uniform dispersal in the biological invasions of two naturalized North American bird species. *Global Ecol. Biogeogr.* 11:155–61
Gaston KJ. 1988. Patterns in the local and regional dynamics of moth populations. *Oikos* 53:49–57
Gaston KJ. 1994. *Rarity*. London: Chapman & Hall
Gaston KJ, Blackburn TM. 2000. *Pattern and Process in Macroecology*. Oxford: Blackwell Sci.
Gilpin M. 1990. Ecological prediction. *Science* 248:88–89
Goldstein H. 1995. *Multilevel Statistical Models*. Oxford: Edward Arnold
Green RE. 1997. The influence of numbers released on the outcome of attempts to introduce exotic bird species to New Zealand. *J. Anim. Ecol.* 66:25–35
Greenberg RS, Mettke-Hofmann C. 2001. Ecological aspects of neophobia and neophilia in birds. In *Current Ornithology*, ed. V Nolan Jr, ED Ketterson, CF Thompson, 16:119–78. New York: Kluwer
Gregory RD, Gaston KJ. 2000. Explanations of commonness and rarity in British breeding birds: separating resource use and resource availability. *Oikos* 88:515–26
Griffith B, Scott JM, Carpenter JW, Reed C. 1989. Translocation as a species tool: status and strategy. *Science* 245:477–80
Guild E. 1938. Tahitian aviculture: acclimation of foreign birds. *Avicultural Magazine* 3:8–11
Guild E. 1940. Western bluebirds in Tahiti. *Avicultural Magazine* 5:284–85
Guix JC, Jover L, Ruiz X. 1997. Muestreos del comercio de psitácidos neotropicales en la ciudad de Barcelona, España: 1991–1996. *Ararajuba* 5:159–67
Hairston NG, Smith FE, Slobodkin LB. 1960. Community structure, population control, and competition. *Am. Nat.* 94:421–25
Hengeveld R. 1989. *Dynamics of Biological Invasions*. London: Chapman & Hall
Herbold B, Moyle PB. 1986. Introduced species and vacant niches. *Am. Nat.* 128:751–60
Hopper KR, Roush RT. 1993. Mate finding, dispersal, number released, and the success of biological control introductions. *Ecol. Entomol.* 18:321–31
Hurlbert SH. 1984. Pseudoreplication and the design of ecological field experiments. *Ecol. Mon.* 54: 187–211
Klopfer PH. 1962. *Behavioral Aspects of Ecology*. London: Prentice-Hall
Koenig WD. 2003. European starlings and their effect on native cavity-nesting birds. *Conserv. Biol.* In press
Kolar CK, Lodge DM. 2001. Progress in invasion biology: predicting invaders. *Trends Ecol. Evol.* 16:199–204
Lack D. 1947. The significance of clutch-size. Part I—Intraspecific variations. *Ibis* 89:302–52
Lack D. 1948. The significance of clutch-size. Part III—Some interspecific comparisons. *Ibis* 90:25–45
Lefebvre L, Reader SM, Sol D. 2003. Brains, innovations and evolution in birds and primates. *Brain Behav. Evol.* In press
Legendre S, Clobert J, Møller AP, Sorci G. 1999. Demographic stochasticity and social mating system in the process of extinction of small populations: the case of passerines introduced to New Zealand. *Am. Nat.* 153:449–63
Lensink R. 1998. Temporal and spatial expansion of the Egyptian goose *Alopochen*

aegyptiacus in The Netherlands, 1967–94. *J. Biogeogr.* 25:251–63

Lever C. 1987. *Naturalized Birds of the World.* New York: Longman Sci. Tech.

Lever C. 1994. *Naturalized Animals: the Ecology of Successfully Introduced Species*. London: T & AD Poyser

Lockwood JL. 1999. Using taxonomy to predict success among introduced avifauna: relative importance of transport and establishment. *Conserv. Biol.* 13:560–67

Lockwood JL, Brooks TM, McKinney ML. 2000. Taxonomic homogeneization of the global avifauna. *Anim. Conserv.* 3:27–35

Lockwood JL, Moulton MP. 1994. Ecomorphological pattern in Bermuda birds: the influence of competition and implications for nature preserves. *Evol. Ecol.* 8:53–60

Lockwood JL, Moulton MP, Anderson SK. 1993. Morphological assortment and the assembly of communities of introduced passeriforms on oceanic islands: Tahiti versus Oahu. *Am. Nat.* 141:398–408

Lodge DM. 1993. Biological invasions: lessons for ecology. *Trends Ecol. Evol.* 8:133–36

Long JL. 1981. *Introduced Birds of the World.* London: David & Charles

Loope LL, Hamman O, Stone CP. 1988. Comparative conservation biology of oceanic archipelagos. Hawaii and the Galápagos. *BioScience* 38:272–82

Mack RN, Simberloff D, Lonsdale WM, Evans H, Clout M, Bazzaz FA. 2000. Biotic invasions: Causes, epidemiology, global consequences, and control. *Ecol. Appl.* 10:689–710

Mayr E. 1965. The nature of colonising birds. In *The Genetics of Colonizing Species*, ed. HG Baker, GL Stebbins, pp. 29–43. New York: Academic

McArdle BH. 1996. Levels of evidence in studies of competition, predation, and disease. *NZ J. Ecol.* 20:7–15

McLain DK, Moulton MP, Redfearn TP. 1995. Sexual selection and the risk of extinction of introduced birds on oceanic islands. *Oikos* 90:599–605

McLain DK, Moulton MP, Sanderson JG. 1999. Sexual selection and extinction: the fate of plumage-dimorphic and plumage-monomorphic birds introduced onto islands. *Evol. Ecol. Res.* 1:549–65

McNeely JA, ed. 2001. *The Great Reshuffling: Human Dimensions of Invasive Alien Species.* Gland, Switz./Cambridge, UK: IUCN

Mielke HW. 1989. *Patterns of Life: Biogeography of a Changing World.* Boston: Unwin Hyman

Mooney HA, Cleland EE. 2001. The evolutionary impact of invasive species. *Proc. Natl. Acad. Sci. USA* 98:5446–51

Morse DH. 1980. *Behavioral Mechanisms in Ecology*. Cambridge, MA: Harvard Univ. Press

Moulton MP. 1985. Morphological similarity and coexistence of congeners: an experimental test with introduced Hawaiian birds. *Oikos* 44:301–5

Moulton MP. 1993. The all-or-none pattern in introduced Hawaiian passeriformes: the role of competition sustained. *Am. Nat.* 141:105–19

Moulton MP, Miller KE, Tillman EA. 2001a. Patterns of success among introduced birds in the Hawaiian islands. *Stud. Avian Biol.* 22:31–46

Moulton MP, Pimm SL. 1983. The introduced Hawaiian avifauna: biogeographic evidence for competition. *Am. Nat.* 121:669–90

Moulton MP, Pimm SL. 1986. Species introductions to Hawaii. In *Ecology of Biological Invasions in North America and Hawaii*, ed. HA Mooney, JA Drake, pp. 231–49. Berlin: Springer-Verlag

Moulton MP, Sanderson JG. 1996. Predicting the fate of Passeriform introductions in oceanic islands. *Conserv. Biol.* 11:552–58

Moulton MP, Sanderson JG. 1999. Fate of Passeriform introductions: reply to Duncan and Young. *Conserv. Biol.* 13:937–38

Moulton MP, Sanderson JG, Labisky RF. 2001b. Patterns of success in game bird (Aves: Galliformes) introductions to the Hawaiian islands and New Zealand. *Evol. Ecol. Res.* 3:507–19

Nee S, Read AF, Greenwood JJD, Harvey PH. 1991. The relationship between abundance and body size in British birds. *Nature* 351:312–13

Newsome AE, Noble IR. 1986. Ecological and physiological characters of invading species. In *Biological Invasions*, ed. RH Groves, JJ Burdon, pp. 1–20. Cambridge, UK: Cambridge Univ. Press

Owens IPF, Bennett PM. 1995. Ancient ecological diversification explains life-history variation among living birds. *Proc. R. Soc. London Ser. B* 261:227–32

Parker IM, Simberloff D, Lonsdale WM, Goodell K, Whonham M, et al. 1999. Impact: toward a framework for understanding the ecological effects of invaders. *Biol. Invasions* 1:3–19

Penny M. 1974. *The Birds of Seychelles and the Outlying Islands*. London: Collins

Perrings C, Williamson M, Dalmazzone S, eds. 2000. *The Economics of Biological Invasions*. Cheltenham, UK: Edward Elgar. 249 pp.

Pimm SL. 1991. *The Balance of Nature?* Chicago: Univ. Chicago Press

Pimm SL, Jones HL, Diamond J. 1988. On the risk of extinction. *Am. Nat.* 132:757–85

Plotkin HC, Odling-Smee FJ. 1979. Learning, change and evolution: an enquiry into the telenomy of learning. *Adv. Stud. Behav.* 10:1–41

Rahbek C. 1997. The relationship among area, elevation, and regional species richness in Neotropical birds. *Am. Nat.* 149:875–902

Reed JM. 1999. The role of behavior in recent avian extinctions and endangerments. *Conserv. Biol.* 13:232–41

Reznick DN, Ghalambor CK. 2001. The population ecology of contemporary adaptations: what empirical studies reveal about the conditions that promote adaptive evolution. *Genetica* 112–13:183–98

Rhymer JM, Simberloff D. 1996. Extinction by hybridization and introgression. *Annu. Rev. Ecol. Syst.* 27:83–109

Ricciardi A, Rasmussen JB. 1998. Predicting the identity and impact of future biological invaders: a priority for aquatic resource management. *Can. J. Fish. Aquat. Sci.* 55:1759–65

Ricciardi A, Steiner WWM, Mack RN, Simberloff D. 2000. Toward a global information system for invasive species. *BioScience* 50:239–44

Richardson DM, Pysek P, Rejmanek M, Barbour MG, Panetta FD, West CJ. 2000. Naturalization and invasion of alien plants: concepts and definitions. *Divers. Distrib.* 6:93–107

Sakai AK, Allendorf FW, Holt JS, Lodge DM, Molofsky J, et al. 2001. The population biology of invasive species. *Annu. Rev. Ecol. Syst.* 32:305–32

Sax DF, Brown JH. 2000. The paradox of invasion. *Global Ecol. Biogeogr.* 9:363–71

Selander RK, Johnston RF. 1967. Evolution in the house sparrow. Intrapopulation variation in North America. *Condor* 69:217–58

Shea K, Chesson P. 2002. Community ecology theory as a framework for biological invasions. *Trends Ecol. Evol.* 17:170–76

Shigesada N, Kawasaki K. 1997. *Biological Invasions: Theory and Practice.* Oxford: Oxford Univ. Press

Sibley CG, Ahlquist JE. 1990. *Phylogeny and Classification of Birds: a Study in Molecular Evolution*. New Haven: Yale Univ. Press

Sibley CG, Monroe BL. 1990. *Distribution and Taxonomy of Birds of the World.* New Haven: Yale Univ. Press

Sibley CG, Monroe BL. 1993. *Supplement to the Distribution and Taxonomy of Birds of the World.* New Haven: Yale Univ. Press

Silva T, Reino LM, Borralho R. 2002. A model for range expansion of an introduced species: the common waxbill *Estrilda astrild* in Portugal. *Divers. Distrib.* 8:319–26

Simberloff D. 1992. Extinction, survival, and effects of birds introduced to the Mascarenes. *Acta Oecol.* 13:663–78

Simberloff D. 1995. Why do introduced species appear to devastate islands more than mainland areas? *Pac. Sci.* 49:87–97

Simberloff D, Boecklen W. 1991. Patterns of extinction in the introduced Hawaiian avifauna: a reexamination of the role of competition. *Am. Nat.* 138:300–27

Smallwood KS. 1994. Site invasibility by exotic birds and mammals. *Biol. Conserv.* 69:251–59

Smith CS, Lonsdale WM, Fortune J. 1999. When to ignore advice: invasion predictions and decision theory. *Biol. Invasion* 1:89–96

Sol D. 2000a. Are islands more susceptible to be invaded than continents? Birds say no. *Ecography* 23:687–92

Sol D. 2000b. *Introduced species: a significant component of the global environmental change.* PhD thesis. Univ. Barcelona. 160 pp.

Sol D. 2003. Behavioural flexibility: a neglected issue in the ecological and evolutionary literature? In *Animal Innovation*, ed. K Laland, SM Reader. Oxford: Oxford Univ. Press. In press

Sol D, Lefebvre L. 2000. Behavioural flexibility predicts invasion success in birds introduced to New Zealand. *Oikos* 90:599–605

Sol D, Timmermans S, Lefebvre L. 2002. Behavioural flexibility and invasion success in birds. *Anim. Behav.* 63:495–502

Sorci G, Møller AP, Clobert J. 1998. Plumage dichromatism of birds predicts introduction success in New Zealand. *J. Anim. Ecol.* 67:263–69

Stevens GC. 1989. The latitudinal gradient in geographical range: how so many species coexist in the tropics. *Am. Nat.* 133:240–56

Thomson GM. 1922. *The Naturalization of Plants and Animals in New Zealand.* Cambridge, UK: Cambridge Univ. Press

Van den Bosch F, Hengeveld R, Metz JAJ. 1992. Analysing the velocity of animal range expansion. *J. Biogeogr.* 19:135–50

van Riper C III, van Riper CG, Goff ML, Laird M. 1986. The epizootiology and ecological significance of malaria in Hawaiian land birds. *Ecol. Monog.* 56:327–44

Veit RR, Lewis MA. 1996. Dispersal, population growth, and the Allee effect: dynamics of the house finch invasion of eastern North America. *Am. Nat.* 148:255–74

Veltman CJ, Nee S, Crawley MJ. 1996. Correlates of introduction success in exotic New Zealand birds. *Am. Nat.* 147:542–57

Vermeij GJ. 1996. An agenda for invasion biology. *Biol. Conserv.* 78:3–9

Vitousek PM. 1988. Diversity and biological invasions of oceanic islands. In *Biodiversity*, ed. EO Wilson, FM Peter, pp. 181–89. Washington, DC: Nat. Acad. Press

Vitousek PM, D'Antonio CM, Loope LL, Rejmánek M, Westbrooks R. 1997. Introduced species: a significant component of human-caused global change. *NZ J. Ecol.* 21:1–16

Williamson M. 1996. *Biological Invasions.* London: Chapman & Hall

Williamson M. 1999. Invasions. *Ecography* 22:5–12

Williamson M. 2001. Can the impacts of invasive species be predicted? In *Weed Risk Assessment*, ed. RH Groves, FD Panetta, JG Virtue, pp. 20–33. Collingwood, Aust.: CSIRO

Williamson M, Fitter A. 1996. The varying success of invaders. *Ecology* 77:1661–66

Wolf CM, Griffith B, Reed C, Temple SA. 1998. Avian and mammalian translocations: update and reanalysis of 1987 survey data. *Conserv. Biol.* 86:1142–54

Annu. Rev. Ecol. Evol. Syst. 2003. 34:99–125
doi: 10.1146/annurev.ecolsys.34.011802.132359

First published online as a Review in Advance on September 29, 2003

THE EFFECTS OF GENETIC AND GEOGRAPHIC STRUCTURE ON NEUTRAL VARIATION

Brian Charlesworth, Deborah Charlesworth, and Nicholas H. Barton
Institute for Cell, Animal, and Population Biology, University of Edinburgh, Edinburgh EH9 3JT, United Kingdom; email: Brian.Charlesworth@ed.ac.uk, Deborah.Charlesworth@ed.ac.uk, N.H.Barton@ed.ac.uk

Key Words genetic drift, effective population size, migration, recombination, coalescent process

■ **Abstract** Variation within a species may be structured both geographically and by genetic background. We review the effects of such structuring on neutral variants, using a framework based on the coalescent process. Short-term effects of sex differences and age structure can be averaged out using fast timescale approximations, allowing a simple general treatment of effective population size and migration. We consider the effects of geographic structure on variation within and between local populations, first in general terms, and then for specific migration models. We discuss the close parallels between geographic structure and stable types of genetic structure caused by selection, including balancing selection and background selection. The effects of departures from stability, such as selective sweeps and population bottlenecks, are also described. Methods for distinguishing population history from the effects of ongoing gene flow are discussed. We relate the theoretical results to observed patterns of variation in natural populations.

INTRODUCTION

The neutral theory of molecular evolution has transformed evolutionary genetics by providing a null model against which alternative hypotheses can be tested (Kimura 1983). There is a rapid accumulation of data on natural variation at the DNA level, from microsatellites to single nucleotide polymorphisms, much of which is presumed to be nearly neutral. The action of selection at a particular site in the genome can cause deviations from the patterns predicted by neutral theory at nearby sites, which are not themselves the direct targets of selection (Kreitman 2000). This is triggering efforts to conduct genome-wide searches for evidence of selection (Akey et al. 2002, Harr et al. 2002). However, many other evolutionary processes can cause departures from the predictions of classical neutral theory. Although this complicates tests for the action of selection, it also provides opportunities to make inferences about such processes.

It is hard to disentangle all the processes that have shaped the properties of samples from natural populations. In particular, neutral variants are affected by both geographic structure and genetic structure (i.e., the division of the gene pool into different genetic backgrounds, defined by loci that influence fitness). Here, we emphasize the close analogy between the different kinds of structures and their similar effects on neutral variation. Rather than reviewing the numerous specific theoretical and empirical results, we aim to provide a general framework for understanding the factors that shape neutral variation. This is essential for correctly interpreting data. There are dangers in merely constructing plausible scenarios that are consistent with what we see, without considering alternative possibilities. Similarly, elaborate statistical methods can be misleading if the basic assumptions (such as a single ancestral population or a lack of selection at linked sites) are incorrect.

MODELS OF GENETIC DRIFT

The process of random genetic drift acting on neutral variants can be viewed in several ways, each appropriate in different contexts. One useful descriptor is the increase in relatedness between alleles in a population, i.e., inbreeding (Wright 1931). The degree of inbreeding can be quantified by the probability of identity by descent, the chance that two distinct alleles trace back to the same ancestral gene (Cotterman 1940; Malécot 1941, 1969). Common ancestry is the inevitable outcome of genetic drift in a finite population because drift consists of fluctuations in the frequencies of alternative alleles, eventually leading to fixation (Fisher 1922, Wright 1931). Neutral variability within a population can be quantified by the probability that a pair of homologous genes or nucleotide sites have the same allelic state (Kimura 1983).

Coalescence Theory

In recent years, genetic drift has increasingly been studied in terms of the coalescence of gene lineages (Donnelly & Tavaré 1995, Hudson 1990, Slatkin & Veuille 2002). If a sample of n homologous genes is taken from a population, we can trace allelic lineages back in time. (By gene we simply mean a defined region of DNA within which recombination is assumed to be negligible.) Consider a standard Wright-Fisher random mating, discrete-generation population of N breeding individuals; the next generation is formed by random draws with replacement from the pool of $2N$ genes of the parental generation (Fisher 1922, Wright 1931). The chance that any pair of genes come from a common ancestral gene in the previous generation (i.e., they coalesce) is $1/(2N)$. If n is small compared with N, at most one coalescent event will occur in any generation. The probability that t generations elapse before the first coalescence is approximated by an exponential distribution with mean $2N/(n[\mathrm{n}-1]/2) = 4N/(n[n-1])$. Once this event has occurred, the time to the next event follows an independent exponential distribution with mean

$4N/([n-1][n-2])$, and so on until all the genes have coalesced into a single common ancestor (Hudson 1990).

Following the ancestry of the genes in samples allows the development of powerful statistical methods for making inferences about evolutionary processes (Donnelly & Tavaré 1995, Hudson 1990, Slatkin & Veuille 2002). These inferences are based on observed variability (i.e., mutations that have occurred since the common ancestor of the sample). Given a distribution of genealogies, it is easy to superimpose any chosen model of mutation and compare the results with data. For example, under the infinite sites model of mutation, where at most a single mutation segregates in the sample at a given nucleotide site (Kimura 1983), the number of differences between a pair of sequences follows a Poisson distribution whose mean is proportional to the time since their last common ancestor (Hudson 1990).

Much classical work on identity probabilities carries over to the coalescent process. For example, under the infinite alleles mutation model (Kimura 1983), each allele is assumed to have a probability μ of mutating to a different allele. We can then determine the probability of identity in allelic state, f, between two genes, from the probability distribution, $\psi(t)$, of their time to coalescence, because they are identical only if neither gene has mutated since they shared a common ancestor (Hudson 1990, Wilkinson-Herbots 1998)

$$f = \sum_{t=1}^{\infty}(1-\mu)^{2t}\psi(t). \quad (1)$$

For the Wright-Fisher model, $\psi(t) \approx 2N \exp -t/(2N)$.

Mathematically, this expression is the moment generating function of the distribution of coalescence times for a pair of genes, from which the mean and higher moments can be obtained by standard methods (Hudson 1990). Thus, classical results concerning f are directly related to this distribution.

The Effects of Population Structure on Neutral Variation

Drift is caused by random variation in the reproductive contributions of individual genes, which depends on the demographic, spatial, and genetic structuring of populations. Alleles at a locus may come from different local populations, or they may differ in their background genotypes (for example, the numbers of deleterious mutations that are present, or the genotypes at a locus subject to balancing selection). They may also be in different sexes, age classes, or stage classes. Multigene families also represent a form of genetic structure (Ohta 2002). Stable genetic or geographic structure sometimes increases variability because lineages may remain associated with different genetic backgrounds or different places for a long time (Wright 1943). Conversely, fluctuations in the structure of a population can greatly reduce variability, as in the case of a population bottleneck (Maruyama & Fuerst 1985) or a selective sweep associated with the spread of a favorable mutation (Maynard Smith & Haigh 1974). The effects of structure depend on how long genes remain associated with particular genetic backgrounds or demes, relative to the timescales of coalescence and fluctuations in structure.

Analysis of movement of lineages between locations or genetic backgrounds requires a theoretical framework known as the structured coalescent (Hey 1991). In the past few years, this has been applied to many problems in evolutionary genetics and widely used to make inferences about evolutionary processes (Slatkin & Veuille 2002). We first describe the effects of different types of structure, treating demographic, geographic, and genetic structure in parallel, and then considering their joint effects.

GENERAL CONSIDERATIONS AND SOME SIMPLIFICATIONS

Much of the power of the coalescent process to generate useful results about samples of alleles from a population comes from the simple properties of the probability distribution of coalescence times for a sample of alleles under the Wright-Fisher model (Donnelly & Tavaré 1995, Hudson 1990). With population structure, the probabilities of coalescence of alleles may depend on their sources, so that there is no longer homogeneity of coalescence time distributions for all alleles. There are two ways to deal with structured populations: one is to find simplifications that avoid some of the complexities, and the other is to look for general but useful properties. These two approaches can, of course, be combined, as we show below.

Short Timescales and the Effective Size of a Population

An important example of the first approach is when alleles at a neutral locus move rapidly between different local populations, age classes, or genotypes at other loci. When viewed over a longer timescale, these rapid movements simply change the rate of genetic drift without causing any significant differentiation between the different compartments that define the population structure (Nagylaki 1980, Nordborg 1997, Nordborg & Krone 2002). This simplifies the study of drift in populations structured by sex, stage, or age, compared with previous methods (Chesser et al. 1993, Wang & Caballero 1999), largely eliminating the need to worry about such structure in the context of a geographically or genetically structured population (Laporte & Charlesworth 2002).

To see this, we note that, to a very good approximation, the rate of genetic drift in more complex situations than the Wright-Fisher model is inversely proportional to the effective population size, N_e (Caballero 1994, Crow & Kimura 1970, Laporte & Charlesworth 2002, Wright 1931). The effects of geographic structuring on genetic differentiation between *demes* (defined as geographical units within which mating is effectively random), depend on both the N_e values of demes and the pattern of gene flow among them (Maruyama 1977, Wright 1951).

It has long been known that subdivision causes little noticeable neutral genetic differentiation between populations, unless the products of migration rates and the effective population sizes of demes are around 1 or less (Maruyama 1977, Nagylaki 1980, Wright 1951). For geographic structure to cause significant genetic differentiation, an allele sampled from deme i with effective population size N_{ei} must have been in this deme for a time on the order of N_{ei} generations. Short-term processes, such as movements of alleles between sexes or age classes, will generally reach their equilibrium states much more quickly (Figure 1). We can then treat the coalescence of alleles within demes (or stably maintained genotypes), and the migration of alleles between demes, as taking place on a long timescale, separate from the short-timescale processes involving age and sex classes (and some others to be described later) (Figure 1). If the discrepancy between these two timescales is large, an allele sampled from deme or genotype i can simply be treated as having a fixed probability α_{ir} of being in class r (defined by age and/or sex), whose value is determined by the matrix describing the flow of genes among classes (Laporte & Charlesworth 2002, Nordborg 1997, Nordborg & Krone 2002).

We can use this principle to determine the probability, P_i, that two alleles sampled from the ith deme coalesced in the previous time period. For a population with nonoverlapping generations, the relevant time period is the previous generation (and the generation time is 1). For age- and stage-structured populations that reproduce over discrete time intervals, the generation time t_i is larger than one time interval (Charlesworth 2001, Nordborg & Krone 2002). In many cases, the chance (P_{ir}) that two alleles which both trace their ancestry back to a particular class r coalesce in this class is independent of the classes from which they were originally sampled. P_i then takes the simple form

$$P_i = \sum_r \alpha_{ir}^2 P_{ir}. \tag{2}$$

The intuitive interpretation of this is that genes can only coalesce once they trace back to the same class (Figure 1); α_{ir}^2 is the equilibrium probability that two genes from deme i both come from class r.

We can then define N_{ei} as the reciprocal of twice the per-generation probability of coalescence, or half the expected number of generations to coalescence: $N_{ei} = 1/(2t_i P_i)$. These expressions can be used to obtain effective population sizes for different modes of inheritance, such as X- versus Y-linked genes, and autosomal versus cytoplasmic genes (Charlesworth 2001, Laporte & Charlesworth 2002, Nordborg & Krone 2002).

Long Timescales and Migration

We can now add the slow process of migration. Migration between demes requires us to specify the probabilities for all possible kinds of migration between classes. We can write m_{irjs} for the probability that a gene sampled from an individual of class r in deme i originated from an individual of class s in deme j in the previous

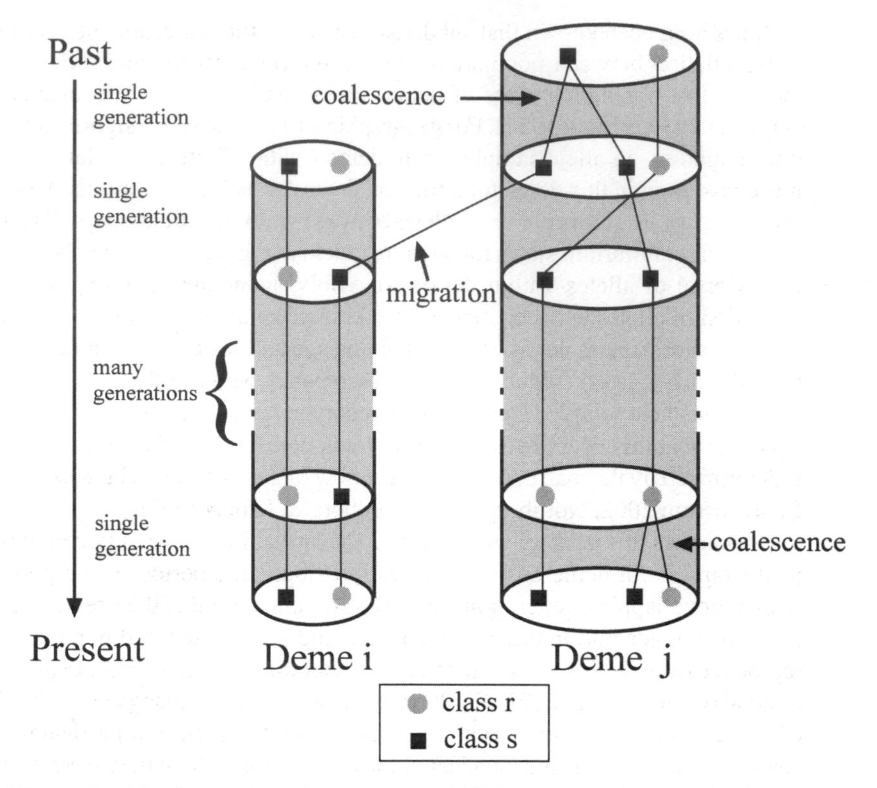

Figure 1 The lines show the ancestry of alleles present in two classes (r and s, indicated by the *square* or *circle* symbols) of individuals in two different demes (i and j, indicated in gray). The classes could be sexes, age classes, or genotypes. An allele in one class can be inherited from a parent in either class, or, more rarely, from a different deme (via migration). The example shows two coalescent events and one migration event; alleles in demes i and j eventually have a common ancestor (coalesce) in deme j. In this example, movement between the classes is much faster than movement between demes (i.e., the migration rate between demes is low). The chance that two alleles from the ith deme, which trace their ancestry back to a given class, coalesce in this class is therefore independent of the classes from which they were found in the sample (i.e., in the present generation of the figure).

time period. Applying the fast timescale approximation, we can average over all possible classes (weighting class r by α_{ir}) and derive a net migration probability m_{ij} for demes i and j (Laporte & Charlesworth 2002). Combining this with Equation 2, the simplification provided by the separation of timescales enables us to obtain a more general version of the standard equilibrium equation for identity probabilities under the infinite alleles model in a geographically structured population (Malécot 1969, Nagylaki 1982).

$$f_{ij} \approx (1-\mu)^2\left(\sum_{kl} m_{ik}m_{jl}f_{kl} + \delta_{ij}(1-f_{ii})P_i\right), \tag{3}$$

where f_{ij} is the probability of identity of alleles sampled from demes i and j, respectively; δ_{ij} is 1 when $i=j$, and zero otherwise.

From Equation 1, this gives the moment generating functions from which the probability distributions of times to coalescences of pairs of alleles can be determined (Wilkinson-Herbots 1998). In particular, the mean coalescence time, T_{ij}, for a pair of alleles sampled from demes i and j (Laporte & Charlesworth 2002, Nagylaki 1998a) is given by

$$T_{ij} \approx 1 + \sum_{kl} m_{ik}m_{jl}T_{kl} - \delta_{ij}P_iT_{ii} \tag{4}$$

However, one can only obtain simple analytic expressions for the T_{ij} with certain specific models of population structure (see below).

The Invariance Principle

There is, however, a powerful general result that can usefully be combined with results derived by this approach. We can define a weighted mean coalescent time for pairs of alleles sampled within demes as

$$T_0 = \sum_i v_i^2 / \sum_i v_i^2 P_i, \tag{5a}$$

where v_i is the ith component of the leading left eigenvector of the migration matrix $\{m_{ij}\}$ (Laporte & Charlesworth 2002, Nagylaki 1998a). (v_i measures the equilibrium probability that an allele sampled from the population will be found in deme i at some time in the past.) Using this definition, Equation 5a gives

$$T_0 = 1/\sum_i v_i^2 P_i. \tag{5b}$$

The weighted mean within-deme expected coalescence time is therefore independent of the migration structure of the population (Laporte & Charlesworth 2002, Nagylaki 1998a). This is a generalization of Maruyama's geographical invariance principle (Maruyama 1971, 1977; Nagylaki 1982).

As discussed above, the expected values of observable properties of genetic variability based on differences between pairs of alleles, such as nucleotide site diversity under the infinite sites model, depend on mean coalescence times (Hudson 1990). It is therefore attractive to use the invariant just derived to characterize the properties of a species with spatially structured populations because it suggests that the appropriately weighted mean diversity among pairs of alleles sampled from the same deme should depend only on the mutation process and T_0.

However, there is no general way to measure v_i or P_i. In the absence of gross asymmetries in the migration process, migration should spread genes fairly evenly among demes, so that v_i should be approximately equal to k_i, the proportion of the total breeding population represented by deme i. When the flow of migrant individuals into each deme exactly balances the outward flow (conservative migration), v_i equals k_i (Maruyama 1971, 1977; Nagylaki 1982). This applies to the most frequently analyzed models of population structure, such as the island and stepping-stone models (Malécot 1969, Weiss & Kimura 1965, Wright 1943). It also applies to "genetic migration" processes such as crossing over and unbiased gene conversion, where the flow is between different genotypes (Nordborg 1997), but it is violated in many biologically significant situations, e.g., with predominantly unidirectional migration from a source to a sink population (Nordborg 1997), or in a metapopulation with frequent local extinction and recolonization of demes (Pannell & Charlesworth 2000, Whitlock & Barton 1997).

If the demographic factors that influence effective population size do not vary among populations, the P_i values for each deme must be approximately inversely proportional to their numbers of breeding individuals. We can then write $P_i = 1/(2c\, k_i N_T)$, where N_T is the total number of breeding individuals in the species and c is the ratio of effective population size to census population size (for ways to calculate c, see Caballero 1994, Crow & Kimura 1970, Laporte & Charlesworth 2002). If these simplifying proportionalities hold, the mean within-deme coalescence time is the same as that for a panmictic population, i.e., $2cN_T$ (Nagylaki 1998a).

Geographic Differentiation: F_{ST}

The most widely used measures of between-population differentiation relative to within-population variation are Wright's F_{ST} (Wright 1943, 1951) and related measures such as G_{ST} (Nei 1973) and θ (Weir & Cockerham 1984). Although F_{ST} has classically been defined in terms of variance in allele frequencies (Nagylaki 1998b; Wright 1943, 1951) or identity probabilities of alleles (Nagylaki 1998b), it can also conveniently be defined in terms of mean coalescence times (Hudson et al. 1992, Slatkin 1991, Takahata 1991). Slightly different measures have been suggested by different authors; these become equivalent for a large number of demes. We shall use the definition

$$F_{ST} = (T_T - T_S)/T_T, \tag{6}$$

where T_T is the expected coalescence time for a pair of alleles sampled randomly from the population as a whole and T_S is the expected time for pairs of alleles sampled within demes (Hudson et al. 1992).

For a given set of demes, T_T involves all possible pairs of demes from which pairs of alleles can be drawn, with a specified weight for each deme pair, and similarly for T_0. If the weights used are the v values in Equation 5a, $T_S = T_0$. The theoretical value of F_{ST} can be related to empirical estimates obtained from genetic data, with appropriate corrections for sampling biases (Hudson et al. 1992; Weir & Cockerham 1984). The relation is particularly close for data on nucleotide site diversity, given the proportionality between coalescence time and divergence in this case (see above). For microsatellite loci that obey the stepwise mutation model, the analogue of F_{ST} based on the variances in number of repeats within and between populations (R_{ST}) can be used with Equation 6 (Goldstein & Schloetterer 1999).

Long Timescales: N_{eT}

Species are often spread over large geographic ranges, or consist of very many demes. Coalescence for a pair of alleles sampled randomly from the species as a whole is usually much slower than the timescale of migration between local subpopulations. The effective size of the whole population, N_{eT}, can be defined by setting the expected time to coalescence for a random pair of genes, T_T, equal to $2N_{eT}$. T_T must obviously exceed T_S because alleles sampled from two different populations can only coalesce once their ancestors find themselves in the same population. Unfortunately, no useful general formula for T_T exists, although some insights can be obtained from calculating the effects of the variance in the long-term contributions of demes to the species' gene pool (Whitlock & Barton 1997). In the following sections, we show that the long timescale approximation provides a powerful way of analyzing models of migration and population history. Over sufficiently long times, ancestral lineages become dispersed uniformly over the species' range, and so the coalescent process for a pair of alleles approximates that in a single panmictic population.

STABLE POPULATION STRUCTURE

The Island Model

The island model assumes d demes, each with the same effective size N_e and a probability m that a gene sampled from a given deme comes from a different, randomly chosen deme in the previous generation. A pair of alleles sampled from different demes waits an average of $(d-1)/(2m)$ generations to trace back to the same deme; from Equation 5, once they are in the same deme, they wait an average of $T_0 = 2dN_e$ generations to coalesce (Slatkin 1991). Their time to coalescence is therefore exponentially distributed, with mean $T_0 + (d-1)/(2m)$. The mean coalescence time of two genes sampled randomly from the species as a whole is

$$T_T = T_0 + \frac{(d-1)^2}{2dm}. \quad (7)$$

This provides a measure of the total amount of genetic diversity in the species as a whole; the difference $T_T - T_0$ measures the absolute amount of population differentiation. The corresponding expression for F_{ST} reduces to Wright's classical formula (Wright 1943, 1951) when there are many demes.

Equations 6 and 7 imply that N_{eT} is increased to $1/(1-F_{ST})$ times the panmictic value (Wright 1943). However, we stress again that N_{eT} is not always increased; reduced T_0, due to departures from the conservative migration assumption, lowers N_{eT} compared with panmixia (Nordborg 1997, Pannell & Charlesworth 2000, Wakeley & Aliacar 2001, Whitlock & Barton 1997). In such situations, diversity is reduced because demes vary greatly in their contributions to the next generation, and alleles coalesce to common ancestors faster than in the above model (Whitlock & Barton 1997). Nonetheless, the results from the island model provide a useful framework for examining the properties of subdivided populations. For example, using N_e values given by Equation 2 for different modes of inheritance, and values of m under different assumptions about sex-specific migration rates, we can derive illuminating conclusions about the expected patterns of genetic differentiation for genes with different modes of inheritance (Laporte & Charlesworth 2002). Because the product $N_e m$ determines the extent of population differentiation, differences in mode of inheritance (e.g., lower N_e for Y-linked versus autosomal genes) and sex differences in migration rates between the sexes (e.g., the possible lower migration rates for human males than females) affect the relative levels of genetic differentiation (Laporte & Charlesworth 2002).

For the island model, Wakeley (1998, 1999) has used the separation of timescales between local migration and drift in the whole population to derive an elegant description of the whole genealogical process for a sample of n alleles. The coalescent process for a set of alleles sampled in an arbitrary way from a large set of demes can be divided into two phases. The first is the scattering phase, during which two or more alleles sampled from the same deme either coalesce within that deme or switch demes. This phase ends when the ancestors of every allele in the original sample are in different demes; from then on, the process enters the collecting phase, which follows a standard coalescent process with effective size N_{eT} and the ancestral allele number remaining when the collecting phase starts (Wakeley 1998, 1999). The scattering phase will usually be short compared to the collecting phase, so that the sampling properties of a collection of alleles can be well approximated by assuming that mutations are placed onto the gene tree during the collecting phase only (Wakeley 1999). This result provides a powerful way of examining models of migration and population history. Below, we show that a large two-dimensional population shows similar qualitative behavior.

Spatial Structure

With arbitrary deme sizes and migration rates, there is no general expression for the distribution of coalescence times. However, numerical values can readily be found by solving recursions such as Equation 3, and there are analytical results for identity probabilities (Malécot 1969, Maruyama 1977, Weiss & Kimura 1965, Wilkinson-Herbots 1998). We next use these to give expressions for the overall effective population size and distribution of coalescence times in spatially structured populations. It is mathematically convenient to work in continuous space, rather than assuming discrete demes as in Equations 3 and 4. There are, however, two difficulties in analyzing spatial continua. First, the interactions between neighboring individuals that are required to regulate population density make truly continuous models intractable (Felsenstein 1975). Second, there is strong genetic differentiation over small spatial scales in a two-dimensional population that must be taken into account. Nevertheless, it is possible to smooth out local fluctuations by defining appropriate effective density and dispersal rates; over all but very local scales, continuous models show the same behavior as their demic counterparts (Barton et al. 2002).

One-Dimensional Habitats

Consider a discrete-generation population of N_T adults with equal probabilities of reproductive success, evenly distributed around a circular habitat of length L; the variance of distance moved in each generation is σ^2. An ancestral lineage makes a random walk through this habitat, with an approximately Gaussian distribution of distance moved and with a variance increasing through time as $\sigma^2 t$. Each gene can be considered to draw its ancestor t generations back from a pool of approximately $2\rho\sigma\sqrt{t}$ individuals, where $\rho = N_T/L$ is the population density (Barton & Wilson 1995, Wright 1943). The probability of coalescence time t for two nearby genes declines with $1/\sqrt{t}$ (Barton & Wilson 1995, Wright 1943).

The mean coalescence time between genes separated by a distance x is

$$T_x = 2N_T + x(L - x)/(2\sigma^2) \tag{8a}$$

(Nagylaki 1978, Wilkinson-Herbots 1998). Two genes sampled from the same point ($x = 0$) have mean coalescence time $T_0 = 2N_T = 2\rho L$, consistent with the invariance principle (Equations 5). However, we emphasize that these results for mean coalescence times may be misleading because the distribution is highly skewed (Barton & Wilson 1995): There is a 1% chance that coalescence occurs more recently than $0.00008\,(4\rho\sigma)^2$ and a 1% chance that it occurs after $3000\,(4\rho\sigma)^2$ generations.

The mean coalescence time for a randomly chosen pair of genes ($2N_{eT}$) is

$$T_T = 2N_{eT} = 2\rho L + L^2/(12\sigma^2). \tag{8b}$$

This expression can be combined with that for T_0 to obtain an analogue of F_{ST} for alleles from populations separated by a given distance (Slatkin 1993, Wilkinson-Herbots 1998). For a pair of alleles sampled at random over all the demes, this approaches 1 for a large number of demes, consistent with the classical result that genetic differentiation in linear habitats can be very high (Maruyama 1977; Weiss & Kimura 1965; Wilkins & Wakeley 2002; Wright 1943, 1946). This implies that species living in long one-dimensional habitats may have much higher total nucleotide sequence diversities than species with comparable population sizes that occupy two-dimensional habitats.

Two-Dimensional Habitats

Most organisms are distributed across a two-dimensional habitat. By the same argument as before, each gene can be considered to draw its ancestor one generation back from a Gaussian distribution covering an area of $4\pi\sigma^2$ and hence from a pool of $4\pi\rho\sigma^2$ individuals. Wright called this the neighborhood size, which we write as N_b (Wright 1946). It can be thought of as the number of individuals within one generation's dispersal range. The probability of coalescence time t for two nearby genes is now expected to decline approximately as $1/(2N_bt)$. There is no explicit formula for the probability distribution, but it can be calculated numerically (Barton & Wilson 1995). There is an appreciable probability, inversely proportional to N_b, that two nearby genes share very recent ancestry; this falls away very rapidly with distance. In contrast, in a one-dimensional population, relatedness decreases smoothly with distance, and genetic differentiation between locations separated by short distances ($<\sigma$) is negligible (Barton et al. 2002).

In a finite range of area L^2, with population density ρ, the mean coalescence time for a pair of randomly sampled genes is

$$T_T = 2\rho L^2 + L^2 \ln(KL/\sigma)/(2\pi\sigma^2), \tag{9}$$

where K is a constant of order 1, which depends on local population structure (Barton et al. 2002). The two terms on the right-hand side of Equation 9 now both increase with L^2 for a given population density. This implies that, even over very large distances initially separating two alleles, there will be significant coalescence over both short and long timescales. The ratio of the two terms is proportional to N_b. For small N_b, the second term dominates, implying that the overall genetic differentiation between populations does not differ greatly from that for the island model, given similar population sizes and migration rates (compare Equations 7 and 9) (Malécot 1969; Maruyama 1977; Wright 1943, 1946).

Stable Genetic Structure: Deleterious Mutations

We now turn from geographic to genetic structure, but for the moment keep the assumption that population structure remains stable. An important type of genetic structuring of a population is caused by the presence of deleterious alleles maintained in the population by recurrent mutation at many loci (Charlesworth et al. 1993). The effects of these deleterious alleles on neutral variability can be investigated using the framework developed above (Nordborg 1997).

If selection against deleterious mutations is sufficiently strong, and if there are no fitness interactions between loci, allele frequencies can be treated by standard deterministic theory: At locus i, the frequency q_i of deleterious alleles is u_i/t_i, where t_i is the selective disadvantage of heterozygotes, and u_i is the mutation rate from the wild-type allele A_i to the mutant allele a_i. We assume that recombination between a given neutral site and locus i takes place at rate r_i. Gametes carrying A_i and a_i can be regarded as two classes; the rate of gene flow between them depends on the mutation and recombination rates. The effect of a single selected locus on the mean time to coalescence of a pair of alleles at the neutral locus can easily be obtained using Equation 2 (Nordborg 1997). This can be extended to m loci, assuming that they are independent of each other (i.e., in linkage equilibrium) (Hudson &

Kaplan 1995, Nordborg et al. 1996, Nordborg 1997, Santiago & Caballero 1998). The mean coalescent time is

$$T \approx 2N_e \exp - \sum_{i=1}^{m} q_i/a_i^2, \qquad (10)$$

where $a_i = 1 + r_i(1 - t_i)/t_i$.

By substituting plausible mutation, selection, and recombination parameters into this equation, it is possible to show that such background selection can account for much of the observed relationship between recombination rate and variability in *Drosophila melanogaster* (Hudson & Kaplan 1995, Charlesworth 1996). This does not, of course, prove that background selection is the sole cause of this relationship, and the issue of how to explain the relationship between recombination rates and levels of genetic diversity remains open (Andolfatto 2001, Lercher & Hurst 2002, Nachman 2001).

More detailed results can be obtained when there is no recombination, as is appropriate for Y chromosomes, asexual species, and highly self-fertilizing species. The flow of neutral alleles between the multilocus genotypic classes can then be described by a matrix similar to that used for flow among sexes and/or age classes (Charlesworth et al. 1995, Gordo et al. 2002). If selection is sufficiently strong relative to $1/N_e$, the system behaves according to the fast timescale approximation. The expected time to coalescence is then $2f_0N_e$, where f_0 is the frequency of the zero-mutation class (Charlesworth et al. 1993, Hudson & Kaplan 1994). If selection is weak, the fast timescale approximation fails, and the equivalent of Equation 4 must be used (Gordo et al. 2002). The effect of background selection on diversity is then much smaller. This approach can even be used to give accurate predictions of expected coalescence times when selection is so weak in relation to drift that *Muller's ratchet* is operating, i.e., the frequency distribution of mutational classes is unstable, with successive losses of the classes with the fewest deleterious mutations (Gordo et al. 2002). It can also be used to construct statistical tests for the departures from neutral expectations of variant frequencies caused by weak background selection, by simulating genealogies of sets of alleles (Charlesworth et al. 1995, Gordo et al. 2002).

Balancing Selection

The flow of genes by recombination between genotypes maintained by long-continued balancing selection can be treated in a similar way to migration between demes but using the long timescale approximation (Hudson 1990, Hudson & Kaplan 1988, Nordborg 1997, Takahata & Satta 1998). For example, consider two selected alleles, A and a, which have been maintained at equal frequencies in a discrete-generation panmictic population with effective size N_e, for much longer than the standard coalescence time. The probability that a neutral allele sampled from a gamete carrying A derives from an a background in the previous generation is then $r/2$, where r is the frequency of recombination between the neutral and selected loci. This is equivalent to the migration rate between two populations, each of size N_e. Substitution into Equation 7 shows that the expected coalescence time between a random pair of alleles at the neutral locus is increased by $1/(2r)$, or by $1/(4N_e r)$ relative to the panmictic coalescence time $2N_e$ (Hudson 1990, Nordborg 1997). Because recombination acts like a conservative migration process, the coalescence time for a pair of neutral alleles sampled from the same allelic class is simply the panmictic value.

We can use Equation 6 to obtain a measure of the extent of genetic differentiation between the allelic classes at the selected locus. This is closely related to measures of linkage disequilibrium between variants at the neutral and selected loci, and between the neutral sites themselves (Charlesworth et al. 1997, Strobeck 1983). For neutral sites for which $4N_e r \ll 1$, the expected diversity and linkage disequilibrium will thus be much higher than for more distant sites. The stochasticity of the coalescent process, however, implies considerable random fluctuation around these expectations, so that peaks of diversity and linkage disequilibrium may be hard to detect in practice (Nordborg 2000).

Sometimes balancing selection acts at several genes [e.g., linked coadapted loci such as the components of the Brassica self-incompatibility system (Sato et al. 2002)], and/or on multiple sites in a gene [e.g., MHC loci (Takahata & Satta 1998)]. Balancing selection at multiple sites can potentially cause greatly increased diversity at tightly linked neutral loci (Barton & Navarro 2002) because many genetic backgrounds are maintained and can diverge from each other. The effect on neutral diversity depends on the number of distinct genotypic classes, which increases exponentially with the number of selected polymorphic sites. But with many selected sites, either linked polymorphisms come into strong linkage disequilibrium, with only two common haplotypes segregating (Kelly & Wade 2000), or random fluctuations reduce variation below the predictions with stable genotype frequencies (Navarro & Barton 2002).

There is evidence for increased neutral diversity due to balancing selection for plant self-incompatibility alleles, a classical example of frequency-dependent selection. *S* alleles are expected to remain polymorphic for long evolutionary times (Vekemans & Slatkin 1994), and very similar alleles have recently been found in related species of Brassica (Sato et al. 2002). *S* alleles indeed have extremely high diversity at synonymous sites (Charlesworth & Awadalla 1998, Richman et al. 1996) and in the introns, which are presumably not under balancing selection (Nishio et al. 1997). High variability is also found in MHC genes (Takahata & Satta 1998) and for some loci encoding allozymes (Filatov & Charlesworth 1999).

The theoretical results can be extended to populations at equilibrium under a system of regular inbreeding, such that the inbreeding coefficient at neutral loci is F. Alleles at the selected locus move between genotypic classes on a fast timescale and so can be treated as if they are in equilibrium. Using this result, the rate of flow between classes at the selected locus is reduced by a factor $(1 - F)$ (Dye & Williams 1997; Nordborg 1997, 2000), enhancing the increase in coalescence time accordingly. We also have to take into account the fact that, by a similar argument, inbreeding decreases the effective population size by a factor of $1/(1 + F)$ (Laporte & Charlesworth 2002, Nordborg 1997, Nordborg & Donnelly 1997). For high inbreeding coefficients, the region over which balancing selection causes increased diversity and linkage disequilibrium, compared with the strictly neutral case, is much wider than with random mating (Charlesworth et al. 1997; Nordborg 1997, 2000).

Combinations of Forces

In the real world, many of the processes we have been discussing act jointly to influence levels of neutral genetic variability within and between demes. These joint effects can be studied with the fast and slow timescale approach. For example, under the island model, background selection reduces expected coalescent times within demes (T_0) without altering the increased expected coalescence time for alleles from different demes (Charlesworth et al. 1997, Nordborg 1997). This reduces the denominator of Equation 6, increasing F_{ST}. The same principle applies to almost any factor that reduces local N_e values while leaving migration rates unchanged, and can be expected to apply to a range of migration models (Charlesworth et al. 1997).

With balancing selection and population subdivision, the increase in mean coalescence time between selected alleles relative to the within-class coalescent time is identical in form to the panmictic case (see above), with the appropriate change in effective population size to N_{eT} (Charlesworth et al. 1997, Nordborg 1997). When background selection acts together with balancing selection, coalescent times are reduced within allelic classes at the selected locus (especially in inbreeding populations, with their low effective recombination rates), but between-class coalescent times are unaffected, so that it should be easier to detect the effects of balancing selection (Charlesworth et al. 1997, Nordborg 1997).

Balancing selection causes reduced differentiation between demes compared with neutral loci (Schierup et al. 2000) because an incoming migrant allele that is not already present in a deme has an increased chance of establishment, so that its effective migration rate is increased. The few available relevant empirical observations are consistent with this prediction. In the fungus *Schizopyllum commune*, the incompatibility loci are much less differentiated than a reference locus (James et al. 1999). Such differences in diversity patterns may allow balancing selection to be detected, particularly now that data from multiple loci can be compared (Akey et al. 2002, Baer 1999, Bamshad et al. 2002).

A different situation exists when different alleles at a locus are favored in different demes (local selection) (Charlesworth et al. 1997). This reduces the effective rate of migration at neutral sites linked to the selected locus because migrants into a population carrying a locally deleterious allele are selected against. For the simple case of a biallelic locus and two demes, with symmetrical migration and strong selection against locally maladapted alleles, the effective migration rate at a neutral site is approximately $mr^*/(s + r^*)$, where m is the migration rate,

s is the selection coefficient against an allele in the "wrong" deme, and r^* is the effective recombination rate between the selected and neutral sites (Charlesworth et al. 1997). This can be substituted into Equations 6 and 7 to find the expected equilibrium level of differentiation among demes. A similar result applies when there is a selective disadvantage to heterozygotes at a selected locus, as can arise with gene flow between partially reproductively isolated populations (Barton & Bengtsson 1986).

The reduced effective migration rate due to these effects leads to increased F_{ST} or other measures of population differentiation. The increase depends on the relative values of the recombination rates and selection coefficients, whereas the effect of balancing selection depends on $N_e r$. Increased variability can therefore extend over a much wider section of genome than with balancing selection (Charlesworth et al. 1997). Examples of increased F_{ST} due to local selection are starting to be discovered (Wilding et al. 2001).

FLUCTUATING STRUCTURE

Bottlenecks and Hitchhiking

A drastic population bottleneck causes an episode of enhanced genetic drift. Seen forward in time, allele frequencies change randomly, and some alleles become fixed in the population. Looking backward in time, we see lineages suddenly coalesce at the bottleneck, such that only a few lineages in the ancestral population contribute to our sample of alleles (Maruyama & Fuerst 1985, Tajima 1989). As variability recovers, each mutation is represented in one lineage, and so there will be an excess of rare variants. However, if several allelic lineages survive the bottleneck, samples could contain several haplotypes, each with different sets of ancestral mutations (Maruyama & Fuerst 1985, Tajima 1989); this produces an excess of high-frequency polymorphisms. This mixture of different patterns, corresponding to mutations that arose before and after the bottleneck, makes it difficult to infer a bottleneck simply from the spectrum of allele frequencies, although one would usually expect some detectable departure from neutral expectation (Charlesworth et al. 1993, Tajima 1989). Population bottlenecks can also dramatically increase levels of linkage disequilibrium, offering a potentially powerful way of detecting their effects (Stumpf & Goldstein 2003).

The fixation of a new favorable mutation has a similar effect to a bottleneck (Barton 2000, Maynard Smith & Haigh 1974). Sites tightly linked to the successful mutation share their ancestry, all tracing back to one founding haplotype. Looking back in time, sites that are not tightly linked may recombine away before coalescence occurs, and so may show a more diverse ancestry. Thus, a hitchhiking event will remove all variation in a narrow region (a selective sweep), whereas the evolution of nearby sites is analogous to a population bottleneck combined with immigration. The region of reduced variation is inversely proportional to the time taken for the mutation to fix in the population, i.e., of order $s/\log(2N)$, where s

is the selection coefficient (Barton 2000, Kim & Stephan 2002). The effects of multiple selective sweeps distributed randomly over the genome have also been modeled (Kaplan et al. 1989, Stephan 1995). These results can be used to ask how frequently selective sweeps must occur in order to account for the observed relations between variability and recombination rate (Nachman 2001, Stephan 1995).

In addition to reduced variability, selective sweeps may be detected from the associated distortions of allele frequency distributions and effects on linkage disequilibrium. Several statistical tests for these have been proposed (Braverman et al. 1995, Fay & Wu 2000, Kim & Stephan 2002, Kreitman 2000, Simonsen et al. 1995). Although some examples of significant departure from neutrality have been detected in regions of low recombination in *Drosophila* (Langley et al. 2000), reduced variability is often observed without such departures (Jensen et al. 2002, Langley et al. 2000).

The contribution of selective sweeps to the association between low recombination and reduced variability thus remains unresolved. However, unexpectedly low variability in regions of normal recombination may indicate the signature of recent adaptive evolution. For example, some microsatellite loci in *D. melanogaster* show reduced variation in non-African populations relative to African populations, suggesting that selective sweeps are associated with adaptation to life outside Africa (Harr et al. 2002). This local reduction was confirmed by finding that sequence variation near these anomalous microsatellites is also reduced (Harr et al. 2002).

General Effects of Fluctuating Structure

We have discussed the most drastic changes in the structure of a population: a sudden reduction in population size or the increase of a single mutant allele. Less severe fluctuations also influence neutral diversity but are harder to describe simply because so many situations are possible. In the context of spatial structure, much attention has been given to metapopulations, in which local populations may go extinct and then be recolonized (Pannell & Charlesworth 2000, Wakeley & Aliacar 2001). More generally, random changes in deme size due to demographic stochasticity, chaotic population dynamics, or extrinsic environmental variability all tend to reduce genetic diversity (Whitlock & Barton 1997).

For similar reasons, balancing selection may sometimes reduce diversity. Imagine a polymorphism maintained by balancing selection for very long periods of time, but with allele frequencies fluctuating as environmental conditions change. Neutral sites very close to sites maintained by selection will have increased diversity (see above). However, neutral sites somewhat farther away will be associated with a given selected allele for less time, and so will be less influenced by the net change in background frequency; the effect is equivalent to a partial selective sweep (Barton 2000). Even if diversity is increased in a narrow genomic region, it may be reduced over a wider region (Sved 1983).

In addition, a locus under balancing selection at a stable frequency may only recently have reached its equilibrium owing to the spread of a new variant to an intermediate frequency. The pattern of variability associated with the new allele will then resemble that for a selective sweep, with greatly reduced variability at closely linked sites and strong linkage disequilibrium, as is observed for the *Sod* gene in *D. melanogaster* (Hudson et al. 1994). Other examples of balancing selection, such as chromosomal inversions in *Drosophila* (Andolfatto et al. 2001) and human loci associated with malaria resistance (Currat et al. 2002, Sabeti et al. 2002), also suggest recent increases in frequency.

Even in constant environmental conditions, deme sizes fluctuate through the random reproductive success of their members, and frequencies of genetic backgrounds fluctuate through genetic drift. These fluctuations can allow weakly selected genes to affect neutral diversity: If $N_e s$ is of order 1, selection is strong enough to have appreciable effects but does not fully determine the genetic structure. The coalescent approach can be extended to model directional selection on a small number of linked selected sites (Fearnhead 2001, Neuhauser & Krone 1997). Simulations show that the joint effects on genealogies of drift and selection are surprisingly small in this case, unless $N_e s$ is very large (Przeworski et al. 1999, Williamson & Orive 2002). There can, however, be large net effects of many weakly selected sites, such as synonymous variants in coding sequences (Comeron & Kreitman 2002, Tachida 2000). Calculations based on the structured coalescent also show that, even with quite strong selection relative to drift (e.g., $N_e s$ of order 10), stochastic fluctuations at a locus subject to balancing selection can substantially reduce its effects on linked neutral diversity (Barton & Etheridge 2003).

Population Structure and Genetic Diversity in Inbreeding Species

Populations reproducing by a regular system of inbreeding, such as self-fertilization, provide a good example of the ways in which multiple forces act on genetic variability within and between populations. As already mentioned, inbreeding reduces local N_e values (Laporte & Charlesworth 2002, Nordborg & Donnelly 1997), the maximal effect being to halve N_e with complete inbreeding. A highly inbred hermaphroditic population should thus have equal N_e values for the nuclear and organelle genomes, assuming maternal transmission of organelles, because both values will be half the nuclear N_e for a random mating hermaphrodite population with a comparable population size. Comparing a dioecious outcrosser with a selfing hermaphrodite with similar population size, the organelle N_e is twice as high in the selfer. Relative neutral diversity values can thus be predicted for panmictic populations.

Another effect of inbreeding is a reduction of the effective frequency of recombination throughout the genome (Dye & Williams 1997, Nordborg 2000). This increases the importance of processes such as background selection and selective sweeps, which will further reduce local N_e (Charlesworth et al. 1993, Hedrick 1980); this even affects the organelle genomes (Charlesworth et al. 1993). Finally, several differences from outcrossers lead to increased isolation between demes in inbreeders (Baker 1953), both because self-pollination prevents female gametes from outcrossing (Lloyd 1979) and because inbreeding species also generally evolve reduced pollen or sperm output (Lloyd 1979), which lowers gene flow. Many inbreeders are weedy colonizing species and may undergo frequent bottlenecks during founder events, further reducing within-deme diversity (Schoen & Brown 1991).

As discussed above, isolation may lead to high between-deme diversity, so that total genetic diversity in a selfing species may be high, possibly even higher than that for an otherwise comparable outcrosser, despite the factors reducing selfers' local N_e values. On the other hand, situations with population turnover, including extinction and recolonization, can greatly reduce species' genetic diversity (Pannell & Charlesworth 2000, Wakeley & Aliacar 2001, Whitlock & Barton 1997), and this may be more prevalent in inbreeding than outcrossing species (Ingvarsson 2002). Thus, no general prediction can be made about relative species-wide diversity in inbreeders and outbreeders.

Adaptation to specific local environmental conditions is also common in plant populations (Baker 1953, Linhart & Grant 1996) and may cause local selective sweeps, although we are not aware of any well-documented example. In addition to reduced within-population diversity, the effect of local adaptation in retarding genetic exchange can be very strong in inbreeders because of their low effective recombination rates. Inbreeding populations are therefore even more isolated (relative to comparable outbreeding ones) than their reduced pollen movement would predict. Large allele frequency differences, and high F_{ST} values, may thus be seen at marker loci (Charlesworth et al. 1997). This may account for the strong differentiation at multiple loci observed between some highly inbreeding populations such as *Hordeum spontaneum* (Volis et al. 2002).

In inbreeders, it may thus be particularly difficult to identify loci that are the targets of selection and distinguish them from genes whose diversity is affected by selection at other loci. If a large proportion of loci are differentiated between populations, this may imply either severe isolation long enough for sequences to have diverged or lesser isolation plus some local adaptation. Haplotype structure in species-wide samples of sequences is therefore not good evidence for balancing selection maintaining variability within demes at the locus, or even at a nearby gene.

Many of the factors just outlined predict higher F_{ST} values in inbreeders than outbreeders. Because many processes (see above) affect variability within demes and differentiation between them, it is impossible to predict the quantitative effect of inbreeding on F_{ST}. However, some qualitative predictions can be tested with data on natural populations. There is clear evidence for reduced allozyme diversity in inbreeding species, both species-wide, and (more markedly), within populations (Hamrick & Godt 1990). Because some allozymes may be subject to balancing selection, silent nucleotide site diversity should also be compared using DNA sequence data. Few comparisons have so far been made for orthologous genes from natural populations of related inbreeding and outcrossing species, but these studies also find low diversity within inbreeding plant populations (Baudry et al. 2001, Charlesworth & Pannell 2001), and inbreeding *Caenorhabditis* species have low diversity species-wide, compared with the dioecious outcrossing congener *C. remanei* (Graustein et al. 2002). The high diversity often found between inbreeding populations suggests that extinction/recolonization cannot explain their low within-deme diversity because this should reduce diversity in the entire set of populations. The allozyme results for other inbreeding plants suggest the same conclusion, but it is difficult to exclude other possibilities (Charlesworth & Pannell 2001).

POPULATION HISTORY VERSUS POPULATION STRUCTURE

Distinguishing Complete Isolation from Partial Gene Flow

An important problem in reconstructing the history of populations within a species, and the history of closely related species, is to distinguish whether populations have been completely isolated from each other since their separation from a common ancestor or whether limited gene flow continues between them (Avise 1999). In either case, substantial genetic differentiation between the populations may coexist with shared polymorphisms. For example, DNA sequence polymorphism data indicate a surprising amount of shared polymorphism among some groups of *Drosophila* sibling species (Kliman et al. 2000, Machado et al. 2002).

Attempts to infer historical events are difficult because of the stochasticity of mutations (which provide the information used) and especially because of the huge number of possible outcomes of the genealogical process (the evolutionary variance) (Hudson 1990). This variance must be taken into account when using genealogies reconstructed from sequences obtained from the same species, which are the basis for inferences about the populations' histories (Slatkin & Veuille 2002). For two populations recently separated from a common ancestor, Wakeley & Hey (1997) provided statistics based on the numbers of fixed nucleotide differences, polymorphisms shared between the two populations, and polymorphisms that are unique to each of the populations. For a single locus, the expectations of these statistics can be related to the rate of gene flow, the time since divergence, and the relative population sizes of the populations and their common ancestor. Tests can be done using simulations of the distribution over loci of test statistics for a null model of no gene flow (Kliman et al. 2000, Wakeley 1996, Wang et al. 1997) or by calculating the likelihood of a set of nucleotide sequence data for a nonrecombining locus in two populations (Nielsen & Wakeley 2001).

Inferring Population History

These considerations imply that we need a large number of markers and a clear pattern across a majority of neutral marker loci to infer the history of a set of populations. With molecular markers, the first requirement can be met for many species in the wild. The second is unlikely to be true within a species, where few variants are diagnostic of populations or races; it may be seen in hybrid zones, where there is a clear geographic pattern so that the hypothesis of isolation is supported by the markers having different alleles at a consistent boundary (Barton & Hewitt 1985). However, the locations of such boundaries are not strong evidence for the locations of ancestral populations; an apparently clear division of a nonrecombining genome into two geographically distinct clades can occur quite frequently by chance in a subdivided population (Irwin 2002). Furthermore, selection may affect some markers in regions of environmental change, and inconsistent marker behavior is indeed often observed in boundary regions (Martinsen et al. 2001, Shaw 2002).

Much of the neutral diversity in outcrossing species is expected to be found within any local population, except for completely isolated populations. In other words, most of the ancestry of a sample of genes is in the distant past and has spread throughout the species' entire ancestral range. Phylogenetic inferences about outcrossing populations will therefore be extremely difficult because shared polymorphisms will be common, and few markers will have patterns that reflect the populations' histories (Wall 2000). A recent simulation study illustrates the low ability to infer even simple historical patterns of population splitting and to test alternative hypotheses (Knowles & Maddison 2002).

In inferring relationships between closely related species, such as humans, chimpanzees and gorillas, multiple independently inherited markers are essential. Indeed, when many genes are studied, variation among gene trees can be used to estimate the effective population size of the common ancestor (Chen & Li 2001, Wakeley & Hey 1997). Nuclear genes have therefore been added to earlier work on mitochondrial sequences on human populations (Chen & Li 2001, Yang 2002). For studying populations within species, mitochondrial sequences (and sometimes chloroplast sequences for plants) have often been used for phylogeographic studies. Because these genomes recombine rarely, or possibly not at all (McVean 2001), they provide only a single outcome of the evolutionary history, with different loci differing only by the mutational variance, so that independent data can be obtained only from nuclear genes.

When population history is simple, or the histories of different species have enough in common, useful inferences may be possible. For example, the general picture of decreasing diversity with latitude suggests recolonization after the last Ice Age in northern Europe (Hewitt 2001), and a similar pattern suggests northward spread of the cactus Lophocereus in Baja, California (Nason et al. 2002). Genetic distances based on studies of multiple allozyme loci are useful in inferring the origins of populations, for example to show multiple recent evolution of inbreeding populations of the plant *Eichhornia paniculata* from outcrossing ones (Husband & Barrett 1993).

The ancestry of human populations is of wide interest and has been inferred in various ways (Harpending & Rogers 2000). One approach uses diversity differences to suggest that Africans form a large and diverse source population from which less diverse populations were derived via founder events and bottlenecks. Although a gradient of human diversity with distance from Africa is well established for many loci (Harpending & Rogers 2000), it is difficult, even with the large datasets now available, to exclude the alternative that low diversity outside Africa reflects a past history of isolated small populations, during which ancestral variants were lost (Takahata & Satta 2002).

LITERATURE CITED

Akey JM, Zhang G, Zhang K, Jin L, Shriver MD. 2002. Interrogating a high-density SNP map for signatures of natural selection. *Genome Res.* 12:1805–14

Andolfatto P. 2001. Adaptive hitchhiking effects on genome variability. *Curr. Opin. Genet. Dev.* 11:635–41

Andolfatto P, Depaulis F, Navarro A. 2001. Inversion polymorphisms and nucleotide variability in *Drosophila. Genet. Res.* 2001:1–8

Avise JC. 1999. *Phylogeography: The History and Formation of Species.* Cambridge, MA: Harvard Univ. Press. 458 pp.

Baer CF. 1999. Among-locus variation in F_{st}: Fish, allozymes and the Lewontin-Krakauer test revisited. *Genetics* 152:653–59

Baker HG. 1953. Race formation and reproductive method in flowering plants. *SEB Symp.* 7:114–45

Bamshad MJ, Mummidi S, Gonzalez E, Ahuja SS, Dunn DM, et al. 2002. A strong signature of balancing selection in the 5′ cis- regulatory region of CCR5. *Proc. Natl. Acad. Sci. USA* 99:10539–44

Barton NH. 2000. Genetic hitchhiking. *Philos. Trans. R. Soc. London Ser. B* 355:1553–62

Barton NH, Bengtsson BO. 1986. The barrier to genetic exchange between hybridising populations. *Heredity* 56:357–76

Barton NH, Depaulis F, Etheridge AM. 2002. Neutral evolution in spatially continuous populations. *Theor. Popul. Biol.* 61:31–48

Barton NH, Etheridge AM. 2003. The effect of selection on genealogies. *Genetics.* In press

Barton NH, Hewitt GM. 1985. Analysis of hybrid zones. *Annu. Rev. Ecol. Syst.* 16:113–48

Barton NH, Navarro A. 2002. Extending the coalescent to multilocus systems: the case of balancing selection. *Genet. Res.* 79:129–39

Barton NH, Wilson I. 1995. Genealogies and geography. *Philos. Trans. R. Soc. London Ser. B* 349:49–59

Baudry E, Kerdelhué C, Innan H, Stephan W. 2001. Species and recombination effects on DNA variability in the tomato genus. *Genetics* 158:1725–35

Braverman JM, Hudson RR, Kaplan NL, Langley CH, Stephan W. 1995. The hitchiking effect on the site frequency spectrum of DNA polymorphism. *Genetics* 140:783–96

Caballero A. 1994. Developments in the prediction of effective population size. *Heredity* 73:657–79

Charlesworth B. 1996. Background selection and patterns of genetic diversity in *Drosophila melanogaster*. *Genet. Res.* 68:131–50

Charlesworth B. 2001. The effect of life-history and mode of inheritance on neutral genetic variability. *Genet. Res.* 77:153–66

Charlesworth B, Morgan MT, Charlesworth D. 1993. The effect of deleterious mutations on neutral molecular variation. *Genetics* 134:1289–303

Charlesworth B, Nordborg M, Charlesworth D. 1997. The effects of local selection, balanced polymorphism and background selection on equilibrium patterns of genetic diversity in subdivided populations. *Genet. Res.* 70:155–74

Charlesworth D, Awadalla P. 1998. The molecular population genetics of flowering plant self-incompatibility polymorphisms. *Heredity* 81:1–9

Charlesworth D, Charlesworth B, Morgan MT. 1995. The pattern of neutral molecular variation under the background selection model. *Genetics* 141:1619–32

Charlesworth D, Pannell JR. 2001. Mating systems and population genetic structure in the light of coalescent theory. In *Integrating Ecology and Evolution in a Spatial Context. British Ecological Society Special Symposium 2000*, ed. J Silvertown, J Antonovics, pp. 73–95. Oxford: Blackwell

Chen FC, Li WH. 2001. Genomic divergences between humans and other hominoids and the effective population size of the common ancestor of humans and chimpanzees. *Am. J. Hum. Genet.* 68:444–46

Chesser RK, Rhodes OE, Sugg DW, Schnabel A. 1993. Effective sizes for subdivided populations. *Genetics* 135:1221–32

Comeron JM, Kreitman M. 2002. Population, evolutionary and genomic consequences of interference selection. *Genetics* 161:389–410

Cotterman CW. 1940. *A calculus for statistico-genetics*. PhD thesis. Ohio State Univ., Columbus. 115 pp.

Crow JF, Kimura M. 1970. *An Introduction to Population Genetics Theory*, New York: Harper & Row. 591 pp.

Currat M, Trabuchet G, Rees D, Perrin P, Harding RM, et al. 2002. Molecular analysis of the beta-globin gene cluster in the Niokholo Mandenka population reveals a recent origin of the beta(S) Senegal mutation. *Am. J. Hum. Genet.* 70:207–23

Donnelly P, Tavaré S. 1995. Coalescents and genealogical structure under neutrality. *Annu. Rev. Genet.* 29:410–21

Dye C, Williams BG. 1997. Multigenic drug resistance among inbred malaria parasites. *Proc. R. Soc. London Ser. B* 264:61–67

Fay JC, Wu CI. 2000. Hitchhiking under positive Darwinian selection. *Genetics* 155:1405–13

Fearnhead P. 2001. Perfect simulation from population genetic models with selection. *Theor. Popul. Biol.* 59:263–79

Felsenstein J. 1975. A pain in the torus: some difficulties with the model of isolation by distance. *Am. Nat.* 109:359–68

Filatov DA, Charlesworth D. 1999. DNA polymorphism, haplotype structure and balancing selection in the Leavenworthia *PgiC* locus. *Genetics* 153:1423–34

Fisher RA. 1922. On the dominance ratio. *Proc. R. Soc. Edinburgh* 52:312–41

Goldstein DB, Schloetterer C, eds. 1999. *Microsatellites. Evolution and Applications.* Oxford: Oxford Univ. Press

Gordo I, Navarro A, Charlesworth B. 2002. Muller's ratchet and the pattern of variation at a neutral locus. *Genetics* 161:835–48

Graustein A, Gaspar JM, Walters JR, Palopoli MF. 2002. Levels of DNA polymorphism vary with mating system in the nematode genus Caenorhabditis. *Genetics* 161:99–107

Hamrick JL, Godt MJ. 1990. Allozyme diversity in plant species. In *Plant Population Genetics, Breeding, and Genetic Resources*, ed. AHD Brown, MT Clegg, AL Kahler, BS Weir, pp. 43–63. Sunderland, MA: Sinauer

Harpending H, Rogers AR. 2000. Genetic perspectives on human origins and differentiation. *Annu. Rev. Genomics Hum. Genet.* 1:361–85

Harr B, Kauer M, Schlötterer C. 2002. Hitchhiking mapping: a population-based finemapping strategy for adaptive mutations in *Drosophila melanogaster*. *Proc. Natl. Acad. Sci. USA* 99:12949–54

Hedrick PW. 1980. Hitch-hiking: a comparison of linkage and partial selfing. *Genetics* 94:791–808

Hewitt GM. 2001. Speciation, hybrid zones and phylogeography—or seeing genes in space and time. *Mol. Ecol.* 10:537–49

Hey J. 1991. A multidimensional coalescent process applied to multiallelic selection models and migration models. *Theor. Popul. Biol.* 39:30–48

Hudson RR. 1990. Gene genealogies and the coalescent process. *Oxford Surv. Evol. Biol.* 7:1–45
Hudson RR, Bailey K, Skarecky D, Kwiatowski J, Ayala FJ. 1994. Evidence for positive selection in the superoxide dismutase (*Sod*) region of *Drosophila melanogaster*. *Genetics* 136:1329–40
Hudson RR, Boos DD, Kaplan NL. 1992. A statistical test for detecting geographic subdivision. *Mol. Biol. Evol.* 9:138–51
Hudson RR, Kaplan NL. 1988. The coalescent process in models with selection and recombination. *Genetics* 120:831–40
Hudson RR, Kaplan NL. 1994. Gene trees with background selection. In *Non-neutral Evolution: Theories and Molecular Data*, ed. B Golding, pp. 140–53. London: Chapman & Hall
Hudson RR, Kaplan NL. 1995. Deleterious background selection with recombination. *Genetics* 141:1605–17
Husband BC, Barrett SCH. 1993. Multiple origins of self-fertilization in tristylous *Eichhornia paniculata* (Pontederiaceae): inferences from style morph and isozyme variation. *J. Evol. Biol.* 6:591–608
Ingvarsson PK. 2002. A metapopulation perspective on genetic diversity and differentiation in partially self-fertilizing plants. *Evolution* 56:2368–73
Irwin DE. 2002. Phylogeographic breaks without geographic barriers to gene flow. *Evolution* 56:2383–94
James TY, Porter D, Hamrick JL, Vilgalys R. 1999. Evidence for limited intercontinental gene flow in the cosmopolitan mushroom *Schizophyllum commune*. *Evolution* 53:1665–77
Jensen MA, Charlesworth B, Kreitman M. 2002. Patterns of genetic variation at a chromosome 4 locus of *Drosophila melanogaster* and *D. melanogaster*. *Genetics* 160:493–507
Kaplan NL, Hudson RR, Langley CH. 1989. The "hitch-hiking" effect revisited. *Genetics* 123:887–99
Kelly JK, Wade MJ. 2000. Molecular evolution near a two-locus balanced polymorphism. *J. Theor. Biol.* 204:83–102
Kim Y, Stephan W. 2002. Detecting a local signature of genetic hitchhiking on a recombining chromosome. *Genetics* 160:765–77
Kimura M. 1983. *The Neutral Theory of Molecular Evolution*. Cambridge, UK: Cambridge Univ. Press. 367 pp.
Kliman RM, Andolfatto P, Coyne JA, Depaulis F, Kreitman M, et al. 2000. The population genetics of the origin and divergence of the *Drosophila simulans* complex of species. *Genetics* 156:1913–31
Knowles LL, Maddison WP. 2002. Statistical phylogeography. *Mol. Ecol.* 11:2623–35
Kreitman M. 2000. Methods to detect selection in populations with applications to humans. *Annu. Rev. Genomics Hum. Genet.* 1:539–59
Langley CH, Lazzaro BP, Phillips W, Heikkinen E, Braverman JM. 2000. Linkage disequilibria and the site frequency spectra in the su(s) and su(wa) regions of the *Drosophila melamogaster* X chromosome. *Genetics* 156:1837–52
Laporte V, Charlesworth B. 2002. Effective population size and population subdivision in demographically structured populations. *Genetics* 162:501–19
Lercher MJ, Hurst LD. 2002. Human SNP variability and mutation rate are higher in regions of high recombination. *Trends. Genet.* 18:337–40
Linhart YB, Grant MC. 1996. Evolutionary significance of local genetic differentiation in plants. *Annu. Rev. Ecol. Syst.* 27:237–77
Lloyd DG. 1979. Some reproductive factors affecting the selection of self-fertilization in plants. *Am. Nat.* 113:67–79
Machado CA, Kliman RM, Markert JA, Hey J. 2002. Inferring the history of speciation from multilocus DNA sequence data: the case of *Drosophila pseudoobscura* and close relatives. *Mol. Biol. Evol.* 19:472–88
Malécot G. 1941. Etudes mathématiques des populations 'mendéliennes.' *Ann. Univ. Lyon Sci. Sect. C* 206:153–55
Malécot G. 1969. *The Mathematics of Heredity*. San Francisco: WF Freeman. 88 pp.

Martinsen GD, Whitham TG, Turek RJ, Keim P. 2001. Hybrid populations selectively filter gene introgression between species. *Evolution* 55:1325–35

Maruyama T. 1971. An invariant property of a subdivided population. *Genet. Res.* 18:81–84

Maruyama T. 1977. *Lecture Notes in Biomathematics. 17. Stochastic Problems in Population Genetics.* Berlin: Springer-Verlag. 245 pp.

Maruyama T, Fuerst PA. 1985. Population bottlenecks and non-equilibrium models in population genetics. II. Number of alleles in a small population that was formed by a recent bottleneck. *Genetics* 111:675–89

Maynard Smith J, Haigh J. 1974. The hitchhiking effect of a favourable gene. *Genet. Res.* 23:23–35

McVean GAT. 2001. What do patterns of genetic variability reveal about mitochondrial recombination? *Heredity* 87:613–20

Nachman MW. 2001. Single nucleotide polymorphisms and recombination rate in humans. *Trends. Genet.* 17:481–84

Nagylaki T. 1978. A diffusion model for geographically structured populations. *J. Math. Biol.* 6:375–82

Nagylaki T. 1980. The strong-migration limit in geographically structured populations. *J. Math. Biol.* 9:101–14

Nagylaki T. 1982. Geographical invariance in population genetics. *J. Theor. Biol.* 99:159–72

Nagylaki T. 1998a. The expected number of heterozygous sites in a subdivided population. *Genetics* 149:1599–604

Nagylaki T. 1998b. Fixation indices in subdivided populations. *Genetics* 148:1325–32

Nason JD, Hamrick JL, Fleming TH. 2002. Historical vicariance and postglacial colonization effects on the evolution of genetic structure in Lophocereus, a sonoran desert columnar cactus. *Evolution* 56:2214–26

Navarro A, Barton NH. 2002. The effects of multilocus balancing selection on neutral variability. *Genetics* 161:849–63

Nei M. 1973. Analysis of gene diversity in subdivided populations. *Proc. Natl. Acad. Sci. USA* 70:3321–23

Neuhauser C, Krone SM. 1997. The genealogy of samples in models with selection. *Genetics* 145:519–34

Nielsen R, Wakeley J. 2001. Distinguishing migration from isolation: a Markov chain Monte Carlo approach. *Genetics* 158:885–96

Nishio T, Kusaba M, Sakamoto K, Ockendon D. 1997. Polymorphism of the kinase domain of the *S*-locus receptor kinase gene (*SRK*) in *Brassica oleracea* L. *Theor. Appl. Genet.* 95:335–42

Nordborg M. 1997. Structured coalescent processes on different timescales. *Genetics* 146:1501–14

Nordborg M. 2000. Linkage disequilibrium, gene trees and selfing: an ancestral recombination graph with partial self-fertilization. *Genetics* 154:923–29

Nordborg M, Charlesworth B, Charlesworth D. 1996. The effect of recombination on background selection. *Genet. Res.* 67:159–74

Nordborg M, Donnelly P. 1997. The coalescent process with selfing. *Genetics* 146:1185–95

Nordborg M, Krone SM. 2002. Separation of time-scales and convergence to the coalescent in structured populations. In *Modern Developments in Population Genetics. The Legacy of Gustave Malécot*, ed. M Slatkin, M Veuille, pp. 194–232. Oxford, UK: Oxford Univ. Press

Ohta T. 2002. Usefulness of the identity coefficients for inferring evolutionary forces. In *Modern Developments in Theoretical Population Genetics*, ed. M Slatkin, M Veuille, pp. 37–51. Oxford, UK: Oxford Univ. Press

Pannell JR, Charlesworth B. 2000. Effects of metapopulation processes on measures of genetic diversity. *Philos. Trans. R. Soc. London Ser. B* 355:1851–64

Przeworski M, Charlesworth B, Wall JD. 1999. Genealogies and weak purifying selection. *Mol. Biol. Evol.* 16:246–52

Richman AD, Uyenoyama MK, Kohn JR. 1996. S-allele diversity in a natural population of ground cherry *Physalis crassifolia* assessed by RT-PCR. *Heredity* 76:497–505

Sabeti PC, Reich DE, Higgins JM, Levine HZ, Richter DJ, et al. 2002. Detecting recent positive selection in the human genome from haplotype structure. *Nature* 419:832–37

Santiago E, Caballero A. 1998. Effective size and polymorphism of linked neutral loci in populations under selection. *Genetics* 149:2105–17

Sato T, Nishio T, Kimura R, Kusaba M, Suzuki G, et al. 2002. Coevolution of the S-locus genes SRK, SLG and SP11/SCR in *Brassica oleracea and B. rapa*. *Genetics* 162:931–40

Schierup MH, Vekemans X, Charlesworth D. 2000. The effect of subdivision on variation at multi-allelic loci under balancing selection. *Genet. Res.* 76:51–62

Schoen DJ, Brown AHD. 1991. Intraspecific variation in population gene diversity and effective population size correlates with the mating system in plants. *Proc. Natl. Acad. Sci. USA* 88:4494–97

Shaw KL. 2002. Conflict between nuclear and mitochondrial DNA phylogenies of a recent species radiation: what mtDNA reveals and conceals about modes of speciation in Hawaiian crickets. *Proc. Natl. Acad. Sci. USA* 99:16122–27

Simonsen KL, Churchill GA, Aquadro CF. 1995. Properties of statistical tests of neutrality for DNA polymorphism data. *Genetics* 141:413–29

Slatkin M. 1991. Inbreeding coefficients and coalescence times. *Genet. Res.* 58:167–75

Slatkin M. 1993. Isolation by distance in equilibrium and nonequilibrium populations. *Evolution* 47:264–79

Slatkin M, Veuille M, eds. 2002. *Modern Developments in Theoretical Population Genetics.* Oxford, UK: Oxford Univ. Press. 264 pp.

Stephan W. 1995. An improved method for estimating the rate of fixation of favorable mutations based on DNA polymorphism data. *Mol. Biol. Evol.* 12:959–62

Strobeck C. 1983. Expected linkage disequilibrium for a neutral locus linked to a chromosomal rearrangement. *Genetics* 103:545–55

Stumpf MPH, Goldstein DB. 2003. Demography, recombination hotspot intensity, and the block structure of linkage disequilibrium. *Curr. Biol.* 13:1–8

Sved JA. 1983. Does natural selection increase or decrease variability at linked loci? *Genetics* 105:239–40

Tachida H. 2000. DNA evolution under weak selection. *Gene* 261:3–9

Tajima F. 1989. The effect of change in population size on DNA polymorphism. *Genetics* 123:597–601

Takahata N. 1991. Genealogy of neutral genes and spreading of selected mutations in a geographically structured population. *Genetics* 129:585–95

Takahata N, Satta Y. 1998. Footprints of intragenic recombination at *HLA* loci. *Immunogenetics* 47:430–41

Takahata N, Satta Y. 2002. Out of Africa with regional interbreeding? Modern human origins. *Bioessays* 24:871–75

Vekemans X, Slatkin M. 1994. Gene and allelic genealogies at a gametophytic self-incompatibility locus. *Genetics* 137:1157–65

Volis S, Mendlinger S, Turuspekov Y, Esnazarov U. 2002. Phenotypic and allozyme variation in mediterranean and desert populations of wild barley, *Hordeum spontaneum* Koch. *Evolution* 56:1403–15

Wakeley J. 1996. The variance of pairwise differences in two populations with migration. *Theor. Popul. Biol.* 49:39–57

Wakeley J. 1998. Segregating sites in Wright's island model. *Theor. Popul. Biol.* 53:166–74

Wakeley J. 1999. Nonequilibrium migration in human history. *Genetics* 153:1863–71

Wakeley J, Aliacar N. 2001. Gene genealogies in a metapopulation. *Genetics* 159:893–905

Wakeley J, Hey J. 1997. Estimating ancestral population parameters. *Genetics* 145:847–55

Wall JD. 2000. Detecting ancient admixture in humans using sequence polymorphism data. *Genetics* 154:1271–79

Wang JL, Caballero A. 1999. Developments in predicting the effective size of subdivided populations. *Heredity* 82:212–26

Wang RL, Wakeley J, Hey J. 1997. Gene

flow and natural selection in the origin of *Drosophila pseudoobscura* and close relatives. *Genetics* 147:1091–106

Weir BS, Cockerham CC. 1984. Estimating F-statistics for the analysis of population structure. *Evolution* 38:1358–70

Weiss GH, Kimura M. 1965. A mathematical analysis of the stepping stone model of genetic correlation. *J. Appl. Prob.* 2:129–49

Whitlock MC, Barton NH. 1997. The effective size of a subdivided population. *Genetics* 146:427–41

Wilding CS, Butlin RK, Grahame J. 2001. Differential gene exchange between parapatric morphs of *Littorina saxatilis* detected using AFLP markers. *J. Evol. Biol.* 14:611–19

Wilkins JF, Wakeley J. 2002. The coalescent in a continuous, finite, linear population. *Genetics* 161:873–88

Wilkinson-Herbots HM. 1998. Genealogy and subpopulation differentiation under various models of population structure. *J. Math. Biol.* 37:535–85

Williamson SM, Orive ME. 2002. The genealogy of a sequence subject to purifying selection at multiple sites. *Mol. Biol. Evol.* 19:1376–84

Wright S. 1931. Evolution in Mendelian populations. *Genetics* 16:97–159

Wright S. 1943. Isolation by distance. *Genetics* 28:114–38

Wright S. 1946. Isolation by distance under diverse systems of mating. *Genetics* 31:39–59

Wright S. 1951. The genetical structure of populations. *Ann. Eugen.* 15:323–54

Yang Z. 2002. Likelihood and Bayes estimation of ancestral population sizes in hominoids using data from multiple loci. *Genetics* 162:1811–23

Annu. Rev. Ecol. Evol. Syst. 2003. 34:127–51
doi: 10.1146/annurev.ecolsys.34.011802.132423

First published online as a Review in Advance on September 29, 2003

DATA, MODELS, AND DECISIONS IN U.S. MARINE FISHERIES MANAGEMENT: Lessons for Ecologists

Kenneth A. Rose and James H. Cowan, Jr.
Coastal Fisheries Institute and Department of Oceanography and Coastal Sciences Energy, Coast and Environment Building, Louisiana State University, Baton Rouge, Louisiana 70803 email: karose@lsu.edu; jhcowan@lsu.edu

Key Words forecasting, communication, uncertainty, Sustainable Fisheries Act, population dynamics

■ **Abstract** Ecological and fisheries approaches to population modeling share many common tools and issues, yet they have developed quite independently over the past decades. The Sustainable Fisheries Act has pushed fisheries modeling into forecasting for management decision-making, which is an area where ecological modeling appears to be headed. We summarize how marine fisheries are managed in the United States, and how data and models are used to make the required forecasts. The recent management deliberations of red grouper in the Gulf of Mexico provide a case study of the sensitive relationship among data, models, and management decisions. We use the U.S. marine fisheries experience and the case study to discuss six lessons that ecologists should consider as they proceed toward forecasting for management. The need for forecasting is accelerating both in ecology and fisheries, while the margin for mistakes is getting smaller.

INTRODUCTION

Applied population ecology and fisheries modeling share many common features, including a focus on predicting the responses of organisms to human activities (e.g., Akcakaya et al. 1997, Hilborn & Walters 1992). Ecological and fisheries population modelers therefore share many common issues related to the development and application of their models. Issues such as model calibration, validation, uncertainty, and density-dependence underlie practically all models of population dynamics. Both applied population ecology and fisheries science also operate in the real world where predictions have societal and economic impacts, and so both must deal with the scrutiny of stakeholders. Both population ecology and fisheries also utilize similar modeling approaches, such as the logistic model and age-structured approaches (Caswell 2001, Hastings 1997, Quinn & Deriso 1999), for simulating population dynamics.

Despite these similarities, ecological and fisheries approaches to population modeling have developed quite independently over the past decades (Frank & Leggett 1994). This separate development has its origins in the early 1950s with the advent of quantitative fisheries analyses (Beverton & Holt 1957, Ricker 1958), and has been reinforced with graduate programs in fisheries and ecology often residing in different colleges or schools at the same university. Whereas we recognize that many population models in fisheries are ecological (e.g., Hermann et al. 2001), we focus here on the special class of stock assessment models that underlie fisheries management.

Over the past 25 years, the implementation of the Sustainable Fisheries Act (SFA) and its predecessors has acted as a major force on fisheries modeling, forcing the models to fulfill the legal requirements and interpretation of the SFA. Whereas legislation, such as the Endangered Species Act, has influenced ecological modeling and invited rigorous evaluation of the population models used (Brook et al. 2000), this influence pales in comparison with the effect of the SFA on fisheries modeling. The influence of the SFA is especially apparent in how data and fishery models are combined and used in management decisions.

Ecological modeling now is being pushed (or pulled) toward forecasting for management (Clark et al. 2002, Ludwig et al. 2001), whereas fisheries modeling has been operating in this capacity for several decades. We have found that many ecologists are unaware of the extent that data, models similar to those they use, and decisions are intertwined in marine fisheries management. Fisheries modeling has dealt with the long-standing issues of calibration, validation, uncertainty, and density-dependence, while producing models that generate very prescribed predictions of the current status of the fish population and optimal harvest rates. The predictions from fisheries models are then used extensively to make management decisions. The fisheries experience is especially enlightening because many fish populations are overexploited and are at low levels worldwide (Alverson 2002, Botsford et al. 1997), which has intensified the scrutiny and criticism both of the data used to configure models and the models themselves. We hope that as ecologists move toward forecasting for management they can learn from fisheries modeling experience about the sometimes treacherous waters where data, models, and decisions converge.

In this review, we discuss the data, models, and decision-making aspects of marine fisheries management in the United States. We first summarize the basic tenants of the SFA, and how these general tenants have been interpreted and implemented for decision-making using stock assessment. The data and models underlying stock assessment are then summarized. We use red grouper (*Epinephelus morio*) in the Gulf of Mexico as a case study to illustrate how model predictions can be very sensitive to the data used, and how changes in the data can greatly affect resulting management decisions. We conclude with a discussion on how ecological modelers can (and must) learn from the fisheries experience, and what ecologists should expect as ecology proceeds toward forecasting.

U.S. MARINE FISHERIES MANAGEMENT

Organizational Structure and the SFA

There are many organizational structures used for the management of marine fisheries worldwide (Buhl-Mortensen & Toresen 2001, OECD 1997). We focus in this review on the structure used by the United States, which originates from the SFA and related legislation. The following description is taken largely from OECD (1997) and Restrepo et al. (1998).

The original legislation that began the development of the present fisheries management organization in the United States was the Fisheries Conservation and Management Act of 1976. This act was renamed the Magnuson-Stevens Fisheries Conservation and Management Act when it was reauthorized and amended by the SFA (Public Law 104–297) in 1996. We refer to the culmination of the standards, guidelines, and implementation of these acts as the SFA. These Acts established the authority for the U.S. federal government to manage fisheries within the U.S. exclusive economic zone (EEZ). Individual states or groups of states are responsible for managing most fisheries within state waters (i.e., shoreward of the EEZ). In some situations, fisheries commissions have been established to deal with fisheries where the majority of fishing takes placc across several states or nations.

The SFA created eight regional management councils that have prepared and amended most of the fishery management plans that are used to manage individual fisheries in the EEZ. Although thc Secretary of Commerce has final management authority for living marine resources in the EEZ, the fishery management plan recommendations of a regional council have generally been approved and implemented by the secretary.

The regional councils are intended to represent diverse interests and generally include stakeholders; most council members are nominated by state governors. Most of the councils make extensive use of industry representatives via advisory panels and scientific groups (e.g., stock assessment panels) that often interpret and summarize stock assessment documents prepared by the National Marine Fisheries Service (NMFS). There are well-established mechanisms in place for public input and review.

The 1996 language in the SFA (Sec. 301, 16 U.S.C. 1851) completed a set of governing guidelines for fishery management plans in the form of ten national standards for fishery conservation and management. While regional councils have great flexibility in the details of how they manage fisheries under their jurisdiction, all must adhere to the national standards. These standards establish the framework within which the modern generation of fishery modelers and stock assessments must operate. The most relevant national standards for our purposes are: (*a*) conservation and management measures shall prevent overfishing while achieving, on a continuing basis, the optimum yield from each fishery for the U.S. fishing industry; and (*b*) conservation and management measures shall be based upon the best scientific information available. The remaining eight national standards deal with related issues, such as consideration of economic impacts, safety, and bycatch.

Criteria for Decisions

Biological reference points and control rules are presently the basis for marine fisheries management in the United States. Biological reference points take the form of target or threshold fishing mortality rates or target or threshold spawning stock biomass levels. Control rules use the values of the biological reference points to determine the status of the stock (e.g., healthy or overfished) and the status of current fishing rates (e.g., optimal, acceptable, causing overfishing). There are two classes of risk or loss: the risk that yields will fall below the maximum sustainable level and the risk that the stock will fall below some minimum population size. Fisheries managers strive to find a fishing strategy that either achieves both the fishing mortality and biomass biological reference points, or strikes a workable compromise between the two (Mace 1994).

Biological reference points have been part of the fisheries vernacular for many years, but have only recently been interpreted, via a series of technical guidelines, that permit implementation of the national standards of the SFA. The revised 1996 SFA resulted in the development of national standard guidelines, which were designed to assist in the development of fishery management plans. Technical guidelines (Restrepo et al. 1998) were then published that translated the national standard guidelines into usable scientific criteria (i.e., biological reference points and control rules) so that scientific advice could be offered to the regional councils.

Key biological reference points arising from the SFA, and the associated national standards and technical guidelines, were: (*a*) that maximum sustainable yield (MSY) is to be viewed as a limit (i.e., a threshold not to be exceeded); (*b*) that two measures are to be used to determine a fish stock's management status, the current fishing mortality rate ($F_{current}$) relative to the fishing mortality rate that would produce MSY (this ratio is denoted as $F_{current}/F_{MSY}$); and the current amount of spawning biomass ($B_{current}$) relative to spawning biomass at MSY (denoted as $B_{current}/B_{MSY}$); (*c*) that there should be maximum standards of fishing mortality rate that should not be exceeded, called the maximum fishing mortality rate threshold (MFMT); (*d*) that there should be a minimum stock size threshold (MSST) under which a stock's spawning biomass is considered depleted; and, (*e*) these biological reference points should be linked together through control rules that specify actions to be taken (i.e., changes in management measures to alter fishing mortality rates) depending upon the current estimate of the spawning biomass relative to spawning biomass at MSY ($B_{current}/B_{MSY}$) and the current estimate of the fishing mortality rate relative to the fishing mortality rate at MSY ($F_{current}/F_{MSY}$) (Restrepo et al. 1998). The actual implementation and selection of biological reference points (e.g., see Smith et al. 1993) varies by council and by stock within each council based on the available data, the status of the stock, how previous analyses were done, and the decisions of the assessment scientists and council on which models to use.

Two types of control rules are specified in the technical guidelines: limit control rules and target control rules. Limit control rules link threshold values of the ratios of current to MSY fishing rate ($F_{current}/F_{MSY}$) and current to MSY spawning biomass ($B_{current}/B_{MSY}$) to statements about overfishing and overfished. The limit control rules establish the limits of overfished (stock too low) and overfishing (fishing rate too high), which should not be surpassed so as to comply with the interpretation of the SFA. These limits are established based upon available scientific data, taking into account uncertainty about the data and its sources. Recognizing that many stocks have only limited data available, the technical guidelines suggest how other, commonly available types of data (e.g., catch and trends in abundance through time), can be used as proxies for spawning biomass at MSY (B_{MSY}) and fishing rate at MSY (F_{MSY}).

Target control rules provide information on the levels of fishing appropriate for a given level of stock biomass. In a sense, the target control rules are a statement of the optimal yield objectives traditionally used in the fishery management plans, and recognition of the uncertainty in estimates of the current biomass and current fishing mortality rate used to compare to the biological reference points. The regional councils have considerable flexibility in determining the target control rules.

Figure 1 shows a default limit control rule (Restrepo et al. 1998). The minimum stock size threshold (MSST) is computed as $(1-M)*B_{MSY}$. We have assumed a natural mortality rate (M) of 0.2 yr^{-1}, so the MSST in Figure 1 is 0.8 (i.e., $0.8*B_{MSY}$ expressed on the x-axis of $B_{current}/B_{MSY}$ becomes 0.8). A stock is considered overfished if the current estimate of spawning biomass falls below the minimum stock size threshold (i.e., $B_{current}$ falls below $0.8*B_{MSY}$ or $B_{current}/B_{MSY}$ falls below 0.8). Overfishing is occurring if a stock is experiencing a current fishing mortality rate that exceeds the MFMT. As with spawning biomass, current and threshold fishing rates can be expressed as mortality rates, or both can be normalized, as in Figure 1, by dividing each by the fishing rate at MSY. Combinations of stock biomass levels and fishing mortality rates that are above or to the left of the limit control rule line should be avoided, as they indicate a need for management actions to reduce fishing pressure to rebuild spawning stock biomass. Combinations to the right and below the limit control rule line are acceptable.

In practice, the MFMT is typically set to fishing mortality at MSY (corresponding to dotted horizontal line at $F_{current}/F_{MSY}$ equal to one in Figure 1). This is because once a stock is overfished (i.e., to the left of 0.8 and where MFMT begins to decrease in the default rule), a rebuilding schedule must be developed to recover the stock to its spawning biomass at MSY (B_{MSY}) within a specified time frame, which dictates the fishing mortality rates (and therefore the MFMT values) that are required. A new control rule line is sometimes developed for stock rebuilding situations that corresponds to the fishing mortality rate required (i.e., the MFMT value) for each specific stock biomass value so that beginning at the particular

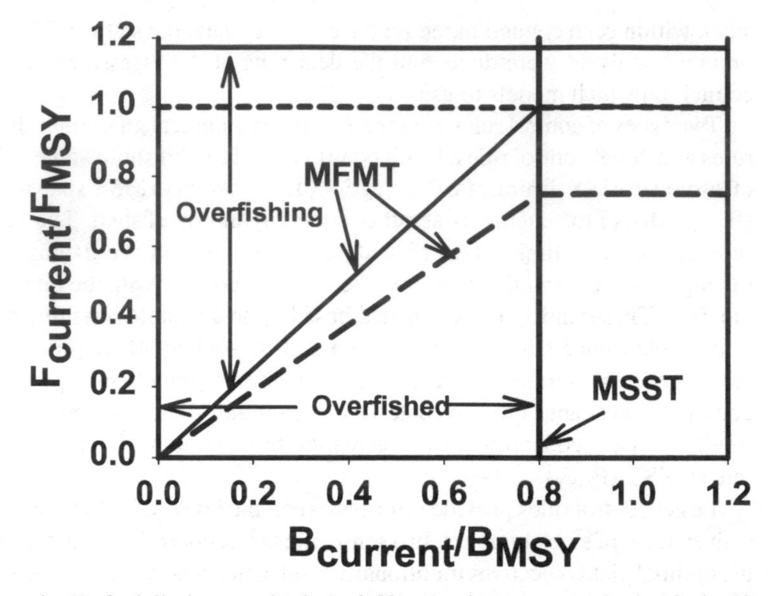

Figure 1 Default limit control rule (*solid line*) and target control rule (*dashed line*) for determining whether a fish stock is overfished and experiencing overfishing (adapted from Restrepo et al. 1998). The x-axis is the ratio of the current spawning biomass to the spawning biomass at maximum sustainable yield (MSY) ($B_{current}/B_{MSY}$), and the y-axis is the ratio of the current fishing mortality rate to the fishing mortality rate at MSY ($F_{current}/F_{MSY}$). MSST is the minimum stock size threshold below which the population is considered overfished. MFMT is the maximum fishing mortality rate above which the population is considered to be experiencing overfishing.

value of spawning stock biomass, spawning biomass at MSY (B_{MSY}) is reached within the specific time period.

The technical guidelines also suggest a default target control rule (Restrepo et al. 1998). The target control rule takes into account the risk of exceeding the limit control rule and also that optimum yield will be less than MSY. A suggested target control rule is a maximum fishing mortality rate set to 75% of limit control rule MFMT, as depicted by the dashed line in Figure 1.

Stock Assessment

The primary tool used for the quantitative evaluation of a fish population's status with respect to exploitation is stock assessment (Hilborn & Walters 1992). Stock assessments can vary in complexity and among stocks, but generally involve a review and synthesis of the available data and the results of a statistical or simulation model that produces the information needed to estimate the biological reference points, and evaluate the control rule. Stock assessments can also provide information on annual yields into the future under different harvest strategies, short-term yield forecasts for formulation of recommendations of allowable biological catch, and identify needed reductions in fishing mortality to rebuild the stock so that the stock is no longer being overfished. Excellent descriptions of fisheries stock assessment methods are provided in Hilborn & Walters (1992) and Quinn & Deriso (1999); many case studies can be found in Funk et al. (1998).

DATA Stock assessment uses data on basic biological information, catch, fishing effort, longevity, and growth, reproductive, and natural mortality rates. Basic biological information consists of a summary of the life cycle and data on geographic distribution, which are needed to define the spatial extent of the stock under management. The assumed relationship between harvest and the definition of the population is important, as the population dynamics and harvest are applied as if all individuals coexisted as a single well-mixed population.

The most common source of data for stock assessment is from the fishery itself. Catch can be numbers or biomass harvested in total or by age, usually reported on an annual basis. The annual catch (by fishing sector if appropriate) and fishing effort are used to derive catch-per-unit-effort (CPUE) indices of population abundance (Harley et al. 2001). Length-at-age (i.e., growth) data are also used to convert measured lengths of individuals in the catch to numbers-at-age (e.g., Goodyear 1997). Length-weight relationships are used to convert between numbers and biomass.

Reproductive rates are especially important for age-structured approaches, and are usually treated as per capita egg production of females by age. Field and laboratory data are used to derive sex ratios and maturity schedules by age, which when combined with fecundity data, result in per capita egg production by age.

Mortality rates of individuals from birth to recruitment (often age one, Houde 1987) are critical to stock assessment. Fisheries science has a long history of encapsulating density-dependent survival in the first-year mortality rate via a spawner-recruit relationship (Needle 2002). The commonly used spawner-recruit relationships can be formulated with two readily interpretable parameters (Haddon 2001): maximum recruitment and steepness. The steepness parameter is the fraction of maximum recruitment expected when spawning biomass is at 20% of its virgin (unfished) biomass. Higher steepness values imply stronger density-dependent survival during the young-of-the-year life stages.

Natural mortality rates of older fish also are very important to stock assessment. Natural mortality is a key element in the limit control rule that determines whether a stock is overfished. Also, in many models, natural mortality influences the relationship between MSY and stock biomass (e.g., Patterson 1992). However, good estimates of natural mortality are very difficult to obtain in situ, so a variety of statistical relationships have been developed to provide estimates of natural mortality rate of adults from information such as longevity and water temperature (Quinn & Deriso 1999).

MODELS Four general types of population-oriented models are used in fisheries stock assessment. These models are surplus production, virtual population analysis, catch-at-age, and matrix projection. All of these approaches are usually formulated on an annual time step. These models use the data and estimate the biological reference points for the current state of the population and fishery, and forecast future population size and catch trajectories for various harvesting or rebuilding scenarios.

Surplus production and matrix projection models are very familiar to ecologists, whereas virtual population analysis and catch-at-age approaches are only loosely related to ecological analogs. Surplus production models are slightly modified logistic population growth models, and matrix projection models are used extensively in ecology. Virtual population analysis and catch-at-age models originated in fisheries and have remained there for decades. Perhaps the closest analog in ecology would be life table analysis, but this does not do justice to the sophisticated extensions of virtual population analysis and catch-at-age models used in fisheries.

Surplus production models are typically a logistic population growth model with an added term for harvest (Hilborn & Walters 1992). Surplus production models are used in situations of very limited data or when aging of individuals is extremely difficult. Surplus production models can be used to generate estimates of spawning biomass and fishing mortality rate at MSY (B_{MSY} and F_{MSY}) and current spawning stock biomass and fishing mortality rate ($B_{current}$ and $F_{current}$), and to predict future stock biomass under assumed harvesting scenarios.

Virtual population analysis is an age-structured approach based upon following year-class cohorts through time. Virtual population analysis is a backward-looking recursive algorithm that calculates numbers alive in each year class for each past year from catch-at-age and natural mortality rates (Hilborn & Walters 1992). Fishing mortality rates are assumed to be separable into the product of an age-specific component (selectivity) and a year-specific component (Megrey 1989). Virtual population analysis is applied to year-class cohorts so that matrices (e.g., years as rows by ages as columns) of estimated numbers of fish and fishing mortality rates are obtained (Hilborn & Walters 1992, Quinn & Deriso 1999). Virtual population analysis can be used to estimate a variety of biological reference points and, with assumptions about initial numbers entering as new cohorts (i.e., recruitment), can be used to make short-term forecasts of population size and catch.

Catch-at-age methods are similar to virtual population analysis, with the additional use of data on fishing effort (Megrey 1989). Catch-at-age attempts to address two weaknesses of virtual population analysis: unreliable estimates of year classes still in the fishery and the requirement of specifying the natural mortality rate (Hilborn & Walters 1992). Catch-at-age analysis uses similar assumptions as virtual population analysis, with an additional assumption of how catch is related to fishing effort (e.g., proportional via a catchability coefficient). These assumptions are used to derive a regression equation relating CPUE by age to terms involving initial year class size and catchability, cumulative fishing mortality (year-specific effort and either constant or age-specific catchabilities), and cumulative natural mortality. Catch-at-age models have been formulated using maximum likelihood and Bayesian methods that enable them to include auxiliary information (Quinn & Deriso 1999).

Matrix projection models are a forward-looking simulation approach that is widely used to simulate animal and plant population dynamics (Caswell 2001). Mortality rates at age are used to project the numbers at age j in year t to the number alive at age j + 1 in year t + 1. Reproductive rates by age are used each year to convert numbers of females at age into a measure of spawning population size. A spawner-recruit relationship then uses the spawning population size to produce the number of age-one recruits that enter the population the following year. This basic model has been modified for use in fisheries stock assessment to allow for complicated treatment of fishing mortality (e.g., multiple fisheries, time-varying mortality with age-specific selectivities) and Bayesian estimation of the many model parameters using age-specific catch, CPUE, and fishery-independent time series (Hilborn et al. 1994).

CASE STUDY: RED GROUPER IN THE GULF OF MEXICO

We use recent deliberations about red grouper in the Gulf of Mexico to illustrate how data and models affect management decisions. Red grouper are part of the temperate reef-fish complex, and are a slow-growing, long-lived species that exhibits high site fidelity and a protogynous hermaphroditic (females become males) life history (Coleman et al. 2000). Because of these life history traits, groupers in general are considered highly susceptible to overfishing.

Red grouper is the most common species in the commercial and recreational grouper catch of the U.S. Gulf of Mexico. Most of the fishery for the species in U.S. waters of the Gulf of Mexico occurs within or immediately to the west of Florida's territorial sea. Although red grouper supports the bulk of the grouper harvest, red grouper received surprisingly little attention in the form of research or management as of the early 1990s when the first formal stock assessments were performed (Goodyear & Schirripa 1993).

Review of the Fishery

Grouper have been harvested off Florida by Cuban vessels since the middle 1800s and by U.S. vessels since about 1880. Starting in 1950, consistent records of Florida west coast grouper landings for the U.S. commercial fleet were kept. However, U.S. commercial landings of red grouper have been separated from other groupers only since 1986. On the basis of initial analyses, estimated total landings of the Cuban fleet peaked at approximately 13 million pounds in the mid-1950s, declined rapidly until 1965, then (perhaps due to Cuba's use of longline fishing gear) increased again until 1976 when the Cuban fleet was expelled from U.S. waters with establishment of the EEZ (Figure 2*a*). Accurate records of U.S. landings begin in 1986, and show a general downward trend through 1998 and then an increase in landings to roughly average levels in the past three years (Figure 2*a*). Historical CPUE indices based on the Cuban fleet and the U.S. handline

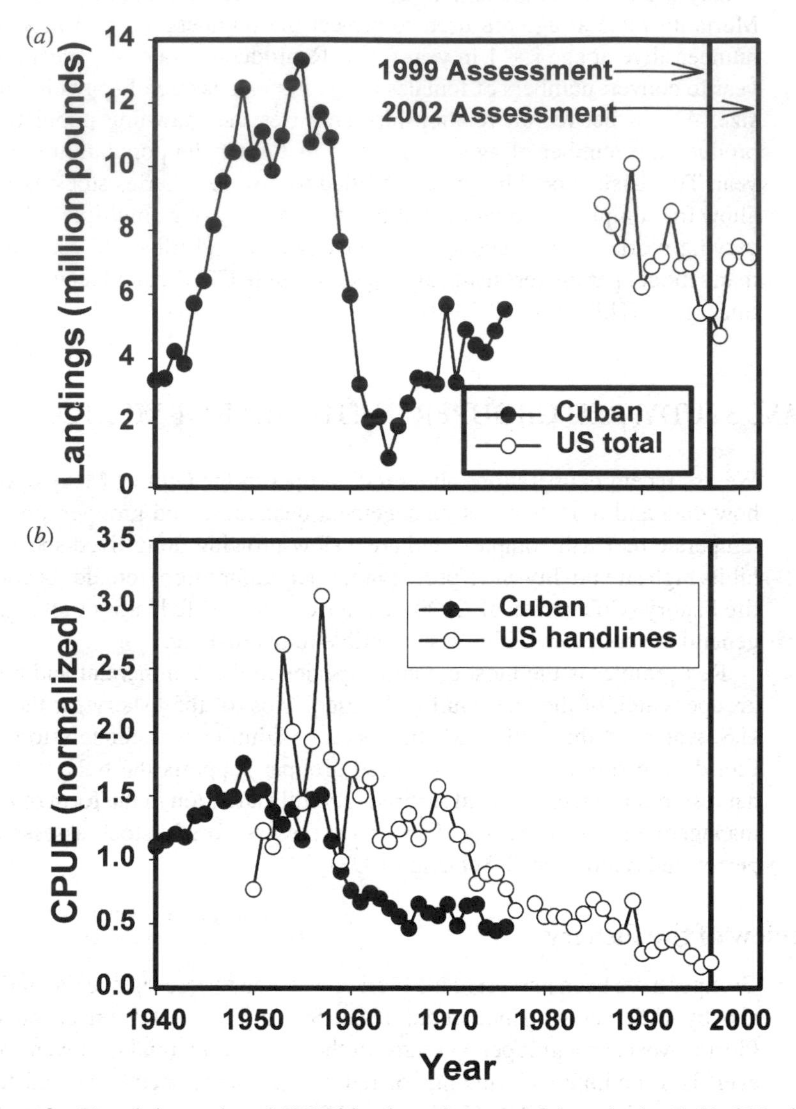

Figure 2 Trends in catch and CPUE for the red grouper fishery in the Gulf of Mexico. (*a*) Total landings by the Cuban fleet and by the U.S. fleet (since 1986 when red grouper were identified by species), (*b*) CPUE of the Cuban catch and the total U.S. catch. Note that subsequent review by an independent consultant showed the Cuban data to be unreliable. Thus, temporal trends in the Cuban CPUE may not be indicative of actual changes in catch rates and should not be compared to U.S. CPUE.

fishery both showed apparent downward trends since the 1950s (Figure 2*b*). Some conservation measures were instituted in Florida state waters in 1985, and in the EEZ in 1990.

The Controversy Begins

A new red grouper stock assessment was performed in 1999 (Schirripa et al. 1999) that included several major changes from the two previous (1991 and 1993) stock assessments, as well as updates of the commercial and recreational catches of red grouper in U.S. waters. Major changes included a new growth curve obtained from a tagging study, inclusion of historical data (e.g., Cuban catch and CPUE since 1940), and the use of a new population dynamics model to evaluate the status of the red grouper stock. There was growing concern about the population, as the historical CPUE time series appeared to show a continuing decline in catches since the peak in the 1950s (Figure 2*b*). Furthermore, with the addition of the data points through 1997, trends in the total U.S. catch showed a continuing decline following the last stock assessment in 1993 (Figure 2*a*).

A new model, called the age-structured matrix projection (ASAP) model, was used in the 1999 stock assessment. ASAP is based upon an age-structured matrix projection approach with a spawner-recruit relationship for egg to age-one survival, and uses a Bayesian parameter estimation routine (Legault & Restrepo 1998). ASAP was selected because it offers advantages over the previously used virtual population analysis model. ASAP enables stock assessment analyses (i.e., estimates of B_{MSY} and F_{MSY}) that are more consistent with the SFA and its associated technical guidelines.

In the 1999 stock assessment, the ASAP model was fit to long or short sets of CPUE time series. The short time series data consisted of eight (four commercial; one recreational; three fishery-independent) CPUE series from 1986 to 1997. The long time series data consisted of nine CPUE time series: seven of which were identical to the short time series; one from the short time series that was extended back before 1986; and the historical Cuban time series. When the long dataset was used the data provided enough contrast in stock sizes so that the parameters of the spawner-recruit relationship (steepness and maximum recruitment) could be estimated from the Bayesian fit of the model to the data. The short time series, while limited in temporal coverage, was considered to be of high quality because it was based on years when the catch was specifically identified as red grouper and the catch was aged for all fishery gears. When the short time series was used to calibrate the model, the parameters of the spawner-recruit relationship had to be specified a priori; a range of steepness values was therefore used because of uncertainty about the steepness value.

When the 1999 stock assessment was completed, the estimated 1997 U.S. landings were the lowest (about 5 million pounds) since 1981 (Figure 2*a*). Since the introduction of bottom longlines to the U.S. commercial fleet in 1979, the estimates of the 1997 U.S. commercial landings were down approximately 55% from their peak. Based upon ASAP model runs utilizing the long time series dataset,

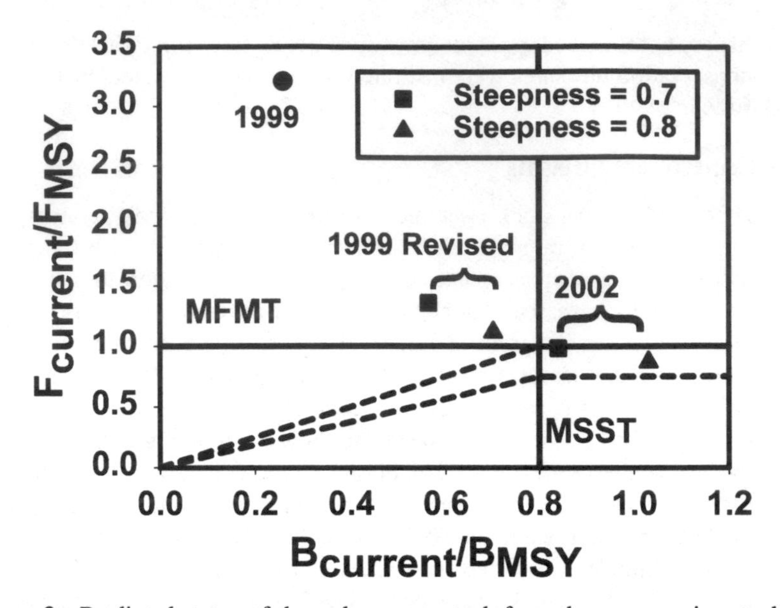

Figure 3 Predicted status of the red grouper stock from three successive stock assessments (1999, 1999 revised, 2002). Stock status is shown as current spawning stock biomass relative to spawning biomass at Maximum Sustainable Yield (MSY) ($B_{current}/B_{MSY}$) and current fishing mortality rate relative to fishing mortality rate at MSY ($F_{current}/F_{MSY}$) on the limit control rule plot. The maximum fishing mortality threshold (MFMT) for overfishing was F_{MSY}; we show the default control rule from Figure 1 for illustrative purposes. The 1999 stock assessment used the long time series data, which included the historical Cuban information. The 1999 revised and 2002 stock assessments only used the short data time series (1986 and later), and thus steepness values of the spawner-recruit relationship (0.7 and 0.8) had to be specified. Current years were 1997 for the 1999 and 1999 revised stock assessments, and 2001 for the 2002 stock assessment.

estimated spawning stock biomass in 1997 relative to the estimate of spawning biomass at MSY ($B_{current}/B_{MSY}$) was 0.26, which was substantially below 1–M or 0.8, indicating that the stock was overfished. The estimated current fishing mortality rate relative to the fish mortality rate at MSY ($F_{current}/F_{MSY}$) was 3.21, which was greater than 1.0, indicating that the stock also was undergoing overfishing (circle in Figure 3).

The 1999 stock assessment was presented to the Reef Fish Stock Assessment Panel (RFSAP), a group of independent scientists appointed by the Gulf Council to evaluate NMFS stock assessments and to suggest a range of allowable biological catch for the species in question. Based upon SFA guidelines, the RFSAP (1999) recommended two rebuilding schedules, one based on a 10-year recovery (spawning biomass reaches spawning biomass at MSY by 2010) and the other based on a recovery to spawning biomass at MSY by 2018. The recommended allowable biological catch range was 0 to 1.5 million pounds for the 10-year recovery plan and 0 to 3.5 million pounds for the 2018 recovery plan. Both of these recommendations required significant reductions with respect to the approximately 5 million pounds landed in 1997, likely imposing significant hardships on Gulf of Mexico red grouper fishers.

Data In Question

Very shortly after the RFSAP completed its report, an oversight advisory committee called the Standing and Special Reef Fish Scientific and Statistical Committee rejected the RFSAP report. The role of the Scientific and Statistical Committee is to ensure that the management advice to the Gulf Council is based upon the best available science (à la National Standard 2). The Scientific and Statistical Committee questioned several portions of the assessment and the RFSAP report, including the validity of the long-term Cuban data upon which the allowable biological catch recommendations were based.

The RFSAP was reconvened in August 2000 and again December 2000 to review updated landings, the Scientific and Statistical Committee report, the NMFS response to the Scientific and Statistical Committee report, and an independent review (Sullivan 1999) of the 1999 red grouper stock assessment. In addition, the RFSAP heard a presentation by an independent consultant hired by a fishers group who raised significant questions about the validity of the long-term Cuban data, and the RFSAP heard comments at the meetings from participants in the grouper fishery.

The RFSAP discussions at the August and December meetings centered around two major issues: (*a*) the reestimation and use of new red grouper discard rates (release of fish too small) and discard mortality rates (subsequent mortality of discarded fish) based upon data not previously known to the assessment scientists and the RFSAP; and (*b*) the validity of the Cuban time series of catches as a long-term index, especially in light of the retooling of the fishing fleet following the Cuban revolution in the early 1960s and uncertainty about changes in how catch and effort were reported.

After much deliberation and many new ASAP model runs (RFSAP 2000), the RFSAP chose to: (*a*) exclude the Cuban catch data altogether in subsequent model runs, thus relying only on the short time series set of CPUE indices to offer management advice; (*b*) use new estimates of depth-specific discard rates for the longline fishery obtained by calibrating simulated discard rates to those estimated from a NMFS field-based observer program; and, (*c*) use lower mortality rates of fish discarded from the longline fishery. Furthermore, because the ASAP model was now using the short CPUE dataset only, the assessment scientists had a priori to specify the parameters of the spawner-recruit relationship. As such, model runs were made with a variety of steepness estimates, ranging from 0.6 to 0.8. The RFSAP ultimately decided that model runs using steepness values of 0.7 and 0.8

(plus excluding the Cuban data and including the calibrated discard rates and lower discard mortality) provided the best scientific advice.

The selection of steepness values of 0.7 and 0.8 was based upon RFSAP members experience with steepness values of other species (see Rose et al. 2001). There is considerable variability in steepness estimates among fish species with periodic life histories roughly similar to red grouper. Rose et al. (2001) reported a median value of 0.68, but a range of 0.31 to 0.96, for 62 species classified as periodic strategists. Furthermore, none of species in the Rose et al. (2001) analysis were groupers and none were protogynous hermaphrodites. On one hand, Huntsman & Schaaf (1994) state that we do not know which density-dependent mechanisms operate in grouper populations, and that the relationship between fishing mortality and reproductive success is very sensitive to the specific mechanisms. On the other hand, Armsworth (2001) compared alternative formulations of age-structured models of a fish species with a protogynous hermaphrodite life history, and concluded that evaluation of harvesting practices did not require detailed understanding of the density-dependent mechanisms.

Given these changes in the data used to generate the stock assessment, the RFSAP again determined that the red grouper stock was overfished and experiencing overfishing. For steepness values of 0.7 and 0.8, the revised estimated spawning stock biomass in 1997 relative to spawning biomass at MSY (0.56 and 0.70, respectively) was still below a MSST of 0.8 (i.e., overfished), and undergoing overfishing because the revised estimated fishing mortality rate for 1997 still exceeded the fishing mortality rate at MSY (square and triangle labeled 1999 in Figure 3). However, model simulations that achieved stock rebuilding within the appropriate time course resulted in the RFSAP recommending an allowable biological catch in year 2000 of 4.3 to 5.2 million pounds. The new allowable biological catch range would require considerably less reduction in red grouper harvest rates than the 1999 assessment, and presumably less hardship on the fishers.

New Data: The Saga Continues

The saga did not end there. The revised 1999 assessment with the changes described above eventually was accepted by the Scientific and Statistical Committee, and was offered as management advice to the Gulf Council. In the intervening two plus years, there was much debate in the Council about how to make the reductions in fishing effort necessary to reduce the fishing mortality rate to the fishing mortality rate at MSY, such that new regulations were not in place before a new red grouper stock assessment was developed by NMFS scientists in 2002 (SEFSC Staff 2002).

As before, new data plugged into the ASAP model for the 2002 assessment greatly changed the perception of stock status, and changed the management advice provided to the Gulf Council. The 2002 assessment used new data with the same ASAP model that resulted in even a brighter picture of the status of the red grouper stock. In the 2002 assessment, catches between 1997 and 2000 (since the revised 1999 assessment) were shown to have increased to around 7 million pounds (near recent historical highs). New data were also available about the reproductive potential-at-age of red grouper. At the RFSAP meeting, three reproduction-at-age relationships were presented to the RFSAP (the original 1999 relationship and two new relationships). The RFSAP and assessment scientists then derived a fourth relationship, which was deemed the best, during the meeting.

The updated catch data and new reproduction-at-age values produced results about stock status that now depended upon whether a steepness value of 0.7 or 0.8 was used (square and triangle labeled 2002 in Figure 3). The stock assessment now showed that for both steepness values red grouper was no longer overfished, as the ratio of spawning stock biomass in 2001 to spawning stock at MSY was 0.84 for steepness of 0.7 and 0.99 for steepness of 0.8. Further, the new stock assessment showed that red grouper were just marginally experiencing overfishing, as the ratio of fishing mortality rate in 2001 to fishing mortality rate at MSY was 1.031 (just above one) for the steepness of 0.7 and the ratio was 0.87 (less than one) for a steepness of 0.8. It should be noted, however, that while the population appeared to be recovering, estimates of spawning stock biomass levels in 1999 (based upon the 2002 assessment) were still less than the required 80% of B_{MSY}. Thus, compliance with the SFA still required the development of a rebuilding schedule.

The RFSAP report (RFSAP 2002) indicated that the RFSAP remained cautiously optimistic about the apparent rapid recovery from the apparent overfished condition. There was some concern that the recent increases in landings were attributable to a single strong year class moving through the fishery; however, the evidence for this was inconclusive. Based on the specified rebuilding schedule for red grouper, the RFSAP recommended an allowable biological catch range of 7.03 to 7.12 million pounds for 2003, which was very similar to recent catches. The Gulf Council finally decided on a total allowable catch of 6.55 million pounds, which was consistent with one of several alternative rebuilding scenarios (all of which were consistent with the SFA) presented in the RFSAP report.

Data, Models, and Decisions

The red grouper case study is by no means unusual, and well illustrates the sensitive relationship between data, models, and decisions in the current world of U.S. marine fisheries management (e.g., see Buhl-Mortensen & Toresen 2001, Wilson & Degnbol 2002). Over the course of about two years, stock assessment model predictions of red grouper stock status went from dire to nearly recovered (Figure 3). Current estimates of spawning stock biomass relative to spawning biomass at MSY went from 26% in the 1999 assessment, to 56–70% in the revised 1999 assessment, to 84–99% in the 2002 assessment. Red grouper fishermen were faced with decisions suggesting that draconian cuts in their livelihoods would be necessary two years ago, to only modest (if any) cuts being required based on the 2002 stock assessment. The dramatic progression of stock status, and consequently the advice to managers, was due to the existing data being rejected and new data becoming available. Whether the fish population changed over this time period, and if so, whether the changes were transient or longer-term, is not clear.

MARINE FISHERIES MANAGEMENT: EXPECTATIONS AND LESSONS FOR ECOLOGISTS

Forecasting is a relevant issue to ecologists, as some think that ecological forecasting is the next frontier for ecological modeling (Clark et al. 2002). As ecological models are pushed toward forecasting and are used for management decisions, there are some lessons and expectations ecologists can take from the fisheries experience.

Unrealistic Demands on the Data and Models

The interpretation and implementation of the SFA has placed unrealistic demands on the data and models underlying the stock assessments (e.g., Hilborn 2002, Wilson & Degnbol 2002). The technical guidelines (biological reference points and decision control rules), which were developed to fulfill the SFA, push the data and models beyond their reasonable capabilities. This is especially true for stock rebuilding situations, where long-term (decades scale) forecasts sometimes are required. These forecasts are typically made assuming that conditions in the future do not change (e.g., constant environment, no change in fishing technology), and then the predictions are being interpreted so strictly by the legal community so that management actions projected to fall 1% or 2% short of achieving the appropriate biological benchmarks decades from now are deemed to be untenable (NRC 2002). We rarely even know what the true baseline or target conditions should be (Jackson et al. 2001), and we are entirely confident that the environment will not remain constant (Chavez et al. 2003, Ruiz et al. 2000, Scavia et al. 2002). Yet, we proceed with long-term forecasts of fish population biomass assuming a constant and well-behaved environment, and make important management decisions based on these forecasts. Strict fulfillment of the SFA results in many situations of "forecasting with disbelief."

Accompanying the use of these fisheries stock assessment models beyond their domain has been intense scrutiny and criticism of the models. There have been numerous peer reviews of individual stock assessments, as well as broad national-in-scope reviews of the stock assessment process (NRC 1998, 1999). While we have not read all of the reviews of specific stock assessments, we can state with some confidence that the reviews generally conclude that stock assessment techniques are sound, and that individual stock assessments are reasonable but have weaknesses that usually are associated with data shortcomings.

In developing models for management, ecologists should be prepared to construct and use models beyond their reasonable domain of applicability. Ecologists will have to knowingly making forecasts decades into the future under assumed conditions we know will not continue. Managers, stakeholders, and especially lawyers, prefer not to deal with the realistic degree of uncertainty and the many caveats that accompany most models of ecological systems. Ecologists have spent much effort on model validation issues (Rykiel 1996), and tend to focus on the uncertainty of model predictions (e.g., Akcakaya et al. 1997, Bartell et al. 1992). As ecological models are used more for management, the focus will likely shift from validation to calibration of the model to historical data, and to presenting forecasts in a simplified format that precludes including all of the uncertainties and caveats.

Ecologists should also be prepared for close scrutiny and review of their data and models. This scrutiny will include reviews according to the conventional peer-review process, and via more directed reviews by consultants hired by various stakeholder groups who do not like the conclusions. The outcome of these reviews will probably not change anyone's mind.

Repeated Calls for Better Data

There have been repeated pleas from many sources for better data upon which to base fishery stock assessments (e.g., NRC 2000), but new information has been slow to arrive. The lack of sufficient data is obvious to all participants in fisheries management. For example, of the 70 species being partially or totally managed by the Gulf Council, 46 species have data that have not been examined beyond simple time series plots or tabulations of catches (Mace et al. 2001). Often, the best data available on population trends are the catch data themselves, and the problems associated with fishery-dependent data have long been discussed (Harley et al. 2001). Requests for better data frequently include a plea for more fishery-independent (e.g., scientific survey) data (NRC 1998), although pragmatic considerations sometimes prevail and improvements in fishery-dependent data are emphasized (NRC 2000). The collection of new data, especially fishery-independent data, costs money and falls to the already underfunded NMFS because many academicians consider such monitoring routine.

General calls for better data may be too vague and seem to go unheeded, while crises get better responses. Although overall progress is slow, there seems to be more success in obtaining support for improvement of the data and models when argued on a case-by-case basis. For example, collapse of the striped bass population in the late 1970s resulted in significant new research by way of congress authorizing the 1979 Emergency Striped Bass Study.

Interestingly, the general reports that include data evaluation and pleas for better data (Mace et al. 2001; NRC 1998, 2000) do not discuss the importance of obtaining better data on density-dependent processes. These pleas make excellent cases for data on abundance trends, and the need for databases to permit sharing of these data. Yet, it is the magnitude of density-dependence (e.g., steepness of the spawner-recruit relationship) that ultimately determines long-term sustainability of a harvested population (Rose et al. 2001, but see Sale & Tolimieri 2000). The difficulties with spawner-recruit relationships have been debated for years (Needle 2002). Krebs (2002) argues that the study of density-dependence is fallow and advocates making short-term predictions that focus on the processes that make populations grow or shrink. However, the current management approach for U.S. marine fisheries requires long-term forecasts.

Ecologists should be prepared to have their pleas for better data ignored, at least initially, and for data collection efforts to be underfunded. All of the fundamental issues that hindered ecological modeling in the past (e.g., density-dependence, Krebs 2002, Rose et al. 2001, Turchin 1995) will again be the Achilles' heel in ecological forecasting. In most cases, long-term forecasts required in the near term will have to be generated knowingly using the inadequate data that are currently available.

Shifting the Focus and Bandwagons

One response to the perceived failure of U.S. marine fisheries management has been calls for major revisions and new approaches (Botsford et al. 1997, Pitcher et al. 2001), followed by a rush to get on the bandwagon. Almost all of U.S. fisheries management takes a single-species approach. The poor state of fish populations has led many to criticize the single species, MSY-oriented approach (Roughgarden & Smith 1996) and has led to calls for ecosystem management (Alaska Sea Grant 1999) and marine reserves (NRC 2001). We find it puzzling how broadening from the single species to the ecosystem, with its exponential increase in data needs, will help with quantitative forecasts. Whereas marine reserves show promise in some specific situations, their general applicability remains to be proven. Indeed, more rational evaluations of the current single-species approach usually leads to a conclusion that the single-species approach, applied in a conservative manner, will be with us for at least a while longer (NRC 1999).

An example of the bandwagon phenomenon is the growing use of the Ecopath modeling software (Christensen & Pauly 1993) in fisheries. We know that no single modeling approach is a panacea, and the Ecopath approach may be appropriate for some, but likely not the vast majority, of situations.

Ecologists should be ready for calls for alternative approaches, and the tendency for scientists and managers, in their desperation if things are not going well, to jump onto "new tools" bandwagons. Consideration of new approaches and new tools is necessary, desirable, and healthy for progress and evolution of the science, but there can be overemphasis on something new that has a nice user-interface, but has not yet undergone trial-by-fire testing. In fisheries, the models and data must be improved; ecosystem approaches, marine reserves, and new sounding models may help, but by themselves, they are not the solution.

Confusion about Uncertainty

The high uncertainty associated with fisheries models has lead to miscommunication, and sometimes misuse, of uncertainty. There have been repeated discussions of the role of risk (Hilborn et al. 1993, Rosenberg & Restrepo 1994) and for application of the precautionary principle (Buhl-Mortensen & Toresen 2001) in fisheries management. Yet, the need for effective communication causes model predictions to be simplified as they are translated from the scientists to the managers and stakeholders (e.g., Buhl-Mortensen & Toresen 2001). The differences between variability resulting from uncertainty (ignorance) versus stochasticity or natural variability, and between precision and accuracy, are usually poorly understood by managers, lawyers, and stakeholders.

When not properly explained or understood, high uncertainty can be used to delay action or to select values at the extreme of the ranges that result in highly risky (or overly conservative) management decisions (Frederick & Peterman 1995). Ludwig et al. (1993) described the ratchet effect: political pressure constantly pushing for increasing fishing mortality because fishery scientists cannot specify with certainty that the next increase will lead to overfishing and population collapse. Ecologists are warned to tread carefully when presenting model forecasts, especially those that are highly uncertain, to managers and stakeholders.

Communication Issues

While not all of the regional councils have adopted the same organizational structure regarding their advisory panels, the need for the results of complex stock assessment models to be translated into a language that the council members can understand is universal. In our opinion, this situation is ripe for controversy because more and more complex models and results, depicting diverse management actions, are being presented to fishery managers who may not understand them, while the criteria for determination of stock status are becoming more exact, time-dependent, and bound by the force of law (e.g., NRC 2000, Wilson & Degnbol 2002). A breakdown in communication, understanding, and trust at any link in the chain (data gatherers ⇒ fishery modelers/assessment scientists in NMFS ⇒ regional council's panels of experts ⇒ regional councils ⇒ NMFS and the secretary of commerce) can debilitate the process (NRC 2002). It is our experience that weaknesses in this chain of communication contribute to difficulties in the decision-making process. Contention can led to management actions that are compromises between conservation and socioeconomic objectives, often resulting in decisions that achieve neither. As of January 2002, the NMFS was faced with over 100 legal cases (NRC 2002).

Many of the council members are stakeholders, representing more often than not, some aspect of the fishing industry. Whereas these individuals usually are quite bright, very few are trained in fisheries mathematics and stock assessment. Rather, they are people who feel grounded by real-world experiences, often at dockside or at sea, and tend to place great measure on what other such stakeholders tell them. People believe in what they know, see, or hear. A common situation in public testimony is that some stakeholders say that the fishing has never been better, while the stock assessment models indicate that the population is being overexploited, that reductions in fishing pressure are required, and that livelihoods will be affected.

Contentious situations can result in second-guessing among the advisory panels and the councils, and compromises that satisfy no one. For effective communication, the advisory panels simplify the model results, often sacrificing comprehensive presentation of the uncertainties inherent in the forecasts. In anticipation of council actions and over concerns that the uncertainty might be misused (highest value used) or used to delay action, the advisory panels may build in their own degree of caution in the results presented to the council. The council then may calibrate their decisions because they know the advisory panels (fishery scientists) have already adjusted the results for uncertainty (Frederick & Peterman 1995).

Communication can also be hampered if one or more of the stakeholders do not like the results of the stock assessment. Not all situations are like the red grouper case study, in which changes in models and data greatly affected management decisions. The red snapper (*Lutjanus campechanus*) fishery in the Gulf of Mexico illustrates a different situation in which, despite significant changes in the models and data through time, the results of the stock assessments consistently indicated that the stock was overfished and undergoing overfishing (e.g., Cummins-Parrack 1986, Goodyear 1995, Schirripa & Legault 1999). Nevertheless, this fishery has been highly controversial. Numerous reviews of the stock assessment data and methodology have been performed (Gulf of Mexico Fishery Management Council 2003). Moreover, the controversy is heightened because of user conflicts between the directed fishery for red snapper and the very large and profitable shrimping industry, which takes juvenile red snapper in large numbers as bycatch. In short, regulation of harvest has been too stringent in the opinion of many, and too lenient in the opinion of others. In our opinion, the issue is controversial only because people do not like the outcome of the stock assessments and the resulting management decisions required.

Ecologists should spend considerable effort on ensuring effective communication of model results to the managers and stakeholders. Complex results will have to be simplified and translated into a language that managers and stakeholders can understand, while still conveying the confidence appropriate for the forecasts. Despite the best efforts, sometimes the results will be rejected based on stakeholder interests, rather than based on the science. Effective communication and honesty (especially being flexible and admitting mistakes) on the part of the ecologists will lead to trust among all parties and ultimately better management decisions.

Improved Data Collection and Modeling Methods

A positive outcome of the fisheries experience has been the stimulus for the development of new methods for data collection and for modeling. While not all of the recent advances in fisheries ecology can be attributed to fisheries management crises, there has been a trickle-down effect of issues and funding that originated in management issues and subsequently entered the research community. Advances in stock identification (Waldman et al. 1997), modeling of early life stages (Hermann et al. 2001), and Bayesian estimation of complex models (Hilborn et al. 1994) are but a few of the many examples.

The fisheries management experience has also emphasized the importance of multidisciplinary studies (e.g., Kendall & Schumacher 1996). Ecological forecasting will require collaboration and multidisciplinary research (Clark et al. 2002).

Ecologists should take note of the fisheries lead on multidisciplinary research, especially the research designed to provide information directly to management. With some patience to let management demands feed back on research, ecologists can expect an exciting period of active research to follow their initial forecasting attempts. Successful multidisciplinary research requires exciting scientific questions and careful planning (Benda et al. 2002), and based on our experiences, some psychology.

CONCLUDING REMARKS

Several recent initiatives are pushing for an increased role of ecological forecasting in management decision making. We have used the U.S. marine fisheries management experience to illustrate what ecologists can expect as they move further into forecasting for management. Our case study of red grouper demonstrated the sensitivity of forecasting to the data and models, and the types of deliberations that can surround highly uncertain, but mandated, forecasting for management. Ecologists can expect:

1. To forecast under knowingly oversimplified conditions, while having to present the results without all of their uncertainties and caveats. There will likely be much greater emphasis on model calibration (i.e., fit to historical data) because that is what convinces people that the model is realistic. Also, there will be close scrutiny and review of the data and models, by both the peer-review process and by stakeholder consultants, that often will not change anyone's opinion.
2. To use existing data and models because general pleas for better data will be ignored until there is a crisis, and then opportunities for better data will be obtained on a case-by-case basis.
3. Calls for alternative approaches and stampedes to jump on "new tools" bandwagons.
4. Difficulties in quantifying the uncertainties in predictions and explaining them to managers and stakeholders.
5. To spend significant time and effort on ensuring effective communication of model results to the managers and stakeholders. Absolute honesty and admission of mistakes will lead to trust among all parties.
6. Ecology to eventually improve as management issues feed back onto the science and ecologists rise to the challenge with better methods and models. Multidisciplinary approaches will no longer be an option, but a requirement.

The need for data and models for forecasting is accelerating both in ecology and fisheries. Many ecosystems are at levels below those which provide optimal ecosystem services, and ecosystems are being subjected to increasing and multiple stressors. The situation for ecology and fisheries is one of forecasting when we are unsure of the baseline and have little idea of what the future will look like. The U.S. marine fisheries management experience provides valuable information on what ecologists can expect. The alternative of no action and walking away from the issues due to limited data or inadequate models is worse.

ACKNOWLEDGMENTS

The authors have served for a combined 15 years (and counting) on the Reef Fish Stock Assessment Panel and related committees within the Gulf of Mexico Fisheries Management Council. Much of the authors' experience with models and marine fisheries management was obtained during various panel and committee meetings. We hope we have described the deliberations of the panel meetings and the time line accurately. Any mistakes or omissions are unintentional and the sole responsibility of the authors. This paper was improved by the comments of an anonymous reviewer and the comments of several of our colleagues (who might prefer to remain anonymous). Partial financial support for the preparation of this paper was provided to KAR by the Electric Power Research Institute, Palo Alto, California.

LITERATURE CITED

Akcakaya HR, Burgman MA, Ginzburg LR. 1997. *Applied Population Ecology*. Setauket, NY: Appl. Biomath. 255 pp.

Alaska Sea Grant. 1999. *Ecosystem approaches for fisheries management. Univ. Alsk. Sea Grant Rep. AK-SG-99-01*, Fairbanks, Alsk.

Alverson DL. 2002. Factors influencing the scope and quality of science and management decisions (The good, the bad and the ugly). *Fish Fish.* 3:3–19

Armsworth PR. 2001. Effects of fishing on a protogynous hermaphrodite. *Can. J. Fish. Aquat. Sci.* 58:568–78

Bartell SM, Gardner RH, O'Neill RV. 1992. *Ecological Risk Estimation*. Boca Raton, FL: Lewis. 252 pp.

Benda LE, Poff NL, Tague C, Palmer MA, Pizzuto J. 2002. How to avoid train wrecks when using science in environmental problem solving. *BioScience* 52:1127–36

Beverton RJH, Holt SJ. 1957. On the dynamics of exploited fish populations. *UK Min. Agric. Food. Fish. Invest. Ser. 2, No. 19*

Botsford LW, Castilla JC, Peterson CH. 1997. The management of fisheries and marine ecosystems. *Science* 277:509–15

Brook BW, O'Grady JJ, Chapman AP, Burgman MA, Akcakaya HR, Frankham R. 2000. Predictive accuracy of population viability analysis in conservation biology. *Nature* 404:385–87

Buhl-Mortensen L, Toresen R. 2001. Fisheries management in a sea of uncertainty: the role and responsibility of scientists in attaining a precautionary approach. *Int. J. Sustain. Dev.* 4:245–64

Caswell H. 2001. *Matrix Population Models: Construction, Analysis, and Interpretation*. Sunderland, MA: Sinauer. 722 pp. 2nd ed.

Chavez FP, Ryan J, Lluch-Cota SE, Niquen CM. 2003. From anchovies to sardines and back: multidecadal change in the Pacific Ocean. *Science* 299:217–21

Christensen V, Pauly D, eds. 1993. *Trophic models of aquatic ecosystems. Proc. ICLARM Conf. 26th*. Manila, Philipp: Int. Cent. Living Aquat. Resour. Manag.

Clark JS, Carpenter SR, Barber M, Collins S,

Dobson A, et al. 2002. Ecological forecasts: an emerging imperative. *Science* 293:657–60

Coleman FC, Koenig CC, Huntsman GR, Musick JA, Eklund AM, et al. 2000. Long-lived reef fishes: the grouper-snapper complex. *Fisheries* 25:14–21

Cummins-Parrack N. 1986. Trends in Gulf of Mexico red snapper population dynamics, 1979–95. *Coast. Resourc. Div. Contrib. No. CRD-86/87-4*, Natl. Mar. Fish. Serv., Miami, FL

Frank KT, Leggett WC. 1994. Fisheries ecology in the context of ecological and evolutionary theory. *Annu. Rev. Ecol. Syst.* 25:401–22

Frederick SW, Peterman RM. 1995. Choosing fisheries harvest policies: when does uncertainty matter? *Can. J. Fish. Aquat. Sci.* 52:291–306

Funk F, Quinn TJ, Heifetz J, Ianelli JN, Powers JE, et al. eds. 1998. *Fishery Stock Assessment Models. Alaska Sea Grant College Prog. Rep. AK-SG-98-01*, Fairbanks, Alsk

Goodyear CP. 1995. *Red Snapper in US Waters of the Gulf of Mexico. Coastal Resources Division Contribution MIA-95/96-05*, Natl. Mar. Fish. Serv., Miami, FL

Goodyear CP. 1997. Fish age determined from length: An evaluation of three methods using simulated red snapper data. *Fish. Bull.* 95:39–46

Goodyear CP, Schirripa MJ. 1993. *The red grouper fishery in the Gulf of Mexico. Miami Lab. Contrib. No. MIA-92/93-75*, Natl. Mar. Fish. Serv., Miami, FL

Gulf Mexico Fish. Manag. Counc. 2003. *Draft environmental impact statement—Appendix A. History of fishery management in the Gulf of Mexico for each fishery management plan. NOAA Rep. Award No. NA17FC1052*, Gulf Mexico Fish. Manag. Counc., Tampa, FL

Haddon M. 2001. *Modelling and Quantitative Methods in Fisheries*. Boca Raton, FL: Chapman & Hall/CRC. 406 pp.

Harley SJ, Myers RA, Dunn A. 2001. Is catch-per-unit-effort proportional to abundance? *Can. J. Fish. Aquat. Sci.* 58:1760–72

Hastings A. 1997. *Population Biology: Concepts and Models*. New York: Springer. 220 pp.

Hermann AJ, Hinckley S, Megrey BA, Napp JM. 2001. Applied and theoretical considerations for constructing spatially explicit individual-based models of marine larval fish that include multiple trophic levels. *ICES J. Mar. Sci.* 58:1030–41

Hilborn R. 2002. The dark side of reference points. *Bull. Mar. Sci.* 70:403–8

Hilborn R, Pikitch EK, Francis RC. 1993. Current trends in including risk and uncertainty in stock assessment and harvest decisions. *Can. J. Fish. Aquat. Sci.* 50:874–80

Hilborn R, Pikitch EK, McAllister MK. 1994. A Bayesian estimation and decision analysis for an age-structured model using biomass survey data. *Fish. Res.* 19:17–30

Hilborn R, Walters CJ. 1992. *Quantitative Fisheries Stock Assessment: Choice, Dynamics and Uncertainty*. New York: Chapman & Hall. 570 pp.

Houde ED. 1987. Fish early life dynamics and recruitment variability. *Am. Fish. Soc. Symp.* 2:17–29

Huntsman GR, Schaaf WE. 1994. Simulation of the impact of fishing on reproduction of a protogynous grouper, the graysby. *N. Am. J. Fish. Manag.* 14:41–52

Jackson JBC, Kriby MX, Berger WH, Bjorndal KA, Botsford LW, et al. 2001. Historical overfishing and the recent collapse of coastal ecosystems. *Science* 293:629–38

Kendall AW, Schumacher JD. 1996. Walleye pollock recruitment in Shelikof Strait: applied fisheries oceanography. *Fish. Oceanogr.* 5(Suppl. 1):4–18

Krebs CJ. 2002. Beyond population regulation and limitation. *Wild. Res.* 29:1–10

Legault CM, Restrepo VR. 1998. A flexible forward age-structured assessment program. *ICCAT Col. Vol. Sci. Pap.* 49:246–53

Ludwig D, Hilborn R, Walters C. 1993. Uncertainty, resource exploitation, and conservation: lessons from history. *Science* 260:17

Ludwig D, Mangel M, Haddad B. 2001. Ecology, conservation, and public policy. *Annu. Rev. Ecol. Syst.* 32:481–517

Mace PM. 1994. Relationships between common biological reference points used as thresholds and targets of fisheries management strategies. *Can. J. Fish. Aquat. Sci.* 51:110–22

Mace PM, Bartoo NW, Hollowed AB, Kleiber P, Methot RD, et al. 2001. Marine fisheries stock assessment improvement plan. *NOAA Tech. Memo. NMFS-F/SPO-56*, Natl. Ocean. Atmos. Adm., Washington, DC

Megrey BA. 1989. Review and comparison of age-structured stock assessment models from theoretical and applied points of view. In *Mathematical Analysis of Fish Stock Dynamics*, ed. EF Edwards, BA Megrey. *Am. Fish. Soc. Symp.* 6:8–48

National Research Council (NRC). 1998. *Improving Fish Stock Assessments*. Washington, DC: Natl. Acad. Press. 77 pp.

National Research Council (NRC). 1999. *Sustaining Marine Fisheries*. Washington, DC: Natl. Acad. Press. 164 pp.

National Research Council (NRC). 2000. *Improving the Collection, Management, and Use of Marine Fisheries Data*. Washington, DC: Natl. Acad. Press. 222 pp.

National Research Council (NRC). 2001. *Marine Protected Areas: Tools for Sustaining Ocean Ecosystems*. Washington, DC: Natl. Acad. Press. 272 pp.

National Research Council (NRC). 2002. *Science and its Role in the National Marine Fisheries Service*. Washington, DC: Natl. Acad. Press. 83 pp.

Needle CL. 2002. Recruitment models: diagnosis and prognosis. *Rev. Fish Biol. Fish.* 11:95–111

Organ. Econ. Co-op. Dev. 1997. *Towards Sustainable Fisheries*. Paris, Fr: OECD. 268 pp.

Patterson K. 1992. Fisheries for small pelagic species: an empirical approach to management targets. *Rev. Fish Biol. Fish.* 2:321–38

Pitcher TJ, Hart PJB, Pauly D, eds. 2001. *Reinventing Fisheries Management*. Boston: Kluwer Acad. 435 pp.

Quinn TJ, Deriso RB. 1999. *Quantitative Fish Dynamics*. New York: Oxford Univ. Press. 542 pp.

Reef Fish Stock Assessment Panel (RFSAP). 1999. *September 1999 report of the reef fish stock assessment panel.* Gulf Mexico Fish. Manag. Counc., Tampa, FL

Reef Fish Stock Assessment Panel (RFSAP). 2000. *December 2000 report of the reef fish stock assessment panel.* Gulf Mexico Fish. Manag. Counc., Tampa, FL

Reef Fish Stock Assessment Panel (RFSAP). 2002. *September 2002 report of the reef fish stock assessment panel.* Gulf Mexico Fish. Manag. Counc., Tampa, FL

Restrepo VR, Thompson GG, Mace PM, Gabriel WL, Low LL, et al. 1998. Technical guidance on the use of precautionary approaches to implementing National Standard 1 of the Magnuson-Stevens Fishery Conservation and Management Act. *NOAA Tech. Memo. NMFS-F/ SPO-31*, Natl. Ocean. Atmos. Adm., Washington, DC

Ricker WE. 1958. Handbook of computations for biological statistics of fish populations. *Bull. Fish. Res. Board Can. 119*

Rose KA, Cowan JH, Winemiller KO, Myers RA, Hilborn R. 2001. Compensatory density dependence in fish populations: importance, controversy, understanding and prognosis. *Fish Fish.* 2:293–327

Rosenberg AA, Restrepo VR. 1994. Uncertainty and risk evaluation in stock assessment advice for U.S. marine fisheries. *Can. J. Fish. Aquat. Sci.* 51:2715–20

Roughgarden J, Smith F. 1996. Why fisheries collapse and what to do about it . *Proc. Natl. Acad. Sci.USA* 93:5078–83

Ruiz GM, Fononoff PW, Carlton JT, Wonham MJ, Hines AH. 2000. Invasion of coastal marine communities in North America: apparent patterns, processes, and biases. *Annu. Rev. Ecol. Syst.* 31:481–531

Rykiel EJ. 1996. Testing ecological models: the meaning of validation. *Ecol. Model.* 90:229–44

Sale PF, Tolimieri N. 2000. Density dependence at some time and place? *Oecologia* 124:166–71

Scavia D, Field JC, Boesch DF, Buddemeier RW, Birkett V, et al. 2002. Climate change

impacts on U.S. coastal and marine ecosystems. *Estuaries* 25:149–64
Schirripa MJ, Legault CM. 1999. *Status of the red snapper in U.S. waters of the Gulf of Mexico: updated through 1998. Sustainable Fisheries Division Contribution SFD-99/1099*, Natl. Mar. Fish. Serv., Miami, FL
Schirripa MJ, Legault CM, Ortiz M. 1999. *The red grouper fishery in the Gulf of Mexico: assessment 3.0. Sustainable Fisheries Division Contribution No. SFD-98/99-56*, Natl. Mar. Fish. Serv., Miami, FL
Smith SJ, Hunt JJ, Rivard D, eds. 1993. Risk evaluation and biological reference points for fisheries management. *Can. Spec. Publ. Fish. Aquat. Sci. 120*
Southeast Fish. Sci. Cent. (SEFSC) Staff. 2002. *Status of red grouper in United States waters of the Gulf of Mexico during 1986–2001. Sustainable Fisheries Division Contribution No. SFD-01/02-175rev*, Natl. Mar. Fish. Serv., Miami, FL
Sullivan PJ. 1999. *Review of Gulf of Mexico red grouper assessment 3.0 with analysis of scientific support for management options.* Rep. Univ. Miami, FL
Turchin P. 1995. Population regulation: old arguments and a new synthesis. In *Population Dynamics: New Approaches and Synthesis*, ed. N Cappuccino, PW Price, pp. 19–40. San Diego: Academic
Waldman JR, Richards RA, Schill WB, Wirgin I, Fabrizio MC. 1997. An empirical comparison of stock identification techniques applied to striped bass. *Trans. Am. Fish. Soc.* 126:369–85
Wilson DC, Degnbol P. 2002. The effects of legal mandates on fisheries science deliberations: the case of Atlantic bluefish in the United States. *Fish. Res.* 58:1–14

Annu. Rev. Ecol. Evol. Syst. 2003. 34:153–81
doi: 10.1146/annurev.ecolsys.34.011802.132435

First published online as a Review in Advance on September 29, 2003

PARTITIONING OF TIME AS AN ECOLOGICAL RESOURCE

Noga Kronfeld-Schor and Tamar Dayan
Department of Zoology, Tel Aviv University, Tel Aviv 69978, Israel;
email: Nogaks@post.tau.ac.il, Dayant@post.tau.ac.il

Key Words competition, circadian rhythms, evolutionary constraints, predation, temporal partitioning

■ **Abstract** Animal species have evolved different diel activity rhythms that are of adaptive value. Theory suggests that diel temporal partitioning may facilitate coexistence between competitors and between predators and prey. However, relatively few studies demonstrate a temporal shift that is predation- or competition-induced. Recorded shifts are usually within the preferred activity phase of animal species (day or night), although there are some inversions to the opposite phase cycle. Temporal partitioning is not perceived as a common mechanism of coexistence. This rarity has been variously ascribed to theoretical considerations and to the rigidity of time-keeping mechanisms, as well as to other physiological and anatomical traits that may constrain activity patterns. Our decade-long study of spiny mice of rocky deserts demonstrates that, while different factors select for activity patterns, endogenous rhythmicity may be an evolutionary constraint.

INTRODUCTION

Different animal species are active during different parts of the diel cycle. Activity patterns have evolved to cope with the time structure of the environment, which changes with a 24 h periodicity (e.g., Daan 1981). These different activity patterns may have ecological implications and evolutionary significance, as well as physiological ramifications. Insight into the interplay between selective forces and evolutionary constraints at these different levels is crucial for understanding the evolution of activity patterns. The selective forces and constraints affecting evolution of activity patterns underlie the partitioning of time as a resource.

How time mediates ecological interactions and shapes the structure of ecological communities is still poorly understood (Jaksic 1982, Schoener 1986, Wiens et al. 1986). Theory postulates that temporal partitioning among competitors and between predators and their prey may promote coexistence in ecological communities (e.g., Schoener 1974a, Richards 2002, Wiens et al.1986). Although the role of temporal partitioning in structuring communities has never been a strong focus of ecology, over the years a number of studies have accumulated that attach ecological significance to activity patterns (e.g., Kenagy 1973, Kunz 1973), as well as studies that record competition- or predation-induced shifts in activity patterns (e.g., Alanara et al. 2001, Fenn & MacDonald 1995).

On the other hand, much physiological, biochemical, and molecular research focuses on the evolution and maintenance of rhythms in activity, physiology, hormone concentrations, biochemistry, and behavior in animals (e.g., De Coursey et al. 1997, Gerkema 1992, Heldmaier et al. 1989, Horton 2001, Refinetti et al. 1992, Turek & Takahashi 2001). These circadian rhythms allow an animal to anticipate environmental changes and to choose the right time for a given response or activity (Aronson et al. 1993). The past decade has seen a surge of research into the nature of circadian rhythms, and, in particular, into mechanisms regulating them. In fact, in 1998 *Science* magazine listed breakthroughs in understanding diel rhythmicity among discoveries transforming our ideas about nature (*Science* Dec. 18:1998:2157–61). Mechanisms regulating circadian rhythmicity may affect the plasticity of response to ecological selective forces and hence the potential for evolving temporal partitioning among animal species.

Our goal here is to review the literature on the ecological significance of temporal partitioning and the physiological literature on the evolution and maintenance of activity rhythms. We aim to gain insight into the ecological significance of temporal activity patterns in light of the physiological literature, and to relate patterns to physiological, morphological, and behavioral adaptations to different activity phases. A close look at the interface between different scientific approaches and disciplines may yield insight into the evolutionary forces and evolutionary constraints at play. We also review a unique case of temporal partitioning among competing rocky desert rodents and point to general inferences that can be drawn from this system.

THE THEORETICAL USE OF TEMPORAL PARTITIONING AT THE DIEL SCALE

The time niche-axis may facilitate niche partitioning between co-occurring organisms. Different diel activity patterns may imply different use of resources or different levels of susceptibility to predation.

Among Competitors

Ecological theory has long considered niche differentiation in heterogeneous environments as a major mechanism of coexistence among competitors (e.g., MacArthur 1958, MacArthur & Levins 1967). Ecological separation is usually considered to involve habitat, food resources, and time axes, or a combination of them. In recent years a growing number of studies have suggested trade-offs in foraging ecology as a mechanism of coexistence (e.g., Brown 1996, Ziv et al. 1993).

Temporal partitioning on the diel scale may facilitate coexistence through avoidance of direct confrontation (interference competition) or through the reduction of resource overlap (resource competition). Temporal partitioning is a viable mechanism for reducing resource competition under either of the following conditions:

- If the shared limiting resources differ between activity times, particularly for predatory species whose prey populations have activity patterns (Schoener 1974a; but see Palomares & Caro 1999); or
- If the limiting resources are renewed within the time involved in the separation (MacArthur & Levins 1967).

A theoretical model suggests that fairly severe resource depletion must occur before it is optimal to cease feeding in a patch frequented by competitors (Schoener 1974b). Based on this model, Carothers & Jaksic (1984) argued that interference competition must be more prevalent than resource competition in driving temporal partitioning. Interference competition and resource competition are not mutually exclusive, however: Temporal partitioning may be generated by interference and yet act to reduce resource overlap, or vice versa (Kronfeld-Schor & Dayan 1999).

Richards (2002) developed a theoretical model to investigate the conditions necessary for temporal partitioning to be an evolutionarily stable strategy. The model highlights how the optimal foraging decision of an individual may depend strongly on the state of the individual and also that of its competitor (Richards 2002).

In Predator-Prey Systems

Some times are more dangerous for activity than others owing to temporal variation in predator activity and predation risk (Lima & Dill 1990). A decrease in activity as well as inactivity in a refuge lower an animal's risk of predation by lowering chances of detection and the probability of encounter (Lima 1998, Skelly 1994, Werner & Anholt 1993). Because many animals experience predictable daily fluctuations in predation risk, they may evolve activity patterns that minimize mortality risks while maximizing foraging. Therefore, it has been suggested that temporal partitioning between predators and their prey at the diel scale may evolve as a mechanism of coexistence (Stiling 1999).

TEMPORAL PARTITIONING IN NATURE

Temporal Partitioning Between Potentially Competing Taxa

Differing activity patterns of sympatric species have been viewed as ways to reduce interspecific resource and interference competition (Johnston & Zucker 1983). Many studies describe different diel activity patterns of potential competitors, many of which date to a time when patterns in ecological communities were not tested (Table 1). A study of four *Acacia* species in a highly seasonal savannah habitat in Tanzania stands out in this respect. Between dawn and dusk, pollen availability maxima (peaks of pollen release) were more regularly spaced than would be predicted by chance alone (Stone et al. 1996). Moreover, the summed

TABLE 1 Temporal partitioning between potentially competing taxa

Species	Observation	Reference
Sympatric congeneric lizards (*Ctenotus*)	The time niche axis is one of three on which these lizards subdivide environment and resources*	Pianka 1969
Great Basin kangaroo rats (*Dipodomys microps*), Merriam's kangaroo rats (*D. merriami*), little pocket mouse, (*Perognathus longimemberis*)	Aggressive interactions with *D. microps* drive *D. merriami* away, and this interaction reduces competition between *D. merriami* and *P. longimemberis*	Kenagy 1973
Twelve species of nocturnal rodents	Differences in time of daily activity between potential competitors	O'Farrell 1974
Hispid cotton rat (*Sigmodon hispidus*), fulvous harvest mouse (*Reithrodontomys fulvescens*)	Temporally partitioned in the Coastal Texas Prairie	Cameron et al. 1979
13 individual cotton rats in captivity	Interpreted as a means for reducing intraspecific competition	Johnston & Zucker 1983
California mouse (*Peromyscus californicus*), cactus mouse (*P. eremicus*), bannertail kangaroo rat (*Dipodomys spectabilis*), Fresno kangaroo rat (*D. nitratoides*)	Differences in response to lunar cycle may reflect temporal segregation among competitors	Lockard & Owings 1974, Owings & Lockard 1971
Cheetahs (*Acinonyx jubatus*), lions (*Felis leo*), spotted hyenas (*Crocuta crocuta*)	The activity pattern of cheetahs reduces kleptoparasitism and interference from lions and spotted hyenas	Mills & Biggs 1993, Palomares & Caro 1999, Schaller 1972
Bat communities	Interspecific differences in onset of activity and species-specific differences in activity times among bat species	Kunz 1973 and references therein
Different species of African dung beetles	Different species emerge from the soil at characteristic times of the day to fly and colonize freshly deposited dung of mammalian herbivores	Caveney et al. 1995

*While ecotothems may reduce food overlap by being active at different times, they may simply be choosing activity periods when their own metabolic performance is "optimal" (Huey & Pianka 1983), so that temporal partitioning of prey may be only an epiphenomenon (Carothers 1983).

activity of pollinators at each *Acacia* species clearly followed the temporal separation between species in pollen release, so the ecological consequences of temporal partitioning have actually been demonstrated.

The interest in the role of competition in evolution and in structuring communities also bears on scientific understanding of the evolution of activity patterns at broad taxonomic scales. For example, Wiens et al. (1986) suggested that mammals were restricted to nocturnal activity because of the largely diurnal activity patterns of dinosaurs, or that flying foxes (*Pteropus*) are constrained to nocturnality by competition with avian and mammalian frugivores. Testing evolutionary-scale patterns is difficult, sometimes impossible, particularly for extinct taxa. However, resource overlap and its partitioning can be studied among temporally partitioned extant taxa, although this has rarely been attempted.

It has often been suggested that raptors reduce competition by differing in activity time (reviewed by Jaksic 1982). Jaksic (1982) and Jaksic et al. (1981) found a high degree of overlap between prey of owls and diurnal raptors and suggested that owls evolved nocturnality in response to interspecific interference (Carothers & Jaksic 1984; Jaksic 1982; but see Simberloff & Dayan 1991). Huey & Pianka (1983), however, found that dietary overlap tended to be lower than expected among pairs of nocturnal and diurnal desert lizards with nonsynchronous activity.

In sum, only in a few cases is there statistical evidence for the occurrence of a meaningful pattern (Stone et al. 1996), or a strong resource-based test of the efficacy of temporal partitioning (Huey & Pianka 1983, Stone et al. 1996). Other studies have failed to demonstrate temporal partitioning or to establish ecological significance relating to interspecific competition (e.g., Cameron et al. 1979, Gabor et al. 2001, Kasoma 2000, Saiful et al. 2001, Saunders & Barclay 1992). However, it would be impossible to assess the prevalence of partitioning on the diel time axis on the basis of current evidence.

Temporal Partitioning Between Predators and Prey

Activity patterns of predators and their prey affect the level of predation risk. It has been suggested that, at the macroevolutionary scale, predator activity patterns track those of their prey. For example, around the Jurassic, insects evolved a waxy epicuticle that enabled them to become day-active; this probably set the stage for an evolutionary boom in diurnal reptiles (Daan 1981).

The bimodal activity pattern of three sympatric species of squirrels in Peninsular Malaysia was interpreted as a means to reduce predation by diurnal raptors that require good light and felids that are not active at these times (Saiful et al. 2001).

Greylag geese (*Anser anser*) on a Danish island responded more strongly to predator-like stimuli (overflying herons and helicopters) during moult when they are flightless, although real predators were absent, and foraged at night, possibly because such stimuli were lowest at night (Kahlert et al. 1996). Before moulting geese remained on the feeding grounds throughout the 24 h.

Fruit bats are active by day and at night on some small, species-poor Pacific islands such as Fiji. Wiens et al. (1986) suggested that they are constrained elsewhere to fly only at night by the presence of predatory diurnal eagles. Although bats are nocturnal, many species emerge from roosts to forage during twilight. Field research on foraging by a maternity colony of Schneider's leafnosed bats (*Hipposideros speoris*) in Sri Lanka suggested that bats captured large numbers of insects that were only available or had marked peaks in abundance during twilight (Pavey et al. 2001).

To test the theory that insectivorous bats have selected for diurnality in earless butterflies, Fullard (2000) compared nocturnal flight patterns of three species of nymphalid butterflies on the bat-free Pacific island of Moorea with those of three nymphalids in the bat-inhabited habitat of Queensland, Australia. No differences were found, however. Fullard (2000) concluded that physiological adaptations constrain the butterflies to diurnal flight. Also, although predation is a very serious threat to microtine rodents (e.g., Jacobs & Brown 2000), studies of their activity patterns suggest that they did not evolve as a means of preventing predator temporal specialization (Halle 1993, Reynolds & Gorman 1994).

In sum, few descriptive studies actually deal with depicting differences in activity patterns between predators and their prey, perhaps because the significance of such separation is perceived as trivial. The scarcity of such studies may also stem from the fact that, during the heydey of descriptive studies of niche partitioning in ecological communities, the focus was on how potential competitors coexist, rather than on predators and their prey.

TEMPORAL SHIFTS IN ECOLOGICAL TIME

Experimental results are the most compelling type of evidence that can be generated in order to test whether differences in activity patterns can actually evolve as a mechanism for coexistence.

Temporal Shifts of Competitors

In the sandy habitats of the western Negev desert of Israel there coexist two species that partition activity times, with *Gerbillus pyramidum* active during the early hours of the night and *G. allenbyi* active during later hours of the night. Upon removal of *G. pyramidum*, the smaller *G. allenbyi* shifted its activity to the earlier hours of the night, suggesting that coexistence between the two species is due to a trade-off between the foraging efficiency of *G. allenbyi* and the dominance of *G. pyramidum* (Ziv et al. 1993). The shared limiting resources, seeds, are renewed daily by afternoon winds redistributing and exposing buried seeds (e.g., Kotler & Brown 1990).

The cue for a temporal shift may be aggressive interference but may also be resource level. Lockard (1978) found that, when food resources were low, bannertail kangaroo rats (*Dipodomys spectabilis*) not only were active under full moonlight but also showed sporadic diurnal activity.

Nectar is renewed within the diel cycle (e.g., Cotton 1998, Craig & Douglas 1984) suggesting that temporal partitioning may facilitate coexistence between nectarivorous species. Cotton (1998) found a hierarchy of dominance over resources between four territorial hummingbird species that is body-size related, with larger species foraging in richer patches. Cotton (1998) suggested that the small body size of some of the species enables them to exploit marginal resources that would not be profitable for individuals of the larger species. In New Zealand bellbirds (*Athornys melanura*), dominant individuals forage in the morning when nectar availability peaks, while subordinate males and females forage in the afternoons, with lower rewards, suggesting intraspecific temporal partitioning (Craig & Douglas 1984).

Arjo & Pletscher (1999) documented interference competition between coyotes and recolonizing wolves in Montana, with records of wolves killing coyotes, and found increased temporal partitioning and changes in coyote behavior after wolf recolonization, in particular in winter (Arjo & Pletscher 1999).

In brown trout (*Salmo trutta*), individuals differ in their daily activity patterns, with dominant individuals feeding mainly at the most beneficial times of dusk and the early part of the night, while more subordinate fish feed at other times (Alanara et al. 2001, Bachmann 1984, Giroux et al. 2000). Moreover, the degree of overlap in foraging times between high-ranking fish depended on energetic demands related to water temperatures. However, predation risk influenced choice of foraging times by dominant individuals, so temporal activity patterns may result from a complex trade-off between ease of access to those resources and diel variation in foraging risk (Alanara et al. 2001).

Temporal Shifts of Predators and Their Prey

Much literature documents how predation risk affects animal activity levels. Predation risk can limit prey activity time. For example, presence of diurnal predatory fish limits activity of large mayfly larvae (*Baetis tricaudatus*), otherwise aperiodic or weakly diurnal, to the night (Culp & Scrimgeour 1993). Hermit crabs (in particular *Coenobita rugosus* and *C. cavipes*) respond to human-induced disturbances in Mozambique mangroves by changing from a 24-h activity cycle to a nocturnal one (Barnes 2001). A diametrically opposite pattern obtains for bank voles (*Clethrionomys glareolus*), which were inactive at night and exhibited a peak of activity at dawn, but in presence of a weasel (*Mustela nivalis*) shifted to being active during both the day and night (Jedrzejewska & Jedrzejewski 1990).

Actual shifts to a different activity pattern are not common (Lima & Dill 1990), but some cases have been reported. Fenn & MacDonald (1995) discovered a population of commensal Norway rats (*Rattus norvegicus*), some members of which were conspicuously diurnal. An experiment revealed that rats shifted to diurnal activity in an area heavily populated by foxes (*Vulpes vulpes*), while in nearby experimental fox-free enclosures, they reverted to nocturnal activity.

Nocturnal Patagonian leaf-eared mice (*Phyllotis xanthophygus*) decreased their activity under high illumination and increased the number of diurnal activity bouts, probably as compensation for reduced foraging (Kramer & Birney 2001). Daly et al. (1992) previously reported a similar pattern of increased crepuscular activity of Merriam's kangaroo rats (*Dipodomys merriami*) in response to full moonlight.

Pheidole titanis, ants that occur in desert and deciduous thorn forest in the southwestern United States and western Mexico, prey on termites during the day in the dry season. During the wet season these ants prey on termites at night, shifting their activity because of a specialist parasitoid fly diurnally active during this season (Feener 1988).

A similar pattern occurs in leaf-cutter ants (*Atta cephalotes*) at the tropical pre-montane wet forest at Parque Nacional Corcovado in Costa Rica (Orr 1992). These ants shift from nocturnal to diurnal activity in the presence of a diurnal parasitoid fly (*Neodohrniphora curvinervis*) (Orr 1992). In the presence of the parasitoid, daytime foraging ants were below the optimal size for foraging efficiency, but also smaller than the minimum size on which *N. curvinervis* will oviposit; nocturnal foragers were larger and within the optimal size range for foraging efficiency (Orr 1992).

Species that can be either nocturnal or diurnal can be used to test whether diel activity patterns respond to variations in predation pressure (Metcalfe et al. 1999). Many fishes can change their activity pattern from nocturnal to diurnal and vice versa, usually on a seasonal basis (Sanchez-Vazquez et al. 1996, Yokota & Oishi 1992). Changes in phasing related to variations in light intensity (Beers & Culp 1990), water temperature (Fraser et al. 1993), social effects (Anras et al. 1997), and nutritional status (Metcalfe & Steele 2001), have been reported. The flexibility in phasing can be an adaptive response to a relatively stable aquatic environment subjected to periodic changes in some biotic factors such as food availability or absence of predators (Sanchez-Vazquez et al. 1996). However, this flexibility in phasing may entail a cost in growth (Baras 2000) and in foraging efficiency (reviewed by Fraser & Metcalfe 1997).

Juvenile Atlantic salmon (*Salmo salar*) exhibit a temperature-dependent shift in the balance of diel activity: At higher temperatures they are found in foraging locations throughout both day and night but acquire most of their food by day when light enables them to forage more efficiently, but they are at greater risk of predation (Fraser & Metcalfe 1997). As temperatures drop they increasingly seek refuge during the day in crevices and emerge at night (Metcalfe et al. 1999). Higher temperatures imply greater energetic requirements, hence the diurnal activity at higher temperatures and more temperate zones. Increased food levels enabled juvenile salmon to minimize their exposure to predators by preferentially reducing daytime (in contrast to nighttime) activity (Metcalfe et al. 1999).

Moreover, the daily timing of activity in juvenile Atlantic salmon is related to the life-history strategy that they have adopted and their current state (body size/relative nutritional state). Salmon preparing to migrate to the sea, which would experience size-dependent mortality during migration, were more diurnal than fish of the same age and size that were delaying migration for a year (Metcalfe et al. 1998).

Dominant brown trout (*Salmo trutta*) preferred to forage during hours that are thought to minimize the predation risk incurred per unit of food obtained (Metcalfe et al. 1999). Thus risk of predation governs not only the activity patterns of dominant individuals, but also those of lower-ranking individuals that they displace (Alanara et al. 2001).

Experimental release from predation risk is a strong test of the significance of predation on modulating activity patterns. The longnose dase (*Rhinichthys cataractae*) is one of the few nocturnal minnow species. Its nocturnal activity pattern is constrained by the risk of predation by diurnally or crepuscularly active predators (Culp 1988). After several months in the laboratory, these fish are asynchronous to each other in their activity pattern with many individuals active during daylight (Culp 1988). Like its nocturnal congeners, the longnose dace have significantly increased encounter rates and reactive distance and significantly decreased search time under twilight conditions compared to starlight (Beers & Culp 1990), so this nocturnal activity clearly has a fitness cost (Calow 1985).

Another interesting case is the diel periodicity of downstream drift of stream invertebrates (Muller 1974). In many invertebrate taxa and among some fishes, drift numbers are low during the day followed by dramatic increases at night (Flecker 1992), a pattern interpreted as an evolutionary response to minimize predation risk by visually hunting fishes. Prey size classes with the greatest risk of predation by size-selective predators exhibit the greatest propensity for nighttime drift (e.g., Allan 1984). A study of mayflies in a series of Andean mountain and piedmont streams revealed that mayfly activity was arrhythmic in fishless streams. A correlation was found between level of predation risk (number of predatory species and their abundance) and propensity toward nocturnal activity (Flecker 1992). Experimental exclusion of fish did not change the nocturnal drift pattern, suggesting that nocturnal activity has evolved as a fixed behavioral response to predation; apparently, this behavior can evolve rapidly because, in formerly fishless streams where trout (*Oncorhynchus mykiss* and *Salvelinus fontinalis*) were introduced, nocturnal peaks in drift were observed for the mayfly *Baetis* (Flecker 1992).

In New Zealand, native common river galaxias (*Galaxias vulgaris*) were replaced in many streams by introduced brown trout (McIntosh & Townsend 1995) that present a higher predation risk to *Deleatidium* mayfly nymphs during the day than at night, while common river galaxias present a similar risk throughout the diel cycle. Mayfly nymphs fed significantly more during the night in streams with introduced brown trout than in streams with native galaxias, a difference that reflects the diel variation in predation risk imposed by the fish (McIntosh & Townsend 1995).

It has been argued that prey activity influences the activity patterns of their predators (e.g., Halle 1993). Zielinski (1988) studied the influence of daily restricted feeding in foraging mustelids, and discovered that while activity of six (of seven) animals shifted in response to modified foraging cost, this shift mostly meant an expansion of foraging period. In only two individuals, both of them mink (*Mustela vison*), was there an increase in activity in the nonpreferred part of the diel cycle (Zielinski 1988).

On the Relative Rarity of Temporal Partitioning at the Diel Scale

Schoener (1974a), reviewing resource partitioning in ecological communities, found temporal partitioning to be significantly less common than habitat or food type partitioning. He argued that "in deciding to omit certain time periods, the consumer is usually trading something—a lowered but positive yield in the time period frequented by competitors—for nothing, no yield at all" (p. 33). Schoener (1974b) developed a theoretical model that predicts that temporal resource partitioning at the diel scale should be relatively rare, requiring severe depletion of resources before it is no longer optimal to feed in a period frequented by competitors.

An alternative hypothesis was presented by Daan (1981), who suggested that diurnal and nocturnal activity require different evolutionary adaptations and therefore closely related species, prime candidates for competition, are usually active during the same part of the diel cycle.

This hypothesis implies that animal species are evolutionarily constrained in their activity patterns (Daan 1981; Kronfeld-Schor et al. 2001a,b,c; Roll & Dayan 2002) and that the plasticity assumed by Schoener (1974a) in adapting to ecological settings is limited. In fact, zoologists recognize that different species tend to have taxon-specific activity patterns. For example, most birds are diurnally active, while most terrestrial amphibians are nocturnal (Daan 1981). A recent quantitative study reveals a strong relationship between between phylogeny (using taxonomic status as a surrogate) and activity patterns in rodents (Roll & Dayan 2002).

These patterns are also in accord with the fact that many recorded shifts in activity times or perceived patterns of temporal partitioning are contained within the preferred part of the diel cycle, whether day or night, although these are detectable only by detailed scientific research. This is in spite of the fact that temporal segregation within the preferred activity phase limits foraging times and hence energy intake of the competing species (see Schoener 1974a). The behavioral response of prey to predation risk is also more commonly manifested in a restriction, rather than shift, of activity times. Actual inversion of activity patterns is not commonly described, although such gross differences in activity patterns (day or night) can be easily discerned. It could be argued that temporal partitioning evolved in response to competitive pressures, but activity patterns may have since become "fixed" and are no longer amenable to manipulation ("ghost of competition past").

Little such general discussion of the potential use of temporal partitioning between predators and their prey is found in the ecological literature, although some authors have raised this issue regarding specific cases (e.g., Flecker 1992, Fullard 2000). For example, Flecker (1992) suggested that some antipredatory behaviors may become "fixed" or "hard-wired," presumably where there is a "prohibitively expensive cost" in assessing risk, meaning a high probability of mortality (Sih 1987). Thus, the general question raised regarding competitors is also relevant to predators and their prey. Specifically, the issue is what is the degree of flexibility of the adaptations to different activity patterns. Are ecological- and evolutionary-level plasticity in the adaptations to activity patterns limited, and hence do they limit the use of the time axis in ecological separation?

In order to explore this issue, we review the literature on adaptations to diurnal and nocturnal activity, in particular that which relates to the evolution and maintenance of diel rhythmicity in animal species.

ADAPTATIONS TO NOCTURNAL AND DIURNAL ACTIVITY

Being active during night or day exposes animals to different challenges; meeting them requires different anatomical, physiological, and behavioral adaptations. For example, environmental conditions affect the activity of animal species, and may even drive them to invert their activity patterns. The bat-eared fox (*Otocyon megalotis*) of the South African deserts forages nightly in the summer when mid-day soil temperatures reach 70°C and diurnally in the winter when night air temperatures drop to −10°C (Lourens & Nel 1990). Desert seed-harvesting ants forage during the day in winter but avoid the heat of summer by foraging crepuscularly, nocturnally, or on cloudy days (Whitford et al. 1981).

Diurnal animals usually use vision for predation and visual pecking, while nocturnal animals use tactile probing, smell, and hearing. Communication is usually vocal and aromatic in nocturnal animals (and also, although rarely, luminescent). Nocturnal animals use camouflage for concealment from their diurnal predators during the day (e.g., moths and owls); diurnal animals use visual signals, e.g., aposematic coloration (Daan 1981).

Retinas of nocturnal and diurnal mammals differ in their photoreceptors; adaptations to vision at a given light level tend to reduce efficiency of activity at other times (Jacobs 1993, Van Schaik & Griffiths 1996). For example, wild-caught antelope ground squirrels (*Ammospermophilus leucurus*) deprived of their suprachiasmatic nuclei [SCN, where the master circadian clock is located (Moore & Eichler 1972, Stephan & Zucker 1972)] behaved normally but became arrhythmic, enabling nocturnal Mojave Desert predators such as rattlesnakes, kit foxes, bobcats, coyotes, and barn owls to take advantage of their limited visual acuity at night and prey upon them (DeCoursey et al. 1997).

In sum, complex adaptations have evolved to accompany diurnal and nocturnal ways of life. Because of the major differences between night and day, particularly in light levels and ambient temperatures (Daan 1981), adaptations to a nocturnal way of life may differ dramatically from those for diurnal activity, and adaptations to a certain mode of activity may be deleterious for another.

DIEL RHYTHMS AND THEIR RELEVANCE TO ECOLOGY

An animal's behavior and physiology result from an integration between its endogenous circadian rhythms generated by an internal clock, direct response to environmental stimuli that mask the expression of the endogenous circadian rhythm independently from the pacemaker, and the influence of the environment on its endogenous circadian clock (entrainment).

The Adaptive Value of Circadian Rhythms

Circadian rhythms allow animals to anticipate environmental changes: Physiological parameters such as body temperature, enzymatic activity, sensitivity of photoreceptors, and storage or mobilization of energy reserves have to be adjusted before the expected environmental changes actually take place. Furthermore, behavioral timing in feeding, reproduction, migration, etc., often precedes the external events (Gerkema 1992). It also allows the animal to choose the right time for a given response or activity without being easily misled by minor environmental disturbances (Aronson et al. 1993), thus contributing to their fitness (Bennet 1987, Daan & Aschoff 1982, Horton 2001).

The internal clock also ensures that internal changes (biochemical and physiological) are coordinated with one another (Horton 2001, Moore-Ede & Sulzman 1981, Turek & Takahashi 2001). In humans under circumstances where social and working routines are disrupted, such as shift-work and jet-lag, the clock may receive photic and nonphotic cues that potentially conflict, leading to circadian dysfunction and poor performance (e.g., Turek & Takahashi 2001).

One way to demonstrate that a character is adaptive is to look for loss or relaxation of this trait in environments lacking diel rhythmicity, where it has no apparent advantage (Horton 2001,Willmer et al. 2000). Subterranean mole rats (*Spalax ehrenbergi*) that live in constant darkness are predominantly diurnal during winter and predominantly nocturnal during summer (Kushnirov et al. 1998). Brazilian cave catfishes (*Pimelodella kronei*, *P. transitoria*, and *Trichomycterus* sp.) showed some degree of rhythmicity (circadian, ultradian, and/or infradian) (Trajano & MennaBarreto 1995, 1996). Most cave-dwelling millipedes (*Glyphiulus cavernicplu sulu*, Cambalidae, Spirostreptida) that occupy the deeper recesses of a cave show circadian rhythmicity (Koilraj et al. 2000). The hypogean loach (*Nemacheilus evezard)*, shows a circadian rhythm of body temperature, with individually varying periods (Pati 2001). Other species, such as the European blind cave salamander (*Proteus anguinus*), have lost their circadian rhythmicity (Hervant et al. 2000). It has been hypothesized that the retention of some rhythmic component reflects the importance of maintaining internal temporal order or the ability to measure seasonal changes in photoperiod (Goldman et al. 1997, Horton 2001).

Plasticity of Diel Rhythms (Laboratory Experiments)

Under constant laboratory conditions, the period of circadian rhythms is usually in the range of 24 h. This fact implies that changes in the physical environment must synchronize or entrain the internal clock system regulating circadian rhythms. In order for the internal timing mechanism to be adaptive, the internal clock should respond to highly predictive environmental cues and not to less predictive ones (Daan & Aschoff 2001).

One of the most important and predictable cues in the environment is the day/night cycle, which allows the development of relatively rigid internal circadian programming of behavior and physiology as an adaptive strategy. The endogenous circadian rhythm is an intrinsic and relatively inflexible component of the organism's physiology and behavior that is common to the species (Daan 1981). At the individual level, this internal component is adjusted to its specific environment based on individual experience with the temporal organization of the environment: Events related to food availability, territory, predation, inter- and intraspecific competition, and temperature affect behavior over the course of the day. Thus, the overt activity rhythms result from the output of the endogenous clock, a direct response to environmental stimuli, and the influence of the environment on the endogenous clock.

The effect of nonphotic stimuli on activity patterns of animals is of special significance to the evolution of temporal partitioning in communities. The ability to respond to nonphotic cues is what provides animal species with some flexibility to respond at the ecological timescale to competition or to predation pressures by shifting diel rhythms.

In the absence of the light-dark cycle (in constant light or dark), nonphotic stimuli such as activity, wheel running, food availability, and social stimuli, can entrain the endogenous circadian clock located in the SCN (e.g., Mistlberger 1991, Mrosovsky 1988) or mask the endogenous circadian system (e.g., Eckert et al. 1986, Gattermann & Weinandy 1997, Refinetti et al. 1992). Masking (Aschoff 1960) is defined as "any process that distorts the original output from the internal clock whether this originates from inside or outside the body" (Minors & Waterhouse 1989). As soon as the masking effect is removed, the underlying circadian rhythm is revealed (Waterhouse et al. 1996).

Under field conditions, light-dark cycles are present for most species. When both light and a nonphotic stimuli are presented, the nonphotic stimulus may be confronted by photic input, so that the perceived shift is a masking effect (Gattermann & Weinandy 1997, Refinetti 1999, Refinetti et al. 1992) or downstream to the clock (Kas & Edgar 1999). Alternatively, it may be blocked by the photic stimulus (Honardo & Mrosovsky 1991, Maywood & Mrosovsky 2001). In any of these cases, the effect of the nonphotic stimulus will disappear when the stimulus is ended (e.g., Blanchong et al. 1999, Kas & Edgar 1999, Kronfeld-Schor et al. 2001a) and will not cause a phase-shift of the circadian system.

A molecular mechanism involving clock gene expression for the interaction between photic and nonphotic circadian clock resetting stimuli was recently suggested. The sensitivity of the circadian pacemaker to light and nonphotic stimuli is phase-dependent. Light causes a phase shift only when given during the subjective night, while nonphotic stimulus causes a phase shift when given during the subjective day (e.g., Hut et al. 1999). Maywood & Mrosovsky (2001) showed in the laboratory that at any phase of the cycle, light and nonphotic stimuli have convergent but opposite effects on the circadian clock gene expression.

According to their model, under natural conditions, the animal is exposed to light only during the day, and any phase shift that may have been caused by a nonphotic stimulus will be blocked by the light stimulus. During the night animals in their natural habitats are not normally exposed to light, and nonphotic stimulus is not expected to cause a phase shift during the night in the absence of light. Interactions between light and nonphotic stimuli that fit this model were reported in several laboratory experiments (e.g., Mistlberger & Antle 1998, Mistlberger & Holmes 1999, Mrosovsky 1991, Ralph & Mrosovsky 1992, Weber & Rea, 1997).

The mechanisms determining diurnal, nocturnal, or crepuscular activity are still unknown (Kas & Edgar 1999, Novak et al. 1999). The internal clock, located in the SCN, is necessary and sufficient for the generation of mammalian circadian rhythms (Moore & Eichler 1972, Stephan & Zucker 1972), but the neuronal and metabolic activity within the SCN and the response to light (Inouye & Kawamura 1979, Sato & Kawamura 1984, Schwartz et al. 1983), and the clock gene expression (Novak et al. 2000, Mrosovsky et al. 2001) are similar in nocturnal and diurnal species, indicating that the center managing activity is located downstream from the core pacemaker.

Unstriped Nile rats (*Arvicanthis niloticus*) in the laboratory show a diurnal rhythm (Blanchong et al. 1999). However, introducing a running wheel to their cage induces an abrupt dramatic change in the basic pattern of activity within one day to a more nocturnal one (Blanchong et al. 1999). A similar effect of a running wheel was described for the diurnal *Octodon degus* (Kas & Edgar 1999), which inverted its activity rhythm in 24 to 48 h (Stevenson et al. 1968). An abrupt change in activity rhythms was also described in cotton rats (*Sigmodon hispidus*), where a variety of distinctly different patterns coexist, and individuals frequently switch from one kind of pattern to another (Johnston & Zucker 1983). Such an abrupt shift without evidence of phase transients (progressive changes in rhythm phase during the course of a phase shift, Kas & Edgar 1999) typical to the process of entrainment suggests that the mechanisms determining the overt diurnal or nocturnal activity rhythm in these species are separate from phase control mechanisms within the circadian pacemaker and that nonphotic stimuli can modulate the mechanisms that determine phase preference (Kas & Edgar 1999). Phase relation of *Fos* expression in the ventrolateral preoptic area (which appears to be the site that integrates circadian and homeostatic signals that influence the sleep-wake cycle) and that of the SCN differ between diurnal and nocturnal rodents exposed to the light-dark cycle, and this fact raises the possibility that the functional outcome of SCN inputs to the ventrolateral preoptic area differs in these two groups of animals, reflecting their activity patterns (Novak et al. 1999).

In nocturnal animals the internal clock period is shorter than 24 h, while in diurnal animals it is longer. Exposure to constant light lengthens and shortens the period of the internal clock of nocturnal and diurnal species, respectively. It was speculated that, as day length changes seasonally, these differences allow diurnal animals to track dawn whereas nocturnal animals track dusk (Pittendrigh & Daan 1976). Several species of rodents expressing shifts from nocturnal to diurnal activity and vice versa have a longer circadian period under constant light conditions than under constant dark, typical of nocturnal animals, even when they are diurnally active [*Arvicanthis ansorgi* (Challet et al. 2002), *Octodon degus* (Lee & Labyak 1997), *Arvicanthis niloticus* (Katona & Smale 1997)]. This fact further supports the hypothesis that the mechanisms determining the overt diurnal or nocturnal activity rhythm in these species are separate from phase control mechanisms within the circadian pacemaker.

Food has a separate entrainable oscillator outside of the light-entrained SCN (reviewed by Stephan 2001, but see Refinetti 1999). Having two separate oscillators with weak coupling may be adaptive, because food sources may shift suddenly, whereas seasonal changes in sunrise and sunset are gradual. This will enable animals to reset the phase of the food entrainable oscillator without shifting the phase of all the circadian system (Stephan 2001). The same reasoning holds for other nonphotic cues affecting activity. At any rate, under natural field conditions the internal circadian clock is primarily entrained by the steadily and reliably occurring light/dark cycle (Gattermann & Weinandy 1997). Thus, it appears that nonphotic stimuli such as food availability, predation, inter- and intraspecific competition, and temperature can cause the animal to shift its activity time without shifting the endogenous circadian clock. In such cases an animal will be active opposite or out of phase to its endogenous circadian rhythm.

For most animals under natural conditions the timing of sleep and wake (rest and activity) is in synchrony with the circadian control of the sleep/wake cycle and all other circadian-controlled rhythms. Humans have the cognitive capacity to override their endogenous circadian clock and its rhythmic output (Turek & Takahashi 2001). We showed here several examples of other species shifting their activity time without shifting their endogenous circadian clock and its rhythmic output. In humans, disturbed circadian rhythmicity has been associated with many mental and physical disorders and can have a negative impact on human safety, performance, and productivity (Turek & Takahashi, 2001). Among the very few studies of animals that shift their activity times, no other differences in diurnal rhythms were found between nocturnally and diurnally active individuals (Blanchong et al. 1999, Kronfeld-Schor et al. 2001a, Weber & Spieler 1987), suggesting that in these cases, the activity-rest cycle is indeed out of synchrony with all other circadian clock–controlled rhythms. Thus the daily optimal temporal arrangement with environmental events and/or the internal temporal order of physiological and biochemical processes may be out of synchrony. As in humans, such a shift may entail severe costs.

In sum, it appears that circadian rhythmicity may limit the response to nonphotic cues such as ecological interactions. The cost of a shift may be considerable in terms of the physiology and ecology of living organisms. Thus the plasticity of use of the time niche-axis at the diel scale may be severely constrained.

SPINY MICE OF ROCKY DESERTS: AN ECOLOGICAL/EVOLUTIONARY CASE STUDY OF TEMPORAL PARTITIONING

Activity Patterns

An excellent model system for the study of the role of temporal partitioning and the evolution of activity patterns is found in a hot rocky desert near the Dead Sea. The common spiny mouse (*Acomys cahirinus*) and the golden spiny mouse (*A. russatus*) coexist in rocky habitats (Kronfeld-Schor et al. 2001a; Shkolnik 1966, 1971) where they overlap in microhabitat use, home ranges, food habits, and reproductive period (Kronfeld et al. 1994, 1996; Kronfeld-Schor & Dayan 1999; Shargal et al. 2000). These species have attracted attention (e.g., Fluxman & Haim 1993; Haim & Borut 1981; Shkolnik 1971; Shkolnik & Borut 1969; Zisapel et al. 1998, 1999) owing to their unique temporal activity patterns: The common spiny mouse is active during the night, as are most desert rodents, whereas the golden spiny mouse is active during the day.

Shkolnik (1966, 1971) repeatedly trapped all individual *A. cahirinus* from a joint habitat, a rock pile, and after several months he began to trap *A. russatus* individuals during the night. This shift implies that the two species compete and that temporal partitioning is a mechanism of coexistence between them (Shkolnik 1971). A recent study with replicated experimental and control enclosures revealed that, while *A. russatus* shifted their activity also into the night in absence of their congener, their diurnal foraging activity remained high (Gutman 2001).

In the past decade we have been investigating the effect of the ecological and environmental challenges of diurnal and nocturnal activity on spiny mouse populations. We studied the costs incurred in diurnal and nocturnal activity and the evolutionary constraints involved in the shift from nocturnal to diurnal activity patterns.

Temporal Partitioning as a Mechanism of Coexistence

Experimental results suggest that temporal partitioning is a mechanism for coexistence between spiny mice (Shkolnik's 1966, 1971), although the actual limiting resource remains to be studied. In the Negev Desert, Abramsky et al. (1992) demonstrated that shelters limit common spiny mice on slopes covered with small stones. However, the Ein Gedi terrain is rich in boulders, and shelter is abundant for the low spiny mouse populations (Kronfeld-Schor & Dayan 1999). In disturbed areas near human settlements where food availability is high, their population densities increase, suggesting that food may be limiting (Kronfeld-Schor & Dayan 1999), as in other desert rodent communities (e.g., Heske et al. 1994, Rosenzweig & Abramsky 1997). Alternatively, the temporal shift results from interference competition but does not actually reduce resource overlap.

The species overlap in food habits, with a preference for arthropods (Kronfeld-Schor & Dayan 1999). In the field, the arthropod component in the diet of both species was low in winter but extremely high in summer (Kronfeld-Schor & Dayan 1999). Thus during winter both species overlap in a largely vegetarian diet.

In winter the two species showed trade-offs in foraging efficiency: The common spiny mouse is a "cream skimmer," a relatively inefficient forager with high giving-up densities, and a habitat generalist; the golden spiny mouse is a habitat specialist that compensates for this restricted niche by foraging very efficiently to low giving-up densities (Jones et al. 2001).

In summer, however, a predation-induced shift in foraging of both species increased the overlap in foraging behaviors between them (Jones et al. 2001). Moreover, an experimental study during summer suggests that foraging trade-offs are not a viable mechanism of coexistence between the two species (Gutman 2001). However, during summer both species turn primarily insectivorous. Because the arthropod prey of *A. cahirinus* and *A. russatus* are likely to show diurnal patterns in availability, temporal partitioning could well promote resource partitioning and coexistence, particularly in summer (Jones et al. 2001, Kronfeld-Schor & Dayan 1999).

Resource competition may be mediated by interference competition, but the evidence for this phenomenon among spiny mice is equivocal (Gutman 2001, Pinter et al. 2002). A possible cue for the displacement of the golden spiny mice from nocturnal to diurnal activity is chemical signals released by common spiny mice (Haim & Fluxman 1996).

Anatomical, Behavioral/Ecological, and Physiological Adaptations

At Ein Gedi, the average maximal temperature in January is 20°C, and the average minimal temperature is 13°C. In July, the average maximal temperature is 38°C, and the average minimal temperature is 28°C (Jaffe 1988). During the day *A. russatus* avoid the heat behaviorally by remaining in the shade (Kronfeld-Schor et al. 2001b, Shkolnik 1971), reducing mid-day activity (Kronfeld-Schor et al. 2001a), and/or using evaporative cooling, which uses water, a scarce resource in the desert. Nevertheless, *A. russatus* has low water requirements owing to their ability to reduce water loss in the feces (Kam & Degen 1993) and to produce highly concentrated urine (Shkolnik 1966, Shkolnik & Borut 1969).

We found no significant differences in water turnover between the species in all seasons, reflecting adaptations of *A. russatus* to water conservation (Kronfeld-Schor et al. 2001c). In summer, energy expenditure of *A. russatus* tended to exceed that of *A. cahirinus*. Energy requirements of *A. cahirinus* in winter were double those of *A. russatus* and may reflect the cost of thermoregulating during cold nights (Kronfeld-Schor et al. 2001c).

A. russatus has evolved some adaptations to diurnal activity, such as dark skin pigmentation and a high concentration of ascorbic acid in its eyes (Koskela et al. 1989). However, it also retained the retinal structure of a nocturnal mammal (Kronfeld-Schor et al. 2001b). Moreover, it has a similar potential for nonshivering thermogenesis (NST) to that of its nocturnal congener (Kronfeld-Schor et al. 2000), which is exposed in winter to much lower ambient temperatures and spends more energy on thermoregulation (Kronfeld-Schor et al. 2001c), suggesting that, in terms of NST *A. russatus* still displays its nocturnal legacy. The degree of NST capacity should be related to the most extreme cold conditions that an animal is expected to encounter. Since reaching maximal NST capacity requires at least several days (Heldmaier et al. 1981), it is crucial that animals not be misled by warm spells within the cold period and reduce their NST capacity. Control of NST capacity by day length obviates this problem, but it also implies that an animal that has shifted to activity in a warmer environment but with the same day length may retain its original cold-adapted NST capacity for an extended period (Kronfeld-Schor et al. 2000).

Ecological Interactions as Selective Forces

The few case studies that concern activity patterns of rodents and their predators take only one predator species or one functional group of predators into account (Halle 1993). General indirect evidence for the evolutionary significance of predation include spines on spiny mouse rumps and a histological mechanism for tail loss (Shargal et al. 1999). In order to gain insight into the evolution of activity patterns of spiny mice, we considered predation risk by owls, snakes, foxes, and diurnal raptors, (Jones et al. 2001).

A. cahirinus reduced their foraging in response to predation risk by owls in open habitats and during moonlit nights (Mandelik et al. 2002). Interestingly, golden spiny mice reduce their daytime foraging following full moon nights, a legacy of their nocturnal activity (Gutman 2001). Also, in response to owl calls, the level of stress hormones of *A. cahirinus* increased (Eilam et al. 1999), and their motor behavior changed with rising illumination levels. Predation risk by owls is a cost during the night, in particular in open habitats, and in particular during moonlit nights.

Nocturnal Blanford's foxes (*Vulpes cana*) prey upon spiny mice, although they constitute only a small portion of their diet (Geffen et al. 1992). Spiny mouse foraging patterns were not clearly affected by the presence of fox feces (Jones & Dayan 2000), and we view risk of predation by foxes as merely reinforcing a pattern driven by risk of owl predation.

The saw-scaled viper (*Echis coloratus*) is active during the summer. *A. cahirinus* have evolved relative immunity to its venom; a single snake bite is not lethal for individuals of this species (Weissenberg et al. 1997). A repeated strike, however, will kill them, so risk of predation by this snake remains a consideration for spiny mice. Predation by vipers is a threat primarily under boulders during the day (where these nocturnal sit-and-wait predators rest curled up) and during the night, both under and between boulders and in open areas, habitats where snakes are either lying still or actively moving at night (H. Hawlena, unpublished data; Mendelssohn 1965). Both *A. cahirinus* and *A. russatus* reduced their foraging in sheltered microhabitats in summer and shifted their foraging activity to more open microhabitats in summer, the viper activity season (Jones et al. 2001). Thus, in summer the response to risk of predation by vipers counters the response to risk of predation by owls during the night (see also Kotler et al. 1992) and that of physiological stress during the day.

In sum, predation pressures clearly affect activity levels of spiny mice. Although they do not appear to cause an inversion in activity patterns, they confer a cost on both diurnal and nocturnal activity that varies seasonally.

Diel Rhythms as an Evolutionary Constraint

In the field spiny mice temperature rhythms are generally compatible with their activity patterns (Elvert et al. 1999). However, immediately upon removal to the laboratory, individuals of both species exhibited typical nocturnal temperature rhythms, and *A. russatus* individuals displayed nocturnal activity rhythms or were active both during the light and dark periods (Kronfeld-Schor et al. 2001a). Immediate inversion of the phase preference without evidence of a phase shift that would be expected in the case of true entrainment (Deacon & Arendt 1996) indicates that the diurnal activity of *A. russatus* in the field and their overt temperature rhythms are merely a masking effect. Furthermore, in the laboratory the presence of *A. cahirinus* provoked a change in daily rhythms of body temperature and urine volume. Lesion of *A. russatus* pineal gland resulted in diminution of urinary 6-sulfatoxymelatonin (6SMT) and modification of body temperature and urine volume rhythms. However, the modifications in body temperature and urine volume provoked by the presence of *A. cahirinus* were similar in pineal-lesioned and sham-operated *A. russatus*, and the presence of *A. cahirinus* did not affect glucose uptake of the SCN in pineal-lesioned and sham-operated *A. russatus*, indicating that the effect of *A. cahirinus* presence on *A. russatus* is a direct, pineal-independent effect (Zisapel et al. 1998, 1999). Thus, many generations of selection for diurnal activity in golden spiny mice have not caused a shift in their underlying rhythmicity. The diel rhythms that normally enable mammals to respond to environmental stimuli appropriately (e.g., Rusak 1981, Ticher et al. 1995) appear to lack the plasticity required to enable *A. russatus* to adapt to community-level interactions, even at this evolutionary scale.

These results suggest that, although the time axis may well be significant for ecological separation, the evolution of temporal partitioning may be severely constrained. If *A. russatus* are indeed constrained in terms of their rhythm biology to their legacy as nocturnal mammals, then they must be paying a price for being active at a phase opposite to their natural rhythm. The rich literature on human shiftworkers suggests that this type of shift, to the diametrically opposite part of the diel cycle, entails severe costs in health and performance (Van Reeth 1998).

In sum, ecological and physiological costs and constraints affect the activity patterns of spiny mice as well as their behavior and their space use. Interspecific competition drives golden spiny mice to invert their activity patterns. This nonphotic cue, however, does not affect circadian rhythmicity, so golden spiny mice in nature are active against their native clock cycle. Circadian rhythms and NST capacity are regulated by photic cues that appear to override nonphotic cues in various biological systems for adaptive purposes; this rigidity implies that animals are not misled by minor environmental disturbances. It also implies that these aspects of animal physiology, as well as morphological traits that relate to their senses, may well act as evolutionary constraints limiting the use of the diel niche axis in structuring ecological communities.

Predation risk of owls, snakes, foxes, and probably also diurnal raptors affects microhabitat use and activity levels and consequently carries a cost, but it does not actually cause an inversion in activity patterns. Moreover, risk of predation by different predators may have opposing effects on activity and foraging and also an affect that opposes that of physiological costs and constraints. Our research so far suggests that, not only are the activity patterns and foraging microhabitat affected by ecological and physiological costs and constraints, but so is the community structure of these rocky desert rodents.

SUMMARY AND PROSPECTS FOR RESEARCH

Temporal partitioning between competitors and between predators and their prey is a significant mechanism of coexistence in some ecological communities. However, relatively few animal species invert their activity patterns as a result of interspecific or intraspecific interactions into the opposite activity phase. Most studies that suggest competition- or predation-induced segregation of activity patterns deal with temporal shifts within the normal nocturnal or diurnal activity time. Such temporal partitioning implies a cost in the overall reduction in activity times (see Schoener 1974a).

Although the time axis may well be significant for ecological separation among competitors and between predators and their prey, the evolution of temporal partitioning may be severely constrained. Physiological adaptations, among which are circadian rhythms, may limit the plasticity of activity patterns of animal species. Research at the interface between chronobiology, animal physiology, and ecological-evolutionary selective forces (see also Marques & Waterhouse 1994) may provide valuable insight into the evolution of activity patterns and of temporal partitioning (Kronfeld-Schor et al. 2001a).

Study of rhythm physiology may provide insight into the conditions under which phase shifts occur and the mechanism involved. In-depth ecological research of communities where temporal partitioning has evolved may provide insight into the selective regimes that have generated temporal segregation as a mechanism of coexistence. Is temporal partitioning a last-resource mechanism of coexistence where other mechanisms fail? Do some taxa have greater evolutionary plasticity than others? Are some environments (e.g., aquatic habitats) more conducive to shifts in activity patterns than others? How does phenotypic plasticity in the response of species to ecological interactions translate into phase shifts and the evolution of different activity patterns? Does history play an important role? Is the time elapsed since selective pressures began a major component in the likelihood of the evolution of temporal segregation?

The majority of animal species have evolved to be active either by day or by night, and it is now difficult to tease apart the evolutionary forces that have selected for their diel activity patterns. Particularly significant for research of the selective forces affecting activity patterns are taxa that can be either nocturnal or diurnal (Metcalfe et al. 1999). These are amenable to manipulative experiments with varying levels of predation risk and interspecific competition.

The remarkable advances in our understanding of the physiology of diel rhythms coupled with a growing understanding of how ecological communities function offers a wonderful opportunity for gaining ecological-evolutionary insight into the role of time as a niche axis.

ACKNOWLEDGMENTS

We are greatly indebted to Daniel Simberloff, Amiram Shkolnik, David Eilam, Gerhard Heldmaier, Abraham Haim, Nava Zisapel, Sara Weissenberg, Amos Bouskila, Aryeh Landsman, Menna Jones, Ralf Elvert, Yael Mandelik, Roee Gutman, Eyal Shargal, Noa Pinter, Einav Vidan, Merav Weinstein, Ornit Hall, and to many undergraduate students who have shared with us the spiny mouse experience. We also thank the many colleagues who have read our previous manuscripts and whose useful comments and criticisms have helped us develop our ideas regarding temporal partitioning. While preparing this manuscript, we were supported by an ISF grant (04320,021).

The *Annual Review of Ecology, Evolution, and Systematics* is online at http://ecolsys.annualreviews.org

LITERATURE CITED

Abramsky Z, Schachak M, Subach A, Brand S, Alfia H. 1992. Predator-prey relationships: rodent-snail interactions in the Central Negev Desert of Israel. *Oikos* 65:128–33

Alanara A, Burns MD, Metcalfe NB. 2001. Intraspecific resource partitioning in brown trout: the temporal distribution of foraging is determined by social rank. *J. Anim. Ecol.* 70:980–86

Allan JD. 1984. The size composition of invertebrate drift in a Rocky Mountain stream. *Oikos* 42:68–76

Anras MLB, Lagardere JP, Lafaye JY. 1997. Diel activity rhythm of seabass tracked in a natural environment: group effect on swimming pattern and amplitude. *Can. J. Fish. Aquat. Sci.* 54:162–68

Arjo WM, Pletscher DH. 1999. Behavioral responses of coyotes to wolf recolonization in northwestern Montana. *Can. J. Zool.* 77:1919–27

Aronson BD, Bell-Pedersen D, Block GD, Bos NP, Dunlap JC, et al. 1993. Circadian rhythms. *Brain Res. Brain Res. Rev.* 18:315–33

Aschoff J. 1960. Exogenous and endogenous components in circadian rhythms. *Cold Spring Harbor Symp. Quant. Biol.* 25:159–82

Aschoff J, ed. 1981. *Handbook of Behavioral*

Neurobiology. Vol. 4: *Biological Rhythms*. New York: Plenum. 562 pp.

Bachmann RA. 1984. Foraging behavior of free-ranging wild and hatchery brown trout in a stream. *Trans. Am. Fish. Soc.* 113:1–32

Baras E. 2000. Day-night alternation prevails over food availability in synchronising the activity of *Piaractus brachypomus* (Characidae). *Aquat. Living Resourc.* 13(2):115–20

Barnes DKA. 2001. Hermit crabs, humans and Mozambique mangroves. *Afr. J. Ecol.* 39(3):241–48

Beers CE, Culp JM. 1990. Plasticity in foraging behavior of a lotic minnow (*Rhinichthys cataractae*) in response to different light intensities. *Can. J. Zool.* 68:101–5

Bennett AF. 1987. The accomplishments of ecological physiology. In *New Directions in Ecological Physiology*,ed. ME Feder, AF Bennett, WW Burggren, RW Huey, pp. 1–8. Cambridge: Cambridge Univ. Press. 376 pp.

Blanchong JA, McElhinny TL, Mahoney MM, Smale L. 1999. Nocturnal and diurnal rhythms in the unstriped Nile rat, *Arvicanthis niloticus*. *J. Biol. Rhythms* 14:364–77

Brown JS. 1996. Coevolution and community organization in three habitats. *Oikos* 75(2):193–206

Calow P. 1985. Adaptive aspects of energy allocation. In *Fish Energetics: New Perspectives,* ed. P Tyler, P Calow, pp. 13–31. Baltimore, MD: Johns Hopkins Univ. 349 pp.

Cameron GN, Kincaid WB, Carnes BA. 1979. Experimental species removal: temporal activity patterns of *Sigmodon hispidus* and *Reithrodontomys fulvescens*. *J. Mammal.* 60(1):195–96

Carothers JH. 1983. Size-related activity patterns in an herbivorous lizard. *Oecologia* 57(1–2):103–6

Carothers JH, Jaksic FM. 1984. Time as a niche difference: the role of interference competition. *Oikos* 42:403–6

Caveney S, Scholtz CH, McIntyre P. 1995. Patterns of daily flight activity in onitine dung beetles (Scarabinae, Onitini). *Oecologia* 103(4):444–52

Challet E, Pitrosky B, Sicard B, Malan A, Pevet P. 2002. Circadian organization in a diurnal rodent: *Arvicanthis ansorgei* Thomas 1910: chronotypes, response to constant lighting conditions, and photoperiodic changes. *J. Biol. Rhythms* 17:52–64

Cotton PA. 1998. Temporal partitioning of a floral resource by territorial hummingbirds. *Ibis* 140(4):647–53

Craig JL, Douglas ME. 1984. Temporal partitioning of a nectar resource in relation to competitive symmetries. *Anim. Behav.* 32:624–25

Culp JM. 1988. Nocturnally constrained foraging of a lotic minnow (*Rhinichthys cataractae*). *Can. J. Zool.* 67:2008–12

Culp JM, Scrimgeour GJ. 1993. Size-dependent diel foraging periodicity of a mayfly grazer in stream with and without fish. *Oikos* 68(2):242–50

Daan S. 1981. Adaptive daily strategies in behavior. See Aschoff 1981, pp. 275–98

Daan S, Aschoff J. 1982. Circadian contribution to survival. In *Vertebrate Circadian System: Structure and Physiology,* ed. J Aschoff, S Daan, GA Groos, pp. 305–21. Berlin: Springer-Verlag. 340 pp.

Daan S, Aschoff J. 2001. The entrainment of circadian rhythms. See Takahashi et al. 2001, pp. 45–57

Daly M, Behrends PR, Wilson MI, Jacobs LF. 1992. Behavioural modulation of predation risk: moonlight avoidance and crepuscular compensation in a nocturnal desert rodent, *Dipodomys merriami*. *Anim. Behav.* 44:1–9

Deacon S, Arendt J. 1996. Adapting to phase shifts, I. an experimental model for jet lag and shift work. *Physiol. Behav.* 59:665–73

DeCoursey PJ, Krulas JR, Mele G, Holley DC. 1997. Circadian performance of suprachiasmatic nuclei (SCN)-lesioned antelope ground squirrels in a desert enclosure. *Physiol. Behav.* 62:1099–108

Duffy JF, Dijk DJ. 2002. Getting through to circadian oscillators: Why use constant routines? *J. Biol. Rhythms* 17:4–13

Eckert HG, Nagel B, Stephani I. 1986. Light and social effects on the free-running circadian activity rhythm in common marmosets

(*Callithrix jacchus:*Primates): Social masking, pseudo-splitting, and relative coordination. *Behav. Ecol. Sociobiol.* 18:443–52

Eilam D, Dayan T, Ben-Eliyahu S, Schulman I, Shefer G, Hendrie C. 1999. Differential behavioural and hormonal response of voles and spiny mice to owl calls: A possible role of individual differences, stimulus interpretation, and habitat structure. *Anim. Behav.* 58:1085–93

Elvert R, Kronfeld N, Dayan T, Haim A, Zisapel N, Heldmaier G. 1999. Telemetric field studies of body temperature and activity rhythms of *Acomys russatus* and *A. cahirinus* in the Judean Desert of Israel. *Oecologia* 119(4):482–92

Feener DH Jr. 1988. Effects of parasites on foraging and defense behavior of a termitophagous ant, *Pheidole titanis* Wheeler (Hymenoptera: Formicidae). *Behav. Ecol. Sociobiol.* 22:421–27

Fenn MGP, MacDonald DW. 1995. Use of middens by red foxes: risk reverses rhythms of rats. *J. Mammal.* 76(1):130–36

Flecker AS. 1992. Fish predation and the evolution of invertebrate drift periodicity: evidence from Neotropical streams. *Ecology* 73:438–48

Fluxman S, Haim A. 1993. Daily rhythms of body-temperature in *Acomys russatus* the response to chemical signals released by *Acomys cahirinus*. *Chronobiol. Int.* 10:159–64

Fraser NHC, Metcalfe NB. 1997. The costs of becoming nocturnal: feeding efficiency in relation to light intensity in juvenile Atlantic salmon. *Funct. Ecol.* 11:385–91

Fraser NHC, Metcalfe NB, Thorp JE. 1993. Temperature-dependent switch between diurnal and nocturnal foraging in salmon. *Proc. R. Soc. London Ser. B* 252:135–39

Fullard JH. 2000. Day-flying butterflies remain day-flying in a Polynesian, bat-free habitat. *Proc. R. Soc. London Ser. B* 267(1459):2295–300

Gabor TM, Hellgren EC, Silvy NJ. 2001. Multi-scale habitat partitioning in sympatric suiforms. *J. Wildl. Manag.* 65:99–110

Gattermann R, Weinandy R. 1997. Lack of social entrainment of circadian activity rhythms in the solitary golden hamster and in the highly social Mongolian gerbil. *Biol. Rhythm Res.* 28:85–93

Geffen E, Hefner R, Macdonald DW, Ucko M. 1992. Diet and foraging behavior of Blanford's foxes, *Vulpes cana*, in Israel. *J. Mammal.* 73:395–402

Gerkema MP. 1992. Biological rhythms: mechanisms and adaptive values. In *Rhythms in Fish*, ed. MA Ali, pp. 27–37. New York: Plenum. 348 pp.

Giroux F, Ovidio M, Philippart J-C, Baras E. 2000. Relationship between the drift of macroinvertebrates and the activity of brown trout in a small stream. *J. Fish Biol.* 56:1248–57

Goldman BD, Goldman SL, Riccio AP, Terkel J. 1997. Circadian patterns of locomotor activity and body temperature in blind mole-rats, *Spalax erenbergi*. *J. Biol. Rhythms* 12:348–61

Gutman R. 2001. *Foraging behavior of spiny mice: a model for testing the role of competition, predation risk and habitat structure.* MSc thesis. Tel Aviv Univ. (From Hebrew, Engl. summ.)

Haim A, Borut A. 1981. Heat production and dissipation in golden spiny mouse, *Acomys russatus*. *J. Comp. Physiol. B* 142:445–50

Haim A, Fluxman S. 1996. Daily rhythms of metabolic rates: role of chemical signals in coexistence of spiny mice of the genus *Acomys*. *J. Chem. Ecol.* 22:223–29

Halle S. 1993. Diel pattern of predation risk in microtine rodents. *Oikos* 68(3):510–18

Heldmaier G, Steinlechner S, Rafael J, Vsiansky P. 1981. Photoperiodic control and effects of melatonin on nonshivering thermogenesis and brown adipose tissue. *Science* 212:917–19

Heldmaier G, Steinlechner S, Ruf T, Wiesinger H, Klingenspor M. 1989. Photoperiod and thermoregulation in vertebrates: Body temperature rhythms and thermogenic acclimation. *J. Biol. Rhythms* 4:251–65

Hervant F, Mathieu J, Durand JP. 2000.

Metabolism and circadian rhythms of the European blind cave salamander *Proteus anguinus* and facultative cave dweller, the Pyrenean newt (*Euproctus asper*). *Can. J. Zool.* 78:1427–32

Heske EJ, Brown JH, Mistry S. 1994. Long-term experimental study of a Chihuahuan desert rodent community: 13 years of competition. *Ecology* 75:438–45

Honardo GI, Mrosovsky N. 1991. Interaction between periodic socio-sexual cues and light-dark cycles in controlling the phasing of activity rhythms in the golden hamsters. *Ethol. Ecol. Evol.* 3:221–31

Horton TS. 2001. Conceptual issues in the ecology and evolution of circadian rhythms. See Takahashi et al. 2001, pp. 45–57

Huey RB, Pianka ER. 1983. Temporal separation of activity and interspecific dietary overlap. In *Lizard Ecology,* ed. RB Huey, ER Pianka, TW Schoener, pp. 281–96. Cambridge, MA: Harvard Univ. Press. 512 pp.

Hut AR, Mrosovsky N, Daan S. 1999. Nonphotic entrainment in a diurnal mammal, the European ground squirrel (*Spermophilus citellus*). *J. Biol. Rhythms* 14:409–19

Inouye ST, Kawamura H. 1979. Persistance of circadian rhythmicity in a mammalian hypothalamic "island" containing the suprachiasmatic nucleus. *Proc. Natl. Acad. Sci. USA* 76:5962–66

Jacobs GH. 1993. The distribution and nature of color vision among the mammals. *Biol. Rev.* 68:413–71

Jacobs J, Brown JS. 2000. Microhabitat use, giving-up densities and temporal activity as short- and long-term anti-predator behaviors in common voles. *Oikos* 91(1):131–38

Jaffe S. 1988. Climate of Israel. In *The Zoogeography of Israel*, ed. Y Yom-Tov, E Tchernov, pp. 79–94. Dordrecht, Neth: Junk. 600 pp.

Jaksic FM. 1982. Inadequacy of activity time as a niche difference: the case of diurnal and nocturnal raptors. *Oecologia* 52:171–75

Jaksic FM, Greene HW, Yanez JL. 1981. The guild structure of a community of predatory verterates in central Chile. *Oecologia* 49:21–28

Jedrzejewska B, Jedrzejewski W. 1990. Antipredatory behaviour of bank voles and prey choice of weasels—enclosure experiments. *Ann. Zool. Fennici* 27:321–28

Johnston PG, Zucker I. 1983. Lability and diversity of circadian rhythms of the cotton rat *Sigmodon hispidus*. *Am. J. Physiol. Regul. Integr. Comp. Physiol.* 244:R338–46

Jones ME, Dayan T. 2000. Foraging behaviour and microhabitat use of spiny mice, *Acomys cahirinus* and *A. russatus*, in the presence of Blandord's fox (*Vulpes cana*) odour. *J. Chem. Ecol.* 26(2):455–69

Jones ME, Mandelik Y, Dayan T. 2001. Coexistence of temporally partitioning spiny mice: roles of habitat structure and foraging behavior. *Ecology* 82(8):2164–76

Kahlert J, Fox AD, Ettrup H. 1996. Nocturnal feeding in moulting greylag geese *Anser anser*—an anti-predator response? *Ardea* 84(1/2):15–22

Kam M, Degen AA. 1993. Effect of dietary preformed water on energy and water budgets of two sympatric desert rodents, *Acomys russatus* and *Acomys cahirinus*. *J. Zool. London* 231:51–59

Kas MJH, Edgar DM. 1999. A nonphotic stimulus inverts the diurnal-nocturnal phase preference in *Octodon degus*. *J. Neurosci.* 19:328–33

Kasoma PMB. 2000. Diurnal activity patterns of three heron species in Queen Elizabeth National Park, Uganda. *Ostrich* 71(1&2):127–30

Katona C, Smale L. 1997. Wheel-running rhythms in *Arvicanthis niloticus*. *Physiol. Behav.* 61:365–72

Kenagy GJ. 1973. Daily and seasonal patterns of activity and energetics in a heteromyid rodent community. *Ecology* 54(6):1201–19

Kohler SL, McPeek MA. 1989. Predation risk and the foraging behavior of competing stream insects. *Ecology* 70:1811–25

Koilraj AJ, Sharma VK, Marimuthu G, Chandrashekaran MK. 2000. Presence of circadian rhythms in the locomotor activity of a cave-dwelling millipede *Glyphiulus cavernicolos sulu* (Cambalidae, Spirostreptida). *Chronobiol. Int.* 17:757–65

Koskela TK, Reiss GR, Brubaker RF, Ellefson RD. 1989. Is the high concentration of Ascorbic Acid in the eye an adaptation to intense solar irradiation? *Invest. Ophthalmol. Vis. Sci.* 30:2265–67

Kotler BP, Blaustein L, Brown JS. 1992. Predator facilitation: the combined effects of snakes and owls on the foraging behavior of gerbils. *Ann. Zool. Fennici.* 29:199–206

Kotler BP, Brown JS. 1990. Rates of seed harvest by two gerbilline rodents. *J. Mammal.* 71:591–96

Kramer KM, Birney EC. 2001. Effect of light intensity on activity patterns of Patagonian leaf-eared mice, *Phyllotis xanthopygus*. *J. Mammal.* 82(2):535–44

Kronfeld N, Dayan T, Zisapel N, Haim A. 1994. Co-existing populations of *Acomys cahirinus* and *A. russatus*: a preliminary report. *Isr. J. Zool.* 40:177–83

Kronfeld N, Shargal E, Dayan T. 1996. Population biology of coexisting *Acomys* species. In *Preservation of our World in the Wake of Change,* ed. Y Steinberger, IV:478–80. Jerusalem, Isr: Isr. Soc. Ecol. Environ. Qual. Sci. (ISEEQS). 886 pp.

Kronfeld-Schor N, Dayan T. 1999. The dietary basis for temporal partitioning: Food habits of coexisting *Acomys* species. *Oecologia* 121(1):123–28

Kronfeld-Schor N, Dayan T, Elvert R, Haim A, Zisapel N, Heldmaier G. 2001a. On the use of the time axis for ecological separation: Diel rhythms as an evolutionary constraint. *Am. Nat.* 158(4):451–57

Kronfeld-Schor N, Dayan T, Jones ME, Kremer I, Mandelik Y, et al. 2001b. Retinal structure and foraging microhabitat use of the golden spiny mouse (*Acomys russatus*). *J. Mamm.* 82(4):1016–25

Kronfeld-Schor N, Haim A, Dayan T, Zisapel N, Klingenspor M, Heldmaier G. 2000. Seasonal thermogenic acclimation of diurnally and nocturnally active desert spiny mice. *Physiol. Biochem. Zool.* 73(1):37–44

Kronfeld-Schor N, Shargal E, Haim A, Dayan T, Zisapel N, Heldmaier G. 2001c. Temporal partitioning among diurnally and nocturnally active desert spiny mice: Energy and water turnover costs. *J. Therm. Biol.* 26:139–42

Kunz TH. 1973. Resource utilization: temporal and spatial components of bat activity in central Iowa. *J. Mamm.* 54(1):14–32

Kushnirov D, Beolchini F, Lombardini F, Nevo E. 1998. Radiotracking studies in the blind mole rat. *Euro-American Mammal Congr., Santiago de Compostela, Spain*, p. 381

Lee TM, Labyak SE. 1997. Free-running rhythms and light- and dark-pulse phase response curves for diurnal *Octodon degus* (Rodentia). *Am. J. Physiol. Regul. Integr. Comp. Physiol.* 273:R278–86

Lima SL. 1998. Nonlethal effects in the ecology of predator-prey interactions. *BioScience* 48(1):25–34

Lima SL, Dill LM. 1990. Behavioral decisions made under the risk of predation: a review and prospectus. *Can. J. Zool.* 68:619–40

Lockard RB. 1978. Seasonal change in the activity pattern of *Dipodomys spectabilis*. *J. Mamm.* 59:563–68

Lockard RB, Owings DH. 1974. Moon-related surface activity of bannertail (*Dipodomys spectabilis*) and Fresno (*D. nitratoides*) kangaroo rats. *Anim. Behav.* 22:262–73

Lourens S, Nel JAJ. 1990. Winter activity of bat-eared foxes *Otocyon megalotis* in the Cape West Coast. *S. Afr. J. Zool.* 25:124–32

MacArthur RH. 1958. Population ecology of some warblers of northeastern coniferous forests. *Ecology* 39:499–519

MacArthur RH, Levins R. 1967. The limiting similarity, convergence and divergence of coexisting species. *Am. Nat.* 101:377–85

Mandelik Y, Jones ME, Dayan T. 2003. Structurally complex habitat and sensory adaptations mediate the behavioral responses of a desert rodent to an indirect cue for increased predation risk. *Evol. Ecol. Res.* 5:501–15

Marques MD, Waterhouse J. 1994. Masking and the evolution of circadian rhythmicity. *Chronobiol. Int.* 11:146–55

Maywood ES, Mrosovsky N. 2001. A molecular explanation of interactions between photic and non-photic circadian clock-resetting stimuli. *Gene Expr. Patterns* 1:27–31

McIntosh AR, Townsend CR. 1995. Impacts of an introduced predatory fish on mayfly grazing in New Zealand streams. *Limnol. Oceanogr.* 40(8):1508–12

Mendelssohn H. 1965. On the biology of the venomous snakes of Israel. *Isr. J. Zool.* 14:185–212

Metcalfe NB, Fraser NHC, Burns MD. 1998. State-dependent shifts between nocturnal and diurnal activity in salmon. *Proc. R. Soc. London Ser. B* 265:1503–7

Metcalfe NB, Fraser NHC, Burns MD. 1999. Food availability and the nocturnal vs. Diurnal foraging trade-off in juvenile salmon. *J. Anim. Ecol.* 68:371–81

Metcalfe NB, Steele GI. 2001. Changing nutritional status causes a shift in the balance of nocturnal to diurnal activity in European Minnows. *Funct. Ecol.* 15:304–9

Mills MGL, Biggs HC. 1993. Prey apportionment and related ecological relationships between large carnivores in Kruger National Park. *Symp. Zool. Soc. London* 65:253–68

Minors DS, Waterhouse JM. 1989. Masking in humans: the problem and some attempts to solve it. *Chronobiol. Int.* 6:29–53

Mistlberger R. 1991. Effect of daily schedules of forced activity on free-running rhythms in the rat. *J. Biol. Rhythms* 6:71–80

Mistlberger RE, Antle MC. 1998. Behavioral inhibition of light-induced circadian phase resetting is phase and serotonin dependent. *Brain Res.* 786:31–38

Mistlberger RE, Holmes MM. 1999. Morphine-induced activity attenuates phase shifts to light in C57BL/6J mice. *Brain Res.* 829:113–19

Moore RY. 1998. Entrainment pathways in the mammalian brain. In *Biological Clocks—Mechanisms and Applications,* ed. Y Touitou, pp. 3–14. Amsterdam: Elsevier Sci. 598 pp.

Moore RY, Eichler VB. 1972. Loss of circadian adrenal corticosterone rhythm following suprachiasmatic lesions in the rat. *Brain Res.* 42:201–6

Moore-Ede MC, Sulzman FM. 1981. Internal temporal order. See Aschoff 1981, pp. 215–41

Mrosovsky N. 1988. Phase response curves for social entrainment. *J. Comp. Physics A* 162:35–46

Mrosovsky N. 1991. Double-pulse experiments with nonphotic and photic phase shifting stimuli. *J. Biol. Rhythms* 6:167–79

Mrosovsky N, Edelstein K, Hastings MH, Maywood ES. 2001. Cycle of period gene expression in a diurnal mammal (*Spermophilus tridecemlineatus*): implications for nonphotic phase shifting. *J. Biol. Rhythms* 16:471–78

Muller K. 1974. Stream drift as a chronobiological phenomenon in running water ecosystems. *Annu. Rev. Ecol. Syst.* 5:309–23

Novak CM, Smale L, Nunez AA. 1999. Fos expression in the sleep-active cell group of the ventrolateral preoptic area in the diurnal murid rodent, *Arvicanthis niloticus*. *Brain Res.* 818:375–82

Novak CM, Smale L, Nunez A. 2000. Rhythms in Fos expression in brain areas related to the sleep-wake cycle in the diurnal *Arvicanthis niloticus. Am. J. Physiol. Regul. Integr. Comp. Physiol.* 278R1267–74

O'Farrell MJ. 1974. Seasonal activity patterns of rodents in a sagebrush community. *J. Mamm.* 55(4):809–23

Orr MR. 1992. Parasitic flies (Diptera: Phoridae) influence foraging rhythms and caste division of labor in the leaf-cutter ant, *Atta cephalotes* (Hymenoptera: Formicidae). *Behav. Ecol. Sociobiol.* 30:395–402

Owings DH, Lockard RB. 1971. Different nocturnal activity patterns of *Peromyscus californicus* and *Peromyscus eremicus* in lunar lighting. *Psychon. Sci.* 22:63–64

Palomares F, Caro TM. 1999. Interspecific killing among mammalian carnivores. *Am. Nat.* 153(5):492–508

Pati AK. 2001. Temporal organization in locomotor activity of the hypogean loach,

Nemacheilus evezardi, and its epigean ancestor. *Environ. Biol. Fish.* 62:119–29
Pavey CR, Burwell CJ, Grunwald JE, Marshall CJ, Neuweiler G. 2001. Dietary benefits of twilight foraging by the insectivorous bat *Hipposideros speoris*. *Biotropica* 33(4):670–81
Pianka ER. 1969. Sympatry of desert lizards (*Ctenotus*) in western Australia. *Ecology* 50(6):1012–30
Pinter N, Dayan T, Eilam D, Kronfeld-Schor N. 2002. Aggressive interactions between two species of spiny mouse: *Acomys russatus* and *Acomys cahirinus*. *Isr. J. Zool.* 48(2):176
Pittendrigh CS. 1993. Temporal organization: Reflections of a Darwinian clock-watcher. *Annu. Rev. Physiol.* 55:17–54
Pittendrigh CS, Daan S. 1976. A functional analysis of circadian pacemakers in nocturnal rodents. IV: Entrainment: pacemaker as a clock. *J. Comp. Physiol. A* 106:291–331
Ralph MR, Mrosovsky N. 1992. Behavioral inhibition of circadian responses to light. *J. Biol. Rhythms* 7:353–59
Refinetti R. 1999. *Circadian Physiology*. Boca Raton: CRC. 200 pp.
Refinetti R, Nelson DE, Menaker M. 1992. Social stimuli fail to act as entraining agents of circadian rhythms in the golden hamster. *J. Comp. Physiol. A* 170:181–87
Reynolds P, Gorman ML. 1994. Seasonal variation in the activity patterns of Orkney vole *Micortus arvalis orcadensis*. *J. Zool. London* 233:605–16
Richards SA. 2002. Temporal partitioning and aggression among foragers; modeling the effects of stochasticity and individual state. *Behav. Ecol.* 13(3):427–38
Roll U, Dayan T. 2002. Family ties and activity time in the order Rodentia. *Isr. J. Zool.* 48(2):177–78
Rosenzweig ML, Abramsky Z. 1997. Two gerbils of the Negev: a long-term investigation of optimal habitat selection and its consequences. *Evol. Ecol.* 11:733–56
Rusak B. 1981. Vertebrate behavioral rhythms. See Aschoff 1981, pp. 183–213
Saiful AA, Rashid YN, Idris AH. 2001. Niche segregation among three sympatric species of squirrels inhabiting a lowland dipterocarp forest, Peninsular Malaysia. *Mamm. Study* 26:133–44
Sanchez-Vazquez FJ, Madrid JA, Zamora S, Iigo M, Tabata M. 1996. Demand feeding and locomotor circadian rhythms in the goldfish, *Carassius auratus*: Dual and independent phasing. *Physiol. Behav.* 60:665–74
Sato T, Kawamura H. 1984. Circadian rhythms in multiple unit activity inside and outside the suprachiasmatic nucleus in the diurnal chipmunk *Eutamias sibiricus*. *Neurosci. Res.* 1:45–52
Saunders MB, Barclay RMR. 1992. Ecomorphology of insectivorous bats: a test of predictions using two morphologically similar species. *Ecology* 73(4):1335–45
Schaller GB. 1972. The Serengeti lion: a study of predator-prey relations. In *Wildlife Behavior and Ecology Series*, ed. GB Schaller. Chicago: Univ. Chicago Press. 480 pp.
Schoener TW. 1974a. Resource partitioning in ecological communities. *Science* 185:27–38
Schoener TW. 1974b. The compression hypothesis and temporal resource partitioning. *Proc. Natl. Acad. Sci. USA* 71(10):4169–72
Schoener TW. 1986. Resource partitioning. In *Community Ecology: Pattern and Processes,* ed. J Kikkawa, DJ Anderson, pp. 91–26. Palo Alto, CA: Blackwell. 432 pp.
Schwartz WJ, Reppert SM, Eagan SM, Moore-Ede MC. 1983. In vivo metabolic activity of the suprachiasmatic nuclei: a comparative study. *Brain Res.* 274:184–87
Shargal E, Kronfeld-Schor N, Dayan T. 2000. Population biology and spatial relationships of coexisting spiny mice of the genus *Acomys*. *J. Mamm.* 81(4):1046–52
Shargal E, Rath-Wolfson L, Kronfeld N, Dayan T. 1999. Ecological and histological aspects of tail-loss in spiny mice (Mammalia; Rodentia; Muridae) with a review of this phenomenon in rodents. *J. Zool.* 249:187–93
Shkolnik A. 1966. *Studies in the comparative biology of Israel's two species of spiny mice*

(*genus* Acomys). PhD thesis. Jerusalem, Isr: Hebrew Univ. 117 pp. (From Hebrew, Engl. summ.).

Shkolnik A. 1971. Diurnal activity in a small desert rodent. *Int. J. Biometeorol.* 15:115–20

Shkolnik A, Borut A. 1969. Temperature and water relation in two species of spiny mice (*Acomys*). *J. Mamm.* 50:245–55

Sih A. 1987. Predators and prey lifestyles: an evolutionary and ecological overview. In *Predation: Direct and Indirect Impacts on Aquatic Communities*, ed. WC Kerfoot, A Sih, pp. 203–24. Hanover, NH: Univ. Press New Engl.

Simberloff D, Dayan T. 1991. Guilds and the structure of ecological communities. *Annu. Rev. Ecol. Syst.* 22:115–43

Skelly DK. 1994. Activity level and the susceptibility of anuran larvae to predation. *Anim. Behav.* 47:465–68

Stephan FK. 2001. Food-entrainable oscillators in mammals. See Takahashi et al. 2001, pp. 223–46

Stephan FK, Zucker I. 1972. Circadian rhythms in drinking behavior and locomotor activity of rats are eliminated by hypothalamic lesions. *Proc. Natl. Acad. Sci. USA* 69:1583–86

Stevenson M, Deatherage G, LaVaque TJ. 1968. Effect of light-dark reversal on the activity cycle of *Sigmodon hispidus. Ecology* 49:1162–63

Stiling PD. 1999. *Ecology: Theories and Applications*. Englewood Cliffs, NJ: Prentice Hall

Stone G, Willmer P, Nee S. 1996. Daily partitioning of pollinators in an African Acacia tree. *Proc. R. Soc. London Ser. B* 263:1389–93

Takahashi JS, Turek FW, Moore RY, eds. 2001. *Handbook of Behavioral Neurobiology.* Vol. 12: *Circadian Rhythms*. New York: Kluwer Acad./ Plenum. 770 pp.

Ticher A, Ashkenazi IE, Reinberg AE. 1995. Preservation of the functional advantage of human time structure. *FASEB J.* 9:269–72

Trajano E, MennaBarreto L. 1995. Locomotor activity pattern of Brazilian cave catfish under constant darkness (Siluriformes, Pimelodidae). *Biol. Rhythm Res.* 26:341–53

Trajano E, MennaBarreto L. 1996. Free-running locomotor activity rhythm in cave dwelling catfishes, *Trichomycterus* sp. from Brazil (*Teleostei, Siluriformes*). *Biol. Rhythm Res.* 27:329–35

Turek FW, Takahashi JS. 2001. Introduction to circadian rhythms. See Takahashi et al. 2001, pp. 3–6

Van Reeth O. 1998. Sleep and circadian disturbances in shiftwork: strategies for their management. *Horm. Res.* 49:158–62

Van Schalk CP, Griffiths M. 1996. Activity periods of Indonesian rain forest mammals. *Biotropica* 28:105–12

Waterhouse J, Minors D, Akerstedt T, Hume K, Kerkhof G. 1996. Circadian rhythms adjustment: difficulties in assessment caused by masking. *Pathol. Biol.* 44:205–7

Weber DN, Spieler RE. 1987. Effect of the light-dark cycle and scheduled feeding on behavioral and reproductive rhythms of the *cyprinodontfish*, medaka, *Oryzias latipes. Experientia* 43:621–24

Weber ET, Rea MR. 1997. Neuropeptide Y blocks light-induced phase advances but not delays of the circadian activity rhythm in hamsters. *Neurosci. Lett.* 231:159–62

Weissenberg S, Bouskila A, Dayan T. 1997. Resistance of the common spiny mouse (*Acomys cahirinus*) to the strikes of the Palestine saw-scaled viper (*Echis coloratus*). *Isr. J. Zool.* 43:119

Werner EE, Anholt BR. 1993. Ecological consequences of the tradeoff between growth and mortality rates mediated by foraging activity. *Am. Nat.* 142:242–72

Whitford WG, Depree DJ, Hamilton P, Ettershank G. 1981. Foraging ecology of seed harvesting ants, *Pheidole* spp. in a Chihuahuan Desert ecosystem. *Am. Midl. Nat.* 105(1):159–67

Wiens JA, Addicot JF, Case TJ, Diamond J. 1986. Overview: the importance of spatial and temporal scale in ecological investigation. In *Community Ecology*, ed. J Diamond,

TJ Case, pp. 145–53. New York: Harper & Row

Willmer PJJ, Johnston IA, Stone G. 2000. *Environmental Physiology of Animals*. Malden, MA: Blackwell. 644 pp.

Yokota T, Oishi T. 1992. Seasonal changes in the locomotor-activity rhythm of the mekada, *Oryzias latipes*. *Int. J. Biometeorol.* 36:39–44

Zielinski WJ. 1988. The influence of daily variation in foraging cost on the activity of small carnivores. *Anim. Behav.* 36:239–49

Zisapel N, Barnea E, Anis Y, Izhaki I, Reiter RJ, Haim A. 1998. Involvement of the pineal gland in daily scheduling of the golden spiny mouse. *Life Sci.* 63:751–57

Zisapel N, Barnea E, Izhaki I, Anis Y, Haim A. 1999. Daily scheduling of the golden spiny mouse under photoperiodic and social cues. *J. Exp. Biol.* 248:100–6

Ziv Y, Abramsky Z, Kotler BP, Subach A. 1993. Interference competition and temporal and habitat partitioning in two gerbil species. *Oikos* 66:237–44

Annu. Rev. Ecol. Evol. Syst. 2003. 34:183–211
doi: 10.1146/annurev.ecolsys.34.011802.132403

First published online as a Review in Advance on July 8, 2003

PERFORMANCE COMPARISONS OF CO-OCCURRING NATIVE AND ALIEN INVASIVE PLANTS: Implications for Conservation and Restoration

Curtis C. Daehler
Department of Botany, University of Hawai'i Manoa, Honolulu, Hawaii 96822; email: daehler@hawaii.edu

Key Words invasive plants, native plants, growth rate, competition, fecundity

■ **Abstract** In the search to identify factors that make some plant species troublesome invaders, many studies have compared various measures of native and alien invasive plant performance. These comparative studies provide insights into the more general question "Do alien invasive plants usually outperform co-occurring native species, and to what degree does the answer depend on growing conditions?" Based on 79 independent native-invasive plant comparisons, the alien invaders were not statistically more likely to have higher growth rates, competitive ability, or fecundity. Rather, the relative performance of invaders and co-occurring natives often depended on growing conditions. In 94% of 55 comparisons involving more than one growing condition, the native's performance was equal or superior to that of the invader, at least for some key performance measures in some growing conditions. Most commonly, these conditions involved reduced resources (nutrients, light, water) and/or specific disturbance regimes. Independently of growing conditions, invaders were more likely to have higher leaf area and lower tissue construction costs (advantageous under high light and nutrient conditions) and greater phenotypic plasticity (particularly advantageous in disturbed environments where conditions are in frequent flux). There appear to be few "super invaders" that have universal performance advantages over co-occurring natives; rather, increased resource availability and altered disturbance regimes associated with human activities often differentially increase the performance of invaders over that of natives.

INTRODUCTION

Invasive plants are nonnative species that have successfully spread outside their native range (Richardson et al. 2000, Williamson 1996). Most invasions over the past several centuries have involved species transported directly or indirectly by humans (McKinney & Lockwood 1999, Pyšek et al. 2002). Invasive plants have attracted much attention because of their economic costs as weeds (Pimentel 2002) and because they may reduce native biodiversity (Daehler & Strong 1994, Wilcove et al. 1998) or alter ecosystem functions (D'Antonio & Vitousek 1992, Vitousek 1990). Because only a small fraction of introduced species become invasive (Williamson 1996), and many invasive species can be considered pests (Daehler 2001), much effort has been focused on understanding what makes some species invasive (Kolar & Lodge 2001, Rejmánek et al. 2003).

A comparative approach has often been useful in helping to understand what makes invasive species so successful (Grotkopp et al. 2002, Mack 1996, Rejmánek 1995). The comparative approach involves pairing invasive species with native species or noninvasive congeners. If a consistent difference can be identified between invader and native, that difference might help explain why an invader has become so successful. Rather than summarizing attributes that might make a species invasive (e.g., Crawley et al. 1996, Pyšek et al. 1995, Williamson & Fitter 1996), this review makes use of published native-invader comparisons to examine the general question "Do invasive plants perform substantially better than co-occurring native plants?" When I refer to "plant performance," I mean a plant's success or aptitude in terms of one or more fitness-related traits. An invader that outperforms co-occurring natives is expected to increase in relative abundance over time, and abundant invaders are expected to have significant impacts on co-occurring native populations (Daehler & Carino 1999). Plant performance can be measured by various traits ranging from competitive ability to fecundity. In any case, whether or not invasive plants substantially outperform co-occurring natives has important consequences for conservation. If most invasive plants substantially and consistently outperform co-occurring natives, then we can expect serious and widespread reductions in global biodiversity as a direct consequence of today's plant invasions; we would be left with few options for preventing this, other than persistent and direct attacks on the invaders. On the other hand, if invaders rarely outperform co-occurring natives, or if their superior performance compared with natives is marginal or dependent on specific environmental conditions, then the possibility remains that the impact of invaders will be strong only under particular environmental circumstances, and these circumstances could be minimized.

Whereas "extreme" environments have often been suggested to be highly resistant to invasion (e.g., Rejmánek 1989, Mueller-Dombois & Loope 1990), Alpert et al. (2000) proposed a specific mechanism that is applicable to a continuum of environmental conditions: The relative performance of native and invasive species could vary depending on the amount of environmental stress. Other environmental variables such as disturbance regime could also differentially increase the performance of native plants relative to co-occurring invaders (Alpert et al. 2000, Hobbs & Huenneke 1992, Mueller-Dombois & Loope 1990). Under these circumstances, land managers could potentially manipulate environmental conditions to thwart invasions or reduce the abundance of unwanted invaders to acceptable levels. Thompson et al. (1995) concluded that at least some native species have the same attributes as invasive species, but they made no attempt to compare the performance of native and invasive plants co-occurring in the same habitats. This review summarizes studies of performance differences between co-occurring native and invasive plants and uses the compiled findings to suggest general strategies for managing invasive plants. Comparisons were not restricted to taxonomically related species pairs; however, in all of the reviewed studies, the invaders and co-occurring natives had the same life form, and/or the study authors had raised concerns that the invader was directly impacting the co-occurring native.

LITERATURE SEARCH FOR PERFORMANCE COMPARISONS

In order to identify a large sample of studies comparing the performance of native and invasive plants, I searched Biological Abstracts (SilverPlatter Information Inc, Norwood, MA) for the publication years 1985–June 2002. The following search was performed for words within the entire database record, including the Abstract: Native *and* (Invasive *or* Invad* *or* Exotic *or* Alien) *and* (Subject = plant *or* plants). For all citations identified using this search filter, the titles and abstracts were read to identify relevant studies. A few additional studies were identified through citations within papers found during the initial search. A number of studies (e.g., Gould & Gorchov 2000, Melgoza et al. 1990) examined the effect of an invader on a native plant but did not provide comparable data on how the native affected the invader; such studies could not be used for the comparative purpose of this review. Some studies compared large groups of invaders and natives (e.g., Crawley et al. 1996, Goodwin et al. 1999, Pyšek et al. 1995), but these studies were not included in this review because they mainly compared general life history traits or biogeographic characteristics, rather than plant performance.

For studies containing appropriate comparative data on one or more native-invasive species pairs, the measure of performance was recorded, and the native's performance was rated as inferior to, equal to, (no statistical difference), or better than that of the invader. In many cases, performance had been measured under more than one condition. If the relative performance of the native depended on the conditions, then the specific conditions under which the native's performance was superior or equal to that of the invader were recorded. Usually, performance assessments (inferior, superior, or equal) were based on the statistical analyses given in the original papers. However, in some cases the authors did not clearly present the statistical comparison for a specific time or condition. In such cases, overlapping standard error bars in graphs were conservatively taken to indicate nonsignificant differences in performance. In some studies, performance data were given for several native and invasive species. Unless the natives were clearly paired with invaders, performance results were recorded using the following rules: If one native was superior to all invaders, then the study was recorded as finding superior performance of a native; if there was no clear pattern (some natives equal to invaders), then the study was recorded as finding equal performance between natives and invaders; if the invaders as a group clearly outperformed the natives as a group, then the study was recorded as finding inferior performance by natives.

A total of 119 published papers containing comparative performance data were identified. Among these, some involved the same species pairs; results from these studies were pooled to create a single summary comparison of that species pair. For four invaders (*Acacia saligna*, *Centaurea maculosa*, *Lythrum salicaria*, and *Tamarix ramosissima*), independent studies compared these invaders to different native species at different locations or times. These cases were considered separate comparisons for purposes of generating summary statistics. Summary statistics on performance results are based on tallies rather than meta-analysis because there was no satisfactory way of weighing context-dependent differences in performance identified within native-invader species pairs. The total number of paired, independent comparisons between native and invasive plants was 79 (Appendix 1, see the Supplemental Material link in the online version of this chapter or at http://www.annualreviews.org/); 13 pairs consisted of congeners and 59 (75%) of the paired comparisons involved data obtained from manipulative experiments. The comparisons included studies from all major geographic regions of the world. North America was the most common geographic region for the native species (46% of comparisons). Asia was underrepresented, with only one study where the native was from Asia (Yamashita et al. 2000). In contrast, Asia and/or Europe (Eurasia) was the origin for over half (54%) of the invaders. The comparisons involved various plant life forms, distributed as follows: 32% grasses, 23% other herbs including vines, 25% shrubs, and 20% trees. Separate analyses by geographic region or life form were not attempted owing to small sample sizes.

PERFORMANCE COMPARISONS

Measures of Performance

The most common measure of performance was growth rate (Figure 1), probably because it is simple and inexpensive to measure in pot and field experiments. Other common measures of performance can be grouped into those related to individual growth rate, at least in theory (e.g., photosynthesis, tissue construction costs, total leaf area) and those related to population spread (e.g., fecundity, dispersal rate, germination rate, survival). Finally, some common measures of performance were not easily placed in either category: competitive ability, standing biomass, and phenotypic plasticity. Specific examples of comparisons involving each of these performance measures are detailed below. Less frequently encountered comparisons of performance between native and invasive plants ($n < 5$ studies) were seed

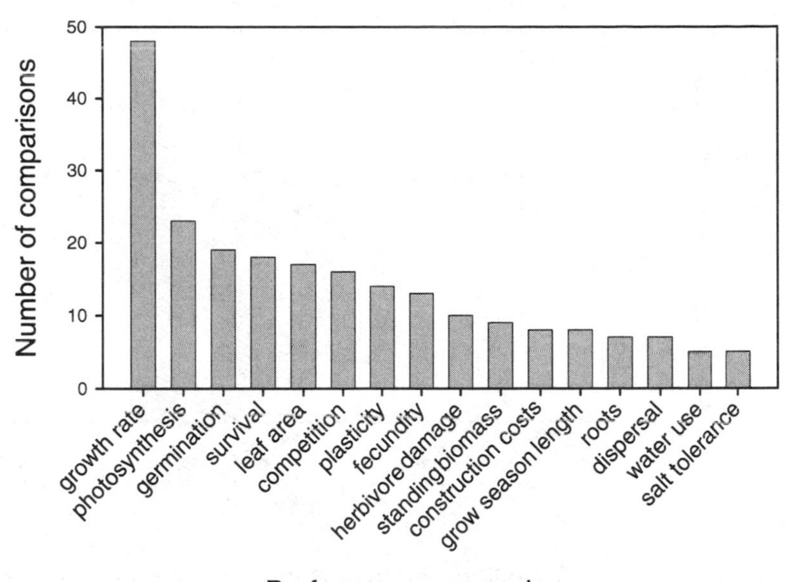

Figure 1 Performance comparisons between native and invasive plants, by frequency of occurrence. Only performance measures for which five or more comparisons were encountered are shown. The total number of comparisons exceeds the total number of studies reviewed because some studies measured multiple aspects of performance.

predation rates, seed longevity, rate of pathogen attack, allelopathic capacity, and environmental breadth.

Context-Dependence of Performance

Among the 79 independent performance comparisons between native and invasive plants, only 10 comparisons (13%) showed consistent performance advantages for the invader for all measured performance variables across all growing conditions (Figure 2). When one considers only the 55 comparisons that involved more than one growing condition, invaders had universally superior performance in only 6% of cases. The most common growing conditions favoring natives over invaders were environments with low resource availability (nutrients, water, or light; Figure 2). Some studies identified a specific disturbance regime, such as periodic flooding (Sher et al. 2000), mowing, or fire that favored the native over the invader (pooled as "special disturb" in Figure 2). On several occasions (e.g., Holmgren et al. 2000), removal of introduced grazers favored the native. Certain invaders consistently had poor performance when directly competing with natives. Presumably, these invaders have become abundant because frequent disturbances reduce the intensity of competition with natives (e.g., *Bunias orientalis*; Dietz et al. 1998).

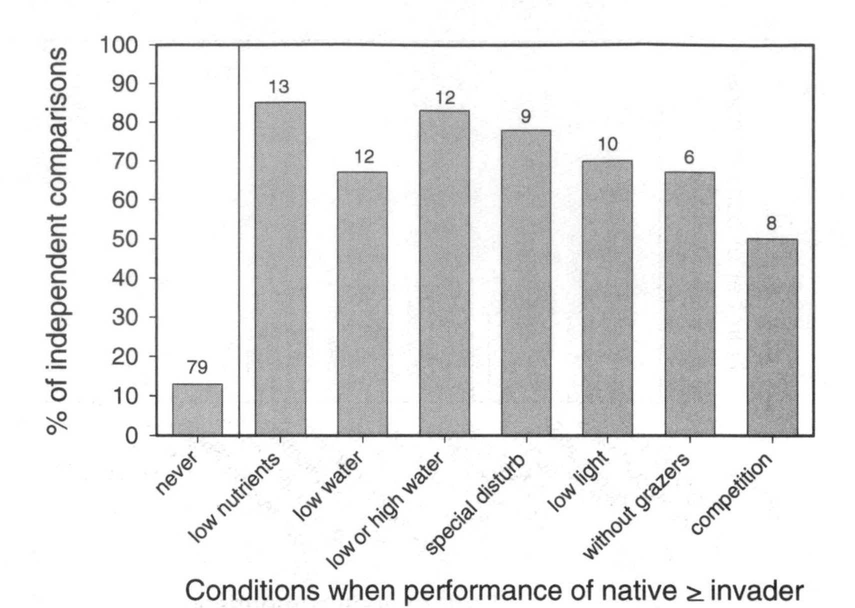

Figure 2 Conditions under which the native's performance was equal to or better than that of the invader. Only conditions favoring natives in three or more independent comparisons are shown. Percentages are based upon the total number of studies that manipulated each condition (given above each bar).

Specific examples of context-dependent performance are detailed in the summaries of individual performance measures below.

Growth-Rate–Related Traits

Invasive plants are often characterized as having unusually rapid individual growth rates, allowing them to outgrow, overgrow, or quickly crowd out natives (Cronk & Fuller 1995). Certainly, examples like "mile-a-minute weed" (*Polygonum perfoliatum*) support the idea that invaders can have very rapid growth rates (Oliver 1996). Nevertheless, among the reviewed comparisons, there was no clear evidence that invaders necessarily grow faster than co-occurring natives (Figure 3). Instead, the more rapid growth rate of invaders appears to be condition- or context-dependent. A study was far more likely to conclude that invaders had a universal growth-rate advantage if the study examined only one growing condition. Among 14 growth-rate studies conducted under only one growing condition, none showed a growth-rate advantage for the native. In contrast, among the 12 comparisons where a growth-rate advantage of the native was observed (Figure 3), all were conducted under more than one growing condition or environment. For example, the growth rate of native plains poplar (*Populus deltoides*) exceeded that of the invasive Russian olive (*Elaeagnus angustifolia*) when plants were grown with a high water table, whereas

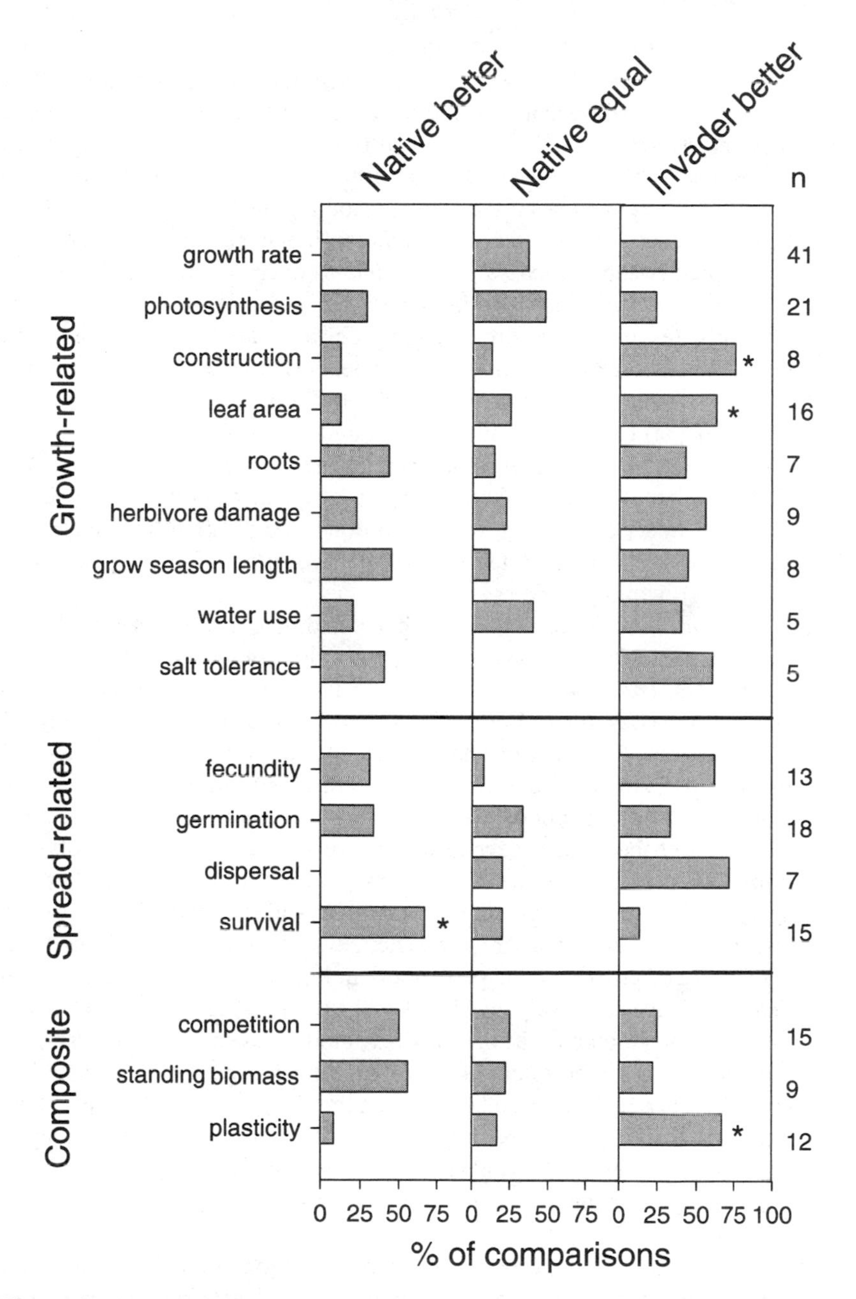

Figure 3 Summary of native versus invader performance, according to whether the native performed better than (*first column*) or as well as (*second column*) the invader under some conditions, or whether the invader always performed better (*third column*). Asterisk indicates significant difference (chi-squared exact test, exact $P < 0.05$). The column to the far right (n) indicates the number of independent comparisons for each measure of performance.

conclusions were different when the table was lower (Shafroth et al. 1995). Under conditions of reduced salinity, the growth rate of native willows (*Salix* spp.) equaled or exceeded that of invasive salt cedar (*Tamarix ramosissima*), a troublesome riverside invader in controlled river courses in the southwestern United States (Cleverly et al. 1997, Glenn et al. 1998). Reduced salinity would be expected along these river courses if the natural flood regime were restored in these areas. In salt marshes of the western United States, higher salinity favored the growth of native pickleweed (*Salicornia subterminalis*) over the invasive European annual beardgrass (*Polypogon monspeliensis*) (Kuhn & Zedler 1997). In the northwestern United States, the native perennial grass *Festuca idahoensis* had a growth advantage over invasive European grasses under lower water conditions, particularly with less winter rain (Borman et al. 1990). In Europe, native *Cystoseira nodicaulis* (a macroalga) had a substantial autumn and winter growth advantage over ecologically similar *Sargassum muticum* introduced from Asia (Arenas et al. 1995). These examples illustrate the importance of making comparisons over different times and under different growing conditions, and the findings from these studies suggest site-specific management strategies for promoting the growth of natives over the growth of invaders.

Similar cases of context-dependent performance were observed for photosynthetic capacity and root growth. Under dry conditions, a native Venezuelan bunchgrass, *Trachypogon plumosus*, had higher photosynthetic rates than invasive African molasses grass (*Melinis minutiflora*) (Baruch 1996). The native California dune grass *Elymus mollis* had higher rates of photosynthesis in the field than the invasive European dune grass, *Ammophila arenaria* (Pavlik 1983a). The same study observed higher photosynthetic rates for the invader in the laboratory, again pointing out the context-dependence of performance. Pattison et al. (1998) reported higher photosynthetic rates among forest invaders when they evaluated several native-invasive species pairs in the Hawaiian Islands. They concluded that "the invasive species appear to be better suited than native (Hawaiian) species to capturing and utilizing light, particularly in high light environments such as those characterized by relatively high levels of disturbance" (Pattison et al. 1998). As the latter statement implies, the invaders' advantage was substantially less or nonexistent under lower light condition (as would be more typical of an undisturbed forest environment). It should also be noted that Pattison et al. (1998) grew all plants in a nutrient-enriched commercial potting medium. It would be interesting to know how photosynthetic rates would compare between invaders and natives growing in a nutrient-poor environment, more similar to soil conditions in most undisturbed Hawaiian rainforest.

Not many studies specifically examined root growth, but among the available studies, there was no consensus advantage for invaders (Figure 3). In New Zealand, the native grass *Festuca novae-zelandiae* had root growth advantages over invasive hawkweeds (*Hieracium* spp.) in certain low-fertility soils (Fan & Harris 1996). Marler et al. (1999) found no difference in fine-root densities of invasive spotted knapweed (*Centaurea maculosa*) and the native bluebunch wheatgrass (*Agropyron spicatum*) when averaged across depths. The invader tended to have deeper roots but the native had denser fine roots at shallower depths. Either rooting strategy could be advantageous, depending on environmental conditions. In comparing invasive ice plant (*Carpobrotus edulis*) with native *Haplopappus* spp., D'Antonio & Mahall (1991) observed the opposite rooting pattern: The invader had denser fine roots near the surface, whereas the natives had deeper roots.

For total leaf area and tissue construction costs, there was a statistically significant trend toward invaders having higher leaf area and lower construction costs than natives (Figure 3). Invaders tended to have greater leaf area, whether measured as absolute area or leaf area ratio (leaf area per total plant mass) and when combined with lower constructions costs (Baruch & Goldstein 1999), this increased area might be expected to give the invaders a growth advantage. The number of studies examining construction costs remains small, and further studies would be useful to determine if a statistical trend toward lower construction costs among invaders is generalizable. One recent study on invasive purple loosestrife (*Lythrum salicaria*) found that the invader did not have lower construction costs relative to some ecologically similar native species in North America (Nagel & Griffin 2001). Furthermore, although low tissue construction costs and higher leaf area might appear to provide a universal growth-rate advantage (but see Poorter & Bergkotte 1992), there are likely to be trade-offs in the form of reduced leaf longevity (Reich et al. 1997). Native plants with lower leaf area ratio and higher construction costs are likely to have longer leaf life spans, which can result in comparable overall above-ground production efficiency per unit foliar biomass, compared with (invasive) plants with larger leaf area ratios and lower construction costs (Reich et al. 1992). In low-nitrogen environments, having a longer leaf life span (with the cost of slower growth) may actually be superior in the long run because plants with longer tissue life spans are more efficient at holding scarce nutrients (Aerts & van der Pejil 1993). If a native plant community is adapted to low-nutrient conditions, then increased nutrient levels are likely to promote the success of invaders adapted to take advantage of these high nutrient levels. Experiments have demonstrated increased success of invaders following nutrient additions alone (Huenneke et al. 1990), although more often nutrient additions interact with physical disturbance to promote invasion (Burke & Grime 1996, Duggin & Gentle 1998, Li & Norland 2001, Weiss 1999). Human-related eutrophication and physical disturbance of the environment can probably explain the success of many invaders (e.g., Allan 1936, D'Antonio et al. 1999, Hobbs & Huenneke 1992).

Competition

Competitive ability or performance can be measured in various ways (Goldberg & Landa 1991). Most of the reviewed studies pitted invaders against ecologically comparable, similar-sized native species. Conclusions about competitive performance were generally based on the final biomass of the invader relative to the native species when grown in competition. As with most other fitness-related traits, the relative competitive performance of native versus invasive species often depended on environmental conditions. Among the 16 studies that experimentally assessed competitive performance, only five (31%) seemed to show universally superior competitive performance by the invader. These five invaders were *Bromus inermis* (Nernberg & Dale 1997), *Dipsacus sylvestris* (Huenneke & Thomson 1995), *Lythrum salicaria* (Mal et al. 1997) *Pennisetum setaceum* (Carino & Daehler 2002), and *Spartina alterniflora* (Callaway & Josselyn 1992). For one of these species (*Dipsacus*), the situation was not clear. This invader seemed to outcompete the native in a greenhouse study, but an accompanying field study was inconclusive (Huenneke & Thomson 1995). The case of *Lythrum* also requires a caveat; the native (*Typha angustifolia*) was clearly the superior competitor in the first year, but by the end of the experiment (fourth year), *Lythrum* appeared to have the competitive advantage (Mal et al. 1997).

Among the remaining 11 comparisons of competitive performance, the native was equivalent or superior to the invader, at least under certain growing conditions. In South Africa, the native woody legume *Virgilia orboides* had superior performance when grown in competition with ecologically similar Australian invaders (*Acacia longifolia* and *Albizia lophantha*) under both high- and low-nutrient conditions (McDowell & Moll 1981). In several comparisons, competitive performance depended on nutrient availability. Under high- and moderate-nutrient conditions, invasive *Acacia saligna* significantly outperformed native *Protea repens*, but when the two were grown together in competition under low-nutrient conditions (native Clovelly soil with acid-washed sand), the performance of the native was comparable to that of the invader (Witkowski 1991). Likewise, when the California native grass *Elymus glaucus* was grown in competition with the European invader *Bromus mollis*, the invader performed better at high soil nitrogen levels, but the outcome was reversed under lower-nitrogen conditions (Claassen & Marler 1998). Similar results were obtained in a comparison of competitive ability between invasive spotted knapweed (*Centaurea maculosa*) and the native perennial grass *Pseudoroegneria* (*Agropyron*) *spicatum* (Herron et al. 2001). In New Zealand, the native grass *Festuca novae-zelandiae* appeared able to suppress an invasive herb (*Hieracium pilosella*) in low-fertility soils if a single clipping treatment was applied, but the native was less successful under higher-nutrient conditions (Fan & Harris 1996). In a study of competitive hierarchies among 20 wetland plants in three environments, Keddy et al. (1994) concluded that invasive purple loosestrife (*Lythrum salicaria*) had strong competitive effects on the phytometer species across all environments; however, it is interesting to note that the phytometer species had strong effects on purple loosestrife in the low-nutrient environment, reducing its biomass by nearly 50%. About half of the native species tested showed lesser reductions in biomass in the low-nutrient environment (Keddy et al. 1994), indicating that many natives tolerated competition better than purple loosestrife did under low-nutrient conditions. Whereas low-nutrient conditions often increased the relative competitive performance of natives, no native species had an increased competitive advantage under high-nutrient conditions.

Another example of an environmental condition that increased the relative competitive performance of a native species was decreased water availability (Smith & Brock 1996). A performance advantage for natives under reduced water conditions was also observed in several other species pairs and for other performance measures (Figure 2). Mesleard et al. (1993) observed superior competitive ability of a native halophytic grass, *Aeluropus littoralis*, over invasive *Paspalum paspalodes* under conditions of increased soil salinity. This pattern has also been reported for other native halophytes in competition with invaders (Kuhn & Zedler 1997, Zedler et al. 1990).

Standing Biomass/Cover Abundance

Several field-based studies measured changes or differences in standing biomass/cover of native versus invasive species under different environmental conditions (Figure 3). Such changes are probably due to a combination of factors, including competition. Working in upland prairies of the western United States, Wilson & Clark (2001) found that a specific mowing regime reduced cover of the invasive Eurasian grass *Arrhenatherum elatius* while increasing cover of native prairie grasses. In the Chillean matorral, Holmgren et al. (2000) observed increased cover of native species, particularly the native grass *Bromus berterianus*, in plots that had been fenced to keep out alien grazers. Stromberg & Griffin (1996) analyzed the vegetation of California grasslands and found that native grasses remained dominant or codominant on lands that had not been historically cultivated. Cultivation presumably altered soil texture and nutrients, favoring invasion by Eurasian grasses. Furthermore, current high levels of gopher disturbance, even on lands no longer in cultivation, seem to favor dominance by the alien grasses (Stromberg & Griffin 1996). Other studies involving seed additions of either natives or aliens (but not both) followed by environmental manipulations (e.g., Hobbs & Atkins 1988) have likewise identified conditions favoring increased cover/biomass of natives, but these studies are not considered here because their differential treatment of natives and invaders precluded direct performance comparisons.

Four studies in the standing biomass category (Figure 3) tested whether invasive species tend to attain higher biomass/productivity than ecologically similar native species. Two of these studies found that the invader attained higher biomass than an ecologically similar native (Pavlik 1983b, Callaway & Josselyn 1992), while the other two studies found no difference in standing biomass between invader and native (Horn & Prach 1994, Smith & Knapp 2001a).

Reproduction and Spread-Related Traits

Some studies examining reproductive ecology have identified spectacular advantages for the invader. For example, invasive smooth cordgrass (*Spartina alterniflora*) in San Francisco Bay had a sevenfold advantage in seed production over native California cordgrass (*Spartina foliosa*), and the germination rate was also higher for the invader (Callaway & Josselyn 1992). Similarly, in the South African fynbos, introduced *Banksia ericifolia* had twofold higher seed production per canopy area compared with native *Leucadendron laureolum* (Honig et al. 1992). The alien also had advantages in germination speed and germination rate (Honig et al. 1992). Despite examples like these, other studies found no clear advantage for the invader. For example, in a glacier foreland, native *Poa kerguelensis* produced nearly three times more seeds per plant than invasive *Poa annua*; percent germination of seeds was similar (80–90%) for both species (Frenot & Gloaguen 1994). The native required a cold pretreatment for germination, but this would not seem to be a disadvantage in a glacial environment. Among 31 comparisons of fecundity and/or germination, the invader had a consistent reproductive advantage in only 14 (45%) of the cases (Figure 3). In the remainder of cases examining fecundity and/or germination, either the invader did not have a clear reproductive advantage (as in *Poa annua* versus *Poa kerguelensis*), or the invader's advantage was context-dependent. For example, although invasive fountain grass (*Pennisetum setaceum*) in Hawai'i had higher seed production per plant than native pili grass (*Heteropogon contortus*) under higher water conditions, the native had the fecundity advantage under drought conditions (Goergen & Daehler 2001b). Likewise, under conditions of low water availability, invasive fountain grass had lower seedling survival compared with the native grass (Goergen & Daehler 2002). An analogous situation was reported from Australia: over 80% of individuals of the native grass *Danthonia richardsonii* flowered and produced seeds under dry field conditions. In comparison, among three co-occurring alien grasses fewer than 5% of individuals flowered under these conditions (Virgona & Bowcher 2000). Survival of the native was also higher in the low water environment (Virgona & Bowcher 2000).

In some cases, fire seemed to give natives a reproductive advantage. Seeds of a native Australian grass (*Austrostipa compressa*) had higher germination rates than an invasive African grass (*Ehrharta calycina*) after exposure to high heat (Smith et al. 1999). The native also had higher densities of germinable seeds in the field after fire (Smith et al. 1999), although germinable seed densities were not compared in unburned areas. Similarly, in Hawai'i, fire caused high seed (and adult) mortality in invasive natal redtop (*Melinis repens*), resulting in low seedling recruitment and cover abundance after fire, compared with the co-occurring fire-tolerant native grass *Heteropogon contortus* (Tunison et al. 1994).

Although most comparisons of fecundity were based on seed production, Aptekar & Rejmánek (2000) compared potential for vegetative reproduction between American beach grass (*Leymus mollis*) and the invasive European beach grass (*Ammophila arenaria*) on U.S. Pacific coasts. They found that the native produced significantly more nodes (potential propagules) per rhizome length than the invader. Rhizome fragments of the native also remained viable in seawater for a longer time (Aptekar & Rejmánek 2000), suggesting a dispersal advantage for the native.

Overall, there was a trend toward a seed dispersal advantage for invaders ($P = 0.06$, Figure 3), but statistical power was limited by the small number of studies. Richardson et al. (1987) concluded that the winged, wind-dispersed seeds of introduced Protaceae (*Hakea* spp.) have a dispersal distance advantage over native South African Protaceae, which are often gravity or ant-dispersed. Also in South Africa, introduced *Banksia ericifolia* was found to have lighter seeds with larger wings compared with native *Leucadendron laureolum*, providing the alien with a dispersal distance advantage (Honig et al. 1992). In California, scats of jackrabbit and deer contained more seeds of invasive *Carpobrotus edulis* than seeds of the presumed native *C. chilensis*, implying a dispersal advantage for the invader (Vila & D'Antonio 1998a). Rejmánek (1996) has proposed that the presence of an efficient bird-disperser may be a key predictor of invasion success among fleshy-fruited plant species, so it is surprising that few studies have compared bird dispersal preferences between native and invasive plants. In a subtropical forest, Montaldo (2000) examined fruit removal rates by birds for two invasive plants (*Rubus ulmifolius* and *Ligustrum lucidum*) and three native plants. There was no clear difference in fruit removal rate between invaders and natives, but the invaders had higher fecundity. Similar findings were reported in a comparison of invasive Oriental bittersweet (*Celastrus orbiculatus*) and native holly (*Ilex opaca*) in a North American temperate forest; fruit removal rates did not differ significantly between the native and invader, although there was a trend toward a higher removal rate for the native (Greenberg et al. 2001). In contrast, fruits of the invasive European hawthorn (*Crataegus monogyna*) were more attractive to American robins than fruits of a native American hawthorn (*C. douglasii*) (Sallabanks 1993). This study was particularly interesting because it showed preference by native birds for fruits of an invasive plant, but so far this situation appears to be the exception rather than the rule (Montaldo 2000).

Among the reproduction and spread-related traits, there was one statistically significant trend: natives tended to have a survival advantage over invaders at some life stage, under at least some environmental conditions (Figure 3). Some of these cases have already been mentioned (e.g., higher relative survival of native *Populus*, *Heteropogon*, and *Danthonia* as well as *Leymus* rhizomes under specific environmental conditions). In Argentinean montane forests, a native tree, *Lithraea ternifolia*, was estimated to have higher seedling and juvenile survival at sites with shallow, rocky soils, compared with an invasive North American competitor, *Gleditsia triacanthos* (Marco & Paez 2000). In New Zealand, although the native vine *Parsonsia heterophylla* had a slow growth rate, it was capable of surviving at lower light levels compared with invasive vines (Baars & Kelly 1996). In woodlands of Ireland, reduced grazing pressure would likely result in higher seedling and juvenile survival of the native understory shrub *Ilex aquifolium* compared with invasive *Rhododendron ponticum*, particularly on sites with an accumulation of leaf litter (Cross 1981). Vila & D'Antonio (1998b) reported a survival advantage for the invader, *Carpobrotus edulis*, over native *C. chilensis* in one environment, but not in another environment. It is surprising that so few demographic studies have compared the survival of native and invasive plants across life stages and environments (e.g., Marco & Paez 2000) as such studies can provide clear insights into management strategies that could differentially promote natives.

Phenotypic Plasticity

Among 12 comparisons of phenotypic plasticity, most concluded that the invader was more plastic than the native (Figure 3). Most observations of greater phenotypic plasticity in invaders involved changes in biomass allocation patterns in response to different environmental conditions (Baruch & Bilbao 1999, Black et al. 1994, Fan & Harris 1996, Luken et al. 1997, Maillet & Lopez 2000, Simoes & Baruch 1991, Yamashita et al. 2000). Other studies reported greater plasticity for the invader in terms of physiological responses (Pattison et al. 1998, Williams & Black 1994), circumnutation (Larson 2000), or germination in response to temperature (Frenot & Gloaguen 1994). It seems that invaders often do have higher phenotypic plasticity than natives, and this plasticity probably allows invaders to succeed in a wider range of environments, but it does not a priori indicate a performance advantage over natives within any single, defined environment. Furthermore, natives probably often have higher genetic variation in comparison with co-occurring invader populations that were established from a small group of founders. It would be interesting to know if the range of phenotypes expressed in an invading population exceeds the range of phenotypes expressed in a native population, since it is this total phenotype range (genetic and plastic) that determines a species' ability to respond to environmental changes over time or space. Kitayama & Mueller-Dombois (1995) found that native species generally had greater overall environmental breadth compared with invaders, and this wide breadth was related to high genetic variation within the natives (Kitayama et al. 1997, Daehler et al. 1999).

Effects of Natural Enemies

If an invader is significantly affected by natural enemies in its native range, then release from natural enemies could give it a fitness advantage in its introduced range. Nine studies were found comparing the effects of herbivory on native and invasive species. In six of these cases, herbivores had a larger impact on the native plants (Figure 3). One of these cases involved introduced grazers (Caldwell et al. 1981) and four involved cases where native vertebrate herbivores likely occurred at unusually high abundances owing to human elimination of their natural predators (Cross 1981, Lesica & Miles 1999, Pyke 1986, Schierenbeck et al. 1994). The one remaining study was largely anecdotal, reporting "heavy nocturnal leaf feeding by crickets" on a native species; the invasive species were "unaffected" by crickets (McDowell & Moll 1981). One of the nine herbivory comparisons found no consistent differences in level of herbivory on native and exotic eucalypts in Australia (Radho et al. 2001). Two studies (22%) found greater herbivore damage on the invasive species than on the native, but in both cases the herbivore was introduced (Bellingham 1998, Gross et al. 2001). Keane & Crawley (2002) summarized eight additional studies examining herbivory on native and invasive plants; five of these showed greater damage among native plants. Based on the studies to date, we cannot generally assume that native plants are more heavily damaged by herbivores than invaders. Successful biocontrol efforts cannot be used as evidence that an invader's advantage was due to release from herbivory because biocontrol agents themselves have been released from their natural predators, pathogens and competitors (Keane & Crawley 2002). There also appears to be no general trend toward greater seed predation rates for natives versus invaders (Blaney & Kotanen 2001b), although individual exceptions have been reported (e.g., Richardson et al. 1987). Likewise, there appears to be no general trend toward greater rates of attack by fungal pathogens among native species than among invaders (Blaney & Kotanen 2001a), although again, exceptions are known (e.g., Goergen & Daehler 2001a). Interestingly, Mitchell & Power (2003) found that native plants had greater absolute numbers of pathogens than invasive pest plants, but unlike Blaney & Kotanen (2001a), they did not make comparisons for co-occurring species and no estimates of pathogen damage were available.

General Conclusions from Performance Comparisons

In the majority of reviewed cases, native plants were equivalent to or had performance advantages over invasive plants under at least some growing conditions (Figure 2). Because the published literature on invasive species is likely to be biased toward studies of the most troublesome invasive pests (Simberloff 1981), it seems safe to conclude that invaders with universal performance advantages over ecologically comparable native species (i.e., super invaders) are quite rare, even among aggressive invaders (Rosenzweig 2001). Trade-offs in physiology and life history (Sinervo & Svensson 1998) are likely important in constraining invaders, but the rare "super invaders" may circumvent some trade-offs (Holway 1999). Assuming sufficient habitat heterogeneity, the performance data imply that most native species should be able to maintain natural populations even as invading plants spread, but the native populations might persist within a narrower range of environmental conditions (narrower realized niche) than before the invaders arrived. Conditions where invaders had the largest performance advantage (high resource availability, high physical disturbance, or departures from the natural disturbance regime; Figure 2) are generally associated with human activities.

STRATEGIES FOR MAXIMIZING PERFORMANCE BY NATIVES

Preserve "Intact" Natural Systems

Preserving "intact" natural systems is an obvious priority for many conservation biologists. From the perspective of invasion biology, natural areas that have experienced minimal human disturbance have long been noted to be less invaded than areas that have been directly disturbed by humans (Allan 1936, Rejmánek

1989). Although this pattern may be partly explained by fewer alien propagules having been introduced to intact natural systems, the studies reviewed here also suggest that the relative performance of most invaders would be reduced in many intact natural systems where key resources are likely to be scarce. Such systems may generally favor the survival, growth, or recruitment of natives. Physical protection of representative samples of intact natural systems also can provide us with invaluable reference information about conditions that are likely to favor local native species over most invaders. Key data that could be obtained from preserved natural systems include measures of nutrient, water, and light availability, assessment of natural disturbance regimes/environmental heterogeneity, and identification of factors associated with crucial population processes such as recruitment.

Focus on the Most Promising Restoration Sites

If data are available from an intact natural system, then promising restoration sites would ideally be identified based on matching resource availabilities and conditions (Critchley et al. 2002). Depending on the system of interest, key resources that maximize native:invader performance ratios will differ, but an obvious strategy for reducing invasive species problems is to avoid restoration sites with unusually high resource availability. For example, in northern California, a former cultivated field is likely to be rich in nutrients from past fertilization and have deeper and richer soil than a sloping, rocky site that has not been cultivated (Stromberg & Griffin 1996). Restoration of the former old field site with native perennial bunch grasses is likely to require substantially more effort and long-term maintenance because of more severe invasive plant problems (Stromberg & Griffin 1996).

In other cases, small differences in rainfall between potential restoration sites may be an important consideration. In the Hawaiian Islands, the native grass *Heteropogon contortus* formerly inhabited large portions of low elevation leeward sides of islands, but in recent years it has nearly disappeared, largely replaced by African grasses like *Pennisetum setaceum* (Daehler & Carino 1998). Under most conditions, invasive *Pennisetum* outcompetes *Heteropogon* (Carino & Daehler 2002); however, under very dry, nutrient poor conditions, *Heteropogon* performs substantially better than the invader (Goergen & Daehler 2002). It is these extremely dry and rocky (poor soil) sites that are the focus of current *Heteropogon* restoration efforts. Although alien grass seedlings are occasionally observed at these sites, they rarely survive owing to inadequate water (C. Daehler, unpublished data). In cases where the native vegetation targeted for restoration is adapted to specific environmental stresses (e.g., halophytes), even a small decrease in environmental stress (salinity) can lead to large increases in the performance of invaders (Mesleard et al. 1993). In such cases, choosing a site with more stressful conditions is likely to minimize invasive plant problems; for halophytic vegetation, salt additions may even be useful means of controlling invaders (Kuhn & Zedler 1997).

Actively Manage Resource Availability and Disturbance—Nutrient Reductions

Reductions in available nitrogen can potentially be achieved either by reducing anthropogenic inputs or by adding a carbon supplement to the soil, which can increase microbial nitrogen uptake (McLendon & Redente 1992, Paschke et al. 2000, Torok et al. 2000). By experimentally manipulating nitrogen inputs to wetlands with controlled hydrology, Green & Galatowitsch (2002) found that increased nitrogen inputs favored dominance of invasive reed canary grass (*Phalaris arundinacea*) over the native plant community. They concluded that reducing nitrate loads to wetland reserves is essential for minimizing declines in native community diversity.

In other systems, such as abandoned agricultural fields, reductions in available nitrogen to natural levels could take decades even without further anthropogenic inputs (Barton et al. 1999, Maron & Jeffries 1999). One strategy for rapidly reducing soil nutrients after anthropogenic disturbance is to remove topsoil. In an effort to restore an abandoned agricultural field to a native fen meadow, Tallowin & Smith (2001) found that removing 15–20 cm of top soil was the most effective strategy. Not only did the lower soil layer contain 85% less available phosphorus and lower nitrogen, it also had a higher calcium content, similar to that of soils in natural fen meadows (Tallowin & Smith 2001).

As a less extreme method of reducing nutrients, available soil nitrogen can be reduced by adding sawdust or sucrose to the soil. Paschke et al. (2000) used sucrose additions to reduce available soil nitrogen on abandoned crop land in Colorado. Over four years, plots that had been dominated by invasive *Bromus tectorum* were converted to a community that closely resembled the natural late seral shortgrass steppe vegetation. There was also less overall recruitment by a variety of weedy species in the reduced-nitrogen plots, even in wet years (Paschke et al. 2000). Interestingly, Evans et al. (2001) found that fires promoted by *Bromus tectorum* decrease nitrogen in the system through volatilization. Over the long run, this change could promote a shift away from dominance by *Bromus* back toward a late seral native community. Zink & Allen (1998) found that available nitrogen was reduced after adding organic mulch to a restoration site, and they suggested that this reduction gave native California sagebrush (*Artemisia californica*) a competitive edge over exotic annuals. Likewise, Alpert & Maron (2000) observed decreased biomass of invaders into nitrogen-rich soil patches after the addition of sawdust, while the sawdust did not reduce the abundance of native species. In contrast, Morghan & Seastedt (1999) did not observe an increase in dominance by natives following addition of sucrose and sawdust to a high elevation Colorado grassland. In that study, the treated plots were small in size (1.5 × 3 m), and these plots may have been swamped by alien seeds originating from outside the plots, allowing the aliens to remain abundant (Morghan & Seastedt 1999). In a mixed prairie grassland, plots receiving sawdust had higher bare ground, but native species responded most strongly to tilling, which presumably created more neighbor-free establishment sites for natives (Wilson & Gerry 1995). The latter two examples point out that additional factors besides nutrient reduction may affect the relative success of natives and invaders.

Other nutrients besides nitrogen could also be profitably manipulated in specific circumstances. For example, the common dandelion, *Taraxicum officinale*, seems to be an unusually poor competitor for potassium, suggesting that simple soil manipulations could be used to control this weed (Tilman et al. 1999). Increases in phosphorus associated with soil disturbance have also been suggested to promote alien invasion in some grasslands (Hobbs & Huenneke 1992), and some specific methods have been proposed for reducing soil phosphorus (Gough & Marrs 1990). Although most nutrient reduction studies to date have involved monitoring herbaceous species in grasslands, the reviewed performance comparisons of native and invasive plants suggest that nutrient reductions would also be beneficial for restorations involving native woody plants. Soil pollution with anthropogenic nitrates is likely to increase as the human population grows to demand higher production agricultural systems (Kawashima et al. 1997, Van Der Voet et al. 1996). Identifying long-term strategies for keeping excess nitrogen and other nutrients out of specific natural or restored native systems will be crucial for maintaining these systems over the long term (Bobbink & Roelofs 1995).

Other resources like light and water could be manipulated locally to promote native species over invasives, but these manipulations may be costly or impractical on a large scale. At the very least, if excesses of certain resources such as water occur at discrete times (e.g., following an unusually heavy rain), then reconnaissance efforts aimed at preventing the establishment and spread of invaders could be focused on these specific times. For example, Burgess et al. (1991) noted that introduced buffel grass (*Cenchrus ciliaris*) failed to spread significantly for many years; however, during two unusually wet years many seedlings became established and later grew to form dense stands. Other studies have also identified the role of brief, discrete periods of increased resource availability in facilitating invasions (Davis & Pelsor 2001).

Manipulate Disturbance

One strategy for promoting natives is to attempt to mimic natural disturbance regimes. For example, restoration of the historic flood regime along rivers in the southwestern United States would probably favor the re-establishment of native vegetation over the currently established invasive *Tamarix ramosissima* (Sher et al. 2000). A similar situation seems to occur with invasive *Elaeagnus angstifolia* along river banks in the western United States; restoration of the historic flood regime would probably favor re-establishment of natives (Shafroth et al. 1995). Maintaining the historic fire regime often favors the growth of native species over invaders (Schultz & Crone 1998, Tveten & Fonda 1999). For example, in California grasslands, warm season burning seems to favor native plants over invasive annual grasses (Meyer & Schiffman 1999). Restoration of historic grazing regimes can also help promote native species diversity (Collins et al. 1998). In other cases, disturbances that do not necessarily match the historic disturbance regime may still favor natives over invaders. For example, early spring mowing alone converted a grassland dominated by an invasive perennial grass (*Arrhenatherum elatius*) to a native perennial grassland (Wilson & Clark 2001). The growth pattern and phenology differed between the invader and natives such that mowing in early spring differentially removed biomass and developing inflorescences of the invader, resulting in increased flowering and growth of the native perennial grass (Wilson & Clark 2001). This example again illustrates how comparative demographic studies of invaders and co-occurring natives could be used to identify restoration strategies.

Further Challenges

Although particular disturbance regimes may tend to favor natives over invaders, maintaining primarily native communities may become increasingly difficult as the pool of introduced species located near a remnant native community increases (Smith & Knapp 2001b). The larger the pool of introduced species nearby, the greater the chance that at least some aliens will be able to tolerate the historic disturbance regime. More generally, the larger the species pool of potential invaders, the greater the chance that some aliens will possess traits that contradict statistical trends among invaders. For example, in a seed addition experiment testing the effects of disturbance and nutrient addition on invasion, Buckland et al. (2001) found that most successful invaders matched the predicted pattern—they became established and abundant in the physical disturbance and/or nutrient addition treatments. But one invasive grass, *Brachypodium pinnatum*, was most successful in the low fertility plots (Buckland et al. 2001). Controlling such exceptional species might require more direct, focused attacks, although the example of *B. pinnatum* is not so clear because in similar grasslands, Bobbink & Willems (1987) concluded that higher nutrients (especially nitrogen) were important in facilitating *B. pinnatum* invasion, matching the predicted general pattern for invaders.

Another problem that becomes apparent as the pool of invaders grows and the size of native habitats decreases is the proportionally greater influx of alien plant seeds into smaller patches of native vegetation (e.g., D'Antonio et al. 2001). A "seed swamping" effect could increase the net establishment of invaders, even if they are unable to develop self-sustaining populations within native habitats. This effect may be partly countered by establishing dense native vegetation around the edges of remnant native habitats (Cadenasso & Pickett 2001).

Perhaps the greatest long-term challenge to maintaining native communities is global climate change associated with urbanization and anthropogenic greenhouse gasses (Walther et al. 2002, White et al. 2002). Increases in global temperature allow warm-climate species to succeed at higher latitudes and elevations while also potentially decreasing the performance of certain native species. Such changes have already been clearly documented (Walther et al. 2002). If anthropogenic climate changes turn out to be large enough to affect vegetation patterns on a global scale, then conservation biologists may need to adopt entirely new objectives and strategies for conserving native biodiversity.

SYNTHESIS

Many conceptual and mathematical models have suggested that increases in resource availability (including space, created by physical disturbance) can increase community susceptibility to invasion (Davis et al. 2000, Fox & Fox 1986, Hobbs 1989, Shea & Chesson 2002, Sher & Hyatt 1999, Tilman 1999). Overall, these models assume that the free resources provide invaders with an opportunity to enter an established community, perhaps by reducing the intensity of competition for a limiting resource (Davis & Pelsor 2001). This idea has been generally supported by studies that manipulate resources (Brooks 1999, Duggin & Gentle 1998, Li & Norland 2001, Weiss 1999, White et al. 1997). However, the performance comparisons reviewed here suggest an additional factor that can help explain these empirical results: the relative performance of invaders versus native species may shift under higher resource conditions. Such shifts in relative performance may then allow invaders to dominate over natives as long as resource availability remains high (increased light, nutrients, or water, usually associated with anthropogenic disturbance). Because resource availability is unlikely to be uniform in any given habitat, patches of natives may be expected to persist in lower resource areas (e.g., on lower quality soil patches, under marginal growing conditions). Theoretically, a return to original (or predisturbance) resource levels would again favor native species over most invaders, but several factors may prevent their re-establishment, including

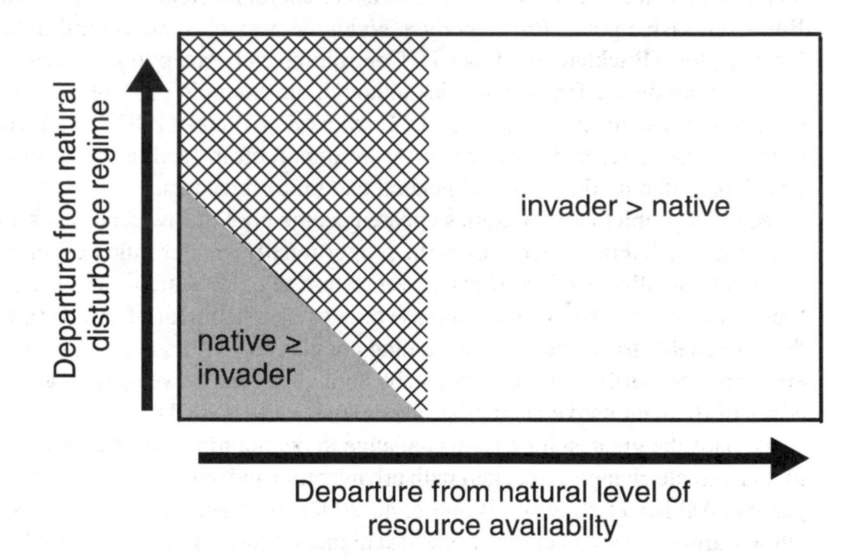

Figure 4 Conceptual model predicting relative performance of native species versus most invaders. The checkerboard region represents possible unnatural disturbance regimes (must be discovered) that might favor the native if resource availability is not extraordinarily high.

inadequate seed input by natives, a priority effect by established invaders, and overwhelming seed inputs by invaders growing in surrounding areas. Furthermore, if invaders are long-lived, then they could potentially prevent significant recruitment of native species for decades without additional intervention to remove the invaders.

Specific disturbance regimes can also favor natives over invaders (Figure 4). In most cases, the natural disturbance regime seems to favor natives (Alpert et al. 2000, Hobbs & Huenneke 1992, Mueller-Dombois & Loope 1990), but other disturbance regimes may also favor natives over specific invaders (Figure 4). Life history and demographic comparisons of co-occurring native and invasive species could aid in identifying artificial disturbance regimes that decrease the performance of invaders relative to that of natives. Manipulation of disturbance, nutrient and hydrological regimes can be considered within the general framework of "ecosystem management" (Christensen et al. 1996, Grumbine 1994), although these strategies for attaining "ecological control" of invaders may also be effective on local scales that do not necessarily extend across an entire ecosystem.

For any given habitat, there are probably a small number of "super invaders" capable of outperforming most co-occurring natives even at natural resource levels and in the presence of natural disturbance regimes. These invaders will require special attention. However, the ability of most invaders to outperform co-occurring natives appears to be context-dependent. Most environments can probably be managed to favor native species by altering resource levels and disturbance regimes so that native species performance is maximized, relative to that of most invaders. This form of "ecological control" is unlikely to eliminate all troublesome invaders from habitats where they already occur; rather, some invaders would probably coexist with natives at lower (acceptable) densities. For most habitats, we have only cursory knowledge, at best, of how environmental manipulations can be used to reduce invasive plant problems and simultaneously to promote natives. The increasing integration of well-planned environmental manipulations into restoration projects promises to provide new insights into managing invasive species problems.

ACKNOWLEDGMENTS

I thank Carla D'Antonio, Dieter Mueller-Dombois, Petr Pyšek, Dave Richardson, and Dan Simberloff for comments on the draft manuscript.

The *Annual Review of Ecology, Evolution, and Systematics* is online at http://ecolsys.annualreviews.org

LITERATURE CITED

Aerts R, van der Pejil MJ. 1993. A simple model to explain dominance of low-productive perennials in nutrient-poor habitats. *Oikos* 66:144–47

Allan HH. 1936. Indigene versus alien in the New Zealand plant world. *Ecology* 17:187–92

Alpert P, Bone E, Holzapfel C. 2000.

Invasiveness, invasibility and the role of environmental stress in the spread of non-native plants. *Perspect. Plant Ecol. Evol. Syst.* 3:52–66

Alpert P, Maron JL. 2000. Carbon addition as a countermeasure against biological invasion by plants. *Biol. Invasions* 2:33–40

Aptekar R, Rejmanek M. 2000. The effect of sea-water submergence on rhizome bud viability of the introduced *Ammophila arenaria* and the native *Leymus mollis* in California. *J. Coast. Conserv.* 6:107–11

Arenas F, Fernandez C, Rico J, Fernandez E, Haya D. 1995. Growth and reproductive strategies of *Sargassum muticum* (Yendo) Fensholt and *Cystoseira nodicaulis* (Whit.) Roberts. *Sci. Mar.* 59(Suppl. 1):1–8

Baars R, Kelly D. 1996. Survival and growth responses of native and introduced vines in New Zealand to light availability. *NZ J. Bot.* 34:389–400

Barton L, McLay CDA, Schipper LA, Smith CT. 1999. Annual denitrification rates in agricultural and forest soils: A review. *Aust. J. Soil Res.* 37:1073–93

Baruch Z. 1996. Ecophysiological aspects of the invasion by African grasses and their impact on biodiversity and function of neotropical savannas. In *Biodiversity and Savanna Ecosystem Processes,* ed. OT Solbrig, E Medina, JF Silva, pp. 79–93. Berlin: Springer-Verlag

Baruch Z, Bilbao B. 1999. Effects of fire and defoliation on the life history of native and invader C4 grasses in a neotropical savanna. *Oecologia* 119:510–20

Baruch Z, Goldstein G. 1999. Leaf construction cost, nutrient concentration, and net CO_2 assimilation of native and invasive species in Hawaii. *Oecologia* 121:183–92

Bellingham PJ. 1998. Shrub succession and invasibility in a New Zealand montane grassland. *Aust. J. Ecol.* 23:562–73

Black RA, Richards JH, Manwaring JH. 1994. Nutrient uptake from enriched soil microsites by three great basin perennials. *Ecology* 75:110–22

Blaney CS, Kotanen PM. 2001a. Effects of fungal pathogens on seeds of native and exotic plants: A test using congeneric pairs. *J. Appl. Ecol.* 38:1104–13

Blaney CS, Kotanen PM. 2001b. Post-dispersal losses to seed predators: An experimental comparison of native and exotic old field plants. *Can. J. Bot.* 79:284–92

Bobbink R, Roelofs JGM. 1995. Nitrogen critical loads for natural and semi-natural ecosystems: The empirical approach. *Water Air Soil Pollut.* 85:2413–18

Bobbink R, Willems JH. 1987. Increasing dominance of Brachypodium-Pinnatum L. Beauv. in chalk grasslands a threat to a species-rich ecosystem. *Biol. Conserv.* 40:301–14

Borman MM, Krueger WC, Johnson DE. 1990. Growth patterns of perennial grasses in the annual grassland type of southwest Oregon (USA). *Agron. J.* 82:1093–98

Brooks ML. 1999. Habitat invasibility and dominance by alien annual plants in the western Mojave desert. *Biol. Invasions* 1:325–37

Buckland SM, Thompson K, Hodgson JG, Grime JP. 2001. Grassland invasions: effects of manipulations of climate and management. *J. Appl. Ecol.* 38:301–9

Burgess TL, Bowers JE, Turner RM. 1991. Exotic plants at the desert laboratory, Tucson, Arizona. *Madroño* 38:96–114

Burke MJW, Grime JP. 1996. An experimental study of plant community invasibility. *Ecology* 77:776–90

Cadenasso ML, Pickett STA. 2001. Effect of edge structure on the flux of species into forest interiors. *Conserv. Biol.* 15:91–97

Caldwell MM, Richards JH, Johnson DA, Nowak RS, Dzurec RS. 1981. Coping with herbivory: photosynthetic capacity and resource allocation in two semiarid *Agropyron* bunchgrasses. *Oecologia* 50:14–24

Callaway JC, Josselyn MN. 1992. The introduction and spread of smooth cordgrass (*Spartina alterniflora*) in South San Francisco Bay. *Estuaries* 15:218–26

Carino DA, Daehler CC. 2002. Can inconspicuous legumes facilitate alien grass invasions?

Partridge peas and fountain grass in Hawaii. *Ecography* 25:33–41

Christensen NL, Bartuska AM, Brown JH, Carpenter S, Francis R, et al. 1996. The report of the Ecological Society of America committee on the scientific basis for ecosystem management. *Ecol. Appl.* 6:665–91

Claassen VP, Marler M. 1998. Annual and perennial grass growth on nitrogen-depleted decomposed granite. *Restor. Ecol.* 6:175–80

Cleverly JR, Smith SD, Sala A, Devitt DA. 1997. Invasive capacity of *Tamarix ramosissima* in a Mojave Desert floodplain: The role of drought. *Oecologia* 111:12–18

Collins SL, Knapp AK, Briggs JM, Blair JM, Steinauer EM. 1998. Modulation of diversity by grazing and mowing in native tallgrass prairie. *Science.* 280:745–47

Crawley MJ, Harvey PH, Purvis A. 1996. Comparative ecology of the native and alien floras of the British Isles. *Philos. Trans. R. Soc. London Ser. B* 351:1251–59

Critchley CNR, Chambers BJ, Fowbert JA, Sanderson RA, Bhogal A, Rose SC. 2002. Association between lowland grassland plant communities and soil properties. *Biol. Conserv.* 105:199–215

Cronk QCB, Fuller JL. 1995. *Plant Invaders.* New York: Chapman & Hall. 241 pp.

Cross JR. 1981. The establishment of *Rhododendron ponticum* in the Killarney oakwoods. *J. Ecol.* 69:807–24

Daehler CC. 2001. Two ways to be an invader, but one is more suitable for ecology. *Bull. Ecol. Soc. Am.* 82:101–2

Daehler CC, Carino DA. 1998. Recent replacement of native pili grass (*Heteropogon contortus*) by invasive African grasses in the Hawaiian Islands. *Pac. Sci.* 52:220–27

Daehler CC, Carino DA. 1999. Threats of invasive plants to the conservation of biodiversity. In *Biodiversity and Allelopathy: From Organisms to Ecosystems in the Pacific,* ed. CH Chou, GR Waller, C Reinhardt, pp. 21–27. Taipei: Acad. Sin.

Daehler CC, Strong DR. 1994. Native plant biodiversity vs. the introduced invaders: status of the conflict and future management options. In *Biological Diversity: Problems and Challenges*, ed. SK Majumdar, FJ Brenner, JE Lovich, JF Schalles, EW Miller, pp. 92–113. Easton, PA: Penn. Acad. Sci.

Daehler CC, Yorkston M, Sun W, Dudley N. 1999. Genetic variation in morphology and growth characters of Acacia koa in the Hawaiian Islands. *Int. J. Plant Sci.* 160:767–73

D'Antonio CM, Dudley TL, Mack M. 1999. Disturbance and biological invasions: direct effects and feedbacks. In *Ecosystems of Disturbed Ground,* ed. LR Walker, New York: Elsevier

D'Antonio CM, Levine JM, Thomson M. 2001. Ecosystem resistance to invasion and the role of propagule supply: a California perspective. *Ecol. Mediterr.* 27:233–45

D'Antonio CM, Mahall BE. 1991. Root profiles and competition between the invasive, exotic perennial, *Carpobrotus edulis*, and two native shrub species in California (USA) coastal scrub. *Am. J. Bot.* 78:885–94

D'Antonio CM, Vitousek PM. 1992. Biological invasions by exotic grasses, the grass/fire cycle, and global change. *Annu. Rev. Ecol. Syst.* 23:63–87

Davis MA, Grime JP, Thompson K. 2000. Fluctuating resources in plant communities: a general theory of invasibility. *J. Ecol.* 88: 528–34

Davis MA, Pelsor M. 2001. Experimental support for a resource-based mechanistic model of invasibility. *Ecol. Lett.* 4:421–28

Dietz H, Steinlein T, Ullmann I. 1998. The role of growth form and correlated traits in competitive ranking of six perennial ruderal plant species grown in unbalanced mixtures. *Acta Oecol.* 19:25–36

Drake JA, Mooney HA, diCastri F, Groves RH, Kruger FJ, et al., eds. 1989. *Biological Invasions: A Global Perspective.* New York: Wiley

Duggin JA, Gentle CB. 1998. Experimental evidence on the importance of disturbance intensity for invasion of *Lantana camara* L. in dry rainforest-open forest ecotones in northeastern NSW, Australia. *Forest Ecol. Manag.* 109:279–92

Evans RD, Rimer R, Sperry L, Belnap J. 2001. Exotic plant invasion alters nitrogen dynamics in an arid grassland. *Ecol. Appl.* 11:1301–10

Fan J, Harris W. 1996. Effects of soil fertility level and cutting frequency on interference among *Hieracium pilosella*, *H. praealtum*, *Rumex acetosella*, and *Festuca novae-zelandiae*. *NZ J. Agric. Res.* 39:1–32

Fox MD, Fox BJ. 1986. The susceptibility of natural communities to invasion. In *Ecology of Biological Invasions: An Australian Perspective*, ed. RH Groves, JJ Burden, pp. 57–66. Canberra: Cambridge Univ. Press

Frenot Y, Gloaguen JC. 1994. Reproductive performance of native and alien colonizing phanerogams on a glacier foreland, Iles Kerguelen. *Polar Biol.* 14:473–81

Glenn E, Tanner R, Mendez S, Kehret T, Moore D, et al. 1998. Growth rates, salt tolerance and water use characteristics of native and invasive riparian plants from the delta of the Colorado River, Mexico. *J. Arid. Environ.* 40: 281–94

Goergen E, Daehler CC. 2001a. Inflorescence damage by insects and fungi in native pili grass (*Heteropogon contortus*) versus alien fountain grass (*Pennisetum setaceum*) in Hawai'i. *Pac. Sci.* 55:129–36

Goergen E, Daehler CC. 2001b. Reproductive ecology of a native Hawaiian grass (*Heteropogon contortus*; Poaceae) versus its invasive alien competitor (*Pennisetum setaceum*; Poaceae). *Int. J. Plant Sci.* 162:317–26

Goergen E, Daehler CC. 2002. Factors affecting seedling recruitment in an invasive grass (*Pennisetum setaceum*) and a native grass (*Heteropogon contortus*) in the Hawaiian Islands. *Plant Ecol.* 161:147–56

Goldberg DE, Landa K. 1991. Competitive effect and response: hierarchies and correlated traits in the early stages of competition. *J. Ecol.* 79:1013–30

Goodwin BJ, McAllister AJ, Fahrig L. 1999. Predicting invasiveness of plant species based on biological information. *Conserv. Biol.* 13:422–26

Gough MW, Marrs RH. 1990. A comparison of soil fertility between semi-natural and agricultural plant communities: implications for the creation of species-rich grassland on abandoned agricultural land. *Biol. Conserv.* 51:83–96

Gould AMA, Gorchov DL. 2000. Effects of the exotic invasive shrub *Lonicera maackii* on the survival and fecundity of three species of native annuals. *Am. Midl. Nat.* 144:36–50

Green EK, Galatowitsch SM. 2002. Effects of *Phalaris arundinacea* and nitrate-N addition on the establishment of wetland plant communities. *J. Appl. Ecol.* 39:134–44

Greenberg CH, Smith LM, Levey DJ. 2001. Fruit fate, seed germination and growth of an invasive vine—an experimental test of "sit and wait" strategy. *Biol. Invasions* 3:363–72

Gross EM, Johnson RL, Hairston NG Jr. 2001. Experimental evidence for changes in submersed macrophyte species composition caused by the herbivore *Acentria ephemerella* (Lepidoptera). *Oecologia* 127: 105–14

Grotkopp E, Rejmánek M, Rost TL. 2002. Toward a causal explanation of plant invasiveness: Seedling growth and life-history strategies of 29 pine (*Pinus*) species. *Am. Nat.* 159:396–419

Grumbine RE. 1994. What is ecosystem management? *Conserv. Biol.* 8:27–38

Herron GJ, Sheley RL, Maxwell BD, Jacobsen JS. 2001. Influence of nutrient availability on the interaction between spotted knapweed and bluebunch wheatgrass. *Restor. Ecol.* 9:326–31

Hobbs RJ. 1989. The nature and effects of disturbance relative to invasion. See Drake et al. 1989, pp. 389–405

Hobbs RJ, Atkins L. 1988. Effect of Disturbance and Nutrient Addition on Native and Introduced Annuals in Plant Communities in the Western Australian Wheatbelt. *Aust. J. Ecol.* 13:171–80

Hobbs RJ, Huenneke LF. 1992. Disturbance, diversity, and invasion: Implications for conservation. *Conserv. Biol.* 6:324–37

Holmgren M, Aviles R, Sierralta L, Segura AM, Fuentes ER. 2000. Why have European herbs so successfully invaded the Chilean matorral? Effects of herbivory, soil nutrients, and fire. *J. Arid Environ.* 44:197–211

Holway DA. 1999. Competitive mechanisms underlying the displacement of native ants by the invasive Argentine ant. *Ecology* 80:238–51

Honig MA, Cowling RM, Richardson DM. 1992. The invasive potential of Australian Banksias in South African fynbos: A comparison of the reproductive potential of *Banksia ericifolia* and *Leucadendron laureolum*. *Aust. J. Ecol.* 17:305–14

Horn P, Prach K. 1994. Aerial biomass of *Reynoutria japonica* and its comparison with that of native species. *Preslia* 66:345–48

Huenneke LF, Hamburg SP, Koide R, Mooney HA, Vitousek PM. 1990. Effects of soil resources on plant invasion and community structure in Californian serpentine grassland. *Ecology* 71:478–91

Huenneke LF, Thomson JK. 1995. Potential interference between a threatened endemic thistle and an invasive nonnative plant. *Conserv. Biol.* 9:416–25

Kawashima H, Bazin MJ, Lynch JM. 1997. A modelling study of world protein supply and nitrogen fertilizer demand in the 21st century. *Environ. Conserv.* 24:50–56

Keane RM, Crawley MJ. 2002. Exotic plant invasions and the enemy release hypothesis. *Trends Ecol. Evol.* 17:164–70

Keddy PA, Twolan SL, Wisheu IC. 1994. Competitive effect and response rankings in 20 wetland plants: Are they consistent across three environments? *J. Ecol.* 82:635–43

Kitayama K, Mueller-Dombois D. 1995. Biological invasion on an oceanic island mountain: Do alien plant species have wider ecological ranges than native species? *J. Veg. Sci.* 6:667–74

Kitayama K, Pattison R, Cordell S, Webb D, Mueller Dombois D. 1997. Ecological and genetic implications of foliar polymorphism in *Metrosideros polymorpha* Gaud. (Myrtaceae) in a habitat matrix on Mauna Loa, Hawaii. *Ann. Bot.* 80:491–97

Kolar CS, Lodge DM. 2001. Progress in invasion biology: Predicting invaders. *Trends Ecol. Evol.* 16:199–204

Kuhn NL, Zedler JB. 1997. Differential effects of salinity and soil saturation on native and exotic plants of a coastal salt marsh. *Estuaries* 20:391–403

Larson KC. 2000. Circumnutation behavior of an exotic honeysuckle vine and its native congener: Influence on clonal mobility. *Am. J. Bot.* 87:533–38

Lesica P, Miles S. 1999. Russian olive invasion into cottonwood forests along a regulated river in north-central Montana. *Can. J. Bot.* 77:1077–83

Li Y, Norland M. 2001. The role of soil fertility in invasion of Brazilian pepper (*Schinus terebinthifolius*) in Everglades National Park, Florida. *Soil Sci.* 166:400–5

Luken JO, Kuddes LM, Tholemeier TC, Haller DM. 1997. Comparative responses of *Lonicera maackii* (amur honeysuckle) and *Lindera benzoin* (spicebush) to increased light. *Am. Midl. Nat.* 138:331–43

Mack RN. 1996. Predicting the identity and fate of plant invaders: emergent and emerging approaches. *Biol. Conserv.* 78:107–21

Maillet J, Lopez GC. 2000. What criteria are relevant for predicting the invasive capacity of a new agricultural weed? The case of invasive American species in France. *Weed Res.* 40:11–26

Mal TK, Lovett-Doust J, Lovett-Doust L. 1997. Time-dependent competitive displacement of *Typha angustifolia* by *Lythrum salicaria*. *Oikos* 79:26–33

Marco DE, Paez SA. 2000. Invasion of *Gleditsia triacanthos* in *Lithraea ternifolia* montane forests of Central Argentina. *Environ. Manag.* 26:409–19

Marler MJ, Zabinski CA, Wojtowicz T, Callaway RM. 1999. Mycorrhizae and fine root dynamics of *Centaurea maculosa* and native bunchgrasses in western Montana. *NW Sci.* 73:217–24

Maron JL, Jeffries RL. 1999. Bush lupine mortality, altered resource availability, and alternative vegetative states. *Ecology* 80:443–54

McDowell CR, Moll EJ. 1981. Studies of seed

germination and seedling competition in *Virgilia oroboides* (Berg.) Salter, *Albizia lophantha* (Willd.) Benth. and *Acacia longifolia* (Andr.) Willd. *J. S. Afr. Bot.* 47:653–85

McKinney ML, Lockwood JL. 1999. Biotic homogenization: a few winners replacing many losers in the next mass extinction. *Trends Ecol. Evol.* 14:450–53

McLendon T, Redente EF. 1992. Effects of nitrogen limitation on species replacement dynamics during early secondary succession on a semiarid sagebrush site. *Oecologia* 91:312–17

Melgoza G, Nowak RS, Tausch RJ. 1990. Soil water exploitation after fire: Competition between *Bromus tectorum* (cheatgrass) and two native species. *Oecologia* 83:7–13

Mesleard F, Ham LT, Boy V, Van Wijck C, Grillas P. 1993. Competition between an introduced and an indigenous species: The case of *Paspalum paspalodes* (Michx) Schribner and *Aeluropus littoralis* (Gouan) in the Camargue (southern France). *Oecologia* 94:204–9

Meyer MD, Schiffman PM. 1999. Fire season and mulch reduction in a California grassland: A comparison of restoration strategies. *Madroño* 46:25–37

Mitchell CE, Power AG. 2003. Release of invasive plants from fungal and viral pathogens. *Nature* 421:625–27

Montaldo NH. 2000. Exito reproductivo de plantas ornitocoras en un relicto de selva subtropical en Argentina. *Rev. Chil. Hist. Nat.* 73:511–24

Morghan RKJ, Seastedt TR. 1999. Effects of soil nitrogen reduction on nonnative plants in restored grasslands. *Restor. Ecol.* 7:51–55

Mueller-Dombois D, Loope LL. 1990. Some unique ecological aspects of oceanic island ecosystems. *Monogr. Syst. Bot. Mo. Bot. Gard.* 32:21–27

Nagel JM, Griffin KL. 2001. Construction cost and invasive potential: Comparing *Lythrum salicaria* (Lythraceae) with co-occurring native species along pond banks. *Am. J. Bot.* 88:2252–58

Nernberg D, Dale MRT. 1997. Competition of five native prairie grasses with *Bromus inermus* under three moisture regimes. *Can. J. Bot.* 75:2140–45

Oliver JD. 1996. Mile-a-minute weed (*Polygonum perfoliatum* L.), an invasive vine in natural and disturbed sites. *Castanea* 61:244–51

Paschke MW, McLendon T, Redente EF. 2000. Nitrogen availability and old-field succession in a shortgrass steppe. *Ecosystems* 3:144–58

Pattison RR, Goldstein G, Ares A. 1998. Growth, biomass allocation and photosynthesis of invasive and native Hawaiian rainforest species. *Oecologia* 117:449–59

Pavlik BM. 1983a. Nutrient and productivity relations of the dune grass *Ammophila arenaria* and *Elymus mollis*. I. Blade photosynthesis and nitrogen use efficiency in the laboratory and field. *Oecologia* 57:227–32

Pavlik BM. 1983b. Nutrient and productivity relations of the dune grass *Ammophila arenaria* and *Elymus mollis*. II. Growth and patterns of dry matter and nitrogen allocation as influenced by nitrogen supply. *Oecologia* 57:233–38

Pimentel D. 2002. *Biological Invasions: Economic and Environmental Costs of Alien Plant, Animal, and Microbe Species*. Boca Raton, FL: CRC Press. 384 pp.

Poorter H, Bergkotte M. 1992. Chemical composition of 24 wild species differing in relative growth rate. *Plant Cell Environ.* 15:221–29

Pyke DA. 1986. Demographic responses of *Bromus tectorum* and seedlings of *Agropyron spicatum* to grazing by small mammals: occurrence. *J. Ecol.* 74:739–54

Pyšek P, Prach K, Smilauer P. 1995. Relating invasion success to plant traits: an analysis of the Czech alien flora. In *Plant Invasions, General Aspects and Special Problems,* ed. P Pyšek, K Prach, M Rejmanek, M Wade, pp. 39–60. Amsterdam, The Netherlands: SPB Acad.

Pyšek P, Sadlo J, Mandak B. 2002. Catalogue of alien plants of the Czech Republic. *Preslia* 74:97–186

Radho TS, Majer JD, Yates C. 2001. Impact

of fire on leaf nutrients, arthropod fauna and herbivory of native and exotic eucalypts in Kings Park, Perth, Western Australia. *Aust. Ecol.* 26:500–6

Reich PB, Walter MB, Ellsworth DS. 1992. Leaf life-span in relation to leaf plant and stand characteristics among diverse ecosystems. *Ecol. Monogr.* 62:365–92

Reich PB, Walters MB, Ellsworth DS. 1997. From tropics to tundra: global convergence in plant functioning. *Proc. Natl. Acad. Sci. USA* 94:13730–34

Rejmánek M. 1989. Invasibility of plant communities. See Drake et al. 1989, pp. 369–88

Rejmánek M. 1995. What makes a species invasive? In *Plant Invasions*, ed. P Pysek, K Prach, M Rejmánek, PM Wade, pp. 1–11. The Hague, Netherlands: SPB Acad.

Rejmánek M. 1996. A theory of seed plant invasiveness: the first sketch. *Biol. Conserv.* 78:171–81

Rejmánek M, Richardson DM, Higgins SI, Pitcairn M, Grotkopp E. 2003. Ecology of invasive plants: state of the art. In *Invasive Alien Species: Searching for Solutions*, ed. HA Mooney, JA McNeely, L Neville, PJ Schei, J Waage. Washington, DC: Island. In press

Richardson DM, Pysek P, Rejmanek M, Barbour MG, Panetta FD, West CJ. 2000. Naturalization and invasion of alien plants: concepts and definitions. *Divers. Distrib.* 6:93–107

Richardson DM, Van Wilgen BW, Mitchell DT. 1987. Aspects of the reproductive ecology of four Australian *Hakea* species Proteaceae in South Africa. *Oecologia* 71:345–54

Rosenzweig ML. 2001. The four questions: what does the introduction of exotic species do to diversity? *Evol. Ecol. Res.* 3:361–67

Sallabanks R. 1993. Fruiting plant attractiveness to avian seed dispersers: native vs. invasive Crataegus in western Oregon. *Madroño* 40:108–16

Schierenbeck KA, Mack RN, Sharitz RR. 1994. Effects of herbivory on growth and biomass allocation in native and introduced species of Lonicera. *Ecology* 75:1661–72

Schultz CB, Crone EE. 1998. Burning prairie to restore butterfly habitat: A modeling approach to management tradeoffs for the Fender's blue. *Restor. Ecol.* 6:244–52

Shafroth PB, Auble GT, Scott ML. 1995. Germination and establishment of the native plains cottonwood (*Populus deltoides* Marshall subsp. *monilifera*) and the exotic Russian-olive (*Elaeagnus angustifolia* L.). *Conserv. Biol.* 9:1169–75

Shea K, Chesson P. 2002. Community ecology theory as a framework for biological invasions. *Trends Ecol. Evol.* 17:170–76

Sher AA, Hyatt LA. 1999. The disturbed resource-flux invasion matrix: A new framework for patterns of plant invasion. *Biol. Invasions* 1:107–14

Sher AA, Marshall DL, Gilbert SA. 2000. Competition between native *Populus deltoides* and invasive *Tamarix ramosissima* and the implications for reestablishing flooding disturbance. *Conserv. Biol.* 14:1744–54

Simberloff D. 1981. Community effects of introduced species. In *Biotic Crises in Ecological and Evolutionary Time*, ed. H Nitecki, pp. 53–81. New York: Academic

Simoes M, Baruch Z. 1991. Responses to simulated herbivory and water stress in two tropical C-4 grasses. *Oecologia* 88:173–80

Sinervo B, Svensson E. 1998. Mechanistic and selective causes of life history trade-offs and plasticity. *Oikos* 83:432–42

Smith MA, Bell DT, Loneragan WA. 1999. Comparative seed germination ecology of *Austrostipa compressa* and *Ehrharta calycina* (Poaceae) in a Western Australian Banksia woodlands. *Aust. J. Ecol.* 24:35–42

Smith MD, Knapp AK. 2001a. Physiological and morphological traits of exotic, invasive exotic, and native plant species in tallgrass prairie. *Int. J. Plant Sci.* 162:785–92

Smith MD, Knapp AK. 2001b. Size of the local species pool determines invasibility of a C4-dominated grassland. *Oikos* 92:55–61

Smith RGB, Brock MA. 1996. Coexistence of *Juncus articulatus* L. and *Glyceria australis* C.E. Hubb. in a temporary shallow wetland in Australia. *Hydrobiologia.* 340:147–51

Stromberg MR, Griffin JR. 1996. Long-term patterns in coastal California grasslands in relation to cultivation, gophers, and grazing. *Ecol. Appl.* 6:1189–1211

Tallowin JRB, Smith REN. 2001. Restoration of a *Cirsio-Molinietum* fen meadow on an agriculturally improved pasture. *Restor. Ecol.* 9:167–78

Thompson K, Hodgson JG, Rich TCG. 1995. Native and alien invasive plants: more of the same? *Ecography* 18:390–402

Tilman D. 1999. The ecological consequences of changes in biodiversity: A search for general principles. *Ecology* 80:1455–74

Tilman EA, Tilman D, Crawley MJ, Johnston AE. 1999. Biological weed control via nutrient competition: Potassium limitation of dandelions. *Ecol. Appl.* 9:103–11

Torok K, Szili Kovacs T, Halassy M, Toth T, Hayek Z, et al. 2000. Immobilization of soil nitrogen as a possible method for the restoration of sandy grassland. *Appl. Veg. Sci.* 3:7–14

Tunison TJ, Leialoha JAK, Loh RH, Pratt LW, Higashino PK. 1994. Fire effects in the coastal lowlands Hawai'i Volcanoes National Park, *CPSU Tech. Rep. 88*. Honolulu: Univ. Hawai'i. 43 pp.

Tveten RK, Fonda RW. 1999. Fire effects on prairies and oak woodlands on Fort Lewis, Washington. *NW Sci.* 73:145–58

Van der Voet E, Kleijn R, De Haes HAU. 1996. Nitrogen pollution in the European Union—An economy-environment confrontation. *Environ. Conserv.* 23:198–206

Vila M, D'Antonio CM. 1998a. Fruit choice and seed dispersal of invasive vs. noninvasive *Carpobrotus* (Aizoaceae) in coastal California. *Ecology* 79:1053–60

Vila M, D'Antonio CM. 1998b. Fitness of invasive *Carpobrotus* (Aizoaceae) hybrids in coastal California. *Ecoscience* 5:191–99

Virgona JM, Bowcher A. 2000. Effects of grazing interval on basal cover of four perennial grasses in a summer-dry environment. *Aust. J. Exp. Agric.* 40:299–311

Vitousek PM. 1990. Biological invasions and ecosystem processes: towards an integration of population biology and ecosystem studies. *Oikos* 57:7–13

Walther G-R, Post E, Convey P, Menzel A, Parmesan C, et al. 2002. Ecological responses to recent climate change. *Nature* 416:389–95

Weiss SB. 1999. Cars, cows, and checkerspot butterflies: Nitrogen deposition and management of nutrient-poor grasslands for a threatened species. *Conserv. Biol.* 13:1476–86

White MA, Nemani RR, Thornton PE, Running SW. 2002. Satellite evidence of phenological differences between urbanized and rural areas of the Eastern United States deciduous broadleaf forest. *Ecosystems* 5:260–73

White TA, Campbel BD, Kemp PD. 1997. Invasion of temperate grassland by a subtropical annual grass across an experimental matrix of water stress and disturbance. *J. Veg. Sci.* 8:847–54

Wilcove DS, Rothstein D, Dubow J, Phillips A, Losos E. 1998. Quantifying threats to imperiled species in the United States. *BioScience* 48:607–15

Williams DG, Black RA. 1994. Drought response of a native and introduced Hawaiian grass. *Oecologia* 97:512–19

Williamson MH. 1996. *Biological Invasions.* London: Chapman & Hall. 244 pp.

Williamson MH, Fitter A. 1996. The characters of successful invaders. *Biol. Conserv.* 78:163–70

Wilson MV, Clark DL. 2001. Controlling invasive *Arrhenatherum elatius* and promoting native prairie grasses through mowing. *Appl. Veg. Sci.* 4:129–38

Wilson SD, Gerry AK. 1995. Strategies for mixed-grass prairie restoration: herbicide, tilling, and nitrogen manipulation. *Restor. Ecol.* 3:290–98

Witkowski ETF. 1991. Growth and competition between seedlings of *Protea repens* (L.) L. and the alien invasive, *Acacia saligna* (Labill.) Wendl. in relation to nutrient availability. *Funct. Ecol.* 5:101–10

Yamashita N, Ishida A, Kushima H, Tanaka N. 2000. Acclimation to sudden increase in light favoring an invasive over native trees in subtropical islands, Japan. *Oecologia* 125: 412–19

Zedler JB, Paling E, McComb A. 1990. Differential responses to salinity help explain the replacement of native *Juncus kraussii* by *Typha orientalis* in Western Australian salt marshes. *Aust. J. Ecol.* 15:57–72

Zink TA, Allen MF. 1998. The effects of organic amendments on the restoration of a disturbed coastal sage scrub habitat. *Restor. Ecol.* 6:52–58

Annu. Rev. Ecol. Evol. Syst. 2003. 34:213–237
doi: 10.1146/annurev.ecolsys.34.030102.151717

First published online as a Review in Advance on July 8, 2003

GENETIC VARIATION IN RARE AND COMMON PLANTS

Christopher T. Cole
Division of Science and Mathematics, University of Minnesota-Morris, Morris, Minnesota 56267; email: colect@mrs.umn.edu

Key Words rare plants, conservation genetics, isozymes, monomorphism

■ **Abstract** Isozyme variation in 247 plant species is summarized as 57 generic-level comparisons of rare and common species. All species-level measures of variation (P_s, A_s, AP_s, H_{es}) and mean population-level measures (P_p, A_p, AP_p, H_{ep}, and H_o) show reductions significant at the $p < 0.001$ level, but F_{IS} and F_{ST} did not differ significantly, reflecting the similarity of breeding system in congeneric species and disparate ranges often sampled for rare and common species. The reduction in gene flow (Nm) among populations of rare species was significant when estimated from F_{ST}, but not when estimated from private alleles. Species monomorphic for isozymes are predominantly endemic and self-fertile. Although census populations of virtually all rare species are higher than levels at which theory would predict genetic erosion, and higher than levels protected by the U.S. Endangered Species Act (ESA), rare plants evidently have more significant reductions in genetic variation and gene flow than have been recognized previously.

INTRODUCTION

When populations become small, genetic drift can lead to a reduction in genetic variation that can be measured in several different ways. As alleles are lost, more loci become monomorphic, and unequal frequencies of alleles remaining at polymorphic loci lead to lower values of expected heterozygosity. The theory behind these predictions for the effects of genetic drift in small populations has been well developed, based on Wright's (1931) work, particularly for the decline in the percent of loci that are polymorphic (P), the number of alleles per locus (A), or per polymorphic locus (AP), and the proportion of individuals or loci that would be expected to be heterozygous (H_e), though these measures of variation may only be affected when populations number less than 100 or so. Also, isolation of remaining individuals can lead to reduced levels of observed heterozygosity (H_o) and increased levels of inbreeding (measured by Wright's F_{IS}), particularly in self-compatible species, whereas decreased migration among populations can lead to greater variance of allele frequencies (Wright's F_{ST} or other indices) and declines in measures of gene flow, the product of population size and migration rate (Nm). Predicting the effects of population size on these latter four measures of variation and differentiation is less certain, especially in plant populations because they are more strongly complicated by factors such as self-incompatiblity, breeding system, and methods of dispersal. H_o, for example, might be stable because self-incompatibility loci, dioecy, optimal outcrossing (sensu Price & Waser 1979) or other breeding system traits prevent selfing or mating among close relatives. The literature predicting and analyzing the effects of these factors on genetic variation and differentiation in plant populations is particularly rich and is not reiterated here; introductions are available in Adams et al. (1990), Ellstrand & Elam (1993), and Falk & Holsinger (1991).

Surveying some of the earlier literature available, Karron (1987, 1991) compared P and AP levels at isozyme loci in 11 genera, finding species with restricted distributions to have lower levels of both measures than congeneric species with widespread distributions. Meta-analyses such as this are particularly important, not only because the specific biology and history of any individual species could overwhelm the effects of drift expected from general theory, but also because an individual study can only assay a tiny and potentially unrepresentative portion of a species' genome; P and A are particularly sensitive to small sample sizes (Nei 1987). More recently, Gitzendanner & Soltis (2000) extended this kind of analysis to more genera (34) and found that H_o was also lower for rare species than for their common congeners, although this relationship did not hold for H_e. Karron's review was limited to reports wherein all the studies reported for a genus were conducted by researchers working in the same lab, and Gitzendanner & Soltis's review was further limited to cases where data on rare and common species were included in the same report. Hamrick & Godt (1989) included a larger number of species and aggregated data by range, from endemic to widespread, as well as examining the roles of other biological factors in determining the levels of genetic variation and differentiation. While this approach is not as powerful for the specific comparison of rare and common species as the paired comparisons used by Karron and by Gitzendanner & Soltis, it did allow the inclusion of a larger number of species. They found that H_e, measured either as a mean across sampled populations or for the species as a whole, was significantly lower for endemic than for widespread species, as were P and A.

The purpose of the present study is to summarize the empirical evidence available comparing genetic variation in rare plant species and closely related common species, extending the approach introduced by Karron to the much more extensive set of data that have become available, as well as to more measures of population variation and differentiation. Specifically, compared to their common congeners, we might expect rare species to have reduced levels of P, A, AP, and H_e, both for the mean values across sampled populations and for the species as a whole. Because of reduced pollinator activity and increased selfing, we might also expect declines in H_o, increases in population differentiation as measured by F_{ST}, and reductions in measures of gene flow (Nm). Also, with declines in H_o we might also see an increase in F_{IS}, though this may be the same for rare and common species since it is most strongly affected by the breeding system, which tends to be consistent within a genus.

MATERIALS AND METHODS

Selection of Taxa

To avoid obvious founder effects that would occur in small populations of introduced species, this review deals with native populations, except in the case of *Aeschyomene indica* (Carulli & Fairbrothers 1988), and *Trifolium hybridum* and *T. pratense* (Hickey et al. 1991), which are common, introduced congeners of rare natives (thus any bias would be conservative). Although "rarity" can take on different meanings (Kruckberg & Rabinowitz 1985, Rabinowitz 1981), I have relied on authors' description of species as "rare" or "endemic" and having limited distributions, but did not include species that are "restricted" but have large populations; with over five million individuals in its small native range, *Pinus radiata* is the most clear example of this (Moran et al. 1988), but other taxa include species of *Eucalyptus* with restricted ranges but large population sizes (e.g., *E. paliformis, E. parvifolia*; Moran 1992). I also limit this review to studies that analyzed eight or more isozyme loci. Studies that only assayed enzymes previously known to be polymorphic were also excluded because that a priori information would bias most of the measures of genetic variation included here. For most genera, the rare and widespread species have been studied by the same people, with at least one author working on both rare and common taxa. This helps consistency because the information comes from studies using similar sampling designs and enzyme assays. Most exceptions are tree genera, most of them economically important, such as *Abies*, *Acacia*, *Eucalyptus*, *Picea*, *Pinus*, and *Salix*.

Because combining data from different subspecies or varieties would exaggerate values of H_{es}, F_{ST}, etc., when data are available for more than one subspecific taxon, only the taxon with the greatest amount of information was used. In virtually all cases this was also the taxon with the widest range, providing a conservative bias in the cases of rare varieties. In some cases, the common species occur across vastly greater ranges than the rare congener(s), and the populations sampled were chosen to represent that greater geographical range. Because this has the potential for biasing measures of genetic variation and structure, when the authors reported data for subsets of the common species, I chose the subset whose range most closely matched that of the rare congener. This also represents a conservative bias in data selection, but was only possible for a few species (*Acacia anomala*, *Achillea millefolium* subsp. *megacephala*, *Deschampsia caespitosa*, *Stellaria longipes*, *Daviesia mimosoides*, and *Erythronium montanum*).

Measures of Genetic Variation and Structure

Several measures of genetic variation and structure are commonly used, but their application has varied among the studies included in this review, as have the symbols used to represent them, so I here define exactly the measures used in this analysis.

PERCENT POLYMORPHIC LOCI (P) This is the percent of sampled loci that are polymorphic. Three criteria are commonly used for scoring a locus as polymorphic: (*a*) counting all loci that have more than one allele at a locus, (*b*) counting all loci where the most common allele has a frequency of less than 99%, or (*c*) counting all loci where the most common allele has a frequency of less than 95%; these criteria are designated herein as P_{100}, P_{99}, and P_{95}, respectively. However, comparing P_{95} values obscures the loss of rare alleles, which is expected to be one of the most immediate results of reduced population size. Consequently, rather than using summary P statistics reported in the original papers, the data analyzed here are taken from allele frequency tables (either published or supplied by the authors) and represent the P_{99} or P_{100} values; for most of the studies summarized here the sample sizes are small enough (i.e., fewer than 50 individuals per population) that the presence of a single copy of an allele represents polymorphism at either the P_{99} or P_{100} criterion. When a study includes multiple populations, the P value can be interpreted in two ways: either as the mean of the values obtained from each population, or when all loci are considered for the species taken as a whole. These two measures are designated here as P_p and P_s, respectively. The same P level is used for all taxa within a genus (with a single exception, where a P_{95} value is used for a common species). In the case of a few species for which P_s was not available, I used the highest value of P reported for the populations studied.

ALLELES PER LOCUS (A) As for P, for each species these measures can be defined either as a total for all of the populations, A_s, or as means for the populations studied, A_p. These indices of variation can be measured either for all loci assayed or just for those loci that are polymorphic, in which case they are designated AP_s and AP_p. In cases where allele frequency data were not available but A and P values (either P_{100} or P_{99}) were provided for each population, I calculated AP values for each population as $AP = (A + P - 1)/P$, with P taken as a proportion, and then took the mean of these population values to calculate AP_p.

MEAN HETEROZYGOSITY (H_{ep}) Several names have been used for this statistic, including "mean gene diversity," "polymorphic index," and "mean expected heterozygosity." This is the proportion of heterozygotes that would be expected under Hardy-Weinberg equilibrium, averaged across populations, where expected heterozygosity H_e is defined as $H_e = 1 - \Sigma p_i^2$, and p_i is the frequency of the ith allele at a locus.

TOTAL HETEROZYGOSITY (H_{es}) This has also received several names in the literature, including "gene diversity," "total genetic diversity," and "mean total limiting variance of allele frequencies." It is defined as $H_{es} = 1 - \Sigma \bar{p}_i^2$, where $\bar{p}_i$ is the frequency of the ith allele, averaged across all sampled populations.

OBSERVED HETEROZYGOSITY (H_o) This is the mean, across populations, of the observed proportion of heterozygotes.

For consistency, the H statistics used in this review are taken across all loci sampled, although this usage varies in the literature, since some authors use H_{ep} and H_{es} measured across all loci, while others measure them only across polymorphic loci. In some cases, where H_{es} or H_{ep} were not reported or were calculated only across polymorphic loci, I estimated one or the other from the equation $H_{es} = H_{ep}/(1 - F_{ST})$. This formula, while correct for parametric values at a single locus, can become skewed when sample sizes available for different loci are greatly different. However, it is more accurate to have this estimate across all loci than to have H_{ep} or H_{es} based on all loci in some taxa and only polymorphic loci in others.

INBREEDING COEFFICIENT (F_{IS}) This is the mean, across loci and populations, of the quantity $F_{IS} = (H_e - H_o)/H_e$.

POPULATION DIFFERENTIATION (F_{ST}, G_{ST}) Although most reports measured differentiation of allozyme frequencies among populations in terms of Wright's F_{ST} (1951), some used Nei's G_{ST}(1987). In practice, though, when these are estimated from genotype data and calculated as weighted averages from different loci, the measures are the same, although the values of these statistics can differ when they are calculated by different algorithms. Most of the values reported here were calculated by Swofford & Selander's (1981) BIOSYS-1 program, although some authors have used other programs. Readers interested in a more thorough discussion of how these measures can differ may want to consult Black & Krafsur (1985), Chakraborty & Danker-Hopfe (1991), Culley et al. (2002), Slatkin (1985a), and Weir (1996). In some cases, original papers compared a few, close populations of a rare species with more populations of a common congener that were widely separated, sometimes by hundreds of kilometers. To avoid this bias when possible, if papers reported data for both rare and common species and identified a set of populations of the common species whose range was similar to that of the rare species, I used that set for calculation of the F_{ST} statistics.

GENE FLOW (Nm) The number of migrants per generation (the product of the population size, N, and the migration rate, m) can be estimated in two ways. Wright (1951) noted that this product is inversely related to the level of population differentiation, measured by F_{ST}. Crow & Aoki (1984) have refined this estimate to reduce the bias resulting from sampling a limited number of populations, as $Nm = (1/4\alpha)[(1/F_{ST}) - 1]$, where $\alpha = [n/(n-1)]^2$ and n is the number of populations. Whereas most of the published estimates of gene flow use Wright's original formula, for the comparisons here I have recalculated Nm (designated Nm-W) from F_{ST} using the Crow & Aoki correction factor (computing Nm separately from the F_{ST} value for each locus, and then taking the mean of those values, can produce a different result; though this latter method has been used by some authors, the values used here provide more consistency for comparisons between taxa). Slatkin (1985a, 1993) and Whitlock & McCauley (1999) provide useful summaries of the relationship between F_{ST} and gene flow, including the pitfalls that can arise from neglecting some of the underlying assumptions. Gene flow (designated N*m*-S) can also be estimated from the mean frequency of private alleles (those found in a single population; Slatkin 1985a). I used published values of N*m*-S if they were included in the original articles; otherwise (the majority of cases), N*m*-S was calculated from private allele frequency data and corrected for sample size as suggested by Slatkin & Barton (1989) using the estimator $\log \bar{p}_{(1)} = a \log (\mathrm{N}m) + b$, where the parameters a and b take values that depend on the sample size.

MONOMORPHIC SPECIES Because genetic variation is strongly influenced by population size and breeding system, and isozyme monomorphism represents the limit in loss of genetic variation, I have summarized breeding system and geographic range information for species that are monomorphic for isozymes.

Several of the *P*, *A*, *H*, and *F* statistics summarized here were not reported in the authors' original publications, but have been calculated from published data either by inspection (*P*, *A*, *AP*) or by using BIOSYS-1 because this program was used most widely in the sources cited. In some cases these data were available in dissertations or by request from the authors. BIOSYS-1 was used to analyze data for *Clarkia amoena*, *C. biloba*, *C. franciscana*, *C. lingulata*, *C. rubicunda*, *Erythronium albidum*, *E. propullens*, *Eucalyptus caesia*, *Gaura demareei*, *G. longiflora*, *Layia discoidea*, *L. glandulosa*, and *Lisianthus skinneri*.

The data available for different species within a genus are based on assays of different numbers of populations, loci, and individuals, as indicated in Table 1. The statistics reported here are weighted by the number of loci sampled; those that are population averages (P_p, A_p, AP_p, H_{ep}, H_o, and F_{IS}) are weighted by the number of populations; and H_o, because it is a quantity measured on individuals, is also weighted by the number of individuals in each sample. However, sample size information was not available for some species in *Acacia*, *Antennaria*, *Eucalyptus*, or *Scutellaria*, so H_o values for those genera are not weighted by number of individuals. In the case of *Pinus*, these data were not available for *P. banksiana* or *P. palustris*, so these species are dropped from the H_o values; sample size data are available for the other *Pinus* species but differ greatly and so should not be omitted from the weighting.

Comparisons of data from rare and common species are illustrated in Figures 1 and 2, and the results were tested for significance using the nonparametric one-tailed Wilcoxon signed-rank test. The various measures of genetic variability all arise from the same ultimate source (isozyme allele frequencies), and thus may be very similar in the information they convey about variation and differentiation of populations, so correlations among the 13 measures of variation and differentiation were calculated and subjected to a principal components analysis of the correlation matrix of the measures. Statistical tests were conducted and figures produced using the S-Plus statistical package.

TABLE 1 Genera analyzed, symbols used in Figures 1 and 2, and number of rare or common species included for each genus

Genus	Symbol	Rare or com.	No. of spp.	Genus	Symbol	Rare or com.	No. of spp.	Genus	Symbol	Rare or com.	No. of spp.
Abies	Abi	R	5	*Delphinium*	Del	R	1	*Myrica*	Myr	R	1
		C	6			C	1			C	1
Acacia	Aca	R	1	*Deschampsia*	Des	R	1	*Oenothera*	Oen	R	2
		C	4			C	1			C	1
Achillea	Ach	R	1	*Echinacea*	Ech	R	1	*Phebalium*	Phe	R	1
		C	1			C	1			C	2
Adenophora	Ade	R	1	*Erythronium*	Ery	R	1	*Picea*	Pic	R	1
		C	1			C	2			C	5
Adenophorus	Ads	R	1	*Eucalyptus*	Euc	R	8	*Pinus*	Pin	R	5
		C	1			C	9			C	25
Aeschyomene	Aes	R	1	*Galvezia*	Gal	R	1	*Polygonella*	Pog	R	5
		C	1			C	4			C	2
Aletes	Ale	R	1	*Gaura*	Gau	R	1	*Polystichum*	Pol	R	1
		C	1			C	1			C	5
Antennaria	Ant	R	1	*Harperocallis*	HT	R	1	*Rhus*	Rhu	R	1
		C	12	Tofieldia		C	1			C	2
Asclepias	Asc	R	1	*Hemerocallis*	Hem	R	1	*Salix*	Sal	R	4
		C	2			C	2			C	5
Aster	Ast	R	3	*Impatiens*	Imp	R	1	*Sarracenia*	Sar	R	2
		C	3			C	1			C	1
Astragalus	Asg	R	1	*Iris*	Iri	R	1	*Scutellaria*	Scu	R	1
		C	2			C	1			C	6
Austromyrtus	Aus	R	2	*Lasthenia*	Las	R	1	*Solanum*	Sol	R	1
		C	1			C	1			C	2
Capsicum	Cap	R	1	*Layia*	Lay	R	4	*Stellaria*	Stl	R	1
		C	3			C	2			C	1
Centaurea	Cen	R	1	*Lespedeza*	Les	R	1	*Stephanomeria*	Ste	R	1
		C	1			C	1			C	1
Cirsium	Cir	R	1	*Limnanthes*	Lim	R	1	*Styrax*	Sty	R	2
		C	1			C	3			C	2
Clarkia	Cla	R	2	*Lindera*	Lin	R	1	*Tricyrtis*	Tri	R	1
		C	4			C	1			C	1
Coreopsis	Cor	R	1	*Lisianthus*	Lis	R	4	*Trifolium*	Trf	R	2
		C	3			C	1			C	2
Cypripedium	Cyp	R	1	*Lomatium*	Lom	R	3	*Widdringtonia*	Wid	R	1
		C	1			C	3			C	1
Daviesia	Dav	R	1	*Lupinus*	Lup	R	1	*Wyethia*	Wye	R	1
		C	1			C	1			C	1

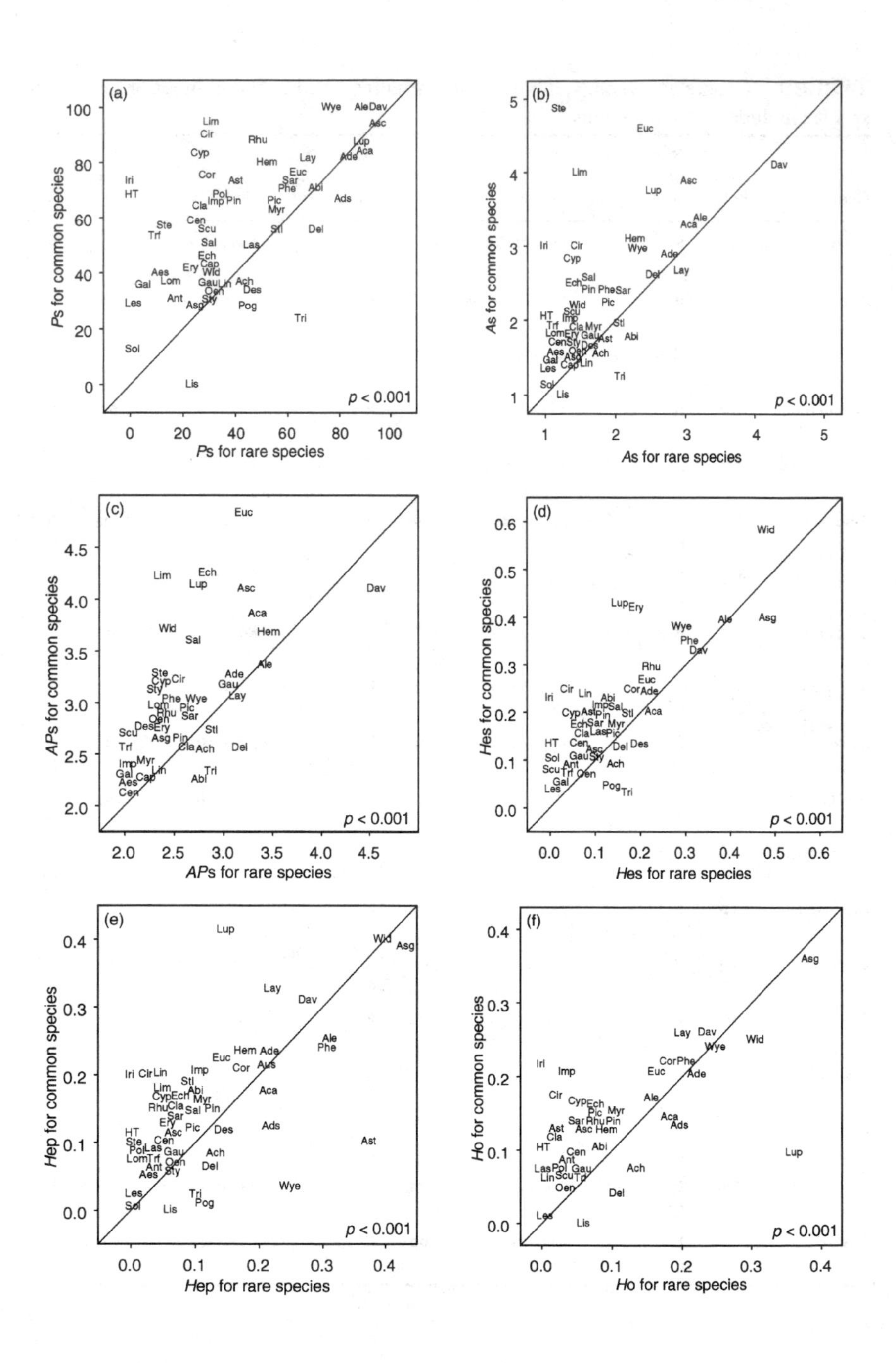
(a)
Ps for common species
Ps for rare species
0
20
40
60
80
100
p < 0.001
(b)
As for common species
As for rare species
1
2
3
4
5
p < 0.001
(c)
APs for common species
APs for rare species
2.0
2.5
3.0
3.5
4.0
4.5
p < 0.001
(d)
Hes for common species
Hes for rare species
0.0
0.1
0.2
0.3
0.4
0.5
0.6
p < 0.001
(e)
Hep for common species
Hep for rare species
0.0
0.1
0.2
0.3
0.4
p < 0.001
(f)
Ho for common species
Ho for rare species
0.0
0.1
0.2
0.3
0.4
p < 0.001

RESULTS

This review summarizes the results from 144 reports for 95 rare taxa and 152 corresponding common taxa. The complete table of the species-specific data on which these results are based is available online, as is the summary table of genus-level values illustrated in Figures 1 and 2; see the Supplemental Material link in the online version of this chapter, or at http://www.annualreviews.org/). On average, 17.8 isozyme loci were analyzed for each species. Data from these 247 species are summarized as 57 generic-level comparisons listed in Table 1 (*Harperocallis* and *Tolfeldia* are compared as rare and common taxa, since *H. flava* and *T. racemosa* are close, confamilial relatives; also, *Achillea millefolium* subspp. *megacephala* and subsp. *lanulosa* are compared as rare and common taxa). The statistical tests and Figures 1 and 2 are based on subsets of these data (as indicated in Table 2) because some reports did not include all of the statistics reported here or the data necessary to calculate them. Figure 1 shows species-level comparisons for P, A, and AP, since the population mean values for these statistics were strongly correlated with the species-wide values; a complete set of graphs for all of the statistics is available on the Supplemental Material link: http://www.annualreviews.org/supmat/supmat.asp. Table 2 records the number of comparisons available for each statistic, the percent that agreed with the theoretical predictions, mean values of each statistic for rare and common species, and the probability measures for each of the Wilcoxon signed-rank tests.

All species-level measures of genetic variation (P_s, A_s, AP_s, and H_s) showed very highly significant decreases in rare species ($p < 0.001$). The percent of loci that are polymorphic in a species (P_s) was lower in 83% of the 56 genera available for comparisons. Fewer alleles were found at all loci (83.7% of 49 A_s values were lower in rare species) and at those loci that remained polymorphic: 76.1% of 44 generic comparisons of AP_s values were lower in rare species. Species-level expected heterozygosity (H_{es}) was also lower in rare species: Of the 46 generic comparisons available, 84.8% showed a decline in rare species.

All of the population-level measures of genetic variation (P_p, A_p, AP_p, H_{ep}, and H_o) also showed very highly significant decreases in rare species ($p < 0.001$). The

---

Figure 1 Levels of genetic (isozyme) variation in rare and common plant species. Abbreviations of genera are listed in Table 1. Some point labels have been moved slightly for legibility. Significance levels are listed in the lower-right corner of each graph. Subscript "s" indicates species-wide values value, "p" indicates mean of population values. (*a*) P, percent polymorphic loci; (*b*) A, alleles per locus; (*c*) AP, alleles per polymorphic locus; (*d,e*) H_e, expected heterozygosity; (*f*) H_o, observed heterozygosity. Results for P_p, A_p, and AP_p (not shown) were highly correlated with those for the corresponding species-wide values. Full-size figures are available in the Supplemental Materials online. (Follow the Supplemental Material link from the Annual Reviews home page at http://www.annualreviews.org.)

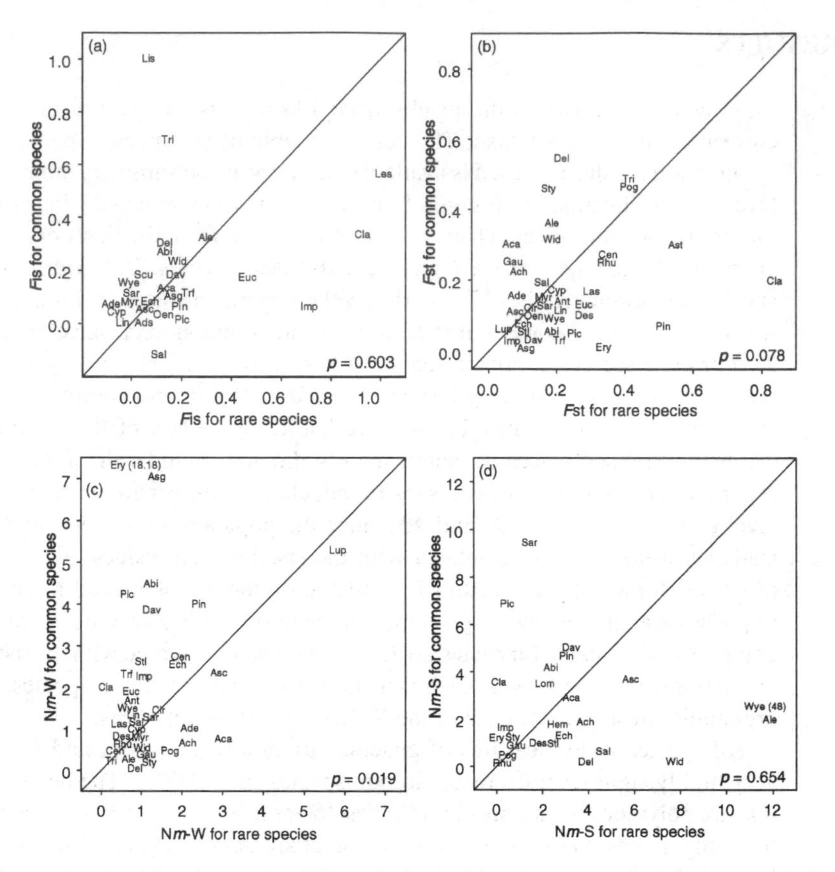

Figure 2 Levels of genetic differentiation; abbreviations as in Figure 1. (*a*) F_{IS}, inbreeding coefficient; (*b*) F_{ST}, population differentiation; (*c*) N*m*-W, Wright's estimate of gene flow based on F_{ST}; (*d*) N*m*-S, Slatkin's estimate of gene flow based on private alleles.

average percent of polymorphic loci in populations (P_p) of rare species was lower in 79.2% of the 48 comparisons available; mean values were 27.6% for rare species and 44.1% for the common species. Both the mean number of alleles per locus (A_p) and the mean number of alleles per polymorphic locus (AP_p) were lower in rare species: 84% of the 50 A_p comparisons and 73.2% of the 43 AP_p comparisons were lower in rare species. Population-level measures of both expected and observed heterozygosity were also lower: 74.1% of the 54 comparisons of H_{ep} available were lower in rare species, and 77.4% of the 41 comparisons of H_o available were lower.

The average value of the inbreeding statistic (F_{IS}), which measures the deviation between the expected and observed levels of heterozygosity within populations, was slightly lower for rare species (0.175) than for common species (0.184),

TABLE 2 Measures of isozyme variation and differentiation from comparisons of rare and common congeneric species. Probability values from Wilcoxon one-tailed signed-rank tests. Abbreviations defined in the text

Statistic	Number of comparisons	Percent of comparisons lower in rare species	Mean for rare species	Mean for common species	*p* value
P_p	48	79.2	27.6	44.1	<0.001
P_s	56	83.0	40.7	58.8	<0.001
A_p	50	84.0	1.42	1.72	<0.001
A_s	49	83.7	1.74	2.34	<0.001
AP_p	43	73.2	2.30	2.49	<0.001
AP_s	44	76.1	2.56	3.01	<0.001
H_{ep}	54	74.1	0.113	0.150	<0.001
H_{es}	46	84.8	0.142	0.199	<0.001
H_o	41	77.4	0.100	0.139	<0.001
F_{is}	28	60.7	0.175	0.184	0.603
F_{st}	38	34.2	0.212	0.198	0.078
Nm-W	37	64.9	1.19	2.24	0.019
Nm-S	25	52.0	4.74	2.09	0.654

but this was not significantly different ($p = 0.603$). For this statistic, 28 comparisons were available, and rare species had lower values in just over half of them (60.7%).

Unlike the measures of genetic variation, the statistics reflecting population differentiation and gene flow showed less difference between rare and common species. Of the 38 comparisons available for F_{ST}, expected to be larger in rare than in common species, only 25 (65.8%) had higher values in the rare species, a difference that was not statistically significant. The average values were nearly identical: 0.212 in rare species and 0.198 in common species. Slatkin's measure of gene flow derived from the frequencies of private alleles, N*m*-S, was lower in 13 of the 25 comparisons available (52.0%), a difference that was not significant. For this estimate of gene flow, rare species had a higher value (4.89) than did common species (2.1). In contrast, Wright's estimate of gene flow based on F_{ST}, N*m*-W, was significantly lower for rare species ($p = 0.019$). This measure of gene flow was lower for rare species in 24 of the 37 comparisons available, with an average value (1.22) was just over half that for the common species (2.35).

Forty-four species found to be monomorphic for isozymes are listed in Table 3, along with their breeding systems and ranges. As illustrated in Figure 3, most of these are rare, self-compatible species: Only five are outcrossers, 36 have some

TABLE 3 Breeding system and ranges for forty-four plant species found to be monomorphic for isozymes ("self" = partial or complete self-fertilization). Numbers indicate sources cited; asterisks indicate systematic studies using small sample sizes

Endemic, Self		**Endemic, Outcross**	
Asteraceae		Asteraceae	
Lactoris fernandeziana	1, 2	*Dendroseris pruinata*	6
Senecio flavus	3	Saxifragaceae	
S. mojavensis	3	*Chrysosplenium iowense*	29
Taraxacum obliquum	4	Scrophulariaceae	
Campanulaceae		*Pedicularis furbishae*	30
Wahlenbergia berteroi	5	*P. dasyantha*	31
W. masafuerae	5		
Chenopodiaceae			
Chenopodium crusoeanum	6		
Ch. sancta-clarae	6		
Fabaceae			
*Cicer bijugum**	7		
*C. chorassanicum**	7		
*C. cuneatum**	7		
*C. echinospermum**	7		
*C. judaicum**	7		
*C. yamashitae**	7		
Trifolium reflexum	8		
Vicia dumetorium	9		
Iridaceae			
Iris lacustris	10, 11		
Lacandoniaceae			
Lacandonia schismatica	12		
Onagraceae			
Oenothera hookeri	13		
Ophioglossaceae			
Ophioglossum vulgatum	14		
Saxifragaceae			
Bensoniella oregona	15		
Sullivantia oregana	16		
Solanaceae			
Solanum fernandezianum	6, 17		
Endemic, Breeding System Unknown			
Asteraceae			
Dendroseris bertoroana	6		
Poaceae			
Glyceria nubigena	18		
Polemoniaceae			
Polemonium occidentale v. *lacustre*	19		

(*Continued*)

TABLE 3 *(Continued)*

Narrow, Self		**Narrow, Outcross**	
Campanulaceae		Alismatidae	
Howelia aquatica	20	*Amphibolis antarctica*	32
Fabaceae			
Lespedeza leptostachya	21		
Broad, Self			
Asteraceae			
Tragopogon praetensis	22		
Cupressaceae			
Thuja plicata	23		
Cyperaceae			
Carex arctogena	24		
Orchidaceae			
Epiactis phyllanthes	25		
E. purpurata	25		
Solanaceae			
*Solanum angustifolium**	26		
*S. davisense**	26		
*S. fructo-tectum**	26		
*S. lumholtzianum**	26		
*S. rostratum**	26		
Typhaceae			
Typha domingensis	27, 28		

1. Bernardello et al. 1999; 2. Crawford et al. 1994; 3. Liston et al. 1989; 4. Van Oostrum et al. 1985; 5. Crawford et al. 1990; 6. Crawford et al. 2001; 7. Kazan & Muehlbauer 1991; 8. Hickey et al. 1991; 9. Black-Samuelsson & Lascoux 1999; 10. Hannan & Orick 2000; 11. Simonich & Morgan 1994; 12. Coello et al. 1993; 13. Levy & Levin 1975; 14. McMaster 1994; 15. Soltis et al. 1992; 16. Soltis 1982; 17. Spooner et al. 1992; 18. Godt & Hamrick 1995; 19. Cole 1998; 20. Lesica et al. 1988; 21. Cole & Biesboer 1992; 22. Roose & Gottlieb 1976; 23. Copes 1981; 24. Reinhammer 1999; 25. Ehlers & Pedersen 2000; 26. Whalen 1979; 27. Mashburn et al. 1978; 28. Sharitz et al. 1980; 29. Schwartz 1985; 30. Waller et al. 1987; 31. Odasz & Savolainen 1996; 32. van Treuren et al. 1991.

level of self-fertilization (data are not available for three species), and all but 11 species have ranges listed as endemic or narrow.

Many of the measures of variation were strongly correlated. Although correlation and principal components analyses were conducted on data sets for the common species alone, the rare species alone, and the combined set, the correlation and PCA results were generally similar for all three; consequently, only the results for the combined data set are discussed here. Only a few of the measures bear mention here, while the full 13×13 correlation matrix is available online. Population mean and species total measures were highly correlated ($P_p/P_s = 0.84$, $A_p/A_s = 0.84$, $AP_p/AP_s = 0.74$). For this reason, only the species' total measures are illustrated in Figure 1. The measures of heterozygosity were not quite as highly correlated ($H_{ep}/H_{es} = 0.68$, $H_{ep}/H_o = 0.61$, and $H_{es}/H_o = 0.75$); accordingly, all

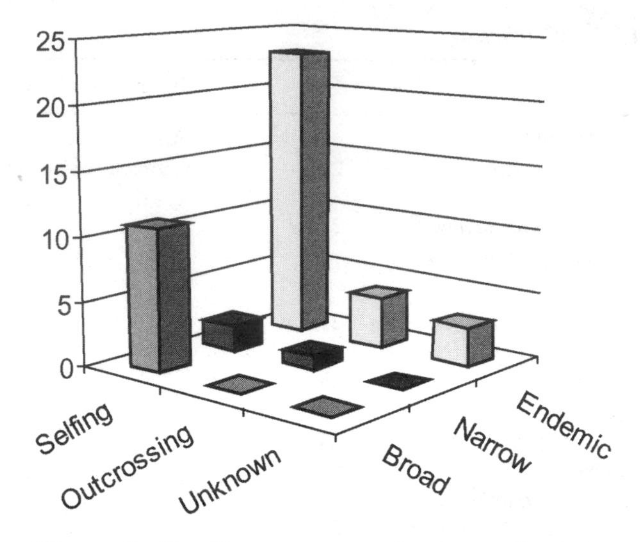

Figure 3 Breeding systems and ranges of species monomorphic for isozymes, listed in Table 3. Vertical axis indicates the number of species in each category.

three of these measures of variation are shown in Figure 1. The strongest correlation was between A_p and P_p, which correlated at the 0.93 level. Both F_{IS} and F_{ST} were negatively correlated with the measures of variation, and with both measures of gene flow. Of these, the strongest correlation was between F_{ST} and Nm-W, which had a correlation coefficient of −0.44. The correlation coefficient between the two fixation indices was 0.52. However, the two measures of gene flow were not correlated, with a value of only −0.07.

The principal components analysis revealed a relatively low dimensionality within the combined data set. Only the first four eigenvalues were greater than one, and together they accounted for 81% of the variation among the measures. The first eigenvalue alone accounted for 46% of the variation. The complete table of correlation coefficients and eigenvectors is available on the Supplemental Materials link online.

DISCUSSION

Although the reports summarized here are mostly from North America, Europe, and Australia, they represent a wide range of taxa, including ferns, gymnosperms, and angiosperms, having a wide variety of breeding systems, pollination, and seed dispersal syndromes, etc., so we may expect the general results to hold in future isozyme studies.

Despite the conservative criteria used for including species in this analysis, data are available from a large number of reports (144) and allow robust comparisons for a large number of rare species (95) and their common congeners (152), providing 57 comparisons of rare and common taxa, based on an average of 17.8 loci per species. Virtually all of these compare rare and common species within a genus, with only two exceptions: *Harperocallis flava* and *Tolfieldia racemosa*, which are each other's closest relatives sampled, and *Achillea millefolium* subspp. *megacephala* and *lanulosa*, which are rare and common conspecific taxa, respectively.

Genetic Variation

From the 57 genera included in this report, all of the measures of genetic variability showed reductions in rare species, compared to their common congeners: The percent of polymorphic loci (*P*), number of alleles per locus (*A*), expected and observed heterozygosity (H_e, H_o), and number of alleles per polymorphic locus (*AP*) was reduced in over 75% of the genera. All of these reductions were very highly significant ($p < 0.001$). As shown in Table 2, P_s and P_p values were roughly one half higher in common species than in rare ones; A_s, H_{es}, H_{ep}, and H_o values were all more than one third higher in common species, and A_p values were higher by one fifth. *AP* values were also higher in common than in rare species, though not as dramatically.

Theoretical work by Wright (1931) and Kimura (1955) on the effects of genetic drift suggests that fluctuating allele frequencies in small populations will result in a decline of expected heterozygosity (also referred to as "gene diversity" by Nei 1987). Subsequent workers (e.g., Allendorf 1986, Denniston 1977, Fuerst & Maruyama 1986, Maruyama & Fuerst 1985) have pointed out that drift should have more immediate effects on the loss of rare alleles, lowering *A* and *P*, than on expected heterozygosity, which can retain high levels even with only two alleles segregating in a population if their frequencies are equal. These theoretical expectations have been strengthened by experiments with model systems, such as Leberg's (1992) work with experimental populations of mosquito fish. However, this sensitivity of *P* and *A* to the stochastic effects of drift also makes their measurement sensitive to error in sampling from natural populations. Nei (1987) points out that the *A* statistic is particularly dependent on sample size because of its sensitivity to the low-frequency alleles in populations, and that *P* is similarly subject to large sampling errors when the number of sampled loci is small. Because the typical research project assays perhaps 0.1% of a species' loci, the results from any one study could be suspect, but the extremely strong pattern shown in the entire set of species studied here indicates that genetic variation is frequently lost in rare species.

These results for *P* and *AP* confirm Karron's earlier summary (1987, 1991) of work comparing genetic variation in rare and common species from 11 genera, showing reduced variation in species with restricted distributions, and Soltis & Soltis's (1991) comparison of *P* and *A* for 14 rare and common species of the Saxifragaceae. They also corroborate Hamrick & Godt's (1989) finding that *P*,

A, and H, measured both as species totals and as means for populations, are significantly reduced in endemic species, compared to widespread (but not necessarily congeneric) species. The present study also confirms some of the results of Gitzendanner & Soltis (2000), whose analysis reviewed 36 genera, finding reductions in P_p, P_s, A_p, and A_s. The results in the present study, drawn from a larger number of genera (57), demonstrate even more significant reductions in genetic variation. Moreover, those authors found no significant reductions in H_{es}, which the present study found to have very highly significant reductions in the rare species. The difference in these results arises from the larger number of species (247 versus about 110) and genera (46 versus 18) available for comparison in the present study, and perhaps from the inclusion of all loci tested, rather than just polymorphic loci, in the measurement of H_{es}.

Studies relating genetic variation to population size within plant species have been reviewed by Ellstrand & Elam (1993) and by Frankham (1996), summarizing data on 10 and 16 species, respectively. Both reviews found a stronger relationship between population size and P than A, and a weaker relationship with H_e. Such a pattern is not universal, however; for example, Ledig et al. (1997) found that while A and H_e were significantly correlated with population size in *Picea chihuahuaensis*, P was not.

Theoretical predictions of the effects of rarity on observed heterozygosity (H_o) are not as clear-cut as they are for the other measures of genetic variation. Mechanisms preventing self-fertilization might protect H_o even when population sizes get small, and Robertson (1965) has even predicted that small populations of sexual species would see an increase in the proportion of heterozygotes, though this prediction would apply only to dioecious species. Alternatively, the loss of pollinator or seed dispersal services in very small populations could lead to higher rates of selfing and mating among close relatives. For example, H_o levels are significantly lower in small populations of *Gentiana pneumonanthae* (Raijmann et al. 1994) and *Eucalyptus albens* (Prober & Brown 1994). The results of the present study, which found that rare species had reduced H_o in 32 of 42 genera studied, indicate that Allee (1949)-type effects such as those documented by Lamont et al. (1993) may commonly have genetic consequences for rare species, and confirm Gitzendanner & Soltis's analysis of a smaller number (24) of genera.

Wright's F_{IS} is affected primarily by the level of inbreeding arising through selfing and mating of close relatives. Because these traits tend to be conserved within genera, we might expect that there would be little difference between the values of F_{IS} we observe in rare and common congeners. In contrast, if we found that rare species have a much higher level of F_{IS} than their common congeners, we might conclude that breeding system differences, rather than small population size, were responsible for the loss of genetic diversity. Of the 28 genera for which comparisons could be made, 11 rare species had higher and 17 had lower levels of F_{IS}, a difference that is not significant. These genetic results corroborate Weller's (1994) conclusion that breeding systems have little effect in determining plant rarity.

Differentiation and Gene Flow

When a species is divided into a series of small populations so isolated from each other that dispersal rarely carries genes from one population to another, genetic drift can lead to divergent allele frequencies in the isolated populations, increasing levels of Wright's F_{ST} statistic. The levels of differentiation among populations of rare and common species were higher in 25 of the 38 genera (65.8%) for which comparisons could be made, a difference that is in the expected direction but is not significant ($p = 0.078$). This result agrees with that of Gitzendanner & Soltis' (2000) analysis, which found even less difference among smaller number of genera (22), and with Hamrick & Godt's (1989) review that found no significant difference among widespread, regional, narrowly distributed, or endemic species. However, all of these analyses are hampered by the sampling design of the original reports on which they are based. While most studies compare similar numbers of loci, individuals, and populations for rare and common species, the geographic dispersion of the samples also usually matches their ranges, whether restricted or widespread. As a result, sampled populations of common species are often much more widely separated than those of rare species, presenting much greater barriers to gene flow and biasing the data set. Unfortunately, there are few studies that sample rare and common species from comparably dispersed populations, and it is not clear what a priori criteria might be used to select appropriate data for comparison without biasing the conclusions.

Nor is it clear that rare plant species will necessarily have lower level of gene flow than common species. Ellstrand (1992) has summarized arguments for expecting that gene flow might increase in rare plant populations, resulting in lower levels of F_{ST}, and refers to some experiments with crop species that support that notion, though the spatial distances between test populations fall within what biologists would consider a single natural population, at least by the criteria used for the studies analyzed here. Very little work has explicitly tested the effect of population size on genetic structure in native populations, and the results also suggest that the effects of rarity on genetic differentiation may be more idiosyncratic than the effects on variation. Van Treuren et al. (1991) found that G_{ST} increased in small populations of *Salvia pratensis* and *Scabiosa columbaria*. On the other hand, two studies comparing genetic differentiation of remnant populations of *Acer saccharum* with differentiation in intact populations (Foré et al. 1992, Young et al. 1993) have shown reduced differentiation among the remnant populations, suggesting that gene flow among small populations may be higher than flow across a continuous population covering a similar area.

Although genetic differentiation among populations was not significantly different for rare and common species, there was a significant difference in levels of gene flow (N*m*) estimated from these F_{ST} values, based on Crow & Aoki's (1984) formula. For the 66 rare species for which N*m*-W values can be calculated, the mean value is $Nm = 1.19$, little over one half that found for 105 common congeners, which had a mean $Nm\text{-W} = 2.24$. The highest N*m*-W value found for any rare species is 8.65 (*Polygonella myriophylla*), whereas four common species have Nm-W values over 10.0 (*Astragalus pattersoni*, *Erythronium albidum*, *Iris cristata*, and *Pinus sylvestris*). The original report for *E. albidum* listed two different G_{ST} values, 0.01 and 0.02; the first value gives the highest level of gene flow in the data set (18.18) and the latter value gives a N*m* value of 9.0, but this does not change the results presented here for either differentiation or gene flow. For both rare and common species, these mean values are high enough that they would prevent differentiation among populations, which is expected to occur if $Nm < 0.5$ (Kimura & Ohta 1971, Slatkin 1987). Of the rare species, 20 (21%) had N*m* values below 0.5, but for common species the N*m* value was lower than 0.5 for 19 species (15.8%). The significance of this comparison was not as strong as for the other statistics ($p = 0.019$) but remains significant ($p = 0.030$) even if *E. albidum* is excluded. For estimation of N*m*, the greater geographic distance that usually occurred between sampled populations of common species than of rare species represents a conservative bias for comparisons of gene flow. Whitlock & McCauley (1999) have cautioned that other factors such as selection or mutation can affect gene frequencies, and hence influence F_{ST} and estimates of gene flow derived from it, as can errors arising from sampling small numbers of loci or individuals. While these concerns apply more to studies of individual species than to a large data set from numerous species, they underscore the caution needed when studying any particular species. These results differ from those of Ellstrand & Elam (1993), whose survey of a slightly smaller number of rare species (32) found the estimates of gene flow to be typical of common species.

Differences in gene flow estimated by Slatkin's private alleles method were not significant. Of 25 comparisons possible, 13 had lower values in the rare species (52%, $p = 0.654$). This may not be surprising, not only because fewer comparisons were available to test the difference than were available for the N*m*-W estimate, but also because this estimator of gene flow, while robust with regards to the number of populations and individuals sampled, is sensitive to the numbers of loci and private alleles available. Even in the common species, relatively few polymorphic loci had private alleles, so the estimates of $p_{(1)}$ would typically be based on about four private alleles from four polymorphic loci, rather than the 20 or so alleles at 10 polymorphic loci typical of the simulations on which the use of this estimator is based (Slatkin 1985b).

Beyond the reductions in genetic variability at these (ostensibly or nearly neutral) marker loci, can we conclude that rare species also have limited genetic variation for adaptively significant traits? And do levels of differentiation for isozyme loci reflect differences for adaptively significant traits? Neither theoretical nor empirical work allow such simple conclusions. Reed & Frankham (2001) have summarized theoretical arguments for and against a correlation between marker variation and quantitative measures of variation; their meta-analysis concluded that population-level variation in molecular markers is only weakly correlated with quantitative variation, and that there was no significant correlation between heterozygosity and life history traits or heritability. Somewhat paradoxically, though, they also found (Reed & Frankham 2003) that marker heterozygosity does correlate significantly with fitness (again as population-level measures), as do population size and quantitative variation. Nor does the differentiation measured by F_{ST} clearly reflect adaptive differences among plant populations, which often are differentiated over distances of just a few meters (Linhart & Grant, 1996). Other species (e.g., Scots pine, *Pinus sylvestris*) have strong quantitative variation across continental-scale ranges but have negligible values of F_{ST} (e.g., Karhu et al. 1996). Reviewing the correlation between F_{ST} and Q_{ST}, a comparable measure of quantitative genetic variation between populations, Lynch et al. (1999), Merilä & Crnokrak (2001), and McKay & Latta (2002) find an increasing number of studies showing higher levels of Q_{ST} than F_{ST} in various species of plants and animals. Especially, as Scots pine exemplifies, populations with low levels of F_{ST} can still have highly significant levels of quantitative differentiation among populations. They also note that the correlation between the two indices is not as strong for life history traits as for quantitative (primarily morphological) traits.

Monomorphism

Although the loss of genetic variation shown here occurs for species with different breeding systems, and rarity has no evident effects on inbreeding, the combination of small population size and self-compatibility can be especially severe for reducing genetic variation. As shown in Table 3, most (56.8%) of the 44 species found to be monomorphic for isozymes are self-compatible species with limited ranges; only 11 have broad ranges, and only five are known to be outcrossers (breeding system information is not available for three other species). This list excludes some reports of monomorphism, e.g., when only isolated, disjunct populations of widespread species have been sampled (e.g., island populations of *Pedicularis* spp.) or some agricultural or horticultural varieties. Some species are included that have limited sample sizes, such as from systematic studies that analyzed small numbers of individuals from a wide range of taxa, and it is possible that more intensive sampling might uncover variation in these species.

Correlation and Principal Components Analysis

As expected, species-wide and population mean measures of variability were highly correlated (especially when only the rare species were analyzed). Direct measures of variability (*P*, *A*, *AP*) also tended to be strongly correlated, in the range of 0.7 to 0.9 (although the correlations between *P* and *AP* values were only 0.55–0.67). Heterozygosity measures of genetic variation, however, were not strongly correlated with the direct measures. For instance, H_{ep} was only correlated with P_p at the 0.22 level, and with P_s at the 0.07 level. It is interesting, then, that observed heterozygosity had a much higher correlation with the percent polymorphic loci than did the expected heterozygosity measures ($P_s = 0.45$, $P_p = 0.67$). The consistently negative correlations between the fixation indices (F_{IS} and F_{ST}) and the measures of variability and of gene flow are to be expected. Perhaps most surprising was the lack of correlation between the two measures of gene flow: N*m*-W and N*m*-S were only correlated at the –0.07 level. However, as mentioned earlier, the number of loci and alleles available for the N*m*-S estimates were consistently below the levels for which the models were developed.

For Future Work

If small populations undergo significant losses of variation for marker genes, what happens to the variation for quantitative traits—especially for adaptive traits? Can we begin to identify those conditions (loss of self-incompatibility alleles being a singularly clear example) in which reduced genetic variation exacerbates demographic problems of rare species? Few studies have examined how fitness, or closely related demographic traits such as survival and reproduction, vary with population size or with levels of genetic variation. Several studies have documented reductions in survival or reproduction in small populations (e.g., DeMauro 1993, Lamont et al. 1993, Menges 1991, Newman & Pilson 1997), although we cannot yet compare their fitness reductions to variation for isozymes or other markers. Even in species that can be grown as explants, and in which population size does have significant genetic effects on adaptive traits, studies such as that by Paschke et al. (2003) indicate how carefully designed such studies must be, to tease out these effects from those of the environment, maternal plant, population identity, and their interactions. Hopefully, such studies may begin to clarify the factors reducing N_e/N ratios in plant populations—an area little studied in plants, in contrast to the rich theory and modicum of animal studies that have been developed.

Similarly, studies of differentiation among populations need to sample roughly equivalent spatial scales for rare and common species, but they also need to examine quantitative and adaptive traits along with markers. Do rare species have higher levels of Q_{ST} than F_{ST}, as seems to be the case for common species? They may have significant differences for quantitative traits even when there are low levels of marker variation, as McKay et al. (2001) found for physiological adaptations to microclimates of the Rocky Mountains in the rare *Arabis fecunda*, although this species has little marker variation. Only a single isozyme locus (of 14 analyzed) was polymorphic, and there was also little variation in chitinase gene sequences analyzed. On the other hand, the limited marker variation we find in rare species may more broadly reflect constraints on differentiation among populations, which is apparently the case that Petit et al. (2001) found in two Mediterranean endemics that have lower levels of differentiation for quantitative traits than for allozymes.

Summary

While not all measures of genetic variation will show reductions in all rare species, this report demonstrates that the reduction in genetic variability in rare plant species is more significant and affects more measures of variation than has previously been realized. Theoretical models (e.g., Allendorf 1986) suggest that even though populations will lose alleles when N is in the range of 10^2 or so, heterozygosity should not be lost until the species declines to a few dozen. Although much more thorough information about census sizes of the species included in this study would be desirable, most rare species exceeded this level by at least two orders of magnitude; in fact, only two species (*Austromyrtus gonoclada* and *Phebalium davesii*) have population sizes down in the range (27 and 43, respectively) where theory predicts that N should affect P, A, H_e, etc. For various reasons, the effective population size (N_e) of many of the rare species studied is probably much smaller than the census population size. In a review of 102 species (mostly animals), Frankham (1995) found that the average N_e value was only 0.11 of the census population size. Moreover, Wilcove et al. (1993) found that the average census size of plant species that are protected under the U.S. Endangered Species Act is only 120 individuals. Clearly, rare plant species lose genetic variation at population sizes far larger than the level usually protected by the ESA.

ACKNOWLEDGMENTS

Many authors generously provided additional information beyond that included in their published reports, which expanded the range of data that could be analyzed and the strength of the conclusions drawn here; this includes those authors whose comments clarified why their work would not be suitable for inclusion in the categories analyzed. I am especially grateful for thoughtful advice from Jeff Karron, Mary Jo Godt, and Brett Purdy, and for conversations with Engin Sungur and Jon Anderson, who helped me wrestle with some of the subtler aspects of the statistics and introduced me to the S-Plus software.

LITERATURE CITED

Adams WT, Strauss SH, Copes DL, Griffin AR, eds. 1990. *Population Genetics of Forest Trees*. Dordrecht: Kluwer

Allee WC. 1949. Group survival value for *Philodina roseola*, a rotifer. *Ecology* 30:395–97

Allendorf FW. 1986. Genetic drift and the loss of alleles versus heterozygosity. *Zoo Biol.* 5:181–90

Bernardello G, Anderson GJ, Lopez P, Cleland MA, Stuessy TF, Crawford DJ. 1999. Reproductive biology of *Lactoris fernandeziana* (Lactoridaceae). *Am. J. Bot.* 86:829–40

Black WC IV, Krafsur ES. 1985. A FORTRAN program for analysis of genotypic frequencies and descriptions of the breeding structure of populations. *Theor. Appl. Genet.* 70:484–90

Black-Samuelsson S, Lascoux M. 1999. Low isozyme diversity in Nordic and central European populations of *Vicia pisiformis* and *V. dumetorum* (Fabaceae). *Nord. J. Bot.* 19: 643–52

Carulli JP, Fairbrothers DE. 1988. Allozyme variation in three eastern United States species of *Aeschyomene* (Fabaceae), including the rare *A. virginica*. *Syst. Bot.* 13:559–66

Chakraborty R, Danker-Hopfe H. 1991. Analysis of population structure: a comparative study of different estimators of Wright's fixation indices. In *Handbook of Statistics, Statistical Methods in Biological and Medical Sciences*, ed. CR Rao, R Chakraborty, 8:203–54. Amsterdam: Elsevier Sci.

Coello G, Escalante A, Soberon J. 1993. Lack of genetic variation in *Lacandonia schismatica* (Lacandoniaceae: Triuridales) in its only known locality. *Ann. Mo. Bot. Gard.* 80:898–901

Cole CT. 1998. Genetic variation and population differentiation in *Polemonium occidentale* var. *lacustre*. *Rep. Wis. Dep. Nat. Resourc. No. 9507713*

Cole CT, Biesboer DD. 1992. Monomorphism, reduced gene flow, and cleistogamy in rare and common species of *Lespedeza* (Fabaceae). *Am. J. Bot.* 79:567–75

Copes DL. 1981. Isoenzyme uniformity in western red cedar seedlings from Oregon and Washington. *Can. J. For. Res.* 11:451–53

Crawford DJ, Ruiz E, Stuessy TF, Tepe E, Aqeveque P, et al. 2001. Allozyme diversity in endemic flowering plant species of the Juan Fernandez archipelago, Chile: ecological and historical factors with implications for conservation. *Am. J. Bot.* 88:2195–203

Crawford DJ, Stuessy TF, Haines DW, Cosner MB, Wiens D, Lopez P. 1994. *Lactoris fernandeziana* in the Juan Fernandez Islands: allozyme uniformity and field observations. *Conserv. Biol.* 8:277–80

Crawford DJ, Stuessy TF, Lammers TG, Silva OM, Pacheco P. 1990. Allozyme variation and evolutionary relationships among three species of *Wahlenbergia* (Campanulaceae) in the Juan Fernandez Islands. *Bot. Gaz.* 151:119–24

Crow JF, Aoki K. 1984. Group selection for a polygenic behavioral trait: estimating the degree of population subdivision. *Proc. Natl. Acad. Sci. USA* 81:6073–77

Culley TM, Wallace LE, Gengler-Nowak KM, Crawford DJ. 2002. A comparison of two methods of calculating G_{ST}, a genetic measure of population differentiation. *Am. J. Bot.* 89:460–66

DeMauro MM. 1993. Relationship of breeding system to rarity in the lakeside daisy (*Hymenoxys acaulis* var. *glabra*). *Conserv. Biol.* 7:542–50

Denniston C. 1977. Small population size and genetic diversity: implications for endangered species. In *Endangered Birds*, ed. SA Temple, pp. 281–89. Madison: Univ. Wis. Press

Ehlers BK, Pedersen HA. 2000. Genetic variation in three species of *Epiactis* (Orchidaceae): geographic scale and evolutionary inferences. *Biol. J. Linn. Soc.* 69:411–30

Ellstrand NC. 1992. Gene flow by pollen: implications for plant conservation genetics. *Oikos* 63:77–86

Ellstrand NC, Elam DR. 1993. Population genetic consequences of small population size: implications for plant conservation. *Annu. Rev. Ecol. Syst.* 24:217–42

Falk DA, Holsinger KE, eds. 1991. *Genetics and Conservation of Rare Plants*. New York: Oxford Univ. Press

Foré SA, Hickey RJ, Vankat JL, Guttman SI, Schaefer RL. 1992. Genetic structure after forest fragmentation: a landscape ecology perspective on *Acer saccharum*. *Can. J. Bot.* 70:1659–68

Frankham R. 1995. Effective population size/adult population size ratios in wildlife: a review. *Genet. Res.* 66:95–107

Frankham R. 1996. Relationship of genetic variation to population size in wildlife. *Conserv. Biol.* 6:1500–8

Fuerst PA, Maruyama T. 1986. Considerations on the conservation of alleles and of genic heterozygosity in small managed populations. *Zoo Biol.* 5:171–79

Gitzendanner MA, Soltis PS. 2000. Patterns of variation in rare and widespread plant congeners. *Am. J. Bot.* 87:783–92

Godt MJW, Hamrick JL. 1995. Allozyme variation in two Great Smoky Mountain endemics: *Glyceria nubigena* and *Rugelia nudicaulis*. *J. Hered.* 86:194–98

Hamrick JL, Godt MJW. 1989. Allozyme diversity in plant species. In *Plant Population Genetics, Breeding, and Genetic Resources*, ed. AHD Brown, MT Clegg, AL Kahler, BS Weir, pp. 43–63. Sunderland, MA: Sinauer

Hannan GL, Orick MW. 2000. Isozyme diversity in *Iris cristata* and the threatened glacial endemic *I. lacustris* (Iridaceae). *Am. J. Bot.* 87:293–301

Hickey RJ, Vincent MA, Guttman SI. 1991. Genetic variation in running buffalo clover (*Trifolium stoloniferum*, Fabaceae). *Conserv. Biol.* 5:309–16

Karhu A, Hurme P, Karjalainen M, Karvonen P, Kärkkäinen K, et al. 1996. Do molecular markers reflect patterns of differentiation in adaptive traits of conifers? *Theor. Appl. Genet.* 93:215–21

Karron JD. 1987. A comparison of levels of genetic polymorphism and self-compatibility in geographically restricted and widespread plant congeners. *Evol. Ecol.* 1:47–58

Karron JD. 1991. Patterns of genetic variation and breeding systems in rare plant species. See Falk & Holsinger 1991, pp. 87–98

Kazan K, Muehlbauer FJ. 1991. Allozyme variation and phylogeny in annual species of *Cicer* (Leguminosae). *Plant Syst. Evol.* 175:11–21

Kimura M. 1955. Stochastic processes and the distribution of gene frequencies under natural selection. *Cold Spring Harbor Symp. Quant. Biol.* 20:33–53

Kimura M, Ohta T. 1971. *Theoretical Aspects of Population Genetics*. Princeton: Princeton Univ. Press

Kruckberg AR, Rabinowitz D. 1985. Biological aspects of endemism in higher plants. *Annu. Rev. Ecol. Syst.* 16:447–79

Lamont BB, Klinkhamer PGL, Witkowksi ETF. 1993. Population fragmentation may reduce fertility to zero in *Banksia goodii*—a demonstration of the Allee effect. *Oecologia* 94:446–50

Leberg PL. 1992. Effects of population bottlenecks on genetic diversity as measured by allozyme electrophoresis. *Evolution* 46:477–94

Ledig FT, Jacob-Cervantes V, Hodgskiss PD, Eguiluz-Piedra T. 1997. Recent evolution and divergence among populations of a rare Mexican endemic, Chihuahua spruce, following Holocene climatic warming. *Evolution* 51:1815–27

Lesica P, Leary RF, Allendorf FW, Bilderback DE. 1988. Lack of genic diversity within and among populations of an endangered plant, *Howelia aquatilis*. *Conserv. Biol.* 2:275–82

Levy M, Levin DA. 1975. Genic heterozygosity and variation in permanent translocation heterozygotes of the *Oenothera biennis* complex. *Genetics* 79:493–512

Linhart YB, Grant MC. 1996. Evolutionary significance of local genetic differentiation in plants. *Annu. Rev. Ecol. Syst.* 27:237–77

Liston A, Rieseberg LH, Elias TS. 1989. Genetic similarity is high between intercontinental disjunct species of *Senecio* (Asteraceae). *Am. J. Bot.* 76:383–88

Lynch M, Pfrender M, Spitze K, Lehman N, Hicks J, et al. 1999. The quantitative and molecular genetic architecture of a subdivided species. *Evolution* 53:100–10

Maruyama T, Fuerst PA. 1985. Population bottlenecks and nonequilibrium models in population genetics. I. Number of alleles in a small population that was formed by means of a recent bottleneck. *Genetics* 111:675–89

Mashburn SJ, Sharitz RR, Smith MH. 1978. Genetic variation among *Typha* populations of the southeastern United States. *Evolution* 32:681–85

McKay JK, Bishop JG, Lin J-Z, Richards JH, Sala A, Mitchell-Olds T. 2001. Local adaptation to climate despite absence of marker diversity in the rare Sapphire Rockcress. *Proc. R. Soc. London Ser. B* 268:1715–21

McKay JK, Latta RG. 2002. Adaptive population divergence: markers QTL and traits. *Trends Ecol. Evol.* 17:285–91

McMaster RT. 1994. Ecology, reproductive biology, and population genetics of *Ophioglossum vulgatum* (Ophioglossaceae) in Massachusetts. *Rhodora* 96:259–86

Menges ES. 1991. Seed germination percentage increases with population size in a fragmented prairie species. *Conserv. Biol.* 5:158–64

Merilä J, Crnokrak P. 2001. Comparison of genetic differentiation at marker loci and quantitative traits. *J. Evol. Biol.* 14:892–903

Moran GF. 1992. Patterns of genetic diversity in Australian tree species. *New For.* 6:49–66

Moran GF, Bell JC, Eldridge KG. 1988. The genetic structure and the conservation status of the five natural populations of *Pinus radiata*. *Can. J. For. Res.* 18:506–14

Nei M. 1987. *Molecular Evolutionary Genetics*. New York: Columbia Univ. Press

Newman D, Pilson D. 1997. Increased probability of extinction due to decreased genetic effective population size: Experimental populations of *Clarkia pulchella*. *Evolution* 51: 354–62

Odasz AM, Savolainen O. 1996. Genetic variation in populations of the arctic perennial *Pedicularis dasyantha* (Scrophulariaceae), on Svalbard, Norway. *Am. J. Bot.* 83:1379–85

Paschke M, Bernasconi G, Schmid B. 2003. Population size and identity influence the reaction norm of the rare, endemic plant *Cochlearia bavarica* across a gradient of environmental stress. *Evolution* 57:496–508

Petit C, Fréville H, Mignot A, Colas B, Riba M, et al. 2001. Gene flow and local adaptation in two endemic plant species. *Biol. Conserv.* 100:21–34

Price MV, Waser NM. 1979. Pollen dispersal and optimal outcrossing in *Delphinium nelsoni*. *Nature* 277:294–97

Prober S, Brown AHD. 1994. Conservation of the grassy white box woodlands: population genetics and fragmentation of *Eucalyptus albens*. *Conserv. Biol.* 8:1003–13

Rabinowitz D. 1981. Seven forms of rarity. In *The Biological Aspects of Rare Plant Conservation*, ed. H Synge, pp. 205–18. Chichester, UK: Wiley

Raijmann LEL, van Leeuwen NC, Kersten R, Oostermeijer JGB, den Nijs HCM, Menken SBJ. 1994. Genetic variation and outcrossing rate in relation to population size in *Gentiana pneumonanthe* L. *Conserv. Biol.* 8:1014–26

Reed DH, Frankham R. 2001. How closely correlated are molecular and quantitative measures of genetic variation? A meta-analysis. *Evolution* 55:1095–103

Reed DH, Frankham R. 2003. Correlation between fitness and genetic diversity. *Conserv. Biol.* 17:230–37

Reinhammar L-G. 1999. Allozyme differentiation between the lowland *Carex capitata* and the alpine *Carex arctogena* (Cyperaceae) in Scandinavia. *Biol. J. Linn. Soc.* 67:377–89

Robertson A. 1965. The interpretation of genotypic ratios in domestic animal populations. *Anim. Prod.* 7:319–24

Roose ML, Gottlieb LD. 1976. Genetic and biochemical consequences of polyploidy in *Tragopogon*. *Evolution* 30:818–30

Schwartz OA. 1985. Lack of protein polymorphism in the endemic relict *Chrysosplenium iowense* (Saxifragaceae). *Can. J. Bot.* 63: 2031–34

Sharitz RR, Wineriter SA, Smith MH, Liu EH. 1980. Comparison of isozymes among *Typha* species in the eastern United States. *Am. J. Bot.* 67:1297–303

Simonich MT, Morgan MD. 1994. Allozymic uniformity in *Iris lacustris* (dwarf lake iris) in Wisconsin. *Can. J. Bot.* 72:1720–22

Slatkin M. 1985a. Gene flow in natural populations. *Annu. Rev. Ecol. Syst.* 16:393–430

Slatkin M. 1985b. Rare alleles as indicators of gene flow. *Evolution* 39:53–65

Slatkin M. 1987. Gene flow and the geographic structure of natural populations. *Science* 236:787–92

Slatkin M. 1993. Gene flow and population structure. In *Ecological Genetics*, ed. LA Real, pp. 3–17. Princeton, NJ: Princeton Univ. Press

Slatkin M, Barton NH. 1989. A comparison of three indirect methods for estimating average levels of gene flow. *Evolution* 43:1349–68

Soltis DE. 1982. Allozymic variability in *Sullivantia* (Saxifragaceae). *Syst. Bot.* 7:26–34

Soltis PS, Soltis DE. 1991. Genetic variation in endemic and widespread plant species: examples from Saxifragaceae and *Polystichum* (Dryopteridaceae). *Aliso* 13:215–23

Soltis PS, Soltis DE, Tucker TL, Lang FA. 1992. Allozyme variability is absent in the narrow endemic *Bensoniella oregona* (Saxifragaceae). *Conserv. Biol.* 6:131–34

Spooner DM, Douches DS, Contreras MA. 1992. Allozyme variation within *Solanum* sect. *Petota*, ser. *Etuberosa* (Solanaceae). *Am. J. Bot.* 79:467–71

Swofford DL, Selander RB. 1981. BIOSYS-1: a FORTRAN program for the comprehensive analysis of electrophoretic data in population genetics and systematics. *J. Hered.* 72:281–83

Van Oostrum H, Sterk AA, Wijsman HJW. 1985. Genetic variation in agamospermous microspecies of *Taraxacum* sect. *Erythrosperma* and sect. *Obliqua*. *Heredity* 55:223–28

van Treuren R, Bijlsma R, van Delden W, Ouborg NJ. 1991. The significance of genetic erosion in the process of extinction. I. Genetic differentiation in *Salvia pratensis* and *Scabiosa columbaria* in relation to population size. *Heredity* 66:181–89

Waller DM, O'Malley DM, Gawler SC. 1987. Genetic variation in the extreme endemic *Pedicularis furbishae* (Scrophulariaceae) *Conserv. Biol.* 1:335–40

Weir BS. 1996. *Genetic Data Analysis*. Sunderland, MA: Sinauer. 2nd ed.

Weller SG. 1994. The relationship of rarity to plant reproductive biology. In *Restoration of Endangered Species*, ed. ML Bowles, CJ Whelan, pp. 90–117. Cambridge, UK: Cambridge Univ. Press

Whalen MD. 1979. Allozyme variation and evolution in *Solanum* section *Androceros*. *Syst. Bot.* 4:203–22

Whitlock MC, McCauley DE. 1999. Indirect measures of gene flow and migration: $F_{ST} \neq 1/(4Nm+1)$. *Heredity* 82:117–25

Wilcove DS, McMillan M, Winston KC. 1993. What exactly is an endangered species? An analysis of the U.S. endangered species list. *Conserv. Biol.* 7:87–93

Wright S. 1931. Evolution in Mendelian populations. *Genetics* 16:91–159

Wright S. 1951. The genetical structure of populations. *Ann. Eugen.* 15:323–54

Young AG, Merriam HG, Warwick SI. 1993. The effects of forest fragmentation on genetic variation in *Acer saccharum* Marsh. (sugar maple) populations. *Heredity* 71:277–89

Annu. Rev. Ecol. Evol. Syst. 2003. 34:239–72
doi: 10.1146/annurev.ecolsys.34.011802.132402

First published online as a Review in Advance on July 11, 2003

THE ECOLOGY AND EVOLUTION OF INSECT BACULOVIRUSES

Jenny S. Cory[1] and Judith H. Myers[2]
[1]*Molecular Ecology and Biocontrol Group, NERC Center for Ecology and Hydrology, Mansfield Road, Oxford, United Kingdom, OX1 3SR; email: jsc@ceh.ac.uk*
[2]*Center for Biodiversity Research, Departments of Zoology and Agricultural Science, University of British Columbia, Vancouver, Canada, V6T 1Z4; email: myers@zoology.ubc.ca*

Key Words virulence, resistance, pathogen, variation, transmission

■ **Abstract** Baculoviruses occur widely among Lepidoptera, and in some species of forest and agricultural insects, they cause epizootics in outbreak populations. Here we review recent developments in baculovirus ecology and evolution, in particular focusing on emerging areas of interest and studies relating to field populations. The expanding application of molecular techniques has started to reveal the structure of baculovirus populations and has highlighted how variable these pathogens are both genotypically and phenotypically at all levels from within individual hosts to among host populations. In addition, the detailed molecular knowledge available for baculoviruses has allowed the interpretation of gene functions across physiological and population levels in a way rarely possible in parasite-host systems and showed the diverse mechanisms that these viruses use to exploit their hosts. Analysis of the dynamic interactions between insects and baculoviruses, and their compatibility for laboratory and field experiments, has formed a basis for studies that have made a significant contribution to unraveling disease interactions in insect populations. In particular, manipulative studies on baculoviruses have been instrumental in developing an understanding of disease transmission dynamics. The results so far indicate that baculoviruses have the potential to be an excellent model for investigations of changes in virulence and resistance in fluctuating and stable host populations.

INTRODUCTION

The role of disease in host populations is a finely balanced interplay between a pathogen's capacity to exploit a host and host susceptibility. Exploitation includes successful transmission, the potential of a disease organism to successfully infect a host, (infectivity), and virulence, the impact of the disease on host fitness from benign to fatal. Host susceptibility is determined by environmental and condition-dependent factors that influence the sensitivity of the host to infection and the ability of the host to modify susceptibility through resistance mechanisms. These factors can change over evolutionary as well as ecological time following fluctuating selection pressures. Baculoviruses and their insect hosts provide a model system for exploring pathogen-host interactions. Baculoviruses are DNA viruses that infect arthropods, mainly insects, in particular Lepidoptera, but also hymenopteran sawflies and some Diptera. Some baculoviruses play a role in the population dynamics of their hosts (Myers 1988) and viruses have also been explored as control agents for insect pests (Moscardi 1999). Here we present an overview of the characteristics that influence the ecology and evolution of baculovirus-host interactions, and identify areas that are ripe for further study. We have not covered all issues; we have focused on new data, emerging areas, and particularly field-related studies. For background on baculovirus biology, molecular biology, and other aspects of their ecology, in particular multi-species interactions, readers are referred to Cory et al. (1997), Entwistle & Evans (1985), Miller (1997), and Rothman & Myers (2000).

The Baculovirus Life Cycle and Infection Process

Baculoviruses have several unique features. First, they only infect the larval feeding stages where they form occlusion bodies (OBs). These are proteinaceous structures that contain the virions (virus particles). The OB is the infectious unit of the baculovirus and is critical for spreading infection between hosts. Second, many baculoviruses contain multiple virions (genomes) in each infective OB. Once in the midgut the virions are rapidly released by the combined action of the alkaline gut pH and proteases, and they then pass through the peritrophic membrane lining the gut (Figure 1). Baculoviruses encode various proteins that enhance the infection process (Peng et al. 1999, Popham et al. 2001). The virions released from the OBs fuse with the plasma membranes of the midgut columnar cells and the DNA-containing nucleocapsids move to the nucleus to initiate infection. In all non-Lepidopteran groups, infection is restricted to the midgut (or its equivalent), however in Lepidoptera most baculoviruses spread to other tissues, initially via the tracheoles (Engelhard et al. 1994, Volkman 1997). The baculovirus replication cycle is biphasic; tissue-to-tissue spread is carried out by nonoccluded budded virus whereas host-to-host transmission is carried out by the OBs. For more detail on baculovirus infection and its molecular basis see Miller (1997) and Volkman (1997).

In Lepidoptera by the end of the infection cycle, larval body tissue is converted into millions of OBs that are released into the environment when the host dies. In host species such as sawflies, baculovirus infections are restricted to the midgut, and infective OBs are shed continually with the feces. Occlusion bodies persist in the environment for considerable periods of time, particularly when protected from degradation by UV irradiation (Carruthers et al. 1988, Thompson et al. 1981). Baculoviruses "sit and wait" until they are ingested by another susceptible host or inactivated. Infection, and thus transmission, expose the baculovirus to two very different environments. Characteristics that promote successful invasion and reproduction within the body of the host may not necessarily maximize transmission to new hosts (DeFillippis & Villarreal 2000). The factors that influence these within and between host processes and shape baculovirus ecology and evolution are discussed in the sections that follow.

THE EVOLUTION OF BACULOVIRUSES

Variation Among Species of Virus

Baculoviruses have been recorded from hundreds of species of insect, however few have been characterized in detail. Baculoviruses are often identified initially by their morphology and their characteristic pathology in their hosts. With the advent of widely available DNA technology, it is now possible to routinely characterize baculoviruses using restriction endonuclease profiles of their DNA, and more recently, DNA sequence data. However, biological characteristics, particularly host range information, are also important for identifying relationships between baculoviruses, but broad-ranging biological data on individual isolates are sparse.

PHYLOGENY Baculoviruses are divided into two groups, the nucleopolyhedroviruses (NPVs) and the granuloviruses (GVs). NPV occlusion bodies are larger (1–5 μm) and contain multiple virions embedded in a protein matrix. Each virion can either contain one nucleocapsid—single NPVs (SNPVs), or many—multiple NPVs (MNPVs) (Figure 1). MNPVs have only been isolated from Lepidoptera and SNPVs from Lepidoptera and non-lepidopteran groups. GVs have only been reported from Lepidoptera and each OB (600–800 nm) contains a single virion enclosing one nucleocapsid. Phylogenetic studies indicate that lepidopteran baculoviruses fall into three groups, the GVs, Group I NPVs, and Group II NPVs,

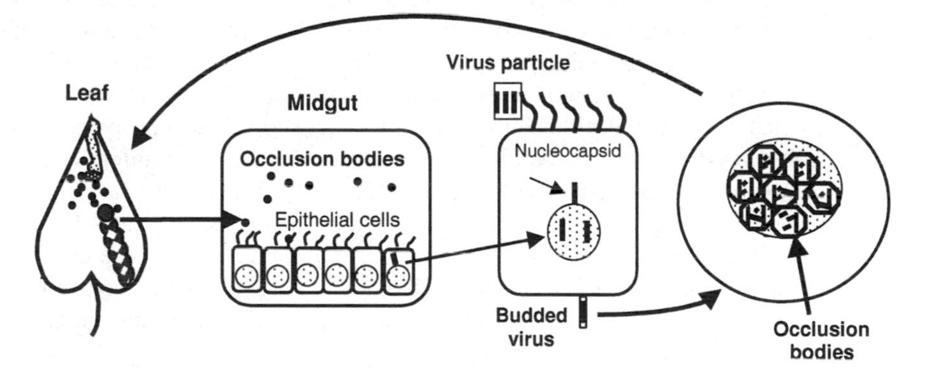

Figure 1 Diagram of the infection cycle of baculoviruses. Caterpillars become infected by ingesting virus that contaminates food plants when an infected larva dies. At the end of the cycle occluded polyhedra are formed, which will eventually be released from the host at death. In NPVs, OBs contain many virions embedded in a protein matrix, each of which may contain one or many nucleocapsids.

with hymenopteran and dipteran NPVs falling outside these clusters (Herniou et al. 2001, 2003; Moser et al. 2001; Zanotto et al. 1993). Analysis of the patterns of gene acquisition and loss among different viruses highlights the very fluid nature of baculovirus genomes (Herniou et al. 2003). A key question is whether baculoviruses have coevolved with insect orders or whether they evolved from one group of hosts. Preliminary indications from wide-ranging sequence analysis indicate that isolates collected from species within an order cluster together, but the relationship between orders is not clear (E.A. Herniou, J.A. Olszewski, D.R. O'Reilly, J.S. Cory, unpublished data).

VIRUS SPECIES Baculoviruses are named after the host from which they were isolated and therefore a virus that infects several host species could be given different names, as is the case with *Anagrapha falcifera* NPV and *Rachiplusia ou* NPV (Harrison & Bonning 1999). Alternatively, isolates from the same host species may represent very different virus species, as has been found with *Orgyia pseudotsugata* MNPVs and SNPVs (Zanotto et al. 1993), *Spodoptera littoralis* NPVs (Kislev & Edelman 1982) and *Mamestra configurata* NPVs (Li et al. 2002a,b). Molecular analysis is helping to clarify this issue but guidelines are needed to determine the criteria for designating different baculovirus isolates as species and the relative importance of molecular and biological data.

HOST RANGE SPECIFICITY The possibility for coevolution between baculoviruses and hosts is closely tied to their host range. Data are limited, but it is clear that different baculovirus species vary in the number of host species they can kill. Some baculovirus species appear to be genuinely host specific, such as the NPVs that infect Lymantriidae (Barber et al. 1993, Cory et al. 2000, Richards et al. 1999a). At the other end of the spectrum, cabbage moth, *Mamestra brassicae* NPV infects species from four families of Lepidoptera (Doyle et al. 1990), and the NPV from the alfalfa looper, *Autographa californica*, (AcMNPV) can infect species within at least 15 families of Lepidoptera (Payne 1986; J.S. Cory, R.D. Possee, M.L. Hirst, unpublished data). Other species of baculovirus (including GVs) have not been studied in such detail, although available evidence suggests that many have narrow host ranges; either monospecific (Gelernter & Federici 1986) or infective within a genus (Allaway & Payne 1984, Carner et al. 1979). Wider host range viruses are not equally infective to all host species. Additionally, studies on AcMNPV indicate that although species closely related to *A. californica* include some of the most highly susceptible species, host range does not closely follow host taxonomy (J.S. Cory, R.D. Possee, M.L. Hirst, unpublished data). This makes host range within the Baculoviridae hard to predict.

All host range data are from laboratory assays. Host usage in the field is likely to be considerably narrower owing to a lack of spatial or temporal synchrony. The trade-offs between generalism and specificity in host range are poorly understood (Woolhouse et al. 2001). Viruses with broader host ranges have the advantage of increased host availability. Specificity may well allow more efficient and less costly host use, however it carries a greater risk of extinction, particularly in unpredictable and patchy environments (Combes 2001). The wider host range viruses tend to be MNPVs isolated from Noctuidae. The M phenotype has been shown to establish irreversible systemic infections more rapidly than SNPVs, and this may make them more efficient at infecting a wider range of hosts (Washburn et al. 1999). Additionally, the high level of genetic variability in baculoviruses combined with the enhanced opportunities for recombination in the multiply occluded morphotypes may predispose MNPVs to wide host ranges.

Variation Within Species of Virus

The potential for selection to act on baculoviruses is related to their levels of genetic variation. Restriction endonuclease (REN) profiles of baculovirus DNA have shown that baculoviruses are extremely variable. Isolates from the same host species in different geographic regions frequently show differences in REN patterns for lepidopteran NPVs (e.g., Gettig & McCarthy 1982, Shapiro et al. 1991, Takatsuka et al. 2003), GVs (Crook et al. 1985, Parnell et al. 2002, Vickers et al. 1991), and hymenopteran sawflies (Brown 1982). However, we know little about the role and maintenance of this variation.

MOLECULAR (GENOTYPIC) VARIATION From an ecological perspective, an infected larva is the relevant unit of inoculum for initiating transmission. However studies of spatial and temporal variation in baculoviruses isolated directly (without amplification) from individual larvae collected in the field are rare (see review in Cooper et al. 2003b). An exception is a recent study that has shown that variation in the NPV of the western tent caterpillar, *Malacosoma californicum pluviale*, exhibits a hierarchical spatial structure. Isolates from individual larvae from within family groups were more likely to be the same as were isolates from within populations compared to those from different island and mainland populations (Cooper et al. 2003b). An exciting area for future work will be to determine if genetic and phenotypic diversity of the NPV (and its host) changes over time and with host density in forest caterpillars with cyclic population dynamics.

Isolates of NPV rarely contain a single genotype or variant. Variants can be separated by in vitro (e.g., Crozier & Ribeiro 1992, Maruniak et al. 1984, Stiles & Himmerich 1998) or in vivo cloning techniques (Jehle et al. 1995, Muñoz et al. 1998, Smith & Crook 1988). In vivo cloning has also been used to separate at least 24 genotypically distinct variants of NPV from a single infected caterpillar of pine beauty moth, *Panolis flammea* (J.S. Cory, B.M. Green, submitted manuscript). Analysis of *Spodoptera exigua* NPV also revealed that mutants with large deletions occurred naturally in wild-type isolates (Muñoz et al. 1998), but these variants were not capable of initiating oral infection alone. They were, however, abundant in the wild-type virus, and more importantly, reduced the pathogenicity of the virus population, indicating that they may well act as parasitic genotypes (Muñoz

& Caballero 2000). More recently Kikhno et al. (2003) found a naturally occurring, plaque-purified mutant of wild-type *Spodoptera littoralis* NPV with a 4.5-kb deletion that was not infective by ingestion, although it was as pathogenic as the wild-type virus by injection. This difference in phenotype was related to the presence of a specific gene, missing from the deletion mutant, which was named "*per os* infectivity factor" (*pif*). The product of this gene appears to be a structural protein associated with occlusion-derived virus that is required for the initial stages of infection within the midgut (Kikhno et al. 2003).

GENERATION OF GENOTYPIC VARIATION The observed high level of genotypic heterogeneity in baculoviruses could result from infection with multiple genotypes, or could be generated during the infection cycle through point mutations, gene duplications, and DNA deletions or insertions. The frequent generation of so-called few plaque mutants during repeated passage in cell culture originates from the insertion of DNA into the genome (Cary et al. 1989). Naturally occurring mutants of codling moth, *Cydia pomonella*, GV also contain host transposable elements that could increase variation (Arends & Jehle 2002, Jehle et al. 1995). Baculoviruses replicate clonally, but co-infection, or the multiple genome packaging found in NPVs, could allow the exchange of genetic material during infection via recombination between virus genotypes, virus and host, or virus and other co-infecting organisms. Recombination occurs in vivo with high frequency (up to 50%) between closely related virus genotypes (Crozier et al. 1988, Crozier & Ribeiro 1992, Hajós et al. 2000), but its importance and frequency in natural populations is unknown. The patterns of insertions and deletions of intraspecific variants indicate that these might involve a limited number of sites within the genome (Muñoz et al. 1999; Stiles & Himmerich 1998; J.S. Cory, B.M. Green, submitted manuscript). Some of the variable regions include long repeat sequences in homologous regions (Garcia-Maruniak et al. 1996, Muñoz et al. 1999), possibly indicating the presence of recombinational hot spots.

Recombination requires coinfection of the same cell and studies using AcMNPV show that the frequency of multiple infection of single cells in vivo is surprisingly high (Bull et al. 2001, 2003; Godfray et al. 1997). Additionally, studies with a recombinant AcMNPV have shown that genetically distinguishable NPVs can be packaged in a single OB (Bull et al. 2001). Thus recombination in vivo as a major source of variation driving the evolution of virulence and host range is highly feasible. If this were the case, NPVs should be more variable than the singly enveloped GVs. However, high levels of genotypic diversity are also found in GVs, and there is indirect evidence for the generation of recombinants in wild-type GVs (Smith & Crook 1993) as well as NPVs (Muñoz et al. 1997). The mechanisms for generating and maintaining genotypic diversity may be different in the two baculovirus groups, and further studies of the mechanisms that generate genotypic heterogeneity in field populations are needed.

PHENOTYPIC VARIATION Whereas genotypic variation may be ubiquitous in baculoviruses, it is of little relevance unless it translates into differences in virus phenotype. Baculoviruses isolated from the same species in different sites frequently vary in their pathogenicity (e.g., Allaway & Payne 1983, Ebling & Kaupp 1995, Hatfield & Entwistle 1988) and their speed of action (e.g., Hughes et al. 1983). Individual variants from within the same virus population obtained by in vitro (Lynn et al. 1993, Ribeiro et al. 1997) and in vivo cloning (Hodgson et al. 2001) can also vary dramatically in virulence. For example, clones of *Anticarsia gemmatalis* NPV, isolated by plaque purification, differed in LD_{50} (number of OBs required to kill 50% of challenged hosts) by over one-hundredfold (Ribeiro et al. 1997). Comparison of four NPV genotypes from a single *P. flammea* larva differed in LD_{50} by a factor of 7, and also showed significant differences in both infection duration and yield of progeny virus (Hodgson et al. 2001). Virus pathology and OB formation can also vary: plaque-purified variants from *Spodoptera frugiperda* NPV differed in the degree to which the infected larvae liquified at death, as well as in the size of the resultant OBs (Hamm & Styer 1985). Lysis of the host and OB size are likely to influence both transmission and persistence of baculoviruses in the wild.

MAINTENANCE OF VARIATION Potential mechanisms for maintaining baculovirus diversity include, (*a*) trade-offs, (*b*) differential selection, (*c*) multiple infection, (*d*) interspecific competition, and (*e*) frequency dependent selection (for a fuller discussion see Hodgson et al. 2001, 2003). Trade-offs occur when negative genetic correlations exist among beneficial traits. For baculoviruses, the most obvious potential trade-off is between the duration of infection and the production of infective OBs, such that the longer a virus takes to kill its host, the more OBs will be produced. Several within genotype comparisons have demonstrated that this is the case, although yield often plateaus (Hernández-Crespo et al. 2001, Hodgson et al. 2001). Intergenotype comparisons are rare. Hodgson et al. (2001) found no significant relationship between yield and infection duration or between the virulence (LD_{50}) and either yield or speed of kill in *P. flammea* NPV genotypes, although the sample size was small. However, Ribiero et al. (1997) found a strong positive relationship between virulence and yield measured as OB number per unit larval weight rather than the absolute yield per cadaver in *Anticarsia gemmatalis* NPV variants. This essentially measures the efficiency of converting insect tissue to virus rather than total production. Thus it is perhaps not surprising that the two are positively correlated.

Differential selection occurs when particular genotypes perform better under different ecological conditions, for example during infection of different host species. Multiple passage of a mixed virus isolate in alternative hosts can change the genotypic structure and biological activity of the virus (Kolodny-Hirsch & van Beek 1997, Tompkins et al. 1988), and individual variants can perform differentially in different hosts (Paul 1997). When AcMNPV was passaged through the diamondback moth, *Plutella xylostella*, 20 times, infectivity toward *P. xylostella* increased 15-fold (Kolodny-Hirsch & van Beek 1997). The changes were accompanied by an increase in virion number per OB, as well as a change in DNA restriction endonuclease profile, but whether they resulted from selection for different genotypes or the production of novel variants is unclear. Abiotic factors could also influence the success of variants through selection on different temperature profiles or for improved persistence against UV (Brassel & Benz 1979). As virus persistence is intimately linked with host plant, it is also possible that adaptation to different plant species could occur. Preliminary support for this idea comes from the pine beauty moth, *P. flammea* NPV system. In Scotland, where this virus was collected, *P. flammea* feeds on two host plants, Scots pine and the introduced lodgepole pine. When insects were infected with two *P. flammea* variants on the two host plants they exhibited differential pathogenicity (Hodgson et al. 2002).

An interesting possibility is that infections by more than one genotype might actually be beneficial to the virus. Conventional theory predicts that mixed infections should be more virulent, as multiple genotypes will increase the rate of host exploitation (Frank 1996). Comparison of two equally infective variants of *P. flammea* NPV alone and in combination showed that mortality was significantly higher in the mixed infection (D.J. Hodgson, R.B. Hitchman, A.J. Vanbergen, et al., submitted manuscript). Unexpectedly, this was not accompanied by a more rapid speed of kill and reduction in overall (combined) yield per insect. The mechanism for increased virulence from mixed infections in mouse malaria was suggested to be the cost to the immune system of having to fight multiple strains (Taylor et al. 1998). However, although differential immunity may be involved in insect defense, other mechanisms, such as variable tissue tropisms, seem more likely causes of enhanced virulence. This has implications for host-pathogen dynamics and virulence management, particularly in the application of baculoviruses as pest control agents. Another possibility is that the structure of the virus population is influenced by interactions with competing natural enemies. Preliminary data suggest that REN profiles of *S. frugiperda* NPV vary depending on whether the host is parasitized or not (Escribano et al. 2000, 2001), however there is little information on this issue.

Coevolutionary processes may also be involved in the maintenance of baculovirus polymorphism. Various coevolutionary scenarios have been proposed for the interaction between parasites and their hosts, in particular escalating arms races between host and parasite and the potential for frequency dependent selection. In antagonistic interactions, such as host-parasite relationships, negative frequency dependent selection may operate whereby dominant genotypes are at a selective disadvantage resulting in fluctuating polymorphisms (Lively 2001). This would provide another route for maintaining baculovirus diversity, however, as yet there are insufficient data on the dynamics of baculovirus (or host) strain structure to address this hypothesis. Frequency dependent selection may also be involved in the maintenance of parasitic genotypes, which potentially have a replication advantage within the host but require an occlusion body for host-to-host transmission (Hodgson et al. 2003). There is already evidence that baculovirus populations show hierarchical spatial structure (Cooper et al. 2003b). Adding a spatial component to all these mechanisms would increase the complexity of the interactions together with the likelihood of increased diversity (Thompson 2001).

Host Manipulation

The manipulation of host behavior by parasites is well studied in vertebrates (Moore 2002), and there is also some information on insect parasitoids (Edwards & Weaver 2001). Pathogens of insects, however, have received considerably less attention, although there is increasing evidence that they may manipulate their hosts in a number of ways both behaviorally and physiologically. Baculoviruses, in particular, offer ideal systems for linking mechanisms of response to underlying genetics.

BEHAVIORAL Some of the earliest descriptions of baculovirus disease mention behavioral changes in infected larvae, in particular, the tendency for infected caterpillars to move up the plant to die, so-called tree-top disease or "wipfelkrankheit" (Steinhaus 1967). However, despite frequent anecdotal observations, this behavior has only recently been quantified. Infection of the cabbage moth, *M. brassicae*, by its NPV produces increased movement (dispersal) and upward migration on cabbage plants (Goulson 1997, Vasconcelos et al. 1996a). It is assumed that this behavior has evolved to enhance the transmission of the virus by either gravity or rainfall (Goulson 1997, D'Amico & Elkinton 1995, Vasconcelos et al. 1996a). Climbing behavior could also make a virus-infected caterpillar more apparent to predators, thus increasing its dispersal rate. A wide range of predators eat infected caterpillars and passively disperse OBs in their feces (Entwistle et al. 1993, Lee & Fuxa 2000, Vasconcelos et al. 1996b). However, although the observed behavioral changes appear to be beneficial for the virus, conclusively demonstrating their adaptive value is more difficult (Moore 2002, Poulin 1998).

From the host's perspective, enhanced mobility may remove infected individuals from the vicinity of their uninfected conspecifics, particularly in gregarious and semigregarious species. In this way, the inclusive fitness of family groups may be increased. Additionally, because baculoviruses are sensitive to sunlight, greater exposure on branch tips may also increase the rate of inactivation. Alternatively, the observed behavioral changes may not be adaptive to either host or virus and may result from pathological changes resulting from virus infection. However, behavioral changes have only been quantified in one species: upward movement may not necessarily enhance virus fitness in all hosts, for example, in cutworms that dwell in the soil, this may not be beneficial. There is thus a need for more studies in host species with diverse life histories.

PHYSIOLOGICAL The function of many baculovirus genes is still not known, but it is evident that not all genes are associated with activities crucial for virus replication or structure. These genes have been termed auxiliary genes (O'Reilly 1997), although when knowledge of their activity increases, it may be more appropriate to ascribe them to functional groups. Baculoviruses can be genetically modified with precision; this allows the function of these nonessential genes to be assessed at both the individual and population level.

The egt gene The only auxiliary gene that has been studied in any detail is the ecdysteroid UDP-glucosyltransferase (*egt*) gene (Cory et al. 2001, O'Reilly & Miller 1989). This gene produces an enzyme that conjugates ecdysteroids (insect molting hormones) with UDP-glucose or galactose (Kelly et al. 1995, O'Reilly et al. 1992). Comparison of wild-type baculoviruses with those without a functional *egt* gene has clearly shown that EGT expression inhibits the larval-larval or larval-pupal molt (O'Reilly 1997, O'Reilly & Miller 1989) and prolongs larval infection by up to 60% (Cory et al. 2001, O'Reilly & Miller 1991, Slavicek et al. 1999). It was originally hypothesized and subsequently shown that the extended infection period would increase production of viral progeny (Slavicek et al. 1999, Wilson et al. 2000). Insect development (rather than time to death) is crucial in determining virus yield in final instar *Heliothis virescens* larvae infected with AcMNPV. Insects for which molting is arrested early produce more than four times the quantity of OBs than insects that go on to the pharate pupal stage (O'Reilly et al. 1998). Thus the *egt* gene manipulates host development, thereby increasing available resources and the quantity of viral progeny produced. As virus productivity is a crucial component of virus fitness, increases in yield should be highly beneficial to the virus.

The *egt* gene might have other potentially beneficial effects. Indirect evidence suggests that insect ecdysteroids could interfere with virus replication (Keeley & Vinson 1975, O'Reilly et al. 1995). However, deletion of the *egt* gene does not reduce pathogenicity (Cory et al. 2001). EGT expression could also influence virus transmission and persistence. A simplified transmission experiment using larvae confined to cut branches showed that gypsy moth NPV without the *egt* gene had a significantly lower transmission rate than the wild-type virus (Dwyer et al. 2002). However, a cost of the greater yield from insects infected with viruses which express EGT is slower death that delays the release of OBs for further rounds of transmission. Theoretical exploration of the dynamics of *egt* plus and minus viruses indicates that the fitness of viruses lacking the *egt* gene would be less than that of the parent wild-type (Dwyer et al. 2002). Extending this analysis to competition between viruses that differ in speed of kill and transmissibility indicates that the faster killing virus requires relatively high levels of transmission to dominate (Dushoff & Dwyer 2001). However, overwinter survival of virus, about which little is known, is crucial in these simulations. The field situation is considerably more complex and it is plausible that baculoviruses lacking a functional *egt* gene could coexist with wild-type virus by essentially "parasitizing" their expression of EGT in the same host.

The chitinase and cathepsin genes The only other auxiliary genes for which a potential function at an organismal level has been identified are the chitinase and cathepsin (cysteine protease) genes. These genes appear to act together to facilitate the release of the virus OBs from the insect cadaver after death by breaking down the chitin and protein present in the insect cuticle. When either gene is deleted, the cuticle of the infected insect fails to liquefy (Hawtin et al. 1997, Slack et al. 1995). The ecological implications of these genes have received little attention. Their most likely function is to enhance horizontal transmission: liquefaction of the infected cadaver will spread the OBs over a greater area thereby potentially increasing the likelihood that a susceptible host will encounter them. However, there is likely to be a trade-off with this strategy as more widely disseminated virus may be more rapidly degraded by sunlight.

Not all baculoviruses contain a chitinase or a cathepsin gene (Herniou et al. 2001, 2003), and liquefaction or "melting" of the cadaver, commonly regarded as a characteristic of baculovirus infection, does not always occur. For example, the Indian mealmoth, *Plodia interpunctella*, does not lyse when infected with its GV, and in this system cannibalism is thought to be the major route of virus transmission (Boots 1998). Whether the lack of liquefaction in this host-virus system is due to a change in chitinase or cathepsin expression is not known. However, the study of the role of genes such as chitinase and cathepsin that appear to exert their effects at an organismal level, would benefit from a comparative approach using species with different ecologies and transmission dynamics.

Variation in Susceptibility

The insect host presents different layers of defense that must be breached for the baculovirus to infect and multiply in the host. Each could impose selectivity on the system and this has recently been described as the "defense component model" (Schmid-Hempel & Ebert 2003). We next discuss some of the factors that influence the susceptibility of insect hosts to baculoviruses.

HOST BEHAVIOR There is no evidence to suggest that insects modify their behavior to reduce the ingestion of baculoviruses. Postingestive behavior is also little studied. In other insect-pathogen systems, fatal infection levels may be reduced through behavioral fever (Blanford & Thomas 1999, Boorstein & Ewald 1987). However, although temperature affects life history parameters that may well have consequences for host-pathogen dynamics such as speed of kill and yield (Kelly & Entwistle 1988, van Beek et al. 2000), there is no evidence that altering temperature can influence fatal infection in baculoviruses (Frid & Myers 2002).

DEVELOPMENTAL RESISTANCE It is well established that susceptibility to baculovirus infection decreases with increasing larval age both within and between instars. Originally this was thought to be related to the increasing weight of the larva. However, comparisons of oral with intrahemocoelic inoculation demonstrated that this resistance did not occur when virus was injected (Teakle et al. 1986); which implies that developmental resistance is related to the infection process in the midgut. Detailed investigations on the development of intrastadial resistance in *Trichoplusia ni* to AcMNPV, showed that fourth instar larvae were most susceptible immediately after molting (Engelhard & Volkman 1995). By using a recombinant AcMNPV that could be tracked in the gut, it was clearly demonstrated that larvae can remove baculovirus infection by sloughing off infected cells lining the midgut (Hoover et al. 2000; Keddie et al. 1989; Washburn et al. 1995, 1998). Thus a decreasingly small window of opportunity exists for the virus to establish infection as the insects age, particularly in later instars.

SYSTEMIC RESISTANCE Although the midgut appears to be the major barrier to baculovirus infection, there is also evidence for systemic resistance in some species. In fourth instar gypsy moth larvae, *Lymantria dispar*, infection levels decreased when the insect was challenged later in the instar and this was not fully reversed when the virus was administered by intrahemocoelic injection, indicating resistance has a systemic component (Hoover et al. 2002). In fifth instar, *H. virescens*, developmental resistance to AcMNPV was partially removed by methoprene, a juvenile hormone analog, suggesting that a component of systemic resistance is mediated by hormones (Hoover et al. 2002, Kirkpatrick et al. 1998).

Insects possess both cellular and humoral mechanisms of immunity and both have been implicated in resistance to baculoviruses. Studies with a recombinant AcMNPV and *Helicoverpa zea* (a less susceptible species) demonstrated that the initial number of infection foci in the midgut was equal to that found in a highly susceptible species (*Heliothis virescens*). However, in *H. zea* the number of infection foci decreased, suggesting that the larvae had somehow removed the infection (Washburn et al. 1996). This was associated with the presence of hemocyte aggregations around infected tracheoles, similar to the type of cellular encapsulation response seen with invading parasitoids. This response was reduced with immunosuppressors confirming the role of the immune system. How systemic resistance is involved between hosts and their homologous viruses is not so clear.

Although links between susceptibility and immunity have received little attention, it has been shown that responses to baculoviruses can vary with rearing density, with insects reared at higher densities less susceptible to infection (Goulson & Cory 1995, Kunimi & Yamada 1990). In the African armyworm, *Spodoptera exempta*, insects reared gregariously were more resistant to NPV infection than those reared solitarily and this was correlated with higher levels of phenoloxidase activity in the hemolymph, indicative of increased immune function (Reeson et al. 1998). Crowding in phase polyphenic Lepidoptera also tends to be related to an increase in melanization. It has been suggested that melanism could indicate enhanced immune activity and disease resistance (Wilson et al. 2001).

FOOD PLANT EFFECTS Herbivorous insects are intimately associated with their food plants and so are their pathogens. Host plant can influence virus interactions in many ways; plant architecture affects virus persistence, palatability modifies host mobility and virus acquisition, plant chemistry modulates infection in the gut and nutrient content determines host survival. The impacts of plant phytochemicals, such as phenolics, on host susceptibility has received most attention and numerous studies have shown that both mortality (Duffey et al. 1995; Farrar & Ridgway 2000; Forschler et al. 1992; Hoover et al. 1998a,b,c; Raymond et al. 2002) and speed of kill vary depending on plant type or allelochemicals (Farrar & Ridgway 2000, Raymond et al. 2002). However the influence of previous herbivory and induced responses on virus susceptibility is equivocal (Ali et al. 1998, Hoover 1998b, Hunter & Schultz 1993). Intraplant variation can also have significant effects. For example, when *H. zea* was infected with its NPV on cotton, susceptibility was lower when the caterpillars were fed on the reproductive rather than the vegetative structures (Ali et al. 1998) and the resulting virus yield was also lower on reproductive tissues (Ali et al. 2002). As heliothines tend to feed cryptically on cotton buds, they are not only avoiding natural enemies but also apparently further reducing their risk of infection.

Finding a mechanism behind the observed differences in plant mediated effects could provide a better framework for understanding and predicting plant-baculovirus interaction. As most of the studies focus on chemical influences at the time of ingestion, alterations in susceptibility are likely to be related to the midgut. In an elegant study, Hoover et al. (2000) compared the susceptibility of *H. virescens* larvae feeding on cotton to those feeding on lettuce. Insects inoculated with AcMNPV on cotton were less susceptible and this was positively correlated with levels of foliar peroxidase. Previous studies suggested that reactive oxygen species associated with peroxidase activity damage the lining of the midgut and cells are sloughed off at early infection (Hoover et al. 1998c). By administering an optical brightener (thought to enhance the retention of gut cells) the effect of feeding on cotton was reversed (Hoover et al. 2000). Reduced susceptibility to NPV on cotton was not observed if the virus was injected directly into the hemocoel. This provides strong support that differences in susceptibility mediated by host plant relate to the effect of phytochemicals on the rate of sloughing gut cells.

Both pre and postingestion nutritional quality of the diet can play significant roles in infection (Duffey et al. 1995). For some host species, epizootics of virus occur following high population densities and the influence of food quantity and quality is relevant. When the susceptibility of the stored product pest, *P. interpunctella* to its GV was assayed, McVean et al. (2002) found that larvae reared on a lower quality diet were less susceptible to virus infection, but died more quickly than larvae on the high quality diet. Recent data, using *Spodoptera littoralis* and its NPV, has shown that ingestion of a high protein: low carbohydrate diet postingestion reduces fatal baculovirus infection (Lee 2002). Additionally, larvae that survived virus challenge increased their intake of protein-biased food after infection when given a choice. Protein may be important in defense against baculoviruses, either via the immune system or by increasing larval development rate (Hoover et al. 1998d, Lee 2002). An intriguing possibility is that insects could alter their dietary intake in response to pathogen challenge by self-medication (Clayton & Wolfe 1993, Lozano 1998). Karban & English-Loeb (1997) report that caterpillars parasitized by a tachinid fly parasitoid adjust their host plant choice to enhance survival. Diet can influence immune function in other groups of insects (Suwanchaichinda & Paskewitz 1998, Vass & Napi 1998) and it has recently been shown that starvation can rapidly decrease phenoloxidase activity in the beetle *Tenebrio molitor* (Siva-Jothy & Thompson 2002). The influence of food plant on virus infection is complex and whether these tritrophic interactions significantly influence host-virus dynamics and evolution in natural populations requires studies of transmission, adaptation and host plant usage in the field.

THE ECOLOGICAL IMPACTS OF BACULOVIRUSES ON HOST POPULATIONS

Theoretical Considerations

Infection by baculoviruses is generally lethal and therefore has the potential to influence host population densities particularly if viral transmission increases with host density. The theoretical relationships of host-microparasite dynamics of insects have been widely explored in mathematical models starting with those of Anderson & May (1980, 1981). More recent models have incorporated modifications such as variation in transmission parameters (Getz & Pickering 1983), vertical transmission (Regniere 1984), both density dependence and vertical transmission (Vezina & Peterman 1985), nonlinear transmission (Hochberg 1991), density dependence (Bonsall et al. 1999, Bowers et al. 1993, White et al. 1996), host stage structure (Briggs & Godfray 1996), heterogeneity in susceptibility (Dwyer et al. 1997), discrete generations and seasonal host reproduction (Dwyer et al. 2000), and sublethal infection (Boots & Norman 2000).

Models of virus versus disease of insect hosts have usually been evaluated by whether or not they generate multigenerational, cyclic population dynamics of the hosts. However, many are unrealistic in other characteristics such as levels of infection (Bowers et al. 1993), e.g., almost 100% infection for three generations of peak host density in the model of Dwyer et al. (2000). Prolonged, high levels of infection do not occur in field populations (Woods & Elkinton 1987, Myers 2000). Whether the outcomes of disease models are realistic depends on how disease transmission is simulated.

Transmission

Transmission in baculoviruses is thought to be primarily horizontal via susceptible larvae ingesting OBs persisting in the environment. However, there is also evidence that baculoviruses can be transmitted vertically from adults to their young. These important processes are discussed in the following sections.

HORIZONTAL TRANSMISSION Transmission depends on the interactions between infected and susceptible individuals, and the rate at which contacts result in new infections. Most models assume that transmission is a mass action process; a density-dependent relationship based on a per capita transmission coefficient, β, the number of susceptible individuals, S, and the number of infected individuals, I, (βSI) although some have argued that this should be related to population size (βSI /N) (reviewed in McCallum et al. 2001).

The transmission process in field populations is unlikely to fit either of these simple representations. For baculoviruses disease transmission depends on susceptible larvae encountering and ingesting a discrete patch of virus that results from the death of an infected individual, and thus depends on the behavior of infected and susceptible larvae. We next consider experimental studies that have estimated the transmission of baculoviruses.

Quantifying transmission In studies that have experimentally estimated the transmission coefficient for baculovirus infection in insect populations, all have found that it declined with increasing inoculum densities (D'Amico et al. 1996, Beisner & Myers 1999, Knell et al. 1998,Vasconcelos 1996). However, the relationship with host density was more variable. Transmission efficiency of NPV increased with increasing density of *M. brassicae* larvae on cabbage (Vasconcelos 1996), and for GV in *P. interpunctella* (Knell et al. 1998). Conversely, in gypsy moth and its NPV, the transmission coefficient declined as the density of susceptible larvae increased (D'Amico et al. 1996). Beisner & Myers (1999) varied both the numbers of susceptible and infected individuals in groups of western tent caterpillar larvae, *M. c. pluviale*, a naturally gregarious species. Again virus transmission was more efficient at lower virus densities, but transmission within groups was not related to density. However, group size did influence infection levels, and increased movement among larger caterpillar groups on individual trees, increased between-group spread of the virus. In a recent analysis, Fenton et al. (2002) explored the influences of transmission functions in a metapopulation model of hosts and microparasites. They suggest that small-scale transmission events can drive large-scale epizootics. This would seem to be the case with the tent caterpillars.

Transmission could vary if the susceptibility of larvae changes at high density, as might result from stress. As mentioned above, crowding has been shown in laboratory studies to reduce the susceptibility of caterpillars to fatal infection. This led Wilson & Reeson (1998) to propose the concept of density-dependent prophylaxis; as the risk of exposure to pathogens increases with density, insects respond physiologically and invest more in resistance. Reeson et al. (2000) compared NPV transmission on maize plants in field cages using *S. exempta* larvae that had been reared either singly or gregariously. Virus transmission was lower in the larvae that had been previously reared gregariously.

These experimental measurements of transmission yield the clear message that simple mass action does not describe baculovirus transmission in the field. More appropriate descriptions of transmission must be found based on a more detailed knowledge of the heterogeneities that influence the infection process.

Spatial distribution and persistence The spatial distribution of virus on the plant, and between plants and the soil, is likely to significantly influence virus transmission at a local scale. Different species of insect, different instars, different host plants, and even different baculovirus-host combinations could influence the pattern of virus distribution. In the first study on the influence of virus distribution on transmission, Dwyer (1991) found that for third instar Douglas fir Tussock moth, *O. pseudotsugata*, transmission was lower when virus distribution was heterogeneous than when it was uniform. For later more mobile instars however, the distribution of the virus was less important. The amount and location of inoculum were varied by Hails et al. (2002) in field experiments with *T. ni*. Whether infected larvae remained on the plant or fell to the ground influenced transmission but not the inoculum (cadaver) size. This suggests that the amount of virus in a patch is less important than the number of patches.

Persistence and thus transmission of virus will also be influenced by the location of the virus OBs. If the virus is on the surface of leaves, it will be exposed to UV and more rapidly inactivated. For example sun exposure is thought to reduce viral infection of forest tent caterpillars, *Malacosoma disstria*, feeding at the edges of forests compared to the interior (Rothman & Roland 1998). The eventual repository for most virus is the soil, where the OBs are protected from exposure to sun but may have reduced opportunity for transmission to new hosts. Hochberg (1989) introduced a pathogen reservoir into a model of host-pathogen dynamics and showed that allowing for translocation of a pathogen from the reservoir to a transmissable surface could result in regulation of the host. A virus reservoir in the soil is crucial to the association between *Wiseana* spp. and baculovirus in New Zealand pastures (Crawford & Kalmakoff 1977). The larvae live in the soil, and grazing animals enhanced the spread of virus, but plowing disrupted the virus reservoir, resulting in outbreaks of the insect. However, Fuxa & Richter (2001) have shown that baculovirus movement from the soil to plants is very low and depends on rainfall. Larval behavior can also have a major impact on virus acquisition from a reservoir. In pine forests in Scotland, early instar larvae of the vapourer moth, *Orygia antiqua*, moved (ballooned) off the trees and into contact with the understory reservoir of NPV where they acquired high levels of infection. When they moved back up to the trees, the virus infection was transmitted to other larvae (Richards et al. 1999b).

Heterogeneity in susceptibility Heterogeneity in host susceptibility related to age structure and variation in resistance within and among populations can also influence transmission. Goulson et al. (1995) investigated the influence of stage-dependent variation in susceptibility on the transmission of NPV among cabbage moth larvae, *M. brassicae*. They predicted that because later instar larvae eat much more leaf material than early instars, transmission should increase for later instars even though they are less susceptible to infection, but this was not supported. In contrast, greater susceptibility of first instar larvae compared to later instars can influence the outcome of models of disease dynamics. Heterogeneity in host susceptibility causes the relationship between virus density and transmission ($\log_n$ proportion susceptible individuals surviving) to be curvilinear with increasingly efficient transmission at lower virus densities (Dwyer et al. 1997). Incorporating heterogeneity in susceptibility improved the fit of the model to the observed NPV infection in low-density gypsy moth populations within generation (Dwyer & Elkinton 1993, Dwyer et al. 1997). Dwyer et al. (2000) extended the model to multiple generations by including a between season component and concluded that having a combination of highly and less susceptible individuals could drive outbreaks of gypsy moth. In this model, virus infection is initiated early and results from overwintering virus contaminating egg masses so that the highly susceptible first instar larvae become infected at hatching. Although, adding heterogeneity to models generally makes dynamics more stable, cyclic dynamics in this model persisted under some conditions. Frid (2002) showed that adding a random variable to the model based on an observed relationship between high levels of sunshine and increased NPV infection (Frid & Myers 2002), expanded the range of conditions under which cycles occurred.

VERTICAL TRANSMISSION Baculoviruses are generally thought to be transmitted horizontally from host to host via OBs persisting in the environment. However, baculoviruses can also be transmitted vertically from the adult to young (Easwaramoorthy & Jayaraj 1989, Fuxa et al. 2002, Kukan 1999, Melamed-Madjar & Raccah 1979, Myers et al. 2000). Vertical transmission encompasses passage of virus from the adult to their progeny by any means. This can be owing to surface contamination of eggs (transovum transmission) or virus passing within the egg (transovarial transmission), including transfer of latent infections; the latter is dealt with below. Infection levels can be as high as 50% in the progeny of adults that were exposed to virus as larvae (Kukan 1999). Surface sterilization usually reduces the level of infection considerably, but rarely eliminates it (Kukan 1999, Myers et al. 2000). Transovum transmission is harder to detect in the field because the eggs could become contaminated when they are laid onto foliage. They could also become contaminated in more subtle ways, e.g., in sawflies the meconia can be infectious, and this could contribute to adult to egg transmission (Olofsson 1989).

Distinguishing between latent and persistent infections The "spontaneous" generation of baculovirus disease from apparently healthy insects has been observed for over 100 years, and has stimulated discussion on possible latent baculoviruses and what might trigger them. Host crowding, fluctuations in temperature or relative humidity, dietary changes, chemical stress, parasitization, and infection by a second pathogen have all been suggested as possible stressors (Fuxa et al. 1999, Podgwaite & Mazzone 1986). Latent virus is a noninfective and nonreplicating form of virus that can be transformed to an infective state by some stressor (Fuxa et al. 1992). Understanding, or even consistently demonstrating, baculovirus latency has proven to be elusive, and triggering an active infection from a latent virus may be an unpredictable, stochastic process. The polymerase chain reaction

(PCR) was used to demonstrate that NPV infection persisted in all life stages of a laboratory culture of *M. brassicae* (Hughes et al. 1993). However, further studies using reverse transcriptase (RT)-PCR showed that a low level of viral transcripts were produced that implied that the virus was replicating and not truly latent (Hughes et al. 1997). More recent controlled experiments with *P. interpunctella* and its GV, demonstrated that a persistent infection was introduced into a high proportion of the survivors of viral challenge, and that this infection was passed via both sexes to the next generation (Burden et al. 2002). Whether this virus persists in later generations is not known. However, these studies suggest that baculoviruses do not persist in a classic latent form, but by a low-level replicating, persistent infection.

Little is known about whether sublethal, persistent virus infection actually occurs in field populations, the rate at which triggering occurs, what causes it, or whether this has any impact on the development of virus epizootics. In a recent study, cross-inoculation with an NPV from the western tent caterpillar (*M. c. pluviale*) was shown to "stress out" a second virus in field populations of the forest tent caterpillar, *M. disstria* (Cooper et al. 2003a). Interestingly, the virus "stressed out" of the population appeared to be a single genotype that differed from the mixed genotypes of horizontally transmitted virus isolated from the same population. Additionally whereas virus was readily activated from high-density forest tent caterpillar populations, no infection resulted in larvae from low-density western tent caterpillar populations in reciprocal cross-infections. This suggests that levels of sublethal infection or the propensity for a covert infection to be reactivated may be related to the history of viral infection in the host population and host density. Interestingly, recent work on the African armyworm, *S. exempta*, using RT-PCR, demonstrated high levels of persistent NPV infection in adults in field populations in Tanzania (L. Vilaplana, E.M. Redman, K. Wilson, J.S. Cory, unpublished data).

The role that vertical transmission plays in virus-host dynamics in field populations is not clear. Transovum transmission appears common in many species, and the emerging picture indicates that persistent, sublethal infections might also be very common in natural populations. The costs of maintaining these persistent infections must be low if they are widespread; a vertically transmitted virus should maximize its fitness by maintaining high host fecundity. It is not known how baculoviruses persist in hosts that have extended periods of low (and unpredictable) density, in situations where the host is highly mobile, or where the host occurs in environments with low environmental persistence of OBs and under these conditions a persistent virus strain may well be able to out-compete a nonpersistant virus. For example, in cyclic populations of tent caterpillars, populations at some sites only occur when densities are high regionally and thus, are initiated by immigration of moths every six to eight years (Myers 2000). However infection from NPV still characterizes these new populations (Cooper et al. 2003b), and vertical transmission on or in eggs must explain the occurrence of virus in these ephemeral populations. The circumstances in which vertical transmission is favored, together with the costs and benefits to both pathogen and host are important for understanding the ecology and evolution of baculoviruses.

Sublethal Effects

Surviving baculovirus challenge is often costly and survivors can have altered development times and reduced size, fecundity, egg viability, and vigor (Duan & Otvos 2001, Matthews et al. 2002, Milks et al. 1998, Myers et al. 2000, Rothman & Myers 1996a). Although the impacts varied with study, on average sublethal effects reduced the reproductive potential (R_0) of the host by 22% (Rothman & Myers 1996a). The causes of these effects are not known, but they could result from fighting off infection or from the continued presence of a persistent virus infection. For example, sloughing off infected gut cells might reduce food intake, or the mobilization of an immune response could be physiologically costly. One interpretation is that infection of late instar larvae is terminated by pupation, and survivors show the influence of early infection as reduced size and fecundity. Infection may be removed at pupation, or could remain as a persistent infection (Burden et al. 2002) and be transmitted to the next generation. Thus baculoviruses can influence host population dynamics through direct mortality and also by delayed effects, such as reduced fecundity.

Resistance

Interactions between parasites and hosts provide an arena in which selection can act. For the host, increased resistance has an obvious benefit but is also expected to have costs (but see Rigby et al. 2002). Baculoviruses are transmitted (horizontally) at the death of the host and transmission is based on the number of new occlusion bodies produced. Therefore, selection, for the virus, should favor (*a*) overcoming host resistance, (*b*) maximizing the conversion of host tissue to OBs, and (*c*) optimizing the time of host death to promote transmission to new hosts. These conflicting selection pressures make theoretical predictions complex. The strength of selection will be influenced by the prevalence of infection, and for many host species the levels of infection by baculoviruses are low.

Variation in resistance to baculoviruses has been identified in several host species (reviewed by Watanabe 1987). However, with the exception of the measurement of a 38-fold increase in resistance of larch budmoth following a GV epizootic (cited in Watanabe 1987), the potential for coevolutionary change in hosts and pathogen following strong selection in the field remains little studied. Resistance has been selected for in the laboratory. Fuxa & Richter (1989) increased resistance threefold in a laboratory colony of fall armyworm, *S. frugiperda*, and resistance was rapidly lost when selection was stopped. After 13 to 15 generations of selection with NPV, colonies of velvetbean caterpillar, *A. gemmatalis*, from Brazil became 1000 times more resistant, whereas those from the United States plateaued at five times more resistant after four generations (Abot et al. 1996). In the U.S. population, resistance was associated with reduced production of young and was rapidly lost (Fuxa & Richter 1998). Milks (1997) compared the susceptibility of 12 lines (8 laboratory strains and 4 wild collections) of *T. ni* and found significant but low (3.5-fold) variation in resistance levels. Females from the most resistant lines laid significantly fewer eggs. Vail & Tebbets (1990) also found variation in resistance among populations of *P. interpunctella* to GV and larval growth was slower in the most resistant populations.

In the only experiment that has selected host resistance and viral virulence, Milks & Myers (2000) increased resistance to *T. ni* SNPV in two laboratory populations of cabbage loopers by 4.4-fold and 22-fold. The virulence of the virus did not change over 26 generations of host selection. Host resistance was stable, and no costs could be identified (Milks et al. 2002). This work agrees with Rigby et al. (2002) who point out that not all mechanisms of resistance are costly. Interestingly, the increased resistance to *T. ni* SNPV in cabbage loopers also conferred resistance to a GV of *Pieris rapae* and a GV of *T. ni* but not to AcMNPV (Milks & Myers 2003). This may indicate more similarity in infection mechanisms between singly embedded viruses and GVs than with (wider host range) multiply embedded NPVs.

The levels of increased host resistance to baculovirus following selection have been low, and the persistence of resistance has varied. Whether the benefits of increased resistance would be sufficiently large to be realized in field populations is impossible to predict. Thus, theoretical considerations assuming patterns of coevolution in regard to baculovirus must remain speculative.

Viral Epizootics

Epizootics are outbreaks of disease in animal populations. An epizootic requires the presence of the pathogen and sufficient numbers of susceptible hosts to allow an increased rate of infection. Mitchell & Fuxa (1990) showed that NPV infection of the fall armyworm, *S. frugiperda*, was only sustainable if host densities averaged one larva per corn plant. Epizootics of baculoviruses occur irregularly in agricultural pests including many species in the genera *Agrotis*, *Feltia*, *Mamestra*, *Plusia*, *Trichoplusia*, *Heliothis*, *Helicoverpa*, *Spodoptera*, and others (Weiser 1987). For example, *Mamestra configurata* periodically outbreaks in canola fields in the prairie provinces of Canada (Mason et al. 1998). NPV infection is widespread among populations (Turnock 1988), and infection of more than 95% of the population has been recorded (Erlandson 1990). Epizootics have been best studied in cyclic species of temperate forest Lepidoptera.

EPIZOOTICS AND FOREST LEPIDOPTERA Baculoviruses infect many species of forest lepidopterans (Cory et al. 1997, Evans & Entwistle 1987, Rothman & Myers 2000, Weiser 1987) and most of the mathematical models of baculovirus-host interactions specifically focus on population outbreaks and cyclic dynamics of forest Lepidoptera. To determine if virus is commonly associated with outbreaks of forest caterpillars, we reviewed information collected by the Forest Insect and Disease Survey of the Canadian Forest Service (Martineau 1984). Of 41 species of native, outbreaking species of Lepidoptera and sawflies (Hymenoptera) in Canada, NPV was recorded associated with the decline of nine species; less than a quarter. Of eight introduced forest outbreak species, virus infection was associated with the declines of four, two sawflies and two Lymantriids.

Other studies of forest caterpillars indicate that epizootics are not the norm. Species for which the population dynamics have been well studied, but for which viral epizootics do not regularly occur are; spruce needleminer, *Epinotia tedella* (Munster-Swendsen 1991), larch budmoth, *Zeiraphera diniana* (Baltensweiler & Fischlin 1988), pine beauty moth, *Panolis flammea* (Watt & Hicks 2000), winter moth and bruces spanworm, *Operophtera brumata*, (Tenow 1972), and *O. bruceata*, (Roland & Embree 1995), *Epirrita autumnata* (*Oporinia autumnata*) (Ruohomäki et al. 2000), jack pine budworm, *Choristoneura pinus pinus* (McCullough 2000), beech caterpillar, *Syntypistis punctatella*, (Kamata 2000), and fall webworm, *Hyphantria cunea* (Weiser 1987). Viral epizootics are largely limited to the families Lymantriidae (gypsy and nun moth, *Lymantria*, tussock moths, *Orgyia*), Lasiocampidae (tent caterpillars, *Malacosoma*), and Diprionidae (sawflies, *Neodiprion*). What is different about these hosts and their NPVs is not readily apparent.

Field tests of the impact of virus have occurred for three species. Spraying virus on western spruce budworm, *Choristoneura occidentalis*, a tortricid moth for which viral infection is not part of the population dynamics, caused high levels of infection (50%) and some vertical transmission. However, the epizootics were not sufficient to control the outbreak (Otvos et al. 1989). Similarly, spraying NPV on introduced populations of the winter moth, *O. brumata*, a geometrid, caused initial infection, but the virus did not persist (Cunningham et al. 1981). Spraying Douglas fir tussock moth, a lymantriid species for which virus is part of the population dynamics, successfully initiated an early epizootic and controlled the populations (Otvos et al. 1987). Thus for species that normally have epizootics, early initiation may be beneficial for control.

Long-term studies of baculoviruses and their hosts are rare. Patterns of infection with population change have been most closely studied in gypsy moth, *L. dispar* (Dwyer et al. 1997, Dwyer & Elkinton 1993, Woods & Elkinton 1987). Long-term patterns of population change and infection have also been monitored for tent caterpillars (Myers 2000). This shows that infection is not always associated with peak population densities, and that infection falls rapidly with population density after the peak (Figure 2). It has been suggested that sublethal effects might explain the reduced fecundity of moths often observed after epizootics of forest caterpillars (Myers 1988). Most of the evidence gathered on tent caterpillars supports this (Myers 2000, Myers & Kukan 1995, Rothman 1997, Rothman & Myers 1996b). An alternative is the "disease-resistance hypothesis" (Myers 2000) in which reduced fecundity is a cost of resistance of insects surviving the epizootic. Forest insect populations show large changes in population density and are frequently found in disjunct populations. This, combined with their considerable genetic variation at all levels; individuals, families, populations, and location, means that this is likely

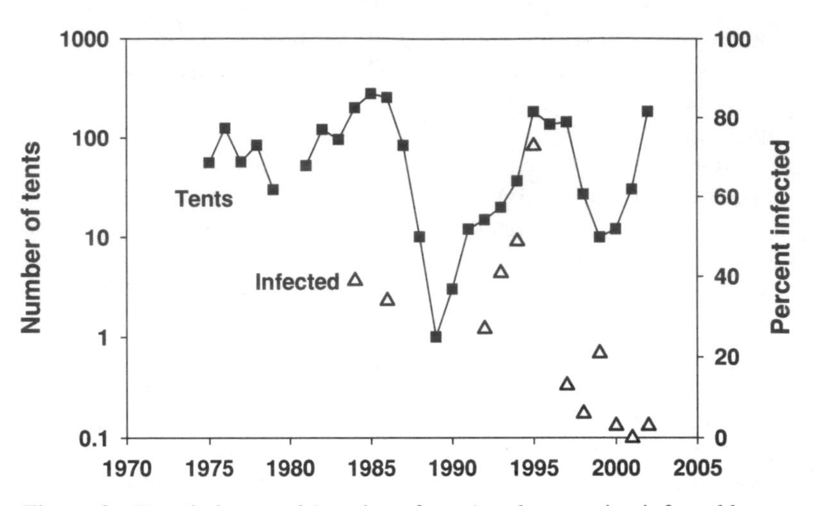

Figure 2 Population trend (number of tents) and proportion infected larvae for western tent caterpillars monitored on Mandarte Island, British Columbia.

to lead to a heterogeneous mixture of host-parasite traits, features underpinning Thompson's (1999) geographic mosaic theory of coevolution. Baculovirus-insect interactions in strongly fluctuating populations have great potential for further field tests of theory of host-parasite interaction.

CONCLUSIONS

Baculoviruses are proving to be a rich source of information on how insects and their pathogens interact, allowing the examination of the relationship from the gene to the population level. The development of increasingly sophisticated molecular tools such as microarray technology, combined with a rapidly expanding database on baculovirus genomes will provide a superb resource for examining issues such as the molecular basis of host range and the genetic basis of persistent infections. They will also provide a framework that will allow the examination of the coevolution of baculoviruses with their hosts and the features that define baculovirus species. A rich theoretical literature exists on the evolution of disease virulence (Day 2001, 2002; Lipsitch & Moxon 1997; Myers & Rothman 1995), and recent studies on baculoviruses have shown they could play a role in furthering our understanding of this key issue. The high level of variation seen in baculovirus populations is particularly fascinating, and it is already contributing to our understanding of what generates and maintains pathogen diversity and the role its plays in disease severity.

Much of the theory on epizootics of baculoviruses and forest Lepidopterans has been in the form of mathematical models that explore the conditions that might yield cyclic host population dynamics. However, many questions still remain about baculoviruses in natural populations. For example, why do baculoviruses occur widely but generally remain rare in most host populations? How is virus maintained when host densities are low? Can tri-trophic interactions influence disease epizootics? If infection is closely associated with host population fluctuations, do virulence and resistance change temporally and spatially? How are the ecology and evolution of baculoviruses influenced by their interaction with other natural enemies? Do baculoviruses (or any insect pathogen) play a role in determining insect community structure? Without learning more about the interactions between baculoviruses and their hosts in field populations, understanding of the evolutionary changes and ecological impacts of these diseases will remain elusive.

ACKNOWLEDGMENTS

We acknowledge funding from NERC, UK, and NSERC, Canada. We thank our graduate students and postdoctoral scientists for all their hard work and valuable contributions to this area of study. We would like to thank Kelli Hoover, David Theilmann, and Greg Dwyer for their comments on all, or part, of the manuscript. We would particularly like to thank Elisabeth Herniou, Lluisa Vilaplana, and Kwang Lee for use of their unpublished data and Jim Bull for access to a manuscript before publication.

LITERATURE CITED

Abot AR, Moscardi F, Fuxa JR, Sosa-Gomez DR, Richter AR. 1996. Development of resistance by *Anticarsia gemmatalis* from Brazil and the United States to a nuclear polyhedrosis virus under laboratory selection pressure. *Biol. Control* 7:126–30

Ali MI, Felton GW, Meade T, Young SY. 1998. Influence of interspecific and intraspecific host plant variation on the susceptibility of heliothines to a baculovirus. *Biol. Control* 12:42–49

Ali MI, Young SY, Felton GW, McNew RW. 2002. Influence of the host plant on occluded virus productivity and lethal infectivity of a baculovirus. *Biol. Control* 81:158–65

Allaway GP, Payne CC. 1983. A biochemical and biological comparison of three European isolates of nuclear polyhedrosis viruses from *Agrotis segetum*. *Arch. Virol.* 75:43–54

Allaway GP, Payne CC. 1984. Host range and virulence of five baculoviruses from lepidopterous hosts. *Ann. Appl. Biol.* 105:29–37

Anderson RM, May RM. 1980. Infectious diseases and population cycles of forest insects. *Science* 210:658–61

Anderson RM, May RM. 1981. The population dynamics of microparasites and their invertebrate hosts. *Phil. Trans. R. Soc. London Ser. B* 291:451–524

Arends HM, Jehle JA. 2002. Homologous recombination between the inverted terminal repeats of defective transposon TCp3.2 causes an inversion in the genome of *Cydia pomonella*. *J. Gen. Virol* 83:1573–78

Baltensweiler W, Fischlin A. 1988. The larch budmoth in the Alps. In *Dynamics of Forest Insect Populations: Patterns, Causes, Implications*. ed. A Berryman, pp. 332–53. New York: Plenum

Barber KN, Kaupp WJ, Holmes SB. 1993.

Specificity testing of the nuclear polyhedrosis virus of the gypsy moth, *Lymantria dispar* (L.) (Lepidoptera: Lymantriidae). *Can. Entomol.* 125:1055–66

Beisner BE, Myers JH. 1999. Population density and transmission of virus in experimental populations of the western tent caterpillar (Lepidoptera: Lasiocampidae). *Environ. Entomol.* 28:1107–13

Blanford S, Thomas M. 1999. Host thermal biology: the key to understanding host-pathogen interactions and microbial pest control? *Agric. For. Entomol.* 1:195–202

Bonsall M, Godfray HCJ, Briggs C, Hassell MP. 1999. Does host self-regulation increase the likelihood of insect-pathogen population cycles? *Am. Nat.* 153:228–35

Boorstein SM, Ewald PW. 1987. Costs and benefits of behavioral fever in *Melanoplus sanguinipes* infected by *Nosema acridophagus*. *Physiol. Zool.* 60:586–95

Boots M. 1998. Cannibalism and the stage-dependent transmission of a viral pathogen of the Indian meal moth, *Plodia interpunctella*. *Ecol. Entomol.* 23:118–22

Boots M, Norman R. 2000. Sublethal infection and the population dynamics of host-microparasite interactions. *J. Anim. Ecol.* 69:517–24

Bowers R, Begon M, Hodgkinson D. 1993. Host-pathogen population cycles in forest insects? Lessons from simple models reconsidered. *Oikos* 67:529–38

Brassel J, Benz G. 1979. Selection of a strain of the granulosis virus of the codling moth with improved resistance against artificial ultraviolet radiation and sunlight. *J. Invertebr. Pathol.* 33:358–63

Briggs CJ, Godfray HCJ. 1996. The dynamics of insect-pathogen interactions in seasonal environments. *Theor. Popul. Biol.* 50:149–77

Brown D. 1982. Two naturally occuring nuclear polyhedrosis virus variants of *Neodiprion sertifer* Geoffr. (Hymenoptera; Diprionidae). *Appl. Environ. Micro.* 43:65–69

Bull JC, Godfray HCJ, O'Reilly DR. 2001. Persistence of an occlusion-negative recombinant nucleopolyhedrovirus in *Trichoplusia ni* indicates high multiplicity of cellular infection. *Appl. Environ. Micro.* 67:5204–9

Bull JC, Godfray HCJ, O'Reilly DR. 2003. A few polyhedra (FP) mutant and wild-type nucleopolyhedrovirus remains as a stable polymorphism during serial coinfection in *Trichoplusia ni*. *Appl. Environ. Micro.* 69:2052–57

Burden JP, Griffiths CM, Cory JS, Smith P, Sait SM. 2002. Vertical transmission of sublethal granulovirus infection in the Indian meal moth, *Plodia interpunctella*. *Mol. Ecol.* 11:547–55

Carner GR, Hudson JS, Barnett OW. 1979. The infectivity of a nuclear polyhedrosis virus of the velvetbean caterpillar for eight noctuid hosts. *J. Invertebr. Pathol.* 33:211–16

Carruthers WR, Cory JS, Entwistle PF. 1988. Recovery of pine beauty moth *Panolis flammea* nuclear polyhedrosis virus from pine foliage. *J. Invertebr. Pathol.* 52:27–32

Cary LC, Goebel M, Corsar BG, Wang H, Rosen E, Fraser MJ. 1999. Transposon mutagenesis of baculoviruses: analysis of *Trichoplusia ni* transposon IFP2 insertions within the FP locus of nuclear polyhedrosis viruses. *Virology* 172:156–69

Clayton DH, Wolfe ND. 1993. The adaptive significance of self-medication. *Trends Ecol. Evol.* 8:60–63

Combes C. 2001. *Parasitism, the Ecology and Evolution of Intimate Interactions*. Chicago: Univ. Chicago Press. 728 pp.

Cooper D, Cory JS, Myers JH. 2003b. Hierarchical spatial structure of genetically variable nucleopolyhedroviruses infecting cyclic populations of western tent caterpillars. *Mol. Ecol.* 12:881–90

Cooper D, Cory JS, Theilmann DA, Myers JH. 2003a. Nucleopolyhedroviruses of forest and western tent caterpillars: cross-infectivity and evidence for activation of latent virus in high density field populations. *Ecol. Entomol.* 28:41–50

Cory JS, Hails RS, Sait SM. 1997. Baculovirus ecology. In *The Baculoviridae*, ed. LK Miller, pp. 301–9. New York: Plenum

Cory JS, Hirst ML, Sterling PH, Speight

MR. 2000. Narrow host range nucleopolyhedrovirus for control of the browntail moth (Lepidoptera: Lymantriidae). *Environ. Entomol.* 29:661–67

Cory JS, Wilson KR, Hails RS, O'Reilly DR. 2001. Host manipulation by insect pathogens: the effect of the baculovirus *egt* gene on host-virus interaction. In *Endocrine Interactions of Insect Parasites and Pathogens*, ed. JP Edwards, RJ Weaver, pp. 233–44, Oxford: BIOS

Crawford AM, Kalmakoff J. 1977. A host virus interaction in a pasture habitat *Wiseana* spp. Lepidoptera Hepialidae and its baculoviruses. *J. Invertebr. Pathol.* 29:81–87

Crook NE, Spencer RA, Payne CC, Leisy DJ. 1985. Variation in *Cydia pomonella* granulosis virus isolates and physical maps of the DNA from three variants. *J. Gen. Virol.* 66:2423–30

Crozier G, Crozier L, Quiot JM, Lereclus D. 1988. Recombination of *Autographa californica* and *Rachiplusia ou* nuclear polyhedrosis viruses in *Galleria mellonella* L. *J. Gen. Virol.* 69:179–85

Crozier G, Ribeiro HCT. 1992. Recombination as a possible major cause of genetic heterogeneity in *Anticarsia gemmatalis* nuclear polyhedrosis virus wild populations. *Virus Res.* 26:183–96

Cunningham JC, Tonks NV, Kaupp WJ. 1981. Viruses to control winter moth *Operophtera brumata* Lepidoptera Geometridae. *J. Entomol. Soc. B. C.* 78:17–24

D'Amico V, Elkinton JS. 1995. Rainfall effects on transmission of gypsy moth (Lepidoptera: Lymantriidae) nuclear polyhedrosis virus. *Environ. Entomol.* 24:1144–49

D'Amico V, Elkinton JS, Dwyer G, Burand JP, Buonaccorsi JP. 1996. Virus transmission in gypsy moths is not a simple mass action process. *Ecology* 77:201–6

Day T. 2001. Parasite transmission modes and the evolution of virulence. *Evolution* 55:2389–400

Day T. 2002. The evolution of virulence in vector-borne and directly transmitted parasites. *Theoret. Popul. Biol.* 62:199–213

DeFillippis VR, Villarreal LP. 2000. An introduction to the evolutionary ecology of viruses. See Hurst 2000, pp. 125–208

Doyle CJ, Hirst ML, Cory JS, Entwistle PF. 1990. Risk assessment studies: detailed host range testing of wild-type cabbage moth *Mamestra brassicae* Lepidoptera Noctuidae nuclear polyhedrosis virus. *Appl. Environ. Micro.* 56:2704–10

Duan L, Otvos IS. 2001. Influence of larval age and virus concentration on mortality and sublethal effects of a nucleopolyhedrovirus on the Western spruce budworm (Lepidoptera: Tortricidae). *Environ. Entomol.* 30:136–46

Duffey SS, Hoover K, Bonning BC, Hammock BD. 1995. The impact of host plant on the efficacy of baculoviruses. *Rev. Pest. Toxicol.* 3:137–275

Dushoff J, Dwyer G. 2001. Evaluating the risks of engineered viruses: modeling pathogen competition. *Ecol. Appl.* 11:1602–9

Dwyer G. 1991. The roles of density, stage, and patchiness in the transmission of an insect virus. *Ecology* 72:559–74

Dwyer G, Dushoff J, Elkinton JS, Burand JP, Levin SA. 2002. Host heterogeneity in susceptibility: lessons from an insect virus. In *Adaptive Dynamics of Infectious Diseases*, ed. U Dieckmann, JAJ Metz, MW Sabelis, K Sigmund, pp. 74–84. Cambridge, UK: Cambridge Univ. Press

Dwyer G, Dushoff J, Elkinton JS, Levin SA. 2000. Pathogen-driven outbreaks in forest defoliators revisited: building models from experimental data. *Am. Nat.* 156:105–20

Dwyer G, Elkinton JS. 1993. Using simple models to predict virus epizootics in gypsy moth populations. *J. Anim. Ecol.* 62:1–11

Dwyer G, Elkinton JS, Buonaccorsi JP. 1997. Host heterogeneity in susceptibility and disease dynamics: tests of a mathematical model. *Am. Nat.* 150:685–707

Easwaramoorthy E, Jayaraj S. 1989. Vertical transmission of granulosis virus of sugarcane shoot borer, *Chilo infuscatellus* Snell. *Trop. Pest Manage.* 35:352–53

Ebling PM, Kaupp WJ. 1995. Differentiation and comparative activity of six isolates of a nuclear polyhedrosis virus from the forest tent caterpillar, *Malacosoma disstria*, Hübner. *J. Invertebr. Pathol.* 66:198–200

Edwards JP, Weaver RJ, eds. 2001. *Endocrine Interactions of Insect Parasites and Pathogens*. Oxford, UK: BIOS

Engelhard EK, Volkman LE. 1995. Developmental resistance in fourth instar *Trichoplusia ni* orally inoculated with *Autographa californica* M nuclear polyhedrosis virus. *Virology* 209:384–89

Engelhard EK, Kam-Morgan LNW, Washburn JO, Volkman LE. 1994. The insect tracheal system: a conduit for the systemic spread of *Autographa californica* M nuclear polyhedrosis virus. *Proc. Natl. Acad. Sci. USA* 91:3224–27

Entwistle PF, Evans HF. 1985. Viral control. In *Comprehensive Insect Physiology, Biochemistry, and Pharmacology*, ed. LI Gilbert, GA Kerkut, 12:347–412. Oxford: Pergamon

Entwistle PF, Forkner AC, Green BM, Cory JS. 1993. Avian dispersal of nuclear polyhedrosis virus after induced epizootics in the pine beauty moth, *Panolis flammea*, (Lepidoptera: Noctuidae). *Biol. Control* 3:61–69

Erlandson MA. 1990. Biological and biochemical comparison of *Mamestra configurata* and *Mamestra brassicae* nuclear polyhedrosis virus isolates pathogenic for the bertha armyworm, *Mamestra configurata* (Lepidoptera: Noctuidae). *J. Invertebr. Pathol.* 56:47–56

Escribano A, Williams T, Goulson D, Cave RD, Chapman JW, Caballero P. 2000. Effect of parasitism on a nucleopolyhedrovirus amplified in *Spodoptera frugiperda* larvae parasitized by *Campoletis sonorensis*. *Entomol. Exp. Appl* 97:257–64

Escribano A, Williams T, Goulson D, Cave RD, Caballero P. 2001. Consequences of interspecific competition on the virulence and genetic composition of a nucleopolyhedrovirus in *Spodoptera frugiperda* larvae parasitized by *Chelonus insularis*. *Biocontrol Sci. Technol.* 11:649–62

Evans HF, Entwistle PF. 1987. Viral diseases. In *Epizootiology of Insect Disease*, ed. JR Fuxa, Y Tanada, pp. 257–322. New York: Wiley & Sons

Farrar RRJ, Ridgway RL. 2000. Host plant effects on the activity of selected nuclear polyhedrosis viruses against the corn earworm and beet armyworm (Lepidoptera: Noctuidae). *Environ. Entomol.* 29:108–15

Fenton A, Fairbairn JP, Norman R, Hudson PJ. 2002. Parasite transmission: reconciling theory and reality. *J. Anim. Ecol.* 71:893–905

Forschler BT, Young SY, Felton GW. 1992. Diet and the susceptibility of *Helicoverpa zea* (Noctuidae: Lepidoptera) to a nuclear polyhedrosis virus. *Environ. Entomol.* 21:1220–23

Fox CW, Roff, DA, Fairbairn DJ, eds. 2001. *Evolutionary Ecology*. Oxford, UK: Oxford Univ. Press

Frank SA. 1996. Models of parasite virulence. *Q. Rev. Biol.* 71:37–78

Frid L. 2002. *Thermal ecology of western tent caterpillars Malacosoma californicum pluviale and infection by nucleopolyhedrovirus*. MSc thesis. Univ. British Columbia, Vancouver. 48 pp.

Frid L, Myers JH. 2002. Thermal ecology of western tent caterpillars *Malacosoma californicum pluviale* and infection by nucleopolyhedrovirus. *Ecol. Entomol.* 27:665–63

Fuxa JR, Richter AR. 1989. Reversion of resistance by *Spodoptera frugiperda* to nuclear polyhedrosis virus. *J. Invertebr. Pathol.* 53:52–56

Fuxa JR, Richter AR. 1998. Repeated reversion of resistance to nucleopolyhedrovirus by *Anticarsia gemmatalis*. *J. Invertebr. Pathol.* 71:159–64

Fuxa JR, Richter AR. 2001. Quantification of soil-to-plant transport of recombinant nucleopolyhedrovirus: effects of soil type and moisture, air currents, and precipitation. *Appl. Environ. Micro.* 67:5166–70

Fuxa JR, Richter AR, Ameen AO, Hammock BD. 2002. Vertical transmission of TnSNPV, TnCPV, AcMNPV, and possibly recombinant NPV in *Trichoplusia ni*. *J. Invertebr. Pathol.* 79:44–50

Fuxa JR, Sun J-Z, Weidner EH, LaMotte LR.

1999. Stressors and rearing diseases of *Trichoplusia ni*: evidence of vertical transmission of NPV and CPV. *J. Invertebr. Pathol.* 74:149–55

Fuxa JR, Weidner EH, Richter AR. 1992. Polyhedra without virions in a vertically transmitted nuclear polyhedrosis virus. *J. Invertebr. Pathol.* 60:53–58

Garcia-Maruniak A, Pavan OHO, Maruniak JE. 1992. A variable region of *Anticarsia gemmatalis* nuclear polyhedrosis virus contains tandemly repeated DNA sequences. *Virus Res.* 41:123–32

Gelernter WD, Federici BA. 1986. Isolation, identification, and determination of virulence of a nuclear polyhedrosis virus from the beet armyworm, *Spodoptera exigua* (Lepidoptera: Noctuidae). *Environ. Entomol.* 15:240–45

Gettig RG, McCarthy WJ. 1982. Genotypic variation among wild isolates of *Heliothis* spp. nuclear polyhedrosis viruses from different geographic regions. *Virology* 117:245–52

Getz W, Pickering J. 1983. Epidemic models: thresholds and population regulation. *Am. Nat.* 121:893–98

Godfray HCJ, O'Reilly DR, Briggs CJ. 1997. A model of nucleopolyhedrovirus (NPV) population genetics applied to co-occlusion and the spread of the few polyhedra (FP) phenotype. *Proc. R. Soc. London Ser. B* 264:315–22

Goulson D. 1997. Wipfelkrankheit: modification of host behavior during baculoviral infection. *Oecologia* 109:219–28

Goulson D, Cory JS. 1995. Responses of *Mamestra brassicae* (Lepidoptera: Noctuidae) to crowding: interactions with disease resistance, color phase and growth. *Oecologia* 104:416–23

Goulson D, Hails RS, Williams T, Hirst ML, Vasconcelos SD, et al. 1995. Transmission dynamics of a virus in a stage-structured insect population. *Ecology* 76:392–401

Hails RS, Hernandez-Crespo P, Sait SM, Donnelly CA, Green BM, Cory JS. 2002. Transmission patterns of natural and recombinant baculoviruses. *Ecology* 83:906–16

Hamm JJ, Styer EL. 1985. Comparative pathology of isolates of *Spodoptera frugiperda* nuclear polyhedrosis virus in *S. frugiperda* and *S. exigua*. *J. Gen. Virol.* 66:1249–62

Harrison RF, Bonning BC. 1999. The nucleopolyhedroviruses of *Rachoplusia ou* and *Anagrapha falcifera* are isolates of the same virus. *J. Gen. Virol.* 80:2793–98

Hajós JP, Pijnenburg J, Usmany M, Zuidema D, Závodszky P, Vlak JM. 2000. High frequency recombination between homologous baculoviruses in cell culture. *Arch. Virol.* 145:159–64

Hatfield PR, Entwistle PF. 1988. Biological and biochemical comparison of nuclear polyhedrosis virus isolates pathogenic for the oriental armyworm, *Mythimna separata* (Lepidoptera: Noctuidae). *J. Invertebr. Pathol.* 52:168–76

Hawtin RE, Zarkowska T, Arnold K, Thomas CJ, Gooday GW, et al. 1997. Liquefaction of *Autographa californica* nucleopolyhedrovirus-infected insects is dependent on the integrity of virus-encoded chitinase and cathepsin genes. *Virology* 238: 243–53

Hernández-Crespo P, Sait SM, Hails RS, Cory JS. 2001. Behavior of a recombinant baculovirus in lepidopteran hosts with different susceptibilities. *Appl. Environ. Micro.* 67: 1140–46

Herniou EA, Luque T, Chen X, Vlak JM, Winstanley D, et al. 2001. Use of whole genome sequence data to infer baculovirus phylogeny. *J. Virol.* 75:8117–26

Herniou EA, Olszewski J, Cory JS, O'Reilly DR. 2003. The genome sequence and evolution of baculoviruses. *Annu. Rev. Entomol.* 48:211–34

Hochberg M. 1991. Nonlinear transmission rates and the dynamics of infectious diseases. *J. Theor. Biol.* 153:301–21

Hochberg ME. 1989. The potential role of pathogens in biological control. *Nature* 337:262–65

Hodgson DJ, Hitchman RB, Vanbergen AJ, Hails RS, Hartley SE, et al. 2003. The existence and persistence of genotypic variation in nucleopolyhedrovirus populations. In *Genes in the Environment, British Ecological Soc. Symp. 15*, ed. RS Hails, JE Beringer, HCJ Godfray, pp. 258–80. Oxford, UK: Blackwell

Hodgson DJ, Vanbergen AJ, Hartley SE, Hails RS, Cory JS. 2002. Differential selection of baculovirus genotypes mediated by different species of host food plant. *Ecol. Letts.* 5:512–18

Hodgson DJ, Vanbergen AJ, Watt AD, Hails RS, Cory JS. 2001. Phenotypic variation between naturally co-existing genotypes of a Lepidopteran baculovirus. *Evol. Ecol. Res.* 3:687–701

Hoover K, Alaniz SA, Yee JL, Rocke DM, Hammock BD, Duffey SS. 1998d. Dietary protein and chlorogenic acid effect on baculoviral disease of noctuid (Lepidoptera: Noctuidae) larvae. *Environ. Entomol.* 27:1264–72

Hoover K, Grove MJ, Su S. 2002. Systemic component to intrastadial developmental resistance in *Lymantria dispar* to its baculovirus. *Biol. Control* 25:92–98

Hoover K, Kishida KT, Digiorgio LA, Workman J, Alaniz SA, et al. 1998c. Inhibition of baculoviral disease by plant-mediated peroxidase activity and free radical generation. *J. Chem. Ecol.* 24:1949–2001

Hoover K, Stout MJ, Alaniz SA, Hammock BD, Duffey SS. 1998b. Influence of induced plant defenses in cotton and tomato on the efficacy of baculoviruses on noctuid larvae. *J. Chem. Ecol.* 24:253–71

Hoover K, Washburn JO, Volkman LE. 2000. Midgut-based resistance of *Heliothis virescens* to baculovirus infection mediated by phytochemicals in cotton. *J. Insect Physiol.* 46:999–1007

Hoover K, Yee JL, Schultz CM, Rocke DM, Hammock BD, Duffey SS. 1998a. Effects of plant identity and chemical constituents on the efficacy of a baculovirus against (*Heliothis virescens*). *J. Chem. Ecol.* 24:221–52

Hughes D, Possee RD, King LA. 1993. Activation and detection of a latent baculovirus resembling *Mamestra brassicae* nuclear polyhedrosis virus in *M. brassicae* insects. *Virology* 194:600–15

Hughes DS, Possee RD, King LA. 1997. Evidence for the presence of a low level, persistent baculovirus infection of *Mamestra brassicae* insects. *J. Gen. Virol.* 78:1801–5

Hughes PR, Gettig RR, McCarthy WJ. 1983. Comparison of the time-mortality response of *Heliothis zea* to 14 isolates of *Heliothis* nuclear polyhedrosis virus. *J. Invertebr. Pathol.* 41:256–61

Hunter MD, Schultz JC. 1993. Induced plant defenses breached? Phytochemical induction protects an herbivore from disease. *Oecologia* 94:195–203

Hurst CJ, ed. 2000. *Viral Ecology*. San Diego: Academic

Jehle JA, Fritsch E, Nickel A, Huber J, Backhaus H. 1995. TCI4.7: a novel lepidopteran transposon found in *Cydia pomonella* granulosis virus. *Virology* 207:369–79

Kamata N. 2000. Population dynamics of the beech caterpillar, *Syntypistis punctatella*, and biotic and abiotic factors. *Popul. Ecol.* 42:267–78

Karban R, English-Loeb G. 1997. Tachinid parasitoids affect host plant choice by caterpillars to increase caterpillar survival. *Ecology* 78:603–11

Keddie BA, Aponte GW, Volkman LE. 1989. The pathway of infection of *Autographa californica* nuclear polyhedrosis virus in an insect host. *Science* 243:1728–30

Keeley LL, Vinson SB. 1975. β-ecdysone effects on the development of nucleopolyhedrosis in *Heliothis* spp. *J. Invertebr. Pathol.* 26:121–23

Kelly PM, Entwistle PF. 1988. In vivo mass production in the cabbage moth (*Mamestra brassicae*) of a heterologous (*Panolis*) and a homologous (*Mamestra*) nuclear polyhedrosis virus. *J. Virol. Meth.* 19:249–56

Kelly TJ, Park EJ, Masler CA, Burand JP. 1995. Characterization of the glycosylated ecdysteroids in the hemolymph of baculovirus-infected gypsy moth larvae and cells in culture. *Eur. J. Entomol.* 92:51–61

Kikhno I, Gutierrez S, Crozier L, Croizier G,

López-Ferber M. 2002. Characterisation of *pif*, a gene required for the per os infectivity of *Spodoptera littoralis* nucleopolyhedrovirus. *J. Gen. Virol.* 82:3013–22

Kirkpatrick BA, Washburn JO, Volkman LE. 1998. AcMNPV pathogenesis and developmental resistance in fifth instar *Heliothis virescens*. *J. Invertebr. Pathol.* 72:63–72

Kislev N, Edelman M. 1982. DNA restriction-pattern differences from geographic isolates of *Spodoptera littoralis* nuclear polyhedrosis virus. *Virology* 119:219–22

Knell RJ, Begon M, Thompson DJ. 1998. Transmission of *Plodia interpunctella* granulosis virus does not conform to the mass action model. *J. Anim. Ecol.* 67:592–599

Kolodny-Hirsch DM, van Beek NAM. 1997. Selection of a morphological variant of *Autographa californica* nuclear polyhedrosis virus with increased virulence following serial passage in *Plutella xylostella*. *J. Invertebr. Pathol.* 69:205–11

Kukan B. 1999. Vertical transmission of nucleopolyhedrovirus in insects. *J. Invertebr. Pathol.* 74:103–11

Kunimi Y, Yamada E. 1990. Relationship of larval phase and susceptibility of the armyworm, *Pseudoletia separata* Walker (Lepidoptera, Noctuidae) to a nuclear polyhedrosis-virus and a granulosis-virus. *Appl. Entomol. Zool.* 25:289–97

Lee KP. 2002. *Ecological factors impacting on the nutritional biology of a generalist and a specialist caterpillar: effects of pathogen and plant structural compound on macronutrient balancing*. DPhil thesis. Univ. Oxford. 168 pp.

Lee Y, Fuxa JR. 2000. Transport of wild-type and recombinant nucleopolyhedroviruses by scavenging and predatory arthropods. *Micro. Ecol.* 39:301–13

Li Q, Donly C, Li L, Willis LG, Theilmann DA, Erlandson M. 2000a. Sequence organization of the *Mamestra configurata* nucleopolyhedrovirus genome. *Virology* 294:106–21

Li L, Donly C, Li Q, Willis LG, Keddie BA, et al. 2002b. Identification and genomic analysis of a second species of nucleopolyhedrovirus isolated from *Mamestra configurata*. *Virology* 297:226–44

Lipsitch M, Moxon ER. 1997. Virulence and transmissibility of pathogens: what is the relationship? *Trends Microbiol.* 5:31–37

Lively CM. 2001. Parasite-host interactions. See Fox et al. 2001, pp. 290–302

Lozano GA. 1998. Parasitic stress and self-medication in wild animals. *Adv. Stud. Behav.* 27:291–317

Lynn DE, Shapiro M, Dougherty EM. 1993. Selection and screening of clonal isolates of the Abington strain of gypsy moth nuclear polyhedrosis virus. *J. Invertebr. Pathol.* 62:191–95

Martineau R. 1984. *Insects Harmful to Forest Trees*. Montreal: Multiscience

Maruniak JE, Brown SE, Knudson DL. 1984. Physical maps of SfMNPV baculovirus DNA and its genomic variants. *Virology* 136:221–34

Mason PG, Arthur AP, Olfert OO, Erlandson MA. 1998. The bertha armyworm (*Mamestra configurata*) (Lepidoptera: Noctuidae) in western Canada. *Can. Entomol.* 130:321–36

Matthews HJ, Smith I, Edwards JP. 2002. Lethal and sublethal effects of a granulovirus on the tomato moth *Lacanobia oleracea*. *J. Invertebr. Pathol.* 80:73–80

McCallum H, Barlow N, Hone J. 2001. How should pathogen transmission be modelled? *Trends Ecol. Evol.* 16:295–300

McCullough DG. 2000. A review of factors affecting the population dynamics of jack pine budworm (*Choristoneura pinus pinus* Freeman). *Popul. Ecol.* 42:243–56

McVean RIK, Sait SM, Thompson DJ, Begon M. 2002. Dietary stress reduces the susceptibility of *Plodia interpunctella* to infection by a granulovirus. *Biol. Control* 25:81–84

Melamed-Madjar V, Raccah B. 1979. The transstadial and vertical transmission of a granulosis virus from the corn borer *Sesamia nonagrioides*. *J. Invertebr. Pathol.* 33:259–64

Milks ML. 1997. Comparative biology and susceptibility of cabbage looper (Lepidoptera:

Noctuidae) lines to a nuclear polyhedrosis virus. *Environ. Entomol.* 26:839–48

Milks ML, Burnstyn I, Myers JH. 1998. Influence of larval age on the lethal and sublethal effects of the nucleopolyhedrovirus of *Trichoplusia ni* the cabbage looper. *Biol. Control* 12:119–26

Milks ML, Myers JH. 2000. The development of larval resistance to a nucleopolyhedrovirus is not accompanied by an increased virulence in the virus. *Evol. Ecol.* 14:645–64

Milks ML, Myers JH. 2003. Cabbage looper resistance to a nucleopolyhedrovirus confers cross-resistance to two granuloviruses. *Environ. Entomol.* 32:286–89

Milks ML, Myers JH, Leptich MK. 2002. Costs and stability of cabbage looper resistance to a nucleopolyhedrovirus. *Evol. Ecol.* 16:369–85

Miller LK, ed. 1997. *The Baculoviridae*. New York: Plenum

Mitchell FL, Fuxa JR. 1990. Multiple regression analysis of factors influencing a nuclear polyhedrosis virus in populations of fall armyworm Lepidoptera Noctuidae in corn. *Environ. Entomol.* 19:260–67

Moore J. 2002. *Parasites and the Behavior of Animals*. Oxford, UK: Oxford Univ. Press. 315 pp.

Moscardi F. 1999. Assessment of the application of baculoviruses for control of Lepidoptera. *Annu. Rev. Entomol.* 44:257–89

Moser BA, Becnel JJ, White SE, Alfonso C, Kutish G, et al. 2001. Morphological and molecular evidence that *Culex nigripalpus* baculovirus is an unusual member of the family Baculoviridae. *J. Gen. Virol.* 82:283–97

Muñoz D, Caballero P. 2000. Persistence and effects of parasitic genotypes in a mixed population of the *Spodoptera exigua* nucleopolyhedrovirus. *Biol. Control* 19:259–64

Muñoz D, Castillejo J, Caballero P. 1998. Naturally occurring deletion mutants are parasitic genotypes in a wild-type nucleopolyhedrovirus population. *Appl. Environ. Micro.* 64:4372–77

Muñoz D, Murillo R, Krell PJ, Vlak JM, Caballero P. 1999. Four genotypic variants of a *Spodoptera exigua* nucleopolyhedrovirus (Se-SP2) are distinguishable by a hypervariable genomic region. *Virus Res.* 59:61–74

Muñoz D, Vlak JM, Caballero P. 1997. In vivo recombination between two strains of the genus *Nucleopolyhedrovirus* in its natural host *Spodoptera exigua*. *Appl. Environ. Micro.* 63:3025–31

Munster-Swendsen M. 1991. The effect of sublethal neogregarine infections in the spruce needleminer, *Epinotia tedella* (Lepidoptera: Tortricidae). *Ecol. Entomol.* 16:211–19

Myers JH. 1988. Can a general hypothesis explain population cycles of forest Lepidoptera? *Adv. Ecol. Res.* 18:179–242

Myers JH. 2000. Population fluctuations of the western tent caterpillar in southwestern British Columbia. *Popul. Ecol.* 42:231–41

Myers JH, Kukan B. 1995. Changes in the fecundity of tent caterpillars: a correlated character of disease resistance or sublethal effects of disease? *Oecologia* 103:475–80

Myers JH, Malakar R, Cory JS. 2000. Sublethal nucleopolyhedrovirus infection effects on female pupal weight, egg mass size, and vertical transmission in gypsy moth (Lepidoptera: Lymantriidae). *Environ. Entomol.* 29:1268–72

Myers JH, Rothman LE. 1995. Virulence and transmission of infectious diseases in humans and insects: evolutionary and demographic patterns. *Trends Ecol. Evol.* 10:194–98

Olofsson E. 1989. Transmission of the nuclear polyhedrosis virus of the European pine sawfly from adult to offspring. *J. Invertebr. Pathol.* 54:322–30

O'Reilly DR. 1997. Auxiliary genes of baculoviruses. In *The Baculoviruses*, ed. LK Miller, pp. 267–300. New York: Plenum

O'Reilly DR, Brown MR, Miller LK. 1992. Alteration of ecdysteroid metabolism due to baculovirus infection of the fall armyworm *Spodoptera frugiperda* host ecdysteroids are conjugated with galactose. *Insect Biochem. Mol. Biol.* 22:313–20

O'Reilly DR, Hails RS, Kelly TJ. 1998. The impact of host developmental status on baculovirus replication. *J. Invertebr. Pathol.* 72:269–75

O'Reilly DR, Kelly TJ, Masler EP, Thyagaraja BS, Robson RM, et al. 1995. Overexpression of *Bombyx mori* prothoracicotropic hormone using baculovirus vectors. *Insect Biochem. Mol. Biol* 25:475–85

O'Reilly DR, Miller LK. 1989. A baculovirus blocks insect molting by producing ecdysteroid UDP-glucosyltransferase. *Science* 245:1110–12

O'Reilly DR, Miller LK. 1991. Improvement of a baculovirus pesticide by deletion of the *egt* gene. *Bio/Technology* 9:1086–89

Otvos IS, Cunningham JC, Alfaro RI. 1987. Aerial application of nuclear polyhedrosis virus against Douglas-fir tussock moth *Orgyia pseudotsugata* Mcdunnough Lepidoptera Lymantriidae ii. Impact 1 and 2 years after application. *Can. Entomol.* 119:707–16

Otvos IS, Cunningham JC, Kaupp WJ. 1989. Aerial application of two baculoviruses against the western spruce budworm *Choristoneura occidentalis* Freeman Lepidoptera Tortricidae in British Columbia Canada. *Can. Entomol.* 121:209–18

Parnell M, Grzywacz D, Jones KA, Brown M, Odour G, Ong'aro J. 2002. The strain variation and virulence of granulovirus of diamondback moth (*Plutella xylostella* Linnaeus, Lep., Yponomeutidae) isolated in Kenya. *J. Invertebr. Pathol.* 79:192–96

Paul RK. 1997. *Evolution and interaction of insect pathogens*. PhD thesis. Univ. Reading. 248 pp.

Payne CC. 1986. Insect pathogenic viruses as pest control agents. In *Biological Plant and Health Protection*, ed. JM Franz. pp. 183–200. Stuttgart: Fischer

Peng J, Zhong J, Granados RR. 1999. A baculovirus enhancin alters the permeability of a mucosal midgut peritrophic matrix from lepidopteran larvae. *J. Insect Physiol.* 45:159–66

Podgwaite JD, Mazzone HM. 1986. Latency of insect viruses. *Adv. Virus Res.* 31:293–320

Popham HJR, Bischoff DS, Slavicek JM. 2001. Both *Lymantria dispar* nucleopolyhedrovirus enhancin genes contribute to viral potency. *J. Virol.* 75:8639–48

Poulin R. 1998. *Evolutionary Ecology of Parasites*. Boca Raton, FL.: Chapman and Hall. 212 pp.

Raymond B, Vanbergen A, Pearce I, Hartley SE, Cory JS, Hails RS. 2002. Host plant species can influence the fitness of herbivore pathogens: the winter moth and its nucleopolyhedrovirus. *Oecologia* 131:533–41

Reeson AF, Wilson K, Cory JS, Hankard P, Weeks JM, et al. 2000. Effects of phenotypic plasticity on pathogen transmission in the field in a Lepidoptera-NPV system. *Oecologia* 124:373–80

Reeson AF, Wilson K, Gunn A, Hails RS, Goulson D. 1998. Baculovirus resistance in the noctuid *Spodoptera exempta* is phenotypically plastic and responds to population density. *Proc. R. Soc. London Ser. B* 265:1787–91

Regniere J. 1984. Vertical transmission of diseases and population dynamics of insects with discrete generations: a model. *J. Theor. Biol.* 107:287–301

Ribeiro HCT, Pavan OHO, Muotri AR. 1997. Comparative susceptibility of two different hosts to genotypic variants of the *Anticarsia gemmatalis* nuclear polyhedrosis virus. *Entomol. Exp. Appl.* 83:233–37

Richards A, Cory J, Speight M, Williams T. 1999b. Foraging in a pathogen reservoir can lead to local host population extinction: A case study of a Lepidoptera-virus interaction. *Oecologia* 118:29–38

Richards A, Speight M, Cory J. 1999a. Characterization of a nucleopolyhedrovirus from the vapourer moth, *Orgyia antiqua* (Lepidoptera Lymantriidae). *J. Invertebr. Pathol.* 74:137–42

Rigby MC, Hechinger, RF, Stevens L. 2002. Why should parasite resistance be costly? *Trends Parasitol.* 18:116–20

Roland JH, Embree DG. 1995. Biological control of the winter moth. *Annu. Rev. Entomol.* 40:475–92

Rothman LD. 1997. Immediate and delayed effects of a viral pathogen and density on tent caterpillar performance. *Ecology* 78:1481–93

Rothman LD, Myers JH. 1996a. Debilitating effects of viral diseases on host Lepidoptera. *J. Invertebr. Pathol.* 67:1–10

Rothman LD, Myers JH. 1996b. Is fecundity correlated with resistance to viral disease in the western tent caterpillar? *Ecol. Entomol.* 21:396–98

Rothman LD, Myers JH. 2000. Ecology of insect viruses. See Hurst 2000, pp. 385–412

Rothman LD, Roland J. 1998. Forest fragmentation and colony performance of forest tent caterpillar. *Ecography* 21:383–91

Ruohomäki K, Tanhuanpää M, Ayres MD, Kaitaniemi P, Tammaru T, Haukioja E. 2000. Causes of cyclicity of *Epirrita autumnata* (Lepidoptera, Geometridae): grandiose theory and tedious practice. *Popul. Ecol.* 42:211–24

Schmid-Hempel P, Ebert D. 2003. On the evolutionary ecology of specific immune defense. *Trends Ecol. Evol.* 18:27–32

Shapiro DI, Fuxa JR, Braymer HD, Pashley DP. 1991. DNA restriction polymorphism in wild isolates of *Spodoptera frugiperda* nuclear polyhedrosis virus. *J. Invertebr. Pathol.* 58:96–105

Siva-Jothy MT, Thompson JJW. 2002. Short-term nutrient deprivation affects immune function. *Physiol. Entomol.* 27:206–12

Slack JM, Kuzio J, Faulkner P. 1995. Characterization of v-cath, a cathepsin L-like proteinase expressed by the baculovirus *Autographa californica* multiple nuclear polyhedrosis virus. *J. Gen. Virol* 76:1091–98

Slavicek JM, Popham HJR, Riegel CI. 1999. Deletion of the *Lymantria dispar* multicapsid nucleopolyhedrosis ecdysteroid UDP-glucosyl transferase gene enhances viral killing speed in the last instar of the gypsy moth. *Biol. Control* 16:91–103

Smith IRL, Crook NE. 1988. In vivo isolation of baculovirus genotypes. *Virology* 166:240–44

Smith IRL, Crook NE. 1993. Characterization of new baculovirus genotypes arising from inoculation of *Pieris brassicae* with granulosis viruses. *J. Gen. Virol.* 74:415–24

Steinhaus EA. 1967. *Principles of Insect Pathology*. New York: McGraw-Hill

Stiles S, Himmerich B. 1998. *Autographa californica* NPV isolates: restriction endonuclease analysis and comparative biological activity. *J. Invertebr. Pathol.* 72:174–77

Suwanchaichinda C, Paskewitz SM. 1998. Effects of larval nutrition, adult body size and adult temperature on the ability of *Anopheles gambiae* (Diptera: Culicidae) to melanise sephadex beads. *J. Med. Entomol.* 35:157–61

Takatsuka J, Okuno S, Nakai M, Kunimi Y. 2003. Genetic and biological comparison of ten geographic isolates of a nucleopolyhedrovirus that infects *Spodoptera litura* (Lepidoptera: Noctuidae). *Biol. Control* 26:32–39

Taylor LH, MacKinnon MJ, Read AF. 1998. Virulence of mixed-clone and single clone infections of the rodent malaria *Plasmodium chabaudi*. *Evolution* 52:583–91

Teakle RE, Jensen JM, Giles JE. 1986. Age-related susceptibility of *Heliothis punctiger* to a commercial formulation of nuclear polyhedrosis virus. *J. Invertebr. Pathol.* 47:82–92

Tenow O. 1972. The outbreaks of *Oporina autumnata* Bkh. and *Operophtera* spp. (Lep. Geometridae) in the Scandinavian mountain chain and northern Finland 1862–1968. *Zool. Bidrag Uppsala, Suppl.* 2:1–107

Thompson CG, Scott DW, Wickman BE. 1981. Long-term persistence of the nuclear polyhedrosis virus of the Douglas-fir tussock moth, *Orgyia pseudotsugata* (Lepidoptera: Lymantriidae), in forest soil. *Environ. Entomol.* 10:254–55

Thompson JN. 1999. Specific hypothesis on the geographic mosaic of coevolution. *Am. Nat.* 153:S1–S14

Thompson JN. 2001. The geographic dynamics of coevolution. See Fox et al. 2001, pp. 331–43

Tompkins GJ, Dougherty EM, Adams JR, Diggs D. 1988. Changes in the virulence of nuclear polyhedrosis viruses when propagated in alternate noctuid (Lepidoptera: Noctuidae) cell lines and hosts. *J. Econ. Entomol.* 81:1027–32

Turnock WJ. 1988. Density, parasitism, and disease incidence of larvae of the bertha armyworm, *Mamestra configurata* Walker (Lepidoptera: Noctuidae), in Manitoba, 1973–86. *Can. Entomol.* 120:401–13

Vail PV, Tebbets JS. 1990. Comparative biology and susceptibility of *Plodia-interpunctella* Lepidoptera Pyralidae populations to a granulosis virus. *Environ. Entomol.* 19:791–94

Van Beek N, Hughes PR, Wood HA. 2000. Effects of incubation temperature on the dose-survival time relationship of Trichoplusia ni larvae infected with *Autographa californica* nucleopolyhedrrovirus. *J. Invertebr. Pathol.* 76:185–90

Vasconcelos SD. 1996. *Studies on the transmission and dispersal of baculoviruses in Lepidopteran populations*. DPhil thesis, Univ. Oxford. 168 pp.

Vasconcelos SD, Cory JS, Wilson KR, Sait SM, Hails RS. 1996a. Modified behavior in baculovirus-infected lepidopteran larvae and its impact on the spatial distribution of inoculum. *Biol. Control* 7:299–306

Vasconcelos SD, Williams T, Hails RS, Cory JS. 1996b. Prey selection and baculovirus dissemination by carabid predators of Lepidoptera. *Ecol. Entomol.* 21:98–104

Vass E, Napi AJ. 1998. The effects of dietary yeast on the cellular immune response of *Drosophila melanogaster* against the larval parasitoid, *Leptopilina boulardi*. *J. Parasitol.* 84:870–72

Vezina A, Peterman R. 1985. Tests of the role of nuclear polyhedrosis virus in the population dynamics of its host, Douglas-fir tussock moth, *Orgyia pseudotsugata* (Lepidpotera: Lymantriidae). *Oecologia* 67:260–66

Vickers JM, Cory JS, Entwistle PF. 1991. DNA characterization of eight geographic isolates of granulosis virus from the potato tuber moth *Phthorimaea operculella* Lepidoptera Gelechiidae. *J. Invertebr. Pathol.* 57:334–42

Volkman LE. 1997. Nucleopolyhedrosis interactions with their insect hosts. *Adv. Virus Res.* 48:313–48

Washburn JO, Kirkpatrick BA, Haas-Stapleton E, Volkman LE. 1998. Evidence that the stilbene-derived optical brightener M2R enhances *Autographa californica* M nucleopolyhedrovirus infection of *Trichoplusia ni* and *Heliothis virescens* by preventing sloughing of infected midgut epithelial cells. *Biol. Control* 11:58–69

Washburn JO, Kirkpatrick BA, Volkman LE. 1995. Comparative pathogenesis of *Autographa californica* M nuclear polyhedrosis virus in larvae of *Trichoplusia ni* and *Heliothis virescens*. *Virology* 209:561–68

Washburn JO, Kirkpatrick BA, Volkman LE. 1996. Insect protection against viruses. *Nature* 383:767

Washburn JO, Lyons EH, Haas-Stapleton EJ, Volkman LE. 1999. Multiple nucleocapsid packaging of *Autographa californica* nucleopolyhedrovirus accelerates the onset of systemic infection in *Trichoplusia ni*. *J. Virol.* 73:411–16

Watanabe H. 1987. The host population. In *Epizootiology of Insect Diseases*, ed. JA Fuxa, Y Tanada, pp. 71–112. New York: Wiley & Sons

Watt AD, Hicks BJ. 2000. A reppraisal of the populations dynamics of the pine beauty moth, *Panolis flammea*, on lodgepole pine, *Pinus contorta*, in Scotland. *Popul. Ecol.* 42:225–30

Weiser J. 1987. Patterns over place and time. In *Epizootiology of Insect Disease*, ed. JR Fuxa, Y Tanada, pp. 215–42. New York: Wiley & Sons

White A, Bowers R, Begon M. 1996. Host-pathogen cycles in self-regulated forest insect systems: resolving conflicting predictions. *Am. Nat.* 148:220–25

Wilson K, Cotter SC, Reeson AF, Pell JK. 2001. Melanism and disease resistance in insects. *Ecol. Letts.* 4:637–49

Wilson K, Reeson AF. 1998. Density-dependent prophylaxis: evidence from Lepidoptera-baculovirus interactions? *Ecol. Entomol.* 23:100–1

Wilson KR, O'Reilly DR, Hails RS, Cory JS. 2000. Age-related effects of the

Autographa californica multiple nucleopolyhedrovirus *egt* gene in the cabbage looper (*Trichoplusia ni*). *Biol. Control* 19:57–63

Woods SA, Elkinton JS. 1987. Bimodal patterns of mortality from nuclear polyhedrosis virus in gypsy moth *Lymantria dispar* populations. *J. Invertebr. Pathol.* 50:151–57

Woolhouse MEJ, Taylor LH, Haydon DT. 2001. Population biology of multihost pathogens. *Science* 292:1109–12

Zanotto PMA, Kessing BD, Maruniak JE. 1993. Phylogenetic interrelationships among baculoviruses: evolutionary rates and host associations. *J. Invertebr. Pathol.* 62:147–62

Annu. Rev. Ecol. Evol. Syst. 2003. 34:273–309
doi: 10.1146/annurev.ecolsys.34.012103.144032

First published online as a Review in Advance on July 11, 2003

LATITUDINAL GRADIENTS OF BIODIVERSITY: Pattern, Process, Scale, and Synthesis

M.R. Willig,[1] D.M. Kaufman,[2] and R.D. Stevens[3]
[1]Ecology Program, Department of Biological Sciences, Texas Tech University, Lubbock, Texas 79409-3131; email: michael.willig@ttu.edu
[2]National Center for Ecological Analysis and Synthesis, University of California, Santa Barbara, California 93101-3351 and Division of Biology, Kansas State University, Manhattan, Kansas 66506-4901; email: dkaufman@ksu.edu
[3]Ecology Program, Department of Biological Sciences, Texas Tech University, Lubbock, Texas 79409-3131 and National Center for Ecological Analysis and Synthesis, University of California, Santa Barbara, California 93101-3351; email: rstevens@nceas.ucsb.edu

Key Words species richness, species diversity, macroecology, geographic ecology, biogeography

■ **Abstract** The latitudinal gradient of decreasing richness from tropical to extra-tropical areas is ecology's longest recognized pattern. Nonetheless, notable exceptions to the general pattern exist, and it is well recognized that patterns may be dependent on characteristics of spatial scale and taxonomic hierarchy. We conducted an extensive survey of the literature and provide a synthetic assessment of the degree to which variation in patterns (positive linear, negative linear, modal, or nonsignificant) is a consequence of characteristics of scale (extent or focus) or taxon. In addition, we considered latitudinal gradients with respect to generic and familial richness, as well as species evenness and diversity. We provide a classification of the over 30 hypotheses advanced to account for the latitudinal gradient, and we discuss seven hypotheses with most promise for advancing ecological, biogeographic, and evolutionary understanding. We conclude with a forward-looking synthesis and list of fertile areas for future research.

INTRODUCTION

The oldest and one of the most fundamental patterns concerning life on earth is the increase in biological diversity from polar to equatorial regions (Brown & Lomolino 1998, Gaston 1996a, Rosenzweig 1995, Willig 2001). Indeed, a cogent statement of the gradient was articulated by von Humboldt early in the nineteenth century (Hawkins 2001). With notable exceptions, the pattern generally holds true, regardless of the biota's taxonomic affiliation (e.g., mammals, fishes, insects, and plants), geographic context (e.g., all continents and oceans), or time domain (e.g., Recent and 70 Mya). It was recognized clearly by the progenitors of evolutionary theory (Darwin 1862, Wallace 1878), and has been integral to the thinking of some of the most influential biologists of the past century (Dobzhansky 1950, Hutchinson 1959, MacArthur 1972). Indeed, a lustful preoccupation with high tropical biodiversity stimulated the conceptual development of much of community ecology. Similarly, serious concerns about the erosion of biodiversity at global, regional, and local scales have catalyzed a considerable body of research in conservation biology (Chown & Gaston 2000).

Broad recognition of the latitudinal gradient occurred in the 1800s, with an emphasis on quantification from the 1950s onward. The number of explanatory hypotheses for the gradient proliferated at the end of the twentieth century, with few attempts to falsify them in the sense of strong inference (Platt 1964). Although, a number of synthetic reviews have set the stage for advancement (especially Rohde 1992, Schemske 2002), rigorous attention to questions of spatial scale only has begun to characterize quantitative research concerning latitudinal gradients of biodiversity. Similarly, the relative importance of causal mechanisms, including stochastic processes, remains a poorly understood yet critical area for future investigation and synthesis. Latitudinal gradients of diversity are ultimately dependent on the historical, geographic, biotic, abiotic, and stochastic forces (Schemske 2002) affecting the geometry, internal structure, and location of species ranges in ecological or evolutionary time. Indeed, latitude is a surrogate for a number of primary environmental gradients (e.g., temperature, insolation, seasonality) that interact and are correlated to each other, making direct tests of hypotheses difficult and controvertible.

Biodiversity: Context and Constraints

Biodiversity became popularized as a term to mean "life on earth" as a consequence of a "National Forum on BioDiversity" that met in Washington D.C. in September 1986. In its full amplification, biodiversity includes variation in life at a spectrum of hierarchical levels, from genes to the biosphere (Swingland 2001). Although biodiversity has genetic, taxonomic, and ecological attributes (Gaston & Spicer 1998), we focus on a taxonomic component—species richness—because that characteristic of biodiversity has been the predominant concern of most research concerning latitudinal gradients. Nonetheless, we examine latitudinal gradients in other aspects of biodiversity (e.g., generic and familial richness, species evenness and diversity), albeit in a more limited fashion, when such patterns have been quantified over broad spatial extents.

Unlike most research in biology and much in ecology, understanding patterns and mechanisms related to the latitudinal gradient is limited severely by practical and ethical concerns. The broad geographic domain of research (countries, continents, and hemispheres) means that the primary data concerning the presence and absence of species for a particular biota are likely compiled by numerous scientists over a large number of years. Moreover, powerful investigative approaches associated with experimental design and manipulation are not feasible or ethical, thereby limiting progress in identifying the mechanistic bases of patterns. Nonetheless, a variety of statistical and simulation approaches provide considerable insight to latitudinal patterns and mechanisms that give rise to them, as they do for other macroecological patterns regarding population density, body mass, or range size (Brown 1995).

Spatial Scale

In general, considerations of spatial scale have become increasingly important in ecology because the detection of pattern and the identification of causal mechanisms critically depend on it (Gardner et al. 2001, Levin 1992, Peterson & Parker 1998). Two attributes of scale, focus and extent (sensu Scheiner et al. 2000), are particularly relevant to studies of gradients regarding species richness. The focus of a research design is defined by the inference space to which each datum applies, whereas the extent of a research design relates to the inference space to which the entire collection of data applies in an analysis. Thus, the focus is intimately associated with the size or dimensions of sampling units, whereas the extent is defined by the geographic or ecological space that has been sampled in the overall analysis. As might be expected, analyses of the latitudinal gradient in richness have involved a variety of different foci and extents (Willig 2001), making direct comparison among studies difficult. Nonetheless, the latitudinal increase in taxonomic richness toward the tropics characterizes studies with a focus on broad climatic zones (e.g., Fischer 1960), assemblages occupying arbitrary geographic subdivisions such as quadrats (e.g., Kaufman 1994, Simpson 1964, Willig & Selcer 1989, Wilson 1974) or bands (e.g., Kaufman & Willig 1998, McCoy & Connor 1980, Willig & Gannon 1997, Willig & Sandlin 1991) and local ecological communities (e.g., Kaufman 1998, Stevens & Willig 2002).

Scale dependence occurs when the form (e.g., linear versus quadratic versus modal) or parameterization (e.g., magnitude and sign of regression coefficients) of a relationship changes with focus or extent. Studies involving species richness, as well as with most other aspects of biodiversity, are expected to be strongly dependent on spatial scale because of the manner in which species richness increases with area. Larger areas contain more individuals, more habitats, and more biomes or biogeographic provinces than do smaller areas (Rosenzweig 1995). As a consequence of all three considerations, species richness increases as well. In any particular study, the exact way in which species richness increases with area determines the impact of scale on biological conclusions (Kolasa & Pickett 1991, Palmer & White 1994). A variety of models have been advanced to characterize the species-area relationship. The two most common are the exponential or semilogarithmic function of Gleason (1922, 1925) and the power function of Arrhenius (1921, 1923a,b), although alternative models based on the random placement of individuals exist as well (Coleman 1981, Coleman et al. 1982).

Regardless of the particular algorithm used to characterize the species-area relationship, three idealized kinds of scale effects (Scheiner et al. 2000,

Figure 1, see color insert) are relevant to the latitudinal gradient in diversity, including scale-invariant, rank-invariant, and scale-dependent patterns. If the forms (e.g., power versus semilogarithmic function) of the species area curves differ at different latitudes, then scale dependence in the latitudinal gradient will be pervasive and difficult to control in comparative analyses. Thus, differences among studies in the form or parameterization of the latitudinal gradient could be a consequence of inherent differences among them with respect to characteristics of scale (i.e., focus or extent) or with respect to biological, historical, or geographical characteristics.

Several studies of the latitudinal gradient have attempted to control for considerations of focal scale (i.e., area) through various kinds of covariance analysis in the context of regression models (e.g., Kaufman & Willig 1998, Mares & Ojeda 1982). Most of these are problematic because of methodological concerns. Such studies have been based on focal units with linear dimensions ranging from 62.5 km to 500 km, or degree-based dimensions ranging from 1° to 10° (Anderson & Marcus 1993). Quadrats defined by meridians and parallels are of variable size because the distance encompassed by 1° of longitude depends on latitudinal position. Longitudinal meridians converge toward the poles so that the area of degree-based quadrats decreases as latitude increases, creating a systematic bias that confounds measures of area and latitude. Even quadrats based on linear dimensions differ in the amount of land they contain because most coastal quadrats will not be full of land (Rosenzweig 1995). Consequently it is difficult to disentangle areal effects from other geographic, historical, or ecological phenomena associated with land-ocean interfaces. Finally, almost all studies of latitudinal gradients fail to take into account the spatially autocorrelated nature of measures of species richness. Because species ranges are idealized as continuous surfaces, proximate sampling units are likely to have more similar magnitudes of richness than would more distant sampling units. Failure to account for autocorrelation can result in unreliable or biased estimates of significance in statistical analyses (Dale & Fortin 2002, Diniz-Filho et al. 2002a, Legendre 1993). For example, Baquero & Telleria (2001) examined patterns of species richness at 289 evenly distributed sampling points in Europe. Species richness, the number of endemic species, and the number of rare species were related negatively to latitude. Nonetheless, latitudinal effects became nonsignificant after accounting for the geographic proximity of points. Caution is needed in interpreting such results as a plethora of statistical approaches exist for dealing with autocorrelation, especially when environmental variables are correlated with spatial distances. Because latitude itself is an explicitly spatial dimension, its treatment in such analyses requires particularly careful consideration. Two general approaches seem appropriate. The first involves using subsets of the data that are sufficiently distant from each other so that autocorrelative effects are eliminated from analyses. The second involves covariance analyses for which distance effects that are independent of latitude (e.g., "pure distance," such as those associated with longitude) are distinguished from distance effects that include differences along environmental gradients that parallel meridia.

Recent studies (Lyons & Willig 2002) overcame many of these problems by examining nested quadrats of fixed area (1000, 10,000, 15,000, 20,000, and 25,000 km^2) that were centered at regular intervals in the continental New World (i.e., at 2.5° latitudinal intervals along each 5° meridian). Following the pioneering work of Pastor et al. (1996), Lyons & Willig (2002) directly evaluated if the parameterization of the power function depended on latitude. Results for both bats and marsupials in the New World documented that the slope (z) of the power function ($S = CA^z$, where S is species richness, A is quadrat area, and $\ln C$ [intercept] and z [slope] are fixed constants derived from least-squares analysis of the log-log transformation of the relationship) systematically decreased with decreased latitude. Lyons & Willig (2002) suggested that some of the very mechanisms (e.g., environmental variation, geometric constraints, Rapoport-rescue effects) thought to promote the latitudinal gradient in richness would also predispose those patterns to be scale sensitive.

SPATIAL DYNAMICS OF DIVERSITY

The richness of any region is a consequence of two factors: the richness of each of the smaller areas that compose it, and the turnover in species composition among them (Whittaker 1960). A variety of terms has been suggested for these three levels of diversity, depending on their spatial scale and ecological characteristics (Whittaker 1977), but their use has not been consistent (Koleff & Gaston 2002). Schneider (2001) provided a reasonable summary of terminology that is of relevance to the latitudinal gradient. Diversity can be characterized at various focal scales including point diversity (richness of a subset of a community), alpha diversity (richness within the full extent of a single community), gamma diversity (richness of a landscape that comprises different communities), and epsilon diversity (richness of a broad geographic area that comprises different landscapes). Although reasonable, these are arbitrary distinctions along a continuum of possible focal scales. In practice, it is difficult to distinguish between gamma and epsilon diversity in a biogeographic context because the boundaries of and distinction between a landscape and a broad region are unclear. Consequently, we distinguish among three focal scales: point diversity, alpha diversity, and gamma diversity (here meaning the richness of an area comprising multiple communities). Three other measures of diversity refer to turnover among focal units (Schneider 2001): local beta diversity (change in species composition among subsets of a community), beta diversity (change in species composition among different communities in a landscape or along an environmental gradient), delta diversity (change in species composition along a climate gradient or among geographic areas). Latitude is clearly a geographic gradient that includes multiple communities, reflecting climatic and environmental characteristics. Again, the distinction between kinds of turnover in a biogeographic context is arbitrary, and we adopt the more popular term, beta diversity, to mean turnover among focal units representing distinct communities.

Beta diversity can be quantified in several interrelated ways. One approach is designed so that the product of beta diversity and alpha diversity equals gamma diversity (Harrison et al. 1992; Whittaker 1960, 1972). An alternative formulation has been constructed so that the sum of alpha and beta diversities yields gamma diversity (e.g., Williams 1996). The additive metric recently has been the focus of a cogent review (Veech et al. 2002), and shows much promise for its application to issues in conservation biology (Gering et al. 2003). Another type of beta diversity is based on consideration of species losses and gains along a gradient or transect (Wilson & Shmida 1984). The selection of a particular index depends on characteristics of the focal samples, the domain or gradients over which they are evaluated, and the hypotheses of interest.

A large body of evidence (see Empirical Patterns, below) (Figure 2) supports an inverse relationship between gamma diversity and latitude. This could be a consequence of underlying gradients in alpha diversity, beta diversity, or both. Beta diversity increases from poles to tropical areas in a number of groups: mammals (Kaufman 1998, Stevens & Willig 2002), birds (Blackburn & Gaston 1996b, Koleff & Gaston 2001), shallow water bryozoans (Clarke & Lidgard 2000), trees (D.F. Sax, unpublished data), and herbaceous plants (Perelman et al. 2001). Moreover, the proportion of the regional biota represented in local communities was equivalent in North American trees and South American herbs, after controlling for hemisphere (Perelman et al. 2001; D.F. Sax, unpublished data). For mammals in the New World, Kaufman (1998) corroborated the poleward decrease in beta diversity but detected a plateau effect in tropical areas, suggesting nonlinearities in the gradient.

At least three studies simultaneously have analyzed latitudinal variation in alpha, beta, and gamma diversity, providing synoptic assessments of latitudinal gradients. Nonetheless, the spatial limits associated with measures of alpha and gamma diversity remain somewhat subjective, albeit informed by the expert opinion of the taxonomic specialists in each study. That is, the spatial limits of a bat community may be larger than those of a rodent community, both of which are considerably larger than those of a community of microarthropods. This transpires because of the different grains at which these taxa likely perceive their environment. For nonvolant species of mammals in North America, latitudinal patterns differed among levels of diversity (Kaufman 1998). Alpha diversity was relatively constant across the Temperate Zone but increased into the tropics, with nearly twice as many species in tropical communities as were found in temperate communities.

→

Figure 2 A broad survey of the published literature (see text for details) illustrates the complexion (*A*, tempo of publication; *B*, taxonomic focus of research; *C*, geographic focus of research) and scale (*D*, geographic extents of analysis; *E*, geographic characteristics) of research on the latitudinal gradient of richness. The conclusion of most studies (*F*) was that the relationship between richness and latitude was negative, although positive, modal, and no patterns were detected as well.

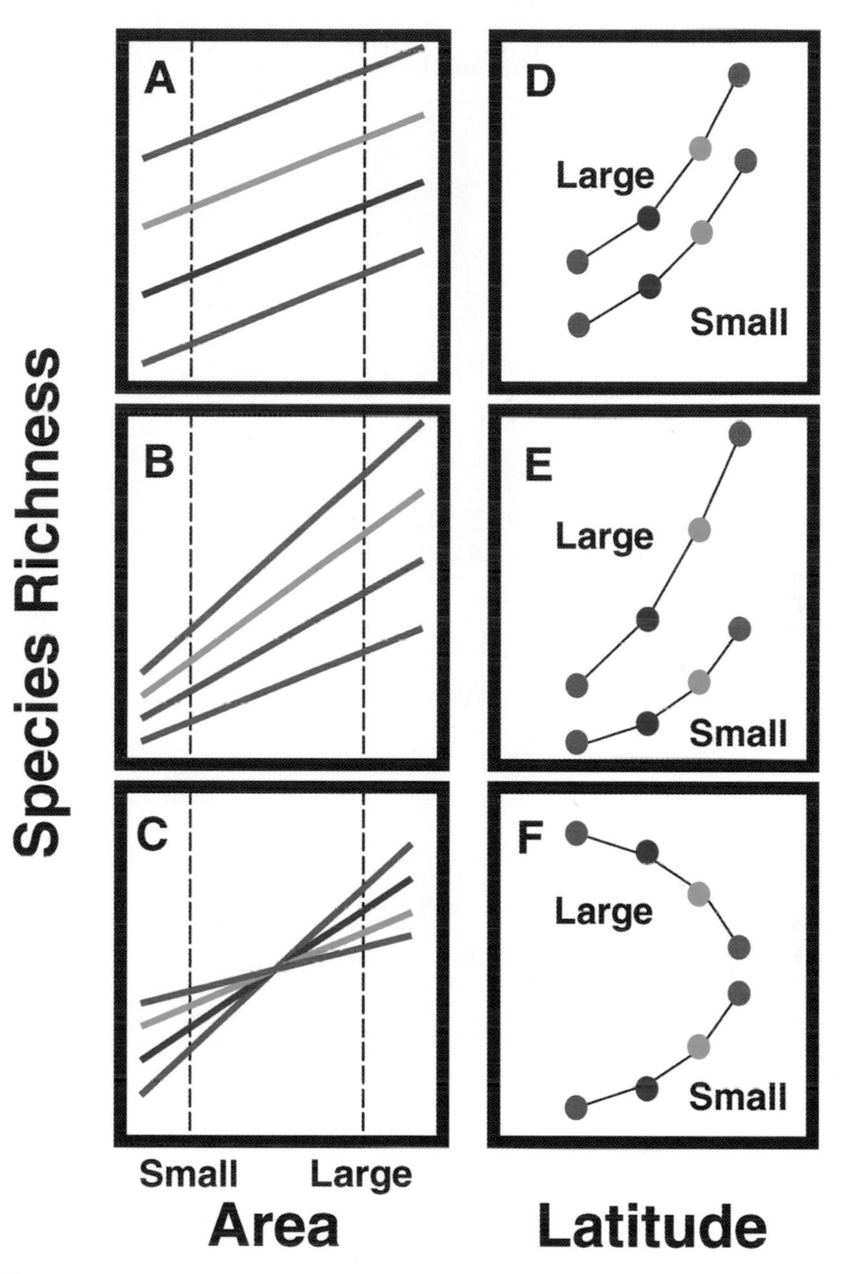

Figure 1 The latitudinal gradient in richness is scale independent when the species-area relationship is parallel at all latitudes (*A* and *D*). The latitudinal gradient is rank-invariant when the species area relationships do not intersect within the domain of areas considered in an analysis (*B* and *E*). The latitudinal gradient is neither scale- nor rank-invariant when the species area relationships intersect within the domain of areas considered in an analysis (*C* and *F*).

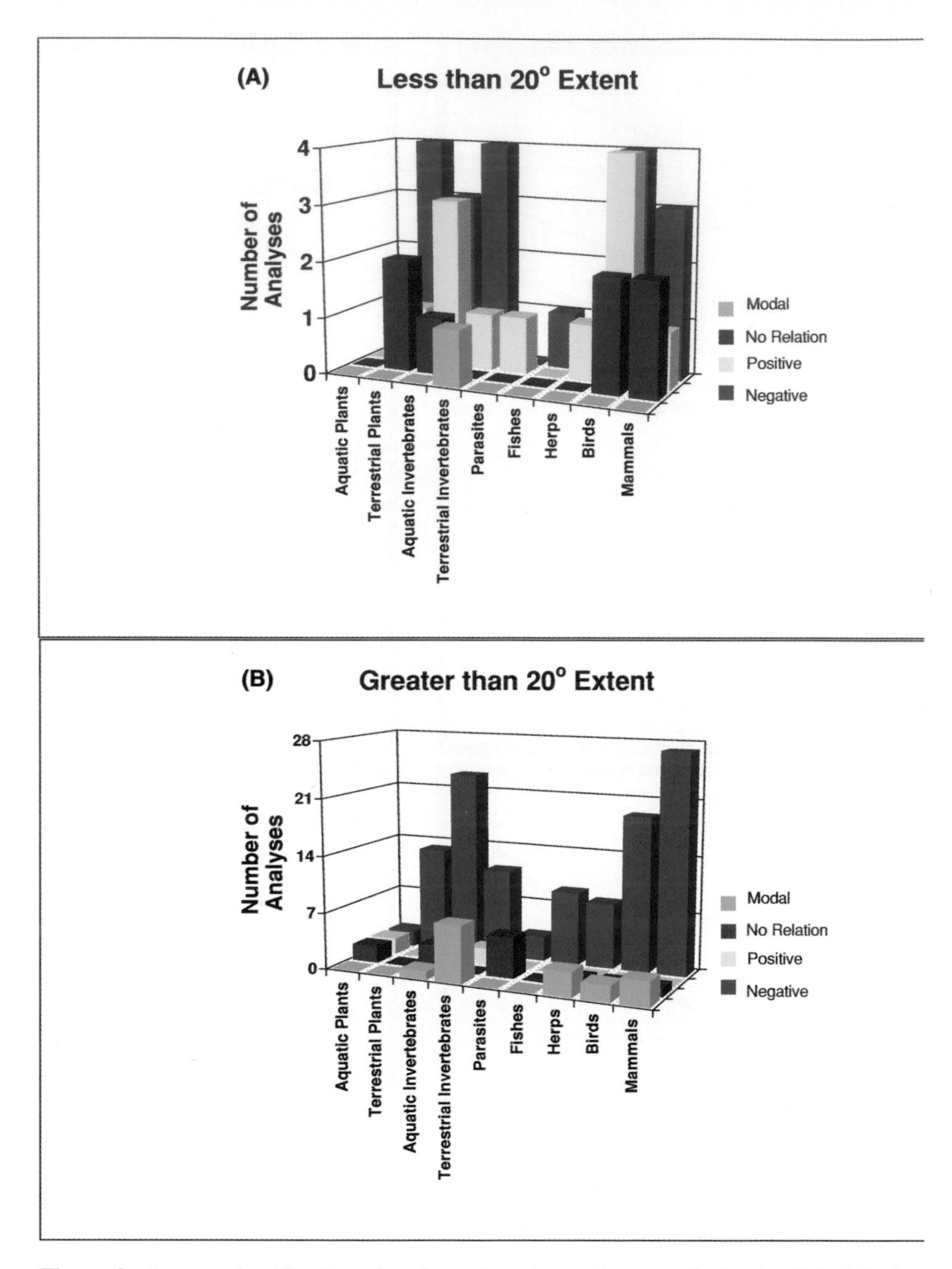

Figure 3 A cross-classification of analyses (see electronic appendix for details) of the latitudinal gradient of richness by latitudinal extent (*A*, narrow [<20°]; *B*, broad [≥20°]) and pattern (increasing toward the tropics [positive], decreasing toward the tropics [negative], modal, and nonsignificant).

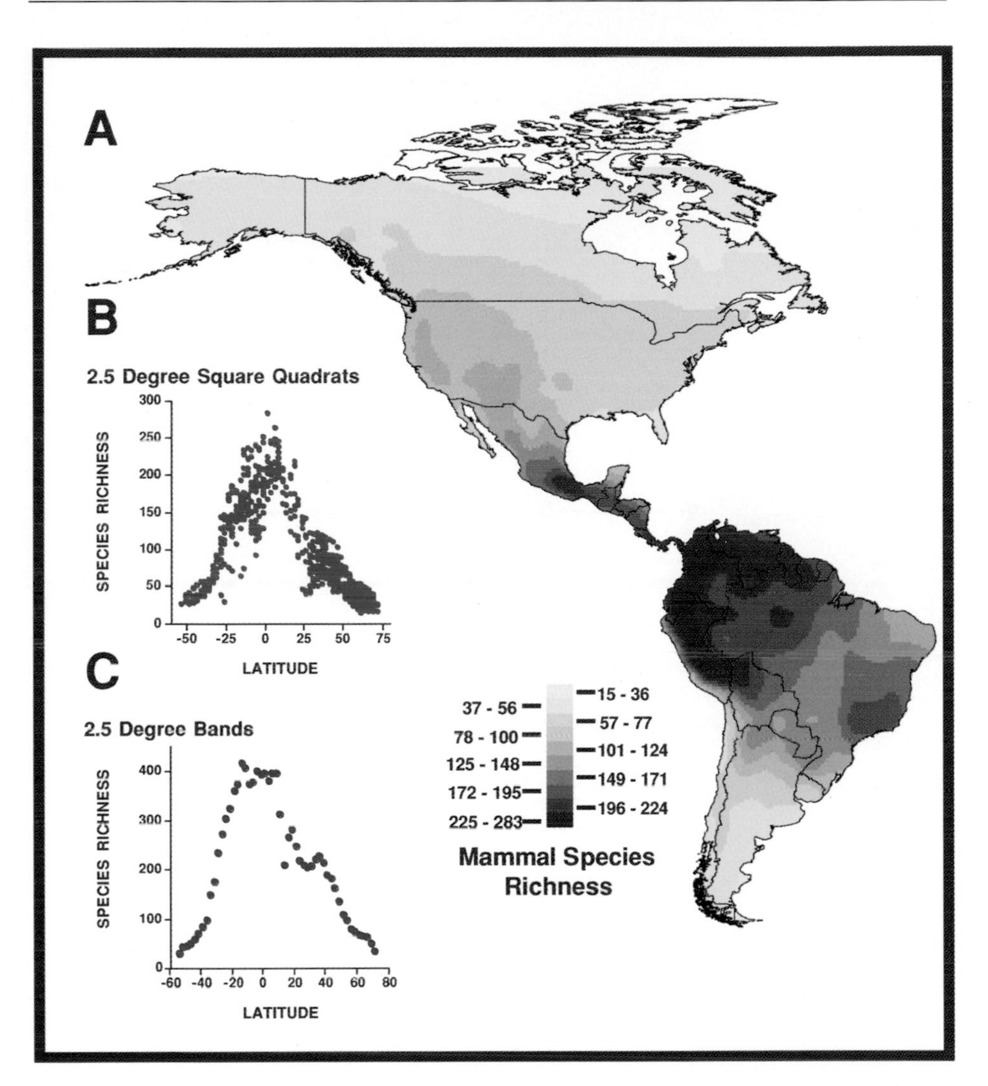

Figure 4 Spatial gradient of mammalian species richness in the continental New World for cells defined by 2.5° parallels and meridians. (*A*). Interpolated richness values in the map were created using the tension spline function in the Spatial Analyst extension to ArcGIS 8.2. Graphic representation of the latitudinal gradient in species richness for those same data (negative values for latitude indicate southern parallels), based on 2.5° cells (*B*) and 2.5° latitudinal bands (*C*). Data from Kaufman & Willig (1998).

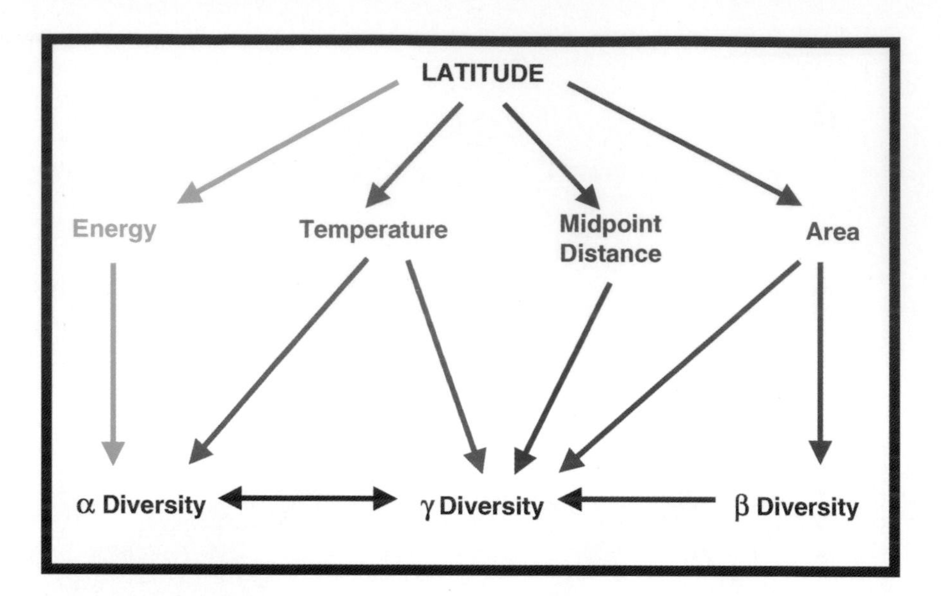

Figure 5 Conceptual model indicating the avenues by which mechanistic factors affect alpha, beta, or gamma diversity in the context of broad latitudinal gradients. Measures of alpha diversity are assumed to be based on areas of equal size or to be determined by asymptotic estimators.

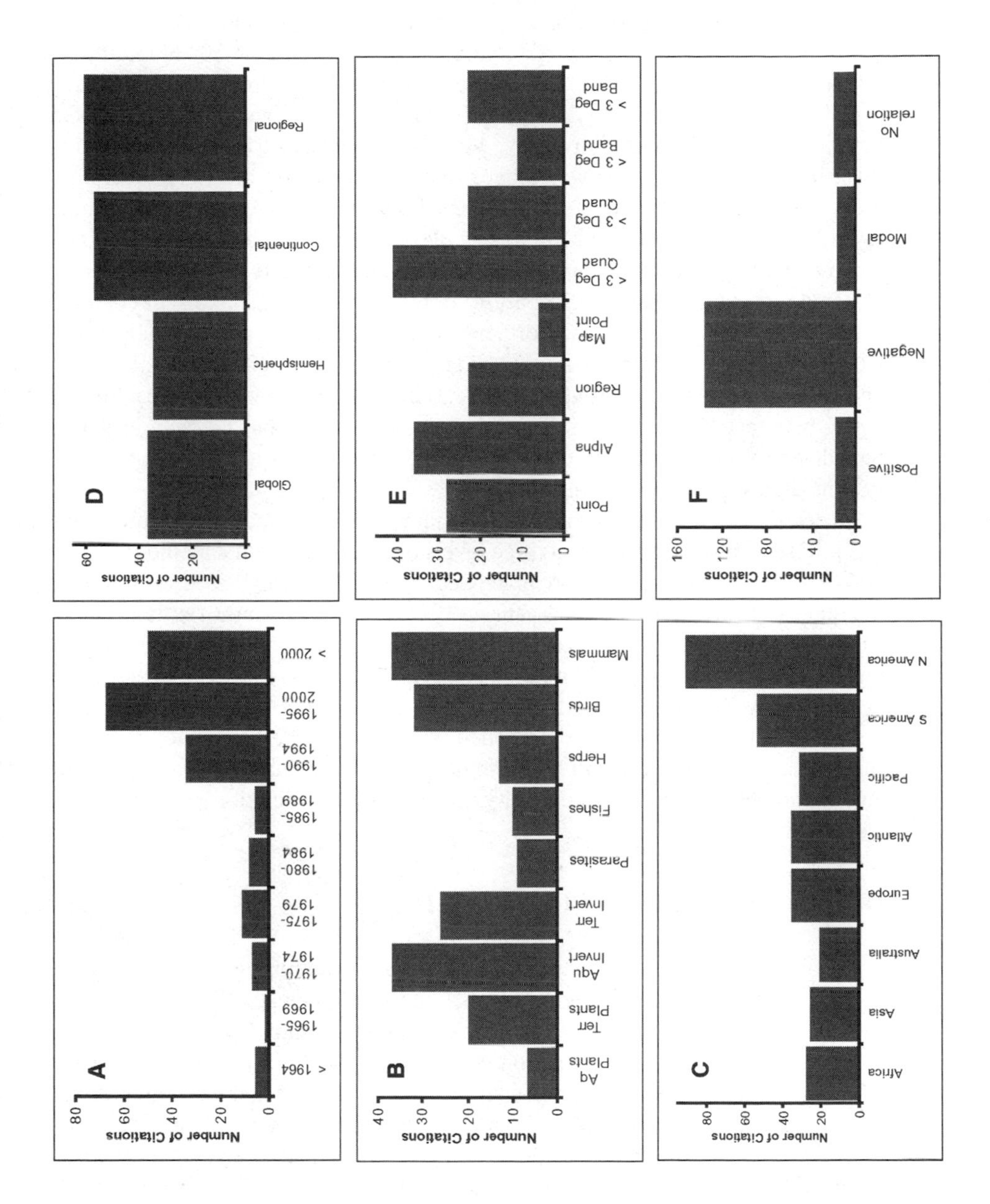
A
Number of Citations
80
60
40
20
0
< 1964
1965-1969
1970-1974
1975-1979
1980-1984
1985-1989
1990-1994
1995-2000
> 2000
B
Number of Citations
40
30
20
10
0
Aq Plants
Terr Plants
Aqu Invert
Terr Invert
Parasites
Fishes
Herps
Birds
Mammals
C
Number of Citations
80
60
40
20
0
Africa
Asia
Australia
Europe
Atlantic
Pacific
S America
N America
D
Number of Citations
60
40
20
0
Global
Hemispheric
Continental
Regional
E
Number of Citations
40
30
20
10
0
Point
Alpha
Region
Map Point
< 3 Deg Quad
> 3 Deg Quad
< 3 Deg Band
> 3 Deg Band
F
Number of Citations
160
120
80
40
0
Positive
Negative
Modal
No relation

In contrast, gamma and beta diversity increased steadily with decreasing latitude throughout the Temperate Zone, but the rate of increase decreased within the tropics (i.e., gamma diversity exhibited a negative sigmoidal relationship). Beta diversity remained more or less constant (i.e., attained a plateau) below 30° N (Kaufman 1998). Stevens & Willig (2002) demonstrated linear relationships between latitude and all three levels of diversity for New World bats. A positive association existed between alpha and gamma diversity. Moreover, the rate of increase in gamma diversity toward the tropics was higher than that for alpha diversity, with both measures approximately equal at the highest latitudes and diverging in magnitude toward the equator. Thus, alpha diversity plays a more dominant role in shaping gamma diversity in extratropical areas, whereas beta diversity plays a more dominant role in shaping gamma diversity in tropical areas in this taxon. In contrast, shallow-water bryozoans in the northern Atlantic Ocean show no evidence of a latitudinal gradient in alpha diversity (Clarke & Lidgard 2000). Moreover, beta and gamma diversity decreased with increased latitude. In conclusion, patterns of alpha, beta, and gamma diversity differed among studies, but the way in which beta diversity interacted with alpha diversity to produce latitudinal gradients in gamma diversity was consistent.

A deep understanding of latitudinal gradients is predicated on simultaneous quantification of gradients for all three levels of diversity. From a methodological perspective, a unified framework is needed to relate patterns based on different measures of beta diversity. This is especially critical because turnover that is measured at different scales often is based on different indexes of beta diversity, and the detection and strength of latitudinal gradients of beta diversity are clearly scale dependent (Arita & Rodriguez 2002, Koleff & Gaston 2002). Until the differences that are due to scale can be distinguished from those related to choice of metric for estimating turnover, it will be difficult to assess the qualitative generality of any latitudinal pattern of beta diversity, much less its association with alpha and gamma diversity.

EMPIRICAL PATTERNS: SPECIES RICHNESS

To obtain information with which to characterize latitudinal gradients in extant species richness, we conducted a suite of literature searches. Our primary endeavor utilized SciSearch, the Science Citation Index Citation Database located on the Institute for Scientific Information website (http://webofscience.com/ciw.cgi). We searched for the key-word combinations "latitud* and diversity and gradient*," "latitud* and diversity and species," "latitud* and richness and gradient*," "latitud* and richness and species," "gradient* and diversity and species," as well as "gradient* and richness and species" up to the end of 2002. The asterisk (*) is a wildcard of undetermined length. By using this wildcard, one can search for a basic root word with different endings. In a second search, we used a variety of search engines including Agricola, ArticleFirst, Biosys, Biological and Agricultural Index, Cambridge Scientific Abstracts, and JSTOR to find additional citations up to 2001. We used the search phrases "diversity and latitude," "richness and latitude," "latitudinal gradient," "diversity gradient," and "richness gradient." Additional journal articles were encountered by perusing the bibliographies of articles on latitudinal gradients in diversity that were identified by this suite of searches.

The publications discovered through this process (N = 162) were perused, categorized, and tallied according to focal scale and latitudinal extent, as well as by the detected latitudinal pattern of species richness. When a single article presented analyses for a number of taxa (e.g., mammals, birds, herptiles) or at a number of scales (e.g., alpha and gamma diversity), results for each combination of analysis were tallied individually [i.e., the number of analyses (191) > the number of articles (162)].

The popularity of the latitudinal gradient as a research topic has grown rapidly over the past 40 years (Figure 2*A*). Moreover, latitudinal patterns have been examined rather thoroughly across taxa, but with a disproportionate emphasis on aquatic invertebrates, birds, and mammals (Figure 2*B*). Similarly, the gradient has been examined broadly from a geographic perspective, covering most continents and a number of oceans (Figure 2*C*). From a methodological perspective, the gradient has been examined at a variety of spatial scales with respect to aspects of extent (Figure 2*D*) and focus (Figure 2*E*). The majority of analyses (135, approximately 71%) have corroborated that the latitudinal gradient is one in which species richness increases toward the tropics (Figure 2*F*). Patterns in which species richness peaked at mid-latitudes or decreased toward the tropics were far less common, only as prevalent as was no relationship at all.

Evidence Corroborating the Classical Pattern

The negative association between taxonomic richness and latitude (i.e., the classical pattern) has been demonstrated for latitudinal extents that are short (i.e., <20° latitude: Cushman et al. 1993, Gotelli & Ellison 2002, Meserve et al. 1991) and long (i.e., ≥20° latitude; Fischer 1960, Rosenzweig 1992), as well as when estimated as point (Gentry 1988, Kocher & Williams 2000, Rex et al. 1993), alpha (Barbour & Brown 1974, Meserve et al. 1991, Stevens & Willig 2002), or gamma diversity (Anderson 1977, Ellison 2002, Gaston 2000). Moreover, this pattern has been described for many major organismal groups including: terrestrial plants (Cowling & Samways 1995, Gentry 1988), mangrove trees (Ellison 2002), marine protists (Culver & Buzas 2000), coral (Harriot & Banks 2002, Reyes-Bonilla & Cruz-Pinon 2000), mollusks (Rex et al. 1993; Roy et al. 1994, 1996, 1998; Stehli et al. 1967; Taylor & Taylor 1977), marine and freshwater arthropods (Dworschak 2000, Steele 1988), terrestrial arthropods (Cushman et al. 1993, Davidowitz & Rosenzweig 1998, Kocher & Williams 2000, Lobo 2000), marine and freshwater fishes (Barbour & Brown 1974, Oberdorff et al. 1995, Stevens 1996), herptiles (Currie 1991, Kiester 1971, Schall & Pianka 1978), birds (Blackburn & Gaston 1996a, 1996b, 1997; Cardillo 2002; Cook 1969; Diniz-Filho et al. 2002b; Rahbek

& Graves 2001), and mammals (Kaufman 1995, Lyons & Willig 2002, McCoy & Connor 1980, Simpson 1964). Although Platnick (1991) warned that global views about patterns of biodiversity may be subject to a "megafauna bias," it is clear that the classical pattern is a pervasive characteristic of life on earth, especially when the gradient comprises broad spatial scales (Figure 3, see color insert).

Some of the best examples of the classical pattern have been based on the gamma diversity of mammals, especially in the New World (Figure 4, see color insert), but also in Africa. In North America, the classical pattern has been corroborated by a multitude of investigators (Anderson 1977, Badgley & Fox 2000, Currie 1991, Davidowitz & Rosenzweig 1998, McCoy & Connor 1980, Pagel et al. 1991, Rosenzweig 1992, Simpson 1964, Wilson 1974). Compared with terrestrial mammals, bats exhibited steeper latitudinal gradients in North America and disproportionately contributed to the tropical increase in species richness of mammals in general (Kaufman 1995; Ruggiero 1994, 1999). In the past decade, the geographic extent of latitudinal studies of New World mammals have expanded to jointly consider North and South America (e.g., Kaufman 1995, Kaufman & Willig 1998, Lyons & Willig 2002, Willig & Gannon 1997, Willig & Sandlin 1991, Willig & Selcer 1989). Although spatial-scale dependence characterized the relationship for bats and marsupials (Lyons & Willig 2002) in the Western Hemisphere, the Mammalia exhibited indistinguishable latitudinal gradients in North and South America once accounting for differences in area of 2.5° and 5° latitudinal bands (Kaufman & Willig 1998). The classical pattern has been demonstrated in Africa as well (Andrews & O'Brien 2000, Cowlishaw & Hacker 1997, Eeley & Foley 1999).

The form of the latitudinal gradient in gamma diversity in the New World differs among mammalian orders (Kaufman 1995, Ruggiero 1994), with maxima in richness occurring at different locations for different taxa. Nonetheless, peaks of species richness for all orders occurred within 30° of the equator, except for the Insectivora, which attained maximum richness in the Temperate Zone of North America. The rate of increase in species richness toward the tropics was steepest for bats and primates compared with other mammalian orders.

Most of the work on latitudinal gradients of bird species richness also has been conducted in the New World. In general, species richness increased toward the equator (Anderson 1984; Blackburn & Gaston 1996a, 1997; Buckton & Ormerod 2002; Cardillo 2002; Cook 1969; Currie 1991; Diniz-Filho et al. 2002b; Gaston 2000; Rabinovich & Rapoport 1975; Rahbek & Graves 2001; Schall & Pianka 1978; Tramer 1974). Although methodological differences existed among studies, especially regarding the inclusion or exclusion of wintering ranges in the determination of species richness, results were correlated (Blackburn & Gaston 1996a, 1996b, 1997) or similar (Tramer 1974).

Latitudinal gradients of species richness for fishes generally corroborated the current paradigm. The tropical increase in richness was true for marine and freshwater taxa as well as for assemblages in lentic and lotic habitats. Although most highly correlated with the surface area of lakes, the species richness of fishes increased with decreased latitude throughout the world (Barbour & Brown 1974). Similarly, the species richness of fish assemblages from 292 rivers throughout the world increased toward the tropics (Oberdorff et al. 1995). For global extents, the number of species of pelagic fishes was lowest at high latitudes and highest at low latitudes, with maximum species richness at approximately 15° N (Rohde 1978a). In addition, fish species richness increased toward the equator in the Atlantic Ocean (Angel 1993, Macpherson 2002, McClatchie et al. 1997), Pacific Ocean (Stevens 1996), and Sea of Japan (Kafanov et al. 2000). Moreover, significant latitudinal gradients were obtained for taxonomic subsets of the fish fauna. For example, species richness of both teleosts and elasmobranch fishes increased toward the equator (Macpherson & Duarte 1994).

Latitudinal gradients in the Mollusca are well established (Crame 2000, 2002; Rex et al. 1993; Roy et al. 1994, 1996, 1998; Stehli et al. 1967). Latitude accounts for significant amounts of variation in the number of species, genera, and families, as well as in the number of species within functional groups, in benthic marine gastropods and bivalves from the eastern Pacific (Roy et al. 1994, 1996, 1998, 2000). Moreover, patterns of generic and familial richness are correlated highly with those of species richness (Roy et al. 1996). For all three levels in the taxonomic hierarchy, richness increased toward the tropics in equivalent fashions. Stehli et al. (1967) reported similar results regarding the species, generic, and familial patterns of taxonomic diversity in Recent bivalves.

Predatory gastropods on the eastern Atlantic Shelf exhibited a latitudinal gradient in which the number of species decreased toward the poles (Taylor & Taylor 1977). Roy et al. (1998) documented a similar gradient in the species richness of prosobranch gastropods. They collected data regarding the geographic ranges of 3916 species from throughout the western Atlantic and eastern Pacific. Gradients of richness were similar in the two oceans despite their marked physical and historical differences. Nonetheless, differences appeared in the comparison of patterns in Northern and Southern Hemispheres. More specifically, latitudinal gradients were significant and steep in the Northern Hemisphere, based on local samples of deep-sea bivalves, gastropods, and isopods, but less strong in the Southern Hemisphere as a result of appreciable interregional variation (Rex et al. 1993). Moreover, younger clades within the Bivalvia exhibited the steepest gradients in richness (Crame 2000).

Exceptions to the Classical Pattern

A number of exceptions exist regarding the strength and direction of latitudinal gradients in richness. In a small proportion of cases (Figure 3), the relationship between species richness and latitude was nonsignificant (10%; e.g., Bolton 1994, Clarke & Lidgard 2000, Dexter 1992, Dingle et al. 2000, Ellingsen & Gray 2002, Mann et al. 1999, Oliva 1999, Reid 1994, Poulin 1995, Poulsen & Krabbe 1997, Pysek 1998, Rohde 1999), modal (9%; e.g., Chown et al. 1998, Davidowitz & Rosenzweig 1998, France 1992, Janzen 1981, Krystufek & Griffiths 2002, Price et al. 1998, Skillen et al. 2000), or positive (10%; e.g., Blaylock et al. 1998, Hawkins & Compton 1992, Heip et al. 1992, Heip & Craeymeersch 1995, Lambshead et al. 2000, Rabenold 1979, Santelices & Meneses 2000). Exceptions primarily were associated with (*a*) narrow latitudinal extents (i.e., a form of scale dependence), (*b*) species with parasitic life histories, or (*c*) aquatic floras.

Many of the exceptions to the classical pattern (36%) were associated with scale and occurred when the gradient in richness was quantified over short latitudinal extents (<20° latitude). Indeed, analyses based on small latitudinal extents had a significantly greater proportion of results that failed to corroborate the classical pattern (Contingency Analysis—$G = 9.39$, d.f. $= 1$, $P = 0.002$). The number of positive or nonsignificant patterns was quite low (Figure 3) when the extent of the analyses was broad (≥20° latitude), with examples scattered among the various organismal groups. In contrast, half of the analyses based on narrow latitudinal extents failed to corroborate the classical pattern. Variation and heterogeneity associated with more local geography, geology, hydrology, or history likely overpowered the effects of causative mechanisms associated with latitude in such cases.

The earliest described exceptions to classical latitudinal pattern involved the Hymenoptera, in particular wasps, and especially members of the parasitoidal Ichneumonidae (Janzen 1981, Price et al. 1998, Skillen et al. 2000). In North America, species richness of ichneumonids peaked at 38°–42° N (Janzen 1981), although some have suggested a broader peak (Skillen et al. 2000). The mechanistic basis of this mid-latitudinal peak in richness remains uncertain, although mid-latitudinal peaks in host density (Janzen 1981) and mid-domain effects (Skillen et al. 2000) have been suggested to be important.

The species richnesses of ecto- and endoparasites do not respond to latitude in a consistent fashion (Blaylock et al. 1998, Poulin 1999, Rohde 1978b). In general, the number of ectoparasites in fish, birds, or mammals was unrelated to latitudinal variation (Poulin 1995). Similarly, the species richness of gastrointestinal helminths of teleost fish did not change in a consistent fashion with latitude (Rohde 1998). Moreover, no significant latitudinal variation existed in the number of species of trematodes parasitizing fish (Rohde 1999) or in the number of species of metazoan parasites found on jack mackerel (Oliva 1999). Symbiotic associations, especially for endoparasites of homeotherms, insulate species from the environmental conditions faced by their hosts, thereby mitigating any latitudinal mechanisms that might influence the species richness of parasites. Nonetheless, even modest host specificity coupled with a latitudinal gradient in host species richness would be expected to effect the classical pattern in parasites as well.

With few exceptions (Pielou 1977, Santelices & Meneses 2000), aquatic plants do not exhibit latitudinal gradients. Bolton (1994) investigated the distribution and abundance of seaweeds from 29 floras throughout the world. Although subtropical and tropical areas exhibited the highest levels of species richness, they also exhibited relatively low levels. No consistent latitudinal trend in species richness was detected, and the highest number of species did not occur in floras situated nearest to the equator. Crow (1993) corroborated this finding for aquatic angiosperms from throughout North and Central America. More specifically, higher levels of angiosperm richness existed at warm temperate sites than at tropical sites. Similarly, a tropical peak in richness was absent for North American charophytes (Mann et al. 1999). Latitude is also unimportant in determining the number of aquatic macrophytes in Fennoscandia (Virola et al. 2001). The reasons remain enigmatic.

Finally, secondary marine birds and mammals (i.e., taxa that have reverted to a marine lifestyle after diverging from terrestrial ancestors) exhibited extratropical peaks in species richness. These taxa include pelagic birds and members of the mammalian suborder Pinnipedia (Chown et al. 1998, Proches 2001). The modal patterns of species richness of secondary marine organisms may be related to an increase in heterogeneity and resources that occurs in association with the Subtropical Convergence and Antarctic Polar Frontal Zone.

EMPIRICAL PATTERNS: SPECIES DIVERSITY AND EVENNESS

Two critical considerations apply when comparing studies of latitudinal gradients between local and regional levels. Although richness at the local and the regional scale is interdependent (Ricklefs & Schluter 1993), these scales represent distinct levels of biological organization (i.e., reflecting ecological mechanisms that operate within communities versus those ecological and historical mechanisms that effect turnover among communities in a region). Accordingly, the importance of particular structuring mechanisms likely differs between such levels of organization. Mechanisms such as interspecific interactions or dispersal limitations are likely more important at the local level, whereas mechanisms related to speciation and extinction are more important at the regional level (Ricklefs & Schluter 1993). Moreover, species abundances can be quantified more readily at the level of local communities than at larger spatial scales. This facilitates consideration of weighted measures of diversity such as species evenness, species dominance, or species diversity. Such measures provide for more comprehensive examinations of latitudinal gradients. Drawbacks to such approaches include the rarity of density data for multiple communities as well as a tendency for such studies to encompass relatively narrow latitudinal gradients when the data do exist.

Porembski et al. (1996) evaluated patterns in the local diversity of inselbergs (i.e., steep ridges or hills that remain after a mountain erodes within an otherwise flat terrain) along a 6° latitudinal transect in the Ivory Coast of Africa. No relationship existed between the number of plant species and latitude. Nonetheless, dominance (Simpson index) was related to latitude and increased from north to south (Porembski et al. 1995). Bowman (1996) investigated the diversity of plant communities along a latitudinal gradient in northern Australia. Although beta diversity among equal-sized quadrats within each of 15 cells increased toward the tropics, species richness and Shannon's diversity exhibited no significant latitudinal gradient.

Marine invertebrates exhibited a variety of different latitudinal gradients with respect to species richness, evenness, and diversity. Rex et al. (2000) found that the number of species of bivalves, gastropods, and isopods as well as species evenness (Pielou's index) and diversity (Shannon's index) in both the Northern and Southern Hemisphere decreased toward the poles. For copepods of the Atlantic Ocean, average evenness was greater and variance of evenness lower at the equator than at higher latitudes (Woodd-Walker et al. 2002). In contrast, Errhif et al. (1997) detected no significant latitudinal relationships for copepods with respect to Shannon's diversity or Shannon's evenness along a transect that spanned 23° of latitude in the Antarctic Ocean.

The richness and evenness of species of wireworms (Collembola: Elateridae) exhibited opposite latitudinal gradients in Central Russia (Penev 1992). More specifically, species richness decreased and evenness increased with increased latitude. Their interaction caused species diversity to exhibit no significant change with latitude.

Vertebrates also exhibited inconsistent latitudinal patterns of taxonomic diversity. The species richness, diversity, and evenness of demersal fishes exhibited positive relationships with latitude in the southern Atlantic Ocean in the vicinity of New Zealand (McClatchie et al. 1997). The species richness and diversity of Finnish birds of prey exhibited positive associations with latitude (Solenon 1994). For waders in Fennoscandia, the latitudinal relationship was positive for species richness, negative for species evenness, and nonsignificant for species diversity (Jarvinen & Vaisanen 1978). Similarly, no significant relationship existed between latitude and species richness, evenness, and diversity of small mammal communities on clearcuts in Sweden between 55° and 65° N (Hansson 1992). In contrast, small mammal communities in North America did exhibit latitudinal relationships in evenness (negative) and dominance (positive), as characterized by rank-abundance relationships (Kaufman 1998). Finally, on the Iberian Peninsula, small mammal diversity (Shannon's index) determined from barn owl pellets increased with increased latitude (Barbosa & Benzal 1996). Although such positive relationships between aspects of mammalian biodiversity and latitude were uncommon, this particular result may have been associated with increased habitat complexity from north to south in this region of Europe.

Although patterns of species evenness and richness can be similar, they may describe two unique aspects of spatial variation among communities (see review by Stirling & Wilsey 2001). Moreover, if communities or assemblages become richer as a consequence of sustaining more rare species, then latitudinal increases in richness should be accompanied by latitudinal decreases in evenness. Stevens & Willig (2002) evaluated the spatial variation of 32 New World bat communities based on 14 different measures of taxonomic diversity including those that were sensitive to changes in richness, evenness, diversity, and dominance. A factor analysis determined that the spatial variation in aspects of richness was independent of spatial variation in aspects of evenness. In fact, the species richness of New World bat communities exhibited significant and strong latitudinal gradients, whereas evenness exhibited no relationship with latitude. Equally important, no two measures of richness, evenness, dominance, or diversity varied spatially in identical fashions. This cautions that comparisons of aspects of biodiversity that are weighted by species abundances may be compromised if they are not based on identical indexes. Moreover, it clearly demonstrates that all measures of diversity do not vary in concert along latitudinal gradients.

EMPIRICAL PATTERNS: ANCIENT ASSEMBLAGES AND HIGHER TAXA

The latitudinal gradient has been demonstrated in contemporary time for a spectrum of taxa across the globe because individual organisms can be observed, identified, and counted at a variety of spatial and temporal scales. A number of methodological issues concerning the assessment of large-scale patterns of diversity characterize studies of extant taxa; however, quantification of latitudinal patterns in the fossil record has additional constraints that constitute a methodological challenge of Herculean dimension. Nonetheless, study of the fossil record reveals that the latitudinal gradient of richness has existed for at least a quarter of a billion years!

Biases in the fossil record have been the subject of much paleontological research (Kidwell & Holland 2002). Preservation potential differences among organisms and fossil samples are biased representations of the communities that existed in the past. These biases may be related to morphology (i.e., only species with skeletons or hard parts are preserved; soft-bodied organisms are preserved only under exceptional circumstances; Briggs & Kear 1993), mineralogy (e.g., phosphatic, aragonitic, or calcitic shells have different preservation potentials; Cherns & Wright 2000), size (i.e., larger fossils are more durable over time and are easier for paleontologists to discover; Kidwell & Holland 2002), habitat (i.e., preservation of marine organisms is more likely than preservation of freshwater or terrestrial organisms; Kidwell & Holland 2002), population size (i.e., common organisms have a higher likelihood of preservation; Kidwell & Holland 2002), and the uneven spatial and temporal distribution of rock (e.g., the absence of appropriately aged rock associated during each mass extinction event; Peters & Foote 2001). Because of these constraints, data relevant to paleolatitudinal gradients are fewer and more scattered than are those for contemporary latitudinal gradients. Nonetheless, evidence from a large span of paleontological times (from deep to shallow) has been used to construct the latitudinal gradient for two major groups: marine invertebrates and terrestrial plants.

Paleolatitudinal gradients of increasing richness toward the tropics are known from three major eras of earth history (i.e., Paleozoic, Mesozoic, and Cenozoic). Currently, the oldest evidence comes from the oceans, in the form of Permian fossils of marine brachiopods that lived 270 Mya in the Northern Hemisphere (7° N–78° N; Stehli et al. 1969). Various marine organisms provide a glimpse of the paleolatitudinal gradient of richness at a variety of different time periods. Jurassic marine bivalves from 150 Ma (sampled over ≤6 Ma) exhibited a paleolatitudinal gradient of richness at specific, generic, and familial levels in both Northern and Southern Hemispheres (33° N–71° N and 30° S–75° S; Crame 2002). Stehli et al. (1969) also documented the paleolatitudinal pattern in Cretaceous planktonic Foraminifera from 70 to 80 Ma in the Northern Hemisphere (10° N–57° N). Finally, Oligocene Foraminifera from the ocean displayed a paleolatitudinal gradient in richness, which began about 35 Ma in the Pacific Ocean (0°–18° N compared with 64° S–65° S; Thomas & Gooday 1996).

The oldest documented terrestrial gradient—one pertaining to vascular plants—originated more than 100 Ma. Terrestrial angiosperms, primarily from Europe and North America, exhibited a latitudinal pattern in richness at a variety of levels in the taxonomic hierarchy for a large portion of the Cretaceous (110 Ma to 65 Ma, in ~5 Ma time bins), with the gradient growing stronger through time (80° N–20° S; Crane & Lidgard 1989). Fossil pollen samples have been used to assess in detail the paleolatitudinal gradient of richness for angiosperms from the end of the Pleistocene through the Holocene to the present. European plant samples evinced a well-defined gradient at 1000-year intervals, from 13,000 onward to 1000 years ago. This included taxonomic richness, primarily at the specific or generic level for woody plants (Silvertown 1985), and familial richness for angiosperms (Haskell 2001). The gradient for woody plants became steeper over time, whereas the gradient for angiosperms was more consistent temporally. In addition, North American plants exhibited a latitudinal gradient in which richness of angiosperm families and woody plant genera increased toward the tropics for time periods from 12,000 or 13,000 years ago, respectively, growing steeper through time, until about 10,000 years ago, at which time the slope became constant (D.M. Kaufman & J.W. Williams, unpublished manuscript).

In summary, a latitudinal gradient in richness has existed for a considerable segment of geologic time, at numerous time steps from 270 Ma to now, and in six periods from the past three eras. Many of the ancient patterns were based on superspecific taxonomic levels, because much of the fossil data, such as those related to pollen or calcareous shells, cannot be resolved to the species level. Higher taxa have been used in numerous studies of latitudinal richness gradients for additional reasons. First, extremely broad geographic coverage of taxon-rich biotas make species-level collection of data impractical. Second, studies based on higher taxa facilitate ecological or evolutionary comparisons across diverse clades or comparisons to species-level patterns.

Gaston et al. (1995) examined familial richness of extant vertebrates, invertebrates, and plants in the same global domain. Strong latitudinal gradients were demonstrated for mammals, reptiles, amphibians, and angiosperms. In contrast, familial richness of beetles showed no relationship to latitude. In addition, latitudinal gradients of plant genera have been demonstrated consistently for North America (Qian 1998; D.M. Kaufman & J.W. Williams, unpublished manuscript). In terms of marine organisms, latitude was related inversely to richness for copepod genera, families, and superfamilies (Atlantic Ocean: Woodd-Walker et al. 2002) as well as for coral genera (Great Barrier Reef: Fischer 1960; western Pacific and Atlantic Oceans: Stehli & Wells 1971).

In an early study, Fischer (1960) showed a latitudinal gradient in specific and generic richness for extant marine tunicates. A latitudinal gradient of generic richness was demonstrated across the globe for extant mammals as well (Gaston et al. 1995). A strong latitudinal gradient in richness existed at specific, generic, familial, and ordinal levels for North and South America, whether considered separately or as a single domain (Kaufman 1994, 1995). Nevertheless, the form of the richness-latitude relationship differed among levels in the taxonomic hierarchy—it was a negative sigmoidal model for specific and generic richness, but linear for familial and ordinal richness (Kaufman 1994, 1995). A recent study of sub-Saharan Africa demonstrated that generic richness is a good predictor of plant species richness, but familial or higher richness is a less suitable indicator (La Ferla et al. 2002). Although patterns based on higher taxa may recapitulate those at the species level, caution must be taken in assuming such a quantitative relationship.

MECHANISTIC BASES OF LATITUDINAL GRADIENTS

As early as 1807, von Humboldt provided the first hypothesis (based on climate) to explain latitudinal gradients of richness (Hawkins 2001). It embodied a general mechanism that was not specific to a particular taxon or to a particular place. In this vein, subsequent authors sought general and pervasive mechanisms that would effect a latitudinal gradient in species richness based on the premise that a single mechanism would be responsible for a pattern that is taxonomically, geographically, and temporally prevalent (Pianka 1966, but see Gaston 2000). Nonetheless, consensus about the identity of "the" mechanism has been elusive, and more synthetic multifactor approaches are emerging.

Quantitative examination of the latitudinal pattern of species richness began in the 1950s, with concurrent speculation as to its possible causes (e.g., Dobzhansky 1950, Hutchinson 1959). Pianka (1966) was the first to provide a comprehensive review of the hypotheses proposed to account for the latitudinal gradient (Table 1), including climatic stability, competition, predation, productivity, spatial heterogeneity, and time hypotheses. In the decades that followed, particular hypotheses were modified and amplified, but efforts to distinguish among them have been few. Ironically, the list of potential mechanisms has grown dramatically, so that by the 1980s, the number of explanatory hypotheses increased to ten (Brown 1988, Brown & Gibson 1983). By the early 1990s, a further threefold increase had occurred (Rohde 1992).

Hypothesized Mechanisms

A review of the potential explanatory power of each of the more than 30 currently posed hypotheses (Table 1) as a cause for the latitudinal gradient is a daunting

TABLE 1 Hypotheses proposed to account for the latitudinal gradient of diversity*

†**Abiotic-biotic**[1]	§**Geographic area**[RI]	§**Rapoport rescue**[4]
§**Ambient energy**[R]	§**Geometric constraints**[3]	Rapoport's rule[RI]
Environmental predictability[RI]	Interspecific interactions[B]	†**Scale hierarchy**[5]
Environmental stability[P, RI]	Competition[P, RC]	**Spatial heterogeneity**[P, B]
Harshness[B, RC]	‡Host diversity[RC]	Biotic spatial heterogeneity[RC]
Seasonality[RI]	Mutualism[RC]	Epiphyte load[RC]
†**Energetic-equivalents**[2]	Niche width[B, RC]	Number of habitats[RI]
Evolutionary rates	Predation[P, RC]	Patchiness[RC]
Extinction rate[B]	**Population dynamics**	Physical heterogeneity[RI]
Origination rate[B]	Epidemics[RC]	‡Solar angle[RI]
§**Evolutionary speed**[R]	Population growth rate[RC]	**Time**[P, B]
Temperature-dependent chemical reactions[R]	Population size[RC]	Abiotic rarefaction[RI]
	§**Productivity**[P, B, RI]	Ecological time[R]
	‡Aridity[RI]	Evolutionary time[R]

*Augmented from Rohde (1992) and modified from D.M. Kaufman & J.H. Brown (in review). Originating authors follow for those hypotheses not included in Rohde; for others, see Rohde (1992).

†Recent hypotheses not yet evaluated thoroughly in the literature; published sources are indicated by numeric superscript: ([1]Kaufman 1995, 1998; [2]Allen et al. 2002; [3]Colwell & Hurtt 1994, Lyons & Willig 1997; and [5]Whittaker et al. 2001).

§Hypotheses discussed in detail in text ([4]Taylor & Gaines 1999).

‡Hypotheses too specific to provide a general mechanism.

[P]Hypotheses included by Pianka (1966).

[B]Hypotheses included by Brown (1988, Brown & Gibson 1983).

[R]Hypotheses included by Rohde (1992; with [C] denoting "circular" hypotheses and [I] for "insufficiently supported" hypotheses).

challenge. Some hypotheses (e.g., aridity, host diversity, solar angle; Table 1) are too specific to explain the ubiquity of the latitudinal gradient. Others are circular and cannot generate the latitudinal gradient in isolation (Rohde 1992). Many are tightly interlinked because of their mechanistic bases. Previous authors (e.g., Schemske 2002) have attempted to narrow the potential mechanisms by sorting hypotheses via contrasting bases (i.e., historical or equilibrial, physical or biotic, evolutionary or ecological). Rohde (1992) offered yet another categorization, perhaps more pragmatic, in his division of hypotheses into two categories: insufficiently supported (those in which the invoked factor has not been shown to vary monotonically with latitude or richness) and circular (those that require latitudinal variation in factors extrinsic to the hypothesis) hypotheses (Table 1). This classification has practical utility as we seek to disentangle the plausibility of and evidence for potential mechanisms. For ease of consideration, rather than examine each hypothesis individually among which there are substantial redundancies, we organized them into five broad categories or themes. Each theme includes hypotheses that are related by similar mechanistic bases or by shared organizational concepts (Table 1). Importantly, many of these themes contain specific hypotheses that have little empirical support, are difficult to test, or could only be secondary explanations for the latitudinal gradient in richness. Nonetheless, the mechanisms that are subsumed by these themes might modify empirical gradients in a substantive manner.

The time theme assumes that older communities are more diverse (Pianka 1966). Ecological forms of this hypothesis involve ecological disturbance and local dispersal (including abiotic rarefaction; Rohde 1992), whereas evolutionary forms entail geological perturbation and speciation or extinction. These hypotheses cannot be tested directly, but Pleistocene glaciations have been used to explain the relative impoverishment of the Temperate Zone as compared with the relative constancy of the tropics and its greater richness (Brown 1988).

The spatial heterogeneity theme assumes an inverse relationship between latitude and environmental complexity (Pianka 1966). Nonetheless, no general gradient in heterogeneity has been demonstrated (Rohde 1992), and any explanation based on biotic heterogeneity would require another underlying mechanism to produce that biotic gradient. Heterogeneity may be an important component of the latitudinal gradient when viewed as an axis of variation along which other factors, such as interspecific interactions, operate to mold patterns of species richness.

The interspecific interaction theme assumes that increased intensity of competition, predation, or mutualism facilitates greater species richness in the tropics. For example, competition may lead to more species per unit habitat space, or predation may reduce prey population levels, thereby allowing more prey species to coexist in the tropics (Pianka 1966). Indeed, interspecific interactions may be more costly at lower latitudes where, for example, organisms invest more energy or resources in combating parasitism (Møller 1998). Moreover, differences in range boundaries at the high- and low-latitude portions of species ranges are consistent with the assumption of greater interspecific effects at low latitudes and greater abiotic controls at high latitudes (Kaufman 1998). That is, range boundaries in extratropical areas were more linear and adherent to latitudinal parallels compared with range boundaries in more equatorial areas. Importantly, none of these mechanisms alone can generate the latitudinal gradient of richness. Each requires an underlying pattern in diversity or the factors that effect it before becoming operational.

The population theme requires a latitudinal pattern in demographic characteristics (e.g., individuals within populations or population dynamics) that must be produced by some other factor, such as species packing. Empirical support for the appropriate latitudinal gradients in population characteristics is not pervasive. For example, monogenean and digenean parasites of marine fishes did not have smaller population size nor were their niches narrower in the tropics (Rohde 1978a).

The evolutionary rate theme has been invoked to produce latitudinal gradients of richness (Rosenzweig 1975). For example, high speciation rates have been linked to the great richness of the lowland wet tropics (Brown 1988). Other mechanisms (such as the larger area, increased productivity, and greater environmental predictability) have been invoked as causes for these purported high speciation rates and low extinction rates in the tropics (Rosenzweig 1975). From a methodological perspective, the accurate measurement of rates is a challenge in the evaluation of these hypotheses (Brown 1988).

We consider six hypotheses for further elaboration. These have the most support and potential, are least easy to refute based on current information, or are general and synthetic. We do not elaborate further on recently advanced hypotheses (Table 1) that have not been evaluated in the literature.

Geographic Area Hypothesis

Although this mechanism was originated by Terborgh (1973), it has been most developed and amplified as a hypothesis by Rosenzweig (1995). The primary tenet is that the tropics support more species than other regions because they comprise more area. A secondary tenet is that elevated productivity in equatorial regions along with enhanced zonal bleeding in the tropics interact with areal effects to produce a gradient of increasing richness with decreasing latitude. Although considerable controversy surrounds the hypothesis (Rohde 1997, 1998; Rosenzweig & Sandlin 1997), it does not question if areal mechanisms increase richness; rather, it focuses on the degree to which area is the dominant factor effecting the latitudinal gradient (Willig 2001).

Two aspects of the earth's geometry predispose area to increase from polar to tropical areas. First, the earth is essentially a sphere, and the circumference of latitudinal parallels is greatest at the equator and becomes progressively smaller toward the poles. Consequently, the area of a latitudinal band of fixed width is greatest at 0° and least at the poles. Second, the distribution of Tropical, Temperate, and Polar Zones is symmetrical around the equator. More specifically, northern and southern Polar Zones are disjunct, separated by the intervening temperate and tropical areas. Similarly, northern and southern Temperate Zones are disjunct, separated by intervening tropical areas. In contrast, northern and southern Tropical Zones are adjacent, straddling the equator to form a contiguous tropics. If the earth's landmasses or bodies of water were randomly distributed across the globe then the area of available land or water would gradually increase from polar to tropical regions. Thus, diminished area and increased isolation of extratropical areas initiates a concomitant gradient of species richness.

More pervasive than the latitudinal gradient is the observation that species richness increases with area. As the area of a region increases, so too does the number of habitats, biomes, or biogeographic provinces within it. Similarly, as the area of a region increases, so too does the area of its constituent habitats, biomes, or provinces. Larger areas support more individuals and populations of a species, and reduce the likelihood of extinction. Larger areas have a higher likelihood of containing geographic barriers to gene flow, which would enhance speciation rates. Larger areas will have more diverse habitats facilitating the development of specialization, adaptation, and speciation. On average then, a larger area potentially will support more richness at a variety of scales (i.e., alpha, beta, and gamma diversity) and will certainly enjoy higher overall species richness.

Evidence regarding this hypothesis is inconclusive. Rosenzweig (1995) and Rosenzweig & Sandlin (1997) provided evidence in support of the hypothesis. Regardless of latitude, larger biotic provinces contain more taxa (species, genera, and families) than do their smaller counterparts. Moreover, species richnesses from the same biome but different continents or provinces vary as a function of their latitudinal extents. Ruggiero (1999) also provided support for the hypothesis in a comprehensive statistical analysis of mammalian gradients in South America in which she controlled for the presence of tropical species in extratropical biomes. In contrast, Rohde (1997, 1998) does not consider area to drive the latitudinal gradient. For example, smaller tropical areas of Eurasia contain far more species of freshwater fishes than do larger temperate regions. In addition, the expansive deep-sea biome with essentially constant temperature supports only a fraction of the species that occur in its much smaller tropical counterpart. In North America, the greatest continental width occurs at relatively high latitudes, and latitude and area are correlated positively, whereas in South America the greatest continental width occurs at equatorial regions, and latitude and area are correlated negatively. Nonetheless, for mammals in North America and in South America, area had little unique explanatory power as compared with latitude, and the latitudinal gradient of richness was significant and steep for each continent, even when the effects of area were removed in analyses of covariance (Kaufman & Willig 1998).

Productivity Hypothesis

The idea that energy limits richness dates back at least as far as Hutchinson (1959). Pianka (1966) wrote that "greater production results in greater richness, everything else being equal." Wright (1983) advanced the species-energy hypothesis as a more general extension of the species-area theory of MacArthur & Wilson (1963, 1967). The productivity hypothesis posits that the annual input of solar radiation determines energy availability, productivity, and biomass, and is tightly related to latitude in an inverse manner (Robinson 1966). Measures of productivity, such as actual evapotranspiration (AET), generally are correlated to species richness, but not in all organisms and generally not as highly as is potential evapotranspiration (PET). For example, in North America, neither bird nor mammal richness exhibited high correlations with AET (Currie 1991). For other organisms, such as North American trees and amphibians, the correlation between AET and species richness is quite high (Currie 1991).

Although a positive relationship between productivity and richness seems likely, at least for an appreciable portion of the gradient, the productivity hypothesis has not been accepted generally as an important cause of geographic patterns of species richness (Brown & Gibson 1983). Currie (1991) identified a critical shortcoming of the productivity hypothesis. It fails to elucidate a mechanism as to how or why species richness would increase to a maximum set by energy availability, as opposed to population densities simply increasing in magnitude. Although highly productive environments can exhibit great richness, they also may exhibit low richness in some situations (MacArthur 1972). Indeed, little generality about the form of the relationship between species richness and productivity (i.e., positive linear, negative linear, modal) exists (Mittelbach et al. 2001, Waide et al. 1999) other than the view that it is scale dependent (Gross et al. 2000, Pastor et al. 1996, Scheiner et al. 2000, Scheiner & Jones 2002). Thus, productivity alone cannot explain latitudinal gradients of richness.

Ambient Energy Hypothesis

This hypothesis considers the input of solar energy to create a physical environment that affects organisms through their physiological responses to temperature. Ambient energy essentially serves as an umbrella hypothesis under which other explanations, such as climatic stability, environmental stability, environmental predictability, seasonality, and harshness, are subsumed. This hypothesis is based on the concept that environments at high latitudes have mean conditions farther from organismal optima (e.g., their thermal neutral zones) than do their low-latitude counterparts. In addition, high-latitude environments are thought to be more variable and seasonal than are those at low latitudes. Based on temperature records from climatological stations throughout the New World, high latitudes evinced lower mean temperature, colder winter extremes, higher annual variability, and shorter growing season as compared with lower latitudes (Kaufman 1998). This is equivalent to the idea that higher latitudes are harsher for organisms than are lower latitudes. Species-poor habitats often have physiologically harsh and unpredictable environmental conditions that require special and costly adaptations (Brown 1988). Rohde (1992) considered the harshness hypothesis to be circular. Nonetheless, physiological characteristics can be used to assess favorableness instead of simply reasoning that low latitudes are favorable because there are many tropical species and high latitudes harsh because there are few polar species. For example, it is more physiologically costly to live at high latitudes than low latitudes because the ambient temperature in polar regions is outside the thermal neutral zone of many organisms for a majority of the year.

Several measures of ambient energy are highly correlated to species richness. For example, sunshine and temperature were the primary underlying factors affecting butterfly richness in Great Britain, as a consequence of basking and physiological requirements (Turner et al. 1987). Moreover, PET, a measure of ambient energy, was generally correlated with species richness of trees, amphibians, reptiles, birds, and mammals in North America (Currie 1991, Currie & Paquin 1987).

Rapoport-Rescue Hypothesis

Rapoport's rule is a pattern in which the size of the distributional ranges of species is related inversely to latitude (Rapoport 1975, Stevens 1989). A diversity of taxa that inhabit aquatic and terrestrial environments exhibited the pattern (e.g., mammals, reptiles, amphibians, fish, crayfish, amphipods, mollusks, and trees). When proposed, Rapoport's rule was designed as a mechanism to underlie the latitudinal gradient of richness. Because seasonal variation at high latitudes is great, organisms with broad climatic tolerances will be favored by natural selection in these areas. Broad tolerance not only allows organisms to persist through time in a particular locale, but also results in the possession of large ranges. In contrast, organisms from more tropical latitudes generally have narrower climatic tolerances and more restricted ranges. Furthermore, the narrower tolerances of tropical organisms cause the environment to be more heterogeneous from their perspective, and they are more likely to disperse or spillover into unfavorable areas. Overall species richness is augmented by the addition of "accidentals," species that normally would not persist but are "rescued" by continual dispersal from nearby favorable areas. That many taxa exhibited both the latitudinal gradient of richness and Rapoport's rule suggested a link between them; furthermore, taxa that did not exhibit a latitudinal gradient were generally exceptions to Rapoport's rule.

Rapoport's rule is more uniformly exhibited by taxa in the Northern Hemisphere than in the Southern Hemisphere. Moreover, the rule is less substantiated in the tropics compared with the Temperate Zone (Gaston et al. 1998). Indeed, empirical evidence and quantitative modeling (Colwell & Hurtt 1994, Lyons & Willig 1997, Taylor & Gaines 1999, Willig & Lyons 1998) together suggest that the logic underlying the Rapoport-rescue hypothesis is flawed or only applicable under restrictive circumstances (Rohde 1996). Moreover, some taxa that exhibit a marked latitudinal gradient of richness do not exhibit Rapoport's rule (e.g., New World bats and marsupials—Lyons & Willig 1997, Willig & Gannon 1997, Willig & Selcer 1989), diminishing the likelihood that the Rapoport-rescue hypothesis represents a general mechanism producing latitudinal gradients of richness.

Evolutionary Speed Hypothesis

Rohde (1992) proposed the evolutionary speed hypothesis, which holds that species richness increases toward the tropics because of temperature-induced increases in rates of speciation. That is, high temperature enhances evolutionary speed. Latitudinal patterns of temperature result in shorter generation times, higher mutation rates, and accelerated selection pressure in the tropics, which combine to enhance rates of speciation, and as a consequence, species richness (Rohde 1992). His view diverged from those who support the ambient energy hypothesis because only under equilibrial conditions and community saturation would available energy be able to effect a latitudinal gradient. Such conditions have not been shown and would be challenging to document (Rohde 1992).

The number of generations per year is related negatively to latitude in some taxa; however, short generation times do not always lead to faster evolutionary rates (Rohde 1992). Although little actual evidence supports or refutes the mechanistic underpinnings of this explanation, Jablonski (1993) suggested that the tropics have been a source of evolutionary novelty throughout geologic time as evidenced by the higher origination rates of post-Paleozoic marine orders in equatorial areas compared with temperate areas. However, a recent quantitative analysis based on phylogenetically independent contrasts of rates of evolution in latitudinally separated pairs of bird species provided no support for the contention that rates of molecular evolution increase toward the tropics (Bromham & Cardillo 2003). If similar research and data for other taxa fail to support the negative association between evolutionary rates and latitude, the evolutionary speed hypothesis can be removed as a plausible mechanism that effects the latitudinal gradient in species richness.

Geometric Constraints Hypothesis

The geometric constraints hypothesis originated as a consequence of a radically different approach for understanding the latitudinal gradient—one that does not require environmental gradients to be associated with changes in latitude and one that does not require the biota to respond to environmental gradients if they do exist. Colwell & Hurtt (1994) suggested that "nonbiological" latitudinal gradients of species richness could be produced as a consequence of the random placement of species ranges within a bounded domain. A bounded domain is an area circumscribed by a physical (e.g., continental coastlines for terrestrial species) or a physiological barrier (e.g., salinity levels for freshwater fish) that restricts the distribution of species within a taxon to a subset of the earth's surface. If the world is effectively unbounded, the random placement of species ranges fails to produce a gradient (i.e., produces a line with zero slope for the relationship between species richness and latitude). This condition is the equivalent of the null hypothesis, a statement that is inherent to all statistical assessments of latitudinal gradients. If the world is characterized by boundaries, gradients in richness of various types are produced, depending on a number of constraints.

Indeed, the sizes of species ranges as well as their central locations are the proximate determinants of any latitudinal gradient. Consequently, a full spectrum of simulation models has been developed that reflects various constraints on the placement of species ranges within a bounded domain (Colwell & Hurtt 1994, Colwell & Lees 2000). The three general categories of such null models include those that are (*a*) unconstrained, (*b*) constrained by the location of range midpoints, and (*c*) constrained by the distribution of range sizes. In general, all of these models produce latitudinal gradients with peaks in richness at mid-domains. The models differ with respect to predictions concerning the rate of increase in richness toward the mid-domain, the magnitude of maximum richness, and the kurtosis of the latitudinal gradient of richness. In addition to simulation approaches, analytical models also suggest a latitudinal gradient with mid-domain peaks in species richness (e.g., Lees et al. 1999, Willig & Lyons 1998).

Unlike all the other hypotheses concerning the latitudinal gradient in species richness, geometric constraint models make quantitative predictions in terms of the latitudinal location of the richness peak, the form of the latitude-richness relationship, and in the richness of species at any particular latitude. Empirical support for the geometric constraints model is increasing (see Hawkins & Diniz-Filho 2002 for an alternate view). For example, Willig & Lyons (1998) applied a binomial model to the New World and found that it produced a tropical peak in species richness. Moreover, they compared their predictions to empirical data for bats and marsupials and found that much of the latitudinal variation in species richness was related to the random placement of species ranges, depending on the location of hard boundaries. For bats, between 67% and 77% of the variation in richness was accounted for by the model. Similarly for marsupials, between 35% and 94% of the variation in richness was accounted for by the model. Importantly, empirical patterns deviated from predictions for each biota in a different way, providing additional insights not possible with other models. In particular, marsupials deviated from model predictions by being less rich at all latitudes than predicted by the model, whereas bats deviated from model predictions by being richer in the tropics and more depauperate in temperate regions. Other biotas exhibited latitudinal gradients that conform to the predictions of various geometric constraint models, including rodents in the deserts of the American Southwest (McCain 2003) and endemic rainforest butterflies, frogs, rodents, tenrecs, chameleons, and birds in Madagascar (Lees et al. 1999).

Finally, geometric constraint models make a unique prediction about the latitudinal gradient that is not shared by any of the other hypotheses and often is in opposition to them (Willig & Lyons 1998). That is, species whose distributions are wholly contained within any geographic domain—even randomly determined hard boundaries—should exhibit a mid-domain peak in richness. Bats and marsupials in the New World conformed to these predictions for 100% and 95% of 20 random locations that extended at least 20° of latitude and contained at least 20 species in the biota of interest. The application of two-dimensional geometric constraint models has successfully accounted for a significant amount of the variation in bird species richness in southern Africa (Jetz & Rahbek 2001, 2002). It has been less successful for South American raptors (Diniz-Filho et al. 2002b), Nearctic birds (Hawkins & Diniz-Filho 2002), and New World mammals (Bokma et al. 2001).

Predictions from geometric constraint approaches may act as a null model against which empirical data are evaluated regarding the latitudinal gradient. Nonetheless, Hawkins & Diniz-Filho (2002) considered the hypothesis to be flawed because the assumption that ranges exist independently of the environment or that ranges are randomly placed has no theoretical justification. Of course, such assumptions of neutrality are inherent to geometric constraint models and null models in general. Moreover, neutral mechanisms as well as deterministic mechanisms can operate simultaneously to produce patterns, and need not be constrained to represent mutually exclusive phenomena (Hubbell 2001), a view that is gaining appreciation in ecology and biogeography. An alternative view is that each species in a biota is unique by its very nature. It responds to the multivariate environmental template in different ways throughout its distribution and in a manner unlike any other species. Numerous examples concerning the diversity of factors affecting species distribution are the fodder for textbooks in ecology and biogeography (e.g., Krebs 1994). Moreover, studies of the response of species to the last glaciation cycle support this view in that taxa shift their ranges in species-specific and unpredictable manners given present knowledge of their niche dimensions (e.g., Holman 1995 for North American herpetofauna; Valentine & Jablonski 1993 for marine mollusks; Davis & Shaw 2001 for North American trees; Lyons 2003 for North American mammals). Correspondence between empirical gradients and geometric constraint models may not reflect the role of chance in determining the distribution of species per se. Rather, it may reflect the many and unpredictable ways in which evolutionary processes have molded the adaptations of species so that they respond to the environment in a multitude of ways (Lyons & Willig 1997).

Distinguishing Among the Hypotheses

The problem of distinguishing among hypotheses concerning the latitudinal gradient of richness is that most represent conceptual models rather than quantitative models and only make the qualitative prediction that species richness should increase with decreasing latitude. Moreover, many of the mechanistic factors associated with different hypotheses are shared or correlated; consequently, indirect tests are problematic and controvertible. However, progress has occurred in reducing the number of likely causative mechanisms (e.g., Willig 2001; D.M. Kaufman & J.H. Brown, in review). By capturing the fundamental nature of over 30 hypotheses, and categorizing them into themes, we provide a reasonable springboard from which to launch future investigations of the mechanistic bases of the gradient. The detailed review of a number of popular, controversial, or promising hypotheses will hopefully direct future research in ways that will challenge contemporary perspectives or corroborate mechanistic explanations of nature's longest recognized and most pervasive pattern. Nonetheless, advancing ecological and evolutionary understanding of the bases of the latitudinal gradient requires adoption of new or additional conceptual, empirical, and analytical approaches.

CONCLUSIONS

The wide variety of data that have been collected across taxa, space, time, and scale has allowed ecologists to assemble considerable evidence in support of the current paradigm (Figure 1). Scale effects generally are of two types. Those related to extent can result in considerable variation in the form and parameterization of the latitudinal gradient. In contrast, those associated with focus usually result only in variation in the parameterization of the classical pattern of richness increasing toward the equator. Further collection of data will increase the likelihood that we truly understand the extent to which the pattern is universal and the contexts in which exceptions occur. Similarly, expanded focus on causes of elevational or bathymetric gradients, known to be affected by similar environmental parameters as those likely affecting the latitudinal gradient, will lead to a more mature body of theory regarding diversity in general, and the latitudinal gradient in particular.

Although additional pattern analysis will enrich our characterization and knowledge of the latitudinal gradient of diversity, the current challenge is to elucidate a convincing, mechanistically based theory to explain the generation and maintenance of the gradient, including the conditions that account for the current exceptions (e.g., Willig 2001, Whittaker et al. 2001). Clearly, no single mechanism appears adequate to account for the taxonomic, geographic, and temporal ubiquity of the latitudinal gradient. Pianka (1966) believed that combining of hypotheses should be avoided because it would be "less testable and useful." However, if multiple factors vary in concert with latitude and interact to cause the latitudinal gradient of richness, it is doubtful that the evaluation of these factors in isolation will bring about meaningful progress (Rosenzweig 1975; Willig 2001; D.M. Kaufman & J.H. Brown, in review).

Because beta diversity connects levels of richness at local and regional levels, it is a key to understanding scale dependence. Consequently, renewed efforts to understand how the environment molds beta diversity will contribute significantly to advancing a number of subdisciplines in macroecology, especially those dealing with gradients such as latitude, elevation, depth, or productivity. At the same time, a careful consideration of the consequences of stochastic processes operating in a bounded world must be incorporated into hypothesis testing and the construction of predictive models, even those with strong mechanistic underpinning. Indeed, only a few hypotheses have made explicit predictions, about the qualitative or quantitative nature of the richness gradient or about other macroecological characteristics, such as range size distribution, abundance, and diversity.

Recently amplified hypotheses are more synthetic in nature (e.g., Whittaker et al. 2001). Indeed, synthetic hypotheses should employ multiple mechanistic features, such as the incorporation of ambient energy, productivity, and interspecific interactions, and should make explicit predictions, an approach advanced by Kaufman (1995, 1998). As Gaston (2000) stressed, a "predictive theory of species richness" is still distant, but work over the past decade characteristically exhibits a much greater emphasis on predictiveness. Furthermore, research that contributes to an understanding of the latitudinal gradient is departing from a narrow focus on latitude, per se. Two schools of thought are emerging. One continues to examine factors in an explicitly latitudinal framework (e.g., Koleff & Gaston 2001). The other examines richness and causative mechanisms across spatial dimensions without including latitude as a particular factor in the framework (e.g., Allen et al. 2002, Hawkins et al. 2003, Hurlbert & Haskell 2003). As with any complex system (Rosenzweig 1975), it is unlikely that any one variable will entirely account for latitudinal gradients of richness. Instead, several factors will inform a comprehensive mechanism that applies to most biotas in most places during much of evolutionary time (Gaston 2000). As such, we provide a simple conceptual model (Figure 5, see color insert) that links mechanisms from a variety of hypotheses, and suggest how each would affect alpha, beta, and gamma diversity. We end with a litany of recommendations for future research in terms of empirical, statistical, and conceptual approaches.

Recommendations for Conceptual Approaches

- To guide future research, an overarching model with hierarchical representation of particular mechanisms (e.g., Figure 5) should comprehensively reflect the primary and secondary gradients that cause or modify spatial patterns of diversity.

- Models should be developed that make explicit predictions about how various metrics of diversity (e.g., richness, evenness, dominance, turnover) at multiple spatial or ecological scales (e.g., alpha, beta, or gamma diversity) are associated with mechanistic processes.
- Synthetic linkages are needed between models associated with gradients of richness and models associated with other macroecological characteristics (e.g., body size, range size, niche dimensions).
- The ramifications of particular hypotheses should be explored to identify unique quantitative or qualitative predictions that differ from those associated with competing hypotheses.
- Advocates of particular hypotheses should postulate critical tests that can be addressed by empirical evidence (hypotheses that cannot be wounded fatally are not vital or useful).
- The assumption that increased energy leads to elevated species richness rather than simply more individuals needs to be validated from first principles.

Recommendations for Empirical Approaches

- The spatial focus of analyses (e.g., point diversity versus alpha diversity versus gamma diversity) needs to be explicit and clearly linked to mechanisms at relevant spatial scales.
- Empirical evidence for gradients in turnover at a variety of ecological scales (i.e., local beta diversity, beta diversity, and delta diversity) are needed to understand the spatial dynamics of diversity, and how heterogeneity affects broad scale patterns.
- Additional empirical research on latitudinal gradients in alpha diversity are needed to guide investigation of how variation in local conditions affects the richness of communities.
- Research should be amplified to include latitudinal gradients in weighted measures of diversity (e.g., species diversity, evenness, and dominance) and to contrast these to patterns in species richness.

Recommendations for Analytical Approaches

- Analytical methods that minimize the effects of spatial autocorrelation on analyses of latitudinal gradients need to be developed and deployed.
- Broader use of simulation approaches and dynamic modeling is needed to decouple the effects of confounding variables in assessments of causative mechanisms.

- Statistical testing of the mechanistic basis of the latitudinal gradient should be undertaken in the context of geometric constraint models in two dimensions, including more comprehensive analyses of residuals.

ACKNOWLEDGMENTS

This manuscript had its genesis at the National Center for Ecological Analysis and Synthesis [a Center funded by NSF (Grant #DEB-0072909), the University of California, and the Santa Barbara campus], which substantively aided all of the authors by provisioning sabbatical (M.R.W.) or postdoctoral (D.M.K. and R.D.S.) support during extensive periods of manuscript preparation. Additional support to M.R.W. was provided through The Office of the Provost (J.M. Burns) and Department of Biological Sciences (J.C. Zak) at Texas Tech University. The clarity of ideas was improved significantly by the critical reviews of S.J. Andelman, S.K. Lyons, N.E. McIntyre, and D.W. Schemske, as well as by comments from C.P. Bloch, D. Chalcraft, D.W. Kaufman, L.A. Levin, M.A. Rex, and S.M. Scheiner. We gratefully acknowledge S.J. Andelman for crafting the map of Figure 4 in an accurate and appealing manner.

The *Annual Review of Ecology, Evolution, and Systematics* is online at http://ecolsys.annualreviews.org

LITERATURE CITED

Allen AP, Brown JP, Gillooly JF. 2002. Global biodiversity, biochemical kinetics, and the energetic-equivalence rule. *Science* 297: 1545–48

Anderson S. 1977. Geographic ranges of North American Terrestrial Mammals. *Am. Mus. Novit.* 2629:1–15

Anderson S. 1984. Areography of North American fishes, amphibians, and reptiles. *Am. Mus. Novit.* 2802:1–16

Anderson S, Marcus LF. 1993. Effect of quadrat size on measurements of species density. *J. Biogeogr.* 20:421–28

Andrews P, O'Brien EM. 2000. Climate, vegetation, and predictable gradients in mammal species richness in southern Africa. *J. Zool.* 251:205–31

Angel MV. 1993. Biodiversity of the pelagic ocean. *Conserv. Biol.* 7:760–72

Arita HT, Rodriguez P. 2002. Geographic range, turnover rate and the scaling of species diversity. *Ecography* 25:541–50

Arrhenius O. 1921. Species and area. *J. Ecol.* 9: 95–99

Arrhenius O. 1923a. Statistical investigations in the constitution of plant associations. *Ecology* 4:68–73

Arrhenius O. 1923b. On the relation between species and area—a reply. *Ecology* 4:90–91

Badgley C, Fox DL. 2000. Ecological biogeography of North American mammals: species density and ecological structure in relation to environmental gradients. *J. Biogeogr.* 27:1437–67

Baquero RA, Telleria JL. 2001. Species richness, rarity, and endemicity of European mammals: a biogeographical approach. *Biodiv. Conserv.* 10:29–44

Barbosa A, Benzal J. 1996. Diversity and abundance of small mammals in Iberia: peninsular effect or habitat suitability? *Z. Säugetierkd.* 61:236–41

Barbour CD, Brown JH. 1974. Fish species diversity in lakes. *Am. Nat.* 108:473–89

Blackburn TM, Gaston KJ. 1996a. Spatial patterns in the species richness of birds in the New World. *Ecography* 19:369–76

Blackburn TM, Gaston KJ. 1996b. The distribution of bird species in the New World: patterns in species turnover. *Oikos* 77:146–52

Blackburn TM, Gaston KJ. 1997. The relationship between geographic area and the latitudinal gradient in species richness in New World birds. *Evol. Ecol.* 11:195–204

Blaylock RB, Margolis L, Holmes JC. 1998. Zoogeography of the parasites of Pacific halibut (*Hippoglossus stenolepis*) in the northeast Pacific. *Can. J. Zool.* 76:2262–73

Bokma F, Bokma J, Monkkonen M. 2001. Random processes and geographic species richness patterns: why so few species in the north? *Ecography* 24:43–49

Bolton JJ. 1994. Global seaweed diversity: patterns and anomalies. *Bot. Mar.* 37:241–45

Bowman DMJS. 1996. Diversity patterns of woody species on a latitudinal transect from the monsoon tropics to desert in the Northern Territory, Australia. *Aust. J. Bot.* 44:571–80

Briggs DEG, Kear AJ. 1993. Fossilization of soft tissue in the laboratory. *Science* 259:1439–42

Bromham L, Cardillo M. 2003. Testing the link between the latitudinal gradient in species and rates of molecular evolution. *J. Evol. Biol.* 16:200–7

Brown JH. 1988. Species diversity. In *Analytical Biogeography: An Integrated Approach to the Study of Animal and Plant Distributions*, ed. AA Myers, PS Giller, pp. 57–89. London: Chapman & Hall

Brown JH. 1995. *Macroecology*. Chicago, IL: Univ. Chicago Press

Brown JH, Gibson AC. 1983. *Biogeography*. St. Louis: Mosby. 1st ed.

Brown JH, Lomolino MV. 1998. *Biogeography*. Sunderland, MA: Sinauer. 2nd ed.

Buckton ST, Ormerod SJ. 2002. Global patterns of diversity among the specialist birds of riverine landscapes. *Freshw. Biol.* 47:695–709

Cardillo M. 2002. Body size and latitudinal gradients in regional diversity of New World birds. *Glob. Ecol. Biogeogr.* 11:59–65

Cherns L, Wright VP. 2000. Missing molluscs as evidence of early skeletal aragonite dissolution in a Silurian Sea. *Geology* 28:791–94

Chown SL, Gaston KJ. 2000. Areas, cradles and museums: the latitudinal gradient in species richness. *Trends Ecol. Evol.* 15:311–15

Chown SL, Gaston KJ, Williams PH. 1998. Global patterns in species richness of pelagic seabirds: the Procellariiformes. *Ecography* 21:342–50

Clarke A, Lidgard S. 2000. Spatial patterns of diversity in the sea: bryozoan species richness in the North Atlantic. *J. Anim. Ecol.* 69:799–814

Coleman BD. 1981. On random placement and species-area relations. *Math. Biosci.* 54:191–215

Coleman BD, Mares MA, Willig MR, Hsieh Y-H. 1982. Randomness, area, and species richness. *Ecology* 63:1121–33

Colwell RK, Hurtt GC. 1994. Nonbiological gradients in species richness and a spurious Rapoport effect. *Am. Nat.* 144:570–95

Colwell RK, Lees DC. 2000. The mid-domain effect: geometric constraints on the geography of species richness. *Trends Ecol. Evol.* 15:70–76

Cook RE. 1969. Variation in species density of North American birds. *Syst. Zool.* 18:63–84

Cowling RM, Samways MJ. 1995. Predicting global patterns of endemic plant species richness. *Biodivers. Lett.* 2:127–31

Cowlishaw G, Hacker JE. 1997. Distribution, diversity, and latitude in African primates. *Am. Nat.* 150:505–12

Crame JA. 2000. Evolution of taxonomic diversity gradients in the marine realm: evidence from the composition of Recent bivalve faunas. *Paleobiology* 26:188–214

Crame JA. 2002. Evolution of taxonomic diversity gradients in the marine realm: a comparison of Late Jurassic and Recent bivalve faunas. *Paleobiology* 28:184–207

Crane PR, Lidgard S. 1989. Angiosperm diversification and paleolatitudinal gradients in

Cretaceous floristic diversity. *Science* 246: 675–78

Crow GE. 1993. Species diversity in aquatic angiosperms: latitudinal patterns. *Aquat. Bot.* 44:229–58

Culver SJ, Buzas MA. 2000. Global latitudinal species diversity gradient in deep-sea benthic Foraminifera. *Deep-Sea Res.* 47:259–75

Currie DJ. 1991. Energy and large-scale patterns of animal- and plant-species richness. *Am. Nat.* 137:27–49

Curric DJ, Paquin V. 1987. Large-scale biogeographical patterns of species richness of trees. *Nature* 329:326–27

Cushman JH, Lawton JH, Manly BFJ. 1993. Latitudinal patterns in European ant assemblages: variation in species richness and body size. *Oecologia* 95:30–37

Dale MRT, Fortin M-J. 2002. Spatial autocorrelation and statistical tests in ecology. *Ecoscience* 9:162–67

Darwin C. 1862. *The Voyage of the Beagle*. Garden City, NJ: Doubleday

Davidowitz G, Rosenzweig ML. 1998. The latitudinal gradient in species diversity among North American grasshoppers (Acrididae) within a single habitat: a test of the spatial heterogeneity hypothesis. *J. Biogeogr.* 25: 553–60

Davis MB, Shaw RG. 2001. Range shifts and adaptive responses to Quaternary climate change. *Science* 292:673–79

Dexter DM. 1992. Sandy beach community structure: the role of exposure and latitude. *J. Biogeogr.* 19:59–66

Dingle H, Rochester WA, Zalucki MP. 2000. Relationships among climate, latitude, and migration: Australian butterflies are not temperate-zone birds. *Oecologia* 124:196–207

Diniz-Filho JAF, Bini LM, Hawkins BA. 2002a. Spatial autocorrelation and red herrings in geographical ecology. *Global Ecol. Biogeogr.* 12:53–64

Diniz-Filho JAF, de Sant'Ana CER, de Souza MC, Rangel TFLVB. 2002b. Null models and spatial patterns of species in South American birds of prey. *Ecol. Lett.* 5:47–55

Dobzhansky T. 1950. Evolution in the tropics. *Am. Sci.* 38:209–21

Dworschak PC. 2000. Global diversity in the Thalassinidea (Decopoda). *J. Crust. Biol.* 20: 238–45

Eeley HAC, Foley RA. 1999. Species richness, species range size, and ecological specialization among African primates: geographical patterns and conservation implications. *Biodivers. Conserv.* 8:1033–56

Ellingsen KE, Gray JS. 2002. Spatial patterns of benthic diversity: is there a latitudinal gradient along the Norwegian continental shelf? *J. Anim. Ecol.* 71:373–89

Ellison AM. 2002. Macroecology of mangroves: large-scale patterns and processes in tropical coastal waters. *TREES* 16:181–94

Errhif A, Razouls C, Mayzaud P. 1997. Composition and community structure of pelagic copepods in the Indian sector of the Antarctic Ocean during the end of the austral summer. *Polar Biol.* 17:418–30

Fischer AG. 1960. Latitudinal variations in organic diversity. *Evolution* 14:64–81

France R. 1992. The North American latitudinal gradient in species richness and geographical range of freshwater crayfish and amphipods. *Am. Nat.* 139:342–54

Gardner RH, Kemp WM, Kennedy VS, Petersen JE. 2001. *Scaling Relations in Experimental Ecology.* New York: Columbia Univ. Press

Gaston KJ. 1996a. *Biodiversity: A Biology of Numbers and Difference.* Oxford: Blackwell Sci.

Gaston KJ. 1996b. Biodiversity—latitudinal gradients. *Prog. Phys. Geogr.* 20:466–76

Gaston KJ. 2000. Global patterns in biodiversity. *Nature* 405:220–27

Gaston KJ, Blackburn TM, Spicer JI. 1998. Rapoport's rule: time for an epitaph? *Trends Ecol. Evol.* 13:70–74

Gaston KJ, Spicer JI. 1998. *Biodiversity: An Introduction*. Oxford: Blackwell Sci.

Gaston KJ, Williams PH, Eggleton P, Humphries CJ. 1995. Large scale patterns of biodiversity: spatial variation in family richness. *Proc. R. Soc. London Ser. B* 260:149–54

Gentry AH. 1988. Changes in plant community diversity and floristic composition on environmental and geographical gradients. *Ann. Mo. Bot. Gard.* 75:1–34

Gering JC, Crist TO, Veech JA. 2003. Additive partitioning of species diversity across multiple scales: implications for regional conservation of biodiversity. *Conserv. Biol.* 17:488–99

Gleason HA. 1922. On the relation between species and area. *Ecology* 3:158–62

Gleason HA. 1925. Species and area. *Ecology* 6:66–74

Gotelli NJ, Ellison AM. 2002. Biogeography at a regional scale: determinants of ant species density in New England bogs and forests. *Ecology* 83:1604–9

Gross KL, Willig MR, Gough L, Inouye R, Cox SB. 2000. Patterns of species diversity and productivity at different scales in herbaceous plant communities. *Oikos* 89:428–39

Hansson L. 1992. Small mammal communities on clearcuts in a latitudinal gradient. *Acta Oecol.* 13:687–99

Harriot VJ, Banks SA. 2002. Latitudinal variation in coral communities in eastern Australia: a qualitative biophysical model of factors regulating coral reefs. *Coral Reefs* 21: 83–94

Harrison S, Ross SJ, Lawton JH. 1992. Beta diversity on geographic gradients in Britain. *J. Anim. Ecol.* 61:151–58

Haskell JP. 2001. The latitudinal gradient of diversity through the Holocene as recorded by fossil pollen in Europe. *Evol. Ecol. Res.* 3:345–60

Hawkins BA. 2001. Ecology's oldest pattern? *Trends Ecol. Evol.* 16:470

Hawkins BA, Compton SG. 1992. African fig wasp communities: undersaturation and latitudinal gradients in species richness. *J. Anim. Ecol.* 61:361–72

Hawkins BA, Diniz-Filho JAF. 2002. The mid-domain effect cannot explain the diversity gradient of Nearctic birds. *Global Ecol. Biogeogr.* 11:419–26

Hawkins BA, Field R, Cornel HV, Currie DJ, Guégan J-F, et al. 2004. Energy, water, and broad-scale geographic patterns of species richness. *Ecology.* In press

Heip C, Basford D, Craeymeersch JA, Dewarumez J-M, Dorjes J, et al. 1992. Trends in biomass, density, and diversity of North Sea macrofauna. *J. Mar. Sci.* 49:13–22

Heip C, Craeymeersch JA. 1995. Benthic community structure in the North Sea. *Helgol. Wiss. Meeresunters.* 49:313–28

Holman JA. 1995. *Pleistocene Amphibians and Reptiles in North America.* Oxford: Oxford Univ. Press

Hubbell SP. 2001. *The Unified Neutral Theory of Biodiversity and Biogeography.* Princeton: Princeton Univ. Press

Hurlbert AH, Haskell JP. 2003. The effect of energy and seasonality on avian species richness and community composition. *Am. Nat.* 161:83–97

Hutchinson GE. 1959. Homage to Santa Rosalia, or why are there so many kinds of animals? *Am. Nat.* 93:145–59

Jablonski D. 1993. The tropics as a source of evolutionary novelty: the post-Paleozoic fossil record of marine invertebrates. *Nature* 364:142–44

Janzen DH. 1981 The peak in North American ichneumonid species richness lies between 38° and 42° North. *Ecology* 62:532–37

Jarvinen O, Vaisanen RA. 1978. Ecological zoogeography of North European waders, or Why do so many waders breed in the north. *Oikos* 30:496–507

Jetz W, Rahbek C. 2001. Geometric constraints explain much of the species richness pattern of African birds. *Proc. Natl. Acad. Sci. USA* 98:5661–66

Jetz W, Rahbek C. 2002. Geographic range size and determinants of avian species richness. *Science* 297:1548–51

Kafanov AI, Volvenko IV, Fedorov VV, Pitruk DL. 2000. Ichthyofaunistic biogeography of the Japan (East) Sea. *J. Biogeogr.* 27:915–33

Kaufman DM. 1994. *Latitudinal patterns of mammalian diversity in the New World: the role of taxonomic hierarchy.* MS thesis. Texas Tech Univ., Lubbock. 96 pp.

Kaufman DM. 1995. Diversity of New World

mammals: Universality of the latitudinal gradients of species and bauplans. *J. Mammal.* 76:322–34

Kaufman DM. 1998. *The structure of mammalian faunas in the New World: from continents to communities.* PhD thesis. Univ. New Mexico, Albuquerque. 130 pp.

Kaufman DM, Willig MR. 1998. Latitudinal patterns of mammalian species richness in the New World: the effects of sampling method and faunal group. *J. Biogeogr.* 25: 795–805

Kidwell SM, Holland SM. 2002. The quality of the fossil record: implications for evolutionary analyses. *Annu. Rev. Ecol. Syst.* 33:561–88

Kiester AR. 1971. Species density of North American amphibians and reptiles. *Syst. Zool.* 20:127–37

Kocher SD, Williams EH. 2000. The diversity and abundance of North American butterflies vary with habitat disturbance and geography. *J. Biogeogr.* 27:785–94

Kolasa J, Pickett STA, eds. 1991. *Ecological Heterogeneity*. New York: Springer-Verlag

Koleff P, Gaston KJ. 2001. Latitudinal gradients in diversity: real patterns and random models. *Ecography* 24:341–51

Koleff P, Gaston KJ. 2002. The relationships between local and regional species richness and spatial turnover. *Global Ecol. Biogeogr.* 11:363–75

Krebs CJ. 1994. *Ecology: The Experimental Analysis of Distribution and Abundance.* New York: Harper Collins College Publ.

Krystufek B, Griffiths HI. 2002. Species richness and rarity in European rodents. *Ecography* 25:120–28

La Ferla B, Taplin J, Ockwell D, Lovett JC. 2002. Continental scale patterns of biodiversity: can higher taxa accurately predict African plant distribution? *Bot. J. Linn. Soc.* 138:225–35

Lambshead PJD, Tietjen J, Ferrero T, Jensen P. 2000. Latitudinal diversity gradients in the deep sea with special reference to North Atlantic nematodes. *Mar. Ecol. Prog. Ser.* 194: 159–67

Lees DC, Kremen C, Andriamampianina L. 1999. A null model of species richness gradients: bounded range overlap of butterflies and other rainforest endemics in Madagascar. *Biol. J. Linn. Soc.* 67:529–84

Legendre P. 1993. Spatial autocorrelation: Trouble or new paradigm. *Ecology* 74:1659–73

Levin SA. 1992. The problem of pattern and scale in ecology. *Ecology* 73:1943–46

Levin SA, ed. 2001. *Encyclopedia of Biodiversity.* San Diego: Academic

Lobo JM. 2000. Species diversity and composition of dung beetle (Coleoptera: Scarabaeoidea) assemblages in North America. *Can. Entomol.* 132:307–21

Lyons SK. 2003. A quantitative assessment of range shifts of Pleistocene mammals. *J. Mammal.* 84:385–402

Lyons SK, Willig MR. 1997. Latitudinal patterns of range size: methodological concerns and empirical evaluations for New World bats and marsupials. *Oikos* 79:568–80

Lyons SK, Willig MR. 2002. Species richness, latitude, and scale-sensitivity. *Ecology* 83:47–58

MacArthur RH. 1972. *Geographical Ecology: Patterns in the Distribution of Species.* Princeton: Princeton Univ. Press

MacArthur RH, Wilson EO. 1963. An equilibrium theory of insular zoogeography. *Evolution* 17:373–87

MacArthur RH, Wilson EO. 1967. The theory of island biogeography. *Monogr. Popul. Biol.* 1:1–203

Macpherson E. 2002. Large-scale species-richness gradients in the Atlantic Ocean. *Proc. R. Soc. London Ser. B* 269:1715–20

Macpherson E, Duarte CM. 1994. Patterns in species richness, size, and latitudinal range of East Atlantic fishes. *Ecography* 17:242–48

Mann H, Proctor VW, Taylor AS. 1999. Toward a biogeography of North American charophytes. *Aust. J. Bot.* 47:445–58

Mares MA, Ojeda RA. 1982. Patterns of diversity and adaptation in South American hystricognath rodents. In *Mammalian Biology*

in South America, Spec. Publ. Ser. No. 6, Pymatuning Lab Ecol., pp. 393–432. Pittsburgh, PA: Univ. Pittsburgh
McCain CM. 2003. North American desert rodents: a test of the mid-domain effect in species richness. *J. Mammal.* In press
McClatchie S, Millar RB, Webster F, Lester PJ, Hurst R, Bagley N. 1997. Demersal fish community diversity off New Zealand: is it related to depth, latitude, and regional surface phytoplankton? *Deep-Sea Res.* 44:647–67
McCoy ED, Connor EF. 1980. Latitudinal gradients in the species diversity of North American mammals. *Evolution* 34:193–203
Meserve PL, Kelt DA, Martinez DR. 1991. Geographical ecology of small mammals in continental Chile chico, South America. *J. Biogeogr.* 18:179–87
Mittelbach GG, Steiner CF, Scheiner SM, Gross KL, Reynolds HL, et al. 2001. What is the observed relationship between species richness and productivity? *Ecology* 82:2381–96
Møller AP. 1998. Evidence of larger impact of parasites on hosts in the tropics: investment in immune function within and outside the tropics? *Oikos* 82:265–70
Oberdorff T, Guegan J-F, Hugueny B. 1995. Global scale patterns of fish species richness in rivers. *Ecography* 18:345–52
Oliva ME. 1999. Metazoan parasites of the Jack Mackerel *Trachurus murphyi* (Teliostei, Carangidae) in a latitudinal gradient from South America (Chile and Peru). *Parasite* 6:223–30
Pagel MD, May RM, Collie AR. 1991. Ecological determinants of the geographical distribution and diversity of mammalian species. *Am. Nat.* 137:791–815
Palmer MW, White PS. 1994. Scale dependence and the species-area relationship. *Am. Nat.* 144:717–40
Pastor J, Downing A, Erickson HE. 1996. Species-area curves and diversity-productivity relationships in beaver meadows of Voyageurs National Park, Minnesota, USA. *Oikos* 77:399–406
Penev LD. 1992. Qualitative and quantitative spatial variation in soil wire-worm assemblages in relation to climatic and habitat factors. *Oikos* 63:180–92
Perelman SB, León RJC, Oesterheld M. 2001. Cross-scale vegetation patterns of flooding pampa grasslands. *J. Ecol.* 89:562–77
Peters SE, Foote M. 2001. Biodiversity in the Phanerozoic: a reinterpretation. *Paleobiology* 27:583–601
Peterson DL, Parker VT. 1998. *Ecological Scale: Theory and Application.* New York: Columbia Univ. Press
Pianka ER. 1966. Latitudinal gradients in species diversity: a review of concepts. *Am. Nat.* 100:33–46
Pielou EC. 1977. The latitudinal spans of seaweed species and their patterns of overlap. *J. Biogeogr.* 4:299–311
Platt JR. 1964. Strong inference. *Science* 146: 347–53
Porembski S, Brown G, Barthlott W. 1995. An inverted latitudinal gradient of plant diversity in shallow depressions on Ivorian inselbergs. *Vegetation* 117:151–63
Porembski S, Brown G, Barthlott W. 1996. A species-poor tropical sedge community: *Afrotrilepis pilosa* mats on inselbergs in West Africa. *Nord. J. Bot.* 16:239–45
Platnick NI. 1991. Patterns of biodiversity: tropical versus temperate. *J. Nat. Hist.* 25:1083–88
Poulin R. 1995. Phylogeny, ecology, and the richness of parasite communities in vertebrates. *Ecol. Monogr.* 65:283–302
Poulin R. 1999. Speciation and diversification of parasite lineages: an analysis of congeneric parasite species in vertebrates. *Evol. Ecol.* 13:455–67
Poulsen BE, Krabbe N. 1997. The diversity of cloud forest birds on the eastern and western slopes of the Ecuadorian Andes: a latitudinal and comparative analysis with implications for conservation. *Ecography* 20:475–82
Price PW, Fernandes GW, Lara ACF, Brawn J, Barrios H, et al. 1998. Global patterns in local number of insect galling species. *J. Biogeogr.* 25:581–91

Proches S. 2001. Back to the sea: secondary marine organisms from a biogeographical perspective. *Biol. J. Linn. Soc.* 74:197–203

Pysek P. 1998. Alien and native species in Central European urban floras: a quantitative comparison. *J. Biogeogr.* 25:155–63

Qian H. 1998. Large-scale biogeographic patterns of vascular plant richness in North America: an analysis at the generic level. *J. Biogeogr.* 25:829–36

Rabenold KN. 1979. A reversed latitudinal diversity gradient in avian communities of eastern deciduous forests. *Am. Nat.* 114:275–86

Rabinovich JE, Rapoport EH. 1975. Geographical variation in the diversity of Argentine passerine birds. *J. Biogeogr.* 2:141–57

Rahbek C, Graves GR. 2001. Multiscale assessment of patterns of avian species richness. *Proc. Natl. Acad. Sci. USA* 98:4534–39

Rapoport EH. 1975. *Areografia: Estrategias Geograficas de Especies.* Mexico City, DF: Fundo Cult. Econ.

Reid JW. 1994. Latitudinal diversity patterns of continental benthic species assemblages in the Americas. *Hydrobiologia* 292/293:341–49

Rex MA, Stuart CT, Coyne G. 2000. Latitudinal gradients of species richness in the deep-sea benthos of the North Atlantic. *Proc. Natl. Acad. Sci. USA* 97:4082–85

Rex MA, Stuart CT, Hessler RR, Allen JA, Sanders HL, Wilson GDF. 1993. Global-scale latitudinal patterns of species diversity in the deep-sea benthos. *Nature* 365:636–39

Reyes-Bonilla H, Cruz-Pinon G. 2000. Biogeografia de los corales ahermatipocos (Scleractinia) del Pacifico de Mexico. *Cien. Mar.* 26:511–31

Ricklefs RE, Schluter D. 1993. *Species Diversity in Ecological Communities.* Chicago: Univ. Chicago Press

Robinson N, ed. 1966. *Solar Radiation.* New York: Elsevier

Rohde K. 1978a. Latitudinal differences in host-specificity of marine Monogenea and Digenea. *Mar. Biol.* 47:125–34

Rohde K. 1978b. Latitudinal gradients in species diversity and their causes. II. Marine parasitological evidence for a time hypothesis. *Biol. Zbl.* 197:405–18

Rohde K. 1992. Latitudinal gradients in species diversity: the search for the primary cause. *Oikos* 65:514–27

Rohde K. 1996. Rapoport's rule is a local phenomenon and cannot explain latitudinal gradients in species diversity. *Biodivers. Lett.* 3:10–13

Rohde K. 1997. The larger area of the tropics does not explain latitudinal gradients in species diversity. *Oikos* 79:169–72

Rohde K. 1998. Latitudinal gradients in species diversity: area matters, but how much? *Oikos* 82:184–90

Rohde K. 1999. Latitudinal gradients in species diversity and Rapoport's rule revisited: a review of recent work and what can parasites teach us about the causes of the gradients? *Ecography* 22:593–613

Rosenzweig ML. 1975. On continental steady states of species diversity. In *Ecology and Evolution of Communities*, ed. ML Cody, JM Diamond, pp. 121–40. Cambridge, MA: Belknap

Rosenzweig ML. 1992. Species diversity gradients: we know more and less than we thought. *J. Mammal.* 73:715–30

Rosenzweig ML. 1995. *Species Diversity in Space and Time.* Cambridge, MA: Cambridge Univ. Press

Rosenzweig ML, Sandlin EA. 1997. Species diversity and latitudes: listening to area's signal. *Oikos* 80:172–76

Roy K, Jablonski D, Valentine JW. 1994. Eastern Pacific molluscan provinces and latitudinal diversity gradient: no evidence for "Rapoport's rule". *Proc. Natl. Acad. Sci. USA* 91:8871–74

Roy K, Jablonski D, Valentine JW. 1996. Higher taxa in biodiversity studies: patterns from eastern Pacific marine molluscs. *Philos. Trans. R. Soc. London Ser. B* 35:1605–13

Roy K, Jablonski D, Valentine JW. 2000. Dissecting latitudinal diversity gradients: functional groups and clades of marine bivalves. *Proc. R. Soc. London Ser. B* 267: 293–99

Roy K, Jablonski D, Valentine JW, Rosenberg G. 1998. Marine latitudinal diversity gradients: tests of causal hypotheses. *Proc. Natl. Acad. Sci. USA* 95:3699–702

Ruggiero A. 1994. Latitudinal correlates of the sizes of mammalian geographical ranges in South America. *J. Biogeogr.* 21:545–59

Ruggiero A. 1999. Spatial patterns in the diversity of mammal species: a test of the geographic area hypothesis in South America. *Ecoscience* 6:338–54

Santelices B, Meneses I. 2000. A reassessment of the phytogeographic characterization of temperate Pacific South America. *Rev. Chil. Hist. Nat.* 73:605–14

Schall JJ, Pianka ER. 1978. Geographical trends in numbers of species. *Science* 201: 679–86

Scheiner SM, Cox SB, Willig MR, Mittelbach GG, Osenberg C, Kaspari M. 2000. Species richness, species-area curves, and Simpson's paradox. *Evol. Ecol. Res.* 2:791–802

Scheiner SM, Jones S. 2002. Diversity, productivity and scale in Wisconsin vegetation. *Evol. Ecol. Res.* 4:1097–17

Schneider DC. 2001. Concept and effects of scale. See Levin 2001, pp. 70245–54

Schemske DW. 2002. Ecological and evolutionary perspectives on the origins of tropical diversity. In *Foundations of Tropical Forest Biology: Classic Papers with Commentaries*, ed. R Chazdon, T Whitmore, pp. 163–73. Chicago, IL: Univ. Chicago Press

Silvertown J. 1985. History of a latitudinal diversity gradient: woody plants in Europe 13,000-1000 years b.p. *J. Biogeogr.* 12:519–25

Simpson GG. 1964. Species density of North American Recent mammals. *Syst. Zool.* 13: 57–73

Skillen EL, Pickering J, Sharkey MJ. 2000. Species richness of the Campopleginae and Ichneumoninae (Hymenoptera: Ichneumonidae) along a latitudinal gradient in eastern North America old-growth forests. *Environ. Entomol.* 29:460–66

Solenon T. 1994. Factors affecting the structure of Finnish birds of prey. *Ornis Fenn.* 71:156–69

Steele DH. 1988. Latitudinal variations in body size and species diversity in marine decopod crustaceans of the continental shelf. *Int. Rev. Hydrobiol.* 73:235–46

Stehli FG, Douglas RG, Newell ND. 1969. Generation and maintenance of gradients in taxonomic diversity. *Science* 164:947–49

Stehli FG, McAlester AL, Helsley CE. 1967. Taxonomic diversity of recent bivalves and some implications of geology. *Geol. Soc. Am. Bull.* 78:455–66

Stehli FG, Wells JW. 1971. Diversity and age patterns in hermatypic corals. *Syst. Zool.* 20:115–26

Stevens GC. 1989. The latitudinal gradient in geographical range: How so many species coexist in the tropics. *Am. Nat.* 133:240–56

Stevens GC. 1996. Extending Rapoport's rule to Pacific marine salmon. *J. Biogeogr.* 23:149–54

Stevens RD, Willig MR. 2002. Geographical ecology at the community level: perspectives on the diversity of New World bats. *Ecology* 83:545–60

Stirling G, Wilsey B. 2001. Empirical relationships between species richness, evenness, and proportional diversity. *Am. Nat.* 158: 286–99

Swingland IR. 2001. Definitions of biodiversity. See Levin 2001, pp. 377–91

Taylor JD, Taylor CN. 1977. Latitudinal distribution of predatory gastropods on the eastern Atlantic shelf. *J. Biogeogr.* 4:73–81

Taylor PH, Gaines SD. 1999. Can Rapoport's rule be rescued? Modeling causes of the latitudinal gradient in species richness. *Ecology* 80:2474–82

Terborgh J. 1973. On the notion of favourableness in plant ecology. *Am. Nat.* 107:481–501

Thomas E, Gooday AJ. 1996. Cenozoic deep-sea benthic foraminifers: tracers for changes in oceanic productivity? *Geology* 24:355–58

Tramer EJ. 1974. On latitudinal gradients in avian diversity. *Condor* 76:123–30

Turner JRG, Gatehouse CM, Corey CA. 1987. Does solar energy control organic diversity? Butterflies, moths and the British climate. *Oikos* 48:195–205

Valentine JE, Jablonski D. 1993. Fossil communities: compositional variation at many time scales. In *Species Diversity in Ecological Communities: Historical and Geographical Perspectives*, ed. RE Ricklefs, DL Schlutter, pp. 341–48. Chicago, IL: Univ. Chicago Press

Veech JA, Summerville KS, Crist TO, Gering JC. 2002. The additive partitioning of species diversity: recent revival of an old idea. *Oikos* 99:3–9

Virola T, Kaitala V, Lammi A, Siikamäki P, Suhonen J. 2001. Geographical patterns of species turnover in aquatic plant communities. *Freshw. Biol.* 46:1471–78

Waide RB, Willig MR, Steiner CF, Mittelbach GG, Gough L, et al. 1999. The relationship between productivity and species richness. *Annu. Rev. Ecol. Syst.* 30:247–300

Wallace AR. 1878. *Tropical Nature and Other Essays*. London: Macmillan

Whittaker RH. 1960. Vegetation of the Siskiyou Mountains, Oregon and California. *Ecol. Monogr.* 30:279–338

Whittaker RH. 1972. Evolution and measurement of species diversity. *Taxon* 21:213–51

Whittaker RH. 1977. Evolution of species diversity in land-plant communities. *Evol. Biol.* 10:1–67

Whittaker RH, Willis KJ, Field R. 2001. Scale and species richness: towards a general, hierarchical theory of species diversity. *J. Biogeogr.* 28:453–70

Williams PH. 1996. Mapping variations in the strength and breadth of biogeographic transition zones using species turnover. *Proc. R. Soc. London Ser. B* 263:579–88

Willig MR. 2001. Common trends with latitude. See Levin 2001, pp. 701–14

Willig MR, Gannon MR. 1997. Gradients of species density and turnover in marsupials: a hemispheric perspective. *J. Mammal.* 78:756–65

Willig MR, Lyons SK. 1998. An analytical model of latitudinal gradients of species richness with an empirical test for marsupials and bats in the New World. *Oikos* 83:93–98

Willig MR, Sandlin EA. 1991. Gradients of species density and turnover in New World bats: A comparison of quadrat and band methodologies. In *Latin American Mammals: Their Conservation, Ecology, and Evolution*, ed. MA Mares, DJ Schmidley, pp. 81–96. Norman, OK: Univ. Okla. Press

Willig MR, Selcer KW. 1989. Bat species density gradients in the New World: a statistical assessment. *J. Biogeogr.* 16:189–95

Wilson JW III. 1974. Analytical zoogeography of North American mammals. *Evolution* 28:124–40

Wilson MV, Shmida A. 1984. Measuring beta diversity with presence-absence data. *J. Ecol.* 72:1055–64

Woodd-Walker RS, Ward P, Clarke A. 2002. Large-scale patterns in diversity and community structure of surface water copepods from the Atlantic Ocean. *Mar. Ecol. Prog. Ser.* 236:189–203

Wright DH. 1983. Species-energy theory: an extension of species-area theory. *Oikos* 41:496–506

Annu. Rev. Ecol. Evol. Syst. 2003. 34:311–38
doi: 10.1146/annurev.ecolsys.34.011802.132351

First published online as a Review in Advance on September 2, 2003

RECENT ADVANCES IN THE (MOLECULAR) PHYLOGENY OF VERTEBRATES

Axel Meyer
Department of Biology, University of Konstanz, 78457 Konstanz, Germany; email: axel.meyer@uni-konstanz.de

Rafael Zardoya
Museo Nacional de Ciencias Naturales, CSIC, José Gutiérrez Abascal, 2, 28006 Madrid, Spain; email: rafaz@mncn.csic.es

Key Words molecular systematics, Agnatha, Actinopterygii, Sarcopterygii, Tetrapoda

■ **Abstract** The analysis of molecular phylogenetic data has advanced the knowledge of the relationships among the major groups of living vertebrates. Whereas the molecular hypotheses generally agree with traditional morphology-based systematics, they sometimes contradict them. We review the major controversies in vertebrate phylogenetics and the contribution of molecular phylogenetic data to their resolution: (*a*) the mono-paraphyly of cyclostomes, (*b*) the relationships among the major groups of ray-finned fish, (*c*) the identity of the living sistergroup of tetrapods, (*d*) the relationships among the living orders of amphibians, (*e*) the phylogeny of amniotes with particular emphasis on the position of turtles as diapsids, (*f*) ordinal relationships among birds, and (*g*) the radiation of mammals with specific attention to the phylogenetic relationships among the monotremes, marsupial, and placental mammals. We present a discussion of limitations of currently used molecular markers and phylogenetic methods as well as make recommendations for future approaches and sets of marker genes.

INTRODUCTION

All studies in comparative biology depend upon robust phylogenetic frameworks. Although the history of vertebrates is relatively well documented in the fossil record (Carroll 1997), the answers to several major issues in vertebrate systematics are still debated among systematists. Often debates arise because of large gaps in the fossil record, rapid lineage diversification, and highly derived morphologies of the extant lineages that complicate the reconstruction of evolutionary events and the establishment of solidly supported phylogenetic relationships. Traditional approaches to studying the phylogeny of vertebrates such as paleontological and comparative morphological methods were augmented by the advent of molecular sequence data about a decade ago. Here we review the contribution of molecular

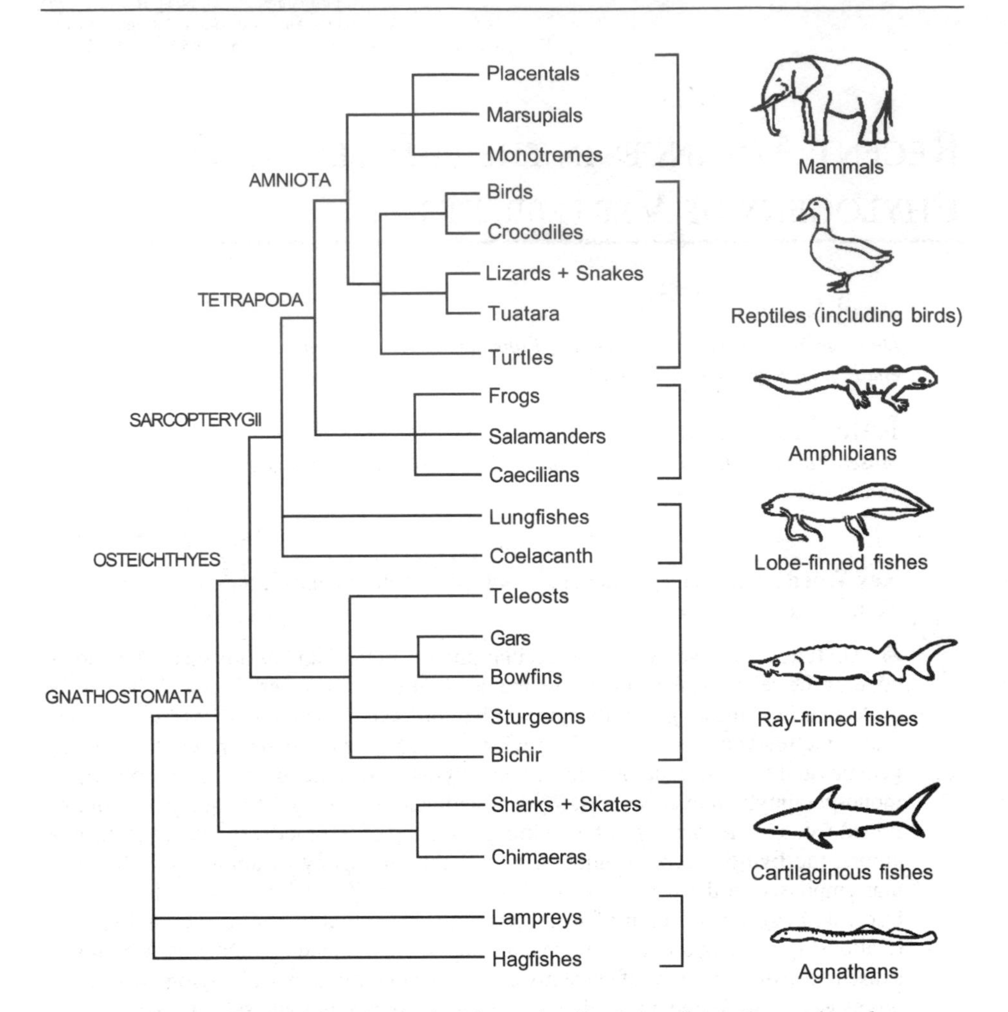

Figure 1 Phylogenetic hypothesis for the major lineages of vertebrates based on morphological, paleontological, and molecular evidence. Disputed relationships are depicted as polytomies.

systematics to the major unresolved questions (polytomies in Figure 1 in the phylogeny of extant vertebrates) (Benton 1990).

One of the first major innovations in the evolution of vertebrates was the origin of jaws in the Cambrian, 540–505 million years ago (mya) (Carroll 1988). Accordingly, vertebrates have been traditionally classified into Agnatha or cyclostomes (hagfishes and lampreys) and Gnathostomata (the jawed vertebrates) (Figure 1). The mono- or paraphyly of living agnathans is the first controversy that we will discuss. Among jawed vertebrates, the major division, based largely on the composition of the skeleton, is between the Chondrichthyes (cartilaginous fishes) and the Osteichthyes (bony fishes) (Figure 1). Bony fishes are further divided into Actinopterygii (ray-finned fishes) and Sarcopterygii (lobe-finned fishes + tetrapods) (Figure 1). The origin and the phylogeny of the major lineages of ray-finned fishes are still subject to debate (Figure 1). The transition to life on land dates back to the Devonian, 408–360 mya (Carroll 1988), and the relationships of the living lobe-finned fishes (lungfishes and coelacanths) to the tetrapods is still actively discussed (Figure 1).

The first lineage of tetrapods that branched off is the Lissamphibia (caecilians, salamanders, and frogs). There is controversy surrounding both their origin(s) in the Permian, 280–248 mya (Carroll 1988), and their interrelationships (Figure 1). One of the major evolutionary novelties of the tetrapods was the origin of the amniote egg that permitted the permanent independence from water for reproduction and ultimately, the colonization of land. Living amniotes are mammals and reptiles (turtles, lizards and snakes, crocodiles and birds) (Figure 1). The origin of amniotes dates back to the Pennsylvanian, 325–280 mya (Carroll 1988). The amniotes have traditionally been divided into three groups based on the fenestration of their skulls. Anapsids (without holes in the skull) are represented by turtles, the diapsids (with two holes in the skull, at least initially) are the tuatara, snakes and lizards, crocodiles, and birds, and the mammals (with one hole in the skull) make up the synapsids. The relationships among the three lineages of amniotes are contended, particularly because of the uncertainty regarding the phylogenetic position of turtles (Figure 1). Molecular phylogenetic data recently shook up the traditional understanding of the ordinal relationships among the birds, as well as those of the placental mammals. This type of data also questioned the previous hypothesis of the evolutionary relationships among the three major groups of mammals, the monotremes, marsupials, and placental mammals (Figure 1).

In the following sections we will review, in some detail, each of the remaining major questions (polytomies in Figure 1) in the evolution of vertebrates. We will discuss the potential causes for the difficulty in resolving these questions, the status of the debate, and the specific contribution that molecular systematics has made to a resolution of these issues.

MONOPHYLY OR PARAPHYLY OF AGNATHANS

Fossils of the earliest vertebrates found in the Chengjiang Lagerstätte from the early Cambrian suggest that vertebrates are part of the Cambrian explosion (Shu et al. 2003). Unlike the chordate filter feeders that had and still have an almost exclusively sedentary or even sessile lifestyle, the earliest vertebrates used their newly evolved head, and sensory organs derived from neural crest tissues, to actively locate and feed on more macroscopic prey. The earliest vertebrates or agnathans were jawless, and reached their peak of species richness in the Devonian. Only a small number of species of hagfishes (Mixiniformes, 43 species) and lampreys (Petromyzontiformes, 41 species) represent the two extant lineages of agnathans.

Jawless vertebrates clearly are a paraphyletic group, at least when extinct lineages are taken into consideration (Janvier 1996). It seems well established that some of the extinct lineages of the Ostracodermata, in particular the Osteostraci, are more closely related to the jawed-vertebrates than they are to other agnathans (Shu et al. 2003). However, whether or not the living hagfishes and lampreys form a monophyletic group, traditionally called the round mouths or cyclostomes (Figure 2*A*), or whether they are paraphyletic, with the lampreys more closely related to the jawed vertebrates (Figure 2*B*), is still debated mostly among molecular phylogeneticists. Most paleontologists now strongly favor the paraphyly hypothesis for the living agnathans (Janvier 1996) (Figure 2*B*).

There are a number of important problems that adversely affect the solution of this problem both for paleontologists and molecular phylogeneticists (Mallat & Sullivan 1998, Mallat et al. 2001, Zardoya & Meyer 2001c). First, the only three surviving lineages of vertebrates (hagfishes, lampreys, and jawed vertebrates) appeared within a time window of less than 40 million years in the Cambrian (Janvier 1996). This allowed only a short time period for the accumulation of diagnostic synapomorphies (both morphological as well as molecular ones) but a long time period of independent evolution and the accumulation of many autapomorphies to overlay the possibly previously existing phylogenetic signal. Second, there are big gaps in the fossil record that can be partly explained by the lack of bone in hagfish and lampreys. Further problems arise because of the rather "featureless" morphology of the earliest vertebrates (Shu et al. 2003).

The traditional classification of vertebrates uniting hagfishes and lampreys as cyclostomes (Figure 2*A*) was supported by a number of morphological traits including the presence of horny teeth, a respiratory velum, and a complex "tongue" apparatus (Delarbre et al. 2000). However, more recent morphological analyses found several apparently shared derived characters between lampreys and jawed vertebrates (Janvier 1981, 1996) (Figure 2*B*). This paraphyletic relationship is even more strongly supported by cladistic analyses of several recent fossil finds from the Chinese Lagerstätten (Shu et al. 1999, 2003).

Several molecular studies have addressed the question of the relationships of the living agnathan lineages to the gnathostomes. Phylogenetic analyses of the nuclear 18S and 28S rRNA genes (Mallat & Sullivan 1998, Mallat et al. 2001, Zardoya & Meyer 2001c) suggested a monophyletic cyclostome clade (Figure 2*A*) with a relatively high support [for the rest of the text we mean a bootstrap value over 70% (Zharkikh & Li 1992)]. The analyses of several other nuclear loci also support the cyclostome hypothesis (Kuraku et al. 1999), and the most recent analyses of the largest data set so far (35 different nuclear markers) also came out in favor of the monophyly hypothesis (Takezaki et al. 2003). By contrast, phylogenetic analyses of mitochondrial protein-coding genes seem to support the paraphyly hypothesis, with lampreys as the closest living sistergroup to jawed vertebrates (Rasmussen et al. 1998) (Figure 2*B*). One of the crucial problems in the reconstruction of early vertebrate phylogeny using molecular data is that the hagfish branch is extremely long (Zardoya & Meyer 2001c). This circumstance could artificially pull the highly divergent hagfish sequence toward the outgroup (Sanderson & Shaffer 2002) and

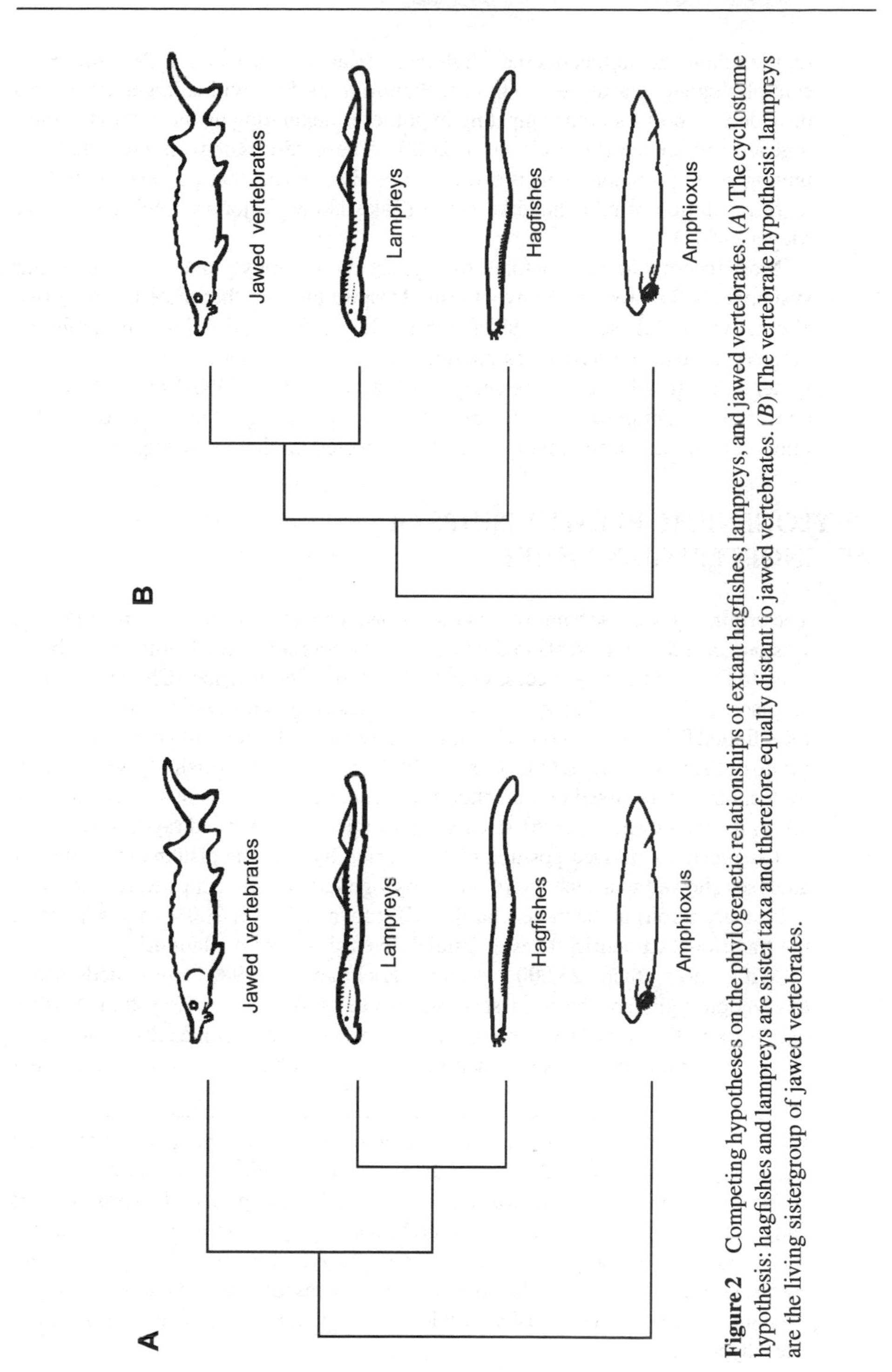

Figure 2 Competing hypotheses on the phylogenetic relationships of extant hagfishes, lampreys, and jawed vertebrates. (*A*) The cyclostome hypothesis: hagfishes and lampreys are sister taxa and therefore equally distant to jawed vertebrates. (*B*) The vertebrate hypothesis: lampreys are the living sistergroup of jawed vertebrates.

may explain the mitochondrial phylogeny (Rasmussen et al. 1998). More recent phylogenetic analyses using mitochondrial markers with a larger set of taxa provided support for both competing hypotheses depending on the method of phylogenetic inference (Delarbre et al. 2000). These results may suggest that this phylogenetic problem involves evolutionary divergences that go beyond the limits of resolution of mitochondrial genes (Takezaki & Gojobori 1999, Zardoya & Meyer 2001c).

Most paleontologists continue to support the paraphyly and most molecular systematists the monophyly hypothesis. Despite the fact that there are only two alternative hypotheses to consider, the phylogenetic relationships of hagfishes, lampreys, and jawed vertebrates still remain an undecided controversy in vertebrate systematics. It will require the analysis of larger (nuclear DNA) data sets with a denser taxon sampling in the lamprey and hagfish lineages that might divide the long branches and lead to more robustly supported phylogenetic hypotheses.

PHYLOGENETIC RELATIONSHIPS OF ACTINOPTERYGIAN FISHES

The origin of jaws, a key innovation that allowed gnathostomes to grasp large prey, was one of the major events in the history of vertebrates. This feature is likely related to the evolutionary success of both the cartilaginous fishes (Chondrichthyes: chimaeras, sharks + skates) and the bony fishes (Osteichthyes: ray-finned fishes, lobe-finned fishes + tetrapods). It is uncontested that Chondrichthyes are the sistergroup of the Osteichthyes (e.g., Carroll 1988) (Figure 3). Surprisingly, some recent molecular studies based on mitochondrial sequence data (Rasmussen & Arnason 1999) recovered sharks as the sistergroup of teleosts (advanced ray-finned fishes) and suggested a derived position of Chondrichthyes in the pisicine tree. Further analyses showed that such unorthodox phylogenetic relationships were caused by noise (saturation) in the molecular data (Zardoya & Meyer 2001c) and supported the traditional Chondrichthyes + Osteichthyes sistergroup relationship.

With more than 25,000 species (Eschmeyer 1998), ray-finned fishes (Actynopterygii) are the most speciose group of vertebrates. Ray-finned fishes date back to the early Devonian (*Dialipina*; Schultze & Cumbaa 2001) and their diversity and number has since then increased steadily. In contrast to most other

---

→

Figure 3 Alternative hypotheses on the phylogeny of the basal lineages of the Actinopterygii (ray-finned fishes). (*A*) Polypteriformes (bichirs and reedfishes) and Acipenseriformes (sturgeons and paddlefishes) are sistergroup taxa (Chondrostei). Gars and bowfins are the sistergroup of teleosts (Neopterygii). (*B*) Acipenseriformes are the sistergroup of Neopterygii to the exclusion of Polypteriformes. (*C*) Polypteriformes are the sistergroup of a clade that includes the Acipenseriformes as the sistergroup of gars and bowfins to the exclusion of teleosts. Most recent molecular data favor this latest hypothesis.

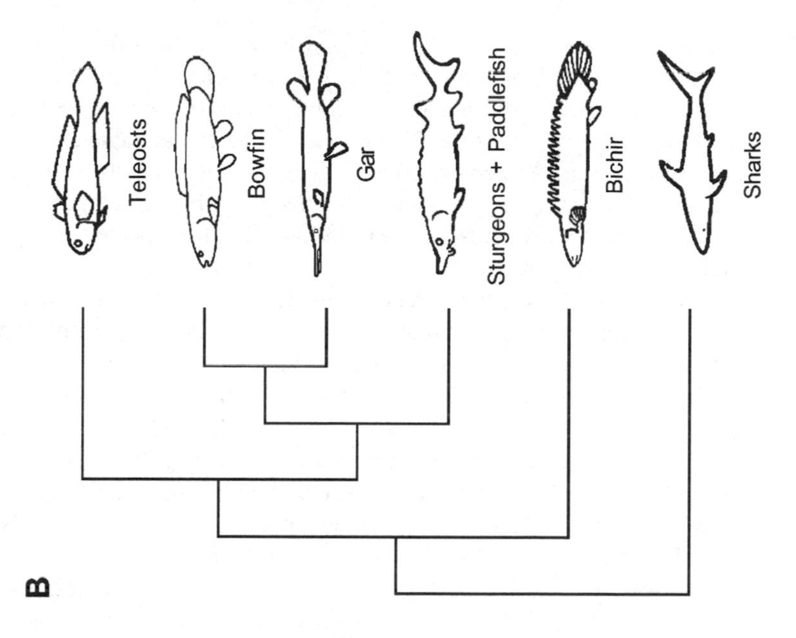
B
Teleosts
Bowfin
Gar
Sturgeons + Paddlefish
Bichir
Sharks

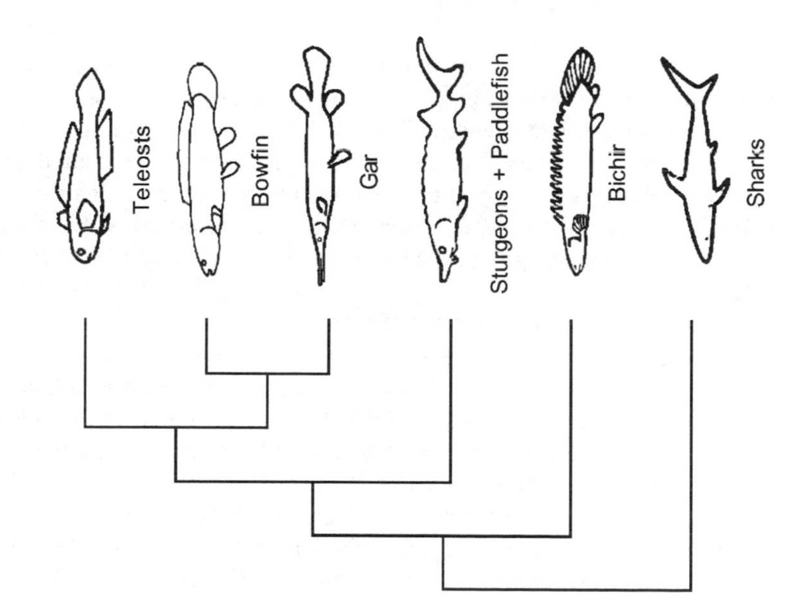
A
Teleosts
Bowfin
Gar
Sturgeons + Paddlefish
Bichir
Sharks

groups of vertebrates, the known diversity of living ray-finned fishes exceeds that of known fossil taxa (Nelson 1994). Ray-finned fish have been traditionally divided into "lower" and "higher" actinopterygians (Gardiner & Schaeffer 1989). The former, also referred to as Chondrostei, includes the Polypteriformes (bichirs and reedfishes) and Acipenseriformes (sturgeons and paddlefishes) (Nelson 1994). The more derived ray-finned fishes, the Neopterygii, include gars, bowfins, and the teleosts that make up more than 96% of all extant species of ray-finned fish (Nelson 1994) (Figure 3*A*).

Some lower actinopterygian relationships are uncertain (Grande & Bemis 1996). Most paleontological and neontological workers place polypteriforms as the most basal lineage of ray-finned fishes (e.g., Gardiner & Schaeffer 1989, Lauder & Liem 1983) (Figures 3*B* and 3*C*). However, because of their peculiar lobed fins, polyteriform fish had even been described as sarcopterygians (Huxley 1861) or classified into their own subclass, the Brachiopterygii (Bjerring 1985, Jessen 1973). All molecular studies bearing on this question agree that polypteriforms are the most basal lineage of the Actinopterygii (Inoue et al. 2003, Le et al. 1993, Noack et al. 1996, Venkatesh et al. 2001).

However, the relative phylogenetic position of Acipenseriformes is debated (Figures 3*B* and 3*C*). Most morphological studies place sturgeons and paddlefishes as the closest living sistergroup of the Neopterygii (Grande & Bemis 1996, Nelson 1969; but see Nelson 1994). This phylogenetic position was further supported by 28S rDNA sequence data (Le et al. 1993) (Figure 3*B*). However, recent molecular studies based on complete mitochondrial genome (Inoue et al. 2003) and nuclear RAG1 (Venkatesh et al. 2001) sequence data favor a close relationship of acipenseriforms to gars and bowfins to the exclusion of teleosts (Figure 3*C*). There is also no consensus on the identity of the closest living sistergroup of teleosts (Arratia 2001). Competing morphological hypotheses suggest that bowfins (Gardiner et al. 1996, Grande & Bemis 1996, Patterson 1973), gars (Olsen 1984), or both bowfins + gars (Holostei; Jessen 1973, Nelson 1969) are the closest relative(s) of teleosts. As mentioned above, recent molecular studies (Inoue et al. 2003, Venkatesh et al. 2001) support that acipenseriforms, bowfins, and gars form a monophyletic group, and therefore, that they are equally related to teleosts (Figure 3*C*).

Because of the large number of taxa involved (there are up to 38 recognized orders of teleosts; Nelson 1994), and the lack of morphological synapomorphies, the higher-level phylogenetic relationships of teleosts have been difficult to resolve (Greenwood et al. 1966, Nelson 1989). An impressive effort was made in a recent study to solve this question by sequencing and analyzing the complete mitochondrial sequence of 100 higher teleosts (Miya et al. 2003). Interestingly, the resulting molecular phylogeny strongly rejected the monophyly of all major groups above the ordinal level as currently defined (Greenwood et al. 1966). Future phylogenetic analyses of nuclear sequences (e.g., from the RAG-1 gene) will be key in resolving the apparent inconsistency between morphological hypotheses and mitochondrial evidence.

THE ORIGIN OF TETRAPODS

The origin of land vertebrates dates back to the Devonian (408–360 mya) (Carroll 1988). The conquest of land by vertebrates was an important evolutionary event that involved morphological, physiological, and behavioral innovations (Clack 2002). A strong paleontological record indicates that early tetrapods evolved from lobe-finned fishes, and recent fossil discoveries have shown that a particular group, the panderichthyids, are the closest relatives of land vertebrates (Ahlberg & Johanson 1998; Ahlberg et al. 1996; Clack 2000, 2002; Cloutier & Ahlberg 1996; Vorobyeva & Schultze 1991). The sistergroup of panderichthyids plus tetrapods are osteolepiforms (Ahlberg & Johanson 1998, Clack 2000, Cloutier & Ahlberg 1996). Dipnomorpha and Actinistia make up the other two major groups of lobe-finned fishes. Dipnomorphs include the extinct porolepiforms, and the air-breathing extant lungfishes (Dipnoi). Actinistia or coelacanths were a highly successful group of lobe-finned fishes during the Devonian that now are represented by only two surviving species (*Latimeria chalumnae* and *L. menadoensis*). Although most recent morphological and paleontological evidence support lungfishes as the closest living sistergroup of tetrapods (Ahlberg & Johanson 1998, Cloutier & Ahlberg 1996) (Figure 4*A*), until recently there was no general agreement regarding which group of living lobe-finned fishes, the Actinistia or the Dipnomorpha, is the one most closely related to the tetrapod lineage (Meyer 1995, Zardoya & Meyer 1997b). There is still disagreement among paleontologists about the homology of some important characters (e.g., the choanae) (Cloutier & Ahlberg 1996) and relevant fossils of intermediate forms connecting the three groups still await discovery.

Significant amounts of molecular phylogenetic data from the living sarcopterygian lineages, lungfishes, coelacanths, and tetrapods have been collected to address this phylogenetic problem. There are three competing phylogenetic hypotheses regarding the relationships among the living lineages of sarcopterygians: lungfishes as the sistergroup to tetrapods (Figure 4*A*), the coelacanth as the sistergroup of tetrapods (Figure 4*B*), and lungfish and coelacanth as a monophyletic sistergroup to tetrapods (Figure 4*C*). The first molecular data set that supported lungfishes as closest living relatives of tetrapods (Figure 4*A*) was based on two fragments of the mitochondrial 12S rRNA and cytochrome *b* genes (Meyer & Wilson 1990). Further support for this hypothesis was obtained from the phylogenetic analysis of complete 12S and 16S rRNA mitochondrial genes (Hedges et al. 1993). However, a reanalysis of this data set with more taxa resulted in an unresolved lungfish + coelacanth + tetrapod trichotomy (Zardoya & Meyer 1997a, Zardoya et al. 1998). Phylogenetic analyses of a data set that combined all mitochondrial protein-coding genes identified lungfishes as the sistergroup of tetrapods (Zardoya & Meyer 1997a, Zardoya et al. 1998) (Figure 4*A*). However, this data set could not statistically reject a lungfish + coelacanth clade (Figure 4*C*) but could reject the coelacanth + tetrapod hypothesis (Figure 4*B*). Phylogenetic analyses of a data set that combined all mitochondrial tRNA genes supported a close relationship between lungfishes and the coelacanth (Zardoya & Meyer 1997a, Zardoya et al.

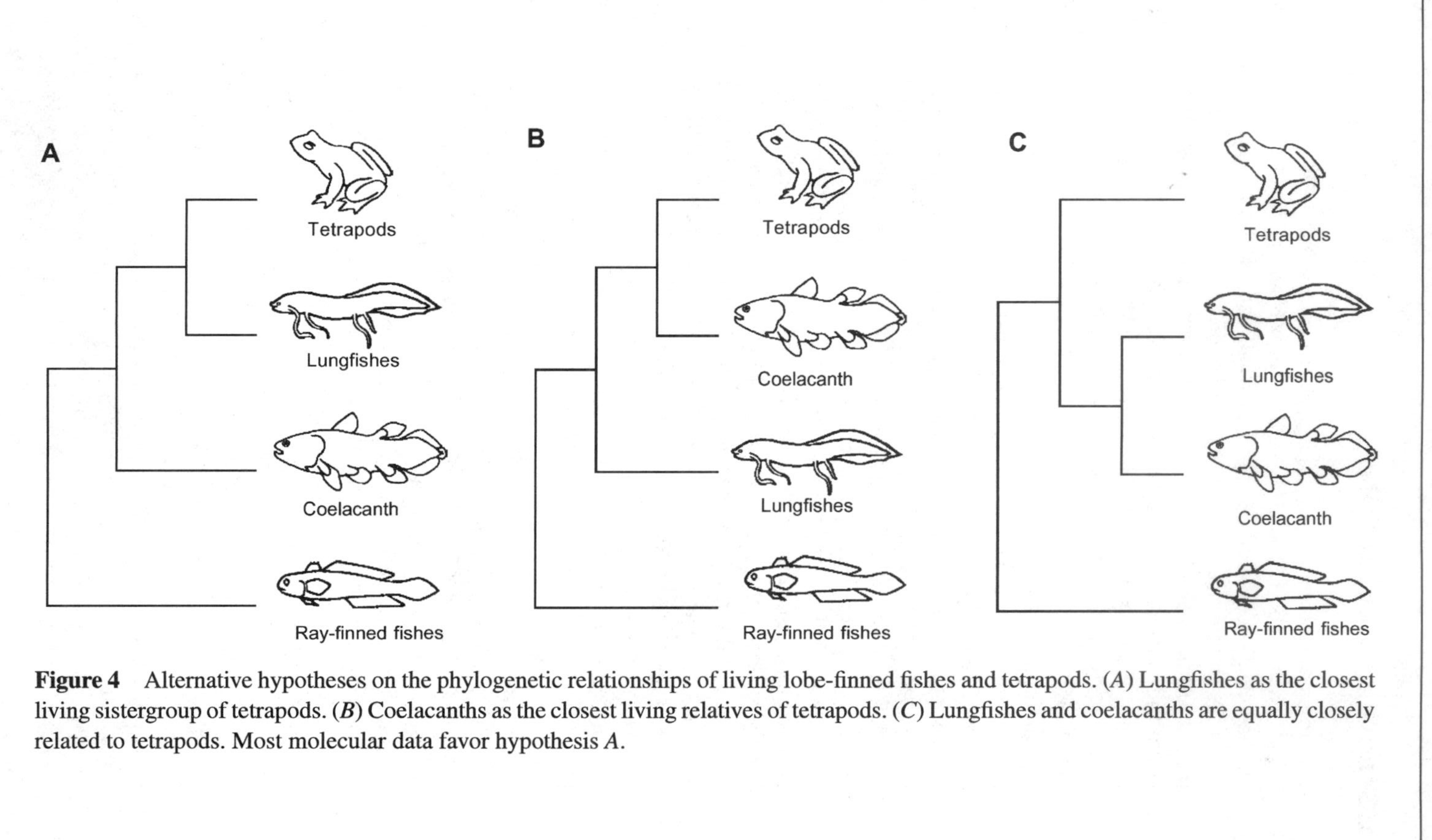

Figure 4 Alternative hypotheses on the phylogenetic relationships of living lobe-finned fishes and tetrapods. (*A*) Lungfishes as the closest living sistergroup of tetrapods. (*B*) Coelacanths as the closest living relatives of tetrapods. (*C*) Lungfishes and coelacanths are equally closely related to tetrapods. Most molecular data favor hypothesis *A*.

1998) (Figure 4*C*). When the mitochondrial protein-coding gene data set was combined with the rest of the mitochondrially encoded (rRNA and tRNA) genes, it also supported lungfishes as the closest living sistergroup of tetrapods (Zardoya et al. 1998). Phylogenetic analyses of nuclear 28S rRNA gene sequences favored a lungfish + coelacanth grouping (Zardoya & Meyer 1996) (Figure 4*C*). The phylogenetic analyses of the combined mitochondrial and 28S rRNA nuclear data sets were not entirely conclusive. Depending on the method of phylogenetic inference used, both a lungfish + tetrapod (Figure 4*A*) or a lungfish + coelacanth clade (Figure 4*C*) were supported (Zardoya et al. 1998). The coelacanth + tetrapod hypothesis (Figure 4*B*) received the least support in all phylogenetic analyses of any molecular data. Recent phylogenetic analyses of a nuclear gene, the myelin DM20 also supported lungfishes as the sistergroup of tetrapods (Tohyama et al. 2000) (Figure 4*A*). The lungfish + tetrapod clade is also supported by a single deletion in the amino acid sequence of a nuclear-encoded gene RAG2 that is shared by lungfishes and tetrapods (Venkatesh et al. 2001). Overall, most molecular and morphological evidence supports lungfishes as the closest living sistergroup of tetrapods (Figure 4*A*) and, albeit cautiously, we conclude that this phylogenetic issue has been solved.

PHYLOGENETIC RELATIONSHIPS AMONG MODERN AMPHIBIANS

Most researchers agree that modern amphibians (Lissamphibia) form a monophyletic group that appeared in the Permian (280–248 mya) (Duellman & Trueb 1994, Parsons & Williams 1963, Szarski 1962). Among paleontologists it is still debated whether the extinct temnospondyls (e.g., Panchen & Smithson 1987, Trueb & Cloutier 1991) or the extinct lepospondyls (Carroll 1995, Laurin 1998, Laurin & Reisz 1997) are their sistergroup. Furthermore, there is no general agreement regarding the phylogenetic relationships among the three living orders of amphibians, the Gymnophiona (caecilians), Caudata (salamanders), and Anura (frogs). Most morphological and paleontological studies suggest that salamanders are the closest relatives of frogs (and form the clade Batrachia) to the exclusion of caecilians (Duellman & Trueb 1994, Milner 1988, Rage & Janvier 1982, Trueb & Cloutier 1991) (Figure 5*A*). Other morphology-based studies suggest that salamanders are the sistergroup of caecilians to the exclusion of frogs (Bolt 1991, Carroll 1995, Laurin 1998) (Figure 5*B*). Because all three lineages of extant amphibians acquired their distinctive body plans early in their evolutionary history, there are few reliable shared derived characters between them. Moreover, a rather poor Permian-Triassic fossil record complicates the determination of the evolutionary relationships among the Lissamphibia (Carroll 2000).

The first phylogenetic studies of this question used nuclear as well as mitochondrial rRNA data and suggested that caecilians are the closest living relatives of salamanders to the exclusion of frogs (Feller & Hedges 1998, Hay et al. 1995,

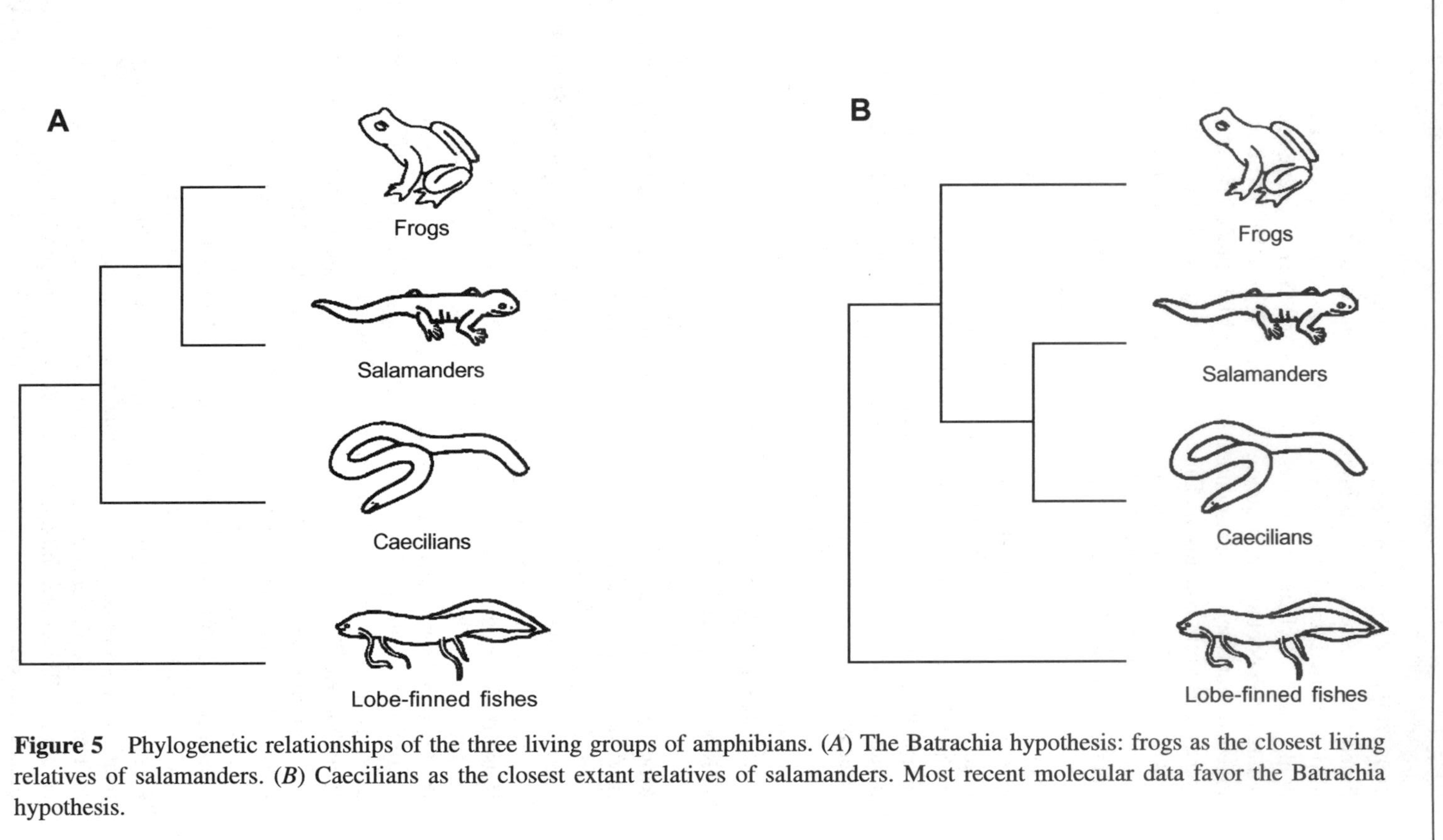

Figure 5 Phylogenetic relationships of the three living groups of amphibians. (*A*) The Batrachia hypothesis: frogs as the closest living relatives of salamanders. (*B*) Caecilians as the closest extant relatives of salamanders. Most recent molecular data favor the Batrachia hypothesis.

Hedges & Maxson 1993, Hedges et al. 1990, Larson & Wilson 1989) (Figure 5*B*). Phylogenetic analyses of complete mitochondrial genomes of a salamander (*Mertensiella luschani*), a caecilian (*Typhlonectes natans*), and a frog (*Xenopus laevis*) supported with high statistical support the Batrachia hypothesis (Zardoya & Meyer 2001b) (Figure 5*A*). This latter result is in agreement with most morphological evidence rather than with earlier molecular studies. The Batrachia hypothesis is currently supported by both morphological and molecular analyses. Yet, more work on nuclear markers and the study of the largely unresolved intraordinal relationships of all three orders possibly also with more complete mitochondrial genomes are expected to settle this long-standing debate in the near future.

AMNIOTE RELATIONSHIPS WITH EMPHASIS ON THE RELATIONSHIPS OF TURTLES

For more than 150 years, the phylogenetic relationships among major amniote lineages have been debated among evolutionary biologists. This phylogenetic problem remains difficult to solve partly because turtles have such a unique morphology and because only few characters can be used to link them with any other group of amniotes. Moreover, different traits provide conflicting phylogenetic signals. Historically, turtles have been considered the only living survivors of anapsid reptiles (those that lack temporal fenestrae in the skull), and the extinct procolophonids (Laurin & Reisz 1995) or pareiasaurs their closest relatives (Gregory 1946; Lee 1995, 1996, 1997). The traditional hypotheses placed turtles (as part of the Anapsida) as sistergroup to all other living amniotes (Gaffney 1980).

More recent phylogenetic analyses based on morphological and fossil data agreed that synapsids—the mammals—(those with a single lower temporal hole in their skulls) are the sistergroup to the remaining amniotes, and they placed anapsids as sistergroup of the diapsids—tuatara, snakes and lizards, crocodiles and birds—(those that have, at least ancestrally, two fenestrae in the temporal region of the skull) (Gauthier et al. 1988, Laurin & Reisz 1995, Lee 1997, Reisz 1997) (Figure 6*A*). However, during the past decade several different amniote phylogenies have been proposed by both paleontologists (Rieppel & Reisz 1999) and molecular phylogeneticists (Zardoya & Meyer 2001a), most of which favor a more derived position for turtles within the reptiles (Figures 6*B*,*C*,*D*).

Recent paleontological analyses reveal that the traditional assignment of turtles to the anapsids may be only weakly supported (deBraga & Rieppel 1997, Rieppel & deBraga 1996, Rieppel & Reisz 1999). Alternatively, turtles have been suggested to be the closest living relatives of the Lepidosauria (tuatara and squamata, i.e., lizards and snakes) (Figure 6*B*) (deBraga & Rieppel 1997, Rieppel & deBraga 1996, Rieppel & Reisz 1999), or the sistergroup of Archosauria (crocodiles and birds) (Figure 6*C*) (Hennig 1983). Both placements imply that the anapsid condition of the turtle skull is a secondary loss or reversal to an ancestral condition.

The first molecular phylogenetic analyses of this issue were based on complete 12S and 16S rRNA mitochondrial gene data sets. They supported a turtle + diapsid

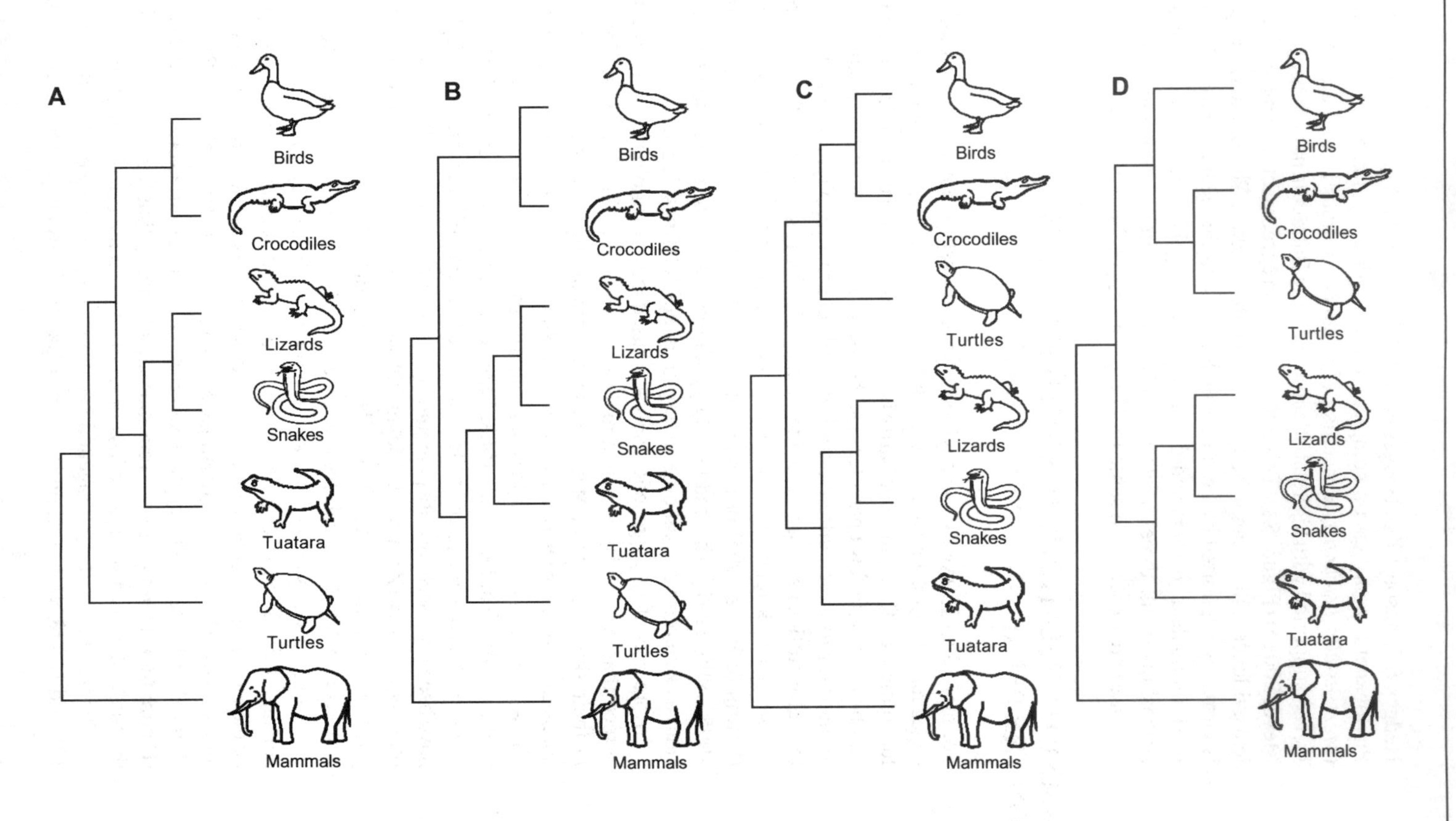
A
Birds
Crocodiles
Lizards
Snakes
Tuatara
Turtles
Mammals
B
Birds
Crocodiles
Lizards
Snakes
Tuatara
Turtles
Mammals
C
Birds
Crocodiles
Turtles
Lizards
Snakes
Tuatara
Mammals
D
Birds
Crocodiles
Turtles
Lizards
Snakes
Tuatara
Mammals

sistergroup relationship to the exclusion of mammals (Cao et al. 1998, Hedges 1994, Strimmer & von Haeseler 1996) (Figure 6*A*). More recent reanalyses of the same genes with additional taxa (including representatives of the two major lineages of turtles, Pleurodira and Cryptodira) recover a turtle + Archosauria clade with moderately high bootstrap support (Zardoya & Meyer 1998) (Figure 6*C*). However, two alternative hypotheses, turtles as anapsids (Figure 6*A*) or turtles as sistergroup of lepidosaurs (Figure 6*B*), could not be statistically rejected based on this data set (Zardoya & Meyer 1998). Recent phylogenetic analyses of relatively large mitochondrial and nuclear sequence data sets further supported the diapsid affinities of turtles, and only differ on their relative position with respect to Lepidosauria and Archosauria. Molecular evidence based on complete mitochondrial protein-coding genes further confirmed the archosaurian affinities of turtles, and statistically rejected alternative hypotheses (Janke et al. 2001, Kumazawa & Nishida 1999) (Figure 6*C*). Phylogenetic analyses of a data set including complete mitochondrial protein-coding, rRNA, and tRNA genes also strongly supported the phylogenetic position of turtles as the sistergroup of archosaurs (Zardoya & Meyer 2001b) (Figure 6*C*). Recent phylogenetic analyses that included the tuatara complete mitochondrial genome firmly support the sistergroup relationship between tuatara and lizards + snakes, and a sistergroup relationship between turtles and archosaurs (Rest et al. 2003). In agreement with mitochondrial evidence, nuclear pancreatic polypeptide data support archosaurs as the living sistergroup of turtles (Platz & Conlon 1997).

Phylogenetic analyses based on eleven nuclear proteins, in addition to the nuclear 18S and 28S rRNA genes, suggested that crocodiles are the closest living relatives of turtles (Hedges & Poling 1999) to the exclusion of birds (Figure 6*D*). Furthermore, a phylogenetic analysis that combined mitochondrial and nuclear data also recovered a crocodile + turtle grouping (Cao et al. 2000). However, morphological data strongly support the monophyly of archosaurs (Gaffney 1980, Gauthier et al. 1988). It is important to note that both crocodiles and turtles show significantly long branches that might introduce biases into the phylogenetic analyses (Sanderson & Shaffer 2002). Hence, the sistergroup relationship of crocodiles and turtles needs to be treated as tentative, and further molecular clarification is needed.

Both recent paleontological and molecular data agree on the more derived position of turtles as diapsids. This new placement of turtles (either as the sistergroup

←

Figure 6 The phylogenetic relationships of turtles to the other groups of living amniotes. (*A*) Turtles as the only living representatives of anapsid reptiles, and as the sistergroup of diapsid reptiles, i.e., the Lepidosauria (the tuatara, snakes, and lizards) + Archosauria (crocodiles and birds). (*B*) Turtles placed as diapsids, and as the sistergroup of the Lepidosauria. (*C*) Turtles as diapsids, and as the sister group of the Archosauria. (*D*) Turtles as diapsids, placed inside the Archosauria, and as the sistergroup of crocodiles. Most recent molecular data favor either hypotheses *C* or *D*.

to Archosaurs, Lepidosaurs, or Crocodilia) has profound implications for the reconstruction of amniote evolution, including, but not limited to, the understanding of the evolution of the fenestration of the skull.

PHYLOGENETIC RELATIONSHIPS OF BIRDS

Ever since the discovery of *Archaeopteryx*, this fossil genus from the Upper Jurassic was recognized as one of the missing links between dinosaurs and birds (Huxley 1868). The sistergroup relationships between theropod dinosaurs (Saurischia) and birds is now firmly established (Ostrom 1975, Xu et al. 2003; but see Feduccia 1996). It is now generally believed by both morphologists and paleontologists that crocodiles are the closest living relatives of birds, and that both groups are the only surviving lineages of the Archosauria (e.g., Gaffney 1980, Gauthier et al. 1988). Most molecular studies based on mitochondrial (e.g., Cao et al. 2000, Hedges 1994, Mindell et al. 1999, Zardoya & Meyer 1998) or nuclear (e.g., Caspers et al. 1996, Platz & Conlon 1997) sequence data agree with this hypothesis. The only exception is the recent molecular work of Hedges and Poling (Hedges & Poling 1999), which supported crocodiles + turtles as the closest living sistergroup of birds (but see above).

Most of the modifications in birds that are associated with powered flight (e.g., feathers, a fully opposable digit for perching, a keeled sternum, and a fused pygostyle that refines flight maneuverability) evolved within a short period of time (less than 10 million years) in the early Cretaceous (Sereno 1999). Adaptation to flight led to a rapid radiation and the origin of the orders of modern birds during the Late Cretaceous (Cooper & Penny 1997; but see Feduccia 1996, 2003), a period in which the fossil record of modern birds is relatively poor (Feduccia 1996). As a result, the phylogenetic relationships of modern avian orders remain unresolved based on paleontological data. Traditionally, extant birds are classified based on the palatal structure into Palaeognathae and Neognathae (Pycraft 1900) (Figure 7*A*). The Palaeognathae include the Struthioniformes (ratites) and Tinamiformes (tinamous). Within the Neognathae, Anseriformes (ducks), Charadriiformes (shorebirds), Gaviiformes (loons), and Procellariiformes (albatrosses) are considered to have diverged early (Feduccia 1996, Mindell et al. 1999).

The classic study of Sibley & Ahlquist (Sibley & Ahlquist 1990) based on DNA-DNA hybridization distances from 1700 species of birds was the first to suggest the palaeognath-neognaths division using molecular data (Figure 7*A*). It was a surprise

→

Figure 7 Major hypotheses about the relationships among the main lineages of birds. (*A*) Basal split between the Palaeognathae (ratites and tinamou) and the Neognathae (the rest). (*B*) Passeriformes (perching birds) are paraphyletic, with oscine passerines (songbirds) as sistergroup of all other birds. Palaeognathae are suggested to be a derived rather than basal group as suggested by the traditional hypothesis (*A*). Most recent molecular data favor hypothesis *A*.

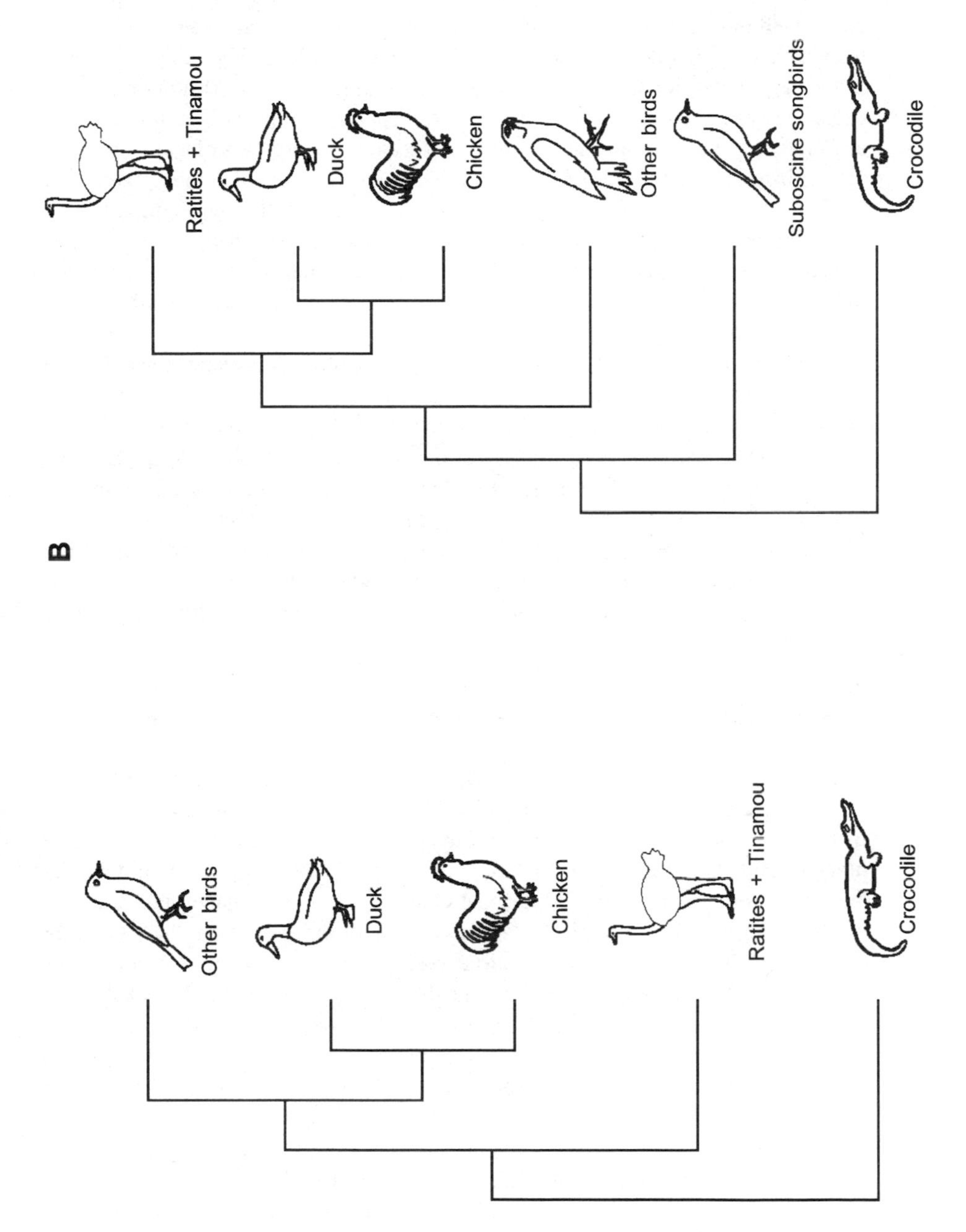
B
Ratites + Tinamou
Duck
Chicken
Other birds
Suboscine songbirds
Crocodile
A
Other birds
Duck
Chicken
Ratites + Tinamou
Crocodile

that more recent work based on complete mitochondrial genome sequence data challenged this traditional view of a basal divergence between Palaeognathae and Neognathae (Härlid & Arnason 1999, Mindell et al. 1999) (Figure 7*B*). In these mitochondrial phylogenies, Passeriformes (perching birds) are paraphyletic, with oscine passerines (songbirds) as the sistergroup of all other birds. Struthioniformes are suggested to be in a rather derived position as sistergroup of galliformes + anseriformes (Galloanserae). However, it has been suggested that these results are likely to be the result of insufficient taxon sampling (van Tuinen et al. 2000). A recent molecular phylogeny of representatives of all modern avian orders based on the complete mitochondrial 12S and 16S rRNA and nuclear 18S genes recovered the basal split between palaeognathans and neognathans and placed Galloanserae as the most basal neognathans (van Tuinen et al. 2000) (Figure 7*A*). The traditional division between palaeognathans and neognathans was also achieved when the mitochondrial DNA data was corrected for rate heterogeneity (Paton et al. 2002). The same results were achieved based on the complete nuclear RAG-1 gene (Groth & Barrowclough 1999). Furthermore, the monophyly of the Passeriformes is supported by several recent molecular studies that are based on nuclear DNA sequences (Barker et al. 2002). It would appear that mitochondrial DNA sequences provide somewhat less reliable phylogenetic information for the question on the ordinal relationships of birds and future studies might need to combine mitochondrial with new nuclear DNA markers to ascertain the relationships among the major ordinal lineages of the Class Aves.

THE SISTERGROUP OF PLACENTAL MAMMALS

The traditional view of the evolution of mammals based on both neontological, morphological, and fossil evidence identified the marsupials as the sistergroup of the eutherians (placental mammals) to the exclusion of the monotremes (the platypus and echidnas) (Carroll 1988) (Figure 8*A*). Many morphological features have been interpreted as shared derived characters between marsupials and placentals (Kermack & Kermack 1984). However, a minority of researchers working on morphological characters advocate a sistergroup relationship of monotremes and marsupials (the Marsupionta hypothesis) based on similar tooth-replacement patterns (Gregory 1947, Kühne 1973), to the exclusion of placentals (Figure 8*B*). The relatively poor fossil record for monotremes (Carroll 1988) complicates the analysis of the phylogenetic relationships among these three living lineages of mammals.

The complete mitochondrial sequences of the platypus and the opossum were determined in an effort to address this debate (Figure 8) (Janke et al. 1994, 1996). Phylogenetic analyses of a data set that combined the inferred amino acid sequences of the mitochondrial protein-coding genes favored, with high statistical support, the monotreme + marsupial clade (Janke et al. 1996) (Figure 8*B*). Follow-up studies based on the same kind of data included the wallaroo, *Macropus robustus* (Janke et al. 2001), the wombat, *Vombatus ursinus*, and the echidna,

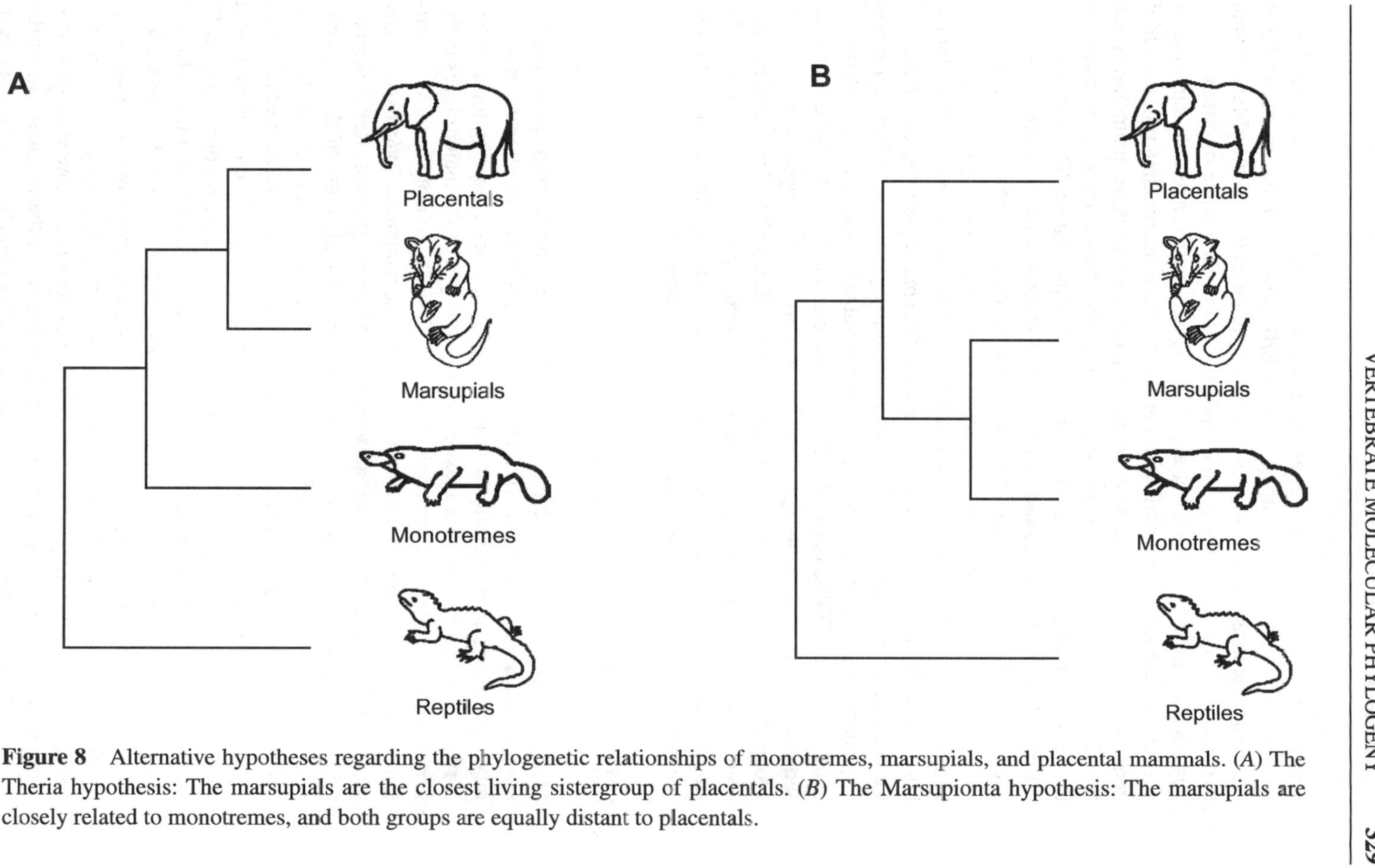

Figure 8 Alternative hypotheses regarding the phylogenetic relationships of monotremes, marsupials, and placental mammals. (*A*) The Theria hypothesis: The marsupials are the closest living sistergroup of placentals. (*B*) The Marsupionta hypothesis: The marsupials are closely related to monotremes, and both groups are equally distant to placentals.

Tachyglossus aculeatus (Janke et al. 2002) and confirmed the mitochondrial support for the Marsupionta hypothesis (Figure 8*B*). However, it has been noted that the strength of the support of the mitochondrial protein data set for the Marsupionta hypothesis varies considerably and depends on both the choice of outgroup and phylogenetic methods (Wadell et al. 1999). Moreover, considerable variation of pyrimidine (cytosine + thymine) frequencies between mammalian mitochondrial genomes seems to affect the recovery of deep divergences in the mammalian tree (Phillips & Penny 2003). Phylogenetic analyses that correct for such bias support the Theria hypothesis (marsupials as sistergroup of placentals) (Phillips & Penny 2003). DNA-DNA-hybridization analyses also supported the monotreme + marsupial clade (Kirsch & Mayer 1998). The validity of these studies was questioned because both monotremes and marsupials show a relatively high GC content in comparison to the placentals (Kirsch & Mayer 1998). Such a base-compositional bias could artificially group the monotremes and the marsupials together. Recently, a nuclear gene, the mannose 6-phosphate/insulin-like growth factor II receptor, was sequenced from representatives of all three mammalian groups in an attempt to clarify this issue (Killian et al. 2001). Phylogenetic analyses of this nuclear gene sequence data favored, with high statistical support, that marsupials are the sistergroup of eutherians to the exclusion of monotremes (Figure 8*A*). These nuclear data seem to corroborate the classical morphology-based hypothesis. Future molecular studies (including, e.g., more nuclear gene sequence data), will certainly improve our understanding of the sistergroup of placental mammals.

CONCLUSIONS AND OUTLOOK

Many of the major events that have occurred throughout the evolution of vertebrates are well documented in the fossil record. Vertebrates therefore offer the opportunity to study long-term evolutionary patterns and processes. However, some nodes, particularly often of those lineages related to the origin of the major clades in the vertebrate tree, remain controversial (Figure 1). This is probably because the origin of the main lineages of vertebrates was often accompanied by/caused by key morphological innovations and subsequent rapid diversification. Rapid origination of lineages, gaps in the fossil record associated with some of these events, and difficulties in the interpretation of synapomorphic character states that were overlaid by long periods of anagenetic changes, hamper the inference of the exact phylogenetic relationships. New vertebrate phylogenies based on molecular data are contributing to the resolution of many of the long-standing problems (Figure 9) (Zardoya et al. 2003). In most cases, molecular data corroborate morphological evidence, but in some cases molecular and morphological signals conflict. Besides the corroboration of many of the traditional morphology-based phylogenetic relationships, new molecular data sets have also been particularly helpful in discerning among competing hypotheses. Examples are (*a*) the now well-supported sistergroup relationships of lungfishes with tetrapods to the exclusion of coelacanths, (*b*) the hypothesis that favors the Batrachia hypothesis (salamanders as

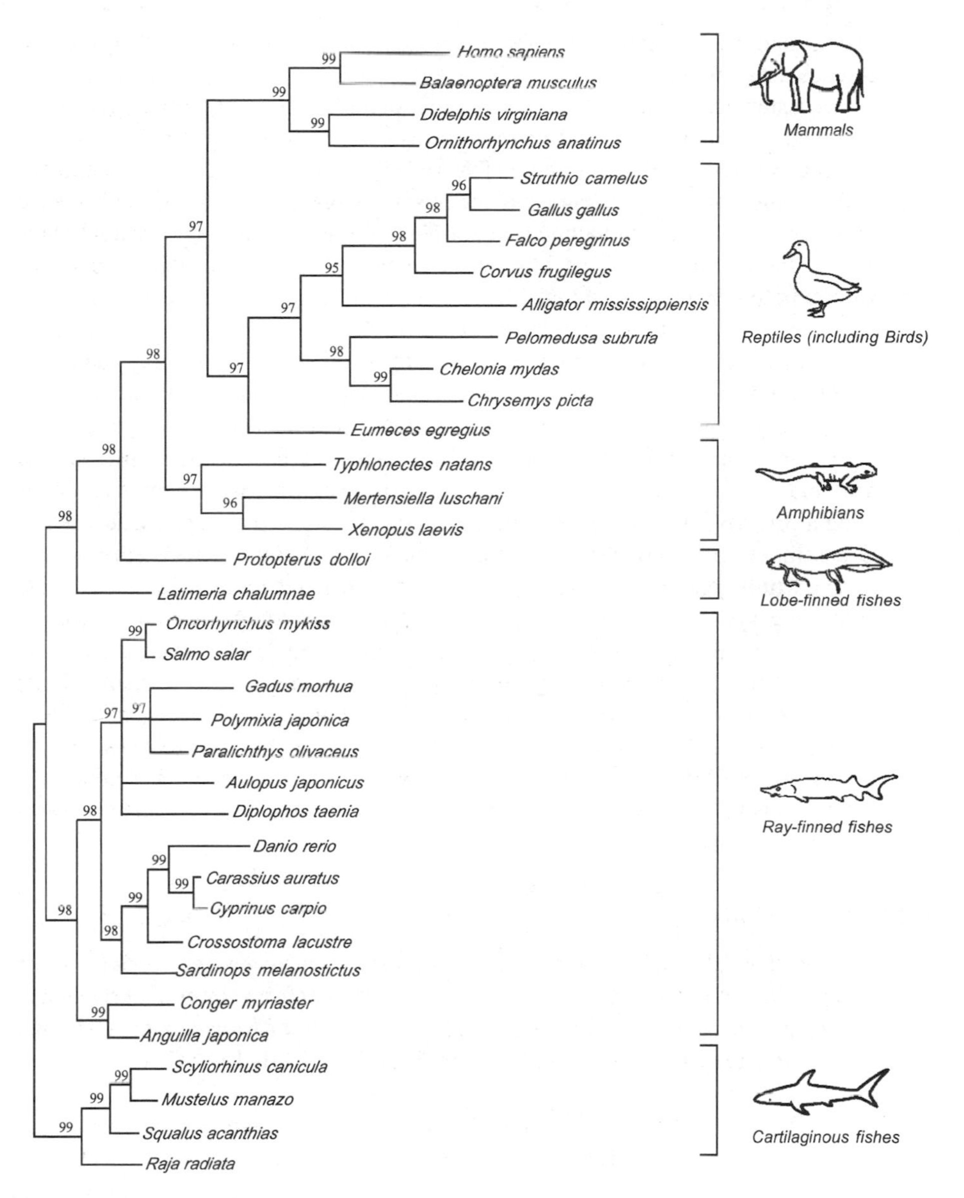

Figure 9 Bayesian phylogeny of the major lineages of vertebrates based on the analysis of complete mitochondrial amino acid data sets (Zardoya et al. 2003). Numbers above branches are posterior probabilities. In this phylogeny the major polytomies of Figure 1 are shown as resolved. Agnathan sequences were not included in this analysis since the phylogenetic limits of mitochondrial genomes are exceeded at this level of phylogenetic inquiry. Note comments in the text for some issues regarding the use of mitochondrial DNA for some phylogenetic questions, e.g., monotreme-marsupial-eutherian relationships and the ordinal phylogeny among birds.

sistergroup of frogs), and (*c*) the placement of turtles as the sistergroup to Archosauria (Figure 9).

Occasionally, when conflicting topologies of molecular and morphological trees are obtained, doubt is raised about the validity of answers to problems that are considered to be settled. This is the case of the recent molecular evidence that supports a sistergroup relationship of hagfishes and lampreys against morphological evidence (Janvier 1996), or the mitochondrial support of a monotreme + marsupial clade against the seemingly well-established Theria hypothesis. Ultimately, comparisons of conflicting signals should enable evolutionary biologists to detect biases that result in misinterpreting one of the two types of data. Understanding the sources of signal conflict will definitively improve phylogenetic inference and may contribute to settling open debates in the systematic relationships among vertebrate lineages.

Two molecular markers, mitochondrial DNA and nuclear rRNA genes, have been widely, and by and large, successfully applied to phylogenetic inference of vertebrate relationships. Several recent advances in molecular techniques such as the development of new nuclear markers (Rag, c-mos, opsins, aquaporin, β-casein, enolase, and creatine kinase, among others) and the possibility of analyzing whole genomes are adding new important insights to the field of vertebrate molecular systematics. More efficient collection techniques for large molecular data sets are already having a major impact on the field. Moreover, more powerful and new phylogenetic algorithms [e.g., Bayesian, Markov Chain, Monte Carlo methods (Huelsenbeck et al. 2001); and the metapopulation genetic algorithm (Lemmon & Milinkovitch 2002)] and alternative new approaches such as reconciled trees (Cotton & Page 2002), as well as faster computers facilitate the estimation of phylogenetic relationships even when using large sequence data sets (Liu et al. 2001).

ACKNOWLEDGMENTS

We thank Brad Shaffer for insightful comments on an earlier version of the manuscript. This work received partial financial support from grants of the Deutsche Forschungsgemeinschaft, the VCI, and the University of Konstanz to A.M., and the Ministerio de Ciencia y Tecnología to R.Z. (REN2001–1514/GLO).

LITERATURE CITED

Ahlberg PE, ed. 2001. *Major Events in Early Vertebrate Evolution. Paleontology, Phylogeny, Genetics, and Development*. London: Taylor & Francis. 418 pp.

Ahlberg PE, Clack JA, Luksevics E. 1996. Rapid braincase evolution between *Panderichthys* and the earliest tetrapods. *Nature* 381:61–63

Ahlberg PE, Johanson Z. 1998. Osteolepiforms and the ancestry of tetrapods. *Nature* 395:792–94

Arratia G. 2001. The sister-group of Teleostei:

consensus and disagreements. *J. Vertebr. Paleontol.* 21:767–73
Barker KF, Barrowclough GF, Groth JG. 2002. A phylogenetic hypothesis for passerine birds: taxonomic and biogeographic implications of an analysis of nuclear DNA sequence data. *Proc. R. Soc. London Ser. B* 269:295–308
Benton MJ. 1990. Phylogeny of the major tetrapod groups: Morphological data and divergence dates. *J. Mol. Evol.* 30:409–24
Bjerring HC. 1985. Facts and thoughts on piscine phylogeny. In *Evolutionary Biology of Primitive Fishes*, ed. RE Foreman, A Gorbman, JM Dodd, R Olsson, pp. 31–57. New York: Plenum
Bolt JR. 1991. Lissamphibian origins. See Schultze & Trueb 1991, pp. 194–222
Cao Y, Adachi J, Hasegawa M. 1998. Comment on the quartet puzzling method for finding maximum-likelihood tree topologies. *Mol. Biol. Evol.* 15:87–89
Cao Y, Sorenson MD, Kumazawa Y, Mindell DP, Hasegawa M. 2000. Phylogenetic position of turtles among amniotes: evidence from mitochondrial and nuclear genes. *Gene* 259:139–48
Carroll RL. 1988. *Vertebrate Paleontology and Evolution.* New York: Freeman
Carroll RL. 1995. Problems on the phylogenetic analysis of paleozoic choanates. *Bull. Mus. Natl. Hist. Nat. Paris 4eme sér.* 17:389–445
Carroll RL. 1997. *Patterns and Processes of Vertebrate Evolution.* Cambridge, UK: Cambridge Univ. Press
Carroll RL. 2000. The Lissamphibian enigma. See Heatwole & Carroll 2000, pp. 1270–73
Caspers GJ, Reinders GJ, Leunissen JA, Watte J, de Jong WW. 1996. Protein sequences indicate that turtles branched off from the amniote tree after mammals. *J. Mol. Evol.* 42:580–86
Clack JA. 2002. *Gaining Ground.* Bloomington: Indiana Univ. Press
Clack JA. 2000. The origin of tetrapods. See Heatwole & Carroll 2000, pp. 980–1029
Cloutier R, Ahlberg PE. 1996. Interrelationships of basal sarcopterygians. See Stiassny et al. 1996, pp. 445–79
Cooper A, Penny D. 1997. Mass survival of birds across the Cretaceous-tertiary boundary: molecular evidence. *Science* 275:1109–13
Cotton JA, Page RDM. 2002. Going nuclear: gene family evolution and vertebrate phylogeny reconciled. *Proc. R. Soc. London Ser. B* 269:1555–61
deBraga M, Rieppel O. 1997. Reptile phylogeny and the interrelationships of turtles. *Zool. J. Linn. Soc.* 120:281–354
Delarbre C, Escriva H, Gallut C, Barriel V, Kourilsky P, et al. 2000. The complete nucleotide sequence of the mitochondrial DNA of the agnathan *Lampetra fluviatilis*: bearings on the phylogeny of cyclostomes. *Mol. Biol. Evol.* 17:519–29
Duellman WE, Trueb L. 1994. *Biology of Amphibians.* Baltimore, MD: Johns Hopkins Univ. Press
Eschmeyer WN. 1998. *Catalog of Fishes.* San Francisco: Cal. Acad. Sci.
Feduccia A. 1996. *The Origin and Evolution of Birds.* New Haven, CT: Yale Univ. Press
Feduccia A. 2003. 'Big bang' for tertiary birds? *Trends Ecol. Evol.* 18:172–76
Feller AE, Hedges SB. 1998. Molecular evidence for the early history of living amphibians. *Mol. Phylogenet. Evol.* 9:509–16
Gaffney ES. 1980. Phylogenetic relationship of the major groups of ammniotes. In *The Terrestrial Enviroment and the Origin of the Land Vertebrates*, ed. AL Panchen, pp. 593–610. London: Academic
Gardiner BG, Maisey JG, Littlewood DT. 1996. Interrelationships of basal neopterygians. See Stiassny et al. 1996, pp. 117–46
Gardiner BG, Schaeffer B. 1989. Interrelationships of lower actinopterygian fishes. *Zool. J. Linn. Soc.* 97:135–87
Gauthier J, Kluge AG, Rowe T. 1988. Amniote phylogeny and the importance of fossils. *Cladistics* 4:105–209
Grande L, Bemis WE. 1996. Interrelationships of Acipenseriformes, with comments

on "Chondrostei." See Stiassny et al. 1996, pp. 85–115

Greenwood PH, Miles RS, Patterson C, eds. 1973. *Interrelationships of Fishes. Zool. J. Linn. Soc.* (Suppl. 1). 536 pp.

Greenwood PH, Rosen DE, Weitzman SH, Myers GS. 1966. Phyletic studies of teleostean fishes, with a provisional classification of living forms. *Bull. Am. Mus. Nat. Hist.* 131:339–456

Gregory WK. 1946. Pareiasaurs versus placodonts as near ancestors to turtles. *Bull. Am. Mus. Nat. Hist.* 86:275–326

Gregory WK. 1947. The monotremes and the palimpsest theory. *Bull. Am. Mus. Nat. Hist.* 88:1–52

Groth JG, Barrowclough GF. 1999. Basal divergences in birds and the phylogenetic utility of the nuclear RAG-1 gene. *Mol. Phylogenet. Evol.* 12:115–23

Härlid A, Arnason U. 1999. Analyses of mitochondrial DNA nest ratite birds within the Neoagnathae: supporting a neotenous origin of ratite morphological characters. *Proc. R. Soc. London Ser. B* 266:305–9

Hay JM, Ruvinsky I, Hedges SB, Maxson LR. 1995. Phylogenetic relationships of amphibian families inferred from DNA sequences of mitochondrial 12S and 16S ribosomal RNA genes. *Mol. Biol. Evol.* 12:928–37

Heatwole H, Caroll RL, eds. 2000. *Amphibian Biology*. Chipping Norton, Aust.: Surrey Beatty & Sons. 1495 pp.

Hedges SB. 1994. Molecular evidence for the origin of birds. *Proc. Natl. Acad. Sci. USA* 91:2621–24

Hedges SB, Hass CA, Maxson LR. 1993. Relations of fish and tetrapods. *Nature* 363:501–2

Hedges SB, Maxson LR. 1993. A molecular perspective on lissamphibian phylogeny. *Herpetol. Monogr.* 7:27–42

Hedges SB, Moberg KD, Maxson LR. 1990. Tetrapod phylogeny inferred from 18S and 28S ribosomal RNA sequences and a review of the evidence for amniote relationships. *Mol. Biol. Evol.* 7:607–33

Hedges SB, Poling LL. 1999. A molecular phylogeny of reptiles. *Science* 283:998–1001

Hennig W. 1983. Testudines. In *Stammesgeschichte der Chordaten*, ed. W Hennig, pp. 132–39. Hamburg: Parey. 208 pp.

Huelsenbeck JP, Ronquist FR, Nielsen R, Bollback JP. 2001. Bayesian inference of phylogeny and its impact on evolutionary biology. *Science* 294:2310–14

Huxley TH. 1861. Preliminary essay upon the systematic arrangement of the fishes of the Devonian epoch. Figures and descriptions illustrative of British organic remains. *Mem. Geol. Surv. UK* 10:1–40

Huxley TH. 1868. On the animals which are most nearly intermediate between the birds and reptiles. *Ann. Mag. Nat. Hist.* 4:66–75

Inoue JG, Miya M, Tsukamoto K, Nishida M. 2003. Basal actinopterygian relationships: a mitogenomic perspective on the phylogeny of the "ancient fish." *Mol. Phylogenet. Evol.* 26:110–20

Janke A, Erpenbeck D, Nilsson M, Arnason U. 2001. The mitochondrial genomes of the iguana (*Iguana iguana*) and the caiman (*Caiman crocodylus*): implications for amniote phylogeny. *Proc. R. Soc. London Ser. B* 268:623–31

Janke A, Feldmaier-Fuchs G, Thomas K, von Haeseler A, Pääbo S. 1994. The marsupial mitochondrial genome and the evolution of placental mammals. *Genetics* 137:243–56

Janke A, Gemmell NJ, Feldmaier-Fuchs G, von Haeseler A, Paabo S. 1996. The mitochondrial genome of a monotreme-the platypus (*Ornithorhynchus anatinus*). *J. Mol. Evol.* 42:153–59

Janke A, Magnell O, Wieczorek G, Westerman M, Arnason U. 2002. Phylogenetic analyses of 18S rRNA and the mitochondrial genomes of the Wombat, *Vombatus ursinus*, and the spiny anteater, *Tachyglossus aculeatus*: increased support for the Marsupionta hypothesis. *J. Mol. Evol.* 54:71–80

Janvier P. 1981. The phylogeny of the Craniata with particular reference to the significance of fossil "agnathans". *J. Vertebr. Paleontol.* 1:121–59

Janvier P. 1996. *Early Vertebrates*. Oxford, UK: Oxford Univ. Press

Jessen HL. 1973. Interrelationships of actinopterygians and brachiopterygians: evidence from pectoral anatomy. See Greenwood et al. 1973, pp. 227–232

Kermack DM, Kermack KA. 1984. *The Evolution of Mammalian Characters*. London: Croom Helm

Killian JK, Buckley TR, Stewart N, Munday BL, Jirtle RL. 2001. Marsupials and Eutherians reunited: genetic evidence for the Theria hypothesis of mammalian evolution. *Mamm. Genome* 12:513–17

Kirsch JAW, Mayer GC. 1998. The platypus is not a rodent: DNA hybridization, amniote phylogeny and the palimpsest theory. *Philos. Trans. R. Soc. London Ser. B* 353:1221–37

Kühne WG. 1973. The systematic position of monotremes reconsidered (Mammalia). *Z. Morph. Tiere* 75:59–64

Kumazawa Y, Nishida M. 1999. Complete mitochondrial DNA sequences of the green turtle and blue-tailed mole skink: statistical evidence for archosaurian affinity of turtles. *Mol. Biol. Evol.* 16:784–92

Kuraku S, Hoshiyama D, Katoh K, Suga H, Miyata T. 1999. Monophyly of lampreys and hagfishes supported by nuclear DNA-coded genes. *J. Mol. Evol.* 49:729–35

Larson A, Wilson AC. 1989. Patterns of ribosomal RNA evolution in salamanders. *Mol. Biol. Evol.* 6:131–54

Lauder GV, Liem KF. 1983. The evolution and interrelationships of the actinopterygian fishes. *Bull. Mus. Comp. Zool.* 150:95–197

Laurin M. 1998. The importance of global parsimony and historical bias in understanding tetrapod evolution. Part I. Systematics, middle ear evolution and jaw suspension. *Ann. Sci. Nat.* 19:1–42

Laurin M, Reisz RR. 1997. A new perspective on tetrapod phylogeny. In *Amniote Origins: Completing the Transition to Land*, ed. SS Sumida, KL Martin, pp. 9–59. New York: Academic. 400 pp.

Laurin M, Reisz RR. 1995. A reevaluation of early amniote phylogeny. *Zool. J. Linn. Soc.* 113:165–223

Le HL, Lecointre G, Perasso R. 1993. A 28S rRNA-based phylogeny of the Gnathostomes: first steps in the analysis of conflict and congruence with morphologically based cladograms. *Mol. Phylogenet. Evol.* 2:31–51

Lee MSY. 1995. Historical burden in systematics and the interrelationships of 'parareptiles'. *Biol. Rev.* 70:459–547

Lee MSY. 1996. Correlated progression and the origin of turtles. *Nature* 379:812–15

Lee MSY. 1997. Pareiasaur phylogeny and the origin of turtles. *Zool. J. Linn. Soc.* 120:197–280

Lemmon A, Milinkovitch MC. 2002. The metapopulation genetic algorithm: an efficient solution for the problem of large phylogeny estimation. *Proc. Natl. Acad. Sci. USA* 99:10516–21

Liu FR, Miyamoto MM, Freire MP, Ong PQ, Tennant MR, et al. 2001. Molecular and morphological supertrees for eutherian (Placental) mammals. *Science* 291:1786–89

Mallat J, Sullivan J. 1998. 28S and 18S rDNA sequences support the monophyly of lampreys and hagfishes. *Mol. Biol. Evol.* 15:1706–18

Mallat J, Sullivan J, Winchell CJ. 2001. The relationship of lampreys to hagfishes: a spectral analysis of ribosomal DNA sequences. See Ahlberg 2001, pp. 106–18

Meyer A. 1995. Molecular evidence on the origin of tetrapods and the relationships of the coelacanth. *Trends Ecol. Evol.* 10:111–16

Meyer A, Wilson AC. 1990. Origin of tetrapods inferred from their mitochondrial DNA affiliation to lungfish. *J. Mol. Evol.* 31:359–64

Milner AR. 1988. The relationships and origin of living amphibians. In *The Phylogeny and Classification of the Tetrapods, Vol. 1: Amphibians, Reptiles, Birds*, ed. MJ Benton, pp. 59–102. Oxford, UK: Clarendon. 392 pp.

Mindell DP, Sorenson MD, Dimcheff DE, Hasegawa M, Ast JC, Yuri T. 1999. Interordinal relationships of birds and other reptiles based on whole mitochondrial genomes. *Syst. Biol.* 48:138–52

Miya M, Takeshima H, Endo H, Ishiguro NB, Inoue JG, et al. 2003. Major patterns of higher teleostean phylogenies: a new perspective based on 100 complete mitochondrial sequences. *Mol. Phylogenet. Evol.* 26:121–38

Nelson GJ. 1989. Phylogeny of major fish groups. In *The Hierarchy of Life*, ed. B Fernholm, K Bremer, H Jörnvall, pp. 325–36. Amsterdam: Elsevier. 488 pp.

Nelson GJ. 1969. Gill arches and the phylogeny of fishes, with notes on the classification of vertebrates. *Bull. Am. Mus. Nat. Hist.* 141:475–552

Nelson GJ. 1994. *Fishes of the World.* New York: Wiley

Noack K, Zardoya R, Meyer A. 1996. The complete mitochondrial DNA sequence of the bichir (*Polypterus ornatipinnis*), a basal ray-finned fish: ancient establishment of the consensus vertebrate gene order. *Genetics* 144:1165–80

Olsen PE. 1984. The skull and pectoral girdle of the parasemiontid fish *Watsonulus eugnathoides* from the early Triassic Sakanema group of Madagascar, with comments on the relationships of the holostean fishes. *J. Vertebr. Paleontol.* 4:481–99

Ostrom JH. 1975. The origin of birds. *Annu. Rev. Earth Planet. Sci.* 3:55–77

Panchen AL, Smithson TR. 1987. Character diagnosis, fossils and the origin of tetrapods. *Biol. Rev.* 62:341–438

Parsons TS, Williams EE. 1963. The relationships of the modern Amphibia: a reexamination. *Q. Rev. Biol.* 38:26–53

Paton T, Haddrath O, Baker AJ. 2002. Complete mitochondrial DNA genome sequences show that modern birds are not descended from transitional shorebirds. *Proc. R. Soc. London Ser. B* 269:839–46

Patterson C. 1973. Interrelationships of holosteans. See Greenwood et al. 1973, pp. 233–305

Phillips MJ, Penny D. 2003. The root of the mammalian tree inferred from complete mitochondrial genomes. *Mol. Phylogenet. Evol.* In press

Platz JE, Conlon JM. 1997. Reptile relationships turn turtle and turn back again. *Nature* 389:246

Pycraft WP. 1900. The morphology and phylogeny of the Palaeognathae (Ratitae and Crypturi) and the Neognathae (Carinatae). *Trans. Zool. Soc. London* 15:149–290

Rage JC, Janvier P. 1982. Le probleme de la monophylie des amphibiens actuels, a la lumiere des nouvelles donnees sur les affinites des tretapodes. *Geobios* 6:65–83

Rasmussen AS, Arnason U. 1999. Molecular studies suggest that cartilagionous fishes have a terminal position in the piscine tree. *Proc. Natl. Acad. Sci. USA* 96:2177–82

Rasmussen AS, Janke A, Arnason A. 1998. The mitochondrial DNA molecule of the hagfish (*Myxine glutionosa*) and vertebrate phylogeny. *J. Mol. Evol.* 46:382–88

Reisz RR. 1997. The origin and early evolutionary history of amniotes. *Trends Ecol. Evol.* 12:218–22

Rest JS, Ast JC, Austin CC, Waddell PJ, Tibbetts EA, et al. 2003. Molecular systematics of primary reptilian lineages and the tuatara mitochondrial genome. *Mol. Phylogenet. Evol.* In press

Rieppel O, deBraga M. 1996. Turtles as diapsid reptiles. *Nature* 384:453–55

Rieppel O, Reisz RR. 1999. The origin and early evolution of turtles. *Annu. Rev. Ecol. Syst.* 30:1–22

Sanderson MJ, Shaffer HB. 2002. Troubleshooting molecular phylogenetic analyses. *Annu. Rev. Ecol. Syst.* 33:49–72

Schultze HP, Cumbaa SL. 2001. *Dialipina* and the characters of basal actinopterygians. See Ahlberg 2001, pp. 315–32

Schultze L, Trueb L, eds. 1991. *Origins of the Major Groups of Tetrapods: Controversies and Consensus*. Ithaca: Cornell Univ. Press. 724 pp.

Sereno PC. 1999. The evolution of dinosaurs. *Science* 2137–47

Shu DG, Conway Morris S, Han J, Zhang ZF, Yasui K, et al. 2003. Head and backbone of the early cambrian vertebrate *Haikouichthys*. *Nature* 421:526–29

Shu DG, Luo HL, Morris SC, Zhang XL, Hus SX, et al. 1999. Lower Cambrian vertebrates from south China. *Nature* 402:42–46

Sibley CG, Ahlquist JE. 1990. *Phylogeny and Classification of Birds: A Study in Molecular Evolution*. New Haven, CT: Yale Univ. Press

Stiassny MLJ, Parent LR, Johnson GD, eds. 1996. *Interrelationships of Fishes*. San Diego, CA: Academic. 496 pp.

Strimmer K, von Haeseler A. 1996. Quartet puzzling: a quartet maximum-likelihood method for reconstructing tree topologies. *Mol. Biol. Evol.* 13:964–69

Szarski H. 1962. The origin of the Amphibia. *Q. Rev. Biol.* 37:189–241

Takezaki N, Figueroa F, Zaleska-Rutczynska Z, Klein J. 2003. Moleular phylogeny of early vertebrates: monophyly of the agnathans as revealed by sequences of 35 genes. *Mol. Biol. Evol.* 20:287–92

Takezaki N, Gojobori T. 1999. Correct and incorrect vertebrate phylogenies obtained by the entire mitochondrial DNA sequences. *Mol. Biol. Evol.* 16:590–601

Tohyama Y, Ichimiya T, Kasama-Yoshida H, Cao Y, Hasegawa M, et al. 2000. Phylogenetic relation of lungfish indicated by the amino acid sequence of myelin DM20. *Brain Res. Mol. Brain Res.* 80:256–59

Trueb L, Cloutier R. 1991. A phylogenetic investigation of the inter- and intrarelationships of the Lissamphibia (Amphibia: Temnospondyli). See Schultze & Trueb 1991, pp. 223–313

van Tuinen M, Sibley CG, Hedges SB. 2000. The early history of modern birds inferred from DNA sequences of nuclear and mitochondrial ribosomal genes. *Mol. Biol. Evol.* 17:451–57

Venkatesh B, Erdmann MV, Brenner S. 2001. Molecular synapomorphies resolve evolutionary relationships of extant jawed vertebrates. *Proc. Natl. Acad. Sci. USA* 98:11382–87

Vorobyeva E, Schultze HP. 1991. Description and systematics of panderichthyid fishes with comments on their relationship to tetrapods. See Schultze & Trueb 1991, pp. 68–109

Wadell PJ, Cao Y, Hauf J, Hasegawa M. 1999. Using novel phylogenetic methods to evaluate mammailan mtDNA, including amino acid-invariant sites-LogDet plus site stripping, to detect internal conflicts in the data, with special reference to the positions of the hedgehog, armadillo, and elephant. *Syst. Biol.* 48:31–53

Xu X, Zhou ZH, Wang XL, Kuang XW, Zhang FC, Du XK. 2003. Four-winged dinosaurs from China. *Nature* 421:335–40

Zardoya R, Cao Y, Hasegawa M, Meyer A. 1998. Searching for the closest living relative(s) of tetrapods through evolutionary analyses of mitochondrial and nuclear data. *Mol. Biol. Evol.* 15:506–17

Zardoya R, Malaga-Trillo E, Veith M, Meyer A. 2003. Complete nucleotide sequence of the mitochondrial genome of a salamander, *Mertensiella luschani*. *Gene*. In press

Zardoya R, Meyer A. 1996. Evolutionary relationships of the coelacanth, lungfishes, and tetrapods based on the 28S ribosomal RNA gene. *Proc. Natl. Acad. Sci. USA* 93:5449–54

Zardoya R, Meyer A. 1997a. The complete DNA sequence of the mitochondrial genome of a "living fossil", the coelacanth (*Latimeria chalumnae*). *Genetics* 146:995–1010

Zardoya R, Meyer A. 1997b. Molecular phylogenetic information on the identity of the closest living relative(s) of land vertebrates. *Naturwissenschaften* 84:389–97

Zardoya R, Meyer A. 1998. Complete mitochondrial genome suggests diapsid affinities of turtles. *Proc. Natl. Acad. Sci. USA* 95:14226–31

Zardoya R, Meyer A. 2001a. The evolutionary position of turtles revised. *Naturwissenschaften* 88:193–200

Zardoya R, Meyer A. 2001b. On the origin of and phylogenetic relationships among living amphibians. *Proc. Natl. Acad. Sci. USA* 98:7380–83

Zardoya R, Meyer A. 2001c. Vertebrate phylogeny: limits of inference of mitochondrial genome and nuclear rDNA squence data due to an adverse phylogenetic signal/noise ratio. See Ahlberg 2001, pp. 106–18

Zharkikh A, Li WH. 1992. Statistical properties of bootstrap estimation of phylogenetic variability from nucleotide sequences: I. Four taxa with a molecular clock. *Mol. Biol. Evol.* 9:1119–47

Annu. Rev. Ecol. Evol. Syst. 2003. 34:339–64
doi: 10.1146/annurev.ecolsys.34.011802.132412

First published online as a Review in Advance on September 29, 2003

THE ROLE OF REINFORCEMENT IN SPECIATION: Theory and Data

Maria R. Servedio[1] and Mohamed A.F. Noor[2]
[1]*Department of Biology, CB# 3280, Coker Hall, University of North Carolina at Chapel Hill, Chapel Hill, North Carolina 27599; email: servedio@email.unc.edu*
[2]*Department of Biological Sciences, Louisiana State University, Baton Rouge, Louisiana 70803; email: mnoor@lsu.edu*

Key Words premating isolation, postmating isolation, sympatric speciation, incompatibilities, divergent selection

■ **Abstract** To assess the frequency and importance of reinforcement in nature we must begin by looking for its signature in the most likely places. Theoretical studies can pinpoint conditions that favor and inhibit reinforcement, and empirical studies can identify both how often these conditions occur and whether reinforcement results. We examine how well these tools have addressed these questions by searching for gaps and mismatches in theoretical and empirical studies of reinforcement. We concentrate on five areas: (*a*) a broad assessment of selection against interspecific mating, (*b*) the mode and genetic basis of nonrandom mating, (*c*) the geography of speciation, (*d*) divergent selection on mating cues, (*e*) and the genetics of reproductive isolation. We conclude that reinforcement has probably not been looked for where it is most likely to occur. We pinpoint however, many further areas of study that may ultimately provide a strong assessment of the importance of reinforcement in speciation.

> "*The grossest blunder in sexual preference, which we can conceive of an animal making, would be to mate with a species different from its own and with which the hybrids are either infertile or, through the mixture of instincts and other attributes appropriate to different courses of life, at so serious a disadvantage as to leave no descendants.*"
>
> —Fisher, 1930 pp. 130

OVERVIEW

Until recently, the primary questions concerning speciation by reinforcement have been whether it can occur and whether any cases can be documented. In the past decade, these questions have been answered. Theoretical studies have shown reinforcement to be feasible under various conditions, and several compelling cases of species that have diverged via reinforcement have been identified (Noor 1999).

In this review, we suggest changing the emphasis of future studies; reinforcement can occur, but what is its frequency and importance in speciation?

Most studies have traditionally required identifying selection against hybrids as a prerequisite for reinforcement (Butlin 1987, Howard 1993). The response to this selection would result in an increase, and possibly eventual completion of, premating isolation between taxa. We refer to selection against hybrids as a "classic" criterion of reinforcement. Butlin (1987) further restricts this definition to cases where hybrids have nonzero fitness. These definitions ignore many processes that are virtually identical to classic reinforcement. To assess how often this whole set of processes occurs in nature, we must think beyond these classic concepts to form a broader definition. Reinforcement in the "broad sense" is an increase in prezygotic isolation between hybridizing populations in response to any type of selection against interspecific matings, regardless of whether hybrids themselves are unfit. In contrast to the apparent division within the literature, we also consider reinforcement to be virtually identical to the latter stage of sympatric speciation, where hybrids are ultimately selected against because of phenotypic differences from diverging parental populations (Kirkpatrick & Ravigné 2002). We therefore consider many models of sympatric speciation to apply to reinforcement. Let us emphasize that we do not suggest that all prezygotic isolation between any two species pairs is a result of reinforcement. We use the term "speciation by reinforcement" as a shorthand to indicate that reinforcement has contributed to prezygotic isolation.

Previous attempts to determine the importance of reinforcement in speciation have concentrated on reviews or reanalyses of empirical studies, using the classic definition of reinforcement as a criterion for inclusion (e.g., Howard 1993, Noor 1997b). We instead attempt to elucidate the role that reinforcement may play in speciation by combining theoretical and empirical approaches, focusing on the broad-sense definition of reinforcement. In this review, we identify conditions theory has predicted may be favorable for reinforcement and evaluate the fit of empirical studies to the assumptions and conclusions of these models. Have researchers even looked for evidence of reinforcement in the most promising systems? We address this question by examining individual components of reinforcement, not in a comprehensive manner (e.g., Kirkpatrick & Ravigné 2002), but with the intention of interpreting the data that are available and identifying gaps in the theoretical and empirical literature.

INCOMPATIBILITIES

The definitions of reinforcement above include some form of selection against interspecific matings. Traditionally this just encompassed low hybrid viability or fertility resulting from intrinsic genetic incompatibilities or interactions with the environment. A number of other mechanisms, however, can promote the evolution of premating isolation in broad-sense reinforcement. If these alternative incompatibilities are common, this would suggest that the frequency of reinforcement may have been underestimated in the past.

Costs Associated with Forming Unfit Hybrids—Intrinsic Versus Extrinsic Postzygotic Incompatibilities

As the classic driving force behind reinforcement, postzygotic incompatibilities have dominated the theoretical and empirical reinforcement literature. Much of this work concentrates on "intrinsic" genetic incompatibilities: hybrids are unfit regardless of their ecological surroundings. Intrinsic incompatibilities can include hybrid inviability, hybrid sterility, or behavioral dysfunctions that prevent hybrids from mating. There are innumerable examples of intrinsic incompatibilities in hybrids across plants, animals, and fungi.

Theoretical studies of reinforcement have incorporated intrinsic incompatibilities in several ways. One of the most basic is to consider a single locus that is underdominant for a character affecting fitness (Balkau & Feldman 1973 in their "disruptive selection" model; Cain et al. 1999; Sanderson 1989). This type of incompatibility is observed, for example, when hybrid dysfunctions result from chromosomal arrangement differences between species (e.g., King 1993, White 1969), which appears to be an uncommon mode of speciation (e.g., Coyne et al. 1993, Navarro & Barton 2003).

Intrinsic incompatibilities commonly arise from sets of loci that interact epistatically to produce low hybrid fitness, as first described by Dobzhansky (1937) and Muller (1942). Many empirical studies have found evidence for these epistatic incompatibilities (e.g., Fishman & Willis 2001, Lamnissou et al. 1996, Orr & Irving 2001, Rawson & Burton 2002) and several theoretical studies have examined the patterns and consequences of the accumulation of these incompatibilities in populations (e.g., Orr 1995, Turelli & Orr 2000). Dobzhansky-Muller incompatibilities have been shown to drive reinforcement in theoretical studies, even when selection against hybrids is relatively weak (Kelly & Noor 1996, Kirkpatrick & Servedio 1999, Servedio 2000, Servedio & Kirkpatrick 1997).

Few theoretical studies of reinforcement consider intrinsic incompatibilities that lead to hybrid sterility, not inviability (Kelly & Noor 1996, Liou & Price 1994). In general, hybrid sterility appears to evolve slightly earlier in evolutionary divergence than hybrid inviability (Presgraves 2002, Price & Bouvier 2002, Sasa et al. 1998, Wu & Davis 1993), so the former is potentially a more common agent of selection for reinforcement. Liou & Price (1994) directly compared the effect on reinforcement of hybrid sterility and hybrid inviability. Reinforcement was found more often with hybrid inviability than sterility, owing largely to an increased probability of extinction when viable but infertile hybrids consume resources. They modeled reduced fertility by removing a fraction of hybrid adults from the mating pool; this is the same as assuming that these individuals chose not to mate, so can also be considered as a model of an intrinsic hybrid mating dysfunction (which has not been explicitly modeled elsewhere).

Whereas such intrinsic incompatibilities appear very common, until recently, few studies showed that hybrids may bear "extrinsic" genetic incompatibilities: lower fitness associated with a specific set of ecological conditions (Coyne & Orr 1998). As such, the frequency of hybrid incompatibilities that could drive reinforcement may be greater than previously perceived. Ecologically dependent isolation may occur when hybrids intermediate to the parental species in phenotype cannot efficiently exploit the available environments used by its parents. This can be tested using reciprocal transplants (Rundle & Whitlock 2001). An example of these incompatibilities has been described elegantly in the literature on benthic and limnetic morphs of the three-spine stickleback (fish) *Gasterosteus aculeatus* (Hatfield & Schluter 1999, Rundle 2002). When raised in the laboratory, hybrids between these morphs, which display intermediate morphology, are fully viable and fertile (Hatfield & Schluter 1999; McPhail 1984, 1992). However, a transplant experiment in the wild revealed that F_1 hybrids grew at lower rates than benthics in the littoral zone and limnetics in the open water (Hatfield & Schluter 1999), despite showing no growth reduction in the laboratory. Similar data have been obtained in birds (see Grant & Grant 1998; Price 2003).

The most comprehensive model of reinforcement that includes extrinsic incompatibilities is Kirkpatrick's (2001) multilocus model. Kirkpatrick considers two traits; one is involved in premating isolation and the other is an ecologically relevant trait under both stabilizing and directional selection. He identifies three major conditions favorable to reinforcement when hybrids are intermediate to parental species in their ecological niche: (*a*) larger differences between the parental mean phenotypes, (*b*) stronger stabilizing natural selection around a local ecological optimum, and (*c*) many loci contributing to the ecological adaptation. Notably, as the parental mean phenotypes diverge, the effect of selection against hybrids on reinforcement increase rapidly, proportional to the square of the difference in parental mean phenotypes. However, this model assumes weak selection, which may not be representative of many speciation events. It would be useful to evaluate the robustness of the results to higher selection intensities.

A gray area in the dichotomy between intrinsic and extrinsic incompatibilities exists when hybrids are unable or unwilling to secure mates through behavioral hybrid dysfunction or have an intermediate phenotype that is not attractive to the choosing sex. Behavioral hybrid dysfunction has been documented in hybrids of a variety of taxa including Lepidoptera (Davies et al. 1997, Pashley & Martin 1987), Drosophila (e.g., Noor 1997a), birds (Buckley 1969), and frogs (Hobel & Gerhardt 2003), but is surely far more widespread. This behavioral anomaly may be intrinsic if the hybrids have a deficiency that prohibits courting or mating, but may be extrinsic if they are selecting mates based on cues that are rare or absent in their population. As pointed out above, the infertility version of the reinforcement model of Liou & Price (1994) can be interpreted as using this type of incompatibility.

In contrast to the situation where the hybrids are behaviorally dysfunctional, they may merely be unattractive to either of the purebred parental populations when the populations are divergent in sexual ornaments (e.g., in *Heliconius*,

Jiggins et al. 2001, Naisbit et al. 2001). Vamosi & Schluter (1999) demonstrated an example of sexual selection against male hybrids between benthic and limnetic threespine sticklebacks. Although there is no sexual selection against hybrids in the lab, male hybrids have difficulty competing in the wild, where the male's choice of microhabitat for nesting among limnetics is crucial to their mating success. Although no reinforcement models have specifically examined "unsexy" hybrids, all models of sympatric speciation by sexual selection (e.g., Higashi et al. 1999, Kawata & Yoshimura 2000) can be considered reinforcement models based on low hybrid attractiveness, if they are examined with the appropriate initial conditions.

Costs to Interspecific Matings—Postmating, Prezygotic Incompatibilities

Selection against hybrids is not the only mechanism that can drive the evolution of premating isolation. The exact patterns of preference divergence that result from classic reinforcement can evolve owing to postmating, prezygotic incompatibilities. Servedio (2001) used a multilocus model to obtain an expression for the expected strength of selection that these incompatibilities would place on mating preferences. She showed that incompatibilities from postmating, prezygotic interactions are expected to be comparable if not greater in strength to those from low hybrid viability, when both types of incompatibilities are caused by pairs of interacting loci.

Postmating, prezygotic incompatibilities can take the form of (*a*) mechanisms that increase female mortality between mating and offspring production or (*b*) mechanisms that reduce female fertility. In the first category, direct injury may result from morphological or behavioral differences during courtship and copulation, as illustrated in the carabid beetles *Carabus* (*Ohmopterus*) *maiyasnus* and *C.* (*O*). *iwawakianus*. Using dissections, Sota & Kubota (1998) found that females of these species involved in heterospecific matings often die owing to rupture of their vaginal membranes, while males of one of the species often had broken genital parts. A similar type of incompatibility may result when costs do not take effect immediately but imperil a female's future reproduction. For example, interspecific hybridization may pose a threat because of the transmission of parasites or diseases to which only one of the taxa has evolved some resistance.

There is much evidence for the second category of postmating, prezygotic incompatibilities, those that reduce female fertility (reviewed by Markow 1997, Howard 1999). Females of the green lacewing species *Crysopa quadripunctata*, for example, showed low fertility when crossed with *C. slossonae* owing to low rates of sperm transfer from the bursa copulatrix to the spermatheca (Albuquerque et al. 1996). This is only one of many specific mechanisms that can lower fertility in an interspecific cross (see Servedio 2001).

Postmating, prezygotic incompatibilities may often evolve alongside postzygotic incompatibilities, although the former is generally not looked for once the latter is discovered. Postmating, prezygotic incompatibilities may also drive broad sense reinforcement when postzygotic isolation is absent or weak.

What is Needed Now?

When biologists try to demonstrate reinforcement they often look only for hybrid inviability or reduced fertility as a driving force. We strongly discourage this practice. Taken together, theoretical and empirical studies have demonstrated that other kinds of postmating isolation exist that can lead to broad sense reinforcement. We hope that this discussion will convince researchers to look more broadly for sources of incompatibilities in future work, expanding our understanding of when reinforcement may occur.

We initially hoped to review the relative frequencies of extrinsic versus intrinsic incompatibilities in various taxa. However, we have concluded that there has been a strong bias in the types of incompatibilities that researchers have investigated: Drosophila researchers have disproportionately sought hybrid sterility/inviability, whereas vertebrate ecologists have disproportionately studied ecological or behavioral differences. Investigators need to look more uniformly at the relative contributions of these different barriers to gene exchange in diverse taxa.

NONRANDOM MATING AND FERTILIZATION

Fundamental issues affecting the probability of reinforcement are the mode and genetic basis of nonrandom mating and fertilization. Theory predicts that certain conditions are particularly favorable to reinforcement. We briefly discuss these below, concentrating on mechanisms at a single locus (nonrandom mating controlled by many loci is discussed further in the section on Genetics of Reproductive Isolation). There is, unfortunately, scant empirical work examining some of the important distinctions that emerge from theory. We point out promising systems in which future studies may find this evidence. Although our examples focus on nonrandom mating, the generalities drawn should apply to nonrandom fertilization as well.

Nonrandom Mating in Reinforcement

In a landmark paper, Felsenstein (1981) presented one of our most valuable insights into the genetics of speciation. He pointed out that assortative mating during speciation can occur by either a one-allele or a two-allele mechanism. In a one-allele process, the substitution of a single allele across both of two diverging populations can reduce interpopulational mating (e.g., an allele that uniformly depresses migration rates). In contrast, in a two-allele system, assortative mating is caused by the substitution of alternate alleles, contributing a distinct preference or fertilization affinity, into each diverging population. Reinforcement is easier to achieve with a one-allele than a two-allele mechanism. Felsenstein noted that, in a two-allele model, recombination can break up beneficial genetic associations (linkage disequilibrium) between the nonrandom mating alleles and the locally adapted allelic combinations in each population; this can not occur in a one-allele model where specific linkage between the mating locus and locally adapted loci is not necessary for speciation. Using Felsenstein's basic two-allele model, Trickett & Butlin (1994) confirmed that a suppressor of recombination would be favored, and consequently speciation would be easier to achieve.

Felsenstein's distinction has been thoroughly discussed in many papers and reviews (e.g., Kirkpatrick & Ravigné 2002). Felsenstein (1981) himself reviewed the major speciation models published to date in the context of his categories. Since then, further one-allele models (reinforcement: Cain et al. 1999, Kelly & Noor 1996, Sanderson 1989; sympatric speciation: Kawecki 1996, 1997) have generally found reinforcement to be less restrictive than have two-allele models (Kirkpatrick 2000, 2001; Payne & Krakauer 1997; Servedio & Kirkpatrick 1997; with multiple loci: Liou & Price 1994; sympatric speciation models: Dieckmann & Doebeli 1999; Higashi et al. 1999; Kawecki 1996, 1997; Kondrashov & Kondrashov 1999). The one-allele/two-allele distinction does not apply with all forms of population structure however (e.g., one-allele mechanisms do not ease conditions for reinforcement in a peripheral isolate; Servedio 2000).

In a paper comparing a one-allele assortative mating model and a two-allele preference model, Servedio (2000) identified an additional factor that may facilitate speciation in many one-allele models. A single allele that causes assortative mating (for example, by inducing females to prefer males that share their body size) may automatically form genetic associations that lead to it being favored by indirect natural and sexual selection in all parts of its range. This occurs because such an allele may naturally form genetic associations with the high fitness trait alleles in each population (because these traits occur at a high frequency). In one-allele systems that face such unopposed selection, reinforcement will always occur. In contrast, in a two-allele model opposing forces of indirect selection will always be acting on the alleles to maintain a polymorphism in the mating system across both populations. Conditions for the maintenance of such a polymorphism will naturally be more restricted.

A detail often ignored in the literature is that many purported one-allele systems must involve an underlying two-allele mechanism to produce two species; the order in which these mechanisms are established may influence the way in which we categorize the system. Take, for example, a hypothetical case where a single allele causes females to prefer males that share their body size. If size had already diverged in two populations, and we were examining the spread of the allele causing assortative mating, we would consider this to be a one-allele system. If, on the other hand, females initially aggregated and mated with males of similar size, and we were examining the spread of an allele for large size in a population with small size, we would view this as a two-allele system. It may sometimes be impossible to determine post facto whether a one-allele or two-allele mechanism was ultimately responsible for the commencement of isolation.

There is very little firm evidence regarding the relative frequency of one-allele versus two-allele reinforcement mechanisms during natural speciation events. However, some candidate systems may be promising places to distinguish these mechanisms. Two-allele systems may be much easier to demonstrate than one-allele ones, and may be common in cases where preferences are for traits not shared by the female herself. For example, different male song features are associated with mating success in *Drosophila pseudoobscura* and *Drosophila persimilis*: interpulse interval is important to females of the former whereas intrapulse frequency is important to those of the latter (Williams et al. 2001). It is hard to imagine this system arising through the fixation of a single allele.

Other general patterns or types of isolation may also be more likely from two-allele systems than one-allele systems, unless asymmetrical effects of genetic backgrounds are imposed on a one-allele system. Divergence in flowering time and allochronic isolation (e.g., Cox & Carlton 1991, Lloyd et al. 1983), for example, would be likely to fit a two-allele system if these characters have a heritable basis. A two-allele system may also be indicated when there are asymmetrical mate preferences: one taxon may exhibit a strong preference for conspecifics while the other taxon displays little or no preference (e.g., Bordenstein et al. 2000, Helbig et al. 2001, Michalak et al. 1997). This suggests that alleles causing a preference have arisen only in the discriminating taxon. This hypothesis can be directly evaluated by genetic manipulations.

The best examples of one-allele systems may be alleles that cause a reduction in migration rate (Balkau & Feldman 1973, Fisher 1930) and alleles that lead to self-pollination. Fishman & Wyatt (1999) showed that selection against crossing with the heterospecific *Arenaria glabra* promoted the evolution of selfing in *Arenaria uniflora* in areas of sympatry; *A. uniflora* is an outcrosser in areas of allopatry. Although in this specific system *A. glabra* only outcrosses, if alleles for selfing were to spread in two incipient species, this would be an example of speciation by a one-allele mechanism.

Another possible example of a one-allele mechanism may be the spread of alleles causing individuals to sexually imprint on parental phenotypes (Irwin & Price 1999, Lorenz 1952, Slabbekoorn & Smith 2002, Vos 1995). If, however, imprinting is an ancestral factor in mate choice and isolation is caused by song divergence (Grant & Grant 1997a,b), then the mechanism is still fundamentally two-allele (different alleles for song must be predominant in each population). This of course assumes that song divergence has a genetic as well as a learned component, although the evidence may be stronger for the latter (e.g., Lynch & Baker 1994, Mundinger 1982). A similar phenomenon occurs when an allele causes individuals to preferentially mate on the host from which they emerged, as is true for many phytophagous insects (e.g., Feder et al. 1994). Although this is a one-allele mechanism, if mating on the host is ancestral and host choice is the divergent trait, then the system is two-allele. Via & Hawthorne (2002), for example, find different quantitative trait loci (QTL) with positive effects for acceptance of alfalfa and of clover, indicating this possible case of host-race divergence can best be explained as a two-allele system.

Direct Selection for Assortative Mating

Although not the scenario considered in reinforcement, premating isolation may also evolve if selection acts on nonrandom mating alleles directly. Such direct selection can act in either one-allele or two-allele systems. Kirkpatrick & Ravigné (2002) point out that direct selection should be the most efficient source of premating isolation, and describe several scenarios in which it may have acted. Even in the context of reinforcement, any direct selection acting on preferences can often overwhelm indirect selection caused by low hybrid fitness to be the primary force driving the evolution of premating isolation (Kirkpatrick & Barton 1997, Servedio 2001). Ecological character displacement and sensory bias can also cause direct selection on mating preferences to produce a pattern identical to reproductive character displacement, and may therefore be confused with reinforcement (Noor 1999, Servedio 2001). Despite the ease with which direct selection is expected to lead to speciation, there are very few empirical examples, partly owing to the difficulty of measuring selection on mating preferences in most systems. It is therefore very difficult to assess how commonly this mechanism occurs.

What is Needed Now?

Whereas theoretical studies have consistently demonstrated the advantages of one-allele over two-allele systems in leading to reinforcement in two populations, empirical studies have not yet begun to evaluate the relative frequency of these mechanisms. Although this will be a challenging task, as discussed above, this information is necessary to determine the ease with which reinforcement can occur in natural systems.

THE GEOGRAPHY OF SPECIATION

The geographic orientation of incipient species can affect both the likelihood of reinforcement and the chance that it will be detected. The most critical aspect of geography is its influence on rates and patterns of gene flow, which are crucial to determining whether reinforcement will occur. Second, the patterns of physical overlap of species can be important in determining whether researchers will search for reinforcement in the species pair.

Gene Flow

One generalization that may be made about divergence and speciation is that they tend to be inhibited by gene flow. No one argues about whether allopatric speciation is a likely scenario. It is helpful to view the geographical settings considered for speciation and reinforcement with this generalization in mind. We can see that, although increasing gene flow tends to inhibit speciation, patterns of gene flow and structure of populations can be crucial as well.

Fairly low gene flow may occur when two populations with non-overlapping ranges exchange long-distance migrants. Several numerical studies have found that, in this case, lowering migration rates increases the chance of reinforcement (Felsenstein 1981, Kelly & Noor 1996, Servedio & Kirkpatrick 1997); less migration decreases the chance that the populations will homogenize.

As gene flow in these two-island models becomes increasingly asymmetric, premating isolation occurs less often (Servedio & Kirkpatrick 1997). At the extreme, when migration is one-way, as in a peripheral isolate receiving migrants from a large parental population, premating isolation may evolve rarely (Servedio & Kirkpatrick 1997). Migration from the "continent" will effectively swamp premating isolation evolution on the "island." The flycatchers *Ficedula albicollis* and *Ficedula hypoleuca*, which show reinforcement in a continental cline (Sætre et al. 1997), demonstrate much less pronounced isolation on recently colonized islands that receive strongly asymmetric migration (Sætre et al. 1999). Although there may be other explanations for this pattern in flycatchers, it is consistent with the theoretical expectations described above. Nosil et al. (2003), using the walking-sticks *Timema cristinae*, examined the effects of migration asymmetries between host races more closely, by determining the amount of female discrimination present in study populations that evolved under different migration regimes. Their results suggest that reinforcement does have the strongest effect when the size of the two populations involved is the most similar; the effect declines as migration becomes very asymmetric.

Kirkpatrick (2000) elegantly demonstrates that migration in a speciation model has dual effects. He considers a population that can be interpreted as an island receiving continental migrants with rate m. He shows that an equation for the change in the mean of a trait used as a cue for assortative mating in the island population, $\bar{T}$, has two terms that contain migration, $m(\bar{T}' - \bar{T})$ and $-m(\bar{T}' - \bar{T})I$, where $\bar{T}'$ is the mean of the assortative mating trait in the foreign population and I is a measure of the intensity of selection against immigrants and hybrids. The first term in this decomposition represents the homogenizing effect of migration, changing the mean of the focal population toward the mean of the migrants. The second term, however, demonstrates that migration can also cause evolution away from the mean of the foreign population; it provides opportunities for selection against hybrids and immigrants, which are important driving forces for divergence. The consistent homogenizing effect of migration found by the other models discussed above indicates that the former effect may generally outweigh the latter. Nosil et al.'s (2003) study of walking-sticks, however, suggests that with low migration rates (and in finite populations), the latter effect may be sometimes more important. If low migration rates can hinder reinforcement due to a lack of selection against hybrids, an intermediate migration rate would therefore be optimal for reinforcement.

Because migration tends to prevent the evolution of premating isolation, sympatry should be a particularly unfriendly environment for speciation. Despite the fact that sympatric conditions have been the most difficult for reinforcement in numerical studies that have made the comparison (when $m = 0.5$, Felsenstein 1981, Kirkpatrick & Ravigné 2002, Liou & Price 1994, Servedio & Kirkpatrick 1997), models of sympatric speciation often imply that it can occur easily (Felsenstein 1981). There is also increasing empirical evidence for sympatric speciation in many natural systems, such as phytophagous insects (reviewed in Berlocher & Feder 2002). As Kirkpatrick & Ravigné (2002) point out, evolution of premating isolation in the last stages of sympatric speciation is conceptually identical to reinforcement; reinforcement may even be easier because it may be driven by forms of intrinsic postmating isolation that cases of sympatric speciation may not possess.

Reinforcement faces several challenges when secondary contact occurs in a hybrid zone (see Howard 1993). Selection against hybrids driving reinforcement, for example, only occurs in the zone (Moore 1957), and premating adaptations that evolve in the zone are likely to be swamped out by migration from the parental populations outside of the zone (Bigelow 1965). Because hybrid zones generally occur on the edge of a species range, migration may also tend toward unidirectionality into the zone originating mainly from the center of the range, biasing against reinforcement. Liou & Price (1994) found, however, that reinforcement occurred more easily in a three-neighborhood stepping-stone model than within a single population. In contrast to a similar model by Sanderson (1989) that reached the opposite conclusion owing to a direct cost of divergence, the pattern Liou & Price (1994) observed was likely determined by lower effective migration in the hybrid zone model than in sympatry. Cain et al. (1999) suggest reinforcement may also occur more easily when hybrid zones exist in a mosaic structure; with a patchy distribution of each species; than in a tension zone where alleles representative of each species change in a monotonic cline (but see Turelli et al. 2001). This could occur if the formation of hybrids over a broader region in a mosaic zone resulted in stronger selection for premating isolation; however, the increased opportunities for introgression in a mosaic zone may also have the tendency to homogenize the population.

In hybrid zones and areas of geographic overlap fitting a stepping-stone model, patterns of premating isolation are often consistent with reinforcement. Although there are several alternative explanations for these patterns (e.g., ecological character displacement), there are many apparent cases of reproductive character displacement, defined as the divergence of mating characters in sympatry but not in allopatry (e.g., Butlin 1987, Howard 1993, Noor 1999). Displacement along clinal hybrid zones has yielded convincing evidence of reinforcement in flycatchers (Sætre et al. 1997), but still remains to be explored further in other taxa (e.g., snails, Schilthuizen & Lombaerts 1995). Comparisons of areas of sympatry and allopatry have yielded evidence of reproductive character displacement and/or reinforcement across many taxa, including snails (e.g., Wullschleger et al.

2002), Drosophila (e.g., Coyne & Orr 1989, 1997; Noor 1995, 1997b), copepods (Holynska 2000), sticklebacks (Rundle & Schluter 1998), and frogs (e.g., Gerhardt 1994, Loftus-Hills & Littlejohn 1992).

Signature of Reinforcement

One noteworthy fact from the section above is that reinforcement per se has only been convincingly demonstrated in a few geographical situations (e.g., sympatry versus allopatry, tension zones). Reinforcement has not been found, for example, where theory has predicted it may occur the most easily: in two populations exchanging long distance migrants (i.e., when migration occurs at a low rate). The simple explanation for this pattern is that reinforcement is commonly looked for only where it can leave the signature of reproductive character displacement (Howard 1993). To identify reinforcement by this signature, species must have both allopatric and sympatric populations or regions. Reinforcement may occur, however, in cases that do not fit this description, for example in two populations exchanging migrants or in populations whose ranges overlap completely.

Even when species do maintain allopatric and sympatric populations, the signature of reinforcement may be easily erased by the spread of premating isolation mechanisms into areas of allopatry (Walker 1974). This may eventually occur whenever there is gene flow maintaining the integrity of each incipient species. It would be especially likely, however, when premating isolation alleles are selectively neutral in allopatry; this is also the most favorable case for reinforcement. Once again we may be missing reinforcement when it is most likely because we generally look for evidence of a signature.

A signature is not necessary to demonstrate reinforcement, however. An alternative, when feasible, would be to collect temporal data. Pfennig (K.S. Pfennig, unpublished manuscript), for example, demonstrated a decreasing frequency of hybridization between *Spea bombifrons* and *S. multiplicata* over 27 years in an area of syntopy. She ruled out several alternative explanations, including a decline of opportunity for hybridization and changes in habitat affecting hybrid production. This type of convincing demonstration is an informative alternative strategy when reinforcement does not leave a signature.

Extinction

Sympatry of incipient species may affect the likelihood of extinction (e.g., van Doorn et al. 1998). Extinction has been found in several reinforcement models. Liou & Price (1994), using an individual-based model, reached the conclusion that extinction was likely particularly when one species greatly outnumbered the other, such that the rarer species frequently produced maladapted hybrids. Deterministic reinforcement models can obtain results that may also be interpreted as extinction, for example the loss of variation of one population's alleles (e.g., Servedio & Kirkpatrick 1997) or population fusion (Kelly & Noor 1996). There are many studies in the ecological literature examining conditions for the coexistence of sympatric species (e.g., Connell 1983, Schoener 1983, Volterra 1926, Yoshimura & Clark 1994); these may be relevant to studies of reinforcement, and should be examined further in this context.

What is Needed Now?

As discussed above, most empirical studies only look for reinforcement where it leaves the signature of reproductive character displacement; creative approaches such as temporal studies have the potential to greatly broaden the situations in which reinforcement is found. Another obvious omission in both theory and empirical work is further consideration of when incipient species can coexist in sympatry. Finally, further empirical studies should be undertaken to confirm and quantify the effects of migration rates and asymmetries on reinforcement.

SELECTION ON MATING CUES

Although ostensibly driven by selection against interspecific matings, reinforcement may include mating cues that are under natural selection, or sexual selection already present at the time of secondary contact. This selection itself can drive premating divergence. It can be a powerful force, capable of eclipsing selection against hybrids (M.R. Servedio, unpublished manuscript). If this type of selection is common, then the number of putative empirical cases of reinforcement may be overestimated. Here we examine how selection on mating cues promotes premating isolation in theoretical studies, and review evidence for this selection in natural systems. This selection can create the appearance of reinforcement, thereby confounding our attempts to determine reinforcement's frequency in nature.

Two Components to Nonrandom Mating

Many models of speciation and reinforcement consider a system of nonrandom mating with two components. The first is a locus or set of loci that causes mate choice (e.g., a mating preference). The second is the cue upon which the first component acts (e.g., a male trait). Because both components are involved in nonrandom mating, linkage disequilibrium builds between them. Any outside source of selection, such as natural or sexual selection, on the cue will therefore cause both components to evolve. An analogous association could evolve between loci conferring habitat preference and performance in that habitat. In reinforcement models, loci that cause selection against hybrids or against interspecific matings also form genetic associations with the loci involved in mate choice. It is difficult to determine whether selection on the mating cue or selection against hybrids is the primary determinant of nonrandom mating evolution in models where both occur. Selection against hybrids may sometimes contribute little to the evolution of premating divergence in the system (Kirkpatrick & Servedio 1999) even though the traditional signature of reproductive character displacement is present. This occurs even in haploid models arranged so that extrinsic selection on hybrids at the mating cue is absent (M.R. Servedio, unpublished manuscript).

Several reinforcement models include natural selection driving the divergence of a mating cue (Cain et al. 1999, Kirkpatrick & Servedio 1999, Servedio 2000, Servedio & Kirkpatrick 1997). This selection alone is capable of driving speciation; selection against hybrids merely adds to this force (Kirkpatrick & Servedio 1999). This can be illustrated by considering sympatric speciation models, where selection against hybrids is initially absent (see Turelli et al. 2001). Dieckmann & Doebeli (1999), for example, find sympatric speciation occurring when the cue for assortment, an ecological trait determining resource use, is under divergent selection owing to competition.

Like natural selection, sexual selection can also favor mating cues during reinforcement (Kirkpatrick & Ravigné 2002, Kirkpatrick & Servedio 1999). Several studies of reinforcement assume some initial sexual selection is already causing a small amount of divergence at the time of secondary contact (Liou & Price 1994, Kelly & Noor 1996). They then examine the spread of alleles that strengthen this nonrandom mating. This evolution will occur in concert with any selection against hybrids also present, thereby amplifying the perceived importance of reinforcement.

These assumptions about natural and sexual selection on mating cues are well met in threespine sticklebacks, where reinforcement may be involved in some of the divergence of benthic and limnetic forms in postglacial lakes (Rundle & Schluter 1998). Limnetic sticklebacks are generally small and narrow bodied, whereas benthics are large and deep bodied. Not only is body size under divergent natural selection in these lakes, but it is an important component of mate choice, and thus may help drive premating divergence (Nagel & Schluter 1998). Divergent sexual selection also operates on male color in sticklebacks, and may play a similar role in the development of premating isolation (Boughman 2001). Selection against hybrids may have only a minor effect on divergence when compared to these other selective forces, and therefore the importance of reinforcement may be minimal in this system.

Divergent natural selection has also been shown to act on mating cues in Heliconius butterflies and in Darwin's finches. *Heliconius melpomene* and *Heliconius cydno* undergo disruptive selection on their color patterns because they each mimic a different model species (Mallet 1999). Males of each species also preferentially court females of their own mimetic pattern (Jiggins et al. 2001). Reinforcement may also be acting in this species pair, which produces hybrids that are not efficient mimics of either model. Podos (2001) similarly demonstrated in Darwin's finches that divergent selection on bill morphology and body size can shape song performance, which may be used in species recognition. While song performance per se may not act as a mating cue, beak shape and size may themselves function in conspecific mate choice (Grant 1986).

A first step to identifying natural selection on mating cues, and therefore assessing the frequency of this force that may amplify the appearance of reinforcement, may be to search for evidence of natural and sexual selection operating in the same direction on a trait in general. Large song repertoire size in birds, for example, may be selectively favored initially through a physiological benefit to varying the syringial muscles used in song production (the anti-exhaustion hypothesis; Lambrechts & Dhondt 1988). If this selection pressure exists it would reinforce the evolution of female preferences for large repertoires, present in a variety of species (Gil & Gahr 2002, Searcy & Yasukawa 1996), until a constraint is reached. Body size may also sometimes be under divergent natural and sexual selection (see Servedio 2000). There may be few known examples of natural and sexual selection acting in the same direction because of a research bias; many studies instead look for evidence of honesty and handicaps in sexually selected traits (Gil & Gahr 2002, Grafen 1990).

One Component to Nonrandom Mating

In other speciation models there is only a single component to mate choice or habitat selection (e.g., Kondrashov 1983, Rice 1987, Drossel & McKane 2000). Selection on this single component would comprise direct selection for divergence, and would thus greatly facilitate speciation (e.g., Kirkpatrick 2000, Kirkpatrick & Ravigné 2002). The clearest examples of this phenomenon are in the habitat choice of phytophagous insects, which may be selected to colonize new hosts, or to specialize from a generalist ancestor (e.g., Berlocher & Feder 2002).

Several models that include a single component to nonrandom mating find speciation without direct selection on mate or habitat choice (reinforcement: Balkau & Feldman 1973, Felsenstein 1981; sympatric speciation: Kawecki 1996, 1997; Kondrashov & Shpak 1998; two-character model of Kondrashov & Kondrashov 1999). In reinforcement and the latter stages of sympatric speciation, selection against hybridization is likely to be the driving force for this divergence.

What is Needed Now?

Divergent selection on mating cues may be a primary driving force during both reinforcement and sympatric speciation, and may greatly enhance ecological speciation mechanisms (e.g., Schluter 2001). Reinforcement, even in its broad sense, may play a small role in speciation if divergent selection on mating cues is also present. It is therefore crucial that more empirical studies search for evidence of this pattern of divergent selection. One place to start this search may be in potential examples of natural and sexual selection acting in the same direction on traits. Evidence of natural selection on cues in single component systems may also provide much needed evidence of speciation by direct selection.

In most taxa, it may not be possible to demonstrate that divergent sexual selection has preceded secondary contact and thus has served as a driving force for initial divergence. This mechanism may be common, however, and warrants both further empirical and theoretical study.

GENETICS OF REPRODUCTIVE ISOLATION

One may envision coexisting species as sets of alleles in linkage disequilibrium (e.g., Dobzhansky 1937, Mallet 1995, Ortíz-Barrientos et al. 2002). Geneticists can conceive of reinforcement as a way to increase the linkage disequilibrium between genes causing premating and postmating isolation (Felsenstein 1981). Recombination among such loci in a two-allele system therefore opposes reinforcement and leads to fusion. Hence, the genetic basis of reproductive isolation can easily impact the probability or speed of attaining reinforcement. Various theoretical models have examined how the genetics of reproductive isolation (both premating and postmating) can impact reinforcement.

Most empirical studies of the genetics of reproductive isolation have been descriptive and focused on specific issues: the number and relative distribution (especially X versus autosome) of reproductive isolation genes, the interactions causing the preferential sterility of the heterogametic sex (Haldane's Rule) and whether the same genes control male and female components of sexual isolation. This abundant data can be applied to evaluate predictions of theoretical models of the reinforcement process.

Numbers and Distribution of Genes

Results of theoretical models are inconsistent in their predictions of how robust the probability of reinforcement may be to the differences in numbers of genes controlling premating isolation. Models where each mating character is expressed by a single locus (Felsenstein 1981, Kelly & Noor 1996, Servedio & Kirkpatrick 1997) and by multiple loci (e.g., Liou & Price 1994) both find reinforcement under certain conditions. Some models obtain results robust to control by any number of loci as long as selection and migration rates are low (Kirkpatrick 2000, 2001; Kirkpatrick & Servedio 1999), but these assumptions may often not be met in nature. In contrast, several studies that varied the numbers of loci controlling mating traits found speciation to be more difficult (or take longer) with increasing numbers of loci controlling mating (Dieckmann & Doebeli 1999, Kondrashov & Kondrashov 1999). Reflecting the inconsistency of the theoretical models, high-resolution empirical studies have estimated dramatically different numbers of loci affecting facets of premating isolation (e.g., Doi et al. 2001, Ting et al. 2001).

Kirkpatrick & Servedio (1999) varied the number of loci interacting to produce hybrid incompatibilities during reinforcement. They found that although having more loci per interaction leads to more preference evolution, this effect might not be very large (Kirkpatrick & Servedio 1999). A similar effect can result from increasing the number of loci involved in ecological selection during sympatric speciation (but see Dieckmann & Doebeli 1999, Kirkpatrick 2001, Kondrashov & Kondrashov 1999). The chromosomal distribution of genes contributing to reproductive isolation may affect the probability of reinforcement more profoundly than the number of loci, however. For example, the placement of genes on sex chromosomes versus autosomes changes the likelihood of reinforcement. Kelly & Noor's (1996) study, based on the biology of *Drosophila*, is unique in considering both the effects of sex linkage and sex-limited expression. They find that a combination of male limited fertility reduction and X-autosome epistasis (but not one or the other characteristic) is favorable to reinforcement. Both of these conditions hold in *Drosophila*, and evidence of reinforcement has been found in *Drosophila* in several studies (Coyne & Orr 1989, 1997; Noor 1997b).

High-resolution genetic studies of hybrid dysfunctions have shown that they result from deleterious interactions between genes on the X-chromosome and autosomes (Noor et al. 2001a, Orr & Irving 2001; see also Turelli & Orr 1995), genes on the Y-chromosome and autosomes (Lamnissou et al. 1996, Pantazidis et al. 1993), nuclear and cytoplasmic genes (Rawson & Burton 2002, Willett & Burton 2001), and genes on the X- and Y-chromosomes (Orr 1987). However, finer-scale studies have illustrated that seemingly simple interactions causing sterility may actually involve three or more loci (e.g., Carvajal et al. 1996, Palopoli & Wu 1994), even in taxa that diverged less than 200,000 years ago (Orr & Irving 2001). This observation is consistent with mathematical models of the accumulation of incompatibilities causing hybrid sterility. Certain paths to the evolution of hybrid sterility are barred because they require intermediate genotypes that are also sterile. However, the proportion of paths that face this impediment decreases with the complexity of the genetic basis of sterility (Orr 1995). Thus, three-gene interactions causing hybrid sterility can evolve more easily than two-gene interactions, and so on.

These empirical observations are consistent with what appear to be "best-case" scenarios for reinforcement based on the models of Kelly & Noor (1996) and Kirkpatrick & Servedio (1999). To some extent, the models' predictions of greater reinforcement may stem from the fraction of F_2 or backcross hybrids that would be sterile or inviable, both increasing with the conditions outlined. As more hybrids are sterile, the selection intensity for reinforcement will necessarily be stronger. Further theoretical analyses are needed to supplement these initial theoretical studies. For example, the Kirkpatrick & Servedio (1999) model assumes that all hybrid mixtures of alleles at sterility-conferring loci produce equivalent hybrid sterility, though empirical data suggest incompatibilities tend to be complex with only certain combinations causing sterility.

Linkage Between Premating Isolation and Postmating Isolation

As discussed extensively above, many theoretical studies have suggested that linkage between premating and postmating isolation can enhance the probability of reinforcement, particularly in "two-allele" models (Felsenstein 1981, Servedio 2000, Trickett & Butlin 1994; see review in Ortíz-Barrientos et al. 2002). Servedio & Sætre (2003) point out that sex linkage will exaggerate this effect, because, in the heterogametic sex, recombination is absent and recessive alleles involved in postmating isolation are expressed immediately. Tighter linkage and sex linkage both not only facilitate the evolution of premating divergence, but can strengthen amounts of intrinsic postzygotic isolation as well (Servedio & Sætre 2003).

Linkage between genes affecting premating and postmating isolation has been identified in some taxa (e.g., Noor et al. 2001b, Sætre et al. 2003). In the case of *D. pseudoobscura* and *D. persimilis*, all forms of reproductive isolation map to three regions that are inverted between the two species (Noor et al. 2001b). Noor et al. (2001b) suggested that chromosomal rearrangements prevent fusion of hybridizing species and facilitate reinforcement through creating this linkage. Similarly, Sætre et al. (2003) find sex linkage on the Z chromosome for both sexual trait and postmating isolation alleles in Ficedula flycatchers. In hybridizations of *D. mojavensis* and *D. arizonae*, the largest effects on both hybrid sterility and hybrid male sexual isolation map to the fourth chromosome (Pantazidis et al. 1993, Zouros 1981, Zouros et al. 1988). Interestingly, all of these species pairs are proposed to have diverged in part via reinforcement (Noor 1995, Sætre et al. 1997, Wasserman & Koepfer 1977). Such linkage is not universal, however, such that sexual isolation and hybrid sterility are not linked in various allopatric species (e.g., Coyne 1992, 1996).

Nonetheless, simple genetic mapping studies of female species preferences and male hybrid sterility are not necessarily comparable. Alleles causing hybrid male sterility often bear recessive effects (Turelli & Orr 1995), making X-chromosomal sterility alleles easier to identify than autosomal ones. Backcross hybrids can be used to compare hemizygous X-chromosomes from one species with those from the other in any possible autosomal background. In contrast, F_1 male sterility precludes forming hybrid females that are homozygous for X-chromosomes from one species and homozygous for autosomes from the other (as could be formed from an F_2 cross), so recessive interaction effects are more difficult to detect. More appropriate comparisons may be hybrid male sterility with preferred male characters or male preferences.

Linkage Between Male and Female Components of Premating Isolation

In classic theoretical studies, Lande (1981) and Kirkpatrick (1982) showed that female preferences could evolve as a correlated response to selection on males, occurring even with free recombination. Decreased recombination between male trait and female preference loci has been shown, however, to increase the likelihood of speciation (Servedio 2000, Trickett & Butlin 1994), hence possibly facilitating reinforcement. While male and female components of sexual isolation are linked in *D. pseudoobscura* and *D. persimilis* (Noor et al. 2001b), this linkage is not observed in *D. simulans* and *D. mauritana* (Coyne 1996), between the races of *D. melanogaster* (Ting et al. 2001), or between the pheromone races of moth *Ostrinia nubilalis* (Lofstedt et al. 1989).

A different form of "preference" and "preferred traits" observed between species is in the selective binding of conspecific sperm or pollen to eggs (see review in Howard 1999). Interestingly, unlike the cases noted above, in some fungi, plants, and sea urchins, cases have been documented where genes that encode sperm bindin are either the same as or linked to genes that encode its egg-surface receptor (reviewed in Swanson & Vacquier 2002).

What is Needed Now?

Both theory and data are often lacking with regard to the genetic basis of species mating preferences, hybrid dysfunctions, and their effects on the probability or progress of reinforcement. Much available data are exclusive to a small number of Drosophila species, so information from unrelated taxa may be especially interesting. More data on the genetic basis of these traits in taxa with female heterogamety may be particularly enlightening. Sætre et al.'s (2003) observation of disproportionate effects of the Z-chromosome is consistent with extensive empirical evidence of sex linkage of species differences in Lepidoptera (Iyengar et al. 2002; see reviews in Sperling 1994, Prowell 1998). Unfortunately, similar genetic studies in birds are rare and often very low in resolution, particularly with regard to hybrid dysfunctions (see review in Grant & Grant 1997).

CONCLUSIONS

The questions of whether reinforcement can or does happen no longer apply—theoretical studies have demonstrated a range of general conditions under which reinforcement may be expected, and empirical studies have yielded numerous examples that it does occur. As in the classic neutralist-selectionist debate, we now face the more daunting challenge of determining the frequency and importance of reinforcement relative to other means of speciation. We may now return to the question posed in the introduction: whether, according to theoretical predictions, we have looked for reinforcement in the most promising systems.

In short, no. There are several areas where reinforcement should be investigated that have not received much attention. First, most empirical studies have focused on intrinsic genetic incompatibilities as being the driving force for reinforcing selection and considered reinforcement to be impossible without overt evidence of hybrid sterility or inviability. We outlined several alternative factors preventing the full success of interspecies matings; these can also drive reinforcement in the broad sense. Potentially harder to deal with is that reinforcement is only studied in systems where it can leave a signature. Because such systems do not necessarily possess the conditions most conducive to reinforcement, we may be missing many cases by this practice.

Beyond this, confounding factors that may reduce the frequency or importance of reinforcement have not been adequately investigated. For example, divergent selection between populations on mating cues can lead to the appearance of reinforcement despite a relative lack of selection against interspecies matings. Direct selection on nonrandom mating genes can have a similar effect. Unfortunately, very little empirical data exists on the frequency of either of these types of selection. Finally, though not discussed above, hybrids may sometimes be more fit than one or more parental types under certain ecological conditions (Levitan 2002, Pfennig & Simovich 2002; see review by Arnold & Hodges 1995).

We note several other considerations that might help to assess the potential for reinforcement to be common. For example, identifying a high frequency of one-allele systems of nonrandom mating would suggest that reinforcement could occur easily in a wide range of taxa. Gaining a better understanding of the genetic architecture of speciation would also help to determine whether conditions favorable to reinforcement, such as sex linkage of postmating and premating isolating factors, are prevalent. However to identify specific cases of reinforcement, or assess its frequency, we must do more than determine when conditions for it are favorable. We must look at a subset of these cases in more detail with selection, temporal and/or biogeographic studies, to identify whether reinforcement has occurred.

Synthesizing theoretical predictions with empirical data is the means of addressing broad questions about the frequency of any evolutionary process. Intrinsic barriers exist between theory and empirical investigations: no model can capture all the complexities of a natural system without making itself so specific as to be uninformative for any other system. Compromises must be made in theoretical studies to capture generalities observed in numerous systems. Similar allowances must be made in the interpretation of empirical studies—investigators must consider the likelihood of the observed outcome in their particular system and use caution in making generalizations from their results. With these considerations in mind, further integrations of future theoretical and empirical work are necessary for understanding the frequency and importance of reinforcement in speciation.

ACKNOWLEDGMENTS

We thank Jerry Coyne, Mark Kirkpatrick, Lindy McBride, Trevor Price, Brad Shaffer, Michael Turelli, and John Wiens for comments on the manuscript. MRS is supported by NSF grant DEB 0234849. MAFN is supported by NSF grants DEB 9980797 and 0211007.

The *Annual Review of Ecology, Evolution, and Systematics* is online at http://ecolsys.annualreviews.org

LITERATURE CITED

Albuquerque GS, Tauber CA, Tauber MJ. 1996. Postmating reproductive isolation between *Chrysopa quadripunctata* and *Chrysopa slossonae*: Mechanisms and geographic variation. *Evolution* 50:1598–606

Arnold ML, Hodges SA. 1995. Are natural hybrids fit or unfit relative to their parents. *Trends Ecol. Evol.* 10:67–71

Balkau BJ, Feldman MW. 1973. Selection for migration modification. *Genetics* 74:171–74

Berlocher SH, Feder JL. 2002. Sympatric speciation in phytophagous insects: Moving beyond controversy? *Annu. Rev. Entomol.* 47:773–815

Bigelow RS. 1965. Hybrid zones and reproductive isolation. *Evolution* 19:449–58

Bordenstein SR, Drapeau MD, Werren JH. 2000. Intraspecific variation in sexual isolation in the jewel wasp *Nasonia. Evolution* 54:567–73

Boughman JW. 2001. Divergent sexual selection enhances reproductive isolation in sticklebacks. *Nature* 411:944–48

Buckley PA. 1969. Disruption of species-typical behavior patterns in F1 hybrid *Agapornis* parrots. *Zeit. Tierpsychologie* 26:737–47

Butlin R. 1987. Speciation by reinforcement. *Trends Ecol. Evol.* 2:8–13

Cain ML, Andreasen V, Howard D. 1999. Reinforcing selection is effective under a relatively broad set of conditions in a mosaic hybrid zone. *Evolution* 53:1343–53

Carvajal AR, Gandarela MR, Naveira HF. 1996. A three-locus system of interspecific incompatibility underlies male inviability in hybrids between *Drosophila buzzatii* and D-koepferae. *Genetica* 98:1–19

Connell JH. 1983. On the prevalence and relative importance of interspecific competition: evidence from field experiments. *Am. Nat.* 122:661–96

Cox RT, Carlton CE. 1991. Evidence of genetic dominance of the 13-year life-cycle in periodical cicadas (Homoptera, Cicadidae, *Magicicada* Spp). *Am. Midl. Nat.* 125:63–74

Coyne JA. 1992. Genetics of sexual isolation in females of the *Drosophila simulans* species complex. *Genet. Res. Camb.* 60:25–31

Coyne JA. 1996. Genetics of sexual isolation in male hybrids of *Drosophila simulans* and *D. mauritiana. Genet. Res. Camb.* 68:211–20

Coyne JA, Meyers W, Crittenden AP, Sniegowski P. 1993. The fertility effects of pericentric inversions in *Drosophila melanogaster. Genetics* 134:487–96

Coyne JA, Orr HA. 1989. Patterns of speciation in *Drosophila. Evolution* 43:362–81

Coyne JA, Orr HA. 1997. "Patterns of speciation in *Drosophila*" revisited. *Evolution* 51:295–303

Coyne JA, Orr HA. 1998. The evolutionary genetics of speciation. *Philos. Trans. R. Soc. Ser. B* 353:287–305

Davies N, Aiello A, Mallet J, Pomiankowski A, Silberglied RE. 1997. Speciation in two neotropical butterflies: extending Haldane's rule. *Proc. R. Soc. London Ser. B* 264:845–51

Dieckmann U, Doebeli M. 1999. On the origin of species by sympatric speciation. *Nature* 400:354–57

Dobzhansky T. 1937. *Genetics and the Origin of Species.* New York: Columbia Univ. Press

Doi M, Matsuda M, Tomaru M, Matsubayashi H, Oguma Y. 2001. A locus for female discrimination behavior causing sexual isolation in *Drosophila. Proc. Natl. Acad. Sci. USA* 98:6714–19

Drossel B, McKane A. 2000. Competitive speciation in quantiative genetic models. *J. Theor. Biol.* 204:467–78

Feder JL, Opp SB, Wlazlo B, Reynolds K, Go W, Spisak S. 1994. Host fidelity is an effective premating barrier between sympatric races of the apple magot fly. *Proc. Natl. Acad. Sci. USA* 91:7990–94

Felsenstein J. 1981. Skepticism towards Santa Rosalia, or why are there so few kinds of animals? *Evolution* 35:124–38

Fisher RA. 1930. *The Genetical Theory of Natural Selection.* Oxford, UK: Clarendon

Fishman L, Willis JH. 2001. Evidence for Dobzhansky-Muller incompatibilities contributing to the sterility of hybrids between *Mimulus guttatus* and *M. nasutus. Evolution* 55:1932–42

Fishman L, Wyatt R. 1999. Pollinator-mediated competition, reproductive character displacement, and the evolution of selfing in *Arenaria uniflora* (Caryophyllaceae). *Evolution* 53:1723–33

Gerhardt HC. 1994. Reproductive character displacement of female mate choice in the grey treefrog, *Hyla chrysoscelis. Anim. Behav.* 47:959–69

Gil D, Gahr M. 2002. The honesty of bird

song: multiple constraints for multiple traits. *Trends Ecol. Evol.* 17:133–41

Grafen A. 1990. Biological signals as handicaps. *J. Theor. Biol.* 144:517–46

Grant BR, Grant PR. 1998. Hybridization and speciation in Darwin's finches: the role of sexual imprinting on a culturally transmitted trait. See Howard & Berlocher 1998, pp. 404–22

Grant PR. 1986. *Ecology and Evolution of Darwin's Finches*. Princeton, NJ: Princeton Univ. Press. 458 pp.

Grant PR, Grant BR. 1997a. Genetics and the origin of bird species. *Proc. Natl. Acad. Sci. USA* 94:7768–75

Grant PR, Grant BR. 1997b. Hybridization, sexual imprinting, and mate choice. *Am. Nat.* 149:1–28

Hatfield T, Schluter D. 1999. Ecological speciation in sticklebacks: Environment-dependent hybrid fitness. *Evolution* 53: 866–73

Helbig AJ, Salomon M, Bensch S, Seibold I. 2001. Male-biased gene flow across an avian hybrid zone: evidence from mitochondrial and microsatellite DNA. *J. Evol. Biol.* 14:277–87

Higashi M, Takimoto G, Yamamura N. 1999. Sympatric speciation by sexual selection. *Nature* 402:523–26

Hobel G, Gerhardt HC. 2003. Reproductive character displacement in the acoustic system of green treefrogs (*Hyla cinerea*). *Evolution.* 57:894–904

Holynska M. 2000. Is the spinule pattern on the leg 4 coxopodite a tactile signal in the specific mate recognition system of *Mesocyclops* (Copepoda, Cyclopidae)? *Hydrobiologia* 417:11–24

Howard DJ. 1993. Reinforcement: Origin, dynamics, and fate of an evolutionary hypothesis. In *Hybrid Zones and the Evolutionary Process*, ed. RG Harrison, pp. 46–69. New York: Oxford Univ. Press

Howard DJ. 1999. Conspecific sperm and pollen precedence and speciation. *Annu. Rev. Ecol. Syst.* 30:109–32

Howard DJ, Berlocher SH. 1998. *Endless Forms, Species and Speciation*. New York: Oxford Univ. Press

Irwin DE, Price T. 1999. Sexual imprinting, learning and speciation. *Heredity* 82:347–54

Iyengar VK, Reeve HK, Eisner T. 2002. Paternal inheritance of a females moth's mating preference. *Nature* 419:830–832

Jiggins CD, Naisbit RE, Coe RL, Mallet J. 2001. Reproductive isolation caused by colour pattern mimicry. *Nature* 411:302–5

Kawata M, Yoshimura J. 2000. Speciation by sexual selection in hybridizing populations without viability selection. *Evol. Ecol. Res.* 2:897–909

Kawecki TJ. 1996. Sympatric speciation driven by beneficial mutations. *Proc. R. Soc. London Ser. B* 263:1515–20

Kawecki TJ. 1997. Sympatric speciation via habitat specialization driven by deleterious mutations. *Evolution* 51:1751–63

Kelly JK, Noor MAF. 1996. Speciation by reinforcement: A model derived from studies of *Drosophila*. *Genetics* 143:1485–97

King M. 1993. *Species Evolution: The Role of Chromosome Change*. Cambridge: Cambridge Univ. Press

Kirkpatrick M. 1982. Sexual selection and the evolution of female choice. *Evolution* 36:1–12

Kirkpatrick M. 2000. Reinforcement and divergence under assortative mating. *Proc. R. Soc. London Ser. B* 267:1649–55

Kirkpatrick M. 2001. Reinforcement during ecological speciation. *Proc. R. Soc. London Ser. B* 268:1259–63

Kirkpatrick M, Barton NH. 1997. The strength of indirect selection on female mating preferences. *Proc. Natl. Acad. Sci. USA* 94:1282–86

Kirkpatrick M, Ravigné V. 2002. Speciation by natural and sexual selection: models and experiments. *Am. Nat.* 159:S22–S35

Kirkpatrick M, Servedio MR. 1999. The reinforcement of mating preferences on an island. *Genetics* 151:865–84

Kondrashov AS. 1983. Multilocus model of sympatric speciation I. One character. *Theor. Popul. Biol.* 24:121–35

Kondrashov AS, Kondrashov FA. 1999. Interactions among quantitative traits in the course of sympatric speciation. *Nature* 400:351–54

Kondrashov AS, Shpak M. 1998. On the origin of species by means of assortative mating. *Proc. R. Soc. London Ser. B* 265:2273–78

Lambrechts M, Dhondt AA. 1988. The anti-exhaustion hypothesis—a new hypothesis to explain song performance and song switching in the great tit. *Anim. Behav.* 36:327–34

Lamnissou K, Loukas M, Zouros E. 1996. Incompatibilities between Y chromosome and autosomes are responsible for male hybrid sterility in crosses between *Drosophila virilis* and *Drosophila texana*. *Heredity* 26:603–9

Lande R. 1981. Models of speciation by sexual selection on polygenic traits. *Proc. Natl. Acad. Sci. USA* 78:3721–25

Levitan DR. 2002. The relationship between conspecific fertilization success and reproductive isolation among three congeneric sea urchins. *Evolution* 56:1599–609

Liou LW, Price TD. 1994. Speciation by reinforcement of premating isolation. *Evolution* 48:1451–59

Lloyd M, Kritsky G, Simon C. 1983. A simple mendelian model for 13-year and 17-year life-cycles of periodical cicadas, with historical evidence of hybridization between them. *Evolution* 37:1162–80

Lofstedt C, Hansson BS, Roelofs W, Bengtsson BO. 1989. No linkage between genes-controlling female pheromone production and male pheromone response in the european corn-borer, Ostrinia-nubilalis Hubner (Lepidoptera, Pyralidae). *Genetics* 123:553–56

Loftus-Hills JJ, Littlejohn ML. 1992. Reinforcement and reproductive character displacement in *Gastrophryne carolinensis* and *G. olivacea* (Anura: Microhylidae): A reexamination. *Evolution* 46:896–906

Lorenz KZ. 1952. *King Solomon's Ring*. New York: Crowell

Lynch A, Baker AJ. 1994. A population mimetics approach to cultural evolution in chaffinch song: differentiation among populations. *Evolution* 48:351–59

Mallet J. 1995. A species definition for the Modern Sythesis. *Trends Ecol. Evol.* 10:294–99

Mallet J. 1999. Causes and consequences of lack of coevolution in Mullerian mimicry. *Evol. Ecol.* 13:777–806

Markow TA. 1997. Assortative fertilization in *Drosophila*. *Proc. Natl. Acad. Sci. USA* 94:7756–60

McPhail JD. 1984. Ecology and evolution of sympatric sticklebacks (Gasterosteus)—morphological and genetic-evidence for a species pair in Enos Lake, British Columbia. *Can. J. Zool. Rev. Can. Zool.* 62:1402–8

McPhail JD. 1992. Ecology and evolution of sympatric sticklebacks (Gasterosteus)—evidence for a species-pair in Paxton Lake, Texada Island, British Columbia. *Can. J. Zool. Rev. Can. Zool.* 70:361–69

Michalak P, Grzesik J, Rafinski J. 1997. Tests for sexual incompatibility between two newt species, *Triturus vulgaris* and *Triturus montandioni*: No-choice mating design. *Evolution* 51:2045–50

Moore JA. 1957. An embryologist's view of the species concept. In *The Species Problem*, ed. E Mayr, pp. 325–38. Washington, D.C.: Am. Assoc. Adv. Sci.

Muller HJ. 1942. Isolating mechanisms, evolution and temperature. *Biol. Symp.* 6:71–125

Mundinger PC. 1982. Microgeographic and macrogeographic variation in acquired vocalizations in birds. In *Acoustic Communication in Birds*, ed. DE Kroodsma, EH Miller, 2:147–208. New York: Academic

Nagel L, Schluter D. 1998. Body size, natural selection, and speciation in sticklebacks. *Evolution* 52:209–18

Naisbit RE, Jiggins CD, Mallet J. 2001. Disruptive sexual selection against hybrids contributes to speciation between *Heliconius cydno* and *Heliconius melpomene*. *Proc. R. Soc. London Ser. B* 268:1849–54

Navarro A, Barton NH. 2003. Accumulating postzygotic isolation genes in parapatry: a new twist on chromosomal speciation. *Evolution* 57:447–59

Noor MAF. 1995. Speciation driven by natural selection in *Drosophila*. *Nature* 375:674–75
Noor MAF. 1997a. Genetics of sexual isolation and courtship dysfunction in male hybrids of *Drosophila pseudoobscura* and *D. persimilis*. *Evolution* 51:809–15
Noor MAF. 1997b. How often does sympatry affect sexual isolation in *Drosophila*? *Am. Nat.* 149:1156–63
Noor MAF. 1999. Reinforcement and other consequences of sympatry. *Heredity* 83:503–8
Noor MAF, Grams KL, Bertucci LA, Almendarez Y, Reiland J, Smith KR. 2001a. The genetics of reproductive isolation and the potential for gene exchange between *Drosophila pseudoobscura* and *D. persimilis* via backcross hybrid males. *Evolution* 55:512–21
Noor MAF, Grams KL, Bertucci LA, Reiland J. 2001b. Chromosomal inversions and the reproductive isolation of species. *Proc. Natl. Acad. Sci. USA* 98:12084–88
Nosil P, Crespi BJ, Sandoval CP. 2003. Reproductive isolation driven by the combined effects of ecological adaptation and reinforcement. *Proc. R. Soc. London Ser. B*. In press
Orr HA. 1987. Genetics of male and female sterility in hybrids of *Drosophila pseudoobscura* and *D. persimilis*. *Genetics* 116:555–63
Orr HA. 1995. The population genetics of speciation: the evolution of hybrid incompatibilities. *Genetics* 139:1805–13
Orr HA, Irving S. 2001. Complex epistasis and the genetic basis of hybrid sterility in the *Drosophila pseudoobscura* Bogota-USA hybridization. *Genetics* 158:1089–100
Ortíz-Barrientos D, Reiland J, Hey J, Noor MAF. 2002. Recombination and the divergence of hybridizing species. *Genetica* 116:167–78
Palopoli MF, Wu C-I. 1994. Genetics of hybrid male sterility between *Drosophila* sibling species: a complex web of epistasis is revealed in interspecific studies. *Genetics* 138:329–41
Pantazidis AC, Galanopoulos VK, Zouros E. 1993. An autosomal factor from *Drosophila arizonae* restores spermatogenesis in *Drosophila mojavensis* males carrying the *D. arizonae* Y chromosome. *Genetics* 134:309–18
Pashley DP, Martin JA. 1987. Reproductive incompatibility between host strains of the fall armyworm (Lepidoptera: Noctuidae). *Ann. Ent. Soc. Amer.* 80:731–33
Payne RJH, Krakauer DC. 1997. Sexual selection, space, and speciation. *Evolution* 51:1–9
Pfennig KS, Simovich MA. 2002. Differential selection to avoid hybridization in two toad species. *Evolution* 56:1840–48
Podos J. 2001. Correlated evolution of morphology and vocal signal structure in Darwin's finches. *Nature* 409:185–88
Presgraves DC. 2002. Patterns of postzygotic isolation in Lepidoptera. *Evolution* 56:1168–83
Price TD. 2003. Causes of post-mating reproductive isolation in birds. *Acta Zoologica Sinica*. In press
Price TD, Bouvier MM. 2002. The evolution of F1 postzygotic incompatibilities in birds. *Evolution* 56:2083–89
Prowell DP. 1998. Sex linkage and speciation in Lepidoptera. See Howard & Berlocher 1998, pp. 309–19
Rawson PD, Burton RS. 2002. Functional coadaptation between cytochrome c and cytochrome c oxidase within allopatric populations of a marine copepod. *Proc. Natl. Acad. Sci. USA* 99:12955–58
Rice WR. 1987. Speciation via habitat specialization: the evolution of reproductive isolation as a correlated character. *Evol. Ecol.* 1:301–14
Rundle HD. 2002. A test of ecologically dependent postmating isolation between sympatric sticklebacks. *Evolution* 56:322–29
Rundle HD, Schluter D. 1998. Reinforcement of stickleback mating preferences: Sympatry breeds contempt. *Evolution* 52:200–8
Rundle HD, Whitlock MC. 2001. A genetic interpretation of ecologically dependent isolation. *Evolution* 55:198–201

Sætre GP, Borge T, Lindroos K, Haavie J, Sheldon BC, et al. 2003. Sex chromosome evolution and speciationin *Ficedula* flycatchers. *Proc. R. Soc. London Ser. B* 270:53–59

Sætre GP, Kral M, Bures S, Ims RA. 1999. Dynamics of a clinal hybrid zone and a comparison with island hybrid zones of flycatchers (*Ficedula hypoleuca* and *F-albicollis*). *J. Zool.* 247:53–64

Sætre G-P, Moum T, Bures S, Kral M, Adamjan M, Moreno J. 1997. A sexually selected character displacement in flycatchers reinforces premating isolation. *Nature* 387:589–92

Sanderson N. 1989. Can gene flow prevent reinforcement? *Evolution* 43:1223–35

Sasa MM, Chippindale PT, Johnson NA. 1998. Patterns of postzygotic isolation in frogs. *Evolution* 52:1811–20

Schilthuizen M, Lombaerts M. 1995. Life on the edge—a hybrid zone in *Albinaria-hippolyti* (Gastropoda, Clausiliidae) from Crete. *Biol. J. Linn. Soc.* 54:111–38

Schluter D. 2001. Ecology and the origin of species. *Trends Ecol. Evol.* 16:372–80

Schoener TW. 1983. Field experiments on interspecific competition. *Am. Nat.* 122:240–85

Searcy WA, Yasukawa K. 1996. The reproductive success of secondary females relative to that of monogamous and primary females in Red-winged Blackbirds. *J. Avian Biol.* 27:225–30

Servedio MR. 2000. Reinforcement and the genetics of nonrandom mating. *Evolution* 54:21–29

Servedio MR. 2001. Beyond reinforcement: The evolution of premating isolation by direct selection on preferences and postmating, prezygotic incompatibilities. *Evolution* 55:1909–20

Servedio MR, Kirkpatrick M. 1997. The effects of gene flow on reinforcement. *Evolution* 51:1764–72

Servedio MR, Saetre GP. 2003. Speciation as a positive feedback loop between post- and prezygotic barriers to gene flow. *Proc. R. Soc. London Ser. B.* In press

Slabbekoorn H, Smith TB. 2002. Bird song, ecology and speciation. *Philos. Trans. R. Soc. London Ser. B* 357:493–503

Sota T, Kubota K. 1998. Genital lock-and-key as a selective agent against hybridization. *Evolution* 52:1507–13

Sperling FAH. 1994. Sex-linked genes and species-differences in Lepidoptera. *Can. Entomol.* 126:807–18

Swanson WJ, Vacquier VD. 2002. The rapid evolution of reproductive proteins. *Nat. Rev. Genet.* 3:137–44

Ting C-T, Takahashi A, Wu C-I. 2001. Incipient speciation by sexual isolation in Drosophila: Concurrent evolution at multiple loci. *Proc. Natl. Acad. Sci. USA* 98:6709–13

Trickett AJ, Butlin RK. 1994. Recombination suppressors and the evolution of new species. *Heredity* 73:339–45

Turelli M, Barton NH, Coyne JA. 2001. Theory and speciation. *Trends Ecol. Evol.* 16:330–43

Turelli M, Orr HA. 1995. The dominance theory of Haldane's Rule. *Genetics* 140:389–402

Turelli M, Orr HA. 2000. Dominance, epistasis and the genetics of postzygotic isolation. *Genetics* 154:1663–79

Vamosi SM, Schluter D. 1999. Sexual selection against hybrids between sympatric stickleback species: Evidence from a field experiment. *Evolution* 53:874–79

van Doorn GS, Noest AJ, Hogeweg P. 1998. Sympatric speciation and extinction driven by environment dependent sexual selection. *Proc. R. Soc. London Ser. B* 265:1915–19

Via S, Hawthorne DJ. 2002. The genetic architecture of ecological specialization: Correlated gene effects on host use and habitat choice in pea aphids. *Am. Nat.* 159:S76–S88

Volterra V. 1926. Variations and flucutations in the numbers of individuals of animal species living together. *Nature* 118:558–60

Vos DR. 1995. The role of sexual imprinting for sex recognition in zebra finches: a difference between males and females. *Anim. Behav.* 50:645–53

Walker TJ. 1974. Character displacement and acoustic insects. *Amer. Zool.* 14:1137–50

Wasserman M, Koepfer HR. 1977. Character displacement for sexual isolation between

Drosophila mojavensis and *Drosophila arizonensis*. *Evolution* 31:812–23

White MJD. 1969. Chromosomal rearrangements and speciation in animals. *Annu. Rev. Genet.* 3:75–98

Williams MA, Blouin AG, Noor MAF. 2001. Courtship songs of *Drosophila pseudoobscura* and *D. persimilis*. II. Genetics of species differences. *Heredity* 86:68–77

Willett CS, Burton RS. 2001. Viability of cytochrome C depends on cytoplasmic backgrounds in *Tigriopus californicus*. *Evolution* 55:1592–99

Wu C-I, Davis AW. 1993. Evolution of postmating reproductive isolation: The composite nature of Haldane's Rule and its genetic bases. *Am. Nat.* 142:187–212

Wullschleger EB, Wiehn J, Jokela J. 2002. Reproductive character displacement between the closely related freshwater snails *Lymnaea peregra* and *L. ovata*. *Evol. Ecol. Res.* 4:247–57

Yoshimura J, Clark CW. 1994. Population-dynamics of sexual and resource competition. *Theor. Popul. Biol.* 45:121–31

Zouros E. 1981. The chromosomal basis of sexual isolation in two sibling species of Drosophila: *D. arizonensis* and *D. mojavensis*. *Genetics* 97:703–18

Zouros E, Lofdahl K, Martin PA. 1988. Male hybrid sterility in *Drosophila*: Interactions between autosomes and sex chromosomes in crosses of *D. mojavensis* and *D. arizonensis*. *Evolution* 42:1321–31

Annu. Rev. Ecol. Evol. Syst. 2003. 34:365–96
doi: 10.1146/annurev.ecolsys.34.011802.132439

First published online as a Review in Advance on September 15, 2003

EXTRA-PAIR PATERNITY IN BIRDS: Causes, Correlates, and Conflict

David F. Westneat and Ian R.K. Stewart
Department of Biology, University of Kentucky, Lexington, Kentucky 40506-0225; email: biodfw@uky.edu

Key Words sexual selection, reproductive strategy, extra-pair copulation, comparative approach, selection theory, game theory

■ **Abstract** Extra-pair paternity (EPP) is extremely variable among species of birds, both in its frequency and in the behavioral events that produce it. A flood of field studies and comparative analyses has stimulated an array of novel ideas, but the results are limited in several ways. The prevailing view is that EPP is largely the product of a female strategy. We evaluate what is known about the behavioral events leading to EPP and find the justification for this view to be weak. Conflict theory (derived from selection theory) predicts that adaptations in all the players involved will influence the outcome of mating interactions, producing complex and often highly variable patterns of behavior and levels of EPP. Data support some of these predictions, but alternative hypotheses abound. Tests of predictions from conflict theory will require better information on how males and females encounter one another, behave once they have met, and influence fertilization once insemination has occurred.

INTRODUCTION

More than 30 years ago, Trivers (1972) made two bold predictions regarding the consequences of anisogamy for the mating behavior of each sex: (*a*) Males should behave in ways that increase their opportunities for additional matings, and (*b*) females should choose a mate that increases the genetic quality of their offspring. At the time, birds seemed a taxon that generally refuted both predictions. After a review of avian breeding systems, Lack (1968) had declared that more than 90% of the 9000+ species were monogamous, with strong social associations between members of a pair and high levels of biparental care. Thus, male birds apparently had little opportunity for additional copulations, and female mating behavior appeared to be directed toward finding a suitable place for breeding or a partner for raising young rather than a high-quality sire. In birds, some of the selective forces arising from anisogamy seemed to be suppressed.

Yet Trivers' predictions could not have been more correct. The past 20 years have seen a revolution in our views of avian reproductive behavior. Advances in genetic techniques have dramatically enhanced the ability to assess the parentage of offspring and thereby uncover actual mating patterns. To date, parentage studies have been published for more than 150 species of birds, and the results have been stunning. In more than 70% of species, at least some offspring are sired by a male other than the social father (reviewed by Griffith et al. 2002).

The genetic studies of extra-pair paternity (EPP) in birds have transformed the study of avian monogamy from what was once termed a "bland" area of research (Mock 1985) into one teeming with intriguing, confusing, and contradictory results. An array of new theoretical and empirical questions has emerged since Gowaty (1985) first outlined the potential impact of EPP, much of which remains controversial. In this review, we summarize some of these new areas of research, critically examine some of the approaches and conclusions to date, and advocate a conceptual approach that emphasizes a balanced perspective of the sexes and stimulates new areas for future research.

COMPARATIVE APPROACHES TO DIVERSITY

One obvious and provocative result emerging from the burgeoning dataset on EPP frequency is the magnitude of interspecific variation. Some of the patterns behind this variation are strikingly evident, whereas others prove more elusive. It is clear, for example, that EPP levels are higher and more variable among passerines than among nonpasserines [15% ± 16% (SD) versus 3% ± 5%, calculated from appendix A of Griffith et al. 2002]. Indeed, interspecific variation in EPP levels clearly has a large phylogenetic component, with more than 55% of the variation occurring at or above the level of family (Arnold & Owens 2002). Nevertheless, substantial variation exists within related birds, as illustrated by the range of EPP found in the swallows, a group of socially monogamous aerial insectivores (Figure 1*a*). Variation in EPP is also inversely related to the estimated branch lengths of the avian phylogeny, indicating more than a phylogenetic influence (Westneat & Sherman

→

Figure 1 Examples of extensive variation in EPP among birds. (*a*) Phylogenetic relationship and EPP in the swallows (Hirundininae). Phylogeny based on Sheldon & Winkler (1993) and depicts relatedness but not likely genetic distance. Abbreviations stand for, in order: *Hirundo pyrrhonota*, *H. rustica*, *Delichon urbica*, *Tachycineta bicolor*, *T. albilinea*, *Riparia riparia*, *Progne subis*. (*b*) Alphabetic listing of cooperative breeders and extra-group paternity (from appendix A in Griffith et al. 2002). Abbreviations stand for, in order: *Acrocephalus sechellensis*, *Aegithalos caudatus*, *Aphelocoma coerulescens*, *Aphelocoma ultramarina*, *Buteo galapagoensis*, *Campylorhynchus griseus*, *C. nuchalis*, *Dacelo novaeguineae*, *Malurus splendens*, *M. cyaneus*, *Manorina melanocephala*, *M. melanophrys*, *Melanerpes formicivorus*, *Merops apiaster*, *M. bullockoides*, *Picoides borealis*, *Porphyrio porphyrio*, *Psaltriparus minimus*, *Sericornus frontalis*, *Sialia mexicana*, *Turdoides squamiceps*. Note: *Aphelocoma ultramarina* has 40% paternity by helpers within the group.

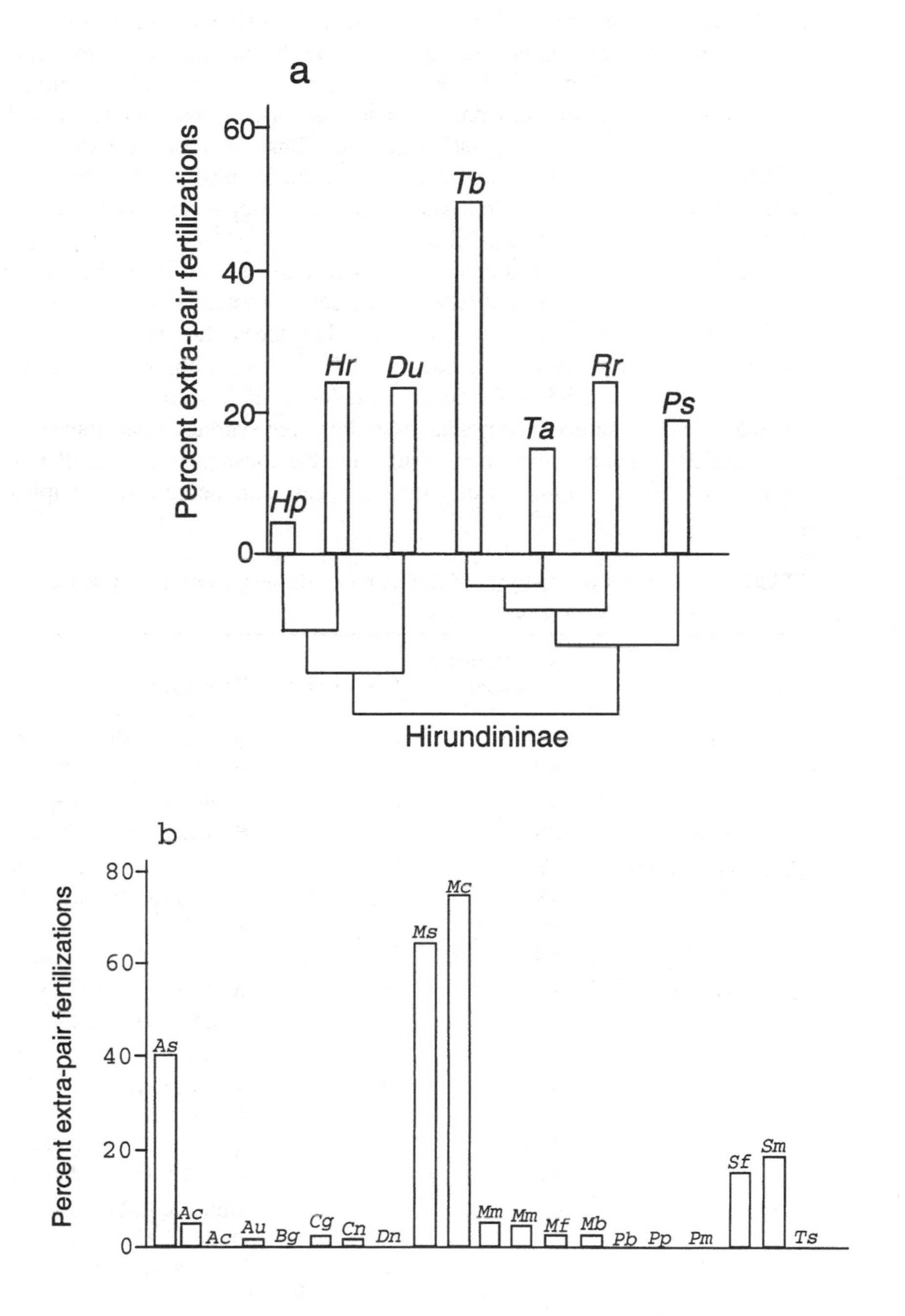
a
Percent extra-pair fertilizations
60
40
20
0
Hp
Hr
Du
Tb
Ta
Rr
Ps
Hirundininae
b
Percent extra-pair fertilizations
80
60
40
20
0
As
Ac
Ac
Au
Bg
Cg
Cn
Dn
Ms
Mc
Mm
Mm
Mf
Mb
Pb
Pp
Pm
Sf
Sm
Ts

1997). Dramatic differences in EPP also exist within groups of species that have similar social mating systems, such as cooperative breeders (Figure 1*b*).

We found 18 comparative studies of potential factors explaining interspecific variation in the level of EPP (Table 1). Several types of variables correlate with EPP across species, ranging from ecological factors, morphological and behavioral traits, and aspects of population genetics. Some of these have very marked effects. Petrie et al. (1998) found that 85% of the interspecific variation in EPP levels was explained by only four variables (study sample size, body size, sexual dichromatism, and genetic variability).

Four points must be considered when interpreting the results of these comparative studies. First, estimates of EPP are subject to substantial sampling variation (Griffith et al. 2002). For example, more than half the studies used by Petrie et al. (1998) have sampling errors that exceed 20%, making it impossible for any set of factors to explain 85% of the actual variation in EPP. Second, it is difficult to separate cause and effect in comparative studies: some variables may have a causal effect on EPP, whereas others may result from the consequences of EPP. For few of the factors listed in Table 1 can one predict, based on theoretical principles, the

TABLE 1 Comparative analyses of EPP in birds and the factors that explain a significant level of variance among species

Factor	% variance explained	N of species	Reference
Phylogeny	21	72	Westneat & Sherman 1997
	55[a]	95	Arnold & Owens 2002
Breeding density	NS	45[b], 64[c]	Westneat & Sherman 1997
	NS	140	Bennett & Owens 2002
Breeding synchrony	30	21	Stutchbury & Morton 1995[d]
	NS	14	Westneat & Sherman 1997
	25	34	Stutchbury 1998b[e]
	NS	140	Bennett & Owens 2002
Adult mortality	47	71	Wink & Dyrcz 1999[f,g]
	25	54	Arnold & Owens 2002[h]
Paternal care	$P < 0.05$[i]	52	Møller & Birkhead 1993[j]
	NS[k]	72	Schwagmeyer et al. 1999
	26	31	Møller 2000
	25	86–112[l]	Møller & Cuervo 2000
	7[m]	87	Arnold & Owens 2002
Sexual dichromatism	9[n]	56	Møller & Birkhead 1994
	49	73	Owens & Hartley 1998
	$p < 0.0001$, r not given	32	Petrie et al. 1998

(*Continued*)

TABLE 1 (*Continued*)

Factor	% variance explained	N of species	Reference
Genetic variability	22	32	Petrie et al. 1998
Sample size	$P < 0.001$, r not given	32	Petrie et al. 1998
Value of male care	30	31	Møller 2000
	22	29	Arnold & Owens 2002
Divorce	35	20	Cezilly & Nager 1995
Clutch size	10[o]	60	Arnold & Owens 2002
Social mating system	26[p], 12[q]	40	Hasselquist & Sherman 2001[r]
Hatching success	NS	58	Morrow et al. 2002
Body mass	$P < 0.0001$, r not given	32	Petrie et al. 1998
Sexual size dimorphism	NS	71	Owens & Hartley 1998
Male plumage brightness	11	56	Møller & Birkhead 1994
Residual testes mass	$P < 0.02$, r not given	53	Møller & Briskie 1995
Sperm length	40	21	Briskie et al. 1997

[a]Nested ANOVA.

[b]Continuous analysis using nearest neighbor distance.

[c]Categorical analysis of colonial versus dispersed nesting.

[d]Comparison of within-genera averages obtained from species-specific data.

[e]See also Stutchbury 1998a, Weatherhead & Yezerinac 1988.

[f]Species-specific data used.

[g]Described by authors as the probability of a pair surviving to breed in the following year.

[h]Result when using molecular phylogeny; 28% when using morphological phylogeny.

[i]From a multiple regression including developmental mode, degree of polygyny and frequency of second broods as nonsignificant variables.

[j]But see Dale 1995, Møller & Birkhead 1995.

[k]No effect of posthatching care, but males tend to undertake a greater share of incubation in species with low EPP.

[l]A composite of four indices of male care. Strongest relationship was with nestling provisioning rate. Degree of nest building, courtship feeding, and incubation all nonsignificant.

[m]Result when using morphological phylogeny. Not significant when using molecular phylogeny.

[n]Male plumage brightness explains 11% of the variation, while female brightness is not significant.

[o]Result when using morphological phylogeny. Marginally nonsignificant when using molecular phylogeny (6%, $P = 0.06$).

[p]When dichotomized as polygynous or monogamous, with EPP rates significantly higher in monogamous species.

[q]When treating mating system as a continuous variable (proportion of males polygynous).

[r]Analysis restricted to passerines.

direction of the causal relationship. Take, for example, the negative association between EPP levels and the amount of male care. A common interpretation of this relationship is that males have a higher benefit to caring if they have higher paternity (e.g., Møller 2000, Møller & Birkhead 1993, Møller & Cuervo 2000). However, increases in the opportunity for gaining EPP during the period of offspring dependency may drive reductions in male care (Westneat et al. 1990). Alternatively, a decline in the value of male care could increase the relative benefits of pursuing EPP to both males and females (Bennett & Owens 2002, Gowaty 1996). Similar problems with cause and effect plague many of the variables related to EPP levels, particularly plumage dichromatism, genetic variability, and duration of the pair bond.

Another reason why so many variables are related to levels of EPP is because many are correlated with each other. For example, species with high levels of male parental care are also likely to have long-term pair bonds, small clutch sizes, low mortality, little sexual dichromatism, and large body size [e.g., many raptors and seabirds (Lack 1968)]. Multiple regression analyses can control for interrelationships between such traits (e.g., Møller 2000, Petrie et al. 1998) and thus uncover independent correlates of EPP. Bennett & Owens (2002) have used such methods to reveal that ancient (>10 mya) phylogenetic diversifications in birds explain a substantial proportion of the variation in interspecific levels of EPP. The evolution of fast life histories (e.g., low adult survival, high fecundity, and little parental care) is also linked to these divisions in phylogeny. Bennett & Owens (2002) suggest that species predisposed to a fast life history necessarily have high levels of EPP and that contemporary ecological factors may have only a minimal effect. However, it is not yet clear how the traits associated with a fast life history (low parental care or high mortality) influence EPP. Some ecological factors could affect both life history and EPP, but how is also not clear. Moreover, such predispositions cannot account for the dramatic level of variation that occurs within families (e.g., swallows, Figure 1*a*) or between populations [e.g., island versus mainland populations (Griffith 2000)]. Also, the variation in which individuals within the same population are involved in EPP cannot be due to general aspects of life history. Bennett & Owens (2002) make the useful suggestion that explanations for EPP may be organized hierarchically, that is, within broad differences in life history (e.g., fast versus long), different types of ecological factors may explain the variation that occurs at more recently separated phylogenetic levels. This is an attractive concept, but it is not yet sufficiently developed to produce testable predictions.

Finally, the frequency of EPP is a population-level variable that emerges from the interactions of the traits of at least three individuals (female, pair male, and extra-pair male; e.g., Lifjeld & Robertson 1992, Lifjeld et al. 1994). The level of EPP can evolve only as those traits change or changes in ecology influence how the players interact. As we show below, EPP is a meta-trait and can arise in a multitude of ways, all of which are glossed over by comparing population EPP frequency. Comparative analyses of the level of EPP per se may therefore be too crude to generate any more novel insights beyond the recent synthesis of Bennett & Owens (2002). Instead, we advocate a closer look at the variation in individual traits associated with EPP across a wide range of species, and subsequent comparative analyses of these traits. This requires us to examine in more detail the within-population variation in the events leading to EPP.

WITHIN-POPULATION VARIATION: AN ANALYSIS OF CONFLICT

The importance of the interactions among the female, the extra-pair male, and the pair male is highlighted by the diversity in behavioral and morphological traits associated with EPP. For example, the frequency of EPP varies from 25% to 35% among populations of red-winged blackbirds (*Agelaius phoeniceus*) (Gray 1996, Westneat 1993a, Weatherhead & Boag 1995), a narrow range in comparison to the variation among species [0%–76% (Griffith et al. 2002)]. Nevertheless, the behavior leading to EPP is strikingly different in these populations, with males initiating extra-pair copulations (EPCs) in eastern populations (Westneat 1992) and females doing so in western ones (Gray 1996). In Canadian populations of common murres (*Uria aalge*), males approach females for EPCs (Birkhead et al. 1985), but in Britain females approach males as well (Hatchwell 1988). Similarly, there is huge variation between species in how males prevent EPCs by their mates (e.g., guard versus copulate frequently) and how they gain them [e.g., advertise versus foray (Birkhead & Møller 1992)].

We suggest that the best approach to understanding this variation is to tease apart the factors influencing the interactions among the three or more parties involved. Our conceptual approach is rooted in sexual conflict theory (Parker 1979, 1984) and extends previous attempts to incorporate conflict into theory about EPP (Gowaty 1996, Lifjeld et al. 1994, Westneat 2000). Sexual conflict has been loosely defined as the differing interests of the sexes (Davies 1992) and modeled using game theory (e.g., Hammerstein & Parker 1987; Parker 1979, 1984). Here we adopt concepts from selection theory (Arnold & Duvall 1994, Arnold & Wade 1984) and extend definitions of conflict (Westneat 2000, Westneat & Sargent 1996) to the interactions producing EPP.

Selection theory focuses on the relationship between trait and fitness. Directional selection on the trait is measured as the slope of that relationship [the selection gradient (Arnold & Wade 1984)]. In social interactions, the trait of one individual is a component of the environment experienced by the other and contributes to fitness variation in that social partner (Rausher 1992, Wolf et al. 1999). Any relationship between one individual's trait and the fitness of its social partner is analogous to the selection gradient (we call this an opportunity gradient) but does not describe selection because the trait affects fitness in an individual who does not necessarily express the trait. Conflict can be quantified as the extent to which an opportunity gradient describing the effect of a trait on a social partner's fitness is of opposite slope from the selection gradient on the trait (Figure 2). Sexual conflict occurs when this interaction is between a male and a female (Westneat 2000).

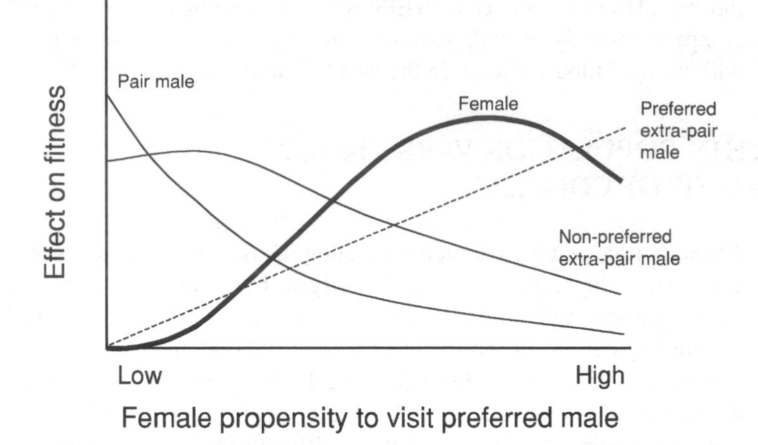

Figure 2 Illustration of sexual conflict arising from female forays. The relationship between female forays and female fitness (*thick line*) is generally increasing except at high values of foraying when benefits may be expected to diminish and costs increase. Preferred extra-pair males experience an increase in fitness as female forays increase (*dashed line*). However, the fitness of non-preferred males or the pair male will decrease (*thin lines*) in different ways as females increase visits to preferred males. Both thin lines depict an opportunity gradient describing sexual conflict on paired or non-preferred extra-pair males due to the propensity of the female to pursue EPCs with preferred extra-pair males.

The opportunity gradient in any conflict, including sexual conflict, produces an opportunity for selection (sensu Crow 1958) favoring any trait that diminishes the impact of the focal trait. A general prediction of this idea is that anytime there is conflict, traits may have arisen that counter traits in social partners. Explicit use of this concept in the case of EPP in birds stimulates a closer focus on the traits of each of the individuals involved and generates an array of novel hypotheses about the factors contributing to EPP.

Female Pursuit

The female is a pivotal player in the interactions leading to EPP. The discovery of frequent EPP in birds and multiple mating by females in other taxa (e.g., Birkhead & Møller 1998) focused much-needed attention on variation in female reproductive behavior and the consequences for female fitness. Unfortunately, there is now a widespread conclusion that EPP is the result of a female mixed reproductive strategy (sensu Trivers 1972), and therefore all adaptive explanations for EPP should focus on female fitness (e.g., Griffith et al. 2002, Petrie & Kempenaers 1998). We review the evidence for this position and explore the implications of sexual conflict for female behavior.

A female can encounter an extra-pair male in two ways: She can pursue EPCs by moving away from the area she normally uses and into areas containing extra-pair males (a foray) or she can stay home and interact with males that initiate encounters by moving to her. To date, there are only 12 species for which we have descriptions of female pursuit (forays leading to encounters with extra-pair males) and genetic evidence of EPP in the same population (Table 2). In several other birds, females initiate EPCs but little or no EPP occurs [e.g., northern fulmar, *Fulmarus glacialis* (Hunter et al. 1992); oystercatcher, *Haematopus ostralegus* (Heg et al. 1993); Humboldt penguin, *Spheniscus humboldti* (Schwartz et al. 1999)]. Radiotracking has been used to document extraterritorial female forays in bluethroats (*Luscinia svecica*) (Smiseth & Amundson 1995) in which EPP is common, although EPCs have only been observed during male forays (Johnsen et al. 1998a,b). Finally, Wagner (1991) observed female razorbills (*Alca torda*), a colonially nesting seabird, moving to traditional mating sites where they encountered extra-pair males. More than 50% of females accepted EPCs (Wagner 1998), but genetic data are not available to document if these result in EPP.

The difficulty with interpreting forays as a female reproductive tactic is that a female could move off territory for reasons other than to copulate with an extra-pair male. For example, female boat-tailed grackles (*Quiscalus major*) leave nesting areas defended by the alpha male and travel to peripheral areas where they gather nesting material. There, males approach them and attempt to mount (Poston 1997). Paternity analyses revealed that often males away from the colony are the sires of offspring (Poston et al. 1999). Given the possible alternative function of foraying for nesting material, it is very difficult in this case to claim that females foray as part of a mixed reproductive strategy. Similar events may occur in other species but could be more cryptic. For example, the demand for specific nutrients during the egg-laying period may prompt females to seek out uncommon food types that are restricted to certain locations (Neudorf et al. 1997). While there, females could encounter extra-pair males. Whether the foray constitutes a mixed reproductive strategy depends on whether benefits of EPP are the main fitness consequence.

No study has explicitly eliminated alternative functions for female forays, but some behavioral observations suggest that their primary function, at least, is to seek copulations. In superb fairy-wrens (*Malurus cyaneus*), for example, radio transmitters were used to track females as they left their mate's territory before dawn and visited other males. Females moved directly to the extra-pair territory and then returned home immediately (Double & Cockburn 2000). Although copulations could not be seen that early in the day, males in the visited territory sired some of the female's offspring. In black-capped chickadees (*Parus atricapillus*) (Smith 1988), blue tits (*Parus caeruleus*) (Kempenaers et al. 1992), and chaffinches (*Fringilla coelebs*) (Sheldon 1994), females make brief forays into other territories, usually those of neighbors, and copulate with the resident male. Females in a Washington population of red-winged blackbirds travel directly to territories of extra-pair males or to sites containing little food or nesting material and solicit copulation (Gray 1997). In razorbills (Wagner 1991), the mating arena is a

TABLE 2 Species of birds for which there is quantitative evidence that EPP is the result of a mixed reproductive strategy by females

Species	Female extra-pair behavior	Male extra-pair behavior	Evidence that females benefit from EPCs	Level of EPP (%)	Reference
Blue tit *Parus caeruleus*	Observations of female forays, 9 EPCs seen	Observations of male forays, 3 EPCs seen	Preferred males are larger, survive better, have longer song strophes, and produce longer-lived offspring	10–15	Kempenaers et al. 1992, 1997
Black-capped chickadee *Parus atricapillus*	Observations of female forays and EPCs[a]	Observations of male forays and EPCs[a]	Extra-pair males of higher dominance rank higher than pair males	11	Otter et al. 1994, 1998; Smith 1988[a]
Bullock's oriole *Icterus galbula*	Not known if females foray, but 3 "solicited EPCs" seen[b]	Males foray to female's territories, >2 forced and 3 unforced EPCs seen[b]	Extra-pair males older than pair males (adult versus subadult)	32	Edinger 1988, Flood 1985, Richardson & Burke 1999
Chaffinch *Fringilla coelebs*	Observations of female forays, 6 EPCs seen. Females also give solicitation calls that may attract extra-pair males.	Observations of male forays, 6 EPCs seen. Some EPCs followed display by intruding male	None to date	17	Sheldon 1994, Sheldon & Burke 1994[c,d]
Common murre *Uria aalge*	Females solicit, 8 EPCs seen[e]	Males foray, 39 UEPCs seen. 508 FEPCs seen, cloacal contact confirmed in >23 (combining populations[e,f])	None to date	8[e]	Birkhead et al. 1985[f], 2001[e], Hatchwell 1988[e]

Great tit *Parus major*	Female forays linked with EPP, females approach male at territory boundary	Males intrude, >3 EPCs seen[g]	None to date	8	Strobach et al. 1998
Hooded warbler *Wilsonia citrina*	Radio-tracked forays by fertile females. Fertile females also give "chip" calls that may attract extra-pair males. No EPCs seen	Observations and radiotracking of male forays. EPCs "rarely observed"	None to date	27	Neudorf et al. 1997; Norris & Stutchbury 2001, 2002; Pitcher & Stutchbury 2000; Stutchbury 1998c; Stutchbury et al. 1994, 1997
Red-winged blackbird (western U.S. population) *Ageliaus phoeniceus*	Observations of forays, 28 EPCs seen in extra pair male's territory, 28 EPCs seen away from marsh	Observations of male forays, 16 EPCs seen	Males allow female to forage on their territory, defend nest against predators. Multiply mated females had higher hatching success and fledged a greater proportion of nestlings	35	Gray 1996, 1997
Superb fairy-wren *Malurus cyaneus*	Observations of female forays. Females radio-tracked during predawn forays, offspring sired by visited males. No EPCs seen	Observations of frequent forays combined with display, 4 EPCs seen	Females visit dominant males who molt earlier	75–85	Double & Cockburn 2000; Dunn & Cockburn 1998, 1999; Green et al. 1995; Mulder 1997
Shag *Phalacrocorax aristotelis*	Female forays in response to male display, 27 EPCs seen	Display to attract females	None to date, although EPCs more likely to be with successful breeders	9	Graves et al. 1992, 1993

(*Continued*)

TABLE 2 (*Continued*)

Species	Female extra-pair behavior	Male extra-pair behavior	Evidence that females benefit from EPCs	Level of EPP (%)	Reference
Starling *Sturnus vulgaris*	Females seen to initiate 5 EPCs[h]	Males move towards females and display, 1 EPC seen[h]	None to date	9	Eens & Pinxten 1990, Smith & von Schantz 1993
Tree swallow *Tachycineta bicolor*	Observations of female forays, 11 EPCs seen	Observations of male forays, >1 possible forced and 24 unforced EPCs seen[i,j]	Clutches containing EPFs had higher hatching success[k]	38–76	Barber & Robertson 1999; Dunn et al. 1994b; Kempenaers et al. 1999, 2001; Morrill & Robertson 1990; Venier et al. 1993

[a]Smith 1988 refers to a separate series of populations in which EPCs were observed although no paternity study was performed. Nine EPCs resulted from female forays, 4 EPCs resulted from male intrusions. Otter et al. 1994, 1998 do not provide data from their population, but they cite unpublished data on extra- and intraterritorial female solicitation of EPCs.

[b]No details on location of solicited EPCs presented in Richardson & Burke 1999. One unforced and at least two forced EPCs were seen by Edinger (1988). Two EPCs were seen by Flood (1985), although both females concerned were widows.

[c]A total of 20 EPCs were observed. Twelve were preceded by female solicitation, of which 6 were in the extra-pair male's territory, and 6 in the female's own territory. The location of the other 8 was unreported.

[d]See Marler (1956) for similar observations in a separate population.

[e]British study site.

[f]Canadian study site.

[g]Björklund et al. (1992) also witnessed male intrusions, one of which resulted in an EPC.

[h]Female behavior reported in Smith & von Schantz 1993, male behavior in Eens & Pinxten 1990.

[i]Unpaired floater males have been observed intruding and copulating with resident females (Barber & Robertson 1999). Seven out of 53 EPCs were achieved by floaters (Kempenaers et al. 2001).

[j]See Lombardo (1986) for observations of single EPCs through male and female forays in a separate population.

[k]The paternity of dead embryos was not assessed in this study. Whittingham & Dunn 2001a accounted for this in a separate population and found no effect of paternity on offspring viability prefledging.

location with no resources (a ledge of bare rock) and so the most likely reason for females to visit is to encounter extra-pair males.

These observations strongly suggest that some females pursue EPCs and that some EPP results from this mixed reproductive strategy. However, how often this occurs is not clear. A striking feature of Table 2 is how few extra-pair events have been witnessed in even the most thoroughly studied species. Only one study has specifically linked forays by particular females with EPP, and even then exactly how copulation occurred was not observed (Double & Cockburn 2000). No study has gathered sufficient data on both female forays and paternity to demonstrate what proportion of EPP occurs through female pursuit. Female pursuit might not be the only route to EPP within a population; males often foray as well in the same populations (Table 2). Hence, even the clearest cases of a mixed reproductive strategy by females should not lead us to conclude that is the basis for all or even most EPP.

If females do pursue EPCs as part of a mixed reproductive strategy, then presumably the benefits of EPP outweigh the costs. Griffith et al. (2002) reviewed the potential benefits to females and concluded that, despite a wealth of mainly observational studies, there is little unambiguous support for any of them. Studies of the costs of female pursuit are rare and even less clear. There are no data on the energetic or time costs of female forays. Aggression by the extra-pair male's mate is also a potential cost (Mays & Hopper 2003) but is poorly documented. The increased risk of acquiring sexually transmitted diseases (STDs) is a likely cost of copulating with additional partners (Lombardo & Thorpe 2000, Sheldon 1993, Westneat & Rambo 2000), but few data are available on STD prevalence and transmission in wild birds. The most widely proposed cost to female pursuit is the retaliatory withholding of parental care by her social mate. A few well-cited studies have found a reduction in paternal care in response to female extra-pair activity (e.g., Chuang-Dobbs et al. 2001a, Dixon et al. 1994, Osorio-Beristain & Drummond 2001, Weatherhead et al. 1994) but many others have not (reviewed by Whittingham & Dunn 2001b). Theory predicts contingent effects of paternity (Westneat & Sherman 1993) and tremendous difficulties in identifying its effect (e.g., Kempenaers & Sheldon 1997). Hence it is premature to cite it as a general cost of EPP for females.

Consequences to Males of Female Pursuit

Selection favoring female pursuit of EPP can have several effects on males. First, female preferences for pursuing EPCs with particular male phenotypes will create or enhance positive selection gradients for those traits. Indeed, several studies have identified aspects of male song or morphology that correlate with success at gaining EPP (Griffith et al. 2002), although there are many studies in which conspicuous male traits do not affect success (e.g., Buchanan & Catchpole 2000, Cordero et al. 1999, Dunn et al. 1994a, Friedl & Klump 2002, Lubjuhn et al. 1999, Rätti et al. 1995, Stutchbury et al. 1997, Weatherhead & Boag 1995). Two studies manipulated male secondary sexual characteristics and found direct effects of plumage on success at gaining extra-pair fertilizations (EPCs) [barn swallow, *Hirundo rustica*) (Saino et al. 1997) and bluethroat, (*Luscinia svecica*) (Johnsen et al. 1998a)]. The results have been interpreted as evidence that these traits serve to increase a male's desirability as an extra-pair mate. Although this may be true, a viable alternative has not been tested. Correlations between male traits and paternity could arise if male traits were correlated with their abilities to repel other males or coerce females into mating. Manipulations might not control for this possibility. Male plumage traits in many species are used as signals of resource holding potential or of status in male-male competition (e.g., Butcher & Rohwer 1988, Evans & Hatchwell 1992, Rohwer 1975). Male signals could also provide females with predictive information about male aggressiveness and hence the costs of resisting copulation attempts by each male. Given that mating interactions in barn swallows can be aggressive (Møller 1985) and have not been seen in bluethroats, it is possible that manipulated males had higher success because females reduced their resistance to them.

A second effect of female pursuit is that it generates sexual conflict for a subset of male-female interactions. Selection favoring female forays produces a negative opportunity gradient for two types of males in the local population, the female's social mate and all extra-pair males that she does not visit (Figure 2). The opportunity gradients experienced by all but the preferred extra-pair males could produce selection for any traits in the female's social mate that reduce her forays or for ways in which nonpreferred males could gain EPP despite her preferences.

Few studies have assessed whether males have evolved counters to the pursuit of EPCs by their social mates. Lifjeld et al. (1994) suggested that males might guard their mates in an attempt to prevent them from making extraterritorial forays, in contrast to the traditional view that mate guarding dissuades other males from approaching (Beecher & Beecher 1979, Birkhead 1979). We might expect behavior that minimizes female forays to differ in some ways from actions that minimize the approach of male intruders. For example, males should follow their mate's extraterritorial forays and force or shepherd them away from other male's territories. We found no published study reporting instances of males herding their mates, although we know of unpublished instances in several species [red-winged blackbirds, (*A. phoeniceus*) (D.F. Westneat, unpublished data) and yellow-breasted chats, *Icteria virens* (Mays 2001)]. Several authors have reported males interrupting EPCs outside their own territory (e.g., Gray 1996, Hatchwell 1988), but it is not clear if this was a response to forays by their social mates. Barash (1976) interpreted several cases of male aggression to his social mate as enforcement of fidelity. Indeed, retaliation by male lesser gray shrikes (*Lanius minor*) to prolonged absences by their mate was recently demonstrated experimentally (Valera et al. 2003). Such behavior may explain the lack of EPP in this species, but few other examples have been reported. Finally, some authors have suggested that song duetting may be a mechanism by which a male can monitor the activities of his mate (e.g., Levin 1996, Sonnenschein & Reyer 1983), but conclusive evidence for this function is lacking.

How nonpreferred extra-pair males could respond to female forays has also not been addressed. Nonpreferred males might be expected to pursue and harass females wherever they encounter them. Wagner (1998) predicted that one potential response of nonpreferred males would be to cluster around preferred males. Such spatial clumping of males for EPCs is an intriguing possibility that has not been tested. Even if extra-pair males do not cluster, they could still focus their forays along the territory boundaries of preferred males where they could intercept and harass prospecting females or even interrupt EPCs. Alternatively, some males could adopt a strategy of floating to maximize their encounters with fertilizable females (e.g., Ewen et al. 1999, Kempenaers et al. 2001). This could lead to higher chances of these males obtaining copulations with foraying females. No study, to our knowledge, has examined this possibility in any depth, although there is some supporting evidence. Even though females foray in superb fairy-wrens, blue tits, chaffinches, black-capped chickadees, and hooded warblers (*Wilsonia citrina*), males of these species also intrude onto the territories of other males, where they approach females, display to them, and sometimes copulate (Table 2). Subordinate male superb fairy-wrens apparently sire 20% of extra-pair offspring by positioning themselves near where the female expects the dominant and presumably preferred male to be (Double & Cockburn 2003). Conflict theory predicts similar behavior by males in many species.

Male Pursuit

Male forays that lead to attempted EPCs do in fact appear common in birds. Males foray into a fertilizable female's area in 39 of 43 species in which EPP is known to occur and observational data have been reported. In 30 of these species, males have attempted EPCs (I.R.K. Stewart & D.F. Westneat, unpublished review). In addition, in the studies (n = 7) in which mate guarding by the pair male was experimentally reduced, all EPCs observed were by intruding males and none occurred through female forays (Björklund & Westman 1983, Björklund et al. 1992, Dickinson 1997, Komdeur et al. 1999, Møller 1987, Sundberg 1994, Westneat 1992). However, we emphasize that female-initiated EPCs could occur and may even predominate in many of these species. Female forays could be infrequent and cryptic, and yet have a disproportionate effect on EPP. Under these conditions, considerable effort may be required to eliminate female pursuit as being involved in EPP. Nevertheless, male-initiated EPCs are the most commonly observed extra-pair events in birds. It is therefore worth examining the impact of these events on both males and females.

Female Responses to Male Pursuit

A central issue when males initiate extra-pair events is whether or not this produces a conflict gradient on females. Females might benefit from male-initiated EPCs or they might not. In no case has the cost of an EPC per se been measured independently of the consequences of how females respond to males. Even in waterfowl, where aggressive copulation attempts by males lead to clear cases of female injury or even death (McKinney & Evarts 1998), it is difficult to ascribe these costs to the attempted copulation itself because in most events both copulation and female resistance occur together. The fact that females resist so vigorously implies that allowing a male to obtain an EPC is costly, but definitive data are needed.

Female responses to male pursuit vary widely within and among species, but can be broadly categorized as cooperation or resistance. For example, in white ibis (*Eudocimus albus*) (Frederick 1987) and common murres (Birkhead et al. 1985), females sometimes cooperate with EPC attempts by holding still and exposing their cloaca during mounting, whereas other times they disrupt mountings by simply standing up. Active female resistance to EPC attempts may also involve moving away from or pecking at the extra-pair male and giving alarm calls (e.g., Westneat 1992).

One explanation for variation in resistance is that females benefit from EPCs with some males but not with others. Differential resistance to an extra-pair male could be a female ploy to manipulate who sires her offspring (Eberhard 1996, Frederick 1987, Gowaty & Buschhaus 1998, McKinney et al. 1983, Wagner et al. 1996, Westneat et al. 1990). A variety of evidence has been marshaled to support the female manipulation hypothesis. Unfortunately, very little of it eliminates the alternative hypothesis that encounters with males result in a net cost to females and variation in female behavior simply minimizes those costs. Because resisting EPC attempts can be costly to females (McKinney & Evarts 1998), cooperating with a male could be a "best of a bad job" scenario for females aiming to reduce the net costs of the encounter (convenience polyandry sensu Thornhill & Alcock 1983). Distinguishing between these two hypotheses is extremely difficult.

Wagner et al. (1996) attempted to do so in purple martins (*Progne subis*), in which older males are rarely cuckolded (4% of offspring are EPP), whereas 43% of the offspring in the nests of younger males result from EPP. Most EPCs occur when males approach females that are on the ground gathering nesting material (Morton 1987, Wagner 1998). Males are aggressive to females, and females display a range of resistance behavior that has been interpreted as a female ploy to manipulate paternity. Wagner et al. (1996) predicted that if females were manipulating EPCs, there would be no relationship between the operational sex ratio (OSR) and EPP, that male age alone would predict the likelihood of cuckoldry and that mate-guarding intensity would not be influenced by OSR. The data fit these predictions, and the authors concluded that females were manipulating EPP through subtle acceptance of EPCs from older males. However, these predictions might also arise from the cost-avoidance hypothesis. Older males may be able to coerce females more effectively, in which case the OSR would have no influence on EPP or on mate-guarding behavior. If older males were also effective at ensuring paternity (either through more effective mate guarding, coercion of their social mates, or better timing and frequency of copulation), then they would also have higher paternity. Although female manipulation is possible in purple martins, the available data are insufficient to eliminate the cost-avoidance hypothesis.

The major difference between the two hypotheses is in the effect on fitness of a female who partially resists EPCs (Figure 3). Manipulation generates higher fitness

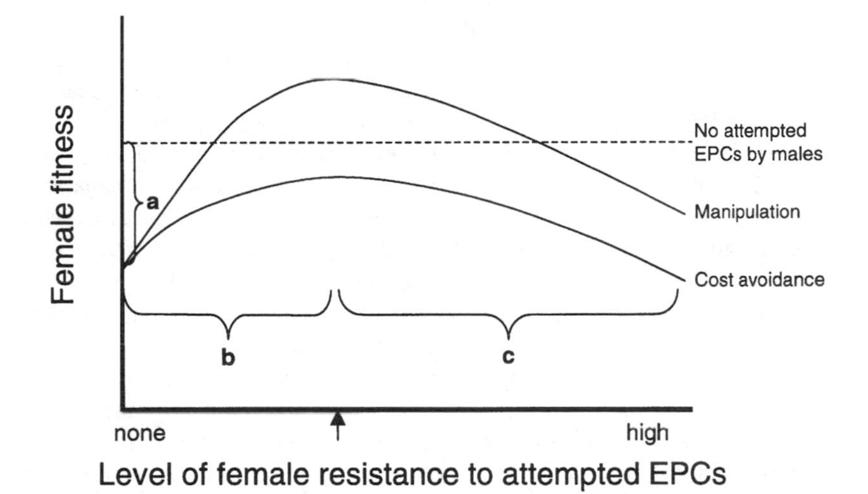

Figure 3 Graphical illustration of the distinction between female resistance as manipulation versus as avoidance-of-costs. The dotted line shows the fitness of a female experiencing no attempted EPCs. A cost of mating with extra-pair males is shown by a decrease in female fitness when they experience EPCs but do not resist them (*a*). Increased resistance reduces that cost and filters out poor-quality males, thereby increasing fitness (*b*). Eventually, the costs of resisting exceed the benefits and the fitness curves drop (*c*). Optimal resistance (*arrow*) is considered manipulation (*upper curve*) if it improves a female's fitness above that obtained if males attempted no EPCs. Resistenace functions as cost-avoidance if the optimal level of resistance does not compensate for the cost of EPCs. Both hypotheses predict limited resistance and involve the same costs and benefits. The essential difference between the two is how much the impact of resistance on mate quality affects female fitness.

than if no EPCs were attempted by males, whereas cost-avoidance minimizes a loss of fitness. Otherwise, the two make identical predictions about the nature of resistance given an encounter and the patterns of copulations and fertilizations that may result, especially if male quality as a sire is correlated with his persistence. However, because manipulation results in higher fitness than could be achieved if there were no extra-pair encounters, females with the potential to manipulate males should encourage extra-pair encounters, whereas if cost limitation were the function of resistance, females should avoid them.

Indeed, some studies suggest that females do encourage extra-pair encounters. Female bearded tits (*Panurus biarmicus*) give a display near a group of males that stimulates them to approach. The female then flies away, eliciting an aggressive chase that results in their copulating with the fastest male (Hoi 1997, Hoi & Hoi-Leitner 1997). Similarly, female chaffinches (Sheldon 1994) and hooded warblers (Neudorf et al. 1997) vocalize loudly and frequently during their fertilizable period. These sounds attract other males and could be a method by which females instigate encounters with potential extra-pair partners without having to foray. However, there also could be other reasons why female chaffinches and hooded warblers emit vocalizations or why female bearded tits display near groups of males [We note that the study reporting these chases was done on a captive population (Hoi & Hoi-Leitner 1997), although chases have been seen in the wild (Hoi 1997).]. Maintaining close proximity to a mate is one possible explanation for both behaviors. In bearded tits, if males are clustered near a resource of value to females, then females may have to approach them and display in order to acquire that resource. None of these studies have explicitly considered alternative hypotheses for female behavior, and so the conclusion that females are manipulating paternity depends partly on the interpretation of the observers.

Lifjeld & Robertson (1992) conducted an experiment testing whether variable female resistance influences paternity. They removed male tree swallows (*Tachycineta bicolor*) midway through the laying period, allowed a replacement male to settle, and determined the paternity of nestlings fertilized both before and after replacement. They found a correlation in behavior and paternity patterns; if the eggs laid before removal were all sired by the female's social mate, then the female was more likely to resist replacement males and produce eggs sired by the original male. If some EPP occurred before the pair male was removed, females were more likely to accept copulations from replacements, although replacements were not more likely to sire eggs fertilized after removal. These results indicate that females can influence fertilization by their behavior but do not distinguish between the two hypotheses for variable resistance. More importantly, the researchers did not manipulate the presence or identity of extra-pair males. The results could be explained by differences in the abilities of nearby males to coerce females or differences in female ability to resist coercive males because these abilities presumably did not change from before to after removal of the pair male. Overtly coercive copulations have not been seen in tree swallows (Barber & Robertson 1999, Venier et al. 1993), but subtle harassment may be difficult to detect.

Male Response to Female Resistance

Resistance to copulation by females, regardless of its function, is costly to males. Coercion is one way in which males could increase the chances they sire offspring when females are not receptive to them. Whether or not males can coerce a resisting female and achieve fertilization is a controversial topic. McKinney & Evarts (1998) reviewed instances of behavior that could be interpreted as sexual coercion. They adopted the definition of sexual coercion given by Smuts & Smuts (1993), which is "use by a male of force, or threat of force, that functions to increase the chances that a female will mate with him at a time when she is likely to be fertile. . .at some cost to the female." McKinney & Evarts (1998) documented numerous accounts of males directing aggressive behavior toward fertile females. The evidence that such behavior can lead to copulation is strongest in waterfowl, partly because copulations occur almost exclusively on open water and are therefore easy to observe. Detailed behavioral observations suggest that, in many species, nearly all EPCs are resisted in some form by females, whereas genetic evidence of mixed paternity shows that at least some can lead to fertilizations (e.g., Birkhead et al. 2001, Burns et al. 1980, Dunn et al. 1999).

However, male waterfowl have intromittent organs, and several authors have maintained that in species lacking such organs, coercion is not possible (e.g., Fitch & Shugart 1984, Gowaty & Buschhaus 1998). Nevertheless, aggressive copulation attempts have been reported in many species without intromittent organs (McKinney & Evarts 1998). Several studies have examined paternity in populations where aggressive copulations occur. In a captive population of zebra finches (*Taeniopygia guttata*), Burley et al. (1994) found that 80% of EPCs were forced (see also Birkhead et al. 1988 for observations of similar events in natural populations). A subsequent paternity analysis revealed a moderate level of EPP [28%, though it is only 4% in free-living populations (Birkhead et al. 1990)]. EPP frequency was directly proportional to the rates at which females engaged in unforced rather than forced EPCs (Burley et al. 1996). In captive Japanese quail (*Coturnix japonica*), forced copulations result in fertilization as often as unforced ones (Adkins-Regan 1995). Most copulations in feral fowl (*Gallus gallus*) are coerced (Pizzari & Birkhead 2000), although females are able to influence insemination by selectively ejecting sperm from subordinate males. However, females do not always do this, nor has it been established that all sperm from an insemination are ejected. Taken together, these data indicate that aggressive copulations can lead to fertilization and that female resistance likely reduces but does not eliminate that possibility.

Gowaty & Buschhaus (1998) present an array of arguments against the hypothesis that coercion functions to acquire fertilizations in species without intromittent organs. These arguments center around the structure of the female cloaca and the myriad ways females could possibly prevent fertilization if inseminated (see also Eberhard 1996 for other possibilities). Gowaty & Buschhaus (1998) suggest that even if males harass females into copulating, then mechanisms for preventing fertilization might be expected to evolve, and males may have little ability to influence those after insemination. Conflict theory predicts female manipulation of male sperm post-insemination in some circumstances, and Pizzari & Birkhead's (2000) results confirm elements of this idea.

However, Gowaty & Buschhaus' (1998) argument that such mechanisms usually have primacy is based on the assumption that the only fitness consequence of extra-pair events involves the genetic quality of the male. There are two flaws with this argument. First, even if females can reduce fertilization of undesirable males, those mechanisms might not be perfect, and males will be selected to counter them (e.g., Rice 2000). Second, fertilization is the last step in a sequence of events. Because the earlier steps of encountering and copulating with females also have unique fitness consequences and affect the set of males that make it to the fertilization stage, it is possible for coercive males that do copulate to be neutral or even attractive sires for a female's offspring (Westneat 2000). Kokko et al. (2003) modeled the effect of female precopulatory resistance when the act of mating is costly. They noted that a consequence of this resistance may be indirect benefits—that is, resistance selects for persistent males because only they will be able to overcome the resistance, and such males may produce sons with similar success. What has not been modeled or addressed empirically in any organism is whether females might be predicted to assist fertilization of successfully coercive males even when the net impact of the whole mating interaction (encounter, copulation, and fertilization) is costly (i.e., on the cost avoidance line in Figure 3).

Conflict theory predicts that male coercion and female resistance will produce a variety of adaptations in each sex that minimize the influence of the other sex on the outcome of these interactions. Males may be selected to narrow a female's options (Gowaty 1996), making copulation the best of a bad situation. Females will be selected to manipulate interactions to limit the influence of males. Thus, encounters between females and extra-pair males may be similar to sequential assessment games (Enquist & Leimar 1983, Maynard Smith 1982) or wars of attrition (Mesterton-Gibbons et al. 1996, Rowe & Arnqvist 2002) in that sometimes the female can escape the constraints imposed by the male and sometimes she cannot. Theory on the impact of sexual conflict has begun to address some of these issues (e.g., Gavrilets et al. 2001, Kokko et al. 2003), but it needs to incorporate the sequential nature of many mating interactions.

Empirical work on avian mating interactions has only begun to consider these ideas. In a detailed study, Dickinson (2001) analyzed sequences of extra-pair interactions and found no effect of male persistence, relative size, or female age on female receptivity but did find a large effect of male age. Whether these patterns are due to female manipulation, male coercion, or a combination of both is unclear because of limited data and an unknown history of interactions before observations. Nevertheless, this level of detail on extra-pair behavior is necessary for designing better tests of ideas about sequential games.

Pair Male's Response to Forays by Extra-Pair Males

Extra-pair events are costly to the male partner of the target female. This should favor traits in the pair male that reduce the chances of being cuckolded (Trivers 1972). Such behavioral and physiological traits appear to be widespread in animals (Birkhead & Møller 1998). Birds have two main paternity assurance behaviors, which are not mutually exclusive. First, males often exhibit noncontact guarding in which they follow their mates closely during their fertilizable period (Birkhead & Møller 1992, 1998). Experiments demonstrate that this does reduce cuckoldry (e.g., Chuang-Dobbs et al. 2001b, Komdeur et al. 1999, Westneat 1994). Second, males copulate with their mates much more frequently than is necessary to fertilize the eggs (Birkhead & Møller 1992). Indeed, testes size and potential ejaculate size are both larger in species where sperm competition is more intense (e.g., Møller & Briskie 1995, Tuttle et al. 1996). Other types of paternity guards are possible [e.g., distracting alarm calls given by pair males (Møller 1990) or antiphonal duetting (e.g., Levin 1996)] but are rarely investigated.

There are, however, important gaps in what we know about both mate guarding and frequent copulation, and these have created some confusing situations. For example, one might predict that if mate guarding reduces cuckoldry, then the individuals who guard more intensely would have higher paternity. This positive relationship was observed in Seychelles warblers (*Acrocephalus sechellensis*) (Komdeur et al. 1999) and black-throated blue warblers (Chuang-Dobbs et al. 2001b). Yet, in bluethroats (Johnsen et al. 1998b), eastern bluebirds (*Sialia sialis*) (Gowaty & Bridges 1991), and purple martins (Wagner et al. 1996) males that have high levels of guarding have lower paternity.

This situation may arise because mate guarding is likely to be shaped by conflicting demands between the benefits of guarding and the costs. When paternity risk was increased (by increasing male density) in Seychelles warblers (Komdeur 2001), males increased guarding effort. Costs of guarding also influence male behavior. In red-winged blackbirds, lapses in guarding occur when males leave their territories to forage (Westneat 1993b); foraging may therefore be sacrificed in order to guard. An experimental increase in food availability on the territories of random males reduced this cost and resulted in increases in both mate guarding and paternity (Westneat 1994). Similarly, in Seychelles warblers, guarding males foraged less and lost weight compared to males induced to reduce guarding (Komdeur 2001). Together, these studies suggest males are sensitive to both the benefits and costs of guarding.

If mate guarding is subject to conflicting demands, then the level of guarding may be contingent upon an individual's circumstances. One explanation for the greater EPP losses found in males that guard more assiduously is that these males have recognized their vulnerability to cuckoldry and have thus diverted more effort into guarding. They may have one or more neighbors that are particularly persistent, or a mate that is particularly prone to foray (see above). Presumably, paternity losses would be even higher if they guarded less, yet no one has manipulated levels of guarding in such individuals with this specific question in mind.

Frequent copulation has also been proposed as a male tactic that increases paternity (Birkhead & Møller 1992, Hunter et al. 1992). Circumstantial evidence indicates multiple copulations do increase paternity (see Hunter et al. 1993), and this is consistent with what is known about the effect of sperm storage on fertilization patterns (Birkhead 1998). However, there has been no experimental work on the costs of frequent copulation to males or its effectiveness.

One difficulty with interpreting frequent copulation as a male tactic for ensuring paternity is that copulation requires some degree of cooperation on the part of the female. Females might cooperate with the pair male if he is of higher quality than the potential extra-pair males. Alternatively, females may benefit from frequent copulations for reasons other than ensuring the pair male's paternity, which could lead to sexual conflict. Negro & Grande (2001) suggested that frequent copulation in raptors may have a signaling function. Lens et al. (1997) proposed that frequent copulation could be a means for a female to assess male vigor. Finally, males may coerce their mates to copulate in a variety of ways. If males can assess their likelihood of paternity via their mate's behavior and reduce care in response to low perceived paternity, that cost may favor female willingness to copulate. Pair males can also harass their mates via excessive courtship or physical attack (Sheldon 1994, Valera et al. 2003, Westneat 1992) or via cloacal pecking (e.g., Davies 1983). In sum, sexual conflict should also influence the copulatory behavior of social mates, potentially in some subtle ways. This possibility emphasizes the need for more studies of the interactions between social mates.

Extra-Pair Male's Response to Paternity Guards

Paternity guards by pair males are costly to extra-pair males because they reduce their paternity and could increase their risk of injury. There has been little empirical or theoretical work exploring the possible counters to paternity guards that may have evolved. Birkhead et al. (1988) suggested that extra-pair males might produce more sperm in their ejaculates than pair males. Indeed, Michl et al. (2002) manipulated male collared flycatchers such that some could release sperm during copulations and others could not. Then, by counting the number of sperm present in the perivitelline layer of different eggs in the laying sequence, they determined the number and timing of EPCs. They found an EPC left >5 times the sperm as a within-pair copulation. The implication is that extra-pair males released more sperm, although the possibility that the female allowed more sperm inside her reproductive tract cannot be eliminated. Other possible adaptations to counter male mate guarding have rarely been addressed.

THE ECOLOGY OF EPP IN BIRDS

We have used conflict theory to explore some of the finer details of mating interactions in birds. Can this approach really lead us to understand the ecological basis for incredible diversity in EPP among species? We claim that it can because ecology will affect the costs and benefits received by each of the players and hence selection acting on their traits, thereby affecting what level of EPP emerges from their interactions. Understanding this will not be easy, however, and to date most hypotheses have focused on the effect of ecology on only one player.

For example, it is clear that neither density nor breeding synchrony affect EPP as strongly or consistently (Bennett & Owens 2002, Griffith et al. 2002, Westneat & Sherman 1997) as initial proponents envisioned (e.g., Birkhead 1979, Stutchbury & Morton 1995) in part because the initial ideas assumed one player's behavior was of most importance. The effects of both density and synchrony are likely to differ depending on the prevalence of female pursuit, the costs and benefits of male mate guarding and aggressiveness, and whether the species is territorial or not (e.g., Westneat & Sherman 1997). These contingencies make simple predictions difficult.

Ecological factors affecting the value of male parental care could also affect EPP either through influencing the allocation of male effort toward the pursuit of

EPCs (Westneat 1988, Westneat et al. 1990) or in its effect on the potential costs of EPP to females (Birkhead & Møller 1996, Gowaty 1996). Comparative tests of this idea have not distinguished between these two mechanisms but do support a general relationship between the value of care and EPP (Table 1). Specific studies indicate that the presence of opportunities to pursue EPCs affect male parental care in some species—indigo bunting, (*Passerina cyanea*) (Westneat 1988); fairy martin (*Hirundo ariel*) (Magrath & Elgar 1997)—but not others [hooded warbler (Stutchbury 1998c)]. Hoi-Leitner et al. (1999a) experimentally manipulated food supply available to particular pairs of the serin (*Serinus serinus*). Supplemented females had higher frequencies of EPP than controls. Given that females visit males at traditional singing perches to copulate (Hoi-Leitner et al. 1999b), this suggests that the increase in resources reduced the potential costs (primarily in amount of male care) of female pursuit of EPCs (Gowaty 1996). However, whether or not food supplements affected male extra-pair behavior was not addressed.

Sherman & Morton (1988) suggested that habitat complexity might influence the success of some types of behavioral events associated with EPP, such as the ability of males to guard their mates. Mays (2001) found that variation in vegetation density affected male proximity to his mate in yellow-breasted chats, but whether this influenced EPP is not yet clear. Davies (1992) noted that certain habitat features influenced the attempts by female dunnocks (*Prunella modularis*) in polyandrous trios to escape the guarding of an alpha male and rendezvous with the beta male. We expect that habitat type will have a large effect on the success of forays by both sexes, particularly those made surreptitiously. However, vegetation density may well be positively related to food or nest site abundance, and therefore territory quality. This interaction leads to some intriguing tradeoffs that surely deserve attention.

The sexual conflict approach to understanding EPP predicts that some very subtle differences in ecology could produce dramatic differences in the behaviors associated with EPP. It is likely as well that individual factors will have contingent effects. For example, male mate-guarding behavior may be influenced by food supply only if females do not foray, and whether females foray or not could depend on both male parental care and the level of genetic variability in viability reliably signaled by male advertisements. Given that we have as yet no coherent theory on what ecological circumstances produce the latter, there is considerable theoretical and empirical work yet to be done on the ecology of EPP.

CONCLUSIONS

Our brief survey of the adaptations associated with EPP in birds stimulates a challenging question. Why is there so much diversity in both EPP frequencies and the behaviors associated with EPP? Why do female bearded tits appear to stimulate males to chase them before copulation, female fairy-wrens travel to a male to copulate, and female mallard ducks (*Anas platyrhynchos*) attempt to avoid all copulations except from their mate? Why do male common murres, a socially monogamous and colonially breeding seabird, engage in coercive attempts to copulate (Birkhead et al. 1985, Hatchwell 1988), whereas male shags (*Phalacrocorax aristotelis*), also a socially monogamous and colonial breeding seabird, do not and instead display to attract females (Graves et al. 1993)? Why do female blue tits, who apparently can pursue EPCs, only produce 10%–15% EPP, whereas female eastern red-winged blackbirds, who often resist EPCs, end up producing 25%–30% EPP? Despite all the work done on EPP in birds over the past 25 years, the answers to such questions remain elusive.

There are two reasons for this state of affairs. The first is the widespread and clearly unsupported assumption that EPP is a female strategy and therefore adaptive explanations for EPP should focus largely or even solely on female fitness (Birkhead 1998, Griffith et al. 2002, Petrie & Kempenaers 1998). Alternative hypotheses are not even being considered, and we suggest the field will not move forward if this narrow view persists. In our opinion, the most fruitful conceptual approach to understanding EPP is to recognize that it arises from a three- (or more) player game in which the fitness of each player depends on the behavioral actions of the others as well as the prevailing ecological and social circumstances. Empirical approaches being used in insects such as water striders (Rowe & Arnqvist 2002) and dung flies (*Sepsis cynipsea*) (Blanckenhorn et al. 2002, Mühlhäuser & Blanckenhorn 2002) have focused directly on conflict between males and females. Although birds clearly have some additional complexities, the approach used in insects could greatly expand our understanding of avian mating behavior.

Interestingly, because EPP emerges from a game, the traits a player does not express are as important to understanding the outcome as those traits players do exhibit. For example, the EPP observed in superb fairy-wrens requires understanding why males apparently do not coerce females (Green et al. 2000) as much as why females foray. Neither theorists nor empiricists have asked these types of questions or confronted the complexity of the three-player game sufficiently to take the field to the next level.

A second reason for the confusion surrounding the causes of EPP is that genetic studies have become more popular than behavioral ones. In contrast to the situation 20 years ago, paternity data are easier to gather and more abundant than information on the behavioral events that affect paternity. These data have spawned the array of comparative studies that have stimulated new questions about paternity patterns within and between populations. Genetic studies have driven the revolution in our understanding of mating patterns in birds. Yet, knowing the mating patterns of birds is not the same as understanding them. Comparative analyses of EPP are reaching their limits, and correlates of paternity, even within populations, are fraught with interpretational difficulties. These will not be solved until we know more about how mating interactions occur. Our review reveals just how little we know about mating interactions, even in the most thoroughly studied species. Conflict theory highlights how critical such information is to understanding the diversity now apparent in birds. We suggest that the integration of conflict theory, genetic analyses, and detailed behavioral data will allow more precise and insightful comparative studies, better tests of hypotheses about the adaptiveness of EPP, and new insights into the ecological causes of the variation we now observe.

ACKNOWLEDGMENTS

We thank the National Science Foundation and the University of Kentucky for support during preparation of this review. We also appreciate the many helpful suggestions on earlier drafts of the manuscript provided by Ann Baker, Peter Dunn, Amanda Ensminger, Herman Mays, Damon Orsetti, Ian Owens, Mike Webster, and an anonymous reviewer.

The *Annual Review of Ecology, Evolution, and Systematics* is online at http://ecolsys.annualreviews.org

LITERATURE CITED

Adkins-Regan E. 1995. Predictors of fertilization in the Japanese quail, *Coturnix japonica*. *Anim. Behav.* 50:1405–15

Arnold KE, Owens IPF. 2002. Extra-pair paternity and egg dumping in birds: life history, parental care and the risk of retaliation. *Proc. R. Soc. London Ser. B* 269:1263–69

Arnold SJ, Duvall D. 1994. Animal mating systems: a synthesis based on selection theory. *Am. Nat.* 143:317–48

Arnold SJ, Wade MJ. 1984. On the measurement of natural and sexual selection: theory. *Evolution* 38:709–19

Barash DP. 1976. The male response to apparent female adultery in the mountain bluebird *Sialia currucoides*: an evolutionary interpretation. *Am. Nat.* 110:1097–101

Barber CA, Robertson RJ. 1999. Floater males engage in extrapair copulations with resident female tree swallows. *Auk* 116:264–69

Beecher MD, Beecher IM. 1979. Sociobiology of bank swallows: reproductive strategy of the male. *Science* 205:1282–85

Bennett PM, Owens IPF. 2002. *Evolutionary Ecology of Birds: Life Histories, Mating Systems and Extinction*. Oxford, UK: Oxford Univ. Press. 278 pp.

Birkhead TR. 1979. Mate guarding in the magpie *Pica pica*. *Anim. Behav.* 27:866–74

Birkhead TR. 1998. Sperm competition in birds: mechanisms and function. See Birkhead & Møller 1998, pp. 579–622

Birkhead TR, Burke T, Zann R, Hunter FM, Krupa AP. 1990. Extra-pair paternity and intraspecific brood parasitism in wild zebra finches *Taeniopygia guttata*, revealed by DNA fingerprinting. *Behav. Ecol. Sociobiol.* 27:315–24

Birkhead TR, Hatchwell BJ, Lindner R, Blomqvist D, Pellatt EJ, et al. 2001. Extra-pair paternity in the common murre. *Condor* 103:158–62

Birkhead TR, Johnson SD, Nettleship DN. 1985. Extra-pair matings and mate guarding in the common murre *Uria aalge*. *Anim. Behav.* 33:608–19

Birkhead TR, Møller AP. 1992. *Sperm Competition in Birds: Evolutionary Causes and Consequences*. London: Academic. 282 pp.

Birkhead TR, Møller AP. 1996. Monogamy and sperm competition in birds. See Black 1996, pp. 323–43

Birkhead TR, Møller AP, eds. 1998. *Sperm Competition and Sexual Selection*. London: Academic. 826 pp.

Birkhead TR, Pellatt JE, Hunter FM. 1988. Extra-pair copulation and sperm competition in the zebra finch. *Nature* 334:60–62

Björklund M, Møller AP, Sundberg J, Westman B. 1992. Female great tits, *Parus major*, avoid extra-pair copulation attempts. *Anim. Behav.* 43:691–93

Björklund M, Westman B. 1983. Extra-pair copulations in the pied flycatcher (*Ficedula hypoleuca*) a removal experiment. *Behav. Ecol. Sociobiol.* 13:271–75

Black JM, ed. 1996. *Partnerships in Birds: The Study of Monogamy*. Oxford, UK: Oxford Univ. Press. 420 pp.

Blanckenhorn WU, Hosken DJ, Martin OY, Reim C, Teuschl Y, Ward PI. 2002. The costs of copulating in the dung fly *Sepsis cynipsea*. *Behav. Ecol.* 13:353–58

Briskie JV, Montgomerie R, Birkhead TR. 1997. The evolution of sperm size in birds. *Evolution* 51:937–45

Buchanan KL, Catchpole CK. 2000. Extra-pair paternity in the socially monogamous sedge warbler *Acrocephalus schoenobaenus* as revealed by multilocus DNA fingerprinting. *Ibis* 142:12–20

Burley NT, Enstrom DA, Chitwood L. 1994. Extra-pair relations in zebra finches: differential male success results from female tactics. *Anim. Behav.* 48:1031–41

Burley NT, Parker PG, Lundy K. 1996. Sexual selection and extrapair fertilization in a socially monogamous passerine, the zebra finch (*Taeniopygia guttata*). *Behav. Ecol.* 7:218–26

Burns JT, Cheng KM, McKinney F. 1980. Forced copulation in captive mallards. I. Fertilization of eggs. *Auk* 97:875–79

Butcher GS, Rohwer S. 1988. The evolution of conspicuous and distinctive coloration for communication in birds. *Curr. Ornithol.* 6:51–108

Cezilly F, Nager RG. 1995. Comparative evidence for a positive association between divorce and extra-pair paternity in birds. *Proc. R. Soc. London Ser. B* 262:7–12

Chuang-Dobbs HC, Webster MS, Holmes RT. 2001a. Paternity and parental care in the black-throated blue warbler, *Dendroica caerulescens*. *Anim. Behav.* 62:83–92

Chuang-Dobbs HC, Webster MS, Holmes RT. 2001b. The effectiveness of mate guarding by male black-throated blue warblers. *Behav. Ecol.* 12:541–46

Cordero PJ, Wetton JH, Parkin DT. 1999. Extra-pair paternity and male badge size in the house sparrow. *J. Avian Biol.* 30:97–102

Crow JF. 1958. Some possibilities for measuring selection intensities in man. *Hum. Biol.* 30:1–13

Dale J. 1995. Problems with pair-wise comparisons: Does certainty of paternity covary with paternal care? *Anim. Behav.* 49:519–21

Davies NB. 1983. Polyandry, cloaca-pecking and sperm competition in dunnocks. *Nature* 302:334–36

Davies NB. 1992. *Dunnock Behaviour and Social Evolution*. Oxford, UK: Oxford Univ. Press. 272 pp.

Dickinson JL. 1997. Male detention affects extra-pair copulation frequency and pair behaviour in western bluebirds. *Anim. Behav.* 53:561–71

Dickinson JL. 2001. Extrapair copulations in western bluebirds (*Sialia mexicana*): Female receptivity favors older males. *Behav. Ecol. Sociobiol.* 50:423–29

Dixon A, Ross D, O'Malley SLC, Burke T. 1994. Paternal investment inversely related to degree of extra-pair paternity in the reed bunting. *Nature* 371:698–700

Double MC, Cockburn A. 2000. Pre-dawn infidelity: Females control extra-pair mating in superb fairy-wrens. *Proc. R. Soc. London Ser.B* 267:465–70

Double MC, Cockburn A. 2003. Subordinate superb fairy-wrens (*Malurus cyaneus*) parasitize the reproductive success of attractive dominant males. *Proc. R. Soc. London Ser. B* 270:379–84

Dunn PO, Afton AD, Gloutney ML, Alisauskas RT. 1999. Forced copulation results in few extrapair fertilizations in Ross's and lesser snow geese. *Anim. Behav.* 57:1071–81

Dunn PO, Cockburn A. 1998. Costs and

benefits of extra-group paternity in superb fairy-wrens. See Parker & Burley 1998, pp. 147–61

Dunn PO, Cockburn A. 1999. Extrapair mate choice and honest signaling in cooperatively breeding superb fairy-wrens. *Evolution* 53:938–46

Dunn PO, Robertson RJ, Michaud-Freeman D, Boag PT. 1994a. Extra-pair paternity in tree swallows: Why do females mate with more than one male? *Behav. Ecol. Sociobiol.* 35:273–81

Dunn PO, Whittingham LA, Lifjeld JT, Robertson RJ, Boag PT. 1994b. Effects of breeding density, synchrony, and experience on extrapair paternity in tree swallows. *Behav. Ecol.* 5:123–29

Eberhard WG. 1996. *Female Control: Sexual Selection by Cryptic Female Choice*. Princeton, NJ: Princeton Univ. Press. 501 pp.

Edinger BB. 1988. Extra-pair courtship and copulation attempts in northern orioles. *Condor* 90:546–54

Eens M, Pinxten R. 1990. Extra-pair courtship in the starling *Sturnus vulgaris*. *Ibis* 132: 618–19

Enquist M, Leimar O. 1983. Evolution of fighting behaviour: decision rules and assessment of relative strength. *J. Theor. Biol.* 102:387–410

Evans MR, Hatchwell BJ. 1992. An experimental study of male adornment in the scarlet-tufted malachite sunbird: I. The role of pectoral tufts in territorial defense. *Behav. Ecol. Sociobiol.* 29:413–19

Ewen JG, Armstrong DP, Lambert DM. 1999. Floater males gain reproductive success through extrapair fertilizations in the stitchbird. *Anim. Behav.* 58:321–28

Fitch MA, Shugart GW. 1984. Requirements for a mixed reproductive strategy in avian species. *Am. Nat.* 124:116–26

Flood NJ. 1985. Incidences of polygyny and extrapair copulation in the northern oriole. *Auk* 102:410–13

Frederick PC. 1987. Extrapair copulations in the mating system of white ibis (*Eudocimus albus*). *Behaviour* 100:170–201

Friedl TWP, Klump GM. 2002. Extrapair paternity in the red bishop (*Euplectes orix*): Is there evidence for the good-genes hypothesis? *Behaviour* 139:777–800

Gavrilets S, Arnqvist G, Friberg U. 2001. The evolution of female mate choice by sexual conflict. *Proc. R. Soc. London Ser. B* 268: 531–39

Gowaty PA. 1985. Multiple parentage and apparent monogamy in birds. See Gowaty & Mock 1985, pp. 11–19

Gowaty PA. 1996. Battles of the sexes and origins of monogamy. See Black 1996, pp. 21–52

Gowaty PA, Bridges WC. 1991. Nestbox availability affects extra-pair fertilizations and conspecific nest parasitism in eastern bluebirds, *Sialia sialis*. *Anim. Behav.* 41:661–75

Gowaty PA, Buschhaus N. 1998. Ultimate causation of aggressive and forced copulation in birds: female resistance, the CODE hypothesis, and social monogamy. *Am. Zool.* 38:207–25

Gowaty PA, Mock DW, eds. 1985. Avian monogamy. *Ornithol. Monogr.* No. 37. Lawrence, KS: Allen. 121 pp.

Graves J, Hay RT, Scallan M, Rowe S. 1992. Extra-pair paternity in the shag, *Phalacrocorax aristotelis*, as determined by DNA fingerprinting. *J. Zool.* 226:399–408

Graves J, Ortega-Rauno J, Slater PJB. 1993. Extra-pair copulations and paternity in shags: Do females choose better males? *Proc. R. Soc. London Ser. B* 253:3–7

Gray EM. 1996. Female control of offspring paternity in a western population of red-winged blackbirds (*Agelaius phoeniceus*). *Behav. Ecol. Sociobiol.* 38:267–78

Gray EM. 1997. Female red-winged blackbirds accrue material benefits from copulating with extra-pair males. *Anim. Behav.* 53:625–39

Green DJ, Cockburn A, Hall ML, Osmond H, Dunn PO. 1995. Increased opportunities for cuckoldry may be why dominant male fairy-wrens tolerate helpers. *Proc. R. Soc. London Ser. B* 262:297–303

Green DJ, Osmond HL, Double MC, Cockburn A. 2000. Display rate by male fairy-wrens

(*Malurus cyaneus*) during the fertile period of females has little influence on extra-pair mate choice. *Behav. Ecol. Sociobiol.* 48:438–46

Griffith SC. 2000. High fidelity on islands: a comparative study of extrapair paternity in passerine birds. *Behav. Ecol.* 11:265–73

Griffith SC, Owens IPF, Thuman KA. 2002. Extra pair paternity in birds: a review of interspecific variation and adaptive function. *Mol. Ecol.* 11:2195–212

Hammerstein P, Parker GA. 1987. Sexual selection: games between the sexes. In *Sexual Selection: Testing the Alternatives*, eds. JW Bradbury, MB Anderson, pp. 119–42. Chichester, UK: Wiley

Hasselquist D, Sherman PW. 2001. Social mating systems and extrapair fertilizations in passerine birds. *Behav. Ecol.* 12:457–66

Hatchwell BJ. 1988. Intraspecific variation in extra-pair copulation and mate defense in common guillemots *Uria aalge*. *Behaviour* 107:157–85

Heg D, Ens BJ, Burke T, Jenkins L, Kruijt JP. 1993. Why does the typically monogamous oystercatcher (*Haematopus ostralegus*) engage in extra-pair copulations? *Behaviour* 126:247–89

Hoi H. 1997. Assessment of the quality of copulation partners in the monogamous bearded tit. *Anim. Behav.* 53:277–86

Hoi H, Hoi-Leitner M. 1997. An alternative route to coloniality in the bearded tit: Females pursue extra-pair fertilizations. *Behav. Ecol.* 8:113–19

Hoi-Leitner M, Hoi H, Romero-Pujante M, Valera F. 1999a. Female extra-pair behaviour and environmental quality in the serin (*Serinus serinus*): a test of the 'constrained female hypothesis.' *Proc. R. Soc. London Ser. B* 266:1021–26

Hoi-Leitner M, Hoi H, Romero-Pujante M, Valera F. 1999b. Multi-male display sites in serins (*Serinus serinus*). *Ecoscience* 6:143–47

Hunter FM, Burke T, Watts SE. 1992. Frequent copulation as a method of paternity assurance in the northern fulmar. *Anim. Behav.* 44:149–56

Hunter FM, Petrie M, Otronen M, Birkhead TR, Møller AP. 1993. Why do females copulate repeatedly with one male? *Trends Ecol. Evol.* 8:21–26

Johnsen A, Andersson S, Ornborg J, Lifjeld JT. 1998a. Ultraviolet plumage ornamentation affects social mate choice and sperm competition in bluethroats (Aves: *Luscinia s. svecica*): a field experiment. *Proc. R. Soc. London Ser. B* 265:1313–18

Johnsen A, Lifjeld JT, Rohde PA, Primmer CR, Ellegren H. 1998b. Sexual conflict over fertilizations: female bluethroats escape male paternity guards. *Behav. Ecol. Sociobiol.* 43:401–8

Kempenaers B, Congdon B, Boag P, Robertson RJ. 1999. Extrapair paternity and egg hatchability in tree swallows: evidence for the genetic compatability hypothesis? *Behav. Ecol.* 10:304–11

Kempenaers B, Everding S, Bishop C, Boag PT, Robertson RJ. 2001. Extra-pair paternity and the reproductive role of male floaters in the tree swallow (*Tachycineta bicolor*). *Behav. Ecol. Sociobiol.* 49:251–59

Kempenaers B, Sheldon BC. 1997. Studying paternity and parental care: pitfalls and problems. *Anim. Behav.* 53:423–27

Kempenaers B, Verheyen GR, Dhondt AA. 1997. Extrapair paternity in the blue tit (*Parus caeruleus*): female choice, male characteristics, and offspring quality. *Behav. Ecol.* 8:481–92

Kempenaers B, Verheyen GR, Vanderbroeck M, Burke T, Vanbroeckoven C, Dhondt AA. 1992. Extra-pair paternity results from female preference for high-quality males in the blue tit. *Nature* 357:494–96

Kokko H, Brooks R, Jennions M, Morley J. 2003. The evolution of mate choice and mating biases. *Proc. R. Soc. London Ser. B* 270:653–64

Komdeur J. 2001. Mate guarding in the Seychelles warbler is energetically costly and adjusted to paternity risk. *Proc. R. Soc. London Ser. B* 268:2103–11

Komdeur J, Kraaijeveld-Smit F, Kraaijeveld K, Edelaar P. 1999. Explicit experimental evidence for the role of mate guarding in minimizing loss of paternity in the Seychelles warbler. *Proc. R. Soc. London Ser. B* 266:2075–81

Lack D. 1968. *Ecological Adaptations for Breeding in Birds*. London: Methuen. 409 pp.

Lens L, Van Dongen S, Van den Broeck M, Van Broeckhoven C, Dhondt AA. 1997. Why female crested tits copulate repeatedly with the same partner: evidence for the mate assessment hypothesis. *Behav. Ecol.* 8:87–91

Levin RN. 1996. Song behaviour and reproductive strategies in a duetting wren, *Thryothorus nigricapillus*: I. Removal experiments. *Anim. Behav.* 52:1093–106

Lifjeld JT, Dunn PO, Westneat DF. 1994. Sexual selection through sperm competition in birds: male-male competition or female choice? *J. Avian Biol.* 25:244–50

Lifjeld JT, Robertson RJ. 1992. Female control of extra-pair fertilization in tree swallows. *Behav. Ecol. Sociobiol.* 31:89–96

Lombardo MP. 1986. Extrapair copulations in the tree swallow. *Wilson Bull.* 98:150–52

Lombardo MP, Thorpe PA. 2000. Microbes in tree swallow semen. *J. Wildl. Dis.* 36:460–68

Lubjuhn T, Strohbach S, Brun J, Gerken T, Epplen JT. 1999. Extra-pair paternity in great tits (*Parus major*)—a long-term study. *Behaviour* 136:1157–72

Magrath MJL, Elgar MA. 1997. Paternal care declines with increased opportunity for extra-pair matings in fairy martins. *Proc. R. Soc. London Ser. B* 264:1731–36

Marler P. 1956. Behaviour of the chaffinch *Fringilla coelebs*. *Behaviour* 5(Suppl.):1–184

Maynard Smith J. 1982. *Evolution and the Theory of Games*. Cambridge, UK: Cambridge Univ. Press. 224 pp.

Mays HL. 2001. *Sexual conflict and constraints on female mating tactics in a monogamous passerine, the yellow-breasted chat*. PhD thesis. Univ. Kentucky, Lexington. 103 pp.

Mays HL, Hopper KR. 2003. Differential responses of yellow-breasted chats (*Icteria virens*) to male and female conspecific model presentations. *Anim. Behav.* In press

McKinney F, Derrickson SR, Mineau, P. 1983. Forced copulation in waterfowl. *Behaviour* 86:250–94

McKinney F, Evarts S. 1998. Sexual coercion in waterfowl and other birds. See Parker & Burley 1998, pp. 163–95

Mesterton-Gibbons M, Marden JH, Dugatkin LA. 1996. On wars of attrition without assessment. *J. Theor. Biol.* 181:65–83

Michl G, Torok J, Griffith SC, Sheldon BC. 2002. Experimental analysis of sperm competition mechanisms in a wild bird population. *Proc. Natl. Acad. Sci. USA.* 99:5466–70

Mock DW. 1985. An introduction to the neglected mating system. In *Ornithological Monographs: Avian Monogamy*, ed. PA Gowaty, DW Mock, 37:1–10. Lawrence, KS: Allen

Møller AP. 1985. Mixed reproductive strategy and mate guarding in a semi-colonial passerine, the swallow *Hirundo rustica*. *Behav. Ecol. Sociobiol.* 17:401–8

Møller AP. 1987. Mate guarding in the swallow *Hirundo rustica*—an experimental study. *Behav. Ecol. Sociobiol.* 21:119–23

Møller AP. 1990. Deceptive use of alarm calls by male swallows *Hirundo rustica*: a new paternity guard. *Behav. Ecol.* 1:1–6

Møller AP. 2000. Male parental care, female reproductive success, and extrapair paternity. *Behav. Ecol.* 11:161–68

Møller AP, Birkhead TR. 1993. Certainty of paternity covaries with paternal care in birds. *Behav. Ecol. Sociobiol.* 33:261–68

Møller AP, Birkhead TR. 1994. The evolution of plumage brightness in birds is related to extrapair paternity. *Evolution* 48:1089–100

Møller AP, Birkhead TR. 1995. Certainty of paternity and parental care in birds: a reply to Dale. *Anim. Behav.* 49:522–23

Møller AP, Briskie JV. 1995. Extra-pair paternity, sperm competition and the evolution of testis size in birds. *Behav. Ecol. Sociobiol.* 36:357–65

Møller AP, Cuervo JJ. 2000. The evolution of

paternity and paternal care in birds. *Behav. Ecol.* 11:472–85
Morrill SB, Robertson RJ. 1990. Occurrence of extra-pair copulation in the tree swallow (*Tachycineta bicolor*). *Behav. Ecol. Sociobiol.* 26:291–96
Morrow EH, Arnqvist G, Pitcher TE. 2002. The evolution of infertility: Does hatching rate in birds coevolve with female polyandry? *J. Evol. Biol.* 15:702–9
Morton ES. 1987. Variation in mate-guarding intensity by male purple martins. *Behaviour* 101:211–24
Mulder RA. 1997. Extra-group courtship displays and other reproductive tactics of superb fairy-wrens. *Aust. J. Zool.* 45:131–43
Mühlhäuser C, Blanckenhorn WU. 2002. The costs of avoiding matings in the dung fly, *Sepsis cynipsea*. *Behav. Ecol.* 13:359–65
Negro JJ, Grande JM. 2001. Territorial signaling: a new hypothesis to explain frequent copulation in raptorial birds. *Anim. Behav.* 62:803–9
Neudorf DL, Stutchbury BJM, Piper WH. 1997. Covert extraterritorial behavior of female hooded warblers. *Behav. Ecol.* 8:595–600
Norris DR, Stutchbury BJM. 2001. Extraterritorial movements of a forest songbird in a fragmented landscape. *Conserv. Biol.* 15:729–36
Norris DR, Stutchbury BJM. 2002. Sexual differences in gap-crossing ability of a forest songbird in a fragmented landscape revealed through radiotracking. *Auk* 119:528–32
Osorio-Beristain H, Drummond H. 2001. Male boobies expel eggs when paternity is in doubt. *Behav. Ecol.* 12:16–21
Otter K, Ratcliffe L, Boag PT. 1994. Extrapair paternity in the black-capped chickadee. *Condor* 96:218–22
Otter K, Ratcliffe L, Michaud D, Boag PT. 1998. Do female black-capped chickadees prefer high-ranking males as extra-pair partners? *Behav. Ecol. Sociobiol.* 43:25–36
Owens IPF, Hartley IR. 1998. Sexual dimorphism in birds: Why are there so many different forms of dimorphism? *Proc. R. Soc. London Ser. B* 265:397–407
Parker GA. 1979. Sexual selection and sexual conflict. In *Sexual Selection and Reproductive Competition in Insects*, ed. MS Blum, NA Blum, pp. 123–66. New York: Academic
Parker GA. 1984. Sperm competition and the evolution of animal mating strategies. In *Sperm Competition and the Evolution of Animal Mating Systems*, ed. RL Smith, pp. 1–60. London: Academic
Parker PG, Burley NT, eds. 1998. *Ornithological Monograph. Avian Reproductive Tactics: Female and Male Perspectives*. Lawrence, KS: Allen, 49:195 pp.
Petrie M, Doums C, Møller AP. 1998. The degree of extra-pair paternity increases with genetic variability. *Proc. Natl. Acad. Sci. USA* 95:9390–95
Petrie M, Kempenaers B. 1998. Extra-pair paternity in birds: explaining variation between species and populations. *Trends Ecol. Evol.* 13:52–58
Pitcher TE, Stutchbury BJM. 2000. Extraterritorial forays and male parental care in hooded warblers. *Anim. Behav.* 59:1261–69
Pizzari T, Birkhead TR. 2000. Female feral fowl eject sperm of subdominant males. *Nature* 405:787–89
Poston JP. 1997. Mate choice and competition for mates in the boat-tailed grackle. *Anim. Behav.* 54:525–34
Poston JP, Wiley RH, Westneat DF. 1999. Male rank, female breeding synchrony, and patterns of paternity in the boat-tailed grackle. *Behav. Ecol.* 10:444–51
Rätti O, Hovi M, Lundberg A, Tegelström H, Alatalo RV. 1995. Extra-pair paternity and male characteristics in the pied flycatcher. *Behav. Ecol. Sociobiol.* 37:419–25
Rausher MD. 1992. The measurement of selection on quantitative traits: biases due to environmental covariances between traits and fitness. *Evolution* 46:616–26
Rice WR. 2000. Dangerous liaisons. *Proc. Natl. Acad. Sci. USA* 97:12953–55
Richardson DS, Burke TA. 1999. Extra-pair paternity in relation to male age in Bullock's orioles. *Mol. Ecol.* 8:2115–26
Rohwer SA. 1975. The social significance of

avian winter plumage variability. *Evolution* 29:593–610

Rowe L, Arnqvist G. 2002. Sexually antagonistic coevolution in a mating system: combining experimental and comparative approaches to address evolutionary processes. *Evolution* 56:754–67

Saino N, Primmer CR, Ellegren H, Møller AP. 1997. An experimental study of paternity and tail ornamentation in the barn swallow (*Hirundo rustica*). *Evolution* 51:562–70

Schwagmeyer PL, St. Clair RC, Moodie JD, Lamey TC, Schnell GD, Moodie MN. 1999. Species differences in male parental care in birds: a reexamination of correlates with paternity. *Auk* 116:487–503

Schwartz MK, Boness DJ, Schaeff CM, Majluf P, Perry EA, Fleischer RC. 1999. Female-solicited extrapair matings in Humboldt penguins fail to produce extrapair fertilizations. *Behav. Ecol.* 10:242–50

Sheldon BC. 1993. Sexually transmitted disease in birds: occurrence and evolutionary significance. *Philos. Trans. R. Soc. London Ser. B* 339:491–97

Sheldon BC. 1994. Sperm competition in the chaffinch: the role of the female. *Anim. Behav.* 47:163–73

Sheldon BC, Burke T. 1994. Copulation behavior and paternity in the chaffinch. *Behav. Ecol. Sociobiol.* 34:149–56

Sheldon FH, Winkler DW. 1993. Intergeneric phylogenetic relationships of swallows estimated by DNA-DNA hybridization. *Auk* 110:798–824

Sherman PW, Morton ML. 1988. Extra-pair fertilizations in mountain white-crowned sparrows. *Behav. Ecol. Sociobiol.* 22:413–20

Smiseth PT, Amundsen T. 1995. Female bluethroats (*Luscinia* s. *svecica*) regularly visit territories of extrapair males before egg laying. *Auk* 112:1049–53

Smith HG, von Schantz, T. 1993. Extra-pair paternity in the European starling: the effect of polygyny. *Condor* 95:1006–15

Smith HG, Wennerberg L, von Schantz T. 1996. Sperm competition in the European starling (*Sturnus vulgaris*): an experimental study of mate switching. *Proc. R. Soc. London Ser. B* 263:797–801

Smith SM. 1988. Extra-pair copulations in black-capped chickadees: the role of the female. *Behaviour* 107:15–23

Smuts BB, Smuts RW. 1993. Male aggression and sexual coercion of females in nonhuman primates and other mammals: evidence and theoretical implications. In *Advances in the Study of Behaviour*, ed. PJB Slater, JS Rosenblatt, CT Snowdon, M Milinski, 22:1–63. New York: Academic

Sonnenschein E, Reyer H-U. 1983. Mate-guarding and other functions of antiphonal duets in the slate-coloured boubou (*Laniarius funebris*). *Z. Tierpsychol.* 63:112–40

Strohbach S, Curio E, Bathen A, Epplen JT, Lubjuhn T. 1998. Extrapair paternity in the great tit (*Parus major*): a test of the "good genes" hypothesis. *Behav. Ecol.* 9:388–96

Stutchbury BJM. 1998a. Female mate choice of extra-pair males: Breeding synchrony is important. *Behav. Ecol. Sociobiol.* 43:213–15

Stutchbury BJM. 1998b. Breeding synchrony best explains variation in extra-pair mating system among avian species. *Behav. Ecol. Sociobiol.* 43:221–22

Stutchbury BJM. 1998c. Extra-pair mating effort of male hooded warblers, *Wilsonia citrina*. *Anim. Behav.* 55:553–61

Stutchbury BJ, Morton ES. 1995. The effect of breeding synchrony on extra-pair mating systems in songbirds. *Behaviour* 132:675–90

Stutchbury BJM, Piper WH, Neudorf DL, Tarof SA, Rhymer JM, et al. 1997. Correlates of extra-pair fertilization success in hooded warblers. *Behav. Ecol. Sociobiol.* 40:119–26

Stutchbury BJ, Rhymer JM, Morton ES. 1994. Extrapair paternity in hooded warblers. *Behav. Ecol.* 5:384–92

Sundberg J. 1994. Paternity guarding in the yellowhammer *Emberiza citrinella*—a detention experiment. *J. Avian Biol.* 25:135–41

Thornhill R, Alcock J. 1983. *The Evolution of Insect Mating Systems*. Cambridge, MA: Harvard Univ. Press. 547 pp.

Trivers RL. 1972. Parental investment and sexual selection. In *Sexual Selection and the Descent of Man, 1871–1971*, ed. B Campbell, pp. 136–79. Chicago: Aldine
Tuttle EM, Pruett-Jones S, Webster MS. 1996. Cloacal protuberances and extreme sperm production in Australian fairy-wrens. *Proc. R. Soc. London Ser. B* 263:1359–64
Valera F, Hoi H, Krištín A. 2003. Male shrikes punish unfaithful females. *Behav. Ecol.* 14:403–8
Venier LA, Dunn PO, Lifjeld JT, Robertson RJ. 1993. Behavioural patterns of extra-pair copulation in tree swallows. *Anim. Behav.* 45:412–15
Wagner RH. 1991. Evidence that female razorbills control extra-pair copulations. *Behaviour* 118:157–69
Wagner RH. 1998. Hidden leks: sexual selection and the clustering of avian territories. See Parker & Burley 1998, pp. 123–45
Wagner RH, Schug MD, Morton ES. 1996. Condition-dependent control of paternity by female purple martins: implications for coloniality. *Behav. Ecol. Sociobiol.* 38:379–89
Weatherhead PJ, Boag PT. 1995. Pair and extra-pair mating success relative to mate quality in red-winged blackbirds. *Behav. Ecol. Sociobiol.* 37:81–91
Weatherhead PJ, Montgomerie R, Gibbs HL, Boag PT. 1994. The cost of extra-pair fertilizations to female red-winged blackbirds. *Proc. R. Soc. London Ser. B* 258:315–20
Weatherhead PJ, Yezerinac SM. 1998. Breeding synchrony and extra-pair mating in birds. *Behav. Ecol. Sociobiol.* 43:217–19
Westneat DF. 1988. Parental care and extra-pair copulations in the indigo bunting. *Auk* 105:149–60
Westneat DF. 1992. Do female red-winged blackbirds engage in a mixed mating strategy? *Ethology* 92:7–28
Westneat DF. 1993a. Polygyny and extra-pair fertilizations in eastern red-winged blackbirds (*Agelaius phoeniceus*). *Behav. Ecol.* 4:49–60
Westneat DF. 1993b. Temporal patterns of within-pair copulations, male mate-guarding, and extra-pair events in eastern red-winged blackbirds (*Agelaius phoeniceus*). *Behaviour* 124:267–90
Westneat DF. 1994. To guard mates or go forage: Conflicting demands affect the paternity of male red-winged blackbirds. *Am. Nat.* 144:343–54
Westneat DF. 2000. Toward a balanced view of the sexes: a retrospective and prospective view of genetics and mating patterns. In *Vertebrate Mating Systems*, ed. M Appolonio, M Festa-Bianchet, D Mainardi, pp. 253–306. Singapore: World Sci.
Westneat DF, Rambo TB. 2000. Copulation exposes female red-winged blackbirds to bacteria in male semen. *J. Avian Biol.* 31:1–7
Westneat DF, Sargent RC. 1996. Sex and parenting: the effects of sexual conflict and parentage on parental strategies. *Trends Ecol. Evol.* 11:87–91
Westneat DF, Sherman PW. 1993. Parentage and the evolution of parental behavior. *Behav. Ecol.* 4:66–77
Westneat DF, Sherman PW. 1997. Density and extra-pair fertilizations in birds: a comparative analysis. *Behav. Ecol. Sociobiol.* 41:205–15
Westneat DF, Sherman PW, Morton ML. 1990. The ecology and evolution of extra-pair copulations in birds. *Curr. Ornithol.* 7:331–70
Whittingham LA, Dunn PO. 2001a. Survival of extrapair and within-pair young in tree swallows. *Behav. Ecol.* 12:496–500
Whittingham LA, Dunn PO. 2001b. Male parental care and paternity in birds. *Curr. Ornithol.* 16:257–98
Wink M, Dyrcz AJ. 1999. Mating systems in birds: a review of molecular studies. *Acta Ornithol.* 34:91–109
Wolf JB, Brodie EB III, Moore AJ. 1999. Interacting phenotypes and the evolutionary process II. Selection resulting from social interactions. *Am. Nat.* 153:254–66

Annu. Rev. Ecol. Evol. Syst. 2003. 34:397–423
doi: 10.1146/annurev.ecolsys.34.011802.132421

First published online as a Review in Advance on August 14, 2003

SPECIES-LEVEL PARAPHYLY AND POLYPHYLY: Frequency, Causes, and Consequences, with Insights from Animal Mitochondrial DNA

Daniel J. Funk[1] and Kevin E. Omland[2]
[1]*Department of Biological Sciences, Vanderbilt University, Nashville, Tennessee 37235; email:daniel.j.funk@vanderbilt.edu*
[2]*Department of Biological Sciences, University of Maryland, Baltimore County, Baltimore, Maryland 21250; email: omland@umbc.edu*

Key Words gene trees, introgression, lineage sorting, species concepts, paraphyletic species

■ **Abstract** Many uses of gene trees implicitly assume that nominal species are monophyletic in their alleles at the study locus. However, in well-sampled gene trees, certain alleles in one species may appear more closely related to alleles from different species than to other conspecific alleles. Such deviations from species-level monophyly have a variety of causes and may lead to erroneous evolutionary interpretations if undetected. The present paper describes the causes and consequences of these paraphyletic and polyphyletic patterns. It also provides a detailed literature survey of mitochondrial DNA studies on low-level animal phylogeny and phylogeography, results from which reveal the frequency of nonmonophyly and patterns of interpretation and sampling. This survey detected species-level paraphyly or polyphyly in 23% of 2319 assayed species, demonstrating this phenomenon to be statistically supported, taxonomically widespread, and far more common than generally recognized. Our findings call for increased attention to sampling and the interpretation of paraphyletic and polyphyletic gene trees in studies of closely related taxa by systematists and population geneticists alike and thus for a new tradition of "congeneric phylogeography."

INTRODUCTION

Intraspecific variation is at the core of modern evolutionary biology, its prevalence and importance having been increasingly documented at the phenotypic and genotypic levels over the course of the twentieth century. Whereas many biological disciplines implicitly adopted a more typological approach—studying the physiology or molecular biology of individuals, then extrapolating to entire species and beyond—evolutionary biology has long emphasized the importance of appropriately sampling any trait or process so as to identify, and thus have the opportunity to interpret, important elements of variation.

Interestingly, however, while an early source of molecular data (allozymes) greatly motivated interest in variation, the more recent introduction of DNA sequences initially reduced the emphasis on certain aspects of variation in studies of evolutionary history. This de-emphasis presumably occurred for the same reason that variation is largely ignored in other fields, namely, constraints on money and effort that restricted the number of individuals that could practically be studied. The sampling traditions of two groups of biologists who embraced these new data reflect different responses to these constraints (Barraclough & Nee 2001, Funk 1999). To caricature these two traditions: Systematists began to use DNA sequences to study the phylogenetic relationships among taxa by sampling a single individual per species, whereas population biologists began to evaluate phylogeographic patterns in DNA sequence variation among many individuals within a single species (Avise 2000, Avise et al. 1987).

In such cases of extremely restricted intraspecific or interspecific sampling, the accuracy of various evolutionary inferences depends on the assumption that individual study species are monophyletic with respect to the alleles at the study locus. That is, they assume that all the DNA sequence alleles that might be collected from individuals of a given species are more closely related to each other than to any alleles that exist in any other species. In turn, this assumption requires that nominal study species represent genetically and reproductively independent lineages whose boundaries have been accurately identified by taxonomists and whose reconstructed gene trees are accurate approximations of organismal history, i.e., species trees. However, only by sampling multiple individuals from each of multiple species can both intraspecific and interspecific variation be assessed, allowing the hypothesis of species-level monophyly to be tested.

The alternatives to species-level monophyly are species-level paraphyly or polyphyly (Figure 1*a*) in which gene trees reveal an allele from one species to be more closely related to particular alleles in a different species than any conspecific allele (but see Wheeler & Nixon 1990). In this review, we use the term polyphyly in referring to both paraphyly—in which all the haplotypes of one or more species are phylogenetically nested within the haplotypes of a second, paraphyletic, species—and narrow-sense polyphyly—in which various haplotypes from the polyphyletic species are phylogenetically interspersed with those of other species such that they are not phylogenetically contiguous with each other on the gene tree. We use polyphyly as our more general term rather than nonmonophyly to avoid awkward prose; we use it rather than paraphyly because polyphyly is the older term and we hope that temporarily expanding its meaning to include paraphyly will be less discordant with past systematics literature than the reverse. We commonly use the term polyphyletic species as convenient shorthand in referring to currently recognized species taxa whose alleles exhibit a polyphyletic pattern in the broad sense outlined above. This pattern is significant both because of what it may reveal about the biology of the polyphyletic species and because of the consequences it may have for evolutionary inference if undetected.

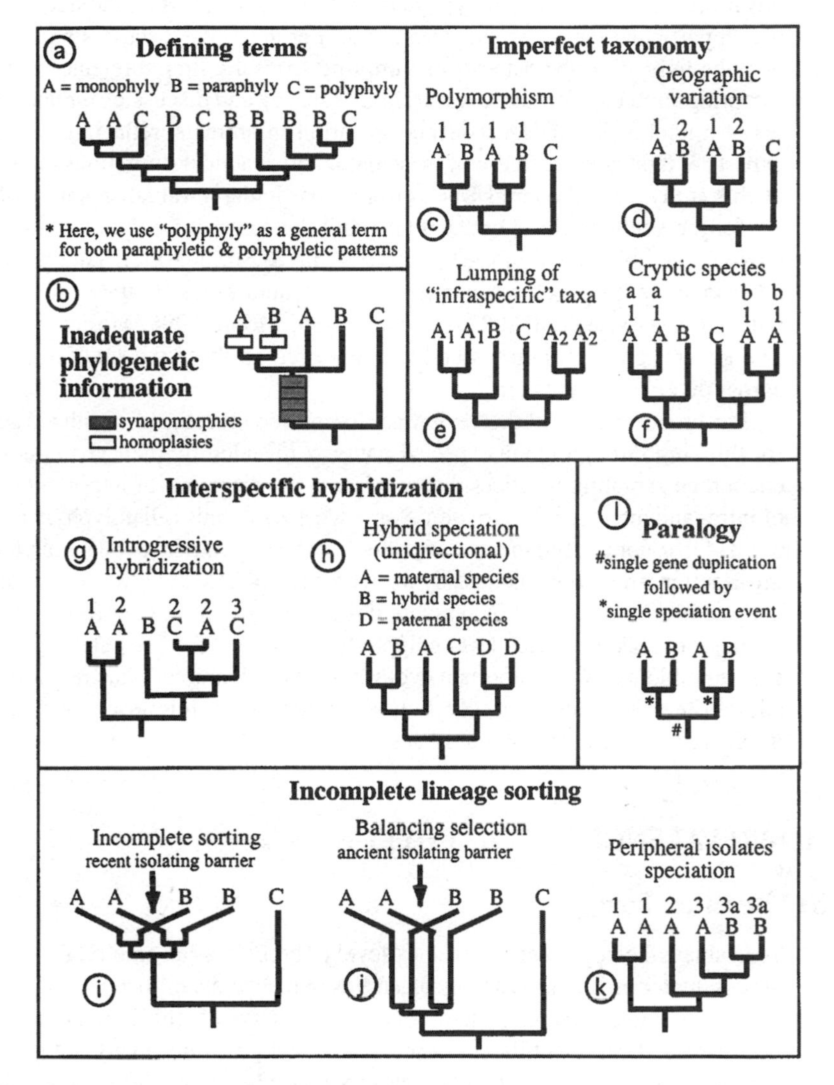

Figure 1 Species-level polyphyly and its causes. This figure illustrates patterns of gene-tree topology that are consistent with various causes of species-level polyphyly. Capital letters represent nominal species; numbers represent geographic regions; the subscripts in (*e*) identify recognized subspecies; the lowercase letters in (*f*) represent subtle phenotypic differences subsequently found to distinguish cryptic species first identified from a gene tree. See text for details.

In this review, we provide the first analysis of the observed frequency and taxonomic distribution of species-level polyphyly. We describe and contrast the various mechanisms that yield polyphyletic patterns and report on the frequency with which particular causes are invoked in the literature. We discuss the implications of polyphyly, describe patterns of sampling from the literature, and recommend sampling strategies for future research. Our survey and discussion emphasize studies of mitochondrial DNA sequence variation in animals, reflecting the authors' expertise, the widespread use of these data, and practical limitations on the scope of this study. Mitochondrial DNA offers a particularly valuable source of markers for the study of closely related taxa and the causes of polyphyly owing to its lack of recombination (but see Maynard Smith & Smith 2002), maternal mode of inheritance (but see below), simple genetic structure, rapid rate of mutation, and reduced N_e (Avise et al. 1987, Harrison 1989, Moore 1995, Moritz et al. 1987). The general principles discussed here, however, apply to the study of gene trees across diverse loci and taxa.

The broadest goals of this review are to provide investigators with a framework for thinking through the unexpected patterns revealed by their gene trees and to encourage sampling practices that maximize the detection of important elements of intra- and interspecific variation. Some workers dismiss all polyphyly as reflecting bad taxonomy. And indeed, imperfect taxonomy and inadequate phylogenetic information are two of the causes we will discuss below. However, we also emphasize introgression and incomplete lineage sorting following recent speciation as major causes of species-level polyphyly that reflect fundamental aspects of organismal biology with important evolutionary implications. This review does not address the observation of polyphyly at higher taxonomic levels or its practical implications for nomenclatural issues.

THE PREVALENCE OF POLYPHYLY

The Literature Survey

To evaluate the importance of species-level polyphyly as an empirical observation, we conducted an intensive survey of studies that evaluate mitochondrial DNA variation in animals in a phylogenetic context. This survey included only those studies with a theoretical possibility of observing polyphyly. Since many of the causes of polyphyly are most likely to affect closely related taxa, we further and arbitrarily limited our data collection to studies of congeners. In order to avoid inflating our estimates of polyphyly, we excluded explicit studies of hybrid zones, where polyphyly would be expected. Included studies were all others that treated at least two congeneric species, at least two individuals from one of these species, and an outgroup. For each species represented by multiple individuals (and thus potentially polyphyletic) we recorded: number of individuals, localities, congeneric species, and congeneric individuals sampled; whether or not polyphyly was observed; and (where presented by the authors, when polyphyly was observed) bootstrap support and possible explanations for the observed polyphyletic pattern. Polyphyly was evaluated using the species-level taxonomy adopted by our studies' authors and their published mtDNA-only phylogenies. Where multiple trees were published, the phylogeny derived from the most data was used. Where unresolved haplotype relationships were consistent with either monophyly or polyphyly, the species was removed from the analysis. When multiple studies treated the same species, the species was recorded as polyphyletic if polyphyly was detected in any study. The large majority of included studies treated mtDNA sequence data, but appropriate mtDNA restriction analysis studies were also evaluated. Surveyed studies were those published between 1990 and 2002 in 14 leading journals: *Annals of the Entomological Society of America, Biological Journal of the Linnaean Society, Copeia, Evolution, Genetics, Heredity, Journal of Evolutionary Biology, Journal of Mammalogy, Journal of Molecular Evolution, Molecular Biology and Evolution, Molecular Ecology, Molecular Phylogenetics and Evolution, Systematic Biology*, and *The Auk*. Citations for all papers treated in our survey are available via the Supplementary Materials link in the online version of this chapter at http://www.annualreviews.org/.

The Distribution of Polyphyletic Species

Our 13-year survey treated 584 studies, 526 genera, and 2319 potentially polyphyletic species (Table 1). Overall, 535 species proved to be polyphyletic, representing 23% of those surveyed. Forty-four percent (44%) of genera included at least one polyphyletic species, with more than half of these study genera including at least two polyphyletic species. A number of studies showed rampant polyphyly involving many (up to 12) congeners, phylogenetically far-flung haplotypes,

TABLE 1 Results of the literature survey

Taxa	Number of: Studies	Genera	Spp.	Percent spp. polyphyletic[a]
Mammals	139	102	469	17.0
Birds	74	87	331	16.7
Reptiles	56	45	147	22.4
Amphibians	35	26	137	21.3
Fishes	100	99	371	24.3
Arthropods	143	126	702	26.5
Other Invertebrates	37	41	162	38.6
TOTAL	584	526	2319	23.1

[a]Percentage of surveyed species observed to exhibit a paraphyletic or polyphyletic pattern of haplotype relationships.

extreme polyphyly, or multiple species emerging from widespread polyphyletic forms (e.g., Crandall & Fitzpatrick 1996, Demboski & Cook 2001, Funk 1999, Porter et al. 2002, Sota & Vogler 2001, van Oppen et al. 2001). The incidence of polyphyly was also taxonomically widespread, observed for at least 15% of species in each evaluated animal class and phylum (Table 1; this is also true of cnidarians, mollusks, insects, crustaceans, arachnids and echinoderms when these invertebrate taxa are considered individually). Interestingly, there seemed to be a negative correlation between intensity of study and proportion of polyphyletic species across taxa, with birds and mammals exhibiting less than half the incidence of polyphyly observed in nonarthropod invertebrates, a pattern than might partly reflect inadequate taxonomy (see below). In sum, these results clearly indicate that species-level monophyly cannot be assumed and that species-level polyphyly is a much more important phenomenon than is generally recognized. To the degree that any bias exists against publishing untidy results, this survey may yet underestimate polyphyly's actual prevalence. Such a bias might explain an apparent recent decline in the reported incidence of polyphyly (1990–1999 = 28.2%, 2000–2002 = 19.7%) that isn't readily explained by changes in sampling or phylogenetic information content. Alternatively, this pattern could reflect a tendency for early studies on a group to focus on problematic taxa.

CAUSES AND INTERPRETATIONS OF POLYPHYLETIC PATTERNS

An observation of polyphyly should prompt a consideration of its particular causes. When interpreting molecular variation, however, it is often tempting to offer ad hoc explanations for unusual patterns without fully considering alternatives. This tendency is exacerbated when certain explanations have achieved wide recognition only recently or by workers in certain fields or students of certain taxa. In this section we try to alert workers to the full range of phenomena that may produce species-level polyphyly and to explain how they do so (Avise 1994, Funk 1996, Slowinski & Page 1999). In some cases, observed polyphyly is an artifact of misidentified specimens, species limits, and study loci, or of inadequate information. In others, it reflects aspects of allelic history that provide important insights into species biology. Where possible, we recommend means of distinguishing among these alternative explanations. Unfortunately, however, clear one-to-one correspondence between specific causes and particular patterns often does not exist so that definitive conclusions may frequently remain elusive.

Inadequate Phylogenetic Information

One potentially quite general cause of observed polyphyly is weak phylogenetic signal, which may result in poor phylogenetic resolution or inaccurate gene trees as an artifact of phylogenetic reconstruction. Phylogenetic algorithms can create topologies regardless of the amount and quality of the data. Thus, if a gene is evolving too slowly relative to the rate of speciation in one's study taxa or if too small a fragment of that gene is analyzed, obtained data may provide too few synapomorphies to robustly recover the underlying gene tree, a challenge that becomes greater if positively misleading homoplasies confound the few variable sites (Figure 1*b*). Although rapidly evolving mitochondrial sequences are less prone to inadequate information than most loci, even mtDNA may exhibit insufficient variation for the accurate reconstruction of very recent phylogenetic radiations. On the other hand, sequences from a gene that evolves very rapidly relative to speciation rates might be saturated and produce an inaccurate gene tree owing to high levels of homoplasy. Thus, even if all study species are in fact monophyletic, a reconstructed gene tree may erroneously exhibit polyphyletic groupings that do not accurately represent the history of the analyzed alleles or species.

The studies in our survey adopted various approaches to assess the likely historical accuracy of polyphyletic gene trees. Some studies tested whether a topology constrained to be monophyletic represented a significantly worse fit to the data than the observed polyphyletic topology using, for example, the method of Kishino & Hasegawa (1989). More commonly, Bremer support (Bremer 1988) and especially bootstrap support (Felsenstein 1985) were offered as estimates of the degree to which the data supported haplotype groupings. Here, we use reported bootstrap values to assess the generality of statistical support for observed polyphyletic patterns. Specifically, for each polyphyletic species (A), we recorded the largest bootstrap value that grouped any haplotypes of A with one or more haplotypes from any other species to the phylogenetic exclusion of some other A haplotypes. This provided a conservative estimate of the support for polyphyly because only one of potentially multiple supporting nodes was considered.

We found that 85% of polyphyletic species were from studies that employed bootstrap proportions, providing a large sample for this analysis. Among these studies, the percentages of polyphyletic species supported by various bootstrap proportions were as follows: $<50 = 17\%$ of species, $50–69 = 15\%$, $70–94 = 22\%$, and $\geq 95 = 46\%$. Thus, in two-thirds of observed cases polyphyly was supported by $\geq 70\%$ of bootstrap replicates. These results provide compelling evidence that the prevalence of polyphyly documented by our survey reflects a common aspect of true mitochondrial gene trees and is not simply a common artifact caused by inadequate data.

The remaining causes of polyphyly result not from imperfect phylogenetic reconstructions, but despite well-supported gene trees with topologies that likely depict the true origins and relationships among sampled alleles. Such gene trees may nonetheless disagree with recognized species boundaries—and produce polyphyly—for a number of reasons. To simplify and separate our discussions of these reasons, we hereafter assume that the phylogenetic patterns invoked are strongly supported, unless stated otherwise.

Imperfect Taxonomy—Inaccurate Species Limits

One important reason for the observation of polyphyly is a failure of the taxonomic circumscription of a nominal species to correspond to patterns of gene flow.

That is, polyphyly sometimes results from "bad taxonomy" when named species fail to identify the genetic limits of separate evolutionary entities. This failure can occur either by underestimating or by overestimating the field of genetic exchange among individuals and populations. In both situations, polyphyly can be validly eliminated simply by changing current taxonomy. More trivially, polyphyly can result from the misidentification of samples, providing a strong argument for maintaining voucher specimens.

SPECIES OVERSPLIT—MISIDENTIFYING INTRASPECIFIC VARIATION AS SPECIES-LEVEL VARIATION Taxonomy underestimates the breadth of species limits when anatomical (or behavioral, ecological, etc.) variants of a single species have erroneously been described as separate nominal species. This may occur, for example, when distinctive variants coexist within individual populations of a single polymorphic species. Under this scenario, no phylogenetic substructure as a function of variant type is expected (e.g., Demastes et al. 2002, Nice & Shapiro 2001, Small & Gosling 2000) (Figure 1*c*). This is because local gene flow among variants should produce a gene tree in which sympatric haplotypes from each variant are cladistically intermingled with those of the other(s), rendering each polyphyletic. Furthermore, levels of genetic variation among these "oversplit" nominal species are expected to be typical of within-species variation in the taxa under study. Incomplete lineage sorting (see below), however, can produce the same patterns.

Species polyphyly may similarly be observed if two nominal species actually represent geographic variants (races, subspecies) of a single species that continue to exchange genes. In this case, the observed phenotypic divergence may be either environmentally induced or genetically based and maintained by strong selection despite gene flow. If haplotypes of these geographic variants do not phylogenetically segregate into separate clades, conspecificity is supported (Figure 1*d*). Unlike the polymorphism example above, however, some degree of phylogenetic substructuring by variant type might be observed if gene flow is geographically restricted, yielding isolation by distance. In such cases, distinguishing between intraspecific variation and interspecific introgression (see below) as a cause of these patterns may be difficult.

SPECIES OVERLUMPED—MISIDENTIFYING SPECIES-LEVEL VARIATION AS INTRASPECIFIC VARIATION Just as intraspecific variants may be mistaken for species, traits diagnostic of species are sometimes assumed to represent intraspecific variation or are simply difficult to detect at all. This may result in the taxonomic "lumping" of multiple species under a single name and the observation of polyphyly when these species are not sister taxa. In such cases, current taxonomy overestimates the breadth of species limits. This is sometimes observed, for example, with respect to subspecies, geographic forms, morphotypes, and other nominally infraspecific taxa that have been recognized on the basis of divergence in particular traits. When one or more infraspecific taxa within a nominal species prove to be mitochondrially monophyletic, a substantial history of genetic isolation of these taxa from other "conspecific" populations is indicated. In the case where distinct clades of this kind separate fully sympatric taxa, reproductive isolation between them is further indicated, as is species status under most species concepts. In the case where distinct clades are also geographically separated (Fukatsu et al. 2001, Kotlik & Berrebi 2002, Riddle et al. 2000), evidence on reproductive compatibility is ambiguous, and the decision of whether or not to recognize these genetically differentiated entities as separate species depends on the species concept applied. In either case, the telltale polyphyletic pattern is caused by the nesting of one or more additional nominal species among the haplotypes of the over-lumped species (A) (Figure 1*e*). When such nesting renders certain infraspecific taxa of A monophyletic, elevating these taxa to species rank is one strategy for taxonomically removing the species-level polyphyly (Omland et al. 1999, Voelker 1999).

Sometimes, clues to lumping may be scarce owing to the highly similar morphologies of unrecognized species. If other described species are more closely related to such "cryptic species" than the cryptic species are to each other, a mitochondrial gene tree might hint at cryptic taxa by revealing polyphyly in the form of two phylogenetically separated clades (Figure 1*f*) (Omland et al. 2000; Williams et al. 2001; D.J. Funk 1998, unpublished data). Such cryptic species might reflect the retention of ancestral morphology (Jarman & Elliott 2000). However, the same polyphyletic pattern would be expected if cryptic species resulted from the convergent evolution of similar morphologies (Kim et al. 2000, Rees et al. 2001, Richmond & Reeder 2002, Su et al. 1996). This might be expected if divergent lineages were responding to similar selection pressures, as in threespine stickleback fishes that have repeatedly evolved complex benthic- and pelagic-adapted morphologies (Bell 1987, Schluter & Nagel 1995). Such convergence creates special problems when traits under selection are also those used by taxonomists to define species.

In the scenarios just reviewed, polyphyly results when the described phenotypic boundaries of nominal species do not adequately or accurately reflect the history of population differentiation and speciation. That is, polyphyly results even if a species tree can be safely assumed to be identical to the gene tree used to infer it. By contrast, the remaining causes of polyphyly generally reflect situations where the history of alleles revealed by a gene tree is incongruent with the actual organismal history embodied by the species tree (but see Doyle 1997, Maddison 1997). This "gene tree/species tree problem" (Avise et al. 1983; Brower et al. 1996; Doyle 1992; Goodman et al. 1979; Maddison 1996, 1997; Nichols 2001; Pamilo & Nei 1988; Slowinski & Page 1999; Wu 1991) represents a major limitation on evolutionary inferences from single-locus (e.g., mitochondrial) gene trees that has not yet been fully incorporated into certain areas of systematic biology.

Interspecific Hybridization

One potential cause of gene tree/species tree discordance and accompanying polyphyly is the occasional mating between otherwise distinct species and resulting transfer of parental alleles to hybrid offspring. Two aspects are worth noting.

INTROGRESSION Alleles from one species may penetrate the gene pool of another through interspecific mating and the subsequent backcrossing of hybrids into parental populations, a process known as introgressive hybridization, introgression, or interspecific gene flow. Introgression yields polyphyly by introducing phylogenetically divergent allelic lineages across species boundaries (e.g., Boyce et al. 1994; Patton & Smith 1994; Shaw 1999, 2002). The phylogenetic effects of mitochondrial introgression are particularly great because a lack of recombination entails that all mitochondrial base positions introgress as a completely linked block (Smith 1992). Thus, any analyzed fragment of introgressed mtDNA will entirely reflect the heterospecific origin of its mitochondrial genome. Furthermore, mitochondrial alleles might be expected to introgress farther, on average, than nuclear loci if their persistence in a foreign gene pool is less constrained by linkage to selected loci than are the alleles of nuclear genes (Barton & Jones 1983, Harrison et al. 1987, Marchant 1988, Tegelström 1987; reviewed in Harrison 1990, Arnold 1993). For these reasons, mitochondrial gene trees could be particularly susceptible to the effects of introgression. An interesting exception is offered by female heterogametic taxa following Haldane's rule, such as birds (Tegelström & Gelter 1990) and butterflies (Sperling 1993). In such cases, female hybrids show reduced viability that might restrict the introgression of maternally inherited mtDNA between species, offering a potential explanation for low mtDNA introgression in several avian hybrid zones (e.g., Allen 2002, Brumfield et al. 2001, Sattler & Braun 2000). More generally, the exposure of haploid mtDNA loci to selection in all individuals may also impede its introgression (Brumfield et al. 2001). The differential introgression of mitochondrial versus nuclear alleles and its effects on polyphyly is an important topic that deserves further attention.

Recognizing mitochondrial introgression requires evaluating a mitochondrial gene tree against a nuclear background that identifies the participating taxa. This background can be provided by gene trees from nuclear loci or simply by consistent taxon-specific phenotypic differences that presumably have a nuclear basis (Smith 1992). The clearest signature of introgression is the sympatric sharing of geographically localized mtDNA sequence haplotypes between otherwise genetically and morphologically divergent species (Figure 1*g*). Such a pattern is hard to interpret as anything but ongoing (or very recent) and geographically localized interspecific gene flow. Importantly, introgression may not be detected in such situations unless populations are indeed sympatrically sampled because species that share haplotypes in regions of geographic overlap may otherwise exhibit reciprocal monophyly in gene trees based on allopatric samples (e.g., Masta et al. 2002, Redenbach & Taylor 2002).

Unfortunately, confidently attributing polyphyly to introgression becomes progressively more difficult the farther in the past that gene flow last occurred. Species that have rather recently ceased exchanging genes may no longer share haplotypes (because of post-introgression mutation) yet still possess very closely related haplotypes that are nested together within the gene tree. However, as the time since last gene flow increases, those introgressed allelic lineages that do persist are more likely to be phylogenetically basal (as a result of the sorting out of allelic polymorphisms dating to the time of introgression) and less likely to show any geographic association with the population from which they introgressed (to the degree that populations change distributions over time).

Other factors can further complicate the recognition of introgression as a cause of polyphyly. If mitochondrial gene flow is bidirectional, very common, or occurs among multiple species, its affect on mitochondrial tree topology may be profound, making it difficult or impossible to confidently infer patterns of genetic exchange or even to determine which mitochondrial clade represents the "native" lineage of particular species. Mitochondrial gene trees may be especially misleading in cases where introgressed haplotype lineages become fixed, leaving no hint that they are of heterospecific origin. The smaller N_e of mtDNA compared with nuclear loci may facilitate this process, such that even low levels of introgression may be sufficient to establish a neutral mitochondrial genotype in a foreign population (Takahata & Slatkin 1984). Patton & Smith (1994), for example, attributed complicated polyphyletic patterns in pocket gophers to sporadic episodes of hybridization combined with small, patchy gopher populations that facilitated the fixation of introgressed alleles. Recurrent hybridization has been similarly invoked to explain rampant polyphyly in a variety of taxa (Freeland & Boag 1999, Funk 1999, Shaw 2002, Sota & Vogler 2001).

HYBRID SPECIATION Polyphyly may also result from the spontaneous formation of a new species through interspecific hybridization, a mechanism that has been demonstrated in various animal taxa (e.g., Moritz et al. 1992; reviewed by Dowling & Secor 1997). In such instances, the initial relationship among parental and a new hybrid species' mitochondrial alleles will depend on the number and symmetry of hybrid speciation events. Most hybrid species appear to originate via asymmetrical hybridization. A hybrid species formed by a single such event will itself be mitochondrially monophyletic, while specifically rendering the mitochondria-contributing maternal species paraphyletic (Figure 1*h*). A hybrid species formed through repeated asymmetric hybridizations (all involving, e.g., a female of species A and a male of species B) will be monophyletic if the participating females have identical mitochondrial haplotypes, polyphyletic otherwise (e.g., Mantovani et al. 2001). A hybrid species formed through symmetric hybridization events would be polyphyletic, as would both parental species. Because hybrid speciation is often associated with polyploidy or asexual reproduction, knowledge of such traits may bolster a suspicion that hybrid speciation is the cause of observed polyphyly (e.g., Johnson & Bragg 1999). However, in several cases of putative hybrid speciation (Hedrick et al. 2002, Wayne & Jenks, 1991; also see Salzburger et al. 2002) alternative explanations have proven difficult to rule out.

Incomplete Lineage Sorting

The incomplete sorting of ancestrally polymorphic allelic lineages represents a very general source of polyphyly, potentially afflicting any single-locus gene tree in any taxon. Within any species, the various alleles at a particular locus have their own history, with some alleles sharing more recent, and others more ancient, coalescent events (Pamilo & Nei 1988). Thus, the random division of allele copies at speciation will generally result in each daughter species possessing certain alleles that are most closely related to those in the other daughter species. For this reason, new species are initially expected to exhibit polyphyletic gene trees (Figure 1*i*). Over time, allelic lineages in each daughter species will be randomly lost by drift, and new alleles will be formed by mutation until eventually only one of the (ancestrally polymorphic) allelic lineages present in the parent species survives in each daughter species and all intraspecific variation reflects post-speciation mutation. At this point, sorting has gone to completion, and alleles in the two daughter species are reciprocally monophyletic. This progression from polyphyly (narrow-sense) to paraphyly to monophyly is expected to take on the order of $4N_e$ generations for mitochondrial loci and ultimately results in a gene tree that accurately reflects the species tree (Avise 1989, Avise & Ball 1990, Harrison 1991, Neigel & Avise 1986, Pamilo & Nei 1988, Tajima 1983, Takahata & Nei 1985).

Because the mitochondrial genome is haploid and maternally inherited, the N_e of mitochondrial loci is generally one-quarter that of nuclear loci (but see Hoelzer 1997), and stochastic lineage sorting is expected to progress more rapidly for mitochondrial alleles. Thus, incomplete sorting is less of a concern for mitochondrial than for nuclear loci, other things being equal, providing one advantage to using mitochondrial gene trees as estimates of species trees for closely related taxa (Hudson & Turelli 2003). Indeed, theory predicts that if one can be 95% certain that an internode in a single mitochondrial gene tree has not been affected by incomplete sorting, 16 independent nuclear gene trees would be required to justify an equal level of confidence (Moore 1995). Nonetheless, incomplete sorting also affects mitochondrial gene trees and can have especially major effects in the case of rapidly radiating taxa, in which succeeding speciation events occur before sorting is completed. This scenario has been invoked to explain the sharing of alleles among multiple species in the rampant polyphyly exhibited by cichlid fishes and other taxa (Moran & Kornfield 1993, 1995; also see Crandall & Fitzpatrick 1996, Goodacre & Wade 2001, Klein & Payne 1998).

Unfortunately, it is impossible to demonstrate conclusively that incomplete sorting explains any particular case of polyphyly. One problem is that because species remain incompletely sorted for a narrow window of evolutionary time, a dearth of accumulated synapomorphies may often make it difficult to distinguish incomplete sorting from inadequate phylogenetic information as a cause of observed polyphyly (Slowinski & Page 1999). Another problem is the difficulty of distinguishing the effects of incomplete sorting and introgression, an issue of considerable interest. A phylogenetically basal position of polyphyly rendering haplotypes hints at retained ancestral polymorphism, while recently introgressed alleles may assume a highly derived position in the gene tree. Also, incomplete sorting is not predicted to promote the geographic proximity of interspecifically shared alleles that may be seen under local introgression (Hare & Avise 1998, Masta et al. 2002). However, these criteria are often inadequate to distinguish ancient mitochondrial introgression from incomplete sorting (Schneider-Broussard et al. 1998). Comparisons with nuclear markers and geography can provide additional insights (e.g., Redenbach & Taylor 2002, Tegelström 1987, Weckstein et al. 2001), as may nested clade analysis (Templeton 1998, Templeton et al. 1995; but see Knowles & Maddison 2002). Moore (1995) suggested comparing maximum intraspecific sequence divergences between a polyphyletic species and related species in an empirical method for evaluating the likelihood of incomplete sorting (see Baker et al. 2003, Holder et al. 2001, Knowles 2000, Mason et al. 1995, Palumbi et al. 2001, Rees et al. 2001). Other, more statistical, methods have also been described (e.g., Nielsen & Wakeley 2001, Sang & Zhong 2000, Wakeley 1996). However, a generally diagnostic and widely agreed-upon approach for documenting incomplete sorting and distinguishing it from introgression has not yet emerged (Holder et al. 2001).

SPECIATION AND SORTING Although the progression of new species from initial polyphyly through paraphyly to monophyly follows quite generally on the heels of speciation, the particular pattern and time course of this progression may be rather distinctive in the case of peripatric, peripheral isolates, or "budding" speciation (Frey 1993; Harrison 1991, 1998; Rieseberg & Brouillet 1994), in which populations along the periphery of a species range become spatially isolated and speciate. To the degree that a "parental" species exhibits geographic substructure and a peripherally speciating population is small and local, this population may be predicted to initially possess a phylogenetically restricted subset of parental alleles and may lose alleles under drift at a faster rate than the larger parental population. For these reasons, peripheral isolates speciation may commonly yield a geographically restricted daughter species whose monophyletic set of haplotypes is embedded within a widely distributed and still paraphyletic parental species (termed a ferespecies by Graybeal 1995; also see Baum & Shaw 1995, Olmstead 1995) (e.g., Avise et al. 1990, Funk et al. 1995a, Hedin 1997, Marko 1998; but see Knowles et al. 1999) (Figure 1*k*). This deep phylogenetic nesting is not expected under large-scale vicariant or parapatric modes of speciation, although it might also be observed (in a different phylogeographic context) in the case of rapid, local sympatric speciation (Harrison 1998). This asymmetrically paraphyletic relationship will persist until sorting renders the parental species monophyletic.

In the case of budding speciation, forcing taxonomy to reflect gene tree monophyly by synonymizing the nested and parent species or by elevating lineages in the paraphyletic lineage to species status ignores the distinctive nature of the nested lineage (de Queiroz & Donoghue 1988; Harrison 1991, 1998; Olmstead 1995; Rieseberg & Brouillet 1994; Rodríguez-Robles & De Jesús-Escobar 2000; Sosef 1997; Wiens & Penkrot 2002). Under budding speciation, the cause of paraphyly is incomplete lineage sorting, yet the gene tree accurately reflects the history of population divergence. Thus, although gene trees from different loci are ordinarily expected, by chance, to be incongruent under incomplete sorting, budding speciation is predicted to produce parallel patterns of paraphyly across nuclear and mitochondrial loci (Hedin 1997, Marko 1998, Petren et al. 1999; but see Ballard 2000, Tosi et al. 2000). Because it reflects population history, this nested pattern is evolutionarily informative, allowing the polarization of the speciation event and of transitions between traits (host plant associations, plumage patterns, geographic ranges, etc.) that accompany and may have promoted speciation (e.g., Brown et al. 1994, 1996; Funk et al. 1995b; Omland 1997).

SELECTION AND SORTING The expected time to complete sorting invoked above, $4N_e$ generations, applies to strictly neutral mitochondrial alleles. However, mtDNA variation may often be subject to selection (Ballard & Kreitman 1995, Hudson & Turelli 2003, Rand 2001), which will affect the rate at which reciprocal monophyly is attained. While positive selection will accelerate allele fixation and sorting, balancing selection may preserve ancestrally polymorphic alleles within a population indefinitely (Figure 1*j*). Polymorphic nuclear MHC alleles, for example, are shared between otherwise genetically divergent species in several animal taxa (Klein et al. 1993, 1998). However, although there is some evidence for balancing selection on mtDNA in animals (James & Ballard 2000) and plants (Städler & Delph 2002), it has not yet been documented as a cause of species-level mitochondrial polyphyly.

Unrecognized Paralogy

Orthologous alleles derive from the same locus whereas paralogous alleles derive from different loci that originated by a gene duplication event. A gene tree that includes paralogous alleles may depict polyphyletic species because its topology reflects gene duplication as well as speciation (Figure 1*l*). The cause of this polyphyly may be misinterpreted if the orthology of alleles is assumed. Because mitochondrial loci are single-copy genes rather than members of multigene families, it was long considered safe to assume the orthology of alleles sequenced with mitochondrial primers. Two phenomena illustrate exceptions to this rule that cause polyphyly.

NUCLEAR PSEUDOGENES It is now well understood that segments of mitochondrial DNA are sometimes transposed into the nucleus where they become functionless pseudogenes (Bensasson et al. 2001, Collura & Stewart 1995, Sorenson & Fleischer 1996, Sunnucks & Hales 1996, Zhang & Hewitt 1996). When such nuclear copies of mtDNA exist, using mitochondrial primers for PCR amplification from whole-genomic DNA extractions (a common approach) may yield sequences of nuclear as well as mitochondrial origin. Indirect evidence for nuclear copies may be provided by unusual patterns of molecular evolution that are consistent with the reduced functional constraint (e.g., elevated frequencies of nonsynonymous substitutions, indels, frameshifts, and stop codons) or nuclear location (slowed rates of substitution) of pseudogenes. Nuclear copies may be more directly detected through the isolation of mtDNA, cloning, and rtPCR (Collura et al. 1996). Nuclear copies of mtDNA and their effects on polyphyly have now been documented in a variety of taxa.

PATERNAL INHERITANCE In a few cases, paternally inherited mitochondrial lineages have been shown to originate from maternally inherited ancestors, much as new loci are formed by a gene duplication event. These divergent maternal and paternal lineages can coexist within species, yielding species-level polyphyly in gene trees that include alleles from both. To date, such instances have generally been taxonomically restricted to various bivalve mollusks (e.g., Rawson & Hilbish 1995), so this phenomenon is not known to present a general cause of polyphyly. Recent results from humans (Bromham et al. 2003), however, illustrate that other taxa may also be affected.

Literature Patterns

Attempting to elucidate the actual causes of polyphyly in the studies from our survey is beyond the scope of the present review. However, some observations on authors' tendencies in reporting potential causes are worth noting. First, 24% of papers with polyphyletic gene trees offered no discussion of this pattern. Second, of those that evaluated polyphyly, 50% specifically suggested faulty taxonomy as one plausible explanation, introgressive hybridization was invoked in 32% of papers, and incomplete lineage sorting was cited in 30%. Inadequate phylogenetic information and unrecognized paralogy received mention in only a few papers each. Third, closer inspection of a subset (~one half) of the polyphyletic papers found that in 56% of these only one or another of three major causes (taxonomy, introgression, sorting) received any mention at all; two causes were mentioned in 25%, and all three in only 16% of the studies.

Although it is encouraging that most authors find polyphyly a worthy subject of comment, these patterns suggest that a fully pluralistic appreciation of its causes has yet to take root. The general disregard of nuclear copies of mtDNA as a possible explanation is especially concerning (but see Weckstein et al. 2001). The equivalent invocation of introgression and incomplete sorting and the considerably greater frequency of taxonomic explanations may reveal the biases of biologists or illuminate the relative importance of different causes. We recommend that future studies seek the most accurate and informative interpretations by systematically considering the full range of alternative explanations in accumulating datasets.

CONSEQUENCES FOR EVOLUTIONARY INFERENCE

Erroneous Estimates

If undetected, species-level polyphyly compromises evolutionary inferences based on gene trees that are erroneously assumed to accurately depict species trees (Funk 1996, 1999) (Figure 2). The extent of these problems will depend on several aspects of polyphyly, among them: (*a*) how commonly polyphyletic species occur in the study taxon, (*b*) how "polyphyletic" a given species is, i.e., how many

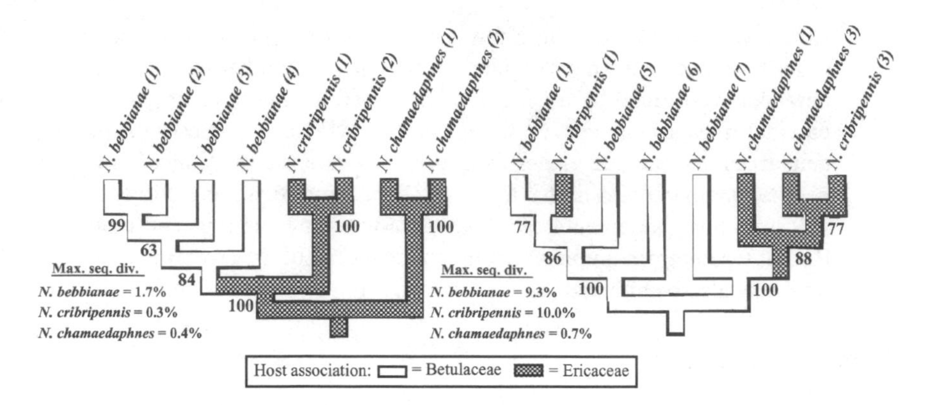

Figure 2 Erroneous evolutionary interpretations due to polyphyly. Each tree depicts the reconstructed relationships among mitochondrial haplotypes collected from different subsets of individuals representing three *Neochlamisus* leaf beetle species (data from Funk 1999). Haplotypes from different individual beetles are indicated by different numbers. Results illustrate how drastically estimates of phylogenetic relationship, character evolution, and genetic divergence can vary as a function of the particular individuals sampled when study species are highly polyphyletic. Bootstrap values indicate strong support of each data set for a different topology.

distinct allelic lineages are represented, (*c*) how genetically and phylogenetically diverse these allelic lineages are, and (*d*) how evenly alleles are distributed among these lineages. The likelihood that evolutionary inferences will vary dramatically according to the individuals sampled will be greatest when polyphyletic species bearing many, diverse, and equally frequent allelic lineages are common. In this context, the consequences of polyphyly may be more or less severe, on average, according to the particular cause. Incomplete lineage sorting, for example, is less likely to involve highly divergent allelic lineages than is an ancient duplication event or introgression between distantly related taxa. Some important inferential problems resulting from polyphyly are described below.

First and most basically, the phylogenetic relationships depicted by a gene tree may vary according to the particular individuals sampled when sequences from one or a few specimens are used as exemplars of polyphyletic species (Ballard 2000, Barraclough & Nee 2001, Funk 1999, Melnick et al. 1993, Omland et al. 1999, Smouse et al. 1991, Zink et al. 1998) (Figure 2). Systematists generally agree that multiple exemplars should be included at the level below the taxonomic rank of interest, and this should also apply to the species level (Ballard 2000, Barraclough & Nee 2001, Graybeal 1995, Omland et al. 1999, Wiens 1999; see also Lanyon 1994). Intensive sampling is more likely to document the underlying polyphyly and alert the systematist that something is amiss. Second, inferred times and rates of evolutionary divergence may be considerably inflated or deflated (depending on the cause of polyphyly) when alleles are sampled from polyphyletic species (Ballard 2000, Funk 1999, Melnick et al. 1993; also see Edwards & Beerli 2000) (Figure 2). Third, when gene trees are used to reconstruct the evolutionary history of particular traits, polyphyly presents especially egregious problems because a single misplaced taxon can have major effects on character transformations throughout the tree (Omland 1997; also see Graybeal 1995) (Figure 2). Fourth, polyphyly may compromise population genetic and phylogeographic studies. This may occur, for example, when sampling of related species is insufficient to identify the heterospecific (e.g., introgressive) origin of divergent haplotypes in the focal species, leading to mistaken conclusions about demography and evolutionary processes that are based on allelic frequencies and relationships (Redenbach & Taylor 2002, Tegelström 1987; also see Ballard 2000). Fifth, mitochondrial polyphyly complicates the identification of species-diagnostic molecular characters for practical issucs of management and conservation, such as identifying endangered species (Baker et al. 1996, Dalebout et al. 1998) and defining evolutionarily significant units (e.g., Moritz 1994, Paetkau 1999).

Intriguing Insights

Although species-level polyphyly can be quite problematic if undetected, when recognized it can provide informative clues that motivate future work (Funk 1996, 1998; Harrison 1998; Omland 1997). For example, appropriately sampled mitochondrial surveys can be an efficient means of detecting initial evidence for gene flow and insights on its direction, geographical and biological correlates, and participating taxa. Such observations can direct workers to informative investigations of hybrid zones, mechanisms of reproductive isolation, or the reexamination of species limits. The revised taxonomic and phylogenetic assessments of species-level taxa that are provoked by polyphyly may provide more stable classifications and more accurate historical inferences, for example, on character evolution (Omland 1997, Omland et al. 2000). The discovery of morphologically cryptic or unusually variable species may prompt studies on the evolutionary causes of convergence/stasis or polymorphism, respectively. Findings consistent with incomplete sorting may lead to studies of demography and speciation rates. Patterns consistent with budding speciation may identify taxa that contribute to ongoing debates on speciation mechanisms (Harrison 1991, Turelli et al. 2001). Even nuclear copies of mtDNA are now being exploited for novel evolutionary insights (Bensasson et al. 2001).

THE IMPORTANCE OF SAMPLING: PATTERNS AND PROSPECTS

Detecting polyphyly requires the sampling of multiple individuals of the polyphyletic species as well as other species with which it shares related alleles. As noted earlier, however, the phylogenetic and phylogeographic traditions initially adopted sampling strategies that sometimes precluded polyphyly detection. These strategies reflect the tendencies of these traditions to deemphasize intraspecific and interspecific sampling, respectively, in their studies of closely related taxa. These different sampling strategies in turn reflect the differing interests of these scientific disciplines—phylogenetic structure versus microevolutionary process—as well as practical constraints on data collection.

Yet sampling is clearly important (Ballard 2000, Funk 1999, Hedin & Wood 2002, Omland et al. 1999). The more individuals sampled, the greater the likelihood of detecting polyphyly and when the interspecifically shared alleles that cause polyphyly are rare, very intensive sampling may be required to document this pattern (Wiens & Servedio 2000). For a given sampling intensity, polyphyly detection should generally be increased by dividing samples across the phenotypic, geographic, and phylogenetic diversity of individuals and species that might plausibly share allelic lineages. An ideal study would thus include all species believed a priori to be closely related (e.g., congeners), maximize the geographic diversity of samples and the number of samples collected from areas of sympatry between study species, and sample broadly from known sources of biological variation (subspecies, ecotypes, morphological variants, etc.).

Our literature survey provided empirical data on the distribution of sampling patterns between 1990 and 2002, beginning with some of the earliest molecular systematic studies using DNA sequence data. Because we included only those studies theoretically capable of detecting polyphyly (those including multiply sampled species), our estimates of sampling intensity may be somewhat upwardly biased. This survey shows a regular increase and possible plateau in the frequency of such studies (yearly frequencies from 1990 to 2002 = 3, 6, 7, 12, 22, 28, 39, 36, 53, 76, 96, 116, and 89, respectively). Two patterns are worth emphasizing here (Figure 3). First, although studies vary greatly in sampling intensity, the majority included no more than a few individuals and sampling localities per study species. Second, median levels of four sampling parameters (total number of congeneric species and individuals sampled, mean number of individuals and localities sampled per study species) that might be expected to correlate positively with polyphyly detection have not notably increased over the 13 years of this survey (Figure 3). While it is certainly true that investigations treating multiple well-sampled species are becoming more common, our survey suggests that many studies continue to adhere to an implicit early established standard of acceptable sampling.

The collection of mtDNA sequences is no longer nearly as onerous as during the early years of our survey. However, the high-volume automatic sequencing and declining costs that have allowed the genomics revolution have not yet translated into generally and considerably improved sampling in mitochondrial studies of closely related animal species. The contribution of improved sampling to the detection of polyphyly is indicated by our survey, which shows that species observed to be polyphyletic were represented, on average, by significantly more conspecific

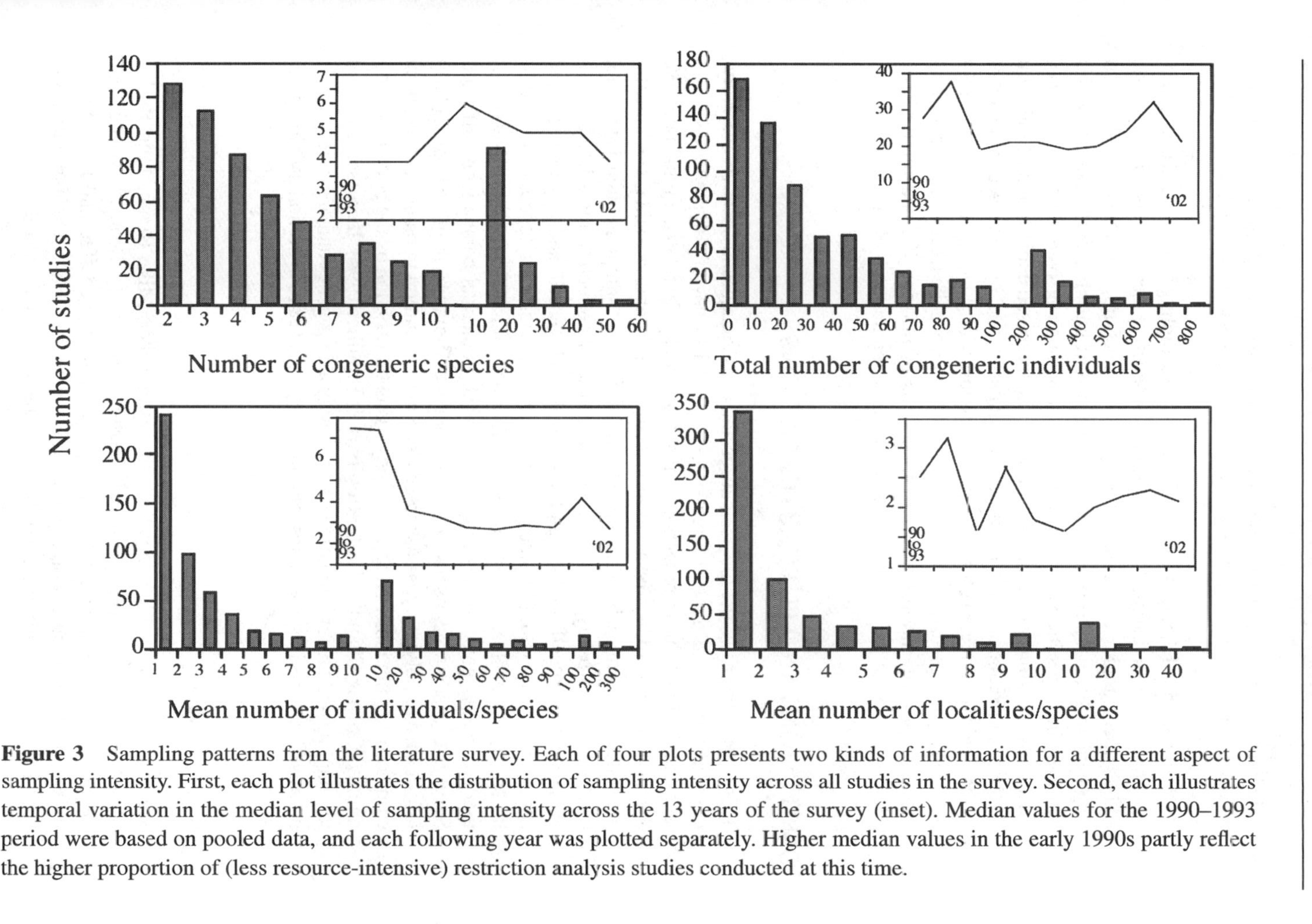

Figure 3 Sampling patterns from the literature survey. Each of four plots presents two kinds of information for a different aspect of sampling intensity. First, each plot illustrates the distribution of sampling intensity across all studies in the survey. Second, each illustrates temporal variation in the median level of sampling intensity across the 13 years of the survey (inset). Median values for the 1990–1993 period were based on pooled data, and each following year was plotted separately. Higher median values in the early 1990s partly reflect the higher proportion of (less resource-intensive) restriction analysis studies conducted at this time.

individuals (22.6 versus 10.7), collection localities (6.1 versus 3.8), and congeneric individuals (49.1 versus 26.9) than other species (one-way ANOVA, $P < 0.001$ in all cases).

We recommend that improved and diversified sampling be embraced as a timely and important goal that might allow a highly productive merger of the phylogenetic and phylogeographic traditions. Specifically, we encourage workers to collect and simultaneously analyze phylogeographic data from multiple closely related species and their infraspecific variants (also see Barraclough & Nee 2001, Hey 1994). These investigations might be facilitated by the increased exploitation of museum material and sequence data from public databases and by increased investment in the field work necessary to obtain diverse material. Investigations in this new tradition of "congeneric phylogeography" will improve the likelihood of detecting and appropriately interpreting critical patterns of intraspecific and interspecific allelic variation to the benefit of systematic and population biology alike.

CONCLUSIONS AND FUTURE DIRECTIONS

This review describes the diverse causes and consequences of species-level polyphyly and represents the first large-scale empirical survey of polyphyletic species and sampling patterns in mitochondrial studies of closely related animal taxa. We demonstrate polyphyly to be a common, statistically supported, and taxonomically general aspect of mitochondrial gene trees. We find that sampling intensity has not increased concurrently with the increasing ease of collecting mitochondrial sequence data. We call for the combination of phylogenetic and phylogeographic approaches to sampling in a new tradition of congeneric phylogeography. Similar surveys of nuclear loci might be informative, although a greater diversity of loci and many fewer studies may as yet make numerical comparisons difficult. Surveys of, and comparisons with, the botanical literature might be particularly informative as polyphyly is suspected to be more common in plants and has been embraced as a fundamental aspect of phylogenetic variation by the botanical community (Crisp & Chandler 1996, Rieseberg & Brouillet 1994). Increased attention to sampling and the interpretation of polyphyly across genes and taxa will provide improved insights in systematics, population genetics, and evolutionary biology in general.

ACKNOWLEDGMENTS

We thank John Avise, William Ballard, Jason Baker, Mike Braun, John Burke, Matt Hare, Rick Harrison, Marshall Hedin, Leo Joseph, Beatrice Kondo, Jeff Peters, Ken Petren, Brad Shaffer, Kerry Shaw, Sonja Scheffer, and Paul Wilson for helpful discussions or comments on this manuscript. Elizabeth Humphries and Paul Prasnik assisted in the compilation of the literature database.

The *Annual Review of Ecology, Evolution, and Systematics* is online at http://ecolsys.annualreviews.org

LITERATURE CITED

Allen ES. 2002. *Long-term hybridization and the maintenance of species identity in orioles (Icterus).* PhD thesis. Indiana Univ., Bloomington. 119 pp.

Arnold J. 1993. Cytonuclear disequilibria in hybrid zones. *Annu. Rev. Ecol. Syst.* 24:521–54

Avise JC. 1989. Gene trees and organismal histories: a phylogenetic approach to population biology. *Evolution* 43:1192–208

Avise JC. 1994. *Molecular Markers, Natural History and Evolution.* New York: Chapman & Hall

Avise JC. 2000. *Phylogeography: The History and Formation of Species.* Cambridge, MA: Harvard Univ. Press

Avise JC, Ankney CD, Nelson WS. 1990. Mitochondrial gene trees and the evolutionary relationship between mallard and black ducks. *Evolution* 44:1109–19

Avise JC, Arnold J, Ball R, Bermingham E, Lamb T, et al. 1987. Intraspecific phylogeography: the mitochondrial DNA bridge between population genetics and systematics. *Annu. Rev. Ecol. Syst.* 18:489–522

Avise JC, Ball RM. 1990. Principles of genealogical concordance in species concepts and biological taxonomy. *Oxf. Surv. Evol. Biol.* 7:45–67

Avise JC, Shapira JF, Daniel SW, Aquadro CF, Lansman RA. 1983. Mitochondrial DNA evolution during the speciation process in *Peromyscus*. *Mol. Biol. Evol.* 1:38–56

Baker CS, Cipriano F, Palumbi SR. 1996. Molecular genetic identification of whale and dolphin products from commercial markets in Korea and Japan. *Mol. Ecol.* 5:671–85

Baker JM, López-Medrano E, Navarro-Sigüenza AG, Rojas-Soto OR, Omland KE. 2003. Recent speciation in the Orchard Oriole group: divergence of *Icterus spurius spurius* and *Icterus spurius fuertesi*. *Auk.* In press

Ballard JWO. 2000. When one is not enough: introgression of mitochondrial DNA in *Drosophila*. *Mol. Biol. Evol.* 17:1126–30

Ballard JWO, Kreitman M. 1995. Is mitochondrial DNA a strictly neutral marker? *Trends Ecol. Evol.* 10:485–88

Barraclough TG, Nee S. 2001. Phylogenetics and speciation. *Trends Ecol. Evol.* 16:391–99

Barton NH, Jones JS. 1983. Mitochondrial DNA: new clues about evolution. *Nature* 306:317–18

Baum DA, Shaw KL. 1995. Geneological perspectives on the species problem. In *Experimental and Molecular Approaches to Plant Biosystematics*, ed. PC Hoch, AG Stephenson, pp. 289–303. Saint Louis, MO: Mo. Bot. Gard.

Bell MA. 1987. Interacting evolutionary constraints in pelvic reduction of threespine stickleback, *Gasterosteus aculeatus* (Pisces, Gasterosteidae). *Biol. J. Linn. Soc.* 31:347–82

Bensasson D, Zhang DX, Hartl DL, Hewitt GM. 2001. Mitochondrial pseudogenes: evolution's misplaced witnesses. *Trends Ecol. Evol.* 16:314–21

Boyce TM, Zwick ME, Aquadro CF. 1994. Mitochondrial DNA in the bark weevils: Phylogeny and evolution in the *Pissodes strobi* species group (Coleoptera: Curculionidae). *Mol. Biol. Evol.* 11:183–94

Bremer K. 1988. The limits of amino acid sequence data in angiosperm phylogenetic reconstruction. *Evolution* 42:795–800

Bromham L, Eyre-Walker A, Smith NH, Smith JM. 2003. Mitochondrial Steve: paternal inheritance of mitochondrial DNA in humans. *Trends Ecol. Evol.* 18:2–4

Brown JM, Abrahamson WG, Way PA. 1996. Mitochondrial DNA phylogeography of host races of the goldenrod gallmaker, *Eurosta solidaginis* (Diptera: Tephritidae). *Evolution* 50:777–86

Brown JM, Pellmyr O, Thompson JN, Harrison RG. 1994. Phylogeny of *Greya* (Lepidoptera: Prodoxidae), based on nucleotide sequence variation in mitochondrial cytochrome oxidase I and II: congruence with morphological data. *Mol. Biol. Evol.* 11:128–41

Brumfield RT, Jernigan RW, McDonald DB, Braun MJ. 2001. Evolutionary implications of divergent clines in an avian (*Manacus*: Aves) hybrid zone. *Evolution* 55:2070–87

Collura RV, Auerbach MR, Stewart CB. 1996. A quick, direct method that can differentiate expressed mitochondrial genes from their nuclear pseudogenes. *Curr. Biol.* 6:1337–39

Collura RV, Stewart C-B. 1995. Insertions and duplications of mtDNA in the nuclear genomes of Old World monkeys and hominids. *Nature* 378:485–89

Crandall KA, Fitzpatrick JF Jr. 1996. Crayfish molecular systematics: using a combination of procedures to estimate phylogeny. *Syst. Biol.* 45:1–26

Crisp MD, Chandler GT. 1996. Paraphyletic species. Telopea 6:813–44

Dalebout ML, Van HA, Van WK, Baker CS. 1998. Molecular genetic identification of southern hemisphere beaked whales (Cetacea: Ziphiidae). *Mol. Ecol.* 7:687–94

Demastes JW, Spradling TA, Hafner MS, Hafner DJ, Reed DL. 2002. Systematics and phylogeography of pocket gophers in the genera *Catogeomys* and *Pappogeomys*. *Mol. Phyl. Evol.* 22:144–54

Demboski JR, Cook JA. 2001. Phylogeography of the dusky shrew, *Sorex monticolus* (Insectivora, Soricidae): insight into deep and shallow history in northwestern North America. *Mol. Ecol.* 10:1227–40

de Queiroz K, Donoghue MJ. 1988. Phylogenetic systematics and the species problem. *Cladistics* 4:317–38

Dowling TE, Secor CL. 1997. The role of hybridization and introgression in the diversification of animals. *Annu. Rev. Ecol. Syst.* 28:593–613

Doyle JJ. 1992. Gene trees and species trees: molecular systematics as one character taxonomy. *Syst. Bot.* 17:144–63

Doyle JJ. 1997. Trees within trees: genes and species, molecules and morphology. *Syst. Biol.* 46:537–53

Edwards SV, Beerli P. 2000. Perspective: gene divergence, population divergence, and the variance in coalescence time in phylogeographic studies. *Evolution* 54:1839–54

Felsenstein J. 1985. Confidence limits on phylogenies: an approach using the bootstrap. *Evolution* 39:783–91

Freeland JR, Boag PT. 1999. The mitochondrial and nuclear genetic homogeneity of the phenotypically diverse Darwin's ground finches. *Evolution* 53:1553–63

Frey JK. 1993. Modes of peripheral isolate formation and speciation. *Syst. Biol.* 42:373–81

Fukatsu T, Shibao H, Nikoh N, Aoki S. 2001. Genetically distinct populations in an Asian soldier-producing aphid, *Pseudoregma bambucicola* (Homoptera: Aphididae), identified by DNA fingerprinting and molecular phylogenetic analysis. *Mol. Phyl. Evol.* 18:423–33

Funk DJ. 1996. *The evolution of reproductive isolation in Neochlamisus leaf beetles: a role for selection*. PhD thesis. State Univ. New York, Stony Brook. 288 pp.

Funk DJ. 1998. Isolating a role for natural selection in speciation: host adaptation and sexual isolation in *Neochlamisus bebbianae* leaf beetles. *Evolution* 52:1744–59

Funk DJ. 1999. Molecular systematics of cytochrome oxidase I and 16S from *Neochlamisus* leaf beetles and the importance of sampling. *Mol. Biol. Evol.* 16:67–82

Funk DJ, Futuyma DJ, Orti G, Meyer A. 1995a. Mitochondrial DNA sequences and multiple data sets: a phylogenetic analysis of phytophagous beetles (*Ophraella*: Chrysomelidae). *Mol. Biol. Evol.* 12:627–40

Funk DJ, Futuyma DJ, Orti G, Meyer A. 1995b. A history of host associations and evolutionary diversification for *Ophraella* (Coleoptera: Chrysomelidae): new evidence from mitochondrial DNA. *Evolution* 49: 1008–17

Goodacre SL, Wade CM. 2001. Patterns of genetic variation in Pacific island land snails: The distribution of cytochrome b lineages among Society Island *Partula*. *Biol. J. Linn. Soc.* 73:131–38

Goodman M, Czelusniak J, Moore GW, Romero-Harrera AE, Matsuda G. 1979. Fitting the gene lineage to its species lineage, a parsimony strategy illustrated by cladograms constructed from globin sequences. *Syst. Zool.* 28:132–63

Graybeal A. 1995. Naming species. *Syst. Biol.* 44:237–50

Hare MP, Avise JC. 1998. Population structure in the American oyster as inferred by nuclear gene genealogies. *Mol. Biol. Evol.* 15:119–28

Harrison RG. 1989. Animal mitochondrial DNA as a genetic marker in population and evolutionary biology. *Trends Ecol. Evol.* 4:6–11

Harrison RG. 1990. Hybrid zones: windows on evolutionary process. *Oxf. Surv. Evol. Biol.* 7:69–128

Harrison RG. 1991. Molecular changes at speciation. *Annu. Rev. Ecol. Syst.* 22:281–308

Harrison RG. 1998. Linking evolutionary patterns and processes: the relevance of species concepts for the study of speciation. In *Endless Forms: Species and Speciation*, ed. DJ Howard, SH Berlocher, pp. 19–31. New York: Oxford Univ. Press

Harrison RG, Rand DM, Wheeler WC. 1987. Mitochondrial DNA variation in field crickets across a narrow hybrid zone. *Mol. Biol. Evol.* 4:144–58

Hedin MC. 1997. Speciational history in a diverse clade of habitat-specialized spiders (Araneae: Nesticidae: *Nesticus*): inferences from geographic-based sampling. *Evolution* 51:1929–45

Hedin M, Wood DA. 2002. Genealogical exclusivity in geographically proximate populations of *Hypochilus thorelli* Marx (Araneae, Hypochilidae) on the Cumberland Plateau of North America. *Mol. Ecol.* 11:1975–88

Hedrick PW, Lee RN, Garrigan D. 2002. Major histocompatibility complex variation in red wolves: evidence for common ancestry with coyotes and balancing selection. *Mol. Ecol.* 11:1905–13

Hey J. 1994. Bridging phylogenetics and population genetics with gene tree models. In *Molecular Ecology and Evolution: Approaches and Applications*, ed. B Schierwater, B Streit, GP Wagner, R DeSalle, pp. 441–45. Basel, Swit.: Birkhauser Verlag

Hoelzer GA. 1997. Inferring phylogenies from mtDNA variation: mitochondrial-gene trees versus nuclear-gene trees revisited. *Evolution* 51:622–26

Holder MT, Anderson JA, Holloway AK. 2001. Difficulties in detecting hybridization. *Syst. Biol.* 50:978–82

Hudson RR. 1990. Gene geneologies and the coalescent process. *Oxf. Surv. Evol. Biol.* 7:1–44

Hudson RR, Turelli M. 2003. Stochasiticity overrules the "three-times rule": genetic drift, genetic draft, and coalescence times for nuclear loci versus mitochondrial DNA. *Evolution* 57:182–90

James AC, Ballard JWO. 2000. Expression of cytoplasmic incompatibility in *Drosophila* simulans and its impact on infection frequencies and distribution of *Wolbachia pipientis*. *Evolution* 54:1661–72

Jarman SN, Elliot NG. 2000. DNA evidence for morphological and cryptic Cenozoic speciations in the Anaspididae, 'living fossils' from the Triassic. *J. Evol. Biol.* 13:624–33

Johnson SG, Bragg E. 1999. Age and polyphyletic origins of hybrid and spontaneous parthenogenetic *Campeloma* (Gastropoda: Viviparidae) from the southwestern United States. *Evolution* 53:1769–81

Kim CG, Zhou HZ, Imura Y, Tominaga O, Su ZH, Osawa S. 2000. Pattern of morphological diversification in the *Leptocarabus* ground beetles (Coleoptera: Carabidae) as deduced from mitochondrial ND5 gene and nuclear 28S rDNA sequences. *Mol. Biol. Evol.* 17:137–45

Kishino H, Hasegawa M. 1989. Evaluation of the maximum likelihood estimate of the evolutionary tree topologies from DNA sequence data, and the branching order in Hominoidea. *J. Mol. Evol.* 29:170–79

Klein J, Sato A, Nagl S, O'hUigín C. 1998. Molecular trans-species polymorphism. *Annu. Rev. Ecol. Syst.* 29:1–21

Klein J, Satta Y, Takahata N, O'hUigín C. 1993. Trans-specific Mhc polymorphism and the origin of species in primates. *J. Med. Primatol.* 22:57–64

Klein NK, Payne RB. 1998. Evolutionary associations of brood parasitic finches (*Vidua*) and their host species: analyses of mitochondrial DNA restriction sites. *Evolution* 52:566–82

Knowles LL. 2000. Tests of Pleistocene speciation in montane grasshoppers (genus *Melanoplus*) from the Sky Islands of western North America. *Evolution* 54:1337–48

Knowles LL, Futuyma DJ, Eanes WF, Rannala B. 1999. Insight into speciation from historical demography in the phytophagous beetle genus *Ophraella*. *Evolution* 53:1846–56

Knowles LL, Maddison WP. 2002. Statistical phylogeography. *Mol. Ecol.* 11:2623–35

Kotlik P, Berrebi P. 2002. Genetic subdivision and biogeography of the Danubian rheophilic barb *Barbus petenyi* inferred from phylogenetic analysis of mitochondrial DNA variation. *Mol. Phyl. Evol.* 24:10–18

Lanyon SM. 1994. Polyphyly of the blackbird genus Agelaius and the importance of assumptions of monophyly in comparative studies. *Evolution* 48:679–93

Maddison WP. 1996. Molecular approaches and the growth of phylogenetic biology. In *Molecular Zoology: Advances, Strategies and Protocols*, ed. D Ferraris, SR Palumbi, pp. 47–63. New York: Wiley-Liss

Maddison WP. 1997. Gene trees in species trees. *Syst. Biol.* 46:523–36

Mantovani B, Passamonti M, Scali V. 2001. The mitochondrial cytochrome oxidase II gene in *Bacillus* stick insects: ancestry of hybrids, androgenesis, and phylogenetic relationships. *Mol. Phyl. Evol.* 19:157–63

Marchant AD. 1988. Apparent introgression of mitochondrial DNA across a narrow hybrid zone in the *Caledia captiva* species complex. *Heredity* 61:39–46

Marko PB. 1998. Historical allopatry and the biogeography of speciation in the prosobranch snail genus *Nucella*. *Evolution* 52: 757–74

Mason DJ, Butlin RK, Gacesa P. 1995. An unusual mitochondrial DNA polymorphism in the *Chorthippus biguttulus* species group (Orthoptera: Acrididae). *Mol. Ecol.* 4:121–26

Masta SE, Sullivan B, Lamb T, Routman EJ. 2002. Phylogeography, species boundaries, and hybridization among toads of the *Bufo americanus* group. *Mol. Phyl. Evol.* 24:302–14

Maynard Smith J, Smith NH. 2002. Recombination in animal mitochondrial DNA. *Mol. Biol. Evol.* 19:2330–32

Melnick DJ, Hoelzer GA, Absher R, Ashley MV. 1993. mtDNA diversity in rhesus monkeys reveals overestimates of divergence time and paraphyly with neighboring species. *Mol. Biol. Evol.* 10:282–95

Moore WS. 1995. Inferring phylogenies from mtDNA variation: mitochondrial-gene trees versus nuclear-gene trees. *Evolution* 49:718–26

Moran P, Kornfield I. 1993. Retention of an ancestral polymorphism in the Mbuna species flock (Teleostei: Cichlidae) of Lake Malawi. *Mol. Biol. Evol.* 10:1015–29

Moran P, Kornfield I. 1995. Were population bottlenecks associated with the radiation of the Mbuna species flock (Teleostei: Cichlidae) of Lake Malawi? *Mol. Biol. Evol.* 12:1085–93

Moritz C. 1994. Applications of mitochondrial DNA analysis in conservation: a critical review. *Mol. Ecol.* 3:401–11

Moritz C, Dowling TE, Brown WM. 1987. Evolution of animal mitochondrial DNA: relevance for population biology and systematics. *Annu. Rev. Ecol. Syst.* 18:269–92

Moritz C, Wright JW, Brown WM. 1992. Mitochondrial DNA analyses and the origin and relative age of parthenogenetic *Cnemidophorus*: phylogenetic constraints on hybrid origins. *Evolution* 46:184–92

Neigel JE, Avise JC. 1986. Phylogenetic relationships of mitochondrial DNA under various demographic models of speciation. In *Evolutionary Processes and Theory*, ed. E Nevo, S Karlin, pp. 515–34. New York: Academic

Nice CC, Shapiro AM. 2001. Patterns of morphological, biochemical, and molecular evolution in the *Oeneis chryxus* complex (Lepidoptera: Satyridae): A test of historical biogeographical hypotheses. *Mol. Phyl. Evol.* 20:111–23

Nichols R. 2001. Gene trees and species trees are not the same. *Trends Ecol. Evol.* 16:358–64

Nielsen R, Wakeley J. 2001. Distinguishing migration from isolation: a Markov chain Monte Carlo approach. *Genetics* :885–96

Olmstead RG. 1995. Species concepts and plesiomorphic species. *Syst. Bot.* 20:623–30

Omland KE. 1997. Examining two standard assumptions of ancestral reconstructions: repeated loss of dimorphism in dabbling ducks (Anatini). *Evolution* 51:1636–46

Omland KE, Lanyon SM, Fritz SJ. 1999. A molecular phylogeny of the New World Orioles (*Icterus*): the importance of dense taxon sampling. *Mol. Phyl. Evol.* 12:224–39

Omland KE, Tarr CL, Boarman WI, Marzluff JM, Fleischer RC. 2000. Cryptic genetic variation and paraphyly in ravens. *Proc. R. Soc. London Ser. B* 267:2475–82

Paetkau D. 1999. Using genetics to identify intraspecific conservation units: a critique of current methods. *Mol. Phyl. Evol.* 13:1507–9

Palumbi SR, Cipriano F, Hare MP. 2001. Predicting nuclear gene coalescence from mitochondrial data: the three-times rule. *Evolution* 55:859–68

Pamilo P, Nei M. 1988. Relationships between gene trees and species trees. *Mol. Biol. Evol.* 5:568–83

Patton JL, Smith MF. 1994. Paraphyly, polyphyly, and the nature of species boundaries in pocket gophers (Genus *Thomomys*). *Syst. Biol.* 43:11–26

Petren K, Grant BR, Grant PR. 1999. A phylogeny of Darwin's finches based on microsatellite DNA length variation. *Proc. R. Soc. London Ser. B* 266:321–29

Porter BA, Cavender TM, Fuerst PA. 2002. Molecular phylogeny of the snubnose darters, subgenus *Ulocentra* (Genus Etheostoma, family Percidae). *Mol. Phyl. Evol.* 22: 364–74

Rand DM. 2001. The units of selection on mitochondrial DNA. *Annu. Rev. Ecol. Syst.* 32: 415–48

Rawson PD, Hilbish TJ. 1995. Evolutionary relationships among the male and female mitochondrial DNA lineages in the *Mytilus edulis* species complex. *Mol. Biol. Evol.* 12:893–901

Redenbach Z, Taylor EB. 2002. Evidence for historical introgression along a contact zone between two species of char (Pisces: Salmonidae) in northwestern North America. *Evolution* 56:1021–35

Rees DJ, Emerson BC, Oromi P, Hewitt GM. 2001a. The diversification of the genus *Nesotes* (Coleoptera: Tenebrionidae) in the Canary Islands: evidence from mtDNA. *Mol. Phyl. Evol.* 21:321–26

Rees DJ, Emerson BC, Oromi P, Hewitt GM. 2001b. Mitochondrial DNA, ecology and morphology: interpreting the phylogeography of the *Nesotes* (Coleoptera: Tenebrionidae) of Gran Canaria (Canary Islands). *Mol. Ecol.* 10:427–34

Richmond JQ, Reeder TW. 2002. Evidence for parallel ecological speciation in scincid lizards of the *Eumeces skiltonianus* species group (Squamata: Scincidae). *Evolution* 56:1498–513

Riddle BR, Hafner DJ, Alexander LF. 2000. Phylogeography and systematics of the *Peromyscus eremicus* species group and the historical biogeography of North American warm regional deserts. *Mol. Phyl. Evol.* 17:145–60

Rieseberg LH, Brouillet L. 1994. Are many plant species paraphyletic? *Taxon* 43:21–32

Rodríguez-Robles JA, De Jesús-Escobar JM. 2000. Molecular systematics of the New World gopher, bull and pinesnakes (Pituophis: Colubridae), a transcontinental species complex. *Mol. Phyl. Evol.* 14:35–50

Salzburger W, Baric S, Sturmbauer C. 2002. Speciation via introgressive hybridization in East African cichlids? *Mol. Ecol.* 11:619–25

Sang T, Zhong Y. 2000. Testing hybridization hypotheses based on incongruent gene trees. *Syst. Biol.* 49:422–34

Sattler GD, Braun MJ. 2000. Morphometric variation as an indicator of genetic interactions between Black-capped and Carolina Chickadees at a contact zone in the Appalachian mountains. *Auk* 117:427–44

Schluter D, Nagel LM. 1995. Parallel speciation by natural selection. *Am. Nat.* 146:292–301

Schneider-Broussard R, Felder DL, Chlan CA, Neigel JE. 1998. Tests of phylogeographic models with nuclear and mitochondrial DNA sequence variation in the stone crabs, *Menippe adina* and *Menippe mercenaria*. *Evolution* 52:1671–78

Shaw KL. 1999. A nested analysis of song groups and species boundaries in the Hawaiian cricket genus *Laupala*. *Mol. Phylogeny Evol.* 11:332–41

Shaw KL. 2002. Conflict between nuclear and mitochondrial DNA phylogenies of a recent species radiation: what mtDNA reveals and conceals about modes of speciation in Hawaiian crickets. *Proc. Natl. Acad. Sci. USA* 99:16122–27

Slowinski JB, Page RDM. 1999. How should species phylogenies be inferred from sequence data. *Syst. Biol.* 48:814–25

Small MP, Gosling EM. 2000. Species relationships and population structure of *Littorina saxatilis* Olivi and L. tenebrosa Montagu in Ireland using single-strand conformational polymorphisms (SSCPs) of cytochrome b fragments. *Mol. Ecol.* 9:39–52

Smith GR. 1992. Introgression in fishes: Significance for paleontology, cladistics, and evolutionary rates. *Syst. Biol.* 41:41–57

Smouse PE, Dowling TE, Tworek JA, Hoeh WR, Brown WM. 1991. Effects of intraspecific variation on phylogenetic inference: a likelihood analysis of mtDNA restriction site data in cyprinid fishes. *Syst. Zool.* 40:393–409

Sorenson MD, Fleischer RC. 1996. Multiple independent transpositions of mitochondrial DNA control region sequences to the nucleus. *Proc. Natl. Acad. Sci. USA* 93:15239–43

Sosef MSM. 1997. Hierarchical models, reticulate evolution and the inevitability of paraphyletic supraspecific taxa. *Taxon* 46:75–85

Sota T, Vogler AP. 2001. Incongruence of mitochondrial and nuclear gene trees in the carabid beetles *Ohomopterus*. *Syst. Biol.* 50:39–59

Sperling FAH. 1993. Mitochondrial DNA variation and Haldane's rule in the *Papilio glaucus* and *Papilio troilus* species groups. *Heredity* 71:227–33

Städler T, Delph LF. 2002. Ancient mitochondrial haplotypes and evidence for intragenic recombination in a gynodioecious plant. *Proc. Natl. Acad. Sci. USA* 99:11730–35

Su ZH, Tominaga O, Ohama T, Kajiwara E, Ishikawa R, et al. 1996. Parallel evolution in radiation of Ohomopterus ground beetles inferred from mitochondrial ND5 gene sequences. *J. Mol. Evol.* 43:662–71

Sunnucks P, Hales DF. 1996. Numerous transposed sequences of mitochondrial cytochrome oxidase I-II in aphids of the genus *Sitobion* (Hemiptera: Aphididae). *Mol. Biol. Evol.* 13:510–24

Tajima F. 1983. Evolutionary relationships of DNA sequences in finite populations. *Genetics* 105:437–60

Takahata N, Nei M. 1985. Gene geneology and variance of interpopulational nucleotide differences. *Genetics* 110:325–44

Takahata N, Slatkin M. 1984. Mitochondrial gene flow. *Proc. Natl. Acad. Sci. USA* 81:1764–67

Tegelström H. 1987. Transfer of mitochondrial DNA from the northern red-backed vole

(*Clethrionomys rutilus*) to the bank vole (*C. glareolus*). *J. Mol. Evol.* 24:218–27

Tegelström H, Gelter HP. 1990. Haldane's rule and sex biased gene flow between two hybridizing flycatcher species (*Ficedula albicollis* and *F. hypoleuca*, Aves: Muscicapidae). *Evolution* 44:2012–21

Templeton AR. 1998. Nested clade analysis of phylogeographic data: testing hypotheses about gene flow and population history. *Mol. Ecol.* 7:381–97

Templeton AR, Routman E, Phillips CA. 1995. Separating population structure from population history: a cladistic analysis of the geographical distribution of mitochondrial DNA haplotypes in the tiger salamander, *Ambystoma tigrinum*. *Genetics* 140:767–82

Tosi AJ, Morales JC, Melnick DJ. 2000. Comparison of Y chromosome and mtDNA phylogenies leads to unique inferences of macaque evolutionary history. *Mol. Phyl. Evol.* 17:133–44

Turelli M, Barton NH, Coyne JA. 2001. Theory and speciation. *Trends Ecol. Evol.* 16:330–42

van Oppen MJH, McDonald BJ, Willis B, Miller DJ. 2001. The evolutionary history of the coral genus *Acropora* (Scleractinia, Cnidaria) based on a mitochondrial and a nuclear marker: reticulation, incomplete lineage sorting, or morphological convergence? *Mol. Biol. Evol.* 18:1315–29

Voelker G. 1999. Molecular evolutionary relationships in the avian genus *Anthus* (Pipits: Motacillidae). *Mol. Phyl. Evol.* 11:84–94

Wakeley J. 1996. Distinguishing migration from isolation using the variance of pairwise differences. *Theor. Popul. Biol.* 49:369–86

Wayne RK, Jenks SM. 1991. Mitochondrial DNA analysis implying extensive hybridization of the endangered red wolf *Canis rufus*. *Nature* 351:565–68

Weckstein JD, Zink RM, Blackwell-Rago RC, Nelson DA. 2001. Anomalous variation in mitochondrial genomes of White-crowned (*Zonotrichia leucophrys*) and Golden-crowned (*Z. atricapilla*) sparrows: pseudogenes, hybridization, or incomplete lineage sorting? *Auk* 118:231–36

Wheeler QD, Nixon KC. 1990. Another way of looking at the species problem: a reply to de Quieroz and Donoghue. *Cladistics* 6:77–81

Wiens JJ. 1999. Polymorphism in systematics and comparative biology. *Annu. Rev. Ecol. Syst.* 30:327–62

Wiens JJ, Penkrot TA. 2002. Deliminating species using DNA and morphological variation and discordant species limits in spiny lizards (*Sceloporus*). *Syst. Biol.* 51:69–91

Wiens JJ, Servedio MR. 2000. Species delimitation in systematics: inferring diagnostic differences between species. *Proc. R. Soc. London Ser. B* 267:631–36

Williams ST, Knowlton N, Weigt LA, Jara JA. 2001. Evidence for three major clades within the snapping shrimp genus *Alpheus* inferred from nuclear and mitochondrial gene sequence data. *Mol. Phyl. Evol.* 20:375–89

Wu C-I. 1991. Inference of species phylogeny in relation to segregation of ancient polymorphism. *Genetics* 127:429–35

Zhang DX, Hewitt GM. 1996. Nuclear integrations: challenges for mitochondrial DNA markers. *Trends Ecol. Evol.* 11:247–51

Zink RM, Weller SJ, Blackwell RC. 1998. Molecular phylogenetics of the avian genus *Pipilo* and a biogeographic argument for taxonomic uncertainty. *Mol. Phyl. Evol.* 10:191–201

Annu. Rev. Ecol. Evol. Syst. 2003. 34:425–53
doi: 10.1146/annurev.ecolsys.34.011802.132410

First published online as a Review in Advance on September 8, 2003

PROTECTIVE ANT-PLANT INTERACTIONS AS MODEL SYSTEMS IN ECOLOGICAL AND EVOLUTIONARY RESEARCH

Martin Heil[1] and Doyle McKey[2]
[1]*Department of Bioorganic Chemistry, Max-Planck-Institute of Chemical Ecology, Beutenberg Campus, Winzerlaer Strasse, D-07745 Jena, Germany; email: Heil_Martin@web.de*
[2]*Céntre d'Ecologie Fonctionelle et Evolutive (CEFE-CNRS, UPR 9056), 1919 Route de Mende, F-34293 Montpellier Cedex 5, France; email: mckey@cefe.cnrs-mop.fr*

Key Words antiherbivore defense, coevolution, myrmecophytism, myrmecophily, plant-animal interaction

■ **Abstract** Protective ant-plant interactions, important in both temperate and tropical communities, are increasingly used to study a wide range of phenomena of general interest. As antiherbivore defenses "worn on the outside," they pose fewer barriers to experimentation than do direct (e.g., chemical) plant defenses. This makes them tractable models to study resource allocation to defense and mechanisms regulating it. As multi-trophic level interactions varying in species specificity and impact on fitness of participants, ant-plant-herbivore associations figure prominently in studies of food-web structure and functioning. As horizontally transmitted mutualisms that are vulnerable to parasites and "cheaters," ant-plant symbioses are studied to probe the evolutionary dynamics of interspecies interactions. These symbioses, products of coevolution between plants and insect societies, offer rich material for studying ant social evolution in novel contexts, in settings where colony limits, resource supply, and nest-site availability are all more easily quantifiable than in the ground-nesting ants hitherto used as models.

INTRODUCTION

In a diverse set of mutualisms, plants are protected, fed, or transported by ants (Beattie 1985). This review deals with ant-plant protection mutualisms, ranging from opportunistic, facultative interactions, in which plants offer food rewards to foraging ant workers, to interactions in which plants also offer hollow structures to nesting ants encouraging more constant association. This latter type of interaction includes a number of specific and obligate symbioses. We cover the spectrum because much has been learned by comparing systems varying in their degree of specialization.

Attracting the attention of naturalists over a century ago (Belt 1874, Rettig 1904, Ridley 1910), these mutualisms have often been considered as interesting, sometimes spectacular, examples of biotic interactions but of limited importance in ecological communities. Recent work, however, suggests that opportunistic ant-plant mutualisms play a key role in structuring food webs in tropical forest canopies. Although restricted to the tropics, symbiotic ant-plant mutualisms involve species of over 100 genera of angiosperms and 40 genera of ants (Davidson & McKey 1993), and they are important components of tropical communities.

Ant-plant protection mutualisms have served as model systems for studying a great range of questions of general interest in ecology and evolutionary biology. In the 1960s, studies of a myrmecophyte symbiosis demonstrated beyond doubt what happened to plants when their defenses against herbivores were removed (Janzen 1966, 1967a,b, 1969). These elegant studies were seminal in stimulating work on plant-animal coevolution in many contexts. Since then, ant-plant protection mutualisms have continued to be tractable systems for testing aspects of plant defense theory (Fonseca 1994, Heil et al. 2002b). Myrmecophyte symbioses have served with increasing frequency as models for examining conditional outcomes in interspecies interactions (Gaume et al. 1998) and understanding the evolutionary stability of mutualisms in the face of potentially destabilizing conflicts (Yu 2001). They have figured prominently in recent work on species coexistence (Palmer et al. 2002, Yu et al. 2001), the structure of food webs (Letourneau & Dyer 1998a,b; Schmitz et al. 2000), and other themes in community ecology. We review recent research, highlight aspects where future research is likely to lead to important advances in our understanding, and show that these systems can serve as material for an even wider range of questions in ecology and evolutionary biology.

NONSYMBIOTIC, "REWARD-BASED" INTERACTIONS (MYRMECOPHILIC INTERACTIONS)

Ants are often among the most important predators of arthropods (Floren et al. 2002, Hölldobler & Wilson 1990, Novotny et al. 1999). Ants attracted by plant-derived food rewards serve as an important indirect defense mechanism of plants in both tropical and temperate communities (Bronstein 1998, Buckley 1982, Davidson & McKey 1993, Huxley & Cutler 1991). These interactions are most commonly facultative, involving opportunistic attraction of ants nesting elsewhere to plant-produced food rewards.

The plant may produce these rewards directly or indirectly. Myrmecophilic ("ant-loving") plants directly produce food rewards such as extrafloral nectar (EFN) or food bodies. Extrafloral nectaries function in protection and not in pollination (Bentley 1977). They are known in at least 66 families of flowering plants (monocots and dicots) and ferns but are absent in "gymnosperms" (Elias 1983). Food bodies are nutrient-rich cellular structures that can easily be removed by foraging ants; they have been described in at least 20 plant families (O'Dowd 1982).

Other plants lack obvious ant-specialized traits but frequently harbor ant-tended hemipteran trophobionts. Recent reviews of ant-hemipteran interactions are provided by Delabie (2001), Gullan (1997), and Gullan & Kosztarab (1997). Most ant-tended hemipterans are phloem-feeders and excrete excess liquid as sugar-rich honeydew. An important resource for ants, honeydew-producing hemipterans are often monopolized by territorial, ecologically dominant ants (Blüthgen et al. 2000). These generalist predators can strongly reduce densities of phytophagous insects.

Protective Effects of Ants

Many studies have demonstrated protection in facultative ant-plant interactions (Bentley 1977, Koptur 1992). Recent work has greatly extended the evidence for protection by ants opportunistically attracted to plant-derived rewards (Costa et al. 1992, de la Fuente & Marquis 1999, delClaro et al. 1996, Koptur et al. 1998, Oliveira et al. 1999, Sobrinho et al. 2002). EFN-consuming ants can even provide protection in plantations of economically important species such as cashew, *Anacardium occidentale* (Rickson & Rickson 1998), showing the potential of such mutualisms in biological control. In addition to herbivores, plant enemies against which nectary-visiting ants have been demonstrated to defend include leaf-cutting ants (Farji Brener et al. 1992) and fungal pathogens (de la Fuente & Marquis 1999). Although herbivores appear to be the general target of both defending ants and scientists studying these interactions, the last cited example shows that we should broaden our attention to other potential interactors.

Other arthropods also exploit these resources and may have diverse effects on ants or plants. Mites and ladybird beetles visit extrafloral nectaries of many plants (Pemberton 1993, Pemberton & Vandenberg 1993, van Rijn & Tanigoshi 1999) and thus might compete with ants, as do stingless bees (O'Dowd 1979) and certain flies (Heil et al. 2003). Other visitors include ichneumonid and braconid wasps (Bugg et al. 1989, Stapel et al. 1997), jumping spiders (Ruhren & Handel 1999), mosquitoes (Foster 1995), and neuropterans (Limburg & Rosenheim 2001). Like ants, some of these EFN consumers can protect plants against herbivores (Pemberton & Lee 1996, Ruhren & Handel 1999, van Rijn & Tanigoshi 1999), whereas others may act as commensals or parasites.

Several studies failed to find protective effects of EFN-consuming ants (Freitas et al. 2000, Mackay & Whalen 1998, O'Dowd & Catchpole 1983, Rashbrook et al. 1992, Tempel 1983). These examples may constitute exceptions or may simply reflect the great variation in both space and time predicted for opportunistic protection mutualisms (Bentley 1976). Variation in the abundance or species composition of ants, phytophagous insects, or other arthropods competing with ants for extrafloral nectar (Heil et al. 2003, O'Dowd 1979) can lead to variation in protective effects (Barton 1986, Di Giusto et al. 2001, Horvitz & Schemske 1984, 1990).

Abiotic conditions influence herbivore pressure, ant visitation rates, or both (de la Fuente & Marquis 1999, Wirth & Leal 2001). The species-richness of assemblages of ants attracted to food rewards varies geographically, seasonally

(Rico-Gray et al. 1998a), and between day and night (Hossaert-McKey et al. 2001). Plant traits also influence ant assemblages. Interspecific variation in EFN structure and placement among *Passiflora* vines may be related to differences in rates of removal of termite baits by nectary-visiting ants (Apple & Feener 2001). In *Triumfetta semitriloba*, the quantity or quality of EFN secretion was considered to dominate over extrinsic factors in determining ant visitation and its effects (Sobrinho et al. 2002). Extrafloral nectary-visiting ants can respond quickly to changes in EFN flow. *Macaranga tanarius* plants treated with jasmonic acid to increase EFN secretion received significantly more nectary visitors for at least 24 h (Heil et al. 2001c). Therefore, EFN-secreting plants can influence the effectiveness of their indirect defense by controlling amount and/or quality of the nectar secreted (see below).

Functional Ecology of EFN Production

Selection should favor production of defense traits in such a way that the protective benefits are maximized and the costs minimized (McKey 1974). Defense should thus be concentrated on plant parts that are most vulnerable or whose loss would be most costly to the plant. Its peak production should be timed to coincide with the greatest risk of herbivore attack, thus showing variation over the 24-h cycle, across developmental stages of plant parts, and over ontogeny of the whole plant. *Dioscorea praehensilis* produces EFN during only one particular phase of its annual cycle, when the single unbranched, leafless shoot climbs from the forest floor to the canopy. Damage to the meristem during this phase would be particularly costly (Di Giusto et al. 2001). EFN production by many plants shows strong diel variation, but in some species, peak production is crepuscular (Heil et al. 2000a, Wickers 1997), in others nocturnal [e.g., *D. praehensilis* (B. DiGiusto, personal communication)] and in still others diurnal [e.g., myrmecophytic *Acacia hindsii*, see Raine et al. 2002; and myrmecophilic *Leonardoxa* (D. McKey, personal observation)]. This variation indicates that timing of production is not the consequence of some simple, physiological (e.g., source-sink) mechanism but could be shaped by selection pressures peculiar to each plant, e.g., the activity rhythms of its particular herbivores. Herbivores of *M. tanarius*, one species with a crepuscular peak of EFN production, show a marked activity peak at dusk and in the first few hours of darkness (M. Heil, personal observation).

In *M. tanarius*, the rate of EFN secretion is responsive to the rate of removal (Heil et al. 2000a), indicating that plants lower costs by reducing their EFN production when EFN is not removed. More studies are required to check whether this regulatory mechanism is a general one.

Does increased investment by the plant in EFN quality or quantity confer increased protective benefit? EFN consists mainly of aqueous solutions of mono- and disaccharides (predominantly sucrose, glucose, and fructose) usually accompanied by much lower concentrations of amino acids and other compounds (Baker et al. 1978, Dress et al. 1997, Galetto & Bernardello 1992, Heil et al. 2000a,

Koptur 1994, Ruffner & Clark 1986, Smith et al. 1990, Stone et al. 1985). Although water in nectar contributes to ant attraction in desert plants (Ruffner & Clark 1986), most ants prefer nectars that are highly concentrated (Galetto & Bernardello 1992) and rich in amino acids or other additional compounds (Koptur 1994, Koptur & Truong 1998, Lanza et al. 1993, Stapel et al. 1997). Such effects can depend on single compounds. Glycine enhances the attractiveness of glucose solutions to *Camponotus* ants (Wada et al. 2001), and fire ants (*Solenopsis invicta*) even distinguish between diastereomers such as D- and L-GLUCOSE (Vandermeer et al. 1995). Although a positive relation between the number of ants attracted and the effectiveness of defense is intuitively evident, few studies, besides the one using inducibility of EFN secretion (Heil et al. 2001c), empirically demonstrate this relation. Even less is known about how differences in plant rewards influence species identity of attracted ants and how this in turn affects protective benefit. More studies are therefore required to determine whether the protective effect of EFN secretion varies with the quantity and/or quality of EFN.

A variety of studies have indicated that EFN secretion, or amino acid concentrations in EFN, may increase in response to herbivory (Koptur 1989, Smith et al. 1990, Stephenson 1982, Swift & Lanza 1993) and that this reaction does not require herbivore-specific elicitors (Heil et al. 2000a, Wäckers & Wunderlin 1999). However, most of these studies suffered from methodological problems (Heil et al. 2000a), and nothing was known about the underlying signaling pathway. In *M. tanarius*, EFN production represents an induced plant defense that is mediated via the octadecanoid signal transduction cascade (Heil et al. 2001c). The earlier reports (see above), along with more recent studies on cotton (*Gossypium hirsutum*, see Wäckers et al. 2001) and several *Acacia* species (M. Heil, S. Greiner, R. Krüger, unpublished manuscript), confirm that inducibility of EFN flow seems to be a taxonomically widespread phenomenon.

Diversity and Abundance of Extrafloral Nectary-Bearing Plants

Important contributions on the taxonomic, floristic, and ecogeographic distribution of extrafloral nectaries are still being made (Dejean et al. 2000; Fiala & Linsenmair 1995; Fonseca & Ganade 1996; O'Brien 1995; Oliveira & Brandão 1991; Rico-Gray et al. 1998a,b). Extrafloral nectary-bearing plant species are diverse and abundant in several different vegetation types. Nearly one-third of 243 plant species surveyed on Barro Colorado Island in Panama (Schupp & Feener 1991), and 12.3% of the 741 plant species surveyed in Pasoh Forest Reserve in West Malaysia (Fiala & Linsenmair 1995), had extrafloral nectaries. Considerable proportions of EFN-producing plants of the total vegetation, or species pool, were also reported for Brazilian cerrado vegetation (Oliveira & Brandão 1991). In contrast, in a study conducted at Los Tuxtlas station in Veracruz, Mexico, only 3% of the 289 tree species investigated possessed extrafloral nectaries (Ibarra-Manríquez & Dirzo

1990). Further studies are required to document and understand such striking variation among plant communities.

Within communities, frequency of extrafloral nectaries appears to vary among plant life forms. In the canopy of an Amazonian rainforest, many species of epiphytes and lianas produced EFN, but a lower proportion of canopy trees did so (Blüthgen et al. 2000). EFN production appears to be disproportionately frequent in vines, most probably because the connectedness of these plants to the surrounding vegetation allows many arboreal ants easy access to nectaries (Bentley 1981, Di Giusto et al. 2001, Hossaert-McKey et al. 2001).

Many ant species visit EFN-producing plants. Twenty-seven ant species were recorded on extrafloral nectaries of *D. praehensilis* (Di Giusto et al. 2001). More than 20 insect species were observed on extrafloral nectaries of *Croton sarcopetalus* (Euphorbiaceae) (Freitas et al. 2000). Thirteen ant species and 42 plant species were involved in 135 pairs of ant-plant associations in the semiarid vegetation of the Zapotitlan valley in Mexico, whereas in the dry coastal tropical lowlands of Veracruz (Mexico), 30 ant species and 102 plant species were involved in 312 associations (Rico-Gray et al. 1998b).

Do Opportunistic Ant-Plant Mutualisms Structure Food Webs?

Recent work in tropical forest ecosystems has suggested a key role of these rewards in shaping the nutritional ecology of tree-dwelling ants and the importance of opportunistic ant-plant mutualisms in structuring entire canopy arthropod communities. Knock-down samples of tree-dwelling arthropods in tropical forests are usually dominated by ants; these predators account for a greater proportion of both individual numbers and biomass than their potential prey (Tobin 1995). This seeming paradox is resolved if many tree-dwelling ants are in fact feeding on plant-derived exudates such as EFN and hemipteran honeydew (Davidson 1997, Delabie 2001, Gullan 1997). A variety of evidence favors this hypothesis, including direct observations and patterns in natural abundance of stable isotopes (Davidson 1997, Davidson et al. 2003) and observations showing that scale insects are far more abundant than indicated by knock-down samples (Dejean et al. 2000).

Davidson (1997) has argued that the abundance of these exudates plays a key role in shaping food-web structure in tropical forest canopies by allowing tree-dwelling ants to evolve energetically costly prey-foraging strategies, in particular when ants are physiologically adapted to feed on resources characterized by a low nitrogen content. This would enable them to reach higher densities, and to maintain prey species at lower densities, than if they depended solely on animal prey. What evidence exists for this last hypothesis? We are aware of no experimental manipulative study of this question. Comparative studies show that ant assemblages vary markedly among plant species and life forms, at least partly in response to types and amounts of resources offered (Blüthgen et al. 2000, Dejean et al. 2000, Hossaert-McKey et al. 2001), but how these differences influence herbivore assemblages remains to be studied.

MYRMECOPHYTIC SYMBIOSES

In over 100 genera of tropical angiosperms, one or more species possess specialized structures for housing ants (Davidson & McKey 1993), encouraging more constant associations. Myrmecophytes offer ants pre-formed nesting sites, or "domatia," in hollow stems (e.g., *Cecropia, Leonardoxa, Macaranga*), thorns (*Acacia*), petioles (*Piper*), or leaf pouches (e.g., *Hirtella, Maieta, Scaphopetalum, Tococa*). The more constant, long-lived, and exclusive association allowed when ants are resident in plants has usually led to specialization of both partners. This specialization may include increased rate of resource supply to ants by plants and increased protective efficacy of ants. According to the coevolutionary scenario first developed by Janzen (1966) and supported by a great body of studies, ants that better protect their host tree, and plants that invest more in maintenance of their protective ant colony, should thereby increase their own survival and reproduction.

True myrmecophytic interactions include a number of highly specific and obligate symbioses. Although, overall much less frequent and widespread than facultative protection mutualisms, myrmecophyte symbioses are often conspicuous and ecologically important components of tropical communities, either as dominant components of forest understories (Fonseca & Ganade 1996, Morawetz et al. 1992), as abundant "weeds" in vast areas (Central American *Acacia* myrmecophytes; see Janzen 1974), or as pioneer trees (*Macaranga* and *Cecropia*; see Davies et al. 1998, Ferguson et al. 1995, Folgarait & Davidson 1994, Whitmore 1967). For all these plants, protection mutualisms are an important ingredient of their ecological success.

Protection by Resident Ants—Recent Findings

In general, ants protect their myrmecophyte hosts against a broad range of herbivores (Bronstein 1998). Whereas herbivores have breached many plant chemical defenses, few seem to have evolved successful counter-adaptations against the resident ants of myrmecophytes. Rates of herbivory on ant-free *Tachigali* trees were ten times higher than on inhabited ones, and both rates of apical growth and leaf longevity were strongly increased by the presence of ants (Fonseca 1994). Ant-free leaves of *Leonardoxa* lost 7 to 12 times more leaf area to chewing insects than ant-tended ones, and the resident *Petalomyrmex* ants also protected against sap-sucking insects that reduced leaf expansion (Gaume et al. 1997). Ant-occupied *M. bancana* increased their total leaf area by about 40% within one year, whereas trees from which ants had been removed experimentally lost 80% of their leaf area on average (Heil et al. 2001a). Defensive efficacy in such myrmecophytic interactions appears to be much greater than in myrmecophilic interactions of other *Macaranga* species (Fiala et al. 1989, 1994). African *A. drepanolobium* effectively combines the effects of indirect, ant-mediated defense (resident *Crematogaster*) with direct defense by thorns (Stapley 1998). *Azteca alfari* ants on *Cecropia* (Vasconcelos & Casimiro 1997), and *Pheidole* ants on *Tococa* (Alvarez et al. 2001), defend their hosts against leaf-cutter ants.

Herbivory to stems may be even more costly to the plant than is destruction of the leaves they bear. Protection against stem herbivores has been reported for *Pheidole* inhabiting Costa Rican *Piper* myrmecophytes (Letourneau 1998), for *Camponotus* in *Endospermum labios* in Papua New Guinea (Letourneau & Barbosa 1999), and for *Crematogaster* in *M. bancana* in West Malaysia (Heil et al. 2001a).

The first hint of protection by ants against pathogens was the observation that food body-producing trichilia of *Cecropia obtusa* were covered by a fungus in the absence of ants (Belin-Depoux et al. 1997). Experimental results on protection of myrmecophytes against fungi have now been presented for *Piper* (Letourneau 1998) and *Macaranga* (Heil et al. 1999, 2001a). Because bacterial and fungal pathogens often gain entry into the plant at wound sites (García-Guzmán & Dirzo 2001), protection against phytophagous insects probably also confers strong indirect protection against pathogen attack in many myrmecophytes.

Plant-ants, inhabiting several genera of myrmecophytes, prune epiphytes and encroaching vines and sometimes neighboring vegetation as well (Renner & Ricklefs 1998, Suarez et al. 1997). This behavior benefits ants directly by reducing access to the plant by competing ants (Davidson et al. 1988, Federle et al. 2002, Yumoto & Maruhashi 1999), and indirectly by its beneficial effects on the host (reduced competition for nutrients, water, and light).

Many studies thus demonstrate that myrmecophytes sustain more herbivory when deprived of their resident ants, and a smaller number document protection against pathogens and against competing plants. However, only a few recent studies join the classical work of Janzen (1966) in clearly demonstrating higher survival rates (Heil et al. 2001a), or higher seed set (Letourneau 1998), when ants are present. Hard evidence for lifetime fitness benefits is therefore still scarce and quantifying such benefits will be even more difficult for these often long-lived woody plants.

Sources of Variation

Providing ants with nesting space and food thus ensures long-term interactions with, in many cases, specialized, plant-adapted ants. However, in some myrmecophytes, such as *Conostegia setosa* (Melastomataceae), ant occupancy varies among sites and depends on clone size and microclimatic conditions (Alonso 1998). Several other myrmecophytes also show striking variation in identity of ant (or even other arthropod) occupants (Bizerril & Vieira 2002, Dejean & Djiéto-Lordon 1996). Protection can differ depending on the occupant ant species (Gaume & McKey 1999, Suarez et al. 1997, Young et al. 1997). Amazonian *Maieta guianensis* and *Tococa bullifera* occupied by different ants differed significantly in size, most probably because of differential effects on plant growth (Vasconcelos & Davidson 2000).

However, mature individuals of most obligate myrmecophytes, such as *Acacia*, *Barteria*, *Cecropia*, *Leonardoxa*, *Macaranga*, and *Piper*, are inhabited by only a restricted number of highly specialized ants. The mechanisms that restrict access, and thereby help to stabilize specific mutualisms, are the subject of active research (see below). However, there may be considerable variation in protective efficacy, due for example to variation in size of the resident colony (Duarte Rocha & Godoy Bergallo 1992, Heil et al. 2001b, Itino et al. 2001b), a factor that itself depends on several internal and external factors (see below).

Trophic Structure of the Symbiosis

Several plant-ants store large amounts of debris (exuviae, dead larvae and workers, remains of arthropod prey) inside domatia, and their host plants often show specific adaptations for efficient nutrient uptake from this debris (Treseder et al. 1995). Although most examples of such "nutritional" mutualisms concern epiphytes, the phenomenon also occurs in some forest-understory treelets, e.g., the melastomes *Tococa* (Alvarez et al. 2001) and *Maieta* (Belin-Depoux & Bastien 2002). However, in most protection mutualisms, the flow of resources appears to be principally from the plant to its resident ants (but see below). Plants feed ants directly by producing extrafloral nectar (*Leonardoxa*, African *Acacia* species), cellular food bodies (*Cecropia*, *Macaranga*, *Piper*), or both (Central American *Acacia* species), or indirectly via hemipteran trophobionts tended by resident ants. Several plant-ants further make use of external, "off-host" food sources. The relative importance of food resources offered directly and indirectly by the plant varies greatly, even among closely related species (Gaume & McKey 2002, Itino et al. 2001b), and can even vary among individuals of a single population.

PLANT-DERIVED FOOD SOURCES EFN is an important plant reward to ants in many myrmecophytes. In *Acacia* (Janzen 1966) and *Leonardoxa* (McKey 2000), nectaries of myrmecophytes are more numerous and more active than those of myrmecophilic congeners. In the most specialized variant of the latter system, EFN is the only known food reward. In other lineages, in contrast, nectar secretion is greatly reduced in myrmecophytes. In *Macaranga*, only the leaf glands of non-myrmecophytic species function as nectaries, and are reduced to hydathodes (i.e., water-secreting glands) in the obligate myrmecophytes (Fiala & Maschwitz 1991).

Food bodies (FB) are ontogenetically derived from pearl body-like emergences (*Cecropia*, *Macaranga*, *Piper*) or leaflet tips (Central American *Acacia*) and can be unicellular (*Piper*) or multicellular. They contain high concentrations of lipids, proteins, and carbohydrates (Fischer et al. 2002, Heil et al. 1998). FBs produced by obligate *Macaranga* (Heil et al. 1998) or *Piper* (Fischer et al. 2002) myrmecophytes are rich in lipids and proteins, whereas those produced by myrmecophilic species mainly contain carbohydrates. FBs produced by myrmecophytes show other striking adaptations for feeding animals, e.g., the presence of glycogen instead of starch in FBs of *Cecropia* (Rickson 1971).

Hemipteran trophobionts are third partners in a large proportion of ant-myrmecophyte mutualisms (Davidson & McKey 1993). Their importance appears to vary among systems. In some cases trophobionts are essential, either because plants appear to produce no direct food rewards or because these rewards do not supply all nutrients required by the ants. Because of their hidden location within domatia, determining the kinds and amounts of resources trophobionts supply to ants is very difficult, and these aspects are much more poorly known than for direct food rewards (McKey & Meunier 1996). Patterns in *Leonardoxa* occupied by *Aphomomyrmex* ants suggested that ants harvested pseudococcid honeydew but consumed coccids as prey (Gaume & McKey 1998). Such differences seem to have important consequences for costs and benefits of the association to the plant (Gaume & McKey 1998, Gaume et al. 1998) and the ant (Gaume & McKey 2002).

Comparative studies suggest that different types of rewards play complementary roles within systems and different roles among systems. For example, both *Macaranga* and neotropical *Acacia* myrmecophytes house their ants in hollow structures and nourish them by FB production. However, the composition of food bodies differs (protein-rich in *Acacia*, containing both lipids and proteins in *Macaranga*), and this is related to differences in other food rewards (M. Heil, B. Baumann, unpublished data). *Acacia* plants provide their ants with abundant extrafloral nectar. In contrast, nectaries are reduced in most myrmecophytic *Macaranga* species (Fiala & Maschwitz 1991), whose *Crematogaster* associates, unlike the ants in *Acacia*, cultivate scale insects (Heckroth et al. 1999) as a source of carbohydrates and probably also proteins. However, *M. puncticulata* is inhabited by a *Camponotus* rather than by a *Crematogaster* species, and this *Camponotus* does not cultivate scale insects. In contrast to other myrmecophytic *Macaranga*, *M. puncticulata* provides its ants with extrafloral nectar (Federle et al. 1998).

FOOD SOURCES EXTERNAL TO THE PLANT Curiously, only a few specialist plant-ants are recorded to eat phytophagous insects that they kill on the plant (Dejean et al. 2001a,b; Gaume et al. 1998; Sagers et al. 2000). Some plant-ants may gather pollen grains or fungal spores that fall onto the host (Davidson et al. 2003), and others gather detritus (Alvarez et al. 2001, Belin-Depoux & Bastien 2002). Such input of externally derived nutrients into the system might benefit the plant partner as well. An estimated 80% of the carbon in *Azteca* workers' bodies was derived from their *Cecropia* host tree, whereas more than 90% of the plant's nitrogen appeared to come from the ants' debris (Sagers et al. 2000). Nutrient flow seems also to be bidirectional in *Tococa* plants defended by *Pheidole* ants, which consume lipid- and sugar-rich trichomes inside some domatia, and deposit detritus in others (Alvarez et al. 2001). These studies suggest that many "protection" mutualisms may also confer nutritional benefits to plants. Further studies are needed to obtain reliable estimates of the complex resource flows between ant-plants and their plant-ants.

Cost-Benefit Relations

Defenses impose costs if the allocation of limited resources to defense entails negative effects on fitness. Allocation costs are generally thought to be an important explanation of both genetic variability in constitutive defense traits and the evolution of induced resistance. However, the allocation of resources to defense is often difficult to quantify, chiefly owing to the multiple functions of many defensive traits. Ant-plants are useful model systems to study such costs. Feeding resident ants, and thereby enabling indirect defense, appears to be the sole function of food rewards produced by ant-plants. Compared with direct chemical defenses these food rewards are easy to remove, quantify, and analyze. For FB production by the myrmecophilic *Ochroma pyramidale*, O'Dowd (1980) estimated costs at about 1% of a leaf's construction costs. In contrast, FB production by saplings of the myrmecophyte *M. bancana* amounted to about 5% of total aboveground biomass production (Heil et al. 1997). FB production by this species is limited by soil nutrient content (Heil et al. 2001b) and responds faster to increased soil nutrient supply than does photosynthesis or plant growth (Heil et al. 2002b).

Production of food rewards for ants can also entail "ecological" costs (Tollrian & Harvell 1999) that result from negative effects on some of the myriad interactions between the plant and its environment (Heil 2002). For example, a defense trait may attract, rather than deter, enemies, or have negative effects on mutualists. Most studies of ecological costs of defense by ants have dealt with specialized ant parasites of mutualisms (see below). However, several studies have hinted at ecological costs generated by other interactions. Some vertebrates are attracted to myrmecophytes as rich sources of ant prey and destroy domatia (Federle et al. 1999). Specialized *Dipoena* spiders and *Phyllobaenus* beetles that exploit *Pheidole* ants on *Piper* effectively control the plant's ants and even the population of their *Piper* hosts (Letourneau & Dyer 1998a,b). *Pheidole* ants detect and avoid leaves carrying *Dipoena* webs, and plants with spiders sustained significantly higher rates of folivory (Gastreich 1999). Ecological costs are even more likely in the less specific interactions among EFN-producing plants and ants, since non-ant EFN consumers can compete with ants for rewards and reduce the effectiveness of defense (Heil et al. 2003).

MAXIMIZING NET BENEFITS—THE PLANT'S POINT OF VIEW The key precondition for the coevolutionary reciprocal intensification of protection and nutrition postulated for myrmecophytes (Janzen 1966) is that by investing more in ants, the plant can increase protective benefits (Fonseca 1993). *M. bancana* can increase the size of their resident ant colony by producing higher amounts of FBs (Heil et al. 2001b, Itino et al. 2001b). The size of ant colonies resident in myrmecophytes might also be limited by nesting space (Fonseca 1999), and there is at least one report that domatia can be induced by plant ants (Blüthgen & Wesenberg 2001).

There is considerable evidence that bigger ant colonies better defend their host (Duarte Rocha & Godoy Bergallo 1992, Gaume et al. 1998, Heil et al. 2001b), although additional ants are less likely to add significantly to a large colony's defensive effect than to that of small colonies (Fonseca 1993). There is thus a theoretical optimum investment in ants. That plant production of food rewards for ants may be sensitive to both costs and benefits is suggested by the observation that

FB production by myrmecophytic *Cecropia* growing in a greenhouse was limited by light and nutrient availability (Folgarait & Davidson 1994, 1995), whereas FB production by *M. bancana* was limited by soil nutrient content at the plant's natural growing site (Heil et al. 2001b, 2002b). Plant investment in ants can respond to information about likely benefits. *Piper* myrmecophytes produce FBs only in the presence of their *Pheidole* ants (Risch & Rickson 1981) or of a parasitic beetle that appears to have broken the code (Letourneau 1990). Ant-free *M. triloba* plants produced fewer FBs than inhabited plants (Heil et al. 1997). Whether this response is triggered by FB removal or by the presence of a specific ant colony is unknown. *C. obtusa* produced more Müllerian bodies when inhabited by ants (Belin-Depoux et al. 1997), and experimental studies have shown that FB production rates by *Cecropia* depend on intensity of mechanical removal (Folgarait et al. 1994).

MAXIMIZING NET BENEFITS—THE ANT'S POINT OF VIEW Ants that are obligate inhabitants of specific host plants have an interest in their host's vigor, growth, and survival. Although ant and plant interests as such converge, they are not entirely congruent. First, in Fonseca's (1993) model, increased investment in ants by the plant reaches a point where the protective benefits the latter receives level off. In contrast, benefits to ants could continue to increase substantially with plant investment in them. Do ants have an interest in pushing their resource demands into the range where they become parasites of the plant? The potential for parasitism may be particularly great in those myrmecophytes in which ants obtain food indirectly from the host via hemipteran trophobionts (Davidson & McKey 1993), a process over which the plant may have limited control. By tending trophobionts at densities above what is optimal for the plant, ants can act as parasites, or as less effective mutualists (Gaume et al. 1998). Although plants might "retaliate" by growing more slowly, thereby reducing the flow of benefits to ants (Fonseca 1993), decreased reproductive effort might be a more frequent response.

Because these associations involve horizontal transmission and are formed anew each generation, neither partner has a short-term interest in the reproduction of the other (Wilkinson & Sherratt 2001, Yu 2001). Selection could favor ants that manipulate plants in ways that cause them to invest more in resources that benefit ants, at the expense of the plant's own reproduction. One demonstrated mechanism is castration (Yu & Pierce 1998), which has now been observed for Peruvian *Cordia nodosa* (Yu & Pierce 1998), African *A. drepanolobium* (Stanton et al. 1999, Young et al. 1997), and Amazonian *Hirtella* (Izzo & Vasconcelos 2002). Flowering and seed set divert resources from vegetative growth and thus are likely to reduce the flow of resources from the plant to its ants. The consequences of castration behavior for the production of ant food or for lifetime fitness of the host plants appear not to have been investigated. In some systems, this behavior appears to be facultative. In *M. bancana*, *Crematogaster* ants often attack their hosts' flowers after experimental removal of FB-producing stipules. The behavior thus occurs under conditions of strongly reduced food production (M. Heil, personal observations).

Immediate "retaliation" against castrating parasites may be difficult. When ant associates castrate, mutualism may be maintained by evolutionary shifts in the location of flowers (Yu & Pierce 1998), or of domatia (Izzo & Vasconcelos 2002), so that ants rarely encounter reproductive structures. Even when ants do not attack flowers, pollinator access might be reduced by aggressive ant-guards (Willmer & Stone 1997). In the African *A. drepanolobium*, ants are deterred from young flowers by a volatile signal, perhaps released by pollen (Willmer & Stone 1997). Similar observations were recently made for the Central American *A. hindsii* (Raine et al. 2002) and might be a general phenomenon because plant-ants inhabiting myrmecophytic *Acacia* plants were even repelled by flowers of several non-myrmecophytes (Ghazoul 2001).

Ants can also selfishly manipulate plants by reducing the quantity of resources plants supply. Competitively inferior *Tetraponera penzigi* ants destroy foliar nectaries of African *Acacia* myrmecophytes, reducing the probability of their being replaced by more aggressive ants that require higher rates of resource supply and are more effective mutualists of the plant (Palmer et al. 2002, Young et al. 1997).

Investment by ants in their own reproduction imposes a cost to the plant but confers no immediate, direct benefit. This aspect of the functional ecology of plant-ants has been virtually ignored. Variation in reproductive effort among plant-ants appears to reflect strategies of ants, not manipulation by their hosts. At the colony level, greater allocation to growth and survival (production of workers and, in secondarily polygynous plant-ants, of supernumerary queens; see Feldhaar et al. 2000, McKey et al. 1999) would be favored when increased colony size and/or longevity are likely to be repaid by increased benefits from better-performing hosts, and when new nest sites are limited (cost of dispersal is high). Greater allocation to reproduction (dispersing males and females) would be favored by high probability of mortality of the ant colony due to factors that it cannot control by more effective protection of the host (Gaume & McKey 1999).

Defense Against "External" Enemies of the Mutualism

"Castration" parasites appear to have evolved from mutualistic partners that started to "cheat." However, parasites can also colonize mutualisms. Janzen (1975) reported that *Pseudomyrmex nigropilosa* inhabits ant-acacias and consumes FBs produced by their hosts without exhibiting the defensive behavior of the plant's mutualistic *Pseudomyrmex* spp. Similarly, *Cataulacus mckeyi* excludes the effective mutualist *Petalomyrmex phylax* from its *Leonardoxa* host trees (Gaume & McKey 1999). Myrmecophilous caterpillars of several lycaenid butterflies have exploited ancestral mutualistic relationships with ants to become parasites of obligate ant-plant interactions. These caterpillars feed on the ants' host plant and are attended and protected by the ants, although they clearly damage the host plant (Forster 2000, Maschwitz et al. 1984). Similarly, beetles of the genus *Coelomera* can live and feed on ant-inhabited *Cecropia* trees without being attacked by the ants (Jolivet 1991). *Phyllobaenus* beetle larvae live in domatia of *Piper obliquum*, where they feed on adult workers and brood of the resident *Pheidole bicornis* ants. If ants are absent, the larvae can also use FBs produced by the plant, and can even induce the production of FBs (Letourneau 1990). A similar system has been reported on *C. obtusa* (Belin-Depoux et al. 1997).

Coevolutionary Specializations

The constancy, intensity and (often) specificity of symbiotic ant-plant associations have led to numerous coevolutionary specializations of each partner. These specializations concern several functional domains.

SPECIALIZATIONS FOR DEFENSE AND ITS ALLOCATION Because the interests of ants and plants are only partly convergent, selection on each may favor traits that enable it to control the flow of resources. However, once the rate of flow is set, selection on both partners favors maximizing protection, because the higher the resource-efficiency of mutualistic benefits, the more each partner can invest in its own current reproduction without reducing its future reproduction by harming the mutualist. Among the ant traits that could be affected by such selection is how the resources invested in the worker force are subdivided among workers, i.e., the evolution of worker size. In some plant-ants, worker size has increased compared with that of their less specialized relatives, whereas in others worker size has decreased (Meunier et al. 1999), and adaptation to the particular suite of herbivores attacking the plant appears to be partly responsible for these divergent specializations (Gaume et al. 1997). Behavioral traits may contribute to the defensive efficacy of the many surprisingly small protective plant-ants. *P. minutula* inhabiting *M. guianensis*, *Allomerus decemarticulatus* ants in *Hirtella physophora*, and *Crematogaster* ants in different *Macaranga* myrmecophytes show very effective mass recruiting systems so that many workers are available when larger herbivores have to be attacked. By focusing their attacks on the most vulnerable parts of the attackers (Fiala & Maschwitz 1990), or by "spread-eagling" of prey (Dejean et al. 2001a,b), these ants can protect despite their small size.

Selection could also favor plant traits that reduce the investment required of the ant colony for functions other than plant defense. Ants inhabiting myrmecophytes must defend not only their host but also their nesting sites and food resources against other ant species. This may require larger worker numbers, or additional resource-consuming activities, such as pruning of encroaching vegetation (Davidson et al. 1988, Federle et al. 2002, Yumoto & Maruhashi 1999) or even more sophisticated ways of "burning bridges" (Palmer et al. 2002) that could be used by competitors. Plant traits that help ants exclude such competitors not only reinforce the specificity of these horizontally transmitted interactions (Davidson & McKey 1993), they could also allow mutualist ants to use limited resources more efficiently in plant defense. Epicuticular waxes produced by some, but not all, myrmecophytic *Macaranga* species form slippery surfaces on which only the adapted specialist ants can walk (Federle et al. 1997). Ant associates of waxy-stemmed species thus require less territorial defense and exhibit reduced pruning activity and lower worker densities (Federle et al. 2002). Another example of filters is the formation by some ant-plants of prostomata (Brouat et al. 2001a, Federle et al. 2001) or membranous or unlignified spots where ants with appropriate behavior and morphology (sometimes coevolved with that of the prostoma; see Brouat et al. 2001a) can easily open entrance holes.

Selection should also favor ant and plant traits that result in concentration of ant effort on sites where defense is most important and most often required. *Azteca* ants living in *Cecropia* plants, and *Crematogaster* on *Macaranga*, can recognize and recruit to damaged sites associated with herbivory and thus represent an effective locally induced resistance (Agrawal & Dubin-Thaler 1999, Fiala & Maschwitz 1990, Fiala et al. 1989). The same response can be elicited by plant sap, particular components of plant extracts, or commercially available green leaf volatiles (Agrawal 1998, Brouat et al. 2000). Many specialized plant-ants, however, patrol even in the complete absence of enemies. This "constitutive" patrolling activity is often concentrated on young leaves, especially in myrmecophytes whose long-lived mature leaves possess direct defenses (Izzo & Vasconcelos 2002, McKey 1984; but see Fonseca 1994). The proximate cues enabling the preferential patrolling of young plant parts remain to be elucidated.

ACQUIRING THE PARTNER: THE BIOLOGY OF JUVENILE MYRMECOPHYTES The association between ant and plant must be established anew in each successive generation, posing several distinct adaptive problems. The first is the orientation to appropriate host plants by mated foundresses of the mutualist ant. The cues used by plant-ant foundresses to locate hosts are only beginning to be studied. Because nuptial flights are often nocturnal [e.g., in *Petalomyrmex* (L. Gaume, personal communication), in *Barteria*-associated *Tetraponera* (C. Djiéto-Lordon, personal communication), and in *Macaranga*-associated *Crematogaster* (B. Fiala, M. Heil, personal observations)], olfactory cues seem most likely. First hints of a role for chemical cues in host plant recognition (whether by olfaction or by contact chemoreception is not yet clear) have been presented (Inui et al. 2001), but more detailed studies are required.

The second problem is survival, which may be a difficult task for juvenile ant-plants and incipient mutualist ant colonies. Interlopers, predators, stochastic mortality factors, and intraspecific competition among incipient colonies of mutualists should all have their greatest impact on ants or plants at this stage (McKey & Meunier 1996). Plants may have limited resources for ants, and ants may be incapable of providing much protection, so that there is little exchange of mutualistic benefits.

Under these circumstances, when in its ontogeny should the plant begin to produce ant-attractant resources? Although some ant-plants already exhibit myrmecophytic traits as seedlings, others become myrmecophytes only when they become small saplings (Fiala & Maschwitz 1992). Brouat & McKey (2000) suggested that the costs of housing and supporting ants are proportionally greater the smaller the plant. This may be true not only of food rewards but also of caulinary domatia, which are especially costly in terms of stem allocation early in plant ontogeny (Brouat & McKey 2001, Gartner 2001). Producing domatia and other ant resources early in ontogeny thus will be favored only when ants possess sufficient mutualistic specializations to provide protection even when both the host plant and the ant colony are small (Brouat & McKey 2000).

This example shows the potential of myrmecophytes as models for exploring the evolution of allometry in modular organisms (Preston & Ackerly, 2003) and for investigating the linkages between coevolutionary interactions and the evolution of life histories, in both plant and ant partners.

REDUCED CHEMICAL DEFENSE IN OBLIGATE MYRMECOPHYTES Plant defenses are assumed to be costly, and plants therefore should avoid redundant defenses. That chemical defense is reduced in ant-plants was originally hypothesized by Janzen (1966), who proposed that chemical defense has been lost in ant-acacias. Indeed, a diet containing leaf powder from a non-ant-acacia (*A. farnesiana*) had a much stronger negative impact on growth of caterpillars than a diet using the ant-acacia, *A. cornigera* (Rehr et al. 1973). Foliage of most ant-acacias contains no or only small amounts of cyanogenic glycosides (Seigler & Ebinger 1987). Recent studies have focused on the enzymatic antifungal defense of ant-plants in the genera *Macaranga* and *Acacia* (Heil et al. 1999, 2000b), on condensed tannin content in *Macaranga* (Eck et al. 2001), and on herbivore-deterrent amides in inhabited and ant-free individuals of *Piper cenocladum* (Dyer et al. 2001). It has also been assumed that plants might switch during ontogeny from biotic to chemical defense or vice versa (Fiala et al. 1994, Nomura et al. 2000).

However, unequivocal patterns have been found only in some studies (Dyer et al. 2001; Heil et al. 1999, 2000b). No trade-offs were found for *Endospermum* (Letourneau & Barbosa 1999), and in a study covering three different classes of defensive phenolics in the genera *Acacia*, *Leonardoxa*, and *Macaranga* (Heil et al. 2002a). In some studies support for trade-offs appears weaker than claimed (Eck et al. 2001, Rehr et al. 1973, Seigler & Ebinger 1987). Although the reduced direct defense of obligate ant-plants has been demonstrated in many studies, most studies focusing on distinct chemical defenses have found no clear evidence of the expected trade-offs. The dramatically increased vulnerability of ant-free myrmecophytes to herbivores and pathogens is thus still in search of a proximate explanation (Heil et al. 2002a).

Phylogenies for Testing Hypotheses About Evolution of Ant-Plant Interactions

The taxonomic and ecological diversity of ant-plant symbioses offers great potential for using the comparative method to test hypotheses about the evolution of mutualism (Davidson & McKey 1993). Realizing this potential requires robust phylogenies of the interacting organisms. Several recent studies confirm earlier conclusions that myrmecophytes have arisen repeatedly (e.g., up to four times in *Macaranga*; see Blattner et al. 2001, Davies et al. 2001). Patterns in the phylogeny of the *Crematogaster* associates of *Macaranga* suggest parallels with those in plants and possible cospeciation (Itino et al. 2001a). In the *Leonardoxa* system, phylogeny of the mutualist ants (Chenuil & McKey 1996) and that of the plants (Brouat et al. 2001b, McKey 2000) suggest a complex history, in which reticulate evolution in plants, local extinctions of ant partners, and colonization of plants by multiple ant lineages all appear to play roles. Application of newly developed microsatellite markers to population genetics and phylogeography of ant associates (Dalecky et al. 2002, Debout et al. 2002) will greatly extend the range of tractable questions about the history of these interactions.

ANT-PLANTS AS MODELS TO STUDY THEORIES ON ANTIHERBIVORE DEFENSE

Empirical tests of theories on plant antiherbivore defense encounter considerable difficulty, because most "defensive" plant traits serve other functions as well, and because precise description of spatial and temporal patterns in the occurrence of direct defenses often requires elaborate chemical analyses. Moreover, low levels of herbivory observed under field conditions may result either from low herbivore pressure or from effective defense, making analyzes of the effectiveness of a given defensive trait extremely difficult. Ants represent an efficient defense mechanism whose location on the plant can be precisely described, allowing studies on the temporal and spatial distribution of defense, and that can be experimentally removed from the plant, allowing tests of their effectiveness. They thus present an elegant model system to test several hypotheses of plant antiherbivore defense.

The optimal defense hypothesis predicts that plant defenses should be concentrated in the most valuable and vulnerable parts of a plant (McKey 1974, Rhoades 1979). In fact, many plant-ants patrol and defend preferably the young leaves, which generally have high potential value to the plant (Harper 1989) and sustain most herbivory (Coley & Barone 1996). In the case of facultative, myrmecophilic interactions, this pattern is mainly caused by patterns in the production of food rewards (such as EFN). However, many ant associates of obligate myrmecophytes patrol preferably the young leaves of their host, often independently of the presence of food rewards (Downhower 1975, Janzen 1972). The concentration of ant defense on young leaves has been reported for *Piper* (Risch 1982), African *Acacia* (Madden & Young 1992), *Tachigali* (Fonseca 1994), *Leonardoxa* (Brouat et al. 2000), *Crypteronia* (Moog et al. 1998), and *Macaranga* (Heil et al. 2001a). There seems to be only one reported exception to the general rule: ants patrol young and mature leaves of *Maieta* with the same intensity, but in this case both age classes were observed to suffer equal herbivore attack (Vasconcelos 1991). In general, plant-ants thus form a defense mechanism whose spatial distribution is consistent with the optimal defense hypothesis.

How should plant defenses respond to variation in resource availability? According to the carbon/nitrogen balance hypothesis (Bryant et al. 1983), the response depends on what particular resources are most limiting. *Cecropia* myrmecophytes produce glycogen-rich "Müllerian bodies" and lipid- and amino acid–rich "pearl bodies." Folgarait & Davidson (1994, 1995) found that the production of pearl bodies increased under conditions of high nutrient level and low light (which should have contributed to relative excess of nitrogen). Although these results supported the C/N balance hypothesis, production of Müllerian bodies increased at high levels of both nutrients and light. Nutrient effects on toughness and leaf expansion rates were also inconsistent with the predictions of this theory (Folgarait & Davidson 1994, 1995). Similarly, production of the (protein-containing, yet lipid-dominated) FBs of *M. triloba* responds quickly and strongly to increased nutrient supply (Heil et al. 2001b, 2002b), and even FBs of *M. tanarius* (containing nearly no proteins) were produced at higher rates when plants received more nitrogen (M. Heil, A. Hilpert, unpublished data). Results from ant-plant studies are thus representative of many other recent studies of chemical defenses that have failed to support the C/N balance hypothesis, which some authors now regard as definitively "rejected" (Hamilton et al. 2001).

According to the resource availability hypothesis (Coley et al. 1985), "mobile" defenses such as alkaloids or other small molecules impose high maintenance costs due to their high turnover rates, but can be reclaimed from leaves before they are shed. Such defenses are predicted to occur mainly in short-lived leaves of fast-growing plants. Ants can move easily over the whole plant surface and are not shed with leaves. They are thus highly "mobile" defenses (McKey 1984). The fact that ant defenses are most spectacularly developed in pioneer trees is consistent with the resource availability hypothesis, as is the frequent restriction of ant defense to young leaves of understory trees whose long-lived mature leaves appear to have immobile direct defenses (Izzo & Vasconcelos 2002, McKey 1984).

PERSPECTIVES

We have focused on ant-plant interactions primarily from the perspective of individuals and populations interacting at local spatial scales. With the exception of a partial treatment of conflicts between mutualists and the evolutionary stability of mutualisms, we have also paid scant attention to the long-term dynamics of these interactions. Examining these interactions at larger scales of space and time would require another review. We indicate what we perceive as some particularly interesting questions in the micro- as well as in the macroecology of ant-plant symbioses.

1. What forces drive the evolution of trophic structure of myrmecophyte symbioses? Are hemipteran trophobionts, apparently essential partners at the outset of many symbioses, sometimes eliminated as plants attempt to gain

control of resource supply to ants? Patterns in *Leonardoxa* strongly suggest such dynamics (Chenuil & McKey 1996, Gaume et al. 1998, Meunier et al. 1999). Does the diversity of trophic structure in ant-plant symbioses reflect the still poorly explored diversity in the nutritional ecology of tree-dwelling ants (Davidson et al. 2003)?

2. Functioning of ant-plant mutualism requires the flow not only of trophic resources but also of information. What mechanisms underlie the flow of information between partners that is required for the exchange of mutualistic benefits? Several mechanisms regulating the provisioning of food rewards by the plants or guiding the ants' patrolling behavior have already been reported, but the nature of the signals appears unclear in most cases. Contemporary methods in chemical and molecular ecology are likely to provide efficient tools to investigate the signals involved in host-finding by founding ant-queens, the mechanisms involved in the restriction of many plant-ants to one or only a few host(s), the regulation of food body production and extrafloral nectar secretion, the diversion of defending ants from flowers, and many other aspects of these interactions.
3. Do conflicts between mutualists drive the evolutionary dynamics of ant-plant symbioses, as current work suggests (Izzo & Vasconcelos 2002, Yu 2001, Yu & Pierce 1998)? Or is evolution of these mutualisms driven by "Red King" mechanisms, in which "the slowest runner wins the coevolutionary race" (Bergstrom & Lachmann 2003)? Interactions of long-lived plants with shorter-lived ants might offer the asymmetry in evolutionary rates that appears to favor such dynamics.
4. How do inclusive-fitness models modify expectations about life-history evolution in plant-ants? How do coevolutionary pressures influence ant social evolution? If selection acts at the colony level, association with a long-lived host that benefits from protection should favor colonies that invest in survival to maximize their future reproduction, and life history of the ant colony should be tied to that of the tree in such a way that their interests tend to converge. However, an ant colony is a society in which individuals both cooperate and conflict. One focus of conflicts is the allocation between growth and reproduction. In long-lived colonies, for example, workers may favor greater allocation to sexuals than does the queen (Bourke & Franks 1995). The outcome may not be that which most favors the plant. The plant may have very little control, even indirectly, over the proportion of plant-derived resources invested in workers (of potential benefit to the plant) and in dispersing sexuals (only costs to the plant).
5. How do species coexist in the "simple" communities represented by a guild of plant-ants that share a population of hosts (Palmer et al. 2002, Young et al. 1997, Yu & Davidson 1997)? Studies thus far have identified a number of niche dimensions where differences among species may facilitate coexistence. Which of these are most important? How do traits of social

organization influence outcomes of ant interaction with plants, and with competing ants? What coevolutionary dynamics characterize spatially structured populations of interacting ants and plants (Yu et al. 2001)?

Studies integrating processes at local scales and at the scale of interacting metapopulations are now required to deepen our understanding of how these systems evolve, and how they can persist faced with human-induced habitat fragmentation and other global change.

The *Annual Review of Ecology, Evolution, and Systematics* is online at http://ecolsys.annualreviews.org

LITERATURE CITED

Agrawal AA. 1998. Leaf damage and associated cues induce aggressive ant recruitment in a neotropical ant-plant. *Ecology* 79:2100–12

Agrawal AA, Dubin-Thaler BJ. 1999. Induced responses to herbivory in the Neotropical ant-plant association between *Azteca* ants and *Cecropia* trees: response of ants to potential inducing cues. *Behav. Ecol. Sociobiol.* 45:47–54

Alonso LE. 1998. Spatial and temporal variation in the ant occupants of a facultative ant-plant. *Biotropica* 30:201–13

Alvarez G, Armbrecht I, Jiménez E, Armbrecht H, Ulloa-Chacón P. 2001. Ant-plant association in two *Tococa* species from a primary rain forest of Colombian Choco (Hymenoptera: Formicidae). *Sociobiology* 38:585–602

Apple JL, Feener DH. 2001. Ant visitation of extrafloral nectaries of *Passiflora*: the effects of nectary attributes and ant behavior on patterns in facultative ant-plant mutualisms. *Oecologia* 127:409–16

Baker HG, Opler PA, Baker I. 1978. A comparison of the amino acid complements of floral and extrafloral nectars. *Bot. Gaz.* 139:322–32

Barton AM. 1986. Spatial variation in the effect of ants on an extrafloral nectary plant. *Ecology* 67:495–504

Beattie AJ. 1985. *The Evolutionary Ecology of Ant-Plant Mutualisms*. Cambridge, UK: Cambridge Univ. Press

Belin-Depoux M, Bastien D. 2002. Glances on the myrmecophily in French Guiana. The devices of *Maieta guianensis* absorption and the triple *Philodendron*-ants-*Aleyrodiciae* association. *Acta Bot. Gall.* 149:299–318

Belin-Depoux M, Solano P-J, Lubrano C, Robin JR, Chteau P, Touzet M-C. 1997. La fonction myrmécophile de *Cecropia obtusa* Trecul (Cecropiaceae) en Guyane française. *Acta Bot. Gall.* 144:289–313

Belt T. 1874. *The Naturalist in Nicaragua*. London: Dent

Bentley BL. 1976. Plants bearing extrafloral nectaries and the associated ant community: interhabitat differences in the reduction of herbivore damage. *Ecology* 57:815–20

Bentley BL. 1977. Extrafloral nectaries and protection by pugnacious bodyguards. *Annu. Rev. Ecol. Syst.* 8:407–27

Bentley BL. 1981. Ants, extrafloral nectaries, and the vine life-form: an interaction. *Trop. Ecol.* 22:127–33

Bergstrom CT, Lachmann M. 2003. The Red King effect: when the slowest runner wins the coevolutionary race. *Proc. Natl. Acad. Sci. USA* 100:593–98

Bizerril MXA, Vieira EM. 2002. *Azteca* ants as antiherbivore agents of *Tococa formicaria* (Melastomataceae) in Brazilian Cerrado. *Stud. Neotrop. Fauna Environ.* 37:145–49

Blattner FR, Weising K, Bänfer G, Maschwitz U, Fiala B. 2001. Molecular analysis of phylogenetic relationships among myrmecophytic *Macaranga* species (Euphorbiaceae). *Mol. Phylogenet. Evol.* 19:331–44

Blüthgen N, Verhaagh M, Goitía W, Jaffé K, Morawetz W, Barthlott W. 2000. How plants shape the ant community in the Amazonian rainforest canopy: the key role of extrafloral nectaries and homopteran honeydew. *Oecologia* 125:229–40

Blüthgen N, Wesenberg J. 2001. Ants induce domatia in a rain forest tree (*Vochysia vismiaefolia*). *Biotropica* 33:637–42

Bourke AF, Franks NR. 1995. *Social Evolution in Ants*. Princeton, NJ: Princeton Univ. Press. 529 pp.

Bronstein JL. 1998. The contribution of ant-plant protection studies to our understanding of mutualism. *Biotropica* 30:150–61

Brouat C, Garcia N, Andary C, McKey D. 2001a. Plant lock and ant key: pairwise coevolution of an exclusion filter in an ant-plant mutualism. *Proc. R. Soc. London Ser. B* 268:2131–41

Brouat C, Gielly L, McKey D. 2001b. Phylogenetic relationships in the genus *Leonardoxa* (Leguminosae: Caesalpinioideae) inferred from chloroplast *trnL* intron and *trnL-trnF* intergenic spacer sequences. *Am. J. Bot.* 88:143–49

Brouat C, McKey D. 2000. Origin of caulinary ant domatia and timing of their onset in plant ontogeny: evolution of a key trait in horizontally transmitted ant-plant symbiosis. *Biol. J. Linn. Soc.* 71:801–19

Brouat C, McKey D. 2001. Leaf-stem allometry, hollow stems, and the evolution of caulinary domatia in myrmecophytes. *New Phytol.* 151:391–406

Brouat C, McKey D, Bessière J-M, Pascal L, Hossaert-McKey M. 2000. Leaf volatile compounds and the distribution of ant patrolling in an ant patroling in an ant-plant protection mutualism: preliminary results on *Leonardoxa* (Fabaceae: Caesalpinioideae) and *Petalomyrmex* (Formicidae: Formicinae). *Acta Oecol.* 21:349–57

Bryant JP, Chapin FS III, Klein DR. 1983. Carbon/nutrient balance of boreal plants in relation to vertebrate herbivory. *Oikos* 40:357–68

Buckley RC. 1982. Ant-plant interactions: a world review. In *Ant-Plant Interactions in Australia*, ed. RC Buckley, pp. 111–62. The Hague/Boston/London: Junk

Bugg RL, Ellis RT, Carlson RW. 1989. Ichneumonidae (Hymenoptera) using extrafloral nectar of faba bean (*Vicia faba* L, Fabaceae) in Massachusetts. *Biol. Agric. Hortic.* 6:107–14

Chenuil A, McKey D. 1996. Molecular phylogenetic study of a myrmecophyte symbiosis: Did *Leonardoxa*-ant associations diversify via cospeciation? *Mol. Phylogenet. Evol.* 6:270–86

Coley PD, Barone JA. 1996. Herbivory and plant defenses in tropical forests. *Annu. Rev. Ecol. Syst.* 27:305–35

Coley PD, Bryant JP, Chapin FS III. 1985. Resource availability and plant antiherbivore defense. *Science* 230:895–99

Costa FBMC, Oliveira-Filho AT, Oliveira PS. 1992. The role of extrafloral nectaries in *Qualea grandiflora* (Vochysiaceae) in limiting herbivory: an experiment of ant protection in cerrado vegetation. *Ecol. Entomol.* 17:363–65

Dalecky A, Debout G, Mondor G, Rasplus J-Y, Estoup A. 2002. PCR primers for polymorphic microsatellite loci in the facultatively polygynous plant-ant *Petalomyrmex phylax* (Formicidae). *Mol. Ecol. Notes* 2:404–7

Davidson DW. 1997. The role of resource imbalances in the evolutionary ecology of tropical arboreal ants. *Biol. J. Linn. Soc.* 61:153–81

Davidson DW, Cook SC, Snelling RR, Chua TH. 2003. Explaining the abundance of ants in lowland tropical rainforest canopies. *Science.* 300:969–72

Davidson DW, Longino JT, Snelling RR. 1988. Pruning of host plant neighbors by ants: an experimental approach. *Ecology* 69:801–8

Davidson DW, McKey D. 1993. The evolutionary ecology of symbiotic ant-plant relationships. *J. Hym. Res.* 2:13–83

Davies SJ, Lum SKY, Chan R, Wng LK. 2001. Evolution of myrmecophytism in western malesian *Macaranga* (Euphorbiaceae). *Evolution* 55:1542–59

Davies SJ, Palmiotto PA, Ashton PS, Lee HS, LaFrankie JV. 1998. Comparative ecology of 11 sympatric species of *Macaranga* in Borneo: tree distribution in relation to horizontal and vertical resource heterogenity. *J. Ecol.* 86:662–73

Debout G, Dalecky A, Mondor G, Estoup A, Rasplus J-Y. 2002. Isolation and characterisation of polymorphic microsatellites in the tropical plant-ant *Cataulacus mckeyi* (Formicidae: Myrmicinae). *Mol. Ecol. Notes* 2:459–61

Dejean A, Djiéto-Lordon C. 1996. Ecological studies on the relationships between ants (Hymenoptera, Formicidae) and the myrmecophyte *Scaphopetalum thonneri* (Sterculiaceae). *Sociobiology* 28:91–102

Dejean A, McKey D, Gibernau M, Belin M. 2000. The arboreal ant mosaic in a Cameroonian rainforest (Hymenoptera: Formicidae). *Sociobiology* 35:403–23

Dejean A, Solano PJ, Belin-Depoux M, Cerdan P, Corbara B. 2001a. Predatory behavior of patrolling *Allomerus decemarticulatus* workers (Formicidae; Myrmicinae) on their host plant. *Sociobiology* 37:571–78

Dejean A, Solano PJ, Orivel J, Belin-Depoux M, Cerdan P, Corbara B. 2001b. The spread-eagling of prey by the obligate plant-ant *Pheidole minutula* (Myrmicinae): Similarities with dominant arboreal ants. *Sociobiology* 38:675–82

Delabie J. 2001. Trophobiosis between Formicidae and Hemiptera (Sternorrhyncha and Auchenorrhyncha): an overview. *Neotrop. Entomol.* 30:501–16

de la Fuente MAS, Marquis RJ. 1999. The role of ant-tended extrafloral nectaries in the protection and benefit of a Neotropical rainforest tree. *Oecologia* 118:192–202

delClaro K, Berto V, Reu W. 1996. Effect of herbivore deterrence by ants on the fruit set of an extrafloral nectary plant, *Qualea multiflora* (Vochysiaceae). *J. Trop. Ecol.* 12:887–92

Di Giusto B, Anstett MC, Dounias E, McKey D. 2001. Variation in the effectiveness of biotic defence: the case of an opportunistic ant-plant protection mutualism. *Oecologia* 129:367–75

Downhower JF. 1975. The distribution of ants on *Cecropia* leaves. *Biotropica* 7:59–62

Dress WJ, Newell SJ, Nastase AJ, Ford JC. 1997. Analysis of amino acids in nectar from pitchers of *Sarracenia purpurea* (Sarraceniaceae). *Am. J. Bot.* 84:1701–6

Duarte Rocha CF, Godoy Bergallo H. 1992. Bigger ant colonies reduce herbivory and herbivore residence time on leaves of an ant-plant: *Azteca muelleri* vs. *Coelomera ruficornis* on *Cecropia pachystachia*. *Oecologia* 91:249–52

Dyer LA, Dodson CD, Beihoffer J, Letourneau DK. 2001. Trade-offs in antiherbivore defenses in *Piper cenocladum*: ant mutualists versus plant secondary metabolites. *J. Chem. Ecol.* 27:581–92

Eck G, Fiala B, Linsenmair KE, Proksch P. 2001. Trade off between chemical and biotic anti-herbivore defense in the South-east Asian plant genus *Macaranga*. *J. Chem. Ecol.* 27:1979–96

Elias TS. 1983. Extrafloral nectaries: Their structure and distribution. In *The Biology of Nectaries*, ed. B Bentley, TS Elias, pp. 174–203. New York: Columbia Univ. Press

Farji Brener AG, Folgarait P, Protomastro J. 1992. Asociatión entre el arbuso *Capparis retusa* (Capparidaceae) y las hormigas *Camponotus blandus* y *Acromyrmex striatus* (Hymenoptera: Formicidae). *Rev. Biol. Trop.* 40:341–44

Federle W, Fiala B, Maschwitz U. 1998. *Camponotus* (*Colobopsis*) (Mayr 1861) and *Macaranga* (Thouars 1806): a specific two-partner ant-plant system from Malaysia. *Trop. Zool.* 11:83–94

Federle W, Fiala B, Zizka G, Maschwitz U. 2001. Incident daylight as orientation cue for hole-boring ants: prostomata in *Macaranga* ant-plants. *Ins. Soc.* 48:165–77

Federle W, Leo A, Moog J, Azarae HI, Maschwitz U. 1999. Myrmecophagy undermines ant-plant mutualisms: ant-eating *Callosciurus* squirrels (Rodentia: Sciuridae) damage ant-plants in Southeast Asia. *Ecotropica* 5:35–43

Federle W, Maschwitz U, Fiala B, Riederer M, Hölldobler B. 1997. Slippery ant-plants and skilful climbers: selection and protection of specific ant partners by epicuticular wax blooms in *Macaranga* (Euphorbiaceae). *Oecologia* 112:217–24

Federle W, Maschwitz U, Hölldobler B. 2002. Pruning of host plant neighbours as defence against enemy ant invasions: *Crematogaster* ant partners of *Macaranga* protected by "wax barriers" prune less than their congeners. *Oecologia* 132:264–70

Feldhaar H, Fiala B, bin Hashim R, Maschwitz U. 2000. Maintaining an ant-plant symbiosis: secondary polygyny in the *Macaranga triloba-Crematogaster* sp association. *Naturwissenschaften* 87:408–11

Ferguson BG, Boucher DH, Maribel Pizzi CR. 1995. Recruitment and decay of a pulse of *Cecropia* in Nicaraguan rain forest damaged by hurricane Joan: relation to mutualism with *Azteca* ants. *Biotropica* 27:455–60

Fiala B, Grunsky H, Maschwitz U, Linsenmair KE. 1994. Diversity of ant-plant interactions: protective efficacy in *Macaranga* species with different degrees of ant association. *Oecologia* 97:186–92

Fiala B, Linsenmair KE. 1995. Distribution and abundance of plants with extrafloral nectaries in the woody flora of a lowland primary forest in Malaysia. *Biodivers. Conserv.* 4:165–82

Fiala B, Maschwitz U. 1990. Studies on the south east Asian ant-plant association *Crematogaster borneensis/Macaranga*: adaptations of the ant partner. *Insect. Soc.* 37:212–31

Fiala B, Maschwitz U. 1991. Extrafloral nectaries in the genus *Macaranga* (Euphorbiaceae) in Malaysia: comparative studies of their possible significance as predispositions for myrmecophytism. *Biol. J. Linn. Soc.* 44:287–305

Fiala B, Maschwitz U. 1992. Domatia as most important adaptions in the evolution of myrmecophytes in the paleotropical tree genus *Macaranga* (Euphorbiaceae). *Plant Syst. Evol.* 180:53–64

Fiala B, Maschwitz U, Tho YP, Helbig AJ. 1989. Studies of a South East Asian ant-plant association: protection of *Macaranga* trees by *Crematogaster borneensis*. *Oecologia* 79:463–70

Fischer RC, Richter A, Wanek W, Mayer V. 2002. Plants feed ants: food bodies of myrmecophytic *Piper* and their significance for the interaction with *Pheidole bicornis* ants. *Oecologia* 133:186–92

Floren A, Biun A, Linsenmair KE. 2002. Arboreal ants as key predators in tropical lowland rainforest trees. *Oecologia* 131:137–44

Folgarait PJ, Davidson DW. 1994. Antiherbivore defenses of myrmecophytic *Cecropia* under different light regimes. *Oikos* 71:305–20

Folgarait PJ, Davidson DW. 1995. Myrmecophytic *Cecropia*: antiherbivore defenses under different nutrient treatments. *Oecologia* 104:189–206

Folgarait PJ, Johnson HL, Davidson DW. 1994. Responses of *Cecropia* to experimental removal of Müllerian bodies. *Funct. Ecol.* 8:22–28

Fonseca CR. 1993. Nesting space limits colony size of the plant-ant *Pseudomyrmex concolor*. *Oikos* 67:473–82

Fonseca CR. 1994. Herbivory and the long-lived leaves of an Amazonian ant-tree. *J. Ecol.* 82:833–42

Fonseca CR. 1999. Amazonian ant-plant interactions and the nesting space limitation hypothesis. *J. Trop. Ecol.* 15:807–25

Fonseca CR, Ganade G. 1996. Asymmetries, compartments and null interactions in an Amazonian ant-plant community. *J. Anim. Ecol.* 65:339–47

Forster PI. 2000. The ant, the butterfly and the ant-plant: Notes on *Myrmecodia beccarii* (Rubiaceae), a vulnerable Queensland endemic. *Haseltonia* (7):2–7

Foster WA. 1995. Mosquito sugar feeding and reproductive energetics. *Annu. Rev. Entomol.* 40:443–74

Freitas L, Galetto L, Bernardello G, Paoli AAS. 2000. Ant exclusion and reproduction of *Croton sarcopetalus* (Euphorbiaceae). *Flora* 195:398–402

Galetto L, Bernardello LM. 1992. Extrafloral nectaries that attract ants in Bromeliaceae: structure and nectar composition. *Can. J. Bot.* 70:1101–6

García-Guzmán G, Dirzo R. 2001. Patterns of leaf-pathogen infection in the understory of a Mexican rain forest: incidence, spatiotemporal variation, and mechanisms of infection. *Am. J. Bot.* 88:634–45

Gartner BL. 2001. Multitasking and tradeoff in stems, and the costly dominion of domatia. *New Phytol.* 151:311–13

Gastreich KR. 1999. Trait-mediated indirect effects of a theridiid spider on an ant-plant mutualism. *Ecology* 80:1066–70

Gaume L, McKey D. 1998. Protection against herbivores of the myrmecophyte *Leonardoxa africana* (Baill.) Aubrev. T3 by its principal ant inhabitant *Aphomomyrmex afer* Emery. *C. R. Acad. Sci. III* 321:593–601

Gaume L, McKey D. 1999. An ant-plant mutualism and its host-specific parasite: Activity rhythms, young leaf patrolling, and effects on herbivores of two specialist plant-ants inhabiting the same myrmecophyte. *Oikos* 84:130–44

Gaume L, McKey D. 2002. How identity of the homopteran trophobiont affects sex allocation in a symbiotic plant-ant: the proximate role of food. *Behav. Ecol. Sociobiol.* 51:197–205

Gaume L, McKey D, Anstett M-C. 1997. Benefits conferred by 'timid' ants: active anti-herbivore protection of the rainforest tree *Leonardoxa africana* by the minute ant *Petalomyrmex phylax*. *Oecologia* 112:209–16

Gaume L, McKey D, Terrin S. 1998. Ant-plant-homopteran mutualism: how the third partner affects the interaction between a plant-specialist ant and its myrmecophyte host. *Proc. R. Soc. London Ser. B* 265:596–75

Ghazoul J. 2001. Can floral repellents preempt potential ant-plant conflicts? *Ecol. Lett.* 4:295–99

Gullan PJ. 1997. Relationships with ants. In *Soft Scale Insects: Their Biology, Natural Enemies and Control*, ed. Y Ben-Dov, CJ Hodgson, pp. 351–73. Amsterdam: Elsevier

Gullan PJ, Kosztarab M. 1997. Adaptations in scale insects. *Annu. Rev. Entomol.* 42:23–50

Hamilton JG, Zangerl AR, DeLucia EH, Berenbaum MR. 2001. The carbon-nutrient balance hypothesis: its rise and fall. *Ecol. Lett.* 4:86–95

Harper JL. 1989. The value of a leaf. *Oecologia* 80:53–58

Heckroth HP, Fiala B, Gullan PJ, Maschwitz U, Azarae HI. 1999. The soft scale (Coccidae) associates of Malaysian ant-plants. *J. Trop. Ecol.* 14:427–43

Heil M. 2002. Ecological costs of induced resistance. *Curr. Opin. Plant Biol.* 5:345–50

Heil M, Delsinne T, Hilpert A, Schürkens S, Andary C, et al. 2002a. Reduced chemical defence in ant-plants? A critical reevaluation of a widely accepted hypothesis. *Oikos* 99:457–68

Heil M, Fiala B, Baumann B, Linsenmair KE. 2000a. Temporal, spatial and biotic variations in extrafloral nectar secretion by *Macaranga tanarius*. *Funct. Ecol.* 14:749–57

Heil M, Fiala B, Boller T, Linsenmair KE. 1999. Reduced chitinase activities in ant plants of the genus *Macaranga*. *Naturwissenschaften* 86:146–49

Heil M, Fiala B, Kaiser W, Linsenmair KE. 1998. Chemical contents of *Macaranga* food bodies: adaptations to their role in ant attraction and nutrition. *Funct. Ecol.* 12:117–22

Heil M, Fiala B, Linsenmair KE, Zotz G, Menke P, Maschwitz U. 1997. Food body production in *Macaranga triloba* (Euphorbiaceae): a plant investment in anti-herbivore defence via mutualistic ant partners. *J. Ecol.* 85:847–61

Heil M, Fiala B, Maschwitz U, Linsenmair KE. 2001a. On benefits of indirect defence: short- and long-term studies in antiherbivore protection via mutualistic ants. *Oecologia* 126:395–403

Heil M, Hilpert A, Fiala B, Kaiser W, bin Hashim R, et al. 2002b. Nutrient allocation of *Macaranga triloba* ant plants to growth, photosynthesis, and indirect defence. *Funct. Ecol.* 16:475–83

Heil M, Hilpert A, Fiala B, Linsenmair KE. 2001b. Nutrient availability and indirect (biotic) defence in a Malaysian ant-plant. *Oecologia* 126:404–8

Heil M, Hilpert A, Krüger R, Linsenmair KE. 2003. Competition among visitors to extrafloral nectaries as a source of ecological costs of an indirect defence. *J. Trop. Ecol.* In press

Heil M, Koch T, Hilpert A, Fiala B, Boland W, Linsenmair KE. 2001c. Extrafloral nectar production of the ant-associated plant, *Macaranga tanarius*, is an induced, indirect, defensive response elicited by jasmonic acid. *Proc. Natl. Acad. Sci. USA* 98:1083–88

Heil M, Staehelin C, McKey D. 2000b. Low chitinase activity in *Acacia* myrmecophytes: a potential trade-off between biotic and chemical defences? *Naturwissenschaften* 87: 555–58

Hölldobler B, Wilson EO. 1990. *The Ants*. Berlin/Heidelberg/New York: Springer. 732 pp.

Horvitz CC, Schemske DW. 1984. Effects of ants and an ant-tended herbivore on seed production of a neotropical herb. *Ecology* 65: 1369–78

Horvitz CC, Schemske DW. 1990. Spatiotemporal variation in insect mutualists of a neotropical herb. *Ecology* 71:1085–95

Hossaert-McKey M, Orivel J, Labeyrie E, Pascal L, Delabie JHC, Dejean A. 2001. Differential associations with ants of three co-occurring extrafloral nectary-bearing plants. *EcoScience* 8:325–35

Huxley CR, Cutler DF, eds. 1991. *Ant-Plant Interactions*. Oxford, UK: Oxford Univ. Press. 601 pp.

Ibarra-Manríquez G, Dirzo R. 1990. Arboreal myrmecophilous plants from Los Tuxtlas biological-station, Veracruz, Mexico. *Rev. Biol. Trop.* 38:79–82

Inui Y, Itioka T, Murase K, Yamaoka R, Itino T. 2001. Chemical recognition of partner plant species by foundress ant queens in *Macaranga-Crematogaster* myrmecophytism. *J. Chem. Ecol.* 27:2029–40

Itino T, Davies SJ, Tada H, Hieda Y, Inoguchi M, et al. 2001a. Cospeciation of ants and plants. *Ecol. Res.* 16:787–93

Itino T, Itioka T, Hatada A, Hamid AA. 2001b. Effects of food rewards offered by ant-plant *Macaranga* on the colony size of ants. *Ecol. Res.* 16:775–86

Izzo TJ, Vasconcelos HL. 2002. Cheating the cheater: domatia loss minimizes the effects of ant castration in an Amazonian ant-plant. *Oecologia* 133:200–5

Janzen DH. 1966. Coevolution of mutualism between ants and acacias in Central America. *Evolution* 20:249–75

Janzen DH. 1967a. Fire, vegetation structure, and the ant x *Acacia* interaction in Central America. *Ecology* 48:26–35

Janzen DH. 1967b. Interaction of the bull's-horn acacia (*Acacia cornigera* L.) with an ant inhabitant (*Pseudomyrmex ferruginea* F. Smith) in eastern Mexico. *Kansas Univ. Sci. Bull.* 47:315–558

Janzen DH. 1969. Allelopathy by myrmecophytes: the ant *Azteca* as an allelopathic agent of *Cecropia*. *Ecology* 50:147–53

Janzen DH. 1972. Protection of *Barteria* (Passifloraceae) by *Pachysima* ants (Pseudomyrmecinae) in a Nigerian rain forest. *Ecology* 53:885–92

Janzen DH. 1974. *Swollen-Thorn Acacias of Central America*. Washington, DC: Smithson. Inst. Press

Janzen DH. 1975. *Pseudomyrmex nigropilosa*: a parasite of a mutualism. *Science* 188:936–37

Jolivet P. 1991. Plante et fourmis: le mutualisme brisé. *La Recherche* 22:792–93

Koptur S. 1989. Is extrafloral nectar an inducible defense? In *The Evolutionary Ecology of Plants*, ed. JH Bock, YB Linhart, pp. 323–29. Boulder, CO: Westview

Koptur S. 1992. Extrafloral nectary-mediated interactions between insects and plants. In *Insect-Plant Interactions*, ed. EA Bernays, 4:81–129. Boca Raton, FL: CRC Press

Koptur S. 1994. Floral and extrafloral nectars of Costa Rican *Inga* trees: a comparison of their constituents and composition. *Biotropica* 26:276–84

Koptur S, Rico-Gray V, Palacios-Rios M. 1998. Ant protection of the nectaried fern *Polypodium plebeium* in Central Mexico. *Am. J. Bot.* 85:736–39

Koptur S, Truong N. 1998. Facultative ant-plant interactions: Nectar sugar preferences of introduced pest ant species in South Florida. *Biotropica* 30:179–89

Lanza J, Vargo EL, Pulim S, Chang YZ. 1993. Preferences of the fire ants *Solenopsis invicta* and *S. geminata* (Hymenoptera: Formicidae) for amino acid and sugar components of extrafloral nectars. *Environ. Entomol.* 22:411–17

Letourneau DK. 1990. Code of ant-plant mutualism broken by parasite. *Science* 248:215–17

Letourneau DK. 1998. Ants, stem-borers, and fungal pathogens: experimental tests of a fitness advantage in *Piper* ant-plants. *Ecology* 79:593–603

Letourneau DK, Barbosa P. 1999. Ants, stem borers and pubescence in *Endospermum* in Papua New Guinea. *Biotropica* 31:295–302

Letourneau DK, Dyer LA. 1998a. Density patterns of *Piper* ant-plants and associated arthropods: top-predator trophic cascades in a terrestrial system. *Biotropica* 30:162–69

Letourneau DK, Dyer LA. 1998b. Experimental test in lowland tropical forest shows top-down effects through four trophic levels. *Ecology* 79:1678–87

Limburg DD, Rosenheim JA. 2001. Extrafloral nectar consumption and its influence on survival and development of an omnivorous predator, larval *Chrysoperla plorabunda* (Neuroptera: Chrysopidae). *Environ. Entomol.* 30:595–604

Mackay DA, Whalen MA. 1998. Associations between ants (Hymenoptera: Formicidae) and *Adriana* Gaudich. (Euphorbiaceae) in East Gippsland. *Aust. J. Entomol.* 37:335–39

Madden D, Young TP. 1992. Symbiotic ants as an alternative defense against giraffe herbivory in spinescent *Acacia drepanolobium*. *Oecologia* 91:235–38

Maschwitz U, Schroth M, Hänel H, Tho YP. 1984. Lycaenids parasitizing symbiotic plant-ant partnerships. *Oecologia* 64:78–80

McKey D. 1974. Adaptive patterns in alkaloid physiology. *Am. Nat.* 108:305–20

McKey D. 1984. Interaction of the ant-plant *Leonardoxa africana* (Caesalpiniaceae) with its obligate inhabitants in a rainforest in Cameroon. *Biotropica* 16:81–99

McKey D. 2000. *Leonardoxa africana* (Leguminosae: Caesalpinioideae): a complex of mostly allopatric subspecies. *Adansonia* 22:71–109

McKey D, Gaume L, Dalecky A. 1999. Les symbioses entre plantes et fourmis arboricoles. *Ann. Biol.* 38:169–94

McKey D, Meunier L. 1996. Evolution des mutualismes plantes-fourmis: quelques éléments de réflexion. *Actes Coll. Insect. Soc.* 10:1–9

Meunier L, Dalecky A, Berticat C, Gaume L, McKey D. 1999. Worker size variation and the evolution of ant-plant mutualisms: Comparative morphometrics of workers of two closely related plant-ants, *Petalomyrmex phylax* and *Aphomomyrmex afer* (Formicinae). *Insect. Soc.* 46:171–78

Moog J, Drude T, Maschwitz U. 1998. Protective function of the plant-ant *Cladomyrma maschwitzi* to its host, *Crypteronia griffithii*, and the dissolution of the mutualism (Hymenoptera: Formicidae). *Sociobiology* 31:105–29

Morawetz W, Henzl M, Wallnöfer B. 1992. Tree killing by herbicide producing ants for the establishment of pure *Tococa occidentalis* populations in the Peruvian Amazon. *Biodivers. Conserv.* 1:19–33

Nomura M, Itioka T, Into T. 2000. Variations in abiotic defense within myrmecophytic and non-myrmecophytic species of *Macaranga* in a Bornean dipterocarp forest. *Ecol. Res.* 15:1–11

Novotny V, Basset Y, Auga J, Boen W, Dal C, et al. 1999. Predation risk for herbivorous insects on tropical vegetation: A search for enemy-free space and time. *Aust. J. Ecol.* 24:477–83

O'Brien SP. 1995. Extrafloral nectaries in *Chamelaucium uncinatum*: A first record in the Myrtaceae. *Aust. J. Bot.* 43:407–13

O'Dowd DJ. 1979. Foliar nectar production and ant activity on a neotropical tree, *Ochroma pyramidale*. *Oecologia* 43:233–48

O'Dowd DJ. 1980. Pearl bodies of a neotropical tree, *Ochroma pyramidale*: ecological implications. *Am. J. Bot.* 67:543–49

O'Dowd DJ. 1982. Pearl bodies as ant food: an ecological role for some leaf emergences of tropical plants. *Biotropica* 14:40–49

O'Dowd DJ, Catchpole EA. 1983. Ants and extrafloral nectaries: no evidence for plant protection in *Helichrysum* spp.—ant interactions. *Oecologia* 59:191–200

Oliveira PS, Brandão CRF. 1991. The ant community associated with extrafloral nectaries in the Brazilian cerrados. See Huxley & Cutler 1991, pp. 198–212

Oliveira PS, Rico-Gray V, Diaz-Castelazo C, Castillo-Guevara C. 1999. Interaction between ants, extrafloral nectaries and insect herbivores in Neotropical coastal sand dunes: herbivore deterrence by visiting ants increases fruit set in *Opuntia stricta* (Cactaceae). *Funct. Ecol.* 13:623–31

Palmer TM, Young TP, Stanton ML. 2002. Burning bridges: priority effects and the persistence of a competitively subordinate acacia-ant in Laikipia, Kenya. *Oecologia* 133:372–79

Pemberton RW. 1993. Observations of extrafloral nectar feeding by predaceous and fungivorous mites. *Proc. Entomol. Soc. Wash.* 95:642–43

Pemberton RW, Lee JH. 1996. The influence of extrafloral nectaries on parasitism of an insect herbivore. *Am. J. Bot.* 83:1187–94

Pemberton RW, Vandenberg NJ. 1993. Extrafloral nectar feeding by ladybird beetles (Coleoptera: Coccinellidae). *Proc. Entomol. Soc. Wash.* 95:139–51

Preston KA, Ackerly DD. 2003. The evolution of allometry in modular organisms. In *Phenotypic Integration: Studying the Ecology and Evolution of Complex Phenotypes*, ed. M Pigliucci, KA Preston. Oxford, UK: Oxford Univ. Press. In press

Raine NE, Willmer P, Stone GN. 2002. Spatial structuring and floral avoidance behavior prevent ant-pollinator conflict in a Mexican ant-acacia. *Ecology* 83:3086–96

Rashbrook VK, Compton SG, Lawton JH. 1992. Ant-herbivore interactions: reasons for the absence of benefits to a fern with foliar nectaries. *Ecology* 73:2167–74

Rehr SS, Feeny PP, Janzen DH. 1973. Chemical defence in Central American non-ant Acacias. *J. Anim. Ecol.* 42:405–16

Renner SS, Ricklefs RE. 1998. Herbicidal activity of domatia-inhabiting ants in patches of *Tococa guianensis* and *Clidemia heterophylla*. *Biotropica* 30:324–27

Rettig E. 1904. Ameisenpflanzen-Pflanzenameisen. *Beih. Bot. Centralblatt* 17:89–122

Rhoades DF. 1979. Evolution of plant chemical defense against herbivores. In *Herbivores: Their Interaction with Secondary Plant Metabolites*, ed. GA Rosenthal, DH Janzen, pp. 4–53. New York/London: Academic

Rickson FR. 1971. Glycogen plastids in Müllerian body cells of *Cecropia peltata*—a higher green plant. *Science* 173:344–47

Rickson FR, Rickson MM. 1998. The cashew nut, *Anacardium occidentale* (Anacardiaceae), and its perennial association with ants: extrafloral nectary location and the potential for ant defense. *Am. J. Bot.* 85:835–49

Rico-Gray V, García-Franco JG, Palacios-Rios M, Díaz-Castelazo C, Parra-Tabla V, Navarro JA. 1998a. Geographical and seasonal variation in the richness of ant-plant interactions in México. *Biotropica* 30:190–200

Rico-Gray V, Palacios-Rios M, García-Franco JG, Mackay WP. 1998b. Richness and seasonal variation of ant-plant associations mediated by plant-derived food resources in

the semiarid Zapotitlan Valley, Mexico. *Am. Midl. Nat.* 140:21–26
Ridley HN. 1910. Symbiosis of ants and plants. *Ann. Bot.* 24:457–83
Risch SJ. 1982. How *Pheidole* ants help *Piper* plants. *Brenesia* 19/20:545–48
Risch SJ, Rickson F. 1981. Mutualism in which ants must be present before plants produce food bodies. *Nature* 291:149–50
Ruffner GA, Clark WD. 1986. Extrafloral nectar of *Ferocactus acanthodes* (Cactaceae): composition and its importance to ants. *Am. J. Bot.* 73:185–89
Ruhren S, Handel SN. 1999. Jumping spiders (Salticidae) enhance the seed production of a plant with extrafloral nectaries. *Oecologia* 119:227–30
Sagers CL, Ginger SM, Evans RD. 2000. Carbon and nitrogen isotopes trace nutrient exchange in an ant-plant mutualism. *Oecologia* 123:582–86
Schmitz OJ, Hamback PA, Beckerman AP. 2000. Trophic cascades in terrestrial systems: a review of the effects of carnivore removals on plants. *Am. Nat.* 155:141–53
Schupp EW, Feener DH. 1991. Phylogeny, life form, and habitat dependence of ant-defended plants in a Panamanian forest. See Huxley & Cutler 1991, pp. 175–97
Seigler DS, Ebinger JE. 1987. Cyanogenic glycosides in ant-acacias of Mexico and Central America. *Southwest. Nat.* 32:499–503
Smith LL, Lanza J, Smith GS. 1990. Amino acid concentrations in extrafloral nectar of *Impatiens sultani* increase after simulated herbivory. *Ecology* 71:107–15
Sobrinho TG, Schoereder JH, Rodrigues LL, Collevatti RG. 2002. Ant visitation (Hymenoptera: Formicidae) to extrafloral nectaries increases seed set and seed viability in the tropical weed *Triumfetta semitriloba*. *Sociobiology* 39:353–68
Stanton ML, Palmer TM, Young TP, Evans A, Turner ML. 1999. Sterilization and canopy modification of a swollen thorn acacia tree by a plant-ant. *Nature* 401:578–81
Stapel JO, Cortesero AM, DeMoraes CM, Tumlinson JH, Lewis WJ. 1997. Extrafloral nectar, honeydew, and sucrose effects on searching behavior and efficiency of *Microplitis croceipes* (Hymenoptera: Braconidae) in cotton. *Environ. Entomol.* 26:617–23
Stapley L. 1998. The interaction of thorns and symbiotic ants as an effective defence mechanism of swollen-thorn acacias. *Oecologia* 115:401–5
Stephenson AG. 1982. The role of the extrafloral nectaries of *Catalpa speciosa* in limiting herbivory and increasing fruit production. *Ecology* 63:663–69
Stone TB, Thompson AC, Pitre HN. 1985. Analysis of lipids in cotton extrafloral nectar. *J. Entomol. Sci.* 20:422–28
Suarez AV, De Moraes C, Ippolito A. 1997. Defense of *Acacia collinsii* by an obligate and nonobligate ant species: the significance of encroaching vegetation. *Biotropica* 30:480–82
Swift S, Lanza J. 1993. How do *Passiflora* vines produce more extrafloral nectar after simulated herbivory? *Bull. Ecol. Soc. Am.* 74:451
Tempel AS. 1983. Bracken fern (*Pteridium aquilinum*) and nectar-feeding ants: a nonmutualistic interaction. *Ecology* 64:1411–22
Tobin JE. 1995. Ecology and diversity of tropical forest canopy ants. In *Forest Canopies*, ed. MD Lowman, NM Nadkarni, pp. 129–47. San Diego, CA: Academic
Tollrian R, Harvell CD. 1999. The evolution of inducible defenses: current ideas. In *The Ecology and Evolution of Inducible Defenses*, ed. R Tollrian, CD Harvell, pp. 306–21. Princeton: Princeton Univ. Press
Treseder KK, Davidson DW, Ehleringer JR. 1995. Absorption of ant-provided carbon-dioxide and nitrogen by a tropical epiphyte. *Nature* 375:137–39
Vandermeer RK, Lofgren CS, Seawright JA. 1995. Specificity of the red imported fire ant (Hymenoptera, Formicidae) phagostimulant response to carbohydrates. *Fla. Entomol.* 78:144–54
van Rijn PCJ, Tanigoshi LK. 1999. The contribution of extrafloral nectar to survival and reproduction of the predatory mite *Iphiseius*

degenerans on *Ricinus communis*. *Exp. Appl. Acarol.* 23:281–96

Vasconcelos HL. 1991. Mutualism between *Maieta guianensis* Aubl., a myrmecophytic melastome, and one of its ant inhabitants: ant protection against insect herbivores. *Oecologia* 87:295–98

Vasconcelos HL, Casimiro AB. 1997. Influence of *Azteca alfari* ants on the exploitation of *Cecropia* trees by a leaf-cutting ant. *Biotropica* 29:84–92

Vasconcelos HL, Davidson DW. 2000. Relationship between plant size and ant associates in two Amazonian ant-plants. *Biotropica* 32:100–11

Wäckers FL, Wunderlin R. 1999. Induction of cotton extrafloral nectar production in response to herbivory does not require a herbivore-specific elicitor. *Entomol. Exp. Appl.* 91:149–54

Wäckers FL, Zuber D, Wunderlin R, Keller F. 2001. The effect of herbivory on temporal and spatial dynamics of foliar nectar production in cotton and castor. *Ann. Bot.* 87:365–70

Wada A, Isobe Y, Yamaguchi S, Yamaoka R, Ozaki M. 2001. Taste-enhancing effects of glycine on the sweetness of glucose: a gustatory aspect of symbiosis between the ant, *Camponotus japonicus*, and the larvae of the lycaenid butterfly, *Niphanda fusca*. *Chem. Senses* 26:983–92

Whitmore TC. 1967. Studies in *Macaranga*, an easy genus of Malayan wayside trees. *Malay. Nat. J.* 20:89–99

Wickers S. 1997. Study of nectariferous secretion in a pioneer plant, *Inga thibaudiana*, in relation with ants. *Acta Bot. Gall.* 144:315–26

Wilkinson DM, Sherratt TN. 2001. Horizontally acquired mutualisms: an unsolved problem in ecology? *Oikos* 92:377–84

Willmer PG, Stone GN. 1997. How aggressive ant-guards assist seed-set in *Acacia* flowers. *Nature* 388:165–67

Wirth R, Leal IR. 2001. Does rainfall affect temporal variability of ant protection in *Passiflora coccinea*? *EcoScience* 8:450–53

Young TP, Stubblefield CH, Isbell LA. 1997. Ants on swollen-thorn acacias: species coexistence in a simple system. *Oecologia* 109:98–107

Yu DW. 2001. Parasites of mutualisms. *Biol. J. Linn. Soc.* 72:529–46

Yu DW, Davidson DW. 1997. Experimental studies of species-specificity in *Cecropia*-ant relationships. *Ecol. Monogr.* 67:273–94

Yu DW, Pierce NE. 1998. A castration parasite of an ant-plant mutualism. *Proc. R. Soc. London Ser. B* 265:375–82

Yu DW, Wilson HB, Pierce NE. 2001. An empirical model of species coexistence in a spatially structured environment. *Ecology* 82:1761–71

Yumoto T, Maruhashi T. 1999. Pruning behavior and intercolony competition of *Tetraponera* (*Pachysima*) *aethiops* (Pseudomyrmecinae, Hymenoptera) in *Barteria fistulosa* in a tropical forest, Democratic Republic of Congo. *Ecol. Res.* 14:393–404

Annu. Rev. Ecol. Evol. Syst. 2003. 34:455–85
doi: 10.1146/annurev.ecolsys.34.011802.132342

First published online as a Review in Advance on August 6, 2003

FUNCTIONAL MATRIX: A Conceptual Framework for Predicting Multiple Plant Effects on Ecosystem Processes

Valerie T. Eviner
Institute of Ecosystem Studies, PO Box AB, Millbrook, New York 12545; email: evinerv@ecostudies.org

F. Stuart Chapin III
Institute of Arctic Biology, University of Alaska, Fairbanks, Alaska 99775; email: terry.chapin@uaf.edu

Key Words biogeochemical cycling, climate feedbacks, ecosystem processes, functional group, functional matrix, litter quality, species effects

■ **Abstract** Plant species differ in how they influence many aspects of ecosystem structure and function, including soil characteristics, geomorphology, biogeochemistry, regional climate, and the activity and distribution of other organisms. Attempts to generalize plant species effects on ecosystems have focused on single traits or suites of traits that strongly covary (functional groups). However, plant effects on any ecosystem process are mediated by multiple traits, and many of these traits vary independently from one another. Thus, most species have unique combinations of traits that influence ecosystems, and there is no single trait or functional-group classification that can capture the effects of these multiple traits, or can predict the multiple functions performed by different plant species.

We present a new theoretical framework, the functional matrix, which builds upon the functional group and single trait approaches to account for the ecosystem effects of multiple traits that vary independently among species. The functional matrix describes the relationship between ecosystem processes and multiple traits, treating traits as continuous variables, and determining if the effects of these multiple traits are additive or interactive. The power of this approach is that the ecosystem effects of multiple traits are the underlying mechanisms determining species effects, how the effects of an individual species change across seasons and under varying environmental conditions, the nonadditive effects of plant species mixtures, and the effects of species diversity.

INTRODUCTION

Plant species can differ in their effects on almost every aspect of ecosystem structure and function. Understanding these effects is critical for predicting the consequences of environmental changes because shifts in vegetation composition can cause ecosystem changes that are larger in magnitude, or even opposite in direction, than the direct effects of environmental change (Kirkby 1995, Hobbie 1996). An in-depth understanding of the multiple ecosystem functions provided by plant species can also be a critical tool for land management, such as restoration, bioremediation, or sustainable agriculture (Eviner & Chapin 2001, Liste & Alexander 2000).

The ecosystem effects of plant species are usually predicted by focusing on one mechanism that is assumed to dominate a plant's effect on a given ecosystem function. One example is the plant functional type approach, which is based on the assumption that there are suites of related plant traits that can generalize how species affect ecosystem processes or respond to environmental changes (Wilson 1999). For example, plants adapted to low-nutrient environments have high nutrient use efficiency, slow growth and photosynthetic rates, and low litter quality, which, in turn, reinforce low nutrient availability (Chapin 1993). Whereas functional grouping is a powerful framework linking plant characteristics and ecosystem processes, it is most appropriate for categorical differences in species functions (e.g., ability to fix N), rather than functions mediated by traits that vary among species in a continuous manner (e.g., litter chemistry). Functional-type generalizations do not adequately account for multiple plant effects because a given species can be classified into multiple functional groupings depending on the function of interest (Wilson 1999). A second approach to predicting the effects of plant composition is to focus on single traits that vary in a continuous manner. This approach is conceptually similar to functional grouping because single traits (e.g., litter C:N) are indicators of a broader suite of traits with similar ecosystem effects.

Both approaches provide large-scale generalizations of the effects of plant species on processes such as N cycling because at a regional scale, environmental conditions select for certain suites of plant traits (Chapin 1993). Thus, the relationships between litter chemistry and biogeochemical cycling tend to be strong at regional scales (Scott & Binkley 1997, Taylor et al. 1989) because nutrient availability at a given site selects for species that reinforce the site's relative level of nutrients. However, traits that strongly covary over environmental gradients can vary independently at local scales. Within a habitat, differences in traits within a species or among species can be as great as trait differences across the full range of an environmental gradient (Fonseca et al. 2000, Wright & Westoby 1999). Thus, relationships between litter quality and N cycling are often weak within a given site (Eviner 2001, Steltzer & Bowman 1998) because the other mechanisms that determine plant effects on N can vary independently from litter chemistry.

This review examines the multiple traits that determine plant species effects on ecosystem processes, the multiple roles that plant species play in ecosystems, and how these multiple roles influence one another. We then present a generalizable, mechanistic framework to improve predictions of plant species effects based on multiple plant traits.

MECHANISMS BY WHICH PLANTS INFLUENCE ECOSYSTEMS

Soil Environment

Plants species can have large effects on the physical structure and chemical properties of soil. These properties strongly influence hydrology, plant growth, biogeochemical cycling, and the activity of soil organisms (reviewed in Angers & Caron 1998, Glinski & Lipiec 1990).

SOIL STRUCTURE Plant roots disaggregate compound rock fragments, resulting in more fine soil particles (<6.3 um), and less medium and coarse silt fractions in the rhizosphere relative to bulk soils (Glinski & Lipiec 1990). The aggregation of these soil particles can vary owing to plant species (Eviner & Chapin 2002, Scott 1998). In general, aggregation is determined by species differences in root characteristics (e.g., biomass and length) (Miller & Jastrow 1990), the quantity and quality of carbon inputs, and associated soil biota (Degens 1997). Plant species also alter soil moisture (Gordon & Rice 1993) and freeze-thawing (Hogg & Lieffers 1991), which can initiate aggregate formation. Grass species often foster greater aggregate stability than other vegetation groups owing to their high root and fungal biomass (Jastrow 1987, Rillig et al. 2002). Other species decrease aggregate stability by exuding chelating agents that bind Fe and Al, destroying the links formed by these metals between organic matter and mineral particles (Reid & Goss 1982).

Soil aggregation and plant rooting characteristics determine the size distribution of pores, and pore volume is enhanced by species with high root turnover, organic matter, and animal activity (Challinor 1968). Soil bulk density decreases with increases in pore volume and soil organic inputs (Troeh & Thompson 1993). High densities of roots displace soil and increase its bulk density, whereas species with taproots penetrate and loosen compact soils (Glinski & Lipiec 1990).

SOIL CHEMISTRY Soil characteristics such as pH, redox, salinity, cation exchange capacity, and water holding capacity are critical determinants of biogeochemical cycling and plant growth. Plant species play active roles in mediating these soil chemical properties.

pH and redox Plants can acidify their rhizosphere soil by as much as two pH units (Glinski & Lipiec 1990). This is largely attributed to release of organic acids in exudates and to higher root uptake of cations than anions, leading to root excretion of H^+ ions. The balance between cation and anion uptake is primarily determined by the form of N taken up. However, even when species take up the same form of N, species differences in other nutrient requirements lead to substantial changes in soil pH (Marschner 1995). These rhizosphere effects often alter bulk soil pH. For example, N-fixing plants decrease bulk soil pH because of their high cation uptake and promotion of nitrification and base cation loss through enhanced NO_3 leaching.

Base cations such as Ca and Mg are key players in the soil's capacity to buffer pH (Marschner 1995). Thus, soil pH is increased by species with high litter calcium inputs (Finzi et al. 1998b) and by species that increase cation exchange capacity or base cation weathering rates (Kelly et al. 1998). These species are particularly important in preventing soil acidity as a result of leaching of base cations over long-term soil development (Ovington 1953). In contrast, base cation leaching and soil acidification are enhanced by species with high litter concentrations of organic acids (particularly tannins) (Finzi et al. 1998b, Ovington 1953). Plant species also influence acid deposition. Deposition is higher in conifers than in hardwood stands owing to the high leaf area of conifers (Ranger & Nys 1994).

Soil redox conditions are strongly influenced by the rhizosphere, where oxygen consumption is high owing to root and microbial respiration. Some plant species maintain high redox potentials by transporting oxygen from shoots to roots through aerenchyma (Marschner 1995).

Salinity When water arrives at the root faster than it can be absorbed, soil evaporation can lead to accumulations of salt that are toxic to plants, particularly in regions with high rates of evaporation or where low rainfall does not leach salts from the rooting zone. Species effects on soil salinity depend on differences in their salt accumulation and effects on hydrology. Iceplant (*Mesembryanthemum crystallinum*) tolerates large accumulations of salt in its tissue, and when it dies, deposits this salt on the soil surface (Kloot 1983). In arid regions, deep-rooted trees such as *Tamarix* draw up large amounts of salt with water (Zavaleta 2000). The presence of trees in arid areas can further enhance salinity because canopy interception and evaporation reduce the water available to flush salts out of the rooting zone (Munzbergova & Ward 2002). In contrast, in more mesic sites, the removal of deep-rooted trees that use large amounts of water can result in waterlogged conditions and soil salinization as this water evaporates (Bell 1999). High soil calcium decreases sodium on cation exchange sites, and thus salinization is combated by planting species that pump calcium from deep in the soil, dissolve $CaCO_3$, or enhance cation exchange capacity (Qadir et al. 2001).

Capacity to retain cations and water Soil cation exchange capacity is critical for the retention of nutrients in soil. Cation exchange sites on organic matter and the edges of clays vary in their capacity to bind cations depending on soil pH, with increases in soil pH by one unit doubling cation exchange capacity in some soils (Camberato 2001). Thus, species effects on soil pH, soil texture, and soil organic matter substantially influence the capacity of soil to retain positively charged nutrients. Organic matter inputs are the primary mechanism by which plants influence cation exchange capacity, and polyphenolic compounds are particularly effective in providing cation exchange sites (Northup et al. 1998). Similarly, water-holding capacity is influenced primarily by organic matter (Troeh & Thompson 1993). Soil porosity can also be important because water is retained more effectively by the large capillary forces of many small pores than in soils with a few large pores.

SOIL DEVELOPMENT Plants have long been recognized as key regulators of soil formation (Jenny 1941), and plant species differentially influence soil structure and processes during primary succession (Lawrence et al. 1967, Matson 1990). Plant species affect weathering through the generation of weathering agents (e.g., CO_2, organic acids, ligands), alteration of hydrology, cycling of cations, and influence on soil pH (Augusto et al. 2000, Kelly et al. 1998). Fulvic and low-molecular-weight humic acids are critical for weathering, and different acids vary in strength and weathering rates of different minerals. Plants also can enhance weathering through the penetration of roots and root-derived acids into the parent material (Viles 1990) explaining why the spread of deeply rooted forests across continental areas resulted in rapid formation of soil (Kelly et al. 1998). Plant species also differentially alter the forms of minerals in the soil. Grass species produce phytoliths, stabilizing soil silica by increasing the fraction of crystalline to noncrystalline soil minerals (Kelly et al. 1998).

In addition to their effects on the weathering of bedrock, plant species alter the accumulation and type of organic matter (Binkley & Giardina 1998). Species differences in the location of organic matter inputs and nutrient redistribution alter the weight, density, and thickness of soil horizons (Boettcher & Kalisz 1990). Podzolic soils typically have a low pH, high C:N and mor-type humus dominated by humic acids. Many conifer and ericaceous species cause acidification and podzolization of the soil largely through inputs of organic acids as a result of the incomplete decomposition of organic matter (Raulund-Rasmussen & Vejre 1995). In contrast, broadleaf trees have a much wider range of effects on pedogenesis, ranging from enhancement to reversal of podzolization (Miles 1985). Herbaceous species often reverse podzolization and acidification of soil through increases in soil base saturation caused by reduced litter acidity and increases in calcium availability (Willis et al. 1997).

LANDSCAPE FORMATION In addition to their effects on soil formation, plant species are also the main determinant of geomorphology. In fact, the indirect effects of climate change through shifts in vegetation composition have larger effects on geomorphology than do the direct effects of changes in the climatic regime (Kirky 1995).

Erosion The force required to disrupt a mass of soil is soil shear strength. Shear strength is critical for preventing multiple types of erosion, ranging from sediment runoff to landslides. Plant species differentially contribute to shear strength through their root strength and effects on soil cohesion (the capacity of particles to bind together) (Angers & Caron 1998). Species enhance soil cohesion through root-soil bonding (Selby 1993), a function of the length and density of fine roots and root hairs (Waldron & Dakessian 1981). Soil cohesion is also enhanced by plant transpiration that dries the soil thereby reducing the buoyancy of soil particles. Erosion through runoff is low in cohesive soils and is further decreased by soil organic matter, which binds soil particles and enhances water infiltration. Surface erosion is also decreased by canopies that intercept rainfall and wind (Viles 1990).

Species that create variable microtopography or have high aboveground cover enhance water infiltration and thus minimize runoff while recapturing sediments in overland flow (Devine et al. 1998). Plant species effects on faunal burrowing, hydrology, and disturbance regime also play large roles in determining erosion rates (Kirkby 1995, Viles 1990).

Root strength contributes to the shear strength of the overall soil profile and prevents larger erosion events (Terwilliger & Waldron 1991). Landslides and soil slips commonly occur when communities dominated by woody species are converted to grasslands (Jones et al. 1983, Prandini et al. 1977). The soil strength provided by deep interlocking roots of woody plants in the upper soil layers can stabilize the overall soil profile, and deep roots can bind the unstable upper soil layer to rocky substrates. These woody roots often provide this protective role even after aboveground woody vegetation has been harvested, and it is not until these roots begin to decompose that substantial soil slips occur. Whereas tree roots significantly stabilize deep soils, on steep slopes with shallow soils, tree root channels can increase landslides by enhancing water infiltration (de Ploey & Cruz 1979). Erosion as a result of runoff rather than landslides also tends to be lower in forests than grasslands because the forest canopy and litter layer better intercept rain and wind, protecting the soil surface particles from transport. In addition, a thick litter layer decreases runoff by increasing water infiltration and retention. Finally, soil stability is enhanced by low soil moisture levels owing to high evapotranspiration in forests (Prandini et al. 1977).

In contrast, in arid systems, the conversion of grasslands to shrublands often results in substantial increases in erosion owing to bare spaces between shrubs (Schlesinger et al. 1990). In these systems, overland flow of water is the major soil transport mechanism. Erosion is decreased by species traits that minimize the physical impact of rainfall and wind on the soil surface (e.g., high cover) (Abrahams et al. 1994), as well as traits that influence hydrology and infiltration (Wainwright et al. 2000).

Sedimentation The architecture of plant shoots and roots determines how well plant species can trap and hold sediments. For example, the dense thickets and extensive root mats formed by *Tamarix* increase sedimentation, resulting in decreases in the width and depth of water channels (Zavaleta 2000). Similarly, invasion of *Spartina alterniflora* into intertidal habitats increases sedimentation, thus increasing the steepness of beaches and decreasing the delivery of sediments to mudflats, converting them to open water habitat (Gleason et al. 1979). Species also produce sediment through the buildup of organic matter. Debris buildup by *Melaleuca quinquenervia* in the Florida everglades leads to the formation of tree islands, converting marshes to swamp forests (Laroche 1994).

Hydrology

Plant species effects on water dynamics range from altering water availability in their rooting zone to stream flow. A number of different plant traits are responsible for these hydrological effects.

WATER AVAILABILITY There are many mechanisms by which plant species alter soil moisture. Species with high litter mass decrease evaporation from the soil (Evans & Young 1970). The amount of water used by species also alters soil water availability, and this effect can be independent of biomass (Gordon & Rice 1993). Root length, water uptake per unit root biomass, and the timing of root development are important influences on the capacity of species to draw down soil water (Gordon & Rice 1993). More extensive water uptake occurs from the fibrous roots of grasses than from the taproots of forbs (Gordon et al. 1989). Deep-rooted species access more water because they tap a larger volume of soil than do shallow-rooted species (Robles & Chapin 1995). Differences in plant phenology can have important effects on ecosystem water fluxes. For example, the invasion of *Andropogon* in Hawaii has lead to boggy conditions because its time of maximum evapotranspiration does not coincide with the rainy season (Mueller-Dombois 1973).

Some species can access unique sources of water and, by tapping into these sources, provide water for other plants in the ecosystem. For example, species with canopies that are tall and have high surface area collect water from fog in many coastal and montane ecosystems (Weathers 1999) providing as much as 34% of the annual ecosystem water input (Dawson 1998). Similarly, a substantial amount of water can be provided by hydraulic lifters, deep-rooted species that take up water from deep layers of soil and passively release it into surface soils at night when transpiration ceases (Caldwell et al. 1998).

WATER MOVEMENT Plant species influence the pathway and fate of precipitation. Canopies of species with a high leaf area intercept precipitation, increasing the amount that evaporates. Infiltration of the remaining water into the soil increases owing to high plant cover (live + dead) (Devine et al. 1998), burrowing soil fauna (Gijsman & Thomas 1996), or increased pore volume (Angers & Caron 1998). Some plant species decrease infiltration by forming a hydrophobic layer of surface soil through inputs of waxes, fatty acids, resins, aromatic oils, and humic materials (Doerr et al. 2000, Spaccini et al. 2002). Infiltration and the path of water flow, can also be influenced by canopy structure. Stem flow concentrates water flow, leading to local water saturation and runoff (Herwitz 1986).

Changes in plant community composition can affect stream flow, flooding regime, and peak storm discharges due to species differences in transpiration rates (Viles 1990), effects on canopy interception and infiltration (Le Maitre et al. 1999), and alterations of water channels through promotion of sedimentation (Laroche 1994). In a Mediterranean climate, runoff in grasslands is 40% higher than in shrublands, leading to higher stream flow (Pitt et al. 1978). In contrast, in semiarid ecosystems, the conversion of grasslands to shrubs increases the amount and rate of water flow because shrubs decrease water infiltration, which increases overland flow. This overland flow is concentrated into fewer and deeper channels through the effects of shrubs on local surface morphology (Wainwright et al. 2000). Conversion of hardwoods to conifers decreases stream flow if conifers have greater canopy interception and transpiration (Swank & Douglass 1974). The opposite might be true in places like Alaska, where conifers have lower transpiration than hardwoods and are associated with boggy soils (Chapin et al. 2000).

Microclimate to Climate

The effects of plant species on soil moisture regimes were discussed above. In this section, we will discuss the effects of plant species on soil temperature, and then extend the local effects of species on soil temperature and moisture to their effects on regional and global climate.

EFFECTS ON SOIL TEMPERATURE Plant species influence local soil temperature primarily through the buffering effects of shoot cover, including live plant tissue and litter (Hogg & Leiffers 1991). The color of litter can also influence soil temperature with darker-colored litter increasing warming by increased absorption of sunlight (Hogg & Leiffers 1991). In California annual grasslands, species differences in litter mass determine winter soil temperature. However, in the spring, as this litter mass decomposes and aboveground growth increases, species effects on soil temperature are mediated primarily by live biomass (Eviner 2001). In boreal forests, thick layers of aboveground tissue can delay soil thaw up to one month, and in the summer, can lead to soils that are 4°C cooler (Hogg & Leiffers 1991). Plant species effects on microclimate are likely to be strongest under extreme environmental conditions. For example, in Mediterranean grasslands, species effects on soil temperature are most pronounced during the cold winters when temperature is most limiting to plant and microbial activity, while species effects on soil moisture become stronger in the warm, dry spring (Eviner 2001).

Species influence not only soil microclimate, but also stand humidity, air temperatures, snow accumulation, and wind speed (Baldocchi & Vogel 1996). Dense canopy cover enhances humidity (Binkley & Giardini 1998) and decreases net canopy energy exchange at the forest floor (Baldocchi & Vogel 1996).

REGIONAL AND GLOBAL CLIMATE Species differences in energy absorption can greatly affect regional temperature. Vegetation dominated by species with tall, dark, complex canopies absorb more solar radiation (i.e., have a lower albedo) than do short, reflective, simple canopies. For this reason, conversion from graminoid tundra to shrub tundra to forest increases radiation absorption and atmospheric heating, just as does succession from deciduous to coniferous forest (Chapin et al. 2000). The atmospheric heating for each of these vegetation changes is similar to that caused by a doubling of atmospheric CO_2 or increased solar input associated with glacial–non-glacial transitions per unit area and time. This vegetation-induced warming of climate occurs only at the scale of the region in which the vegetation changes occur (Chapin et al. 2000, Foley et al. 1994). These vegetation changes and warming can also influence regional and global climate by influencing fluxes of greenhouse gases such as CO_2, CH_4, N_2O, and water vapor (Schlesinger 1997).

Energy absorbed by vegetation can be transferred to the atmosphere as sensible heat or as latent heat (evapotranspiration). Species differences in this energy partitioning have strong effects on moisture transfer to the atmosphere and therefore the moisture available for precipitation (Otterman 1989). The proportion of energy transferred to the atmosphere by evapotranspiration is twice as great in deciduous as in conifer boreal forests (Baldocchi et al. 2000). Simulations with general circulation models suggest that widespread replacement of deep-rooted tropical trees by shallow-rooted pasture grasses would reduce evapotranspiration and lead to a warmer, drier climate (Shukla et al. 1990). In semiarid regions of Australia, cloud accumulation occurs over areas of land dominated by native vegetation with high sensible heat flux but not over adjacent agricultural areas dominated by vegetation with high latent heat flux. Despite the higher atmospheric moisture inputs from agricultural areas, the low surfacc roughncss of this system provides insufficient mixing to allow for the condensation necessary for cloud formation. In contrast, the high sensible heat flux and surface roughness of native vegetation promotes atmospheric mixing and thus cloud formation (Lyons et al. 1993).

General Introduction to Species Effects on Biogeochemical Cycling

Thus far, we have discussed how species influence a number of ecosystem properties that are critical determinants of biogeochemical processes. In the following sections, we discuss how species effects on soil properties and microclimate interact with other species traits (e.g., litter chemistry, exudation rates) to determine rates of biogeochemical cycling.

CARBON CYCLING Plant species differ in their effects on almost every aspect of C cycling, from C fixation into biomass, to the extent to which this C is released from the ecosystem. In many cases, species effects on one aspect of C cycling are independent from their effects on other C dynamics.

Net primary production The amount of carbon fixed by plants through photosynthesis (NPP) provides the energy that drives most biotic processes, and plant species growing at the same site can greatly differ in their NPP. The unique ability of a sedge species in the tundra to access nutrients from ground water leads to a tenfold increase in productivity compared with other species (Chapin et al. 1988), while replacement of native perennials by the annual *Bromus tectorum* decreases productivity by up to 90% in some years in California rangelands (DiTomaso 2000). Species differences in NPP result from a wide range of plant traits, including growth rate, allocation, phenology, nutrient use efficiency, resource requirements, traits that influence access to resource pools (e.g., root depth, symbioses with mycorrhizae or N fixing microorganisms), and traits that influence conditions that limit growth (e.g., temperature, moisture, nutrient pools) (reviewed in Chapin & Eviner 2003).

Decomposition Litter C:N ratios reflect the relative amount of energy (C) and N available to microbes and are often negatively associated with rates of decomposition (Melillo et al. 1982, Taylor et al. 1989). Litter C quality can be a more important determinant of decomposition than litter C:N (Hobbie 1996). Lignin generally inhibits decomposition because it is highly resistant to enzymatic attack and can physically interfere with the decay of other chemical fractions (Gloaguen & Touffet 1982), whereas tannins vary substantially in their resistance to decomposition and in the degree to which they decrease decomposition of other compounds (Lewis & Starkey 1968). The ratio of labile C to these recalcitrant compounds is another key determinant of decomposition (Gillon et al. 1994). Plant species also influence decomposition rates through their allocation patterns (Hobbie 1996), because litter chemistry differs among tissue types within the same species, and species effects on root and shoot decomposition are not correlated (Wardle et al. 1998). Whereas litter C:N and lignin:N are frequently used as indicators of decomposition, they do not always correlate with species differences in decomposition rates because the traits that best predict species effects on decomposition can vary across groups of species (Cornelissen & Thompson 1997).

While litter chemistry is a major determinant of decomposition rates, the presence of growing plants significantly alters decomposition dynamics, both decreasing (Dormaar 1990) and increasing (van der Krift et al. 2002) rates of decomposition. Living plants can decrease decomposition rates because (*a*) microbes preferentially use the labile material provided by living roots rather than more recalcitrant litter, (*b*) roots release compounds that inhibit microbial activity, (*c*) plants compete with microbes for uptake of nutrients or organic compounds, and/or (*d*) exudates stimulate predation on microbes and thus decrease microbial populations. In contrast, growing plants can stimulate decomposition through inputs of labile C that increase the activity and turnover of microbes (Cheng & Coleman 1990, Sallih & Bottner 1988). Plants can also influence decomposition through their effects on soil temperature, moisture (Mack 1998, van der Krift et al. 2002), or O_2 concentrations (Allen et al. 2002). Decomposition of a common substrate can differ substantially in the presence of different plant species, highlighting the importance of plant traits other than litter quality (Eviner 2001, van der Krift et al. 2002).

The relative importance of these different plant effects can vary through the decomposition process. In the early stage of decomposition, microbes degrade primarily labile litter constituents, and rates are determined largely by the stimulating effects of litter nutrients (Berg 2000a, Couteaux et al. 1998), and are often predictable based on litter C:N (Taylor et al. 1989). Once labile compounds are exhausted, decomposition is determined by degradation of more recalcitrant compounds, and, at this stage, litter lignin and phenolic concentrations are the main determinants of decomposition rates (Cortez et al. 1996, Taylor et al. 1989). Microbial degradation of the recalcitrant substrates is often energy limited and thus labile C inputs from growing plants stimulate the later stages of decomposition

(Bottner et al. 1999). The presence of N often inhibits later stages of degradation of high-lignin litter (Heal et al. 1997; but see Melillo et al. 1989) because N and lignin form stable complexes (Couteaux et al. 1998).

Soil organic matter In some species, litter decomposition halts at a later stage, leaving as much as 55% of the litter mass in a stable form that directly contributes to the buildup of humic soil organic matter (Berg 2000a). Species effects on soil organic matter can be predicted based on the quantity of litterfall and the fraction of litter that does not decompose ("limit value"). Lignin (Melillo et al. 1982) and polyphenols (Howard et al. 1998) have been particularly linked with the formation of stable soil organic matter. Interestingly, species with high initial litter quality tend to have higher accumulations of recalcitrant, stabilized litter in the later stages of decomposition, and thus higher limit values (Berg 2000b). For example, legumes often are associated with higher soil organic matter contents than other species (Binkley & Giardini 1998, Cole 1995). This may be a result of the rapid decomposition of the more labile fractions of litter, leaving little labile C to fuel decomposition of the more recalcitrant substrates. This can be exacerbated because high-N tissue supports a higher microbial biomass, and, as microbes turnover, there is an accumulation of their recalcitrant residues (Mueller et al. 1998). In some cases, these limit values appear in lab incubations, but, when this same litter is incubated in the field, the litter continues to decompose (Couteaux et al. 1998, Eviner 2001), likely owing to the effects of growing plants, field variations in microclimate, and exposure to soil fauna.

Species can also influence soil organic matter through their tissue allocation. Soil organic matter is often derived primarily from root litter (Balesdent & Balabane 1996, Puget & Drinkwater 2001). This is likely because root litter is more recalcitrant (higher lignin content), is often physically protected within aggregates, and root-derived C forms the most stable aggregates (Gale et al. 2000). Thus the quantity (Christian & Wilson 1999) and quality (Urquiaga et al. 1998) of root litter can determine plant species differences in accumulation of soil C.

Species also affect the vertical distribution of organic matter (Finzi et al. 1998b), %C and %N in soil organic matter (Wardle 2002), composition of soil humic acids (Howard et al. 1998), and the type of organic matter they foster (Northup et al. 1998). Soil organic matter can be physically protected from degradation in soil macro-aggregates (0.25–2 mm), allowing even labile C pools to have long residence times in the soil (van Veen & Kuikman 1990). Species can also increase C sequestration of labile C through inputs of hydrophobic compounds such as humic materials (Spaccini et al. 2002).

Given that plant species differ in the quality of organic matter, it is not surprising that they also differ in the turnover rates of soil organic matter (Scheu 1997, Wedin et al. 1995). Although labile C pools are relatively small, they are one of the main drivers of soil C dynamics (Updegraff et al. 1995). Just as with recalcitrant litter, degradation of organic matter can be influenced by labile C inputs or inorganic

N availability (Bottner et al. 1999, Mueller et al. 1998), with the effects of these differing depending on the pools of C being degraded (Neff et al. 2002).

Methane dynamics Plant species influence methane fluxes in water-logged soils through the amount and quality of C substrates provided to methanogens, effects on soil oxygen levels, and methane transport from the soil to the atmosphere through aerenchyma of roots (Verville et al. 1998). The abundance of species with aerenchymous roots is one of the best predictors of ecosystem methane flux (Torn & Chapin 1993). Plant species that enhance soil O_2 decrease CH_4 formation (Grosse et al. 1996), and methane flux is negatively correlated with recalcitrant compounds (Updegraff et al. 1995) and positively correlated with root exudation (Marschner 1995). Plant species can also influence microbial methane consumption owing to inhibition of CH_4 uptake by NH_4 (Epstein et al. 1998).

N CYCLING There are many processes involved in terrestrial N cycling, and there are multiple mechanisms by which plant species influence these processes.

N inputs N-fixing plants rely on a plant-microbial symbiosis to fix atmospheric N_2 into reduced form. The presence of N-fixing plants can substantially increase soil N supply and recycling rates (Vitousek & Walker 1989). Legume species differ in the amount of N they fix (Franco & de Faria 1997) and the form (exudates versus litter) of N inputs (Ta & Farris 1987). N fixation by legumes can be altered by neighboring species through shading, effects on nutrients, or the presence of secondary compounds such as polyphenolics that inhibit N fixation (Ta & Faris 1987, Wardle et al. 1994). N fixation by free-living heterotrophic bacteria is common in the rhizosphere of a number of plant species (Noskop et al. 1994) and can be decreased through lignin inputs, low soil pH, or high soil N availability (DeLuca et al. 1996, Vitousek et al. 2002). Species with high leaf surface area can foster wet and dry N deposition, which is delivered to the soil in throughfall (Weathers et al. 2001).

Cycling of N The net conversion of N from organic to inorganic forms through the processes of mineralization and immobilization can be strongly affected by plant species (Hobbie 1992, Wedin & Tilman 1990). While similar mechanisms determine patterns of decomposition and N cycling, species effects on rates of net N mineralization often do not correlate with rates of decomposition (Hart et al. 1994, Prescott et al. 2000). For example, moss decomposes slowly but promotes rapid rates of net N mineralization (Hobbie 1996) because the recalcitrant carbon leads to C starvation among microbes.

The C:N or lignin:N ratios of litter or soil are often good indicators of plant species patterns of N cycling (Finzi et al. 1998a, Scott & Binkley 1997). When comparing net N mineralization among species, litter C:N has a negative correlation with net N mineralization up to a threshold C:N ratio. Above this threshold, no net mineralization occurs, indicating microbial N immobilization (Prescott et al. 2000). However, the C:N ratio that determines the threshold between net mineralization versus immobilization is variable and differs among species (Berg & Ekbohm 1993). This is likely because N mineralization is determined both by N concentrations and C quality (decomposability), and there is a linear relationship between N mineralization and C:N only when there is a uniform C recalcitrance among substrates (Jannsen 1996). The presence of secondary compounds can depress N cycling by serving as a C source for microbial immobilization (Schimel et al. 1996), by inhibiting microbial activity (Lodhi & Killingbeck 1980), or by binding labile compounds or mineral N (Hattenschwiler & Vitousek 2000).

Species effects on N cycling are also mediated by litter quantity, with high litter biomass enhancing N fluxes (Aerts et al. 1992, Hobbie 1992). Plant species can supply as much as 87% or as little as 23% of total plant N inputs to soil through root turnover (Aerts et al. 1992), and fine root production positively correlates with rates of net mineralization and nitrification (Aerts et al. 1992). Species differences in labile C inputs can have strong effects on N cycling (Flanagan & Van Cleve 1983), stimulating release of N from litter and soil organic matter (van der Krift et al. 2001) or decreasing rates of net N cycling by fostering microbial immobilization of N (Hobbie 1992). Even though labile C is a relatively small component of the total soil C pool, it is an extremely active pool, and species effects on labile C are responsible for up to tenfold differences in N cycling (Wedin & Pastor 1993). During some seasons, plant labile C inputs have stronger effects on N cycling than does litter chemistry (Eviner 2001).

Because nitrification is often limited by NH_4 availability, net nitrification rates are often proportional to net mineralization rates (Finzi et al. 1998a, van Vuuren et al. 1993), and species effects on NH_4 availability can be a key determinant of their effects on nitrification. However, the percent of mineralized N that is nitrified varies with species (Wedin & Tilman 1990). Nitrification can be extremely low or absent in the presence of species with high concentrations of phenolics (Cortez et al. 1996, Eviner 2001). While it is often assumed that these phenolics inhibit nitrifiers, different phenolics can have distinct effects on N cycling (Horner et al. 1988), and decreases in nitrification rates can also be due to decreases in NH_4 availability imposed by phenolics, or increased microbial immobilization of N through use of phenolics as C substrates (Schimel et al. 1996).

Plant composition can also influence N cycling by altering local conditions such as temperature, pH, soil moisture, and O_2 concentrations (Mack 1998, van Vuuren et al. 1993). These species effects on microclimate are often more important than litter chemistry or labile C inputs in determining species effects on N dynamics when temperature, moisture, or O_2 limits microbial activity (Engelaar et al. 1995, Eviner 2001, van Vuuren et al. 1993). Ultimately, multiple traits best explain species effects on N cycling (Eviner 2001, Mack 1998, Steltzer & Bowman 1998, Wedin & Tilman 1990).

N loss High levels of N retention are associated with plant species that enhance microbial immobilization and plant uptake, or minimize soil NO_3 (Epstein et al. 2001). Nitrate loss into streams from forested watersheds can be determined by plant species effects on soil C:N ratios (Lovett et al. 2002). High rates of DON leaching are associated with species with high concentrations of tannins and phenolics that bind proteins (Hattenschwiler & Vitousek 2000), but these species may decrease overall N leaching by minimizing NO_3 loss (Northup et al. 1998). Leaching loss can also be influenced by species effects on hydrology (Schlesinger et al. 2000).

Plant species also alter fluxes of NO and N_2O (Bronson & Mosier 1993, Woldendrop 1962). Species effects on nitrification are a key determinant of N gas loss because gaseous N forms are released during nitrification (Firestone & Davidson 1989) and because NO_3 availability often limits denitrification. Species influence denitrification through their C inputs, and effects on soil oxygen, NO_3, temperature, and pH (Hume et al. 2002). The relative amount of acid soluble C to NO_3 is the strongest determinant of species effects on denitrification (Hume et al. 2002), and phenolics are an important C source for denitrifying bacteria (Siqueira et al. 1991). Plant leaves can also directly assimilate and emit N compounds (NO, NO_2, HNO_3, organic NO_3), and species differences in leaf NO_2 flux rates relate to leaf N, nitrate reductase activity, and stomatal conductance (Sparks et al. 2001).

OTHER NUTRIENTS The effect of a plant species on any element cycle can be independent of its effects on other elements (Eviner & Chapin 2003). Fluxes of N, Ca, Mg, Mn, and S are mediated by biological mechanisms, while physical processes (e.g., leaching, atmospheric deposition) dominate fluxes of K, Na, Pb, Cd, and Zn, and chemical processes (e.g., interactions of metals with humic substrates) dominate fluxes of P, Fe, Zn, Pb, and Cd (Laskowski et al. 1995). Thus, plant species have different mechanisms by which they influence these cycles.

P availability is largely controlled by chemical interactions with the soil, such as through complexes with other elements (e.g., Ca, Al, Fe). Plant species influence P availability largely by influencing P solubility (Easterwood & Sartain 1990) through exudation of charged organic compounds that compete with PO_4 for binding surfaces on other elements, alteration of soil pH that influences P-binding with elements, or accumulation of calcium oxalates that enhance P availability by decreasing soil calcium (reviewed in Chapin & Eviner 2003, Eviner & Chapin 2003). Because P is bonded to organic compounds through ester bonds rather than a direct bond to C, mineralization of P does not depend on microbial utilization of C, and plants can release organic P on demand with enzymes (Hinsinger 2001). P inputs from root turnover vary with species, supplying between 21% and 84% of total plant P input to soil (Aerts et al. 1992). Cluster, or proteoid roots are particularly efficient at accessing P because they release large amounts of organic acids that solubilize soil PO_4 in the rhizosphere (Yan et al. 2002). The main controls over P availability change between systems. The main source of available P is from organic matter in the tundra, litter decomposition in forests, weathering of $CaPO_4$ in deserts, and atmospheric deposition in Mediterranean climates (Gressel &

McColl 1997). In waterlogged soils or sediments, plant species that enhance soil O_2 concentrations increase P availability by lowering consumption of SO_4 leading to decreased mobilization of Fe (Christensen 1999).

The control of sulfur dynamics is intermediate between P and N since C-S compounds are mineralized depending on microbial demand for C, and ester-bonded S is released through enzymes. S mineralization rarely occurs in the absence of growing plants (Chowdhury et al. 2000). Mineralization of ester-bonded S compounds is enhanced by plant SO_4 uptake, which decreases soil SO_4, thus stimulating the production of sulfohydrolases (Maynard et al. 1992). Root turnover is an important source of S and K (Burke & Raynal 1994). K dynamics are primarily determined by plant traits that affect K leaching from leaves (e.g., leaf area, leaf thickness, waxes) and soil (e.g., high fine root biomass in the surface soil) (Challinor 1968).

Species differences in Ca dynamics are primarily related to their rates of calcium release from litter (Dijkstra & Smits 2002), uptake, and allocation to biomass pools with different turnover times (Eriksson & Rosen 1994). In general, species that promote faster mineral weathering and have higher uptake rates enhance accumulations of base cations (Bergkvist & Folkeson 1995) and Ca availability (Challinor 1968, Finzi et al. 1998b) in both plant tissue and soil. Herbaceous dicots have a much higher base content than monocots (Cornelissen & Thompson 1997). Deciduous trees are particularly effective in enhancing Ca cycling (Willis et al. 1997) through cation pumping, where species with high fine root biomass in the lower soil profiles take up calcium from deep in the soil, and deposit it onto the surface soil through litter fall (Dijkstra & Smits 2002).

Species differ in their effects on Mg, Mn, Cu, Fe, Zn, and Al through organic acid input, cation uptake, and effects on soil pH and redox (Finzi et al. 1998b, Chapin & Eviner 2003). In general, plant species influence base cation availability by exuding low-molecular-weight organic acids. These acids act as ligands and, depending on the type exuded, have differential effects on metal solubility and speciation (Jones & Darrah 1994). Species that decrease pH increase Fe and Al solubility, which then out-compete base cations on cation exchange sites (Finzi et al. 1998b) and can lead to leaching of base cations. Organic acids rich in humic acid form water-soluble complexes with Fe, Mn, and P increasing Mn solubility 10- to 50-fold (Marschner 1995). Plant species can also enhance retention of nutrients by binding them with phenolics (Mn, Fe, Cu, Ca, Mg, and K) (Hattenschwiler & Vitousek 2000), or root mucilage (Pb, Cu, Zn, Cd) (Glinski & Lipiec 1990). Plant species differentially influence atmospheric deposition of elements through species differences in canopy structure (Weathers et al. 2001).

Disturbance

Species effects on disturbance regimes can dominate many plant effects on ecosystem processes (Mack et al. 2001). Species effects on flooding and erosion were discussed earlier in this paper. Species susceptibility to wind throw depends on their morphology, age, size, rooting conditions, and tissue constituents, with taller trees being more susceptible to wind damage than shorter-statured species with high root allocation. Species with full canopies have increased susceptibility to damage from wind as well as snow and ice accumulation (Boose et al. 1994, Foster 1988). Plant species composition has large impacts on the frequency, intensity, and spatial extent of fires (Billings 1991, D'Antonio & Vitousek 1992). Traits that influence flammability include (from increasing to decreasing importance): litter moisture content, biomass, the fineness of the litter, canopy structure (continuous fuel layer), silica-free ash content, and tissue chemistry (e.g., resins, caloric content). Conifers are more susceptible to high intensity fire than deciduous trees, owing to their lower moisture content, and thinner branches (Johnson 1992). Vegetation can also alter fire regime through its effects on insect outbreaks (Furyaev et al. 1983).

Plant Species Interactions with Other Organisms

Plant species are frequently associated with a specific suite of other organisms that play key roles in ecosystems. Selective feeding is mediated by plant size (Mittelbach & Gross 1984), energetic content (May 1992), and concentrations of nutrients (Tian et al. 1993) and secondary compounds (Slansky 1997). Secondary compounds not only deter feeding of herbivores, but also can be used selectively by herbivores for protection. For example, crabs utilize unpalatable algae as decoration to deter predatory fish (Stachowicz & Hay 1999).

Plant architecture and litter persistence can alter the microclimate and habitat specificity of organisms. Termite populations increase in the favorable microclimate provided by recalcitrant litter (Tian et al. 1993), while the distribution of grazing animals may focus around sources of shade during warm parts of the day (Sanchez & Febles 1999). Plant architecture can also play a role in the distribution of organisms by altering access of prey to predators (Clark & Messina 1998), the energetics of foraging (Whelan 2001), or pollinator access to flowers (Lortie & Aarssen 1999). Leaf characteristics can deter organisms, with plant cuticles inhibiting microbes and viruses, and spines and leaf hairs irritating herbivores (Gutschick 1999). In many cases, organisms prefer multiple plant species for different reasons (Stachowicz & Hay 1999; V.T. Eviner, F.S. Chapin III, submitted manuscript).

In order to determine the ecosystem impact of a plant species, it is vital to include the impact of plants on these other organisms (Hobbie 1995). For example, gophers selectively build large mounds in association with plant species that increase soil cohesion through their high root surface area. This soil disturbance increases rates of N cycling, and consideration of selective gopher disturbance greatly changes the relative effects of plant species on N cycling (V.T. Eviner, F.S. Chapin III, submitted manuscript). Similarly, plant species differences in C inputs are responsible for distinct microbial communities (Grayston et al. 1998) that vary in their ability to degrade specific substrates (Sharma et al. 1998). Plant species rooting patterns and canopy architecture affect the distribution and activity of earthworms (Zaller & Arnone 1999), which have large effects on nutrient cycling (Cortez et al. 2000, Thompson et al. 1993).

THE OVERALL IMPACT OF PLANT SPECIES ON ECOSYSTEMS

Several clear messages emerge from this review.

1. Multiple traits determine plant species effects on any ecosystem process.
2. Many plant traits influence a number of ecosystem properties and processes.
3. Plants have multiple, simultaneous effects on ecosystems.
4. Each ecosystem process is influenced by a different suite of traits or, if regulated by the same suite of traits, may be differentially influenced by these traits.

Most plant-effect studies focus on one type of ecosystem process and often on a subset of the key traits involved in that process. It is vital to consider the overall impact of a plant species on its ecosystem. In order to do this, it is crucial to understand how these traits and multiple effects interact. In this section of the review, we explore the relationships between multiple traits and ecosystem effects of plant species, and we propose a mechanistic framework to improve predictions of plant species effects on ecosystem processes.

Suites of Related Traits or Unique Combinations of Independent Traits?

As discussed in the introduction, there is a solid framework linking a specific plant trait to an ecosystem process. The strength of these correlations results from suites of covarying plant traits that influence plant effects on N cycling (Chapin 1993). However, many ecosystem processes cannot be adequately predicted with a single trait because they are mediated by multiple plant traits that do not covary (Verville et al. 1998, Wardle et al. 1998). For example, species effects on N cycling and decomposition are mediated by litter chemistry, labile C inputs, and effects on soil microclimate (Bottner et al. 1999, Cheng & Coleman 1990, Eviner 2001), and species differences in litter C:N are independent of their differences in labile C inputs and effects on soil moisture (Eviner 2001). Many plant traits vary independently among species (Wardle et al. 1998) and across groups of species (Garnier 1991), and the relative importance of different traits in determining ecosystem processes can change across plant groups (Cornelissen & Thompson 1997). Even plants within the same functional group can have up to a tenfold difference in other traits (Wardle et al. 1998), explaining why the same plant species can be grouped into many different functional groups (Wilson 1999), and why a single plant trait often cannot adequately predict the ecosystem effects of plant species.

Relationship Between Multiple Ecosystem Effects

Species play multiple, simultaneous roles in ecosystems, and these functions are often distributed independently among species. For example, plant species effects on N cycling are unrelated to their effects on P cycling (Hooper & Vitousek 1998), C cycling (Thomas & Prescott 2000), water fluxes, soil stability, and interactions with pests and beneficial organisms (Eviner & Chapin 2001). This is because many ecosystem processes are mediated by a distinct suite of traits. Just as multiple traits determine a plant species effect on any ecosystem process, the multiple effects of plant species can feedback to alter other ecosystem processes. For example, species effects on water availability can have large impacts on N cycling (Eviner 2001).

FUNCTIONAL MATRIX: PREDICTING PLANT SPECIES ECOSYSTEM IMPACTS BASED ON MULTIPLE TRAITS

Single traits such as litter chemistry are often inadequate predictors of plant species effects on ecosystem processes because species exhibit unique combinations of traits that influence ecosystems, and multiple traits are responsible for the ecosystem effect of a plant species. The functional matrix approach describes the relationship between ecosystem processes and multiple traits that vary independently from one another. This approach builds upon the functional group and single trait approaches to consider how different traits of functional significance are distributed among species (Figure 1) and the relationship between multiple traits and an ecosystem function (Figure 2). Consideration of multiple plant traits greatly enhances our ability to account for plant species effects, particularly because this approach can account for changes in the relationship between multiple traits and ecosystem processes by highlighting when the effects of multiple traits are additive or interactive (e.g., the strength of N inhibition of lignin degradation increases with higher lignin concentrations). The variables on the axes can be selected based on the ecosystem process and season of interest and by considering attributes

Figure 1 Existing plant functional type data can be used to form the basis of a functional matrix. Plant species from California annual grasslands are plotted according to their litter quality on the x-axis, and effects on soil moisture on the y-axis. The solid boxes represent functional groups based on litter chemistry, while the dashed boxes represent functional groups based on effects of soil moisture (determined by plant phenology). Both moisture and litter quality are important controllers of species effects on N cycling in California grasslands (Eviner 2001), and these vary independently and continuously among species. Data are from: Brown (1998), Eviner (2001), Franck et al. (1997), Gordon & Rice (1993), Hooper & Vitousek (1998), V.T. Eviner (unpublished manuscript). Species include: *Lupinus bicolor, Trifolium microcephalum, Lupinus sublanatus, Lotus purshianus, Plantago erecta, Erodium botrys, Amsinckia douglasiana, Lasthenia californica, Hemizonia congesta luzulaefolia, Lessingia micradenia, Vulpia microstachys, Avena barbata, Taeniatherum caput-medusae, Hordeum brachyantherum, Poa secunda, Bromus hordeaceous, Aegilops triuncialis, Nasella pulchra, and Elymus glaucus.*

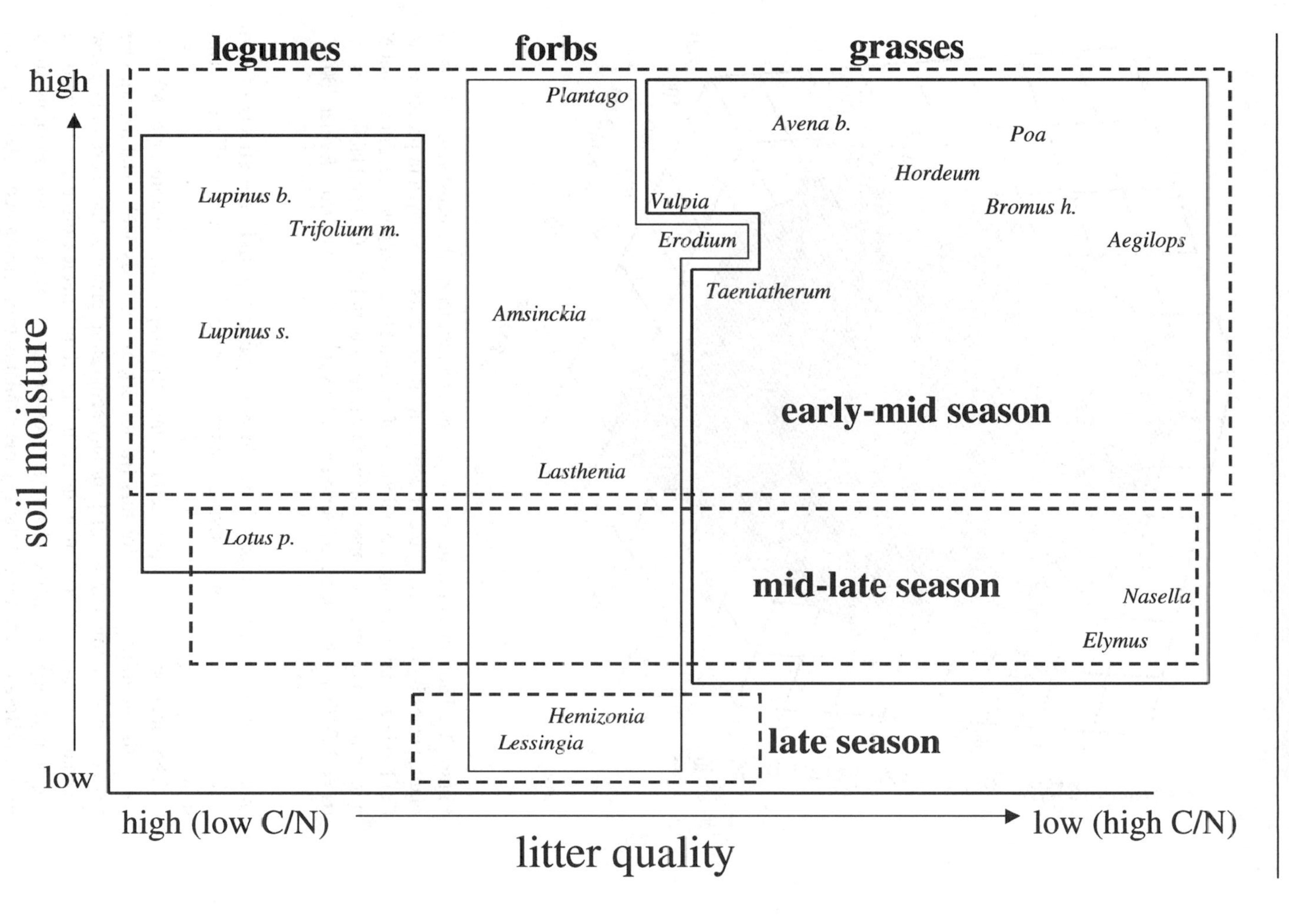

legumes
forbs
grasses
Plantago
Avena b.
Poa
Hordeum
Bromus h.
Aegilops
Lupinus b.
Trifolium m.
Vulpia
Erodium
Taeniatherum
Amsinckia
Lupinus s.
early-mid season
Lasthenia
Lotus p.
mid-late season
Nasella
Elymus
Hemizonia
Lessingia
late season
soil moisture
high
low
high (low C/N)
low (high C/N)
litter quality

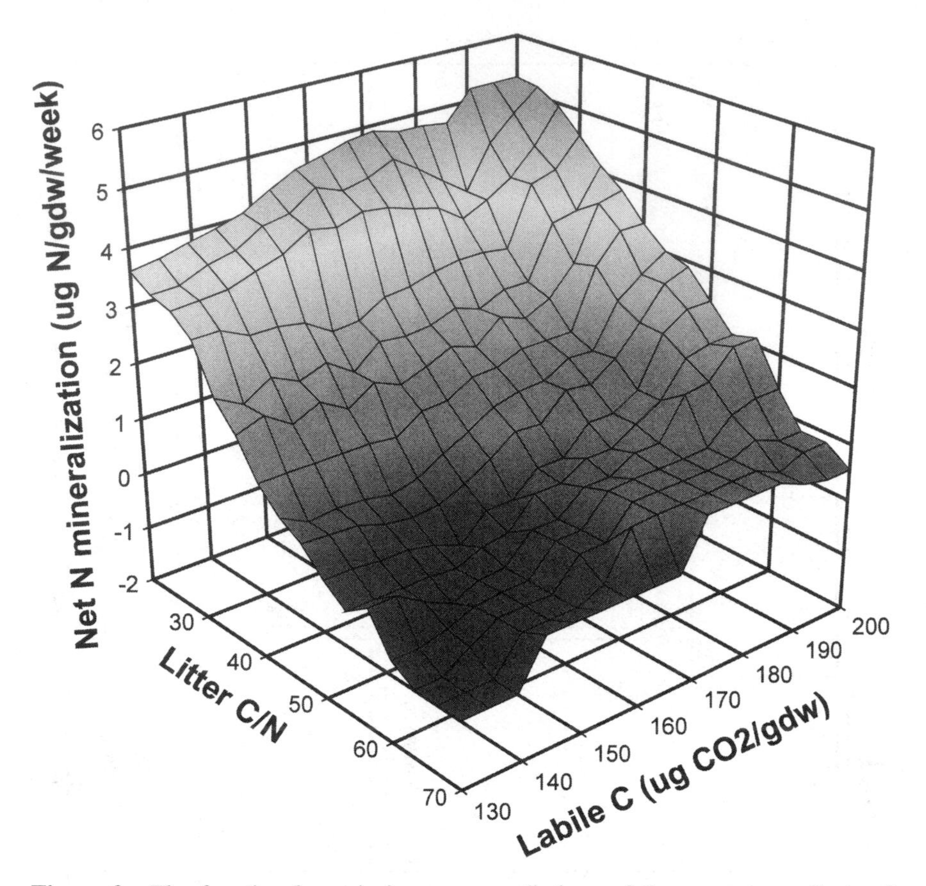

Figure 2 The functional matrix improves predictions of the ecosystem effects of plant species by considering the effects of multiple traits on ecosystem processes. In California annual grasslands, predictions of species effects on potential rates of net N mineralization are greatly improved by considering both litter C:N and labile C inputs ($r^2 = 0.7$), than when just considering litter C:N ($r^2 = 0.4$) (data from Eviner 2001).

of the species being compared. While our figures highlight the effects of two traits for the sake of simplicity, this approach is not limited to the effects of two traits.

The functional matrix approach is relatively simple and feasible especially because the building blocks for this framework are already well established. Our review clearly demonstrates that for any ecosystem process, there are only a few mechanisms that are critical in determining plant species effects. The traits of importance depend on the characteristics of the ecosystem, and previous studies have already determined the key mechanisms influencing ecosystem function in most ecosystems. In many ecosystems, the species-level information is readily available (Figure 1), and the data we collect to determine species effects on one ecosystem function can also enhance our ability to predict multiple ecosystem impacts of plant species because the same plant traits are involved in multiple processes. These multiple impacts can influence one another, and the functional matrix approach also accounts for this (e.g., by using soil moisture as a trait axis to determine how it interacts with litter chemistry to influence nutrient cycling).

Strength of this Approach

The power of this approach is that the ecosystem effects of multiple traits are the underlying mechanisms determining many types of vegetation effects on ecosystem processes. For example, the ranking of species effects on an ecosystem process varies with changing conditions (Cote & Fyles 1994). This approach allows us to consider the context dependence of species effects because it is based on the premise that differences in plant traits have consistent impacts on ecosystems, whether they occur among different plant species, or within a species grown under different conditions (Eviner 2001). Thus, we are able to use this multiple trait approach to account for:

- changes in the relative importance of traits with the stage of the process, season or over environmental gradients
- changes in plant species traits with season, phenology or over environmental gradients
- non-additive effects of plant species mixtures on ecosystem processes.

CHANGES IN THE RELATIVE IMPORTANCE OF TRAITS The traits that determine plant species effects on an ecosystem process can change owing to environmental conditions, plant age, growing season, and stage of a process. The functional matrix approach allows us to respond to these changes by selecting different axes or weighting them differently under changing conditions. For example, when environmental factors are the most limiting to ecosystem processes, the effects of plant species on these environmental factors become a key determinant of ecosystem processes (Eviner 2001, Mack 1998). Changes in the functional axes selected can account for changes in the mechanisms influencing species effects at a number of scales, ranging from local environmental gradients to global biomes.

CHANGES IN PLANT TRAITS Plant traits can change substantially because of different environmental conditions, seasons, and plant age (Austin & Vitousek 2000, Marschner 1995). Plant traits respond independently to environmental variables (Fonseca et al. 2000, Wright & Westoby 1999), and species differ in the responses of their traits to shifts in the environment (Hobbie & Gough 2002, Shierlaw & Alston 1984). These variations of traits within a species due to environmental change can be larger than interspecific differences (Olff 1992). Because the functional matrix approach determines plant species effects based on traits, it can account for changes in the ecosystem effects of a given plant species under changing conditions.

NONADDITIVE EFFECTS OF SPECIES MIXTURES Although the patterns of species effects in monoculture often can't predict the patterns of mixtures, the same plant traits that determine plant species effects on ecosystem processes also determine the ecosystem effect of plant species mixtures. The multiple trait approach can account for these nonadditive effects because it considers the effects of changes in a species trait due to neighbors (Gordon & Rice 1993, Tremmel & Bazzaz 1995). For example, differences in litter C:N ratios and labile C in species grown in mixtures versus monocultures account for the nonadditive effects of mixtures on N cycling (Eviner 2001). In addition, this approach determines the combined effects of multiple traits, whether these multiple traits occur within a given species, or are associated with different plant species growing together. The stimulating effects of labile C on decomposition of recalcitrant C should occur whether these traits are expressed within one plant, or in neighboring plants.

The functional matrix approach, by focusing on the multiple mechanisms underlying plant species effects on ecosystems, greatly improves our predictions of plant effects over a wide range of conditions. It also encompasses the multiple effects of plant species on ecosystems. The ability to predict the overall effects of plants on ecosystems under a variety of conditions can be a powerful management tool, allowing us to simultaneously provide multiple ecosystem services, or address multiple environmental problems through vegetation manipulations (Eviner & Chapin 2001).

ACKNOWLEDGMENTS

Thanks to Sarah Hobbie and Gaius Shaver for comments on this manuscript. This is a contribution to the program of the Institute of Ecosystem Studies.

The *Annual Review of Ecology, Evolution, and Systematics* is online at http://ecolsys.annualreviews.org

LITERATURE CITED

Abrahams AD, Parsons AJ, Wainwright J. 1994. Resistance to overland-flow on semi-arid grassland and shrubland hillslopes, Walnut Gulch, southern Arizona. *J. Hydrol.* 156: 431–46

Aerts R, Bakker C, De Caluwe H. 1992. Root turnover as determinant of the cycling of C, N, and P in a dry heathland ecosystem. *Biogeochemistry* 15:175–90

Aerts R, De Caluwe H. 1997. Nutritional and plant-mediated controls on leaf litter decomposition of Carex species. *Ecology* 78:244–60

Allen WC, Hook PB, Biederman JA, Stein OR. 2002. Temperature and wetland plant species effects on wastewater treatment and root zone oxidation. *J. Environ. Qual.* 31:1010–16

Angers D, Caron J. 1998. Plant-induced changes in soil structure: processes and feedbacks. *Biogeochemistry* 42:55–72

Augusto L, Turpault MP, Ranger J. 2000. Impact of forest tree species on feldspar weathering rates. *Geoderma.* 96:215–37

Austin A, Vitousek P. 2000. Precipitation, decomposition and litter decomposability of Meterosideros polymorpha in native forests on Hawaii. *J. Ecol.* 88:129–38

Baldocchi D, Kelliher FM, Black TA, Jarvis P. 2000. Climate and vegetation controls on boreal zone energy exchange. *Glob. Change Biol.* 6:69–83

Baldocchi D, Vogel C. 1996. Energy and CO_2 flux densities above and below a temperate broad-leaves forest and a boreal pine forest. *Tree Physiol.* 16:5–16

Balesdent J, Balabane M. 1996. Major contribution of roots to soil carbon storage inferred from maize cultivated soils. *Soil Biol. Biochem.* 28:1261–63

Bell D. 1999. Australian trees for the rehabilitation of waterlogged and salinity-damaged landscapes. *Aust. J. Bot.* 47:697–716

Berg B. 2000a. Initial rates and limit values for decomposition of Scots pine and Norway spruce needle litter: a synthesis for N-fertilized forest stands. *Can. J. For. Res.* 30: 122–35

Berg B. 2000b. Litter decomposition and organic matter turnover in northern forest soils. *For. Ecol. Manage.* 133:13–22

Berg B, Ekbohm G. 1993. Decomposing needle litter in a Pinus contorta (Lodgepole pine) and Pinus sylvestris (Scots pine) monoculture systems—is there a maximum mass loss. *Scand J. For. Res.* 8:457–65

Bergkvist B, Folkeson L. 1995. The influence of tree species on acid deposition, proton budgets and element fluxes in south Swedish forest ecosystems. *Ecol. Bull.* 44:90–99

Billings W. 1991. Bromus tectorum, a biotic cause of ecosystem impoverishment in the Great Basin. In *The Earth in Transition: Patterns and Processes of Biotic Impoverishment*, ed. G Woodwell, pp. 301–22. New York: Cambridge Univ. Press

Binkley D, Giardina C. 1998. Why do tree species affect soils? The warp and woof of tree-soil interactions. *Biogeochemistry* 42: 89–106

Blair J. 1988. Nitrogen, sulfur and phosphorus dynamics in decomposing deciduous leaf litter in the southern Appalachians. *Soil Biol. Biochem.* 20:693–701

Boettcher S, Kalisz P. 1990. Single-tree influence on soil properties in the mountains of eastern Kentucky. *Ecology* 71:1365–72

Boose E, Foster D, Fluet M. 1994. Hurricane impacts to tropical and temperate forest landscapes. *Ecol. Monogr.* 64:369–400

Bottner P, Pansu M, Sallih Z. 1999. Modelling the effect of active roots on soil organic matter turnover. *Plant Soil.* 2166:15–25

Bronson K, Mosier A. 1993. Nitrous oxide emissions and methane consumption in wheat and corn-cropped systems in northeastern Colorado. In *Agricultural Ecosystem Effects on Trace Gases and Global Climate Change*, ed. L Harper, A Mosier, J Duxbury, D Rolston, pp. 133–44. Madison, WI: ASA, CSSA, SSSA

Brown C. 1998. *Restoration of California central valley grasslands: applied and theoretical approaches to understanding interactions among prairie species.* PhD thesis. Univ. California, Davis. 177 pp.

Burke M, Raynal D. 1994. Fine root growth phenology, production and turnover in a northern hardwood forest ecosystem. *Plant Soil.* 162:135–46

Caldwell M, Dawson T, Richards J. 1998. Hydraulic lift: consequences of water efflux from the roots of plants. *Oecologia* 113:151–61

Camberato J. 2001. Cation exchange capacity-everything you want to know and much more. *S.C. Turfgrass Found. News* Oct–Dec 2001

Challinor D. 1968. Alteration of surface soil characteristics by four tree species. *Ecology* 49:286–90

Chapin FS III. 1993. Functional role of growth forms in ecosystem and global processes. In *Scaling Physiological Processes: Leaf to Globe*, ed. J Ehleringer, C Field, pp. 287–312. San Diego: Academic

Chapin FS III, Eviner VT. 2003. Biogeochemistry of terrestrial net primary production. In *Treatise on Geochemistry. Volume 8: Biogeochemistry*, ed. WH Schlesinger. In press

Chapin FS III, Fetcher N, Kielland K, Everett KR, Linkins AE. 1988. Productivity and nutrient cycling of Alaskan tundra-enhancement by flowing soil water. *Ecology* 69:693–702

Chapin FS III, McGuire AD, Randerson J, Pielke R, Baldocchi D, et al. 2000. Arctic and boreal ecosystems of western North America as components of the climate system. *Global Change Biol.* 6:211–23

Cheng W, Coleman D. 1990. Effect of living roots on soil organic matter decomposition. *Soil Biol. Biochem.* 22:781–87

Chowdhury MAH, Kouno K, Ando T, Nagaoka T. 2000. Microbial biomass, S mineralization and S uptake by African millet from soil amended with various composts. *Soil Biol. Biochem.* 32:845–52

Christensen KK. 1999. Comparison of iron and phosphorus mobilization from sediments inhabited by Littorella uniflora and Sphagnum sp. at different sulfate concentrations. *Arch. Hydrobiol.* 145:257–75

Christian JM, Wilson SD. 1999. Long-term ecosystem impacts of an introduced grass in the northern Great Plains. *Ecology* 80:2397–407

Clark T, Messina F. 1998. Foraging behavior of lacewing larvae (Neuroptera: Chrysopidae) on plants with divergent architectures. *J. Insect Behav.* 11:303–17

Cole D. 1995. Soil nutrient supply in natural and managed forests. *Plant Soil.* 169:43–53

Cornelissen J, Thompson K. 1997. Functional leaf attributes predict litter decomposition rate in herbaceous plants. *New Phytol.* 135:109–14

Cortez J, Billes G, Bouche M. 2000. Effect of climate, soil type and earthworm activity on nitrogen transfer from a nitrogen-15-labelled decomposing material under field conditions. *Biol. Fert. Soils.* 30:318–27

Cortez J, Demard J, Bottner P, Monrozier L. 1996. Decomposition of mediterranean leaf litters: a microcosm experiment investigating relationships between decomposition rates and litter quality. *Soil Biol. Biochem.* 28:443–52

Cote B, Fyles J. 1994. Leaf litter disappearnce of hardwood species of southern Quebec: interaction between litter quality and stand type. *Ecoscience* 1:322–28

Couteaux M, NcTiernan K, Berg B, Szuberla D, Dardenne P, Bottner P. 1998. Chemical composition and carbon mineralization potential of scots pine needles at different stages of decomposition. *Soil Biol. Biochem.* 30:583–95

D'Antonio C, Vitousek P. 1992. Biological invasions by exotic grasses, the grass/fire cycle, and global change. *Annu. Rev. Ecol. Syst.* 23:63–87

Dawson T. 1998. Fog in the California redwood forest: ecosystem inputs and use by plants. *Oecologia.* 117:476–85

Degens BP. 1997. Macro-aggregation of soils by biological bonding and binding mechanisms and the factors affecting these: a review. *Aust. J. Soil. Res.* 35:431–59

Deluca TH, Drinkwater LE, Wiefling BA, DeNicola DM. 1996. Free-living nitrogen-fixing bacteria in temperate cropping systems: Influence of nitrogen source. *Biol. Fert. Soils.* 23:140–44

De Ploey Y, Cruz O. 1979. Lanslides in the serra de Mar, Brazil. *Catena.* 6:111–22

Devine DL, Wood M, Donart GB. 1998. Runoff and erosion from a mosaic tobosagrass and burrograss community in the northern Chihuahuan Desert grassland. *J. Arid. Environ.* 39:11–19

Dijkstra F, Smits M. 2002. Tree species effects on calcium cycling: the role of calcium uptake in deep soils. *Ecosystems* 5:385–98

DiTomaso JM. 2000. Invasive weeds in rangelands: Species, impacts and management. *Weed Sci.* 48:255–65

Doerr SH, Shakesby RA, Walsh RPD. 2000. Soil water repellency: its causes, characteristics and hydro-geomorphological significance. *Earth Sci. Rev.* 51:33–65

Dormaar J. 1990. Effect of active roots on

the decomposition of soil organic materials. *Biol. Fert. Soils.* 10:121–26

Easterwood G, Sartain J. 1990. Clover residue effectiveness in reducing orthophosphate sorption on ferric hydroxide coated soil. *Soil Sci. Soc. Am. J.* 54:1345–50

Engelaar W, Symens J, Laanbroek H, Blom C. 1995. Preservation of nitrifying capacity and nitrate availability in waterlogged soils by radial oxygen loss from roots of wetland plants. *Biol. Fert. Soils.* 20:243–48

Epstein HE, Burke IC, Mosier AR. 2001. Plant effects on nitrogen retention in shortgrass steppe 2 years after N-15 addition. *Oecologia.* 128:422–30

Epstein H, Burkey I, Mosier A. 1998. Plant effects on spatial and temporal patterns of nitrogen cycling in shortgrass steppe. *Ecosystems.* 1:374–85

Eriksson H, Rosen K. 1994. Nutrient distribution in a Swedish tree species experiment. *Plant Soil.* 164:51–59

Evans R, Young J. 1970. Plant litter and establishment of alien annual species in rangeland communities. *Weed Sci.* 18:697–702

Eviner VT. 2001. *Linking plant community composition and ecosystem dynamics: interactions of plant traits determine the ecosystem effects of plant species and plant species mixtures.* PhD thesis. Univ. California, Berkeley. 404 pp.

Eviner VT, Chapin FI. 2001. The effects of California grassland species on their ecosystems: implications for sustainable agriculture and rangeland management. *Calif. Agr.* 55:254–59

Eviner VT, Chapin FS III. 2002. The influence of plant species, fertilization and elevated CO2 on soil aggregate stability. *Plant Soil.* 246:211–19

Eviner VT, Chapin FS III. 2003. Selective gopher disturbance influences plant species effects on nitrogen cycling. *Oikos.* Submitted

Eviner VT, Chapin FS III. 2003. Biogeochemical interactions and biodiversity. In *Element Interactions: Rapid Assessment Project of SCOPE*, ed. J Melillo, C Field, M Moldan. Washington, DC: Island. In press

Finzi A, Canham C, Van Breemen N. 1998b. Canopy tree-soil interactions within temperate forests: species effects on pH and cations. *Ecol. Appl.* 8:447–54

Finzi A, Van Breeman N, Canham C. 1998a. Canopy tree-soil interactions within temperate forests: species effects on soil carbon and nitrogen. *Ecol. Appl.* 8:440–46

Firestone MK, Davidson E. 1989. Microbiological basis of NO and N_2O production and consumption in soils. In *Exchanges of Trace Gases Between Terrestrial Ecosystems and the Atmosphere*, ed. M Andreae, D Schimel, pp. 7–21. New York: Wiley

Flanagan P, Van Cleve K. 1983. Nutrient cycling in relation to decomposition and organic matter quality in taiga ecosystems. *Can. J. For. Res.* 13:795–817

Foley J, Kutzbach J, Coe M, Lewis S. 1994. Feedbacks between climate and boreal forests during the Holocene epoch. *Nature.* 371:52–54

Fonseca C, Overton J, Collins B, Westoby M. 2000. Shifts in trait combinations along rainfall and phosphorus gradients. *J. Ecol.* 88:964–77

Foster D. 1988. Species and stand response to catastrophic wind in central New England. *J. Ecol.* 76:135–51

Franck VM, Hungate BA, Chapin FS III, Field CB. 1997. Decomposition of litter produced under elevated CO_2: Dependence on plant species and nutrient supply. *Biogeochemistry.* 36:223–37

Franco A, De Faria S. 1997. The contribution of N2-fixing tree legumes to land reclamation and sustainability in the tropics. *Soil Biol. Biochem.* 29:897–903

Furyaev V, Wein R, McLean D. 1983. Fire influences in Abies-dominated forests. In *The role of fire in northern circumpolar ecosystems*, ed. R Wein, D McLean, pp. 221–34. New York: Wiley

Gale WJ, Cambardella CA, Bailey TB. 2000. Surface residue- and root-derived carbon in stable and unstable aggregates. *Soil Sci. Soc. Am. J.* 64:196–201

Garnier E. 1991. Resource capture biomass

allocation and growth in herbaceous plants. *Trends Ecol. Evol.* 6:126–31

Gijsman A, Thomas R. 1996. Evaluation of some physical properties of an oxisol after conversion of native savanna into legume-based or pure grass pastures. *Tropical Grasslands.* 30:237–48

Gillon D, Joffre R, Ibrahima A. 1994. Initial litter properties and decay rate: a microcosm experiment on Mediterranean species. *Can. J. Bot.* 72:946–54

Gleason M, Elmer D, Pien N, Fisher J. 1979. Effects of stem density upon sediment retention by salt marsh cord grass, *Spartina alterniflora* Loisel. *Estuaries.* 2:271–73

Glinski J, Lipiec J. 1990. *Soil Physical Conditions and Plant Roots.* Boca Raton, FL.: CRC. 250 pp.

Gloaguen J, Touffet J. 1982. C-N evolution in the leaves and during litter decomposition under Atlantic climate—the beech and some conifers. *Ann. Sci. Forest.* 39:219–30

Gordon D, Rice K. 1993. Competitive effects of grassland annuals on soil water and blue oak (Quercus douglasii) seedlings. *Ecology* 74:68–82

Gordon D, Welker J, Menke J, Rice K. 1989. Competition for soil water between annual plants and blue oak (*Quercus douglasii*) seedlings. *Oecologia.* 79:533–41

Grayston S, Wang S, Campbell C, Edwards A. 1998. Selective influence of plant species on microbial diversity in the rhizosphere. *Soil Biol. Biochem.* 30:369–78

Gressel N, McColl J. 1997. Phosphorus mineralization and organic matter: a critical review. In *Driven by Nature: Plant Litter Quality and Decomposition*, ed. G Cadish, K Giller, pp. 297–309. Wallington, UK: CAB

Grosse W, Jovy K, Tiebel H. 1996. Influence of plants on redox potential and methane production in water-saturated soil. *Hydrobiologia.* 340:93–99

Gutschick V. 1999. Biotic and abiotic consequences of differences in leaf structure. *New Phytol.* 143:3–18

Hart S, Nason G, Myrold D, Perry D. 1994. Dynamics of gross nitrogen transformations in an old-growth forest—the carbon connection. *Ecology* 75:880–91

Hattenschwiler S, Vitousek PM. 2000. The role of polyphenols in terrestrial ecosystem nutrient cycling. *Trends Ecol. Evol.* 15:238–43

Heal O, Anderson J, Swift M. 1997. Plant litter quality and decomposition: an historical overview. In *Driven by nature: plant litter quality and decomposition*, ed. G Cadish, K Giller, 3–30. Wallington, UK: CAB

Herwitz S. 1986. Infiltration-excess caused by stemflow in a cyclone-prone tropical rainforest. *Earth Surf. Process.* 11:401–12

Hinsinger P. 2001. Bioavailability of soil inorganic P in the rhizosphere as affected by root-induced chemical changes: a review. *Plant Soil.* 237:173–95

Hobbie SE. 1992. Effects of plant-species on nutrient cycling. *Trends Ecol. Evol.* 7:336–39

Hobbie SE. 1995. Direct and indirect effects of plant species on biogeochemical processes in arctic ecosystems. In *Arctic and Alpine Biodiversity*, ed. FS Chapin III, C Korner, pp. 213–24. Berlin: Springer-Verlag

Hobbie SE. 1996. Temperature and plant species control over litter decomposition in Alaskan tundra. *Ecol. Monogr.* 66:503–22

Hobbie SE, Gough L. 2002. Foliar and soil nutrients in tundra on glacial landscapes of contrasting ages in northern Alaska. *Oecologia.* 131:453–62

Hogg E, Lieffers V. 1991. The impact of *Calamagrostis canadensis* on soil thermal regimes after logging in northern Alberta. *Can. J. For. Res.* 21:

Hooper D, Vitousek P. 1998. Effects of plant composition and diversity on nutrient cycling. *Ecol. Monogr.* 68:121–49

Horner J, Gosz J, Cates R. 1988. The role of carbon-based plant secondary metabolites in decomposition in terrestrial ecosystems. *Am. Nat.* 132:869–83

Howard P, Howard D, Lowe L. 1998. Effects of tree species and soil physiochemical conditions on the nature of soil organic matter. *Soil Biol. Biochem.* 30:285–97

Hume N, Fleming M, Horne A. 2002. Denitrification potential and carbon quality of four aquatic plants in wetland microcosms. *Soil Sci. Soc. Am. J.* 66:1706–12

Jannsen B. 1996. Nitrogen mineralization in relation to C:N ratio and decomposability of organic materials. *Plant Soil.* 181:39–45

Jastrow J. 1987. Changes in soil aggregation associated with tallgrass prairie restoration. *Am. J. Bot.* 74:1656–64

Jenny H. 1941. *Factors of Soil Formation.* New York: McGraw-Hill

Johnson D. 1992. Effects of forest management on soil carbon storage. *Water Air Soil Poll.* 64:83–0

Jones DL, Darrah PR. 1994. Role of root derived organic-acids in the mobilization of nutrients from the rhizosphere. *Plant Soil.* 166:247–57

Jones M, Koenigs R, Vaughn C, Murphy A. 1983. Converting chaparral to grassland increases soil fertility. *Calif. Agr.* 37:23–24

Kelly E, Chadwick O, Hilinski T. 1998. The effect of plants on mineral weathering. *Biogeochemistry* 42:21–53

Kirkby M. 1995. Modeling the links between vegetation and landforms. *Geomorphology* 13:319–35

Kloot PM. 1983. The role of common iceplant (Mesembryanthemum-crystallinum) in the deterioration of medic pastures. *Aust. J. Ecol.* 8:301–6

Laroche F. 1994. *Melaleuca management plan for Florida.* Florida: Exotic Pest Plant Council. 88 pp.

Laskowski R, Niklinska M, Maryanski M. 1995. The dynamics of chemical elements in forest litter. *Ecology* 76:1393–406

Lawrence D, Schoenike R, Quispel A, Bond G. 1967. The role of *Dryas drummondii* in vegetation development following ice regression at Glacier Bay, Alaska, with special reference to nitrogen fixation by root nodules. *J. Ecol.* 55:793–813

Le Maitre DC, Scott DF, Colvin C. 1999. A review of information on interactions between vegetation and groundwater. *Water SA.* 25:137–52

Lewis J, Starkey R. 1968. Vegetable tannins, their decomposition and effects of decomposition of some organic compounds. *Soil Sci.* 106:241–47

Liste HH, Alexander M. 2000. Plant-promoted pyrene degradation in soil. *Chemosphere.* 40:7–10

Lodhi M, Killingbeck K. 1980. Allelopathic inhibition of nitrification and nitrifying bacteria in ponderosa pine (*Pinus ponderosa* Dougl.) community. *Am. J. Bot.* 67:1423–29

Lortie CJ, Aarssen LW. 1999. The advantage of being tall: Higher flowers receive more pollen in Verbascum thapsus L-(Scrophulariaceae). *Ecoscience* 6:68–71

Lovett GM, Weathers KC, Arthur MA. 2002. Control of nitrogen loss from forested watersheds by soil carbon: Nitrogen ratio and tree species composition. *Ecosystems* 5:712–18

Lyons T, Schwerdtfeger P, Hacker J, Foster I, Smith R, Huang X. 1993. Land atmosphere interaction in a semiarid region—the bunny fence experiment. *Bull. Am. Meteorol. Soc.* 74:1327–34

Mack M. 1998. *Effects of exotic grass invasion on ecosystem nitrogen dynamics in a Hawaiian woodland.* PhD thesis. Univ. California, Berkeley. 209 pp.

Mack M, D'Antonio C, Ley R. 2001. Alteration of ecosystem nitrogen dynamics by exotic plants: a case study of C4 grasses in Hawaii. *Ecol. Appl.* 11:1323–35

Marschner H. 1995. *Mineral Nutrition of Higher Plants.* London: Academic

Matson P. 1990. Plant-soil interactions during primary succession at Hawaii Volcanoes National Park. *Oecologia.* 85:241–46

May P. 1992. Flower selection and the dynamics of lipid reserves in two nectarivorous butterflies. *Ecology* 73:2181–91

Maynard DG, Stadt JJ, Mallett KI, Volney WJA. 1992. A comparison of sulfur impacted and non-impacted lodgepole pine stands in west central Alberta. *Can. J. Soil Sci.* 72:327–

McClaugherty C, Pastor J, Aber J, Melillo J. 1985. Forest litter decomposition in relation to soil nitrogen dynamics and litter quality. *Ecology* 66:266–75

Melillo J, Aber J, Linkins A, Ricca A, Perry B, Nadelhoffer K. 1989. Carbon and nitrogen dynamics along the decay continuum: Plant litter to soil organic matter. *Plant Soil.* 115: 189–98

Melillo J, Aber J, Muratore J. 1982. Nitrogen and lignin control of hardwood leaf litter decomposition dynamics. *Ecology* 63:621–26

Miles J. 1985. The pedogenic effects of different species and vegetation types and the implications of succession. *J. Soil Sci.* 36:571–84

Miller R, Jastrow J. 1990. Hierarchy of root and mycorrhizal fungal interactions with soil aggregation. *Soil Biol. Biochem.* 22:579–84

Mittelbach G, Gross K. 1984. Experimental studies of seed predation in old-fields. *Oecologia.* 65:7–13

Mueller-Dombois D. 1973. A non-adapted vegetation interferes with water removal in a tropical rainforest area in Hawaii. *Tropical Ecol.* 14:1–18

Mueller T, Jensen L, Nielsen E, Magid J. 1998. Turnover of carbon and nitrogen in a sandy loam soil following incorporation of chopped maize plants, barley straw and blue grass in the field. *Soil Biol. Biochem.* 30:561–71

Munzbergova Z, Ward D. 2002. Acacia trees as keystone species in Negev desert ecosystems. *J. Veg. Sci.* 13:227–36

Neff JC, Townsend AR, Gleixner G, Lehman SJ, Turnbull J, Bowman WD. 2002. Variable effects of nitrogen additions on the stability and turnover of soil carbon. *Nature* 419:915–17

Northup R, Dahlgren R, McColl J. 1998. Polyphenols as regulators of plant-litter-soil interactions in northern California's pygmy forest: a positive feedback? *Biogeochemistry* 42:189–20

Noskop Bliss LC, Cook FD. 1994. The association of free-living nitrogen-fixing bacteria with the roots of high arctic graminoids. *Arctic Alpine Research* 26:180–86

Olff H. 1992. Effects of light and nutrient availability on dry-matter and N-allocation in 6 successional grassland speciestesting for resource ratio effects. *Oecologia.* 89:412–21

Otterman J. 1989. Enhancement of surface-atmosphere fluxes by desert fringe vegetation through reduction of surface albedo and of soil heat flux. *Theor. Appl. Climatol.* 40:67–79

Ovington J. 1953. Studies of the development of woodland conditions under different trees. I. Soils pH. *J Ecol.* 41:13–34

Pitt M, Burgy R, Heady H. 1978. Influences of brush conversion and weather patterns on runoff from a northern California watershed. *J. Range Manage.* 31:23–27

Prandini L, Guidicini G, Bottura J, Poncano W, Santos A. 1977. Behavior of the vegetation in slope stability: a critical review. *B. Int. Assoc. Eng. Geol.* 16:51–55

Prescott C, Chappell H, Vesterdal L. 2000. Nitrogen turnover in forest floors of coastal Douglas-fir at sites differing in soil nitrogen capital. *Ecology* 81:1878–86

Puget P, Drinkwater LE. 2001. Short-term dynamics of root- and shoot-derived carbon from a leguminous green manure. *Soil Sci. Soc. Am. J.* 65:771–79

Qadir M, Schubert S, Ghafoor A, Murtaza G. 2001. Amelioration strategies for sodic soils: A review. *Land Degrad. Dev.* 12:357–86

Ranger J, Nys C. 1994. The effect of spruce (Picea-abies karst) On soil development—an analytical and experimental approach. *Europ. J. Soil Sci.* 45:193–204

Raulund-Rasmussen K, Vejre H. 1995. Effect of tree species and soil properties on nutrient immobilization in the forest floor. *Plant Soil.* 168–169:345–52

Reid J, Goss M. 1982. Interactions between soil drying due to plant water use and decreases in aggregate stability caused by maize roots. *J. Soil Sci.* 33:47–53

Rillig M, Wright S, Eviner VT. 2002. The role of arbuscular mycorrhizal fungi and glomalin in soil aggregation: comparing effects of five plant species. *Plant Soil.* 238:325–33

Robles M, Chapin FS III. 1995. Comparison of the influence of two exotic species on ecosystem processes in the Berkeley hills. *Madrono.* 42:349–57

Sallih Z, Bottner P. 1988. Effect of wheat (*Triticum aestivum*) roots on mineralization rates of soil organic matter. *Biol. Fert. Soils.* 7:67–70

Sanchez R, Febles I. 1999. Behaviour of grazing Holstein cows in natural shade. *Cuban J. Agric. Sci.* 33:241–46

Scheu S. 1997. Effects of litter (beech and stinging nettle) and earthworms (Octolasion lacteum) on carbon and nutrient cycling in beech forests on a basalt-limestone gradient: A laboratory experiment. *Biol. Fert. Soils.* 24:384–93

Schimel J, Van Cleve K, Cates R, Clausen T, Reichardt P. 1996. Effects of balsam poplar (*Populus balsamifera*) tannins and low molecular weight phenolics on microbial activity in taiga floodplain soil: implications for changes in N cycling during succession. *Can. J. Bot.* 74:84–90

Schlesinger W, Ward T, Anderson J. 2000. Nutrient losses in runoff from grassland and shrubland habitats in southern New Mexico: II. Field plots. *Biogeochemistry* 49:69–86

Schlesinger WH. 1997. *Biogeochemistry*. San Diego: Academic

Schlesinger WH, Reynolds J, Cunningham G, Huennecke L, Jarrel W, et al. 1990. Biological Feedbacks in Global Desertification. *Science.* 247:1043–48

Scott N. 1998. Soil aggregation and organic matter mineralization in forests and grasslands: plant species effects. *Soil Sci. Soc. Am. J.* 62:1081–89

Scott N, Binkley D. 1997. Foliage litter quality and annual net N mineralization: Comparison across North American forest sites. *Oecologia* 111:151–59

Selby MJ. 1993. *Hillslope Materials and Processes*. Oxford: Oxford Univ. Press

Sharma S, Rangger A, von Lutzow M, Insam H. 1998. Functional diversity of soil bacterial community increases after maize litter amendment. *Eur. J. Soil Biol.* 34:53–60

Shierlaw J, Alston A. 1984. Effect of soil compaction on root growth and uptake of phosphorus. *Plant Soil.* 77:15–28

Shukla J, Nobre C, Sellers P. 1990. Amazon deforestation and climate change. *Science* 247:1322–25

Siqueira J, Muraleedharan G, Hammerschmidt R, Safir G. 1991. Significance of phenolic compounds in plant soil-microbial systems. *CRC Crit. Rev. Plant Sci.* 10:63–121

Slansky F Jr. 1997. Allelochemical-nutrient interactions in herbivore nutritional ecology. In *Herbivores: Their interactions with secondary plant metabolites*, ed. G Rosenthal, M Berenbaum, pp. 135–75. San Diego: Academic

Spaccini R, Piccolo A, Conte P, Haberhauer G, Gerzabek M. 2002. Increased soil organic carbon sequenstration through hydrophobic protection by humic substances. *Soil Biol. Biochem.* 34:1839–51

Sparks J, Monson R, Sparks K, Lerdau M. 2001. Leaf uptake of nitrogen dioxide (NO_2) in a tropical wet forest: implications for tropospheric chemistry. *Oecologia* 127:214–21

Stachowicz J, Hay M. 1999. Reducing predation through chemically mediated camouflage: indirect effects of plant defenses on herbivores. *Ecology* 80:495–509

Steltzer H, Bowman W. 1998. Differential influence of plant species on soil nitrogen transformations with moist meadow alpine tundra. *Ecosystems* 1:464–74

Swank W, Douglass J. 1974. Streamflow greatly reduced by converting deciduous hardwood stands to pine. *Science* 185:857–59

Ta T, Faris M. 1987. Species variation in the fixation and transfer of nitrogen from legumes to associated grasses. *Plant Soil.* 98:265–74

Taylor B, Parkinson D, Parsons W. 1989. Nitrogen and lignin content as predictors of litter decay rates: a microcosm test. *Ecology* 70:97–104

Terwilliger V, Waldron L. 1991. Effects of root reinforcement on soil-slip patterns in the Transverse Ranges of southern California. *Geol. Soc. Am. Bull.* 103:775–85

Thomas K, Prescott C. 2000. Nitrogen availability in forest floors of three tree species on the same site: the role of litter quality. *Can. J. For. Res.* 30:1698–706

Thompson L, Thomas C, Radley J, Williamson S, Lawton J. 1993. The effect of earthworms and snails in a simple plant community. *Oecologia* 95:171–78

Tian G, Brussaard L, Kang B. 1993. Biological effects of plant residues with contrasting chemical composition under humid tropical conditions: effects on soil fauna. *Soil Biol. Biochem.* 25:731–37

Torn MS, Chapin FS III. 1993. Environmental and biotic controls over methane flux from arctic tundra. *Nato Advanced Research Workshop On Atmospheric Methane: Sources, Sinks And Role In Global Change, Mount Hood, Oregon, Usa, October 7–11, 1991. Chemosphere.* 26:357–68

Tremmel D, Bazzaz F. 1995. Plant architecture and allocation in different neighborhoods: implications for competitive success. *Ecology* 76:262–71

Troeh F, Thompson L. 1993. *Soil and soil fertility*. New York: Oxford Univ. Press

Updegraff K, Pastor J, Bridgham S, Johnston C. 1995. Environmental and substrate controls over carbon and nitrogen mineralization in northern wetlands. *Ecol. Appl.* 5:151–63

Urquiaga S, Cadisch G, Alves BJR, Boddey RM, Giller KE. 1998. Influence of decomposition of roots of tropical forage species on the availability of soil nitrogen. *Soil Biol. Biochem.* 30:2099–106

Van der Krift TAJ, Gioacchini P, Kuikman PJ, Berendse F. 2001. Effects of high and low fertility plant species on dead root decomposition and nitrogen mineralisation. *Soil Biol. Biochem.* 33:2115–24

Van der Krift TAJ, Kuikman PJ, Berendse F. 2002. The effect of living plants on root decomposition of four grass species. *Oikos.* 96:36–45

Van Veen JA, Kuikman PJ. 1990. Soil structural aspects of decomposition of organic matter by microorganisms. *Biogeochemistry* 11:213–34

Van Vuuren MMI, Berendse F, de Visser W. 1993. Species and site differences in the decomposition of litters and roots from wet heathlands. *Can. J. Bot.* 71:167–73

Verville J, Hobbie S, Chapin FI, Hooper D. 1998. Response of tundra CH_4 and CO_2 flux to manipulation of temperature and vegetation. *Biogeochemistry* 41:215–35

Viles H. 1990. The agency of organic beings: a selective review of recent work in biogeomorphology. In *Vegetation and Erosion*, ed. J Thornes, pp. 5–24. New York: Wiley

Vitousek PM, Cassman K, Cleveland C, Crews T, Field CB, et al. 2002. Towards an ecological understanding of biological nitrogen fixation. *Biogeochemistry* 57:1–45

Vitousek PM, Walker L. 1989. Biological invasion by *Myrica fayan* Hawai'i: plant demography, nitrogen fixation, ecosystem effects. *Ecol. Monogr.* 59:247–65

Wainwright J, Parsons AJ, Abrahams AD. 2000. Plot-scale studies of vegetation, overland flow and erosion interactions: case studies from Arizona and New Mexico. *Hydrol. Process.* 14:2921–43

Waldron L, Dakessian S. 1981. Soil reinforcement by roots: calculation of increased soil shear resistance from root properties. *Soil Sci.* 132:427–35

Wardle D. 2002. *Communities and Ecosystems: Linking the Aboveground and Belowground Components*. Princeton, NJ: Princeton Univ. Press. 392 pp.

Wardle D, Barker G, Bonner K, Nicholson K. 1998. Can comparative approaches based on plant ecophysiological traits predict the nature of biotic interactions and individual plant species in ecosystems? *J. Ecol.* 86:405–20

Wardle D, Nicholson K, Ahmed M, Rahman A. 1994. Interference effects of the invasive plant Carduus nutans L. against the nitrogen fixation ability of Trifolium repens L. *Plant Soil.* 163:287–97

Weathers KC. 1999. The importance of cloud and fog in the maintenance of ecosystems. *Trends Ecol. Evol.* 14:214–15

Weathers KC, Cadenasso ML, Pickett STA. 2001. Forest edges as nutrient and pollutant concentrators: Potential synergisms between fragmentation, forest canopies, and the atmosphere. *Conserv. Biol.* 15:1506–14

Wedin D, Pastor J. 1993. Nitrogen mineralization dynamics in grass monocultures. *Oecologia* 96:186–92

Wedin D, Tieszen L, Dewey B, Pastor J. 1995. Carbon isotope dynamics during grass decomposition and soil organic matter formation. *Ecology* 76:1383–92

Wedin D, Tilman D. 1990. Species effects on nitrogen cycling: a test with perennial grasses. *Oecologia* 84:433–41

Whelan C. 2001. Foliage structure influences foraging of insectivorous forest birds: an experimental study. *Ecology* 82:219–31

Willis K, Braun M, Sumegi P, Toth A. 1997. Does soil change cause vegetation change or vice versa? A temporal perspective from Hungary. *Ecology* 78:740–50

Wilson J. 1999. Guilds, functional types and ecological groups. *Oikos.* 86:507–22

Woldendrop J. 1962. The quantitative influence of the rhizosphere on denitrification. *Plant Soil.* 17:267–70

Wright I, Westoby M. 1999. Differences in seedling growth behavior among species: trait correlations across species, and trait shifts along nutrient compared to rainfall gradients. *J. Ecol.* 87:85–97

Yan F, Zhu YY, Muller C, Zorb C, Schubert S. 2002. Adaptation of H+-pumping and plasma membrane H+ ATPase activity in proteoid roots of white lupin under phosphate deficiency. *Plant Physiol.* 129:50–63

Zaller J, Arnone JI. 1999. Interactions between plant species and earthworm casts in a calcareous grassland under elevated CO_2. *Ecology* 80:873–81

Zavaleta E. 2000. Valuing ecosystem services lost to *Tamarix* invasion in the United States. In *Invasive Species in a Changing World*, ed. H Mooney, R Hobbs, pp. 261–302. Washington, DC: Island

Annu. Rev. Ecol. Evol. Syst. 2003. 34:487–515
doi: 10.1146/annurev.ecolsys.34.011802.132419

First published online as a Review in Advance on August 14, 2003

EFFECTS OF HABITAT FRAGMENTATION ON BIODIVERSITY

Lenore Fahrig
Ottawa-Carleton Institute of Biology, Carleton University, Ottawa, Ontario, Canada K1S 5B6; email: Lenore_Fahrig@carleton.ca

Key Words habitat loss, landscape scale, habitat configuration, patch size, patch isolation, extinction threshold, landscape complementation

■ **Abstract** The literature on effects of habitat fragmentation on biodiversity is huge. It is also very diverse, with different authors measuring fragmentation in different ways and, as a consequence, drawing different conclusions regarding both the magnitude and direction of its effects. Habitat fragmentation is usually defined as a landscape-scale process involving both habitat loss and the breaking apart of habitat. Results of empirical studies of habitat fragmentation are often difficult to interpret because (*a*) many researchers measure fragmentation at the patch scale, not the landscape scale and (*b*) most researchers measure fragmentation in ways that do not distinguish between habitat loss and habitat fragmentation per se, i.e., the breaking apart of habitat after controlling for habitat loss. Empirical studies to date suggest that habitat loss has large, consistently negative effects on biodiversity. Habitat fragmentation per se has much weaker effects on biodiversity that are at least as likely to be positive as negative. Therefore, to correctly interpret the influence of habitat fragmentation on biodiversity, the effects of these two components of fragmentation must be measured independently. More studies of the independent effects of habitat loss and fragmentation per se are needed to determine the factors that lead to positive versus negative effects of fragmentation per se. I suggest that the term "fragmentation" should be reserved for the breaking apart of habitat, independent of habitat loss.

INTRODUCTION

A recent search of the Cambridge Scientific Abstracts database revealed over 1600 articles containing the phrase "habitat fragmentation." The task of reviewing this literature is daunting not only because of its size but also because different authors use different definitions of habitat fragmentation, and they measure fragmentation in different ways and at different spatial scales.

This diversity of definitions of habitat fragmentation can be readily seen in the titles of some articles. For example, "Impacts of habitat fragmentation and patch size..." (Collingham & Huntly 2000) suggests that habitat fragmentation and patch size are two different things. However, other authors actually use patch size to measure habitat fragmentation (e.g., Golden & Crist 2000, Hovel & Lipicus 2001). "The effects of forest fragmentation and isolation..." (Goodman & Rakotodravony 2000) suggests that forest fragmentation and isolation are different, in contrast to authors who use forest isolation as a measure of forest fragmentation (e.g., Mossman & Waser 2001, Rukke 2000). "Effect of land cover, habitat fragmentation, and..." (Laakkonen et al. 2001) contrasts with many authors who equate landscape fragmentation with land cover (e.g., Carlson & Hartman 2001; Fuller 2001; Gibbs 1998, 2001; Golden & Crist 2000; Hargis et al. 1999; Robinson et al. 1995; Summerville & Crist 2001; Virgós 2001). "The influence of forest fragmentation and landscape pattern..." (Hargis et al. 1999) contrasts with researchers who define fragmentation as an aspect of landscape pattern (e.g., Wolff et al. 1997, Trzcinski et al. 1999). As a final example, "Effects of experimental habitat fragmentation and connectivity..." (Ims & Andreassen 1999) suggests that habitat fragmentation and connectivity can be examined independently, whereas some researchers actually define fragmentation as "a disruption in landscape connectivity" (With et al. 1997; see also Young & Jarvis 2001).

My goal in this review is to discuss the information available on the effects of habitat fragmentation on biodiversity. To meet this objective I first need to examine the different ways in which habitat fragmentation is conceptualized and measured. Of course, the concept of biodiversity is probably at least as wide-ranging as the concept of habitat fragmentation. However, I do not deal with the issues surrounding the concept of biodiversity. Instead, I include any ecological response variable that is or can be related to biological diversity (see Table 1).

To determine current usage of the term habitat fragmentation, I conducted a search of the Cambridge Scientific Abstracts (Biological Sciences) database on 11 April 2002 for papers containing either "habitat fragmentation," "forest fragmentation," or "landscape fragmentation" in the title of the paper. I reviewed in detail the most recent 100 resulting papers, irrespective of the journal in which they appeared. I limited this search to papers containing "fragmentation" in the title to ensure that my sample included only papers that are directly on the subject of habitat fragmentation. The results are summarized in Table 1.

I then surveyed the broader ecological literature to ask the following: How strong are the effects of habitat fragmentation on biodiversity, and are the effects negative or positive? Habitat fragmentation is generally thought to have a large, negative effect on biodiversity and is therefore widely viewed as an aspect of habitat degradation (Haila 2002). However, as I show, this conclusion is generally valid only for conceptualizations of fragmentation that are inseparable from habitat loss. Other ways of conceptualizing habitat fragmentation lead to other conclusions. I end the paper with recommendations.

TABLE 1 Summary of 100 recent fragmentation studies*

Fragmentation (predictor) variables	Biodiversity (response) variables									
	Abundance/ density (35)	Richness/ diversity (28)	Presence/ absence (26)	Fitness measures (15)	Genetic variability (12)	Species interactions (10)	Extinction/ turnover (8)	Individual habitat use (5)	Movement/ dispersal (4)	Population growth (3)
Patch size[a] (63)	26	21	20	11	3	7	3	3	3	3
Habitat loss/amount (60)	21	17	13	9	8	5	5	3	2	1
Patch isolation[a] (35)	14	7	11	2	6	3	0	0	1	0
Edge[a] (22)	11	5	3	2	0	4	1	2	0	1
Number of patches (10)	2	1	0	2	0	2	3	1	0	1
Structural connectivity[b] (8)	3	1	1	2	0	2	1	0	1	1
Matrix quality (7)	3	2	1	1	2	0	0	2	0	0
Patch shape[a] (4)	0	1	2	0	0	1	0	0	0	0
Qualitative only (28)	13	9	7	10	4	0	1	3	0	1
Patch scale[c] (42)	17	14	16	6	7	4	0	2	1	1
Landscape scale[d] (37)	7	7	4	4	3	3	8	2	3	1
Patch and landscape scales (21)	10	6	6	5	2	3	0	1	0	1

[a]Predictor variables that can be measured at either the patch scale (individually for each patch) or at the landscape scale (averaged or summed across all patches in the landscape).

[b]Includes both connectivity studies and corridor studies.

[c]Each data point in the analysis represents information from a single patch.

[d]Each data point in the analysis represents information from a single landscape.

*Table entries are the numbers of papers that studied the given combination of predictor (fragmentation) variable or scale and response (biodiversity) variable. Numbers in parentheses after variable names are the total number of papers (of 100) using that variable. Columns and rows do not add to 100 because each study may contain more than one fragmentation variable and more than one biodiversity variable.

CONCEPTUALIZATION AND MEASUREMENT OF HABITAT FRAGMENTATION

Fragmentation as Process

Habitat fragmentation is often defined as a process during which "a large expanse of habitat is transformed into a number of smaller patches of smaller total area, isolated from each other by a matrix of habitats unlike the original" (Wilcove et al. 1986) (Figure 1). By this definition, a landscape can be qualitatively categorized as either continuous (containing continuous habitat) or fragmented, where the fragmented landscape represents the endpoint of the process of fragmentation.

Many studies of the effect of habitat fragmentation on biodiversity conform to this definition by comparing some aspect(s) of biodiversity at "reference" sites within a continuous landscape to the same aspect(s) of biodiversity at sites within a fragmented landscape (e.g., Bowers & Dooley 1999, Cascante et al. 2002, Diaz et al. 2000, Groppe et al. 2001, Laurance et al. 2001, Mac Nally & Brown 2001, Mahan & Yahner 1999, Morato 2001, Mossman & Waser 2001, Renjifo 1999, Walters et al. 1999). From my sample of 100 recent studies, 28% conducted such comparisons of continuous versus fragmented landscapes (Table 1). In these studies, the continuous landscape represents a landscape before fragmentation (time 1 in Figure 1) and the fragmented landscape represents a landscape following fragmentation (time 2 or time 3 in Figure 1).

Although this approach conforms to the definition of fragmentation as a process, it has two inherent weaknesses. First, because habitat fragmentation is a landscape-scale process (McGarigal & Cushman 2002), the sample size in such studies, for questions about the effects of habitat fragmentation on biodiversity, is typically

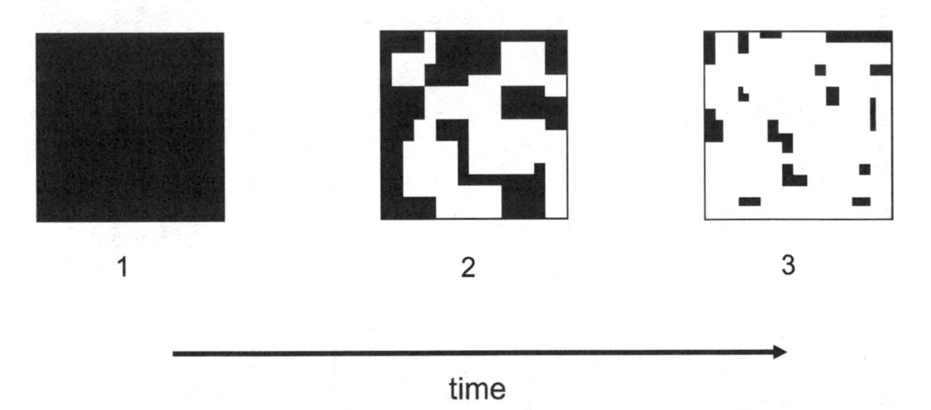

Figure 1 The process of habitat fragmentation, where "a large expanse of habitat is transformed into a number of smaller patches of smaller total area, isolated from each other by a matrix of habitats unlike the original" (Wilcove et al. 1986). Black areas represent habitat and white areas represent matrix.

only two, i.e., one continuous landscape and one fragmented landscape. With such a design, inferences about the effects of fragmentation are weak. Apparent effects of fragmentation could easily be due to other differences between the landscapes. For example, Mac Nally et al. (2000) found consistent vegetation differences between fragments and reference sites and concluded that apparent effects of fragmentation on birds could be due to preexisting habitat differences between the two landscapes.

Second, this characterization of habitat fragmentation is strictly qualitative, i.e., each landscape can be in only one of two states, continuous or fragmented. This design does not permit one to study the relationship between the degree of habitat fragmentation and the magnitude of the biodiversity response. Quantifying the degree of fragmentation requires measuring the pattern of habitat on the landscape. The diversity of approaches in the fragmentation literature arises mainly from differences among researchers in how they quantify habitat fragmentation. These differences have significant implications for conclusions about the effects of fragmentation on biodiversity.

Fragmentation as Pattern: Quantitative Conceptualizations

The definition of habitat fragmentation above implies four effects of the process of fragmentation on habitat pattern: (*a*) reduction in habitat amount, (*b*) increase in number of habitat patches, (*c*) decrease in sizes of habitat patches, and (*d*) increase in isolation of patches. These four effects form the basis of most quantitative measures of habitat fragmentation. However, fragmentation measures vary widely; some include only one effect (e.g., reduced habitat amount or reduced patch sizes), whereas others include two or three effects but not all four.

Does it matter which fragmentation measure a researcher uses? The answer depends on whether the different effects of the process of fragmentation on habitat pattern have the same effects on biodiversity. If they do, we can draw general conclusions about the effects of fragmentation on biodiversity even though the different studies making up the fragmentation literature measure fragmentation in different ways. As I show in Effects of Habitat Fragmentation on Biodiversity, the different effects of the process of fragmentation on habitat pattern do not affect biodiversity in the same way. This has led to apparently contradictory conclusions about the effects of fragmentation on biodiversity. In this section, I review quantitative conceptualizations of habitat fragmentation. This is an important step toward reconciling these apparently contradictory results.

FRAGMENTATION AS HABITAT LOSS The most obvious effect of the process of fragmentation is the removal of habitat (Figure 1). This has led many researchers to measure the degree of habitat fragmentation as simply the amount of habitat remaining on the landscape (e.g., Carlson & Hartman 2001, Fuller 2001, Golden & Crist 2000, Hargis et al. 1999, Robinson et al. 1995, Summerville & Crist 2001, Virgós 2001). If we can measure the level of fragmentation as the amount of habitat, why do we call it "fragmentation"? Why not simply call it habitat loss? The reason is that when ecologists think of fragmentation, the word invokes more than habitat removal: "fragmentation ... not only causes loss of the amount of habitat, but by creating small, isolated patches it also changes the properties of the remaining habitat" (van den Berg et al. 2001).

Habitat can be removed from a landscape in many different ways, resulting in many different spatial patterns (Figure 2). Do some patterns represent a higher degree of fragmentation than others, and does this have implications for biodiversity? If the answer to either of these questions is "no," then the concept of fragmentation is redundant with habitat loss. The assertion that habitat fragmentation means something more than habitat loss depends on the existence of effects of fragmentation on biodiversity that can be attributed to changes in the pattern of habitat that are independent of habitat loss. Therefore, many researchers define habitat fragmentation as an aspect of habitat configuration.

FRAGMENTATION AS A CHANGE IN HABITAT CONFIGURATION In addition to loss of habitat, the process of habitat fragmentation results in three other effects: increase in number of patches, decrease in patch sizes, and increase in isolation of patches. Measures of fragmentation that go beyond simply habitat amount are generally derived from these or other strongly related measures (e.g., amount of edge). There are at least 40 such measures of fragmentation (McGarigal et al. 2002), many of which typically have strong relationships with the amount of habitat as well as with each other (Bélisle et al. 2001, Boulinier et al. 2001, Drolet et al. 1999, Gustafson 1998, Haines-Young & Chopping 1996, Hargis et al. 1998, Robinson et al. 1995, Schumaker 1996, Trzcinski et al. 1999, Wickham et al. 1999) (Figure 3).

The interrelationships among measures of fragmentation are not widely recognized in the current fragmentation literature. Most researchers do not separate the effects of habitat loss from the configurational effects of fragmentation. This leads to ambiguous conclusions regarding the effects of habitat configuration on biodiversity (e.g., Summerville & Crist 2001, Swenson & Franklin 2000). It is also common for fragmentation studies to report individual effects of fragmentation measures without reporting the relationships among them, which again makes the results difficult to interpret.

THE PATCH-SCALE PROBLEM Similar problems arise when fragmentation is measured at the patch scale rather than the landscape scale. Because fragmentation is a landscape-scale process (Figure 1), fragmentation measurements are correctly made at the landscape scale (McGarigal & Cushman 2002). As pointed out by Delin & Andrén (1999), when a study is at the patch scale, the sample size at the landscape scale is only one, which means that landscape-scale inference is not possible (Figure 4; see Brennan et al. 2002, Tischendorf & Fahrig 2000). However, in approximately 42% of recent fragmentation studies, individual data points represent measurements on individual patches, not landscapes (Table 1). Similarly, using a different sample of the literature, McGarigal & Cushman (2002)

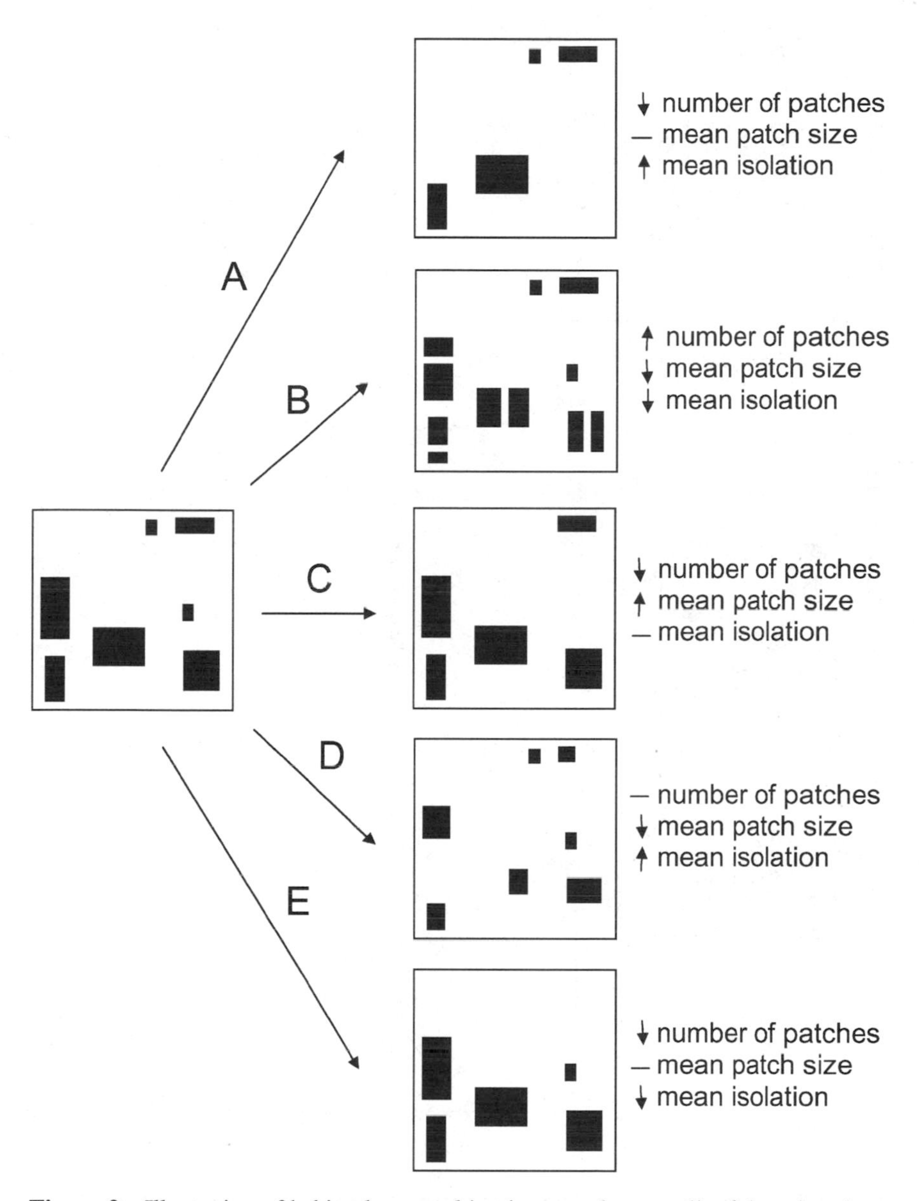

Figure 2 Illustration of habitat loss resulting in some, but not all, of the other three expected effects of habitat fragmentation on landscape pattern. Expected effects are (*a*) an increase in the number of patches, (*b*) a decrease in mean patch size, and (*c*) an increase in mean patch isolation (nearest neighbor distance). Actual changes are indicated by arrows.

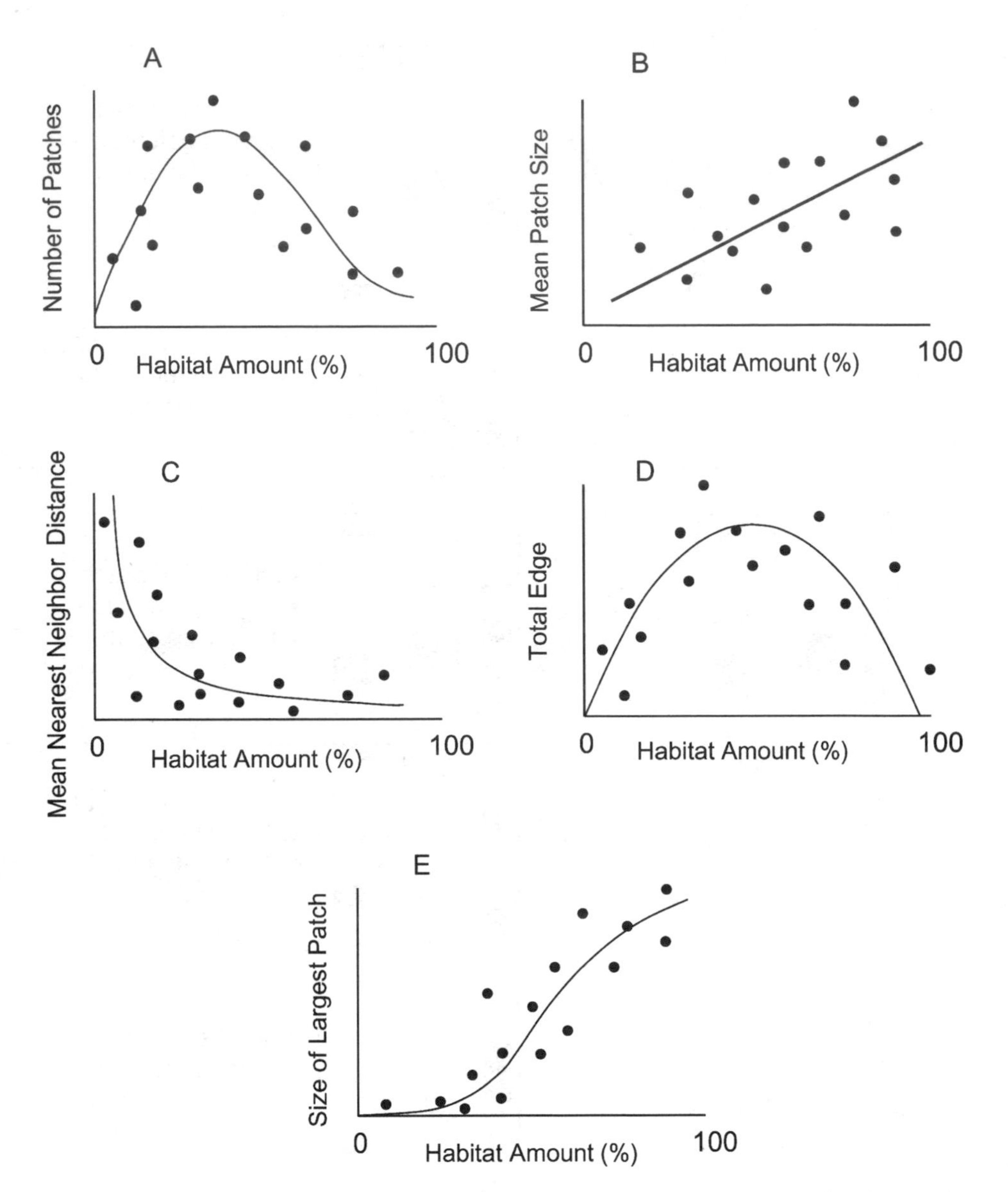

Figure 3 Illustration of the typical relationships between habitat amount and various measures of fragmentation. Individual data points correspond to individual landscapes. Based on relationships in Bélisle et al. (2001), Boulinier et al. (2001), Drolet et al. (1999), Gustafson (1998), Haines-Young & Chopping (1996), Hargis et al. (1998), Robinson et al. (1995), Schumaker (1996), Trzcinski et al. (1999), and Wickham et al. (1999).

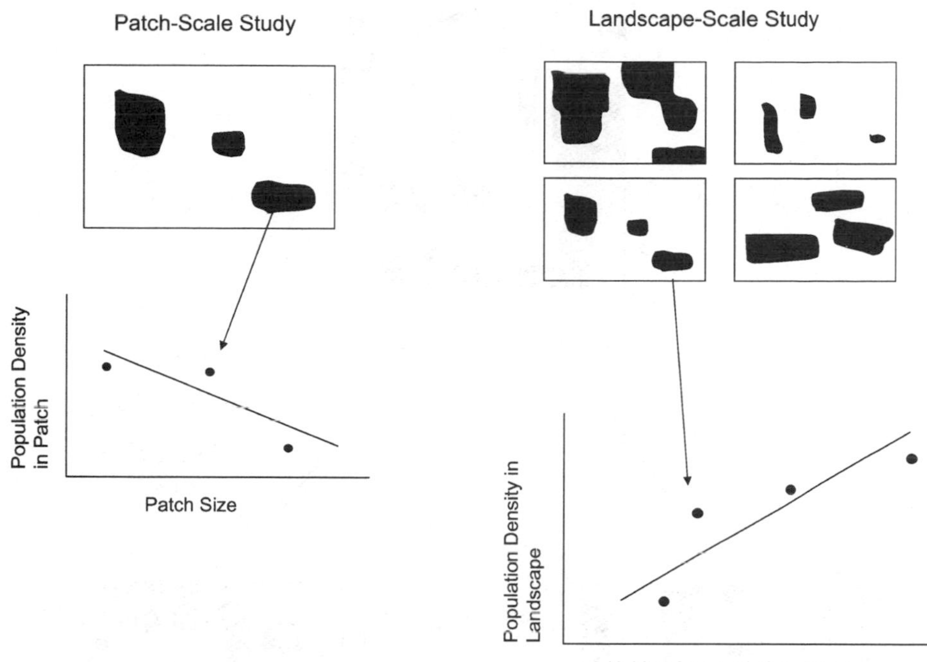

Figure 4 *(A)* Patch-scale study. Each observation represents the information from a single patch. Only one landscape is studied, so sample size for landscape-scale inferences is one. (*B*) Landscape-scale study. Each observation represents the information from a single landscape. Multiple landscapes, with different structures, are studied. Here, sample size for landscape-scale inferences is four.

estimated that more than 57% of all fragmentation studies are at the patch scale. Some researchers even refer to patch-scale measures as landscape features (e.g., Fernandez-Juricic 2000, Schweiger et al. 2000).

Patch size: an ambiguous measure of fragmentation The relationship between patch size and fragmentation is ambiguous because both habitat loss and habitat fragmentation per se (i.e., the breaking apart of habitat, controlling for changes in habitat amount) result in smaller patches (Figure 5). Using patch size as a measure of habitat fragmentation per se implicitly assumes that patch size is independent of habitat amount at the landscape scale (e.g., Niemelä 2001). However, regions where patches are large often correspond to regions where there is more habitat (Fernandez-Juricic 2000, McCoy & Mushinsky 1999) (Figure 6). Ignoring potential relationships between a patch-scale measure (e.g., patch size) and landscape-scale habitat amount does not control for this relationship; it can lead to misinterpretation of results.

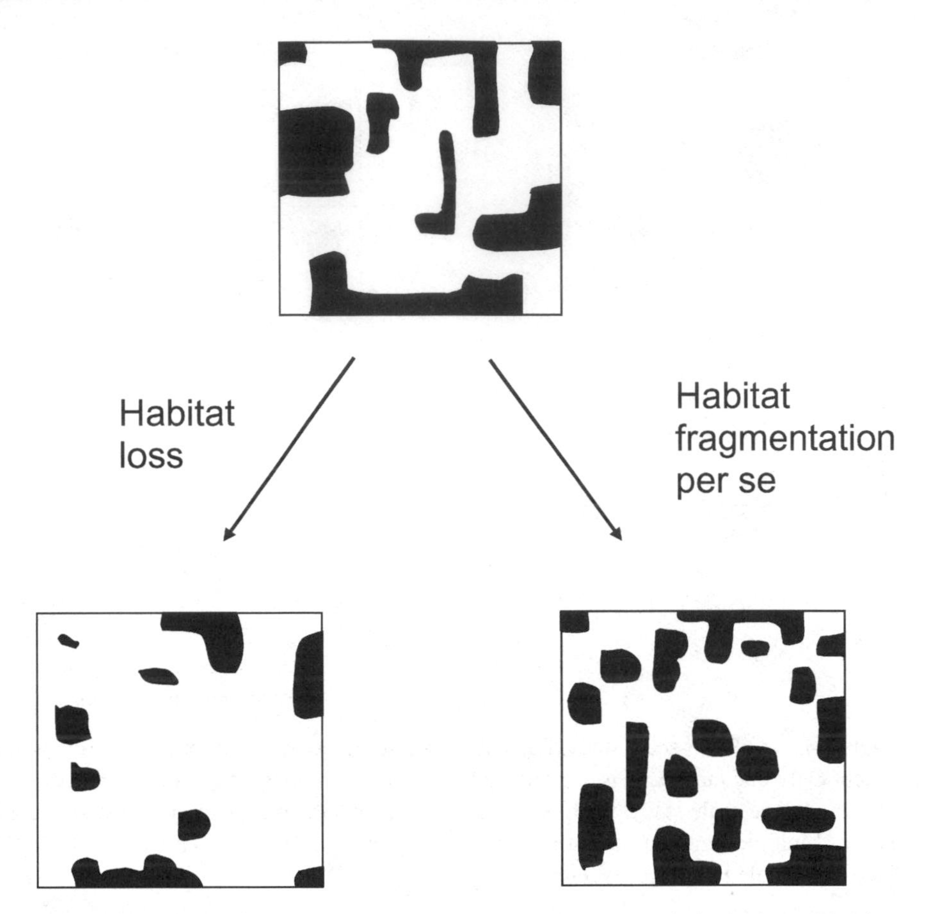

Figure 5 Both habitat loss and habitat fragmentation per se (independent of habitat loss) result in smaller patches. Therefore, patch size itself is ambiguous as a measure of either habitat amount or habitat fragmentation per se. Note also that habitat fragmentation per se leads to reduced patch isolation.

Patch isolation: a measure of habitat amount In the fragmentation literature, patch isolation is almost universally interpreted as a measure of habitat configuration. However, patch isolation is more accurately viewed as a measure of the lack of habitat in the landscape surrounding the patch. The more isolated a patch is, generally speaking, the less habitat there is in the landscape that surrounds it (Figure 7). Therefore, when translated to the landscape scale, isolation of a patch is a measure of habitat amount in the landscape, not configuration of the landscape.

Bender et al. (2003) reviewed measures of patch isolation. All measures are strongly negatively related to habitat amount in the surrounding landscape. The most common measure of patch isolation is the distance to the next-nearest patch,

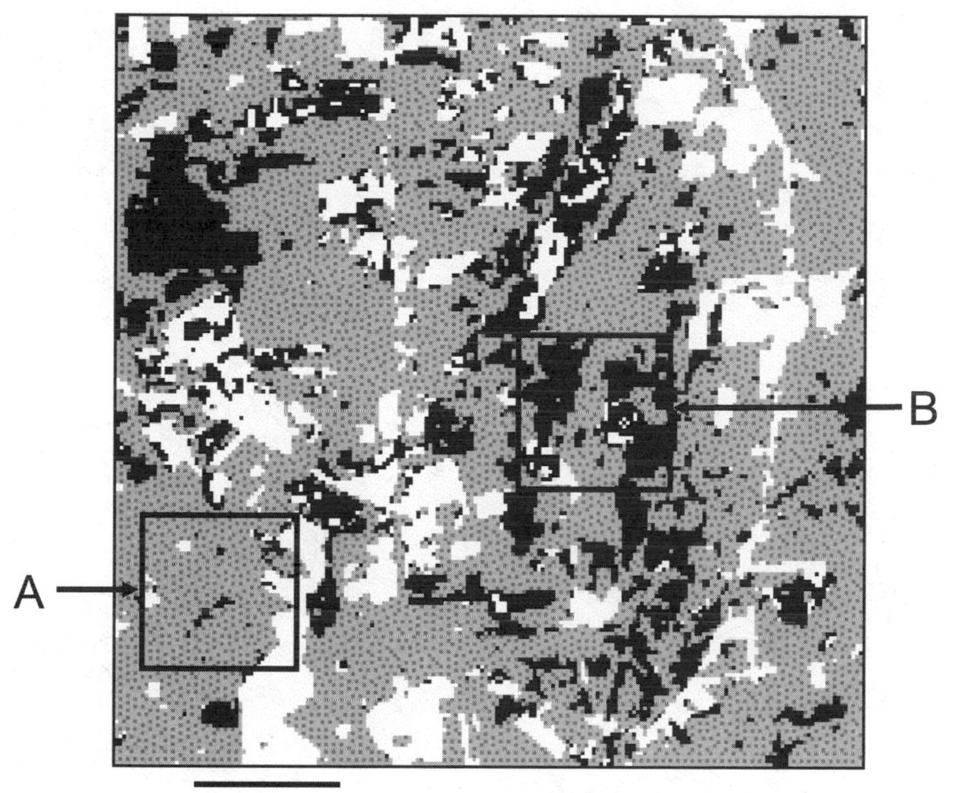

Figure 6 Landscape in southern Ontario (from Tischendorf 2001) showing that regions where forest patches (black areas) are small typically correspond to regions where there is little forest. Compare (*A*) and (*B*), where (*A*) has small patches and less than 5% forest and (*B*) has larger patches and approximately 50% forest.

or "nearest-neighbor distance" (e.g., Delin & Andrén 1999, Haig et al. 2000, Hargis et al. 1999). Patches with small nearest-neighbor distances are typically situated in landscapes containing more habitat than are patches with large nearest-neighbor distances (Figure 7), so in most situations this measure of isolation is related to habitat amount in the landscape. Another common measure of patch isolation is the inverse of the amount of habitat within some distance of the patch in question (e.g., Kinnunen et al. 1996, Magura et al. 2001, Miyashita et al. 1998). In other words, patch isolation is measured as habitat amount at the landscape scale. All other measures of patch isolation are a combination of distances to other patches and sizes of those patches (or the populations they contain) in the surrounding landscape (reviewed in Bender et al. 2003). As such they are all measures of the amount of habitat in the surrounding landscape.

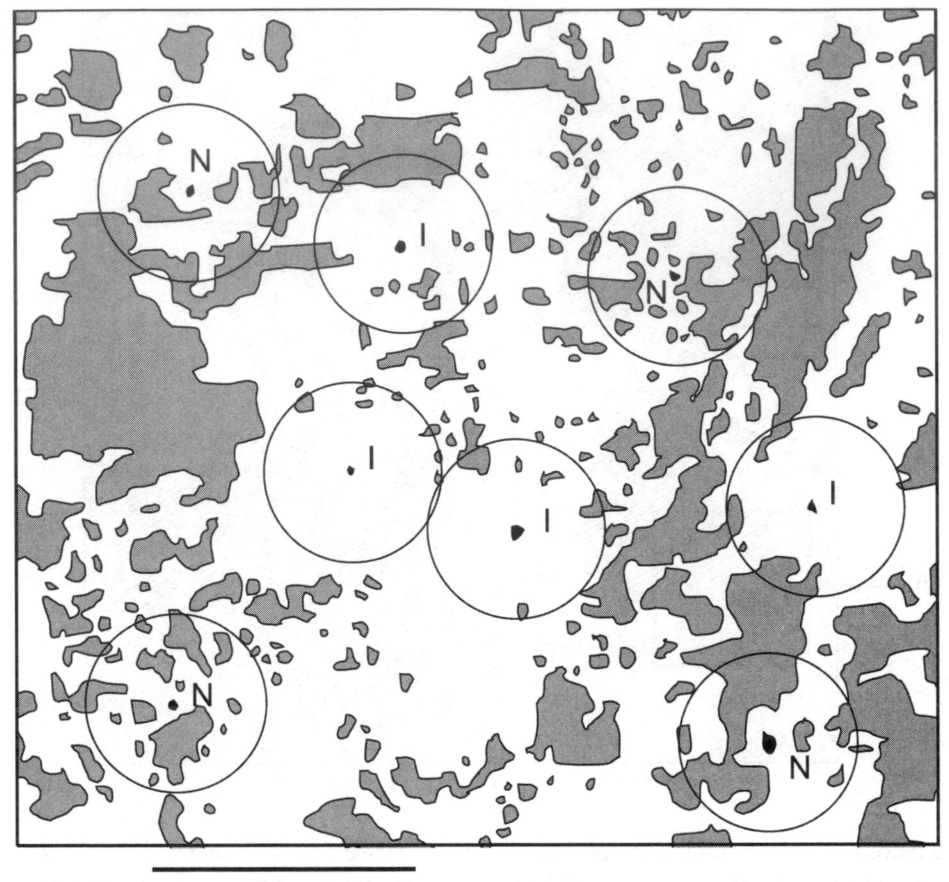

Figure 7 Illustration of the relationship between patch isolation and amount of habitat in the landscape immediately surrounding the patch. Gray areas are forest. Isolated patches (*black patches labeled "I"*) are situated in landscapes (*circles*) containing less forest than are less isolated patches (*black patches labeled "N"*).

MEASURING HABITAT FRAGMENTATION PER SE How can we measure habitat fragmentation independent of habitat amount? Some researchers have constructed landscapes in which they experimentally controlled habitat amount while varying habitat fragmentation per se (e.g., Caley et al. 2001, Collins & Barrett 1997). Researchers studying real landscapes have used statistical methods to control for habitat amount. For example, McGarigal and McComb (1995) measured 25 landscape indices for each of 30 landscapes. They statistically corrected each index for its relationship to habitat amount and then entered the corrected variables into a PCA. Each axis of the resulting PCA represented a different component of landscape configuration. In a similar approach, Villard et al. (1999) measured the number of forest patches, total length of edge, mean nearest-neighbor distance, and percent of forest cover on each of 33 landscapes. They used the residuals of the statistical models relating each of the first three variables to forest amount as measures of fragmentation that have been controlled for their relationships to habitat amount.

EFFECTS OF HABITAT FRAGMENTATION ON BIODIVERSITY

In this section I review the empirical evidence for effects of habitat fragmentation on biodiversity. This review is not limited to the 100 papers summarized in Table 1. The fragmentation literature can be distilled into two major effects: the generally strong negative effect of habitat loss on biodiversity, and the much weaker, positive or negative effect of fragmentation per se on biodiversity. Because the effect of fragmentation per se is weaker than the effect of habitat loss, to detect the effect of fragmentation per se, the effect of habitat loss must be experimentally or statistically controlled.

Effects of Habitat Loss on Biodiversity

Habitat loss has large, consistently negative effects on biodiversity, so researchers who conceptualize and measure fragmentation as equivalent to habitat loss typically conclude that fragmentation has large negative effects. The negative effects of habitat loss apply not only to direct measures of biodiversity such as species richness (Findlay & Houlahan 1997, Gurd et al. 2001, Schmiegelow & Mönkkönen 2002, Steffan-Dewenter et al. 2002, Wettstein & Schmid 1999), population abundance and distribution (Best et al. 2001, Gibbs 1998, Guthery et al. 2001, Hanski et al. 1996, Hargis et al. 1999, Hinsley et al. 1995, Lande 1987, Sánchez-Zapata & Calvo 1999, Venier & Fahrig 1996) and genetic diversity (Gibbs 2001), but also to indirect measures of biodiversity and factors affecting biodiversity. A model by Bascompte et al. (2002) predicts a negative effect of habitat loss on population growth rate. This is supported by Donovan & Flather (2002), who found that species showing declining trends in global abundance are more likely to occur in areas with high habitat loss than are species with increasing or stable trends. Habitat loss has been shown to reduce trophic chain length (Komonen et al. 2000), to alter species interactions (Taylor & Merriam 1995), and to reduce the number of specialist, large-bodied species (Gibbs & Stanton 2001). Habitat loss also negatively affects breeding success (Kurki et al. 2000), dispersal success (Bélisle et al. 2001, Pither & Taylor 1998, With & Crist 1995, With & King 1999), predation rate (Bergin et al. 2000, Hartley & Hunter 1998), and aspects of animal behavior that affect foraging success rate (Mahan & Yahner 1999).

INDIRECT EVIDENCE OF EFFECTS OF HABITAT LOSS Negative effects of habitat loss on biodiversity are also evident from studies that measure habitat amount indirectly, using measures that are highly correlated with habitat amount. For example, Robinson et al. (1995) found that reproductive success of forest nesting bird species was positively correlated with percentage of forest cover, percentage of forest interior, and average patch size in a landscape. Because the latter two variables were highly correlated with percentage of forest cover, these all represent positive effects of habitat amount on reproductive success. Boulinier et al. (2001) found effects of mean patch size on species richness, local extinction rate, and turnover rate of forest birds in 214 landscapes. Because mean patch size had a 0.94 correlation with forest amount in their study, this result most likely represents an effect of habitat amount.

Patch isolation effects Patch isolation is a measure of the lack of habitat in the landscape surrounding the patch (Figure 7). Therefore, the many studies that have shown negative effects of patch isolation on species richness or presence/absence represent further evidence for the strong negative effect of landscape-scale habitat loss on biodiversity (e.g., McCoy & Mushinsky 1999, Rukke 2000, Virgós 2001).

Bender et al. (2003) and Tischendorf et al. (2003) conducted simulation analyses to determine which patch isolation measures are most strongly related to movement of animals between patches. They found that the "buffer" measures, i.e., amount of habitat within a given buffer around the patch, were best. This suggests a strong effect of habitat amount on interpatch movement. It also suggests, again, that effects of patch isolation and landscape-scale habitat amount are equivalent.

Patch size effects Individual species have minimum patch size requirements (e.g., Diaz et al. 2000). Therefore, smaller patches generally contain fewer species than larger patches (Debinski & Holt 2000), and the set of species on smaller patches is often a more-or-less predictable subset of the species on larger patches (e.g., Ganzhorn & Eisenbeiß 2001, Kolozsvary & Swihart 1999, Vallan 2000). Similarly, the amount of habitat on a landscape required for species occurrence there differs among species (Gibbs 1998, Vance et al. 2003), so landscapes with less habitat should contain a subset of the species found in landscapes with more habitat.

Despite this apparent correspondence between patch- and landscape-scale effects, the landscape-scale interpretation of patch size effects depends on the landscape context of the patch. For example, Donovan et al. (1995) found that forest birds had lower reproductive rates in small patches than in large patches. If small patches occur in areas with less forest, the reduced reproductive rate may not be the result of patch size, but may result from larger populations of nest predators and brood parasites that occur in landscapes with more open habitat (Hartley & Hunter 1998, Robinson et al. 1995, Schmiegelow & Mönkkönen 2002).

EXTINCTION THRESHOLD The number of individuals of any species that a landscape can support should be a positive function of the amount of habitat available to

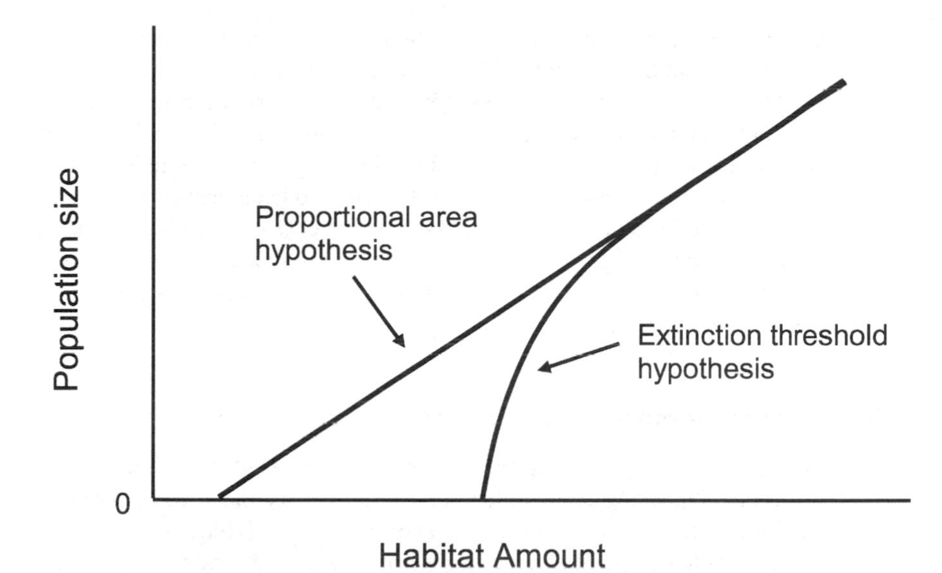

Figure 8 Illustration of the extinction threshold hypothesis in comparison to the proportional area hypothesis.

that species in the landscape. However, several theoretical studies suggest that the relationship is not proportional; they predict a threshold habitat level below which the population cannot sustain itself, termed the extinction threshold (Bascompte & Solé 1996, Boswell et al. 1998, Fahrig 2001, Flather & Bevers 2002, Hill & Caswell 1999, Lande 1987, With & King 1999; Figure 8). There have been very few direct empirical tests of the extinction threshold hypothesis (but see Jansson & Angelstam 1999).

Note that the predicted occurrence of the extinction threshold results from habitat loss, not habitat fragmentation per se. Theoretical studies suggest that habitat fragmentation per se can affect where the extinction threshold occurs on the habitat amount axis. Also, the effects of habitat fragmentation per se are predicted to increase below some level of habitat loss (see The 20–30% Threshold, below). However, the occurrence of the extinction threshold is a response to habitat loss, not fragmentation per se. This has led to some ambiguity in interpretation of empirical literature. For example, Virgós (2001) found that patch isolation affects badger density only for patches in landscapes with <20% forest cover. As explained above, patch isolation is typically an index of habitat amount at the landscape scale. Therefore, this result probably suggests a threshold effect of forest loss on badger density. This conclusion is different from that of the author, who interpreted the isolation effect as an effect of habitat configuration. The interpretation is ambiguous because the relationship between habitat amount and patch isolation was not statistically controlled in this study. Similarly, Andrén (1994) reviewed patch size and patch isolation effects on population density and concluded that these effects increase below a threshold amount of habitat in the landscape. Because patch size and isolation can be indicators of habitat amount at a landscape scale (see Patch size: An Ambiguous Measure of Fragmentation and Patch Isolation: A Measure of Habitat Amount, above), this result could be interpreted as an intensification of the effects of habitat loss at low habitat levels, i.e., it supports the extinction threshold hypothesis (Figure 8). This result has also been viewed as evidence for configuration effects below a threshold habitat level (e.g., Flather & Bevers 2002, Villard et al. 1999). Again, the interpretation is ambiguous because the relationships between patch size and isolation and amount of habitat surrounding each patch were not controlled for.

Effects of Habitat Fragmentation per se on Biodiversity

In this section I review the empirical evidence for fragmentation effects per se, i.e., for effects of "breaking apart" of habitat on biodiversity, that are independent of or in addition to the effects of habitat loss. The 17 studies in Table 2 represent all of the empirical studies of fragmentation per se of which I am aware. Some theoretical studies suggest that the effect of habitat fragmentation per se is weak relative to the effect of habitat loss (Collingham & Huntley 2000, Fahrig 1997, Flather & Bevers 2002, Henein et al. 1998), although other modeling studies predict much larger effects of fragmentation per se (Boswell et al. 1998, Burkey 1999, Hill & Caswell 1999, Urban & Keitt 2001, With & King 1999; reviewed in Fahrig 2002). All these recent models predict negative effects of habitat fragmentation per se, in contrast with some earlier theoretical work (see Reasons for Positive Effects of Fragmentation, below). The empirical evidence to date suggests that the effects of fragmentation per se are generally much weaker than the effects of habitat loss. Unlike the effects of habitat loss, and in contrast to current theory, empirical studies suggest that the effects of fragmentation per se are at least as likely to be positive as negative.

The 17 empirical studies on the effects of habitat fragmentation per se (Table 2) range from small-scale experimental studies to continental-scale analyses. They cover a range of response variables, including abundance, density, distribution, reproduction, movement, and species richness. About half of the studies are on forest birds; other taxa include insects, small mammals, plants, aquatic invertebrates, and a virus, and other habitats include grasslands, cropland, a coral reef, and an estuary.

The 17 studies used a variety of approaches for estimating the effect of fragmentation per se. In five of them, experimental landscapes were constructed to independently control the levels of habitat amount and fragmentation per se. Four of these varied both habitat amount and fragmentation per se, and one varied only fragmentation, holding the amount of habitat constant. Three of the 12 studies in real landscapes compared the response variable in one large patch versus several small patches (i.e., holding habitat amount constant). In the remaining nine studies in real landscapes, the effect of fragmentation per se was estimated by statistically controlling for the effect of habitat amount.

TABLE 2 Summary of empirical studies that examined the effects of habitat fragmentation per se, i.e., controlling for effects of habitat amount on biodiversity

Study	Taxa and response variable(s)	Relative effects of habitat loss versus habitat fragmentation per se	Direction of effect(s) of fragmentation per se on biodiversity
Studies in real landscapes			
Middleton & Merriam 1983	111 forest taxa (various): distribution	n.a.[a]	No effect
McGarigal & McComb 1995	15 late-seral forest bird species: abundance	Amount ≫ fragmentation	6 positive, 1 negative
Meyer et al. 1998	Northern spotted owl: presence/absence, persistence, reproduction	Amount ≫ fragmentation	No effect
Rosenberg et al. 1999	6 Tanager species and populations: presence/absence	Amount ≫ fragmentation	2 negative
Trzcinski et al. 1999	31 forest bird species: presence/absence	Amount ≫ fragmentation	2 positive, 4 negative
Drolet et al. 1999	14 forest bird species: presence/absence	Amount ≫ fragmentation	No effect
Villard et al. 1999	15 forest bird species: presence/absence	Amount ≅ fragmentation	4 positive, 2 negative
Bélisle et al. 2001	3 forest bird species: homing time and homing success	Amount ≫ fragmentation	Positive
Langlois et al. 2001	Hanta virus: incidence	Amount ≫ fragmentation	Positive

(*Continued*)

TABLE 2 (*Continued*)

Study	Taxa and response variable(s)	Relative effects of habitat loss versus habitat fragmentation per se	Direction of effect(s) of fragmentation per se on biodiversity
Hovel & Lipcius 2001	Blue crab: juvenile survival, adult (predator) density	n.a.[a]	1 positive (juvenile survival), 1 negative (adult density)
Tscharntke et al. 2002	Butterflies: species richness, endangered species richness	n.a.[a]	2 positive (total and endangered species richness)
Flather et al. 1999	Forest birds: abundance	Amount $\gg$ fragmentation	No effect
Studies in experimental landscapes:			
Collins & Barrett 1997	Meadow vole: density	n.a.[a]	Positive
Wolff et al. 1997	Gray-tailed vole: abundance, density, reproductive rate, recruitment	Fragmentation > amount	Positive
Collinge & Forman 1998	Grassland insects: abundance, species richness	Not stated	Positive
Caley et al. 2001	8 coral commensals: species richness and abundance	Amount $\gg$ fragmentation	1 positive
With et al. 2002	Clover insects: spatial aggregation	Amount $\gg$ fragmentation	n.a.[b]

[a]Habitat amount was held constant; only fragmentation was varied.

[b]Response variable was spatial distribution of the insects.

The overall result from these studies is that habitat loss has a much larger effect than habitat fragmentation per se on biodiversity measures (Table 2). When fragmentation per se did have an effect, it was at least as likely to be positive as negative (Table 2). Given the relatively small number of studies and the large variation in conditions among studies, it is not possible to tease apart the factors that lead to positive versus negative effects of fragmentation per se. However, the positive effects of fragmentation can not be explained as merely responses by "weedy," habitat generalist species. For example, the results reported from McGarigal & McComb (1995) are specifically limited to late-seral forest species, and Tscharntke et al. (2002) found a positive effect of fragmentation per se on butterfly species richness, even when they only included endangered butterfly species.

THE 20–30% THRESHOLD Some theoretical studies suggest that the effects of fragmentation per se should become apparent only at low levels of habitat amount, below approximately 20–30% habitat on the landscape (Fahrig 1998, Flather & Bevers 2002). To date, there is no convincing empirical evidence for this prediction. If the threshold does occur, it should result in a statistical interaction effect between habitat amount and habitat fragmentation per se; such an interaction would indicate that the effect of fragmentation per se depends on the amount of habitat in the landscape. Trzcinski et al. (1999) tested for this interaction effect but found no evidence for it. The hypothesis that fragmentation effects increase below a threshold of habitat amount has not yet been adequately tested.

REASONS FOR NEGATIVE EFFECTS OF FRAGMENTATION PER SE Negative effects of fragmentation are likely due to two main causes. First, fragmentation per se implies a larger number of smaller patches. At some point, each patch of habitat will be too small to sustain a local population or perhaps even an individual territory. Species that are unable to cross the nonhabitat portion of the landscape (the "matrix") will be confined to a large number of too-small patches, ultimately reducing the overall population size and probability of persistence.

The second main cause of negative effects of fragmentation per se is negative edge effects; more fragmented landscapes contain more edge for a given amount of habitat. This can increase the probability of individuals leaving the habitat and entering the matrix. Overall the amount of time spent in the matrix will be larger in a more fragmented landscape, which may increase overall mortality rate and reduce overall reproductive rate of the population (Fahrig 2002). In addition, there are negative edge effects due to species interactions. Probably the most extensively studied of these is increased predation on forest birds at forest edges (Chalfoun et al. 2002).

REASONS FOR POSITIVE EFFECTS OF FRAGMENTATION PER SE More than half of the effects of fragmentation per se that have been documented are positive (Table 2). Some readers will find this surprising, probably because habitat loss is inextricably included within their conceptualization of habitat fragmentation. In this case even if fragmentation per se has a positive effect on biodiversity, this effect will be masked by the large negative effect of habitat loss.

Haila (2002) describes how the current concept of habitat fragmentation emerged from the theory of island biogeography (MacArthur & Wilson 1967). The two predictor variables in this theory are island size and island isolation, or distance of the island from the mainland. When this theory was conceptually extended from island archipelagos to terrestrial systems of habitat patches, the concept of isolation changed; isolation was now the result of habitat loss, and it represented the distance from a patch to its neighbor(s), not the distance to a mainland. Because of its roots in island biogeography, isolation was viewed as representing habitat subdivision even though it was inextricably linked to habitat loss.

However, a parallel research stream, which arose independently of the theory of island biogeography, suggested that habitat fragmentation could have positive effects on biodiversity. Huffaker's (1958) experiment suggested that subdivision of the same amount of habitat into many smaller pieces can enhance the persistence of a predator-prey system. He hypothesized that habitat subdivision provides temporary refugia for the prey species, where they can increase in numbers and disperse elsewhere before the predator or parasite finds them. The plausibility of this mechanism was supported by early theoretical studies (Hastings 1977, Vandermeer 1973). Early theoretical studies also suggested that habitat fragmentation enhances the stability of two-species competition (Levin 1974, Shmida & Ellner 1984, Slatkin 1974), and in an empirical study, Atkinson & Shorrocks (1981) found that coexistence of two competing species could be extended by dividing the habitat into more, smaller patches. Enhanced coexistence resulted from a trade-off between dispersal rate and competitive ability. This trade-off, along with asynchronous disturbances that locally removed the superior competitor, allowed the inferior competitor (but superior disperser) to colonize the empty patches first, before being later displaced by the superior competitor (Chesson 1985). Other researchers suggested that habitat subdivision could even stabilize single-species population dynamics when local disturbances are asynchronous by reducing the probability of simultaneous extinction of the whole population (den Boer 1981; Reddingius & den Boer 1970; Roff 1974a,b).

Why has this early work, suggesting positive effects of habitat fragmentation per se, been largely ignored in the more recent habitat fragmentation literature? One reason is that later theoretical and empirical studies (reviewed in Kareiva 1990) demonstrated that the predicted positive effects of fragmentation per se depend strongly on particular assumptions about the relative movement rates of predator versus prey (or host versus parasite), the trade-off between competitive ability and movement rate, and the asynchrony of disturbances. It seems that the sensitivity to these assumptions, along with the misrepresentation of patch isolation as a measure of habitat subdivision, led researchers to ignore the possibility that fragmentation per se could have a positive effect on biodiversity.

There are at least four additional possible reasons for positive effects of habitat fragmentation per se on biodiversity. First, Bowman et al. (2002) argued that, for many species, immigration rate is a function of the linear dimension of a habitat patch rather than the area of the patch. For these species, overall immigration rate should be higher when the landscape is comprised of a larger number of smaller patches (higher fragmentation per se) than when it is comprised of a smaller number of larger patches. In situations where immigration is an important determinant of population density, this could result in a positive effect of fragmentation per se on density.

Second, if habitat amount is held constant, increasing fragmentation per se actually implies smaller distances between patches (Figure 5). Therefore, a positive effect of fragmentation per se could be due to a reduction in patch isolation.

Third, many species require more than one kind of habitat (Law & Dickman 1998). For example, immature insects and amphibians often use different habitats than those they use as adults. A successful life cycle requires that the adults can move away from the habitat where they were reared to their adult habitats and then back to the immature habitat to lay eggs. The proximity of different required habitat types will determine the ease with which individuals can move among them. For example, Pope et al. (2000) showed that the proximity of feeding habitat to breeding ponds affected the abundance of leopard frog populations. Pedlar et al. (1997) found that raccoon abundance was highest in landscapes with intermediate amounts of forest. They suggested that this level of forest maximized the accessibility to the raccoons of both feeding areas (grain fields) and denning sites in forest.

The degree to which landscape structure facilitates movement among different required habitat types was labeled "landscape complementation" by Dunning et al. (1992). For the same amount of habitat, a more fragmented landscape (more, smaller patches, and more edge) will have a higher level of interdigitation of different habitat types. This should increase landscape complementation, which has a positive effect on biodiversity (Law & Dickman 1998, Tscharntke et al. 2002).

Finally, it seems likely that positive edge effects are a factor. Some species do show positive edge effects (Carlson & Hartman 2001, Kremsater & Bunnell 1999, Laurance et al. 2001). For a given amount of habitat, more fragmented landscapes contain more edge. Therefore, positive edge effects could be responsible for positive effects of fragmentation per se on abundance or distribution of some species.

CONCLUSIONS AND FUTURE DIRECTIONS

Habitat Loss Versus Fragmentation

Most researchers view habitat fragmentation as a process involving both the loss of habitat and the breaking apart of habitat. The fact that most fragmentation research does not differentiate between these two effects has led to several problems. First, the apparent inconsistency in the effects of a single process (fragmentation) gives the impression that fragmentation effects are difficult to generalize. In fact, generalization is possible, but only for the separate components of fragmentation, not for the combined concept of loss and breaking apart of habitat. Empirical evidence to date suggests that the loss of habitat has large negative effects on biodiversity. On the other hand, the breaking apart of habitat, independent of habitat loss, has rather weak effects on biodiversity, which are as likely to be positive as negative.

Second, the merging of these two aspects of fragmentation has obscured the fact that the effects of habitat loss outweigh the effects of habitat fragmentation per se. In fact, the effects of fragmentation per se are absent or too small to be detected in most empirical tests to date. This is in contrast to several theoretical predictions (Burkey 1999, Hill & Caswell 1999, Urban & Keitt 2001, With & King 1999) and has important implications for conservation. It suggests that conservation efforts should focus on habitat preservation and restoration. It also suggests that research in support of particular conservation problems should focus on determining the amount of habitat required for conservation of the species of concern. The fact that effects of fragmentation per se are usually small and at least as likely to be positive as negative suggests that conservation actions that attempt to minimize fragmentation (for a given habitat amount) may often be ineffectual.

Note, however, that this conclusion is preliminary because there are still only a small number of relevant empirical studies. To my knowledge there are, to date, no studies in tropical regions of the effects of forest fragmentation per se (controlling for habitat loss). Laurance et al. (2002) concluded that in Brazilian tropical forest there are strong negative effects of forest edge on several taxa. These effects are apparently much stronger than negative edge effects in temperate systems (Kremsater & Bunnell 1999). Negative edge effects could translate into a negative effect of fragmentation per se at the landscape scale because fragmentation per se increases the amount of edge on the landscape. This suggests that effects of fragmentation per se may be greater in tropical systems than in temperate systems. This prediction remains to be tested.

Third, ambiguous empirical results could lead to errors in modeling studies. For example, Donovan & Lamberson (2001) constructed a model to look at the effects of habitat fragmentation on population growth rate. They held amount of habitat constant and varied mean patch size. For input parameters they used empirical work suggesting that reproductive success increases with increasing patch size. However, as they point out, in these empirical studies patch size was highly correlated with habitat amount in the surrounding landscape. It is not known whether reproductive success increases with increasing patch size when habitat amount in the landscape is held constant. It could be that reproductive success increases with amount of habitat on the landscape, independent of habitat fragmentation per se. If this is true, the results of the simulation may be misleading.

These conclusions are based on the relatively small, but growing, number of empirical studies that separate the effects of habitat loss and fragmentation per se. So far these studies have been conducted on a limited set of taxa primarily within

North America. More research is needed to determine how general the conclusions are (Harrison & Bruna 1999).

IS "FRAGMENTATION" A USEFUL TERM? The term "fragmentation" is quickly losing its usefulness as more and more effects of human activities are incorporated into this single term. Some authors have even suggested that some species are "indicators of fragmentation" (e.g., Hager 1998, Niemelä 2001). The implication that fragmentation can be indicated by the decline of some species or species group suggests that the term is becoming a catchall for human-caused habitat changes that have negative effects on biodiversity. As questioned by Haila (2002), "Is a conceptually ambiguous and empirically multifaceted term fruitful as a generic description of human effects on landscapes?"

I suggest that the term "fragmentation" should be limited to the breaking apart of habitat. Habitat loss should be called habitat loss; it has important effects on biodiversity that are independent of any effects of habitat fragmentation per se. Habitat fragmentation should be reserved for changes in habitat configuration that result from the breaking apart of habitat, independent of habitat loss.

Implications for Biodiversity Conservation

Does our knowledge about fragmentation effects have general implications for conservation of biodiversity, particularly simultaneous conservation of multiple species? The fragmentation literature provides strong evidence that habitat loss has large, consistently negative effects on biodiversity. This implies that the most important question for biodiversity conservation is probably "How much habitat is enough?" Different species use different kinds of habitat, and different species require different amounts of habitat for persistence. Therefore, conservation of all species in a given region requires identifying which species in that region are most vulnerable to habitat loss (Fahrig 2001, With & King 1999) and estimating the minimum habitat required for persistence of each of these most vulnerable species. This determines the minimum habitat amounts for each kind of habitat in the region. In addition, many species require more than one kind of habitat within a life cycle. Therefore, landscape patterns that maintain the required habitat amounts, but intersperse the different habitat types as much as possible, should produce the largest positive biodiversity response (Law & Dickman 1998).

ACKNOWLEDGMENTS

I thank Jeff Bowman, Julie Brennan, Dan Bert, Tormod Burkey, Neil Charbonneau, Kathryn Freemark, Audrey Grez, Jeff Houlahan, Jochen Jaeger, Maxim Larrivée, Lutz Tischendorf, Rebecca Tittler, Paul Zorn, and members of the Landscape Ecology Laboratory at Carleton for comments on an earlier version of this manuscript. This work was supported by a grant from the Natural Sciences and Engineering Research Council of Canada.

The *Annual Review of Ecology, Evolution, and Systematics* is online at http://ecolsys.annualreviews.org

LITERATURE CITED

Andrén H. 1994. Effects of habitat fragmentation on birds and mammals in landscapes with different proportions of suitable habitat: a review. *Oikos* 71:355–66

Atkinson WD, Shorrocks B. 1981. Competition on a divided and ephemeral resource: a simulation model. *J. Anim. Ecol.* 50:461–71

Bascompte J, Solé RV. 1996. Habitat fragmentation and extinction thresholds in spatially explicit models. *J. Anim. Ecol.* 65:465–73

Bascompte J, Possingham H, Roughgarden J. 2002. Patchy populations in stochastic environments: critical number of patches for persistence. *Am. Nat.* 159:128–37

Bélisle M, Desrochers A, Fortin M-J. 2001. Influence of forest cover on the movements of forest birds: a homing experiment. *Ecology* 82:1893–904

Bender DJ, Tischendorf L, Fahrig L. 2003. Evaluation of patch isolation metrics for predicting animal movement in binary landscapes. *Landsc. Ecol.* 18:17–39

Bergin TM, Best LB, Freemark KE, Koehler KJ. 2000. Effects of landscape structure on nest predation in roadsides of a midwestern agroecosystem: a multiscale analysis. *Landsc. Ecol.* 15:131–43

Best LB, Bergin TM, Freemark KE. 2001. Influence of landscape composition on bird use of rowcrop fields. *J. Wildl. Manage.* 65:442–49

Boswell GP, Britton NF, Franks NR. 1998. Habitat fragmentation, percolation theory and the conservation of a keystone species. *Proc. R. Soc. London Ser. B* 265:1921–25

Boulinier T, Nichols JD, Hines JE, Sauer JR, Flather CH, Pollock KH. 2001. Forest fragmentation and bird community dynamics: inference at regional scales. *Ecology* 82:1159–69

Bowers MA, Dooley JL. 1999. A controlled, hierarchical study of habitat fragmentation: responses at the individual, patch, and landscape scale. *Landsc. Ecol.* 14:381–89

Bowman J, Cappuccino N, Fahrig L. 2002. Patch size and population density: the effect of immigration behavior. *Conserv. Ecol.* 6:9. http://www.consecol.org/vol6/iss1/art9

Brennan JM, Bender DJ, Contreras TA, Fahrig L. 2002. Focal patch landscape studies for wildlife management: optimizing sampling effort across scales. In *Integrating Landscape Ecology into Natural Resource Management*, ed. J Liu, WW Taylor, pp. 68–91. Cambridge, MA: Cambridge Univ. Press. 480 pp.

Burkey TV. 1999. Extinction in fragmented habitats predicted from stochastic birth-death processes with density dependence. *J. Theor. Biol.* 199:395–406

Caley MJ, Buckley KA, Jones GP. 2001. Separating ecological effects of habitat fragmentation, degradation, and loss on coral commensals. *Ecology* 82:3435–48

Carlson A, Hartman G. 2001. Tropical forest fragmentation and nest predation—an experimental study in an Eastern Arc montane forest, Tanzania. *Biodivers. Conserv.* 10:1077–85

Cascante A, Quesada M, Lobo JJ, Fuchs EA. 2002. Effects of dry tropical forest fragmentation on the reproductive success and genetic structure of the tree *Samanea saman*. *Conserv. Biol.* 16:137–47

Chalfoun AD, Thompson FR, Ratnaswamy MJ. 2002. Nest predators and fragmentation: a review and meta-analysis. *Conserv. Biol.* 16:306–18

Chesson PL. 1985. Coexistence of competitors in spatially and temporally varying environments: a look at the combined effects of different sorts of variability. *Theor. Popul. Biol.* 28:263–87

Collinge SK, Forman RTT. 1998. A conceptual model of land conversion processes: predictions and evidence from a microlandscape experiment with grassland insects. *Oikos* 82:66–84

Collingham YC, Huntley B. 2000. Impacts of habitat fragmentation and patch size upon migration rates. *Ecol. Appl.* 10:131–44

Collins RJ, Barrett GW. 1997. Effects of habitat fragmentation on meadow vole (*Microtus pennsylvanicus*) population dynamics in experiment landscape patches. *Landsc. Ecol.* 12:63–76

Debinski DM, Holt RD. 2000. A survey and overview of habitat fragmentation experiments. *Conserv. Biol.* 14:342–55

Delin AE, Andrén H. 1999. Effects of habitat fragmentation on Eurasian red squirrel (*Sciurus vulgaris*) in a forest landscape. *Landsc. Ecol.* 14:67–72

den Boer PJ. 1981. On the survival of populations in a heterogeneous and variable environment. *Oecologia* 50:39–53

Diaz JA, Carbonell R, Virgos E, Santos T, Telleria JL. 2000. Effects of forest fragmentation on the distribution of the lizard *Psammodromus algirus*. *Anim. Conserv.* 3:235–40

Donovan TM, Flather CH. 2002. Relationships among North American songbird trends, habitat fragmentation, and landscape occupancy. *Ecol. Appl.* 12:364–74

Donovan TM, Lamberson RH. 2001. Area-sensitive distributions counteract negative effects of habitat fragmentation on breeding birds. *Ecology* 82:1170–79

Donovan TM, Thompson FR, Faaborg J, Probst J. 1995. Reproductive success of migratory birds in habitat sources and sinks. *Conserv. Biol.* 9:1380–95

Drolet B, Desrochers A, Fortin M-J. 1999. Effects of landscape structure on nesting songbird distribution in a harvested boreal forest. *Condor* 101:699–704

Dunning JB, Danielson BJ, Pulliam HR. 1992. Ecological processes that affect populations in complex landscapes. *Oikos* 65:169–75

Fahrig L. 1997. Relative effects of habitat loss and fragmentation on species extinction. *J. Wildl. Manage.* 61:603–10

Fahrig L. 1998. When does fragmentation of breeding habitat affect population survival? *Ecol. Model.* 105:273–92

Fahrig L. 2001. How much habitat is enough? *Biol. Conserv.* 100:65–74

Fahrig L. 2002. Effect of habitat fragmentation on the extinction threshold: a synthesis. *Ecol. Appl.* 12:346–53

Fernandez-Juricic E. 2000. Forest fragmentation affects winter flock formation of an insectivorous guild. *Ardea* 88:235–41

Findlay CS, Houlahan J. 1997. Anthropogenic correlates of species richness in southeastern Ontario wetlands. *Conserv. Biol.* 11:1000–9

Flather CH, Bevers M. 2002. Patchy reaction-diffusion and population abundance: the relative importance of habitat amount and arrangement. *Am. Nat.* 159:40–56

Flather CH, Bevers M, Cam E, Nichols J, Sauer J. 1999. Habitat arrangement and extinction thresholds: do forest birds conform to model predictions? *Landscape Ecology: the Science and the Action. 5th World Congr. Int. Assoc. Landsc. Ecol., Snowmass, Col.* 1:44–45 (Abstr.)

Fuller DO. 2001. Forest fragmentation in Loudoun County, Virginia, USA evaluated with multitemporal Landsat imagery. *Landsc. Ecol.* 16:627–42

Ganzhorn JU, Eisenbeiß B. 2001. The concept of nested species assemblages and its utility for understanding effects of habitat fragmentation. *Basic Appl. Ecol.* 2:87–95

Gibbs JP. 1998. Distribution of woodland amphibians along a forest fragmentation gradient. *Landsc. Ecol.* 13:263–68

Gibbs JP. 2001. Demography versus habitat fragmentation as determinants of genetic variation in wild populations *Biol. Conserv.* 100:15–20

Gibbs JP, Stanton EJ. 2001. Habitat fragmentation and arthropod community change: carrion beetles, phoretic mites, and flies. *Ecol. Appl.* 11:79–85

Golden DM, Crist TO. 2000. Experimental effects of habitat fragmentation on rove beetles

and ants: patch area or edge? *Oikos* 90:525–38
Goodman SM, Rakotondravony D. 2000. The effects of forest fragmentation and isolation on insectivorous small mammals (Lipotyphla) on the Central High Plateau of Madagascar. *J. Zool.* 250:193–200
Groppe K, Steinger T, Schmid B, Baur B, Boller T. 2001. Effects of habitat fragmentation on choke disease (*Epichloe bromicola*) in the grass *Bromus erectus*. *J. Ecol.* 89:247–55
Gurd DB, Nudds TD, Rivard DH. 2001. Conservation of mammals in Eastern North American wildlife reserves: How small is too small? *Conserv. Biol.* 15:1355–63
Gustafson EJ. 1998. Quantifying landscape spatial pattern: What is the state of the art? *Ecosystems* 1:143–56
Guthery FS, Green MC, Masters RE, DeMaso SJ, Wilson HM, Steubing FB. 2001. Land cover and bobwhite abundance on Oklahoma farms and ranches. *J. Wildl. Manage.* 65:838–49
Hager HA. 1998. Area-sensitivity of reptiles and amphibians: Are there indicator species for habitat fragmentation? *EcoScience* 5:139–47
Haig AR, Matthes U, Larson DW. 2000. Effects of natural habitat fragmentation on the species richness, diversity, and composition of cliff vegetation. *Can. J. Bot.* 78:786–97
Haila Y. 2002. A conceptual genealogy of fragmentation research: from island biogeography to landscape ecology. *Ecol. Appl.* 12: 321–34
Haines-Young R, Chopping M. 1996. Quantifying landscape structure: a review of landscape indices and their application to forested landscapes. *Prog. Phys. Geogr.* 20:418–45
Hanski I, Moilanen A, Gyllenberg M. 1996. Minimum viable metapopulation size. *Am. Nat.* 147:527–41
Hargis CD, Bissonette JA, David JL. 1998. The behavior of landscape metrics commonly used in the study of habitat fragmentation. *Landsc. Ecol.* 13:167–86
Hargis CD, Bissonette JA, Turner DL. 1999. The influence of forest fragmentation and landscape pattern on American martens. *J. Appl. Ecol.* 36:157–72
Harrison S, Bruna E. 1999. Habitat fragmentation and large-scale conservation: What do we know for sure? *Ecography* 22:225–32
Hartley MJ, Hunter ML. 1998. A meta-analysis of forest cover, edge effects, and artificial nest predation rates. *Conserv. Biol.* 12:465–69
Hastings A. 1977. Spatial heterogeneity and the stability of predator-prey systems. *Theor. Popul. Biol.* 12:37–48
Henein K, Wegner J, Merriam G. 1998. Population effects of landscape model manipulation on two behaviourally different woodland small mammals. *Oikos* 81:168–86
Hill MF, Caswell H. 1999. Habitat fragmentation and extinction thresholds on fractal landscapes. *Ecol. Lett.* 2:121–27
Hinsley SA, Bellamy PE, Newton I, Sparks TH. 1995. Habitat and landscape factors influencing the presence of individual breeding bird species in woodland fragments. *J. Avian. Biol.* 26:94–104
Hovel KA, Lipcius RN. 2001. Habitat fragmentation in a seagrass landscape: Patch size and complexity control blue crab survival. *Ecology* 82:1814–29
Huffaker CB. 1958. Experimental studies on predation: dispersion factors and predator-prey oscillations. *Hilgardia* 27:343–83
Ims RA, Andreassen HP. 1999. Effects of experimental habitat fragmentation and connectivity on root vole demography. *J. Anim. Ecol.* 68:839–52
Jansson G, Angelstam P. 1999. Threshold levels of habitat composition for the presence of the long-tailed tit (*Aegithalos caudatus*) in a boreal landscape. *Landsc. Ecol.* 14:283–90
Kareiva P. 1990. Population dynamics in spatially complex environments: theory and data. *Philos. Trans. R. Soc. London Ser. B* 330:175–90
Kinnunen H, Jarvelainen K, Pakkala T, Tiainen J. 1996. The effect of isolation on the occurrence of farmland carabids in a fragmented landscape. *Ann. Zool. Fenn.* 33:165–71
Kolozsvary MB, Swihart RK. 1999. Habitat

fragmentation and the distribution of amphibians: patch and landscape correlates in farmland. *Can. J. Zool.* 77:1288–99

Komonen A, Penttilae R, Lindgren M, Hanski I. 2000. Forest fragmentation truncates a food chain based on an old-growth forest bracket fungus. *Oikos* 90:119–26

Kremsater L, Bunnell FL. 1999. Edge effects: theory, evidence and implications to management of western North American forests. In *Forest Fragmentation: Wildlife and Management Implications*, ed. JA Rochelle, LA Lehmann, J Wisniewski, pp. 117–53. Boston, MA: Brill. 301 pp.

Kurki S, Nikula A, Helle P, Linden H. 2000. Landscape fragmentation and forest composition effects on grouse breeding success in boreal forests. *Ecology* 81:1985–97

Laakkonen J, Fisher RN, Case TJ. 2001. Effect of land cover, habitat fragmentation and ant colonies on the distribution and abundance of shrews in southern California. *J. Anim. Ecol.* 70:776–88

Lande R. 1987. Extinction thresholds in demographic models of territorial populations. *Am. Nat.* 130:624–35

Langlois JP, Fahrig L, Merriam G, Artsob H. 2001. Landscape structure influences continental distribution of hantavirus in deer mice. *Landsc. Ecol.* 16:255–66

Laurance WF, Delamonica P, D'Angelo S, Jerozolinski A, Pohl L, et al. 2001. Rain forest fragmentation and the structure of Amazonian liana communities. *Ecology* 82:105–16

Laurance WF, Lovejoy TE, Vasconcelos HL, Bruna EM, Didham RK, et al. 2002. Ecosystem decay of Amazonian forest fragments: a 22-year investigation. *Conserv. Biol.* 16:605–18

Law BS, Dickman CR. 1998. The use of habitat mosaics by terrestrial vertebrate fauna: implications for conservation and management. *Biodivers. Conserv.* 7:323–33

Levin SA. 1974. Dispersion and population interactions. *Am. Nat.* 108:207–28

MacArthur RH, Wilson EO. 1967. *The Theory of Island Biogeography*. Princeton, NJ: Princeton Univ. Press

Mac Nally R, Brown GW. 2001. Reptiles and habitat fragmentation in the box-ironbark forests of central Victoria, Australia: predictions, compositional change and faunal nestedness. *Oecologia* 128:116–25

Mac Nally R, Bennett AF, Horrocks G. 2000. Forecasting the impacts of habitat fragmentation. Evaluation of species-specific predictions of the impact of habitat fragmentation on birds in the box-ironbark forests of central Victoria, Australia. *Biol. Conserv.* 95:7–29

Magura T, Koedoeboecz V, Tothmeresz B. 2001. Effects of habitat fragmentation on carabids in forest patches. *J. Biogeogr.* 28:129–38

Mahan CG, Yahner RH. 1999. Effects of forest fragmentation on behaviour patterns in the eastern chipmunk (*Tamias striatus*). *Can. J. Zool.* 77:1991–97

McCoy ED, Mushinsky HR. 1999. Habitat fragmentation and the abundances of vertebrates in the Florida scrub. *Ecology* 80:2526–38

McGarigal K, Cushman SA. 2002. Comparative evaluation of experimental approaches to the study of habitat fragmentation effects. *Ecol. Appl.* 12:335–45

McGarigal K, McComb WC. 1995. Relationships between landscape structure and breeding birds in the Oregon Coast Range. *Ecol. Monogr.* 65:235–60

McGarigal K, Cushman SA, Neel MC, Ene E. 2002. *FRAGSTATS: Spatial pattern analysis program for categorical maps.* Comp. software prog. Univ. Mass., Amherst. www.umass.edu/landeco/research/fragstats/fragstats.html

Meyer JS, Irwin LL, Boyce MS. 1998. Influence of habitat abundance and fragmentation on northern spotted owls in western Oregon. *Wildl. Monogr.* 139:1–51

Middleton J, Merriam G. 1983. Distribution of woodland species in farmland woods. *J. Appl. Ecol.* 20:625–44

Miyashita T, Shinkai A, Chida T. 1998. The effects of forest fragmentation on web spider communities in urban areas. *Biol. Conserv.* 86:357–64

Morato EF. 2001. Effects of forest fragmentation on solitary wasps and bees in Central Amazonia. II. Vertical stratification. *Rev. Brasileira Zool.* 18:737–47

Mossman CA, Waser PM. 2001. Effects of habitat fragmentation on population genetic structure in the white-footed mouse (*Peromyscus leucopus*). *Can. J. Zool.* 79:285–95

Niemelä J. 2001. Carabid beetles (Coleoptera: Carabidae) and habitat fragmentation: A review. *Eur. J. Entomol.* 98:127–32

Pedlar JH, Fahrig L, Merriam HG. 1997. Raccoon habitat use at two spatial scales. *J. Wildl. Manage.* 61:102–12

Pither J, Taylor PD. 1998. An experimental assessment of landscape connectivity. *Oikos* 83:166–74

Pope SE, Fahrig L, Merriam HG. 2000. Landscape complementation and metapopulation effects on leopard frog populations. *Ecology* 81:2498–508

Reddingius J, den Boer PJ. 1970. Simulation experiments illustrating stabilization of animal numbers by spreading of risk. *Oecologia* 5:240–84

Renjifo LM. 1999. Composition changes in a Subandean avifauna after long-term forest fragmentation. *Conserv. Biol.* 13:1124–39

Robinson SK, Thompson FR, Donovan TM, Whitehead DR, Faaborg J. 1995. Regional forest fragmentation and the nesting success of migratory birds. *Science* 267:1987–90

Roff DA. 1974a. Spatial heterogeneity and the persistence of populations. *Oecologia* 15:245–58

Roff DA. 1974b. The analysis of a population model demonstrating the importance of dispersal in a heterogeneous environment. *Oecologia* 15:259–75

Rosenberg KV, Lowe JD, Dhondt AA. 1999. Effects of forest fragmentation on breeding tanagers: a continental perspective. *Conserv. Biol.* 13:568–83

Rukke BA. 2000. Effects of habitat fragmentation: increased isolation and reduced habitat size reduces the incidence of dead wood fungi beetles in a fragmented forest landscape. *Ecography* 23:492–502

Sánchez-Zapata JA, Calvo JF. 1999. Rocks and trees: habitat response of Tawny Owls *Strix aluco* in semiarid landscapes. *Ornis Fenn.* 76:79–87

Schmiegelow FKA, Mönkkönen M. 2002. Habitat loss and fragmentation in dynamic landscapes: avian perspectives from the boreal forest. *Ecol. Appl.* 12:375–89

Schumaker NH. 1996. Using landscape indices to predict habitat connectivity. *Ecology* 77:1210–25

Schweiger EW, Diffendorfer JE, Holt RD, Pierotti R, Gaines MS. 2000. The interaction of habitat fragmentation, plant, and small mammal succession in an old field. *Ecol. Monogr.* 70:383–400

Shmida A, Ellner S. 1984. Coexistence of plant species with similar niches. *Vegetatio* 58:29–55

Slatkin M. 1974. Competition and regional coexistence. *Ecology* 55:128–34

Steffan-Dewenter I, Münzenberg U, Bürger C, Thies C, Tscharntke T. 2002. Scale-dependent effects of landscape context on three pollinator guilds. *Ecology* 83:1421–32

Summerville KS, Crist TO. 2001. Effects of experimental habitat fragmentation on patch use by butterflies and skippers (Lepidoptera). *Ecology* 82:1360–70

Swenson JJ, Franklin J. 2000. The effects of future urban development on habitat fragmentation in the Santa Monica Mountains. *Landsc. Ecol.* 15:713–30

Taylor PD, Merriam G. 1995. Habitat fragmentation and parasitism of a forest damselfly. *Landsc. Ecol.* 11:181–89

Tischendorf L. 2001. Can landscape indices predict ecological processes consistently? *Landsc. Ecol.* 16:235–54

Tischendorf L, Fahrig L. 2000. On the usage and measurement of landscape connectivity. *Oikos* 90:7–19

Tischendorf L, Bender DJ, Fahrig L. 2003. Evaluation of patch isolation metrics in mosaic landscapes for specialist vs. generalist dispersers. *Landsc. Ecol.* 18:41–50

Trzcinski MK, Fahrig L, Merriam G. 1999.

Independent effects of forest cover and fragmentation on the distribution of forest breeding birds. *Ecol. Appl.* 9:586–93

Tscharntke T, Steffan-Dewenter I, Kruess A, Thies C. 2002. Contribution of small habitat fragments to conservation of insect communities of grassland-cropland landscapes. *Ecol. Appl.* 12:354–63

Urban D, Keitt T. 2001. Landscape connectivity: a graph-theoretic perspective. *Ecology* 82:1205–18

Vallan D. 2000. Influence of forest fragmentation on amphibian diversity in the nature reserve of Ambohitantcly, highland Madagascar. *Biol. Conserv.* 96:31–43

Vance MD, Fahrig L, Flather CH. 2003. Relationship between minimum habitat requirements and annual reproductive rates in forest breeding birds. *Ecology.* In press

van den Berg LJL, Bullock JM, Clarke RT, Langston RHW, Rose RJ. 2001. Territory selection by the Dartford warbler (*Sylvia undata*) in Dorset, England: the role of vegetation type, habitat fragmentation and population size. *Biol. Conserv.* 101:217–28

Vandermeer JH. 1973. On the regional stabilization of locally unstable predator-prey relationships. *J. Theor. Biol.* 41:161–70

Venier L, Fahrig L. 1996. Habitat availability causes the species abundance-distrubution relationship. *Oikos* 76:564–70

Villard M-A, Trzcinski MK, Merriam G. 1999. Fragmentation effects on forest birds: relative influence of woodland cover and configuration on landscape occupancy. *Conserv. Biol.* 13:774–83

Virgós E. 2001. Role of isolation and habitat quality in shaping species abundance: a test with badgers (*Meles meles* L.) in a gradient of forest fragmentation. *J. Biogeogr.* 28:381–89

Walters JR, Ford HA, Cooper CB. 1999. The ecological basis of sensitivity of brown treecreepers to habitat fragmentation: a preliminary assessment. *Biol. Conserv.* 90:13–20

Wettstein W, Schmid B. 1999. Conservation of arthropod diversity in montane wetlands: Effect of altitude, habitat quality and habitat fragmentation on butterflies and grasshoppers. *J. Appl. Ecol.* 36:363–73

Wickham JD, Jones KB, Riitters KH, Wade TG, O'Neill RV. 1999. Transitions in forest fragmentation: implications for restoration opportunities at regional scales. *Landsc. Ecol.* 14:137–45

Wilcove DS, McLellan CH, Dobson AP. 1986. Habitat fragmentation in the temperate zone. In *Conservation Biology*, ed. ME Soulé, pp. 237–56. Sunderland, MA: Sinauer

With KA, Crist TO. 1995. Critical thresholds in species' responses to landscape structure. *Ecology* 76:2446–59

With KA, Gardner RH, Turner MG. 1997. Landscape connectivity and population distributions in heterogeneous environments. *Oikos* 78:151–69

With KA, King AW. 1999. Dispersal success on fractal landscapes: a consequence of lacunarity thresholds. *Landsc. Ecol.* 14:73–82

With KA, Pavuk DM, Worchuck JL, Oates RK, Fisher JL. 2002. Threshold effects of landscape structure on biological control in agroecosystems. *Ecol. Appl.* 12:52–65

Wolff JO, Schauber EM, Edge WD. 1997. Effects of habitat loss and fragmentation on the behavior and demography of gray-tailed voles. *Conserv. Biol.* 11:945–56

Young CH, Jarvis PJ. 2001. Measuring urban habitat fragmentation: an example from the Black Country, UK. *Landsc. Ecol.* 16:643–58

Annu. Rev. Ecol. Evol. Syst. 2003. 34:517–47
doi: 10.1146/annurev.ecolsys.34.030102.151725

First published online as a Review in Advance on July 30, 2003

SOCIAL ORGANIZATION AND PARASITE RISK IN MAMMALS: Integrating Theory and Empirical Studies

Sonia Altizer,[1] Charles L. Nunn,[2] Peter H. Thrall,[3] John L. Gittleman,[4] Janis Antonovics,[4] Andrew A. Cunningham,[5] Andrew P. Dobson,[6] Vanessa Ezenwa,[6,7] Kate E. Jones,[4] Amy B. Pedersen,[4] Mary Poss,[8] and Juliet R.C. Pulliam[6]

[1]Department of Environmental Studies, Emory University, Atlanta, Georgia 30322; email: saltize@emory.edu

[2]Section of Evolution and Ecology, University of California, Davis, California 95616; email: cnunn@ucdavis.edu

[3]CSIRO-Plant Industry, Center for Plant Biodiversity Research, GPO Box 1600, Canberra ACT 2601, Australia; email: Peter.Thrall@csiro.au

[4]Department of Biology, University of Virginia, Charlottesville, Virginia 22904; email: ja8n@virginia.edu, JLGittleman@virginia.edu, kate.jones@virginia.edu, abp3a@virginia.edu

[5]Institute of Zoology, Zoological Society of London, London, United Kingdom, NW1 4RY; email: Andrew.Cunningham@ioz.ac.uk

[6]Department of Ecology and Evolutionary Biology, Princeton University, Princeton, New Jersey 08544; email: andy@eno.princeton.edu, pulliam@princeton.edu

[7]Present address: U.S. Geological Survey, Reston, Virginia 20192; email: vezenwa@usgs.gov

[8]Division of Biological Sciences, University of Montana, Missoula, Montana 59812; email: mposs@selway.umt.edu

Key Words infectious disease, social structure, mating system, host behavior, transmission mode, biodiversity, conservation

■ **Abstract** Mammals are exposed to a diverse array of parasites and infectious diseases, many of which affect host survival and reproduction. Species that live in dense populations, large social groups, or with promiscuous mating systems may be especially vulnerable to infectious diseases owing to the close proximity and higher contact rates among individuals. We review the effects of host density and social contacts on parasite spread and the importance of promiscuity and mating structure for the spread and evolution of sexually transmitted diseases. Host social organization and mating system should influence not only parasite diversity and prevalence but may also determine the fitness advantages of different transmission strategies to parasites.

Because host behavior and immune defenses may have evolved to reduce the spread and pathogenicity of infectious diseases, we also consider selective pressures that parasites may exert on host social and mating behavior and the evolutionary responses of hosts at both the immunological and behavioral levels. In examining these issues, we relate modeling results to observations from wild populations, highlighting the similarities and differences among theoretical and empirical approaches. Finally, the epidemiological consequences of host sociality are very relevant to the practical issues of conserving mammalian biodiversity and understanding the interactions between extinction risk and infectious diseases.

INTRODUCTION

Social organization, including the size and composition of social groups, and mating systems, including partner exchange rates and variance in male and female mating success, should directly influence host proximity and the number and duration of contacts in a population. These behaviors are therefore expected to have major effects on parasite spread within host species and should influence the distribution of parasites among host species. This point is illustrated vividly in the case of sexually transmitted diseases (STDs), in which the expected risk of infection increases with the number of mating partners (e.g., Anderson & May 1992, Thrall et al. 2000). Consequently, differences in promiscuity among host species should influence the prevalence and diversity of STDs. Social interactions also provide key opportunities for parasite spread, and parasites transmitted by social contacts are expected to be more common in larger groups and in denser populations (Arneberg et al. 1998, Freeland 1976, Loehle 1995). In fact, parasites may ultimately represent a major cost to both social organization and promiscuity, and changes in host behavior may arise from increased parasite prevalence or severity. A cogent example is the shift in human sexual behavior resulting from the AIDS pandemic (e.g., Anderson et al. 1989, Mills et al. 1997).

General expectations for how host behavior might affect parasite spread may seem relatively straightforward (Figure 1), but questions remain to be addressed at two levels: patterns of parasitism within populations, and comparative patterns among host and parasite communities. For example, do highly social hosts harbor greater parasite diversity, and do observed patterns depend on parasite characteristics? Do promiscuous host species experience a greater risk of STDs, and how does host mating behavior affect the evolution of parasite transmission? Have behavioral or immune defenses evolved in highly social species to minimize disease risk? Answering such questions provides an important step toward predicting patterns of disease incidence and will aid in identifying the potential for new parasites to emerge and spread.

This review discusses recent advances in this field and raises new questions for understanding the links between host sociality and the transmission of infectious disease. Two complementary approaches are needed to address these questions: theoretical studies that examine how key variables influence parasite spread, and empirical studies that assess infection risk within populations and the distribution

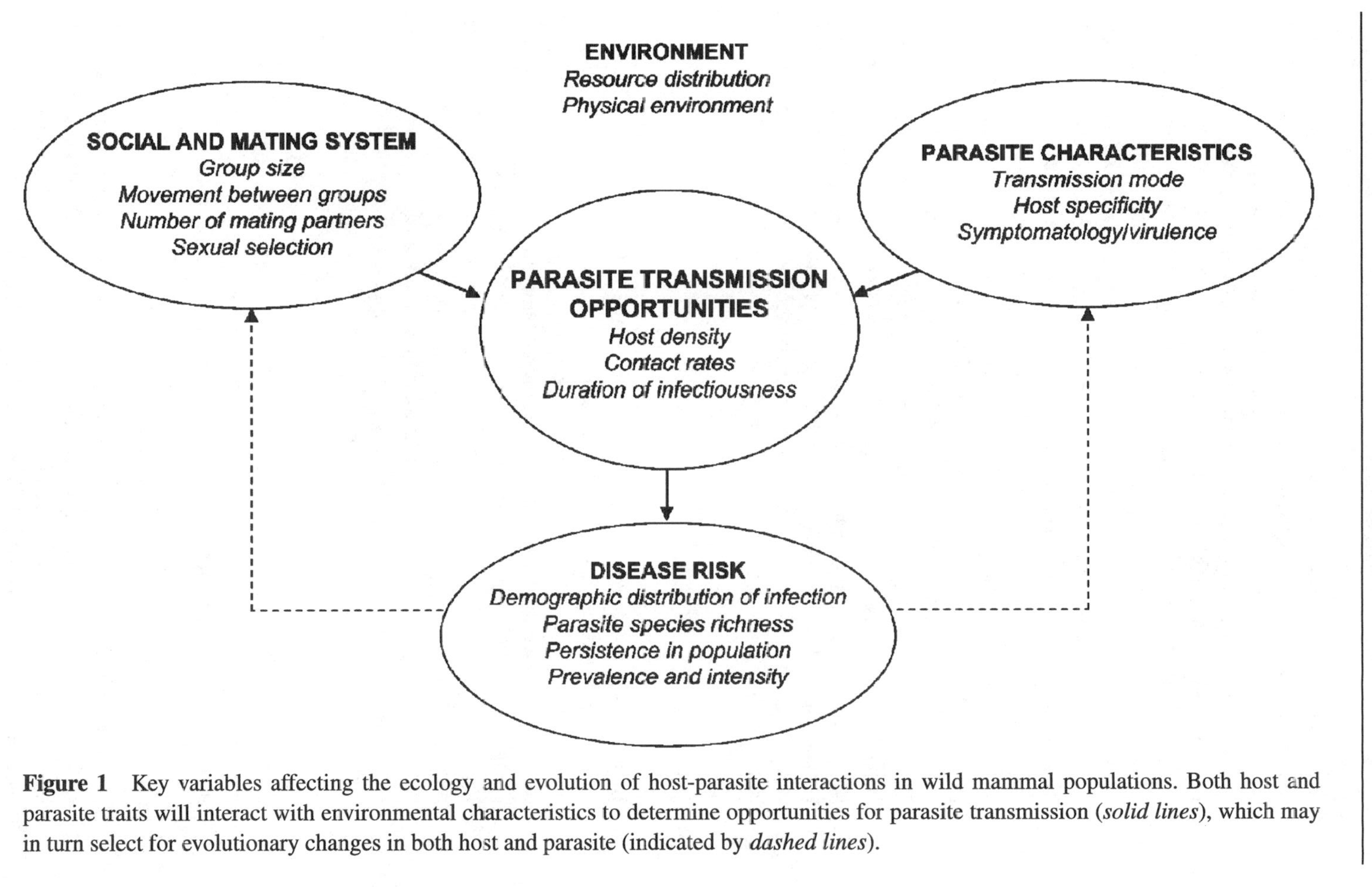

Figure 1 Key variables affecting the ecology and evolution of host-parasite interactions in wild mammal populations. Both host and parasite traits will interact with environmental characteristics to determine opportunities for parasite transmission (*solid lines*), which may in turn select for evolutionary changes in both host and parasite (indicated by *dashed lines*).

of parasites across host populations or species. Throughout this review, we apply the term parasite to any infectious organism capable of colonizing a host, utilizing host resources, and spreading to new hosts. Our conclusions apply to both microparasites (such as viruses, bacteria, and protozoa) and macroparasites (such as helminths and arthropods; Anderson & May 1992).

Integrating theoretical results with empirical approaches is best accomplished using a model system, or a group of related species in which studies have addressed basic natural history and conceptual issues (cf. Dugatkin 2001). We focus on mammals because comprehensive data are available on social, ecological, life history, and biogeographic parameters for a large proportion of species (e.g., Geffen et al. 1996, Gittleman 1996, Smuts et al. 1987). Moreover, because of their uses in farming and biomedical research, a great deal of information is available on the parasites of wild and captive mammal populations, making them particularly well suited for comparative studies (e.g., Arneberg et al. 1998, Morand & Poulin 1998). The evolutionary history of mammals is becoming increasingly well known (e.g., Bininda-Emonds et al. 1999, Jones et al. 2002, Liu et al. 2001) enabling examination of questions in a phylogenetic context. Finally, mammals are important foci for conservation efforts, and understanding the role of parasites in wild populations will become vital for future conservation and management decisions (e.g., Dobson & Lyles 2000, Funk et al. 2001).

GENERAL BACKGROUND

In pioneering research, Freeland (1976, 1979) suggested that primate social interactions and behavior have evolved to reduce the spread and pathogenicity of new and existing parasites. Assuming that larger social groups experience increased disease risk, selection for pathogen avoidance should influence social group size, composition, and intergroup movements. For example, high parasite pressure may lead to increased rates of juvenile dispersal, and activities that induce behavioral or nutritional stress should increase the susceptibility of particular classes of individuals such as newcomers trying to enter a group or individuals vying for potential mates. Despite clear intuitive links and a growing number of theoretical studies that address host social and mating systems and infectious diseases, empirical studies that examine parasite spread and host sociality face many challenges. These include the need to control for a large number of potentially correlated host and parasite traits (Figure 1) and a lack of correspondence between methods for quantifying host behavior and parameters from epidemiological models, in addition to difficulties associated with population-level experimental work with mammalian species.

Social and Mating Systems in Mammals

Social organization, defined by the size and composition of social groups and patterns of intergroup dispersal, should directly influence host density and the number and duration of contacts within a population with important consequences for parasite transmission (Figure 1). For example, monogamous species with strictly defended territories are expected to exhibit fewer parasites because their number of intraspecific contacts is small, whereas social mammals that live in multi-male multi-female groups may afford greater opportunities for parasite spread. Mammalian social systems are distinguished primarily by the temporal and spatial interactions between adults and by the genetic relatedness among individuals (e.g., Alexander 1974). For example, solitary species often come together only to mate, whereas gregarious species may breed colonially in family groups or form herds or packs with either stable or variable composition. Social interactions can also be influenced by territoriality or exclusion of other groups and the presence of substructuring or dominance within groups (e.g., Dunbar 1988, Eisenberg 1981, Smuts et al. 1987).

Mating systems can be defined by variance in male and female mating success within a population, and vary dramatically among mammalian species. Observed patterns range from monogamy to polygynandry (where both sexes mate with multiple partners during a breeding season; e.g., Clutton-Brock 1989, Eisenberg 1981). Mating systems also exist in which males, but not females, have multiple partners (polygyny) and vice versa (polyandry), and such mating bonds can last throughout life or be limited to a few reproductive events.

Several ecological factors are thought to influence mammalian mating and social systems (Clutton-Brock 1989, Emlen & Oring 1977). Both sexes require resources and access to mates. However, sex differences in parental investment may skew the operational sex ratio toward males (Clutton-Brock & Parker 1992), and the relative importance of resources versus mating opportunities affect the two sexes differently (Trivers 1972). In mammals, females often invest more in offspring than do males, leading to the prediction that female social strategies should reflect environmental risks and resource availability, whereas male social strategies should reflect the opportunity to control groups of females (Emlen & Oring 1977). In microtine rodents, for example, female distributions are linked to environmental heterogeneities, and male distributions depend on female densities (e.g., Ostfeld 1985). In polygynous species such as red deer, clumped resources result in males defending areas that attract females, whereas dispersed resources lead to direct male defense of roving females (Carranza et al. 1995). Finally, intersexual conflict may influence social and mating systems. In primates, for example, stable male-female relationships and home range defense may protect against infanticide by extragroup males (as in gibbons; van Schaik & Dunbar 1990), and female promiscuity may confuse paternity in larger social groups (as in baboons and macaques; van Schaik et al. 1999).

A particular challenge arises from the need to quantify mating and social structure in ways that are meaningful to host-parasite dynamics. As a case in point, mating systems defined categorically (as monogamous or polygynous) only poorly capture characteristics that are important from an epidemiological perspective. To assess STD risk in different mating systems, it is necesary to quantify variance in male and female mating contacts, how mating contacts vary with age and social status, the duration and fidelity of contacts within groups, and rates of intergroup migration (Figure 1). Many of these variables are difficult to measure directly in wild populations, and indirect measures, including sexual dimorphism in body size or other traits under sexual selection, may be used to infer the degree of reproductive skew (Andersson 1994). For example, it has been shown that relative testes mass (after controlling for body mass) can indirectly reflect variation in mating promiscuity among host species (Nunn et al. 2000). Similarly, grooming rates with different partners or social group size can be used to quantify social contacts. Finally, intergroup migration is rarely measured in wild populations of social mammals, yet movement among groups should have major consequences for parasite spread (Thrall et al. 2000).

Parasite Ecology and Epidemiological Parameters

A general understanding of parasite ecology and epidemiology provides a set of predictions regarding ecological factors that influence parasite spread and persistence. In simple host-parasite models with direct transmission, the probability that most parasites will spread in a host population is an increasing function of host density and longevity and a decreasing function of parasite-induced mortality and recovery (Anderson & May 1992). New infections usually depend on host contact rates and per contact probabilities of successful infection. This leads to the straightforward prediction that hosts living at high density or with frequent intraspecific contacts will increase the spread and prevalence of any given parasite species, and, by extension, the number of parasite species harbored by a host population (Figure 1; Anderson & May 1979, Arneberg 2002, Roberts et al. 2002).

Parasites exhibit an impressive variety of transmission modes, with a major dichotomy between direct transmission (where parasites are spread directly from host to host by sexual, social, or other close contact) and indirect transmission (where hosts encounter parasites in the environment, or through intermediate hosts or vectors). Different transmission modes should interact with host traits to influence parasite spread and persistence (Figures 1 and 2). The establishment of an STD, for example, depends on both host sexual behavior and parasite adaptations to increase infection probability. Increased sociality and greater host population density are predicted to increase the transmission of parasites spread through direct contact (Thrall & Antonovics 1997), whereas parasites spread by biting vectors or exposure to contaminated soil or water may be less sensitive to changes in host contacts or density (e.g., Anderson & May 1992). Interestingly, sexual (between mating partners) and vertical (parent to offspring) transmission have been suggested as parasite strategies for persistence in low density or solitary host species, as sexual reproduction is one of the few times that conspecifics come into contact in these species (Smith & Dobson 1992).

One useful epidemiological perspective is to characterize conditions for initial spread and persistence in a host population among parasites that vary in their transmission mode. Much recent work has considered the differences between STDs

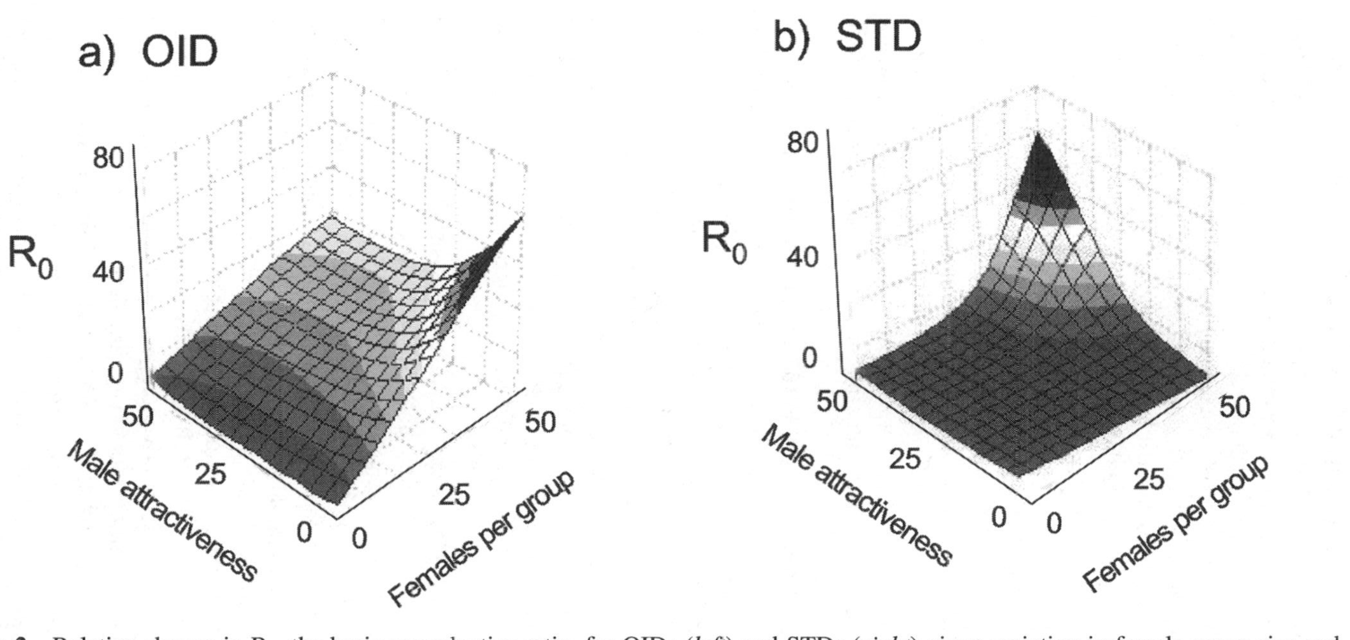

Figure 2 Relative change in R_0, the basic reproductive ratio, for OIDs (*left*) and STDs (*right*) given variation in female group size and male mating success. Here, R_0 provides a measure of the number of secondary infections produced by a single infected host introduced to an entirely susceptible host population. In each figure, the relative magnitude of R_0 is plotted as a function of the number of females in each social group and "male attractiveness," an index of the number of females each breeding male mates with in an annual mating season. For STDs, initial spread is maximized by large female group size and high male attractiveness, which leads to relatively few breeding males mating with a large number of females. For OIDs, R_0 is maximized for large female group size but low male attractiveness, so that many breeding males each mate with a small number of females (A. Dobson et al. unpublished manuscript).

and ordinary infectious diseases (OIDs) that are transmitted by nonsexual direct contact. The temporal dynamics and criteria for establishment of STDs and OIDs are expected to differ, as are their regulatory effects on host abundance (Figure 2). For example, the number of "effective contacts" leading to parasite transmission should increase directly as a function of host density for parasites spread by social contact. However, for STDs, the effective number of sexual contacts is expected to saturate quite rapidly with increasing host density. Therefore, the rate of spread of STDs should depend more strongly on the proportion of infected hosts rather than on total host density (Getz & Pickering 1983, Thrall et al. 1993). This dichotomy has been captured mathematically as the difference between density-dependent transmission (where the change in the number of infected hosts, I, depends on βSI, or the product of the transmission parameter, β, and the number of susceptible and infected hosts) versus frequency-dependent transmission (where the change in the number of infected hosts depends on $\beta SI/N$, where N is the total host population size). Recent studies suggest that parasite transmission is neither purely frequency nor density dependent but is a complex function of both (Antonovics et al. 1995, Begon et al. 1999, Knell et al. 1996).

Another parasite characteristic relevant to successful maintenance of parasites involves the degree of host specificity (Figure 1). Parasites may be classified as generalists that infect many host species, or specialists that infect only one or a few host species. Host specificity can be measured by the actual number of susceptible host taxa (Poulin 1998) but is more accurately measured relative to host taxonomy or phylogeny. The capability and opportunity to infect multiple host taxa should interact with parasite transmission mode as some transmission routes (e.g., sexual, close contact) provide almost no opportunities for cross-species transfers (Figure 1). By comparison, vector transmission or transmission through contaminated soil, water, or intermediate hosts can expose multiple host species to the same parasite (Woolhouse et al. 2001).

Finally, parasite virulence plays a key role in host-parasite dynamics and may coevolve with transmission mode and host behavior (Figure 1; Levin 1996, Messenger et al. 1999). Although precise definitions of virulence vary (e.g., Bull 1994, Ewald 1994, O'Keefe & Antonovics 2002), this term usually refers to reductions in host survival or fecundity stemming from parasite replication or damage to host tissues. Counter to the traditional wisdom that parasites should evolve to cause minimal harm to hosts, virulence may provide a selective advantage to parasites if disease symptoms increase transmission to new hosts, or when multiple strains compete within the same individuals (Bull 1994). Parasites transmitted by sexual or vertical routes require that hosts survive long enough to mate or reproduce, so that in these cases, reductions in host longevity may have strong negative effects on parasite transmission (Lockhart et al. 1996, Thrall et al. 1993). Interestingly, STDs are more likely to induce host sterility than OIDs, an effect that may enhance their transmission if infected (and hence sterile) females undergo more frequent reproductive cycles and mate more often (Lockhart et al. 1996, Nunn & Altizer 2003).

SOCIALITY AND PARASITE SPREAD

If close proximity or contact among host individuals increases parasite transmission, then greater degrees of host sociality or gregariousness should translate to higher parasite prevalence, intensity, and diversity (Møller et al. 1993). Here, prevalence refers to the proportion of infected or diseased hosts, intensity refers to the average number of parasites within infected hosts, and diversity includes the total number of parasite species documented in host populations. Thus, social hosts are predicted to suffer greater exposure to parasites (Brown & Brown 1986, Møller et al. 2001), experience increased selection for innate or acquired immune defenses, and evolve behavioral defenses against parasites (Freeland 1976, Loehle 1995).

A large number of epidemiological models, supported by data from several empirical and comparative studies, point to strong links between host density or local group size and the spread and diversity of directly transmitted parasites (Figures 2 and 3; Anderson & May 1979, Arneberg 2002). For example, Dobson & Meagher (1996) summarized evidence that brucellosis in North American bison has a host density threshold for establishment, and Packer et al. (1999) showed that the incidence of infection with four different viruses in African lions increased with the estimated number of previously unexposed individuals. A comparative study of parasites in wild primates showed that host density was the most consistent factor predicting increases in the diversity of both macro- and microparasite communities (Nunn et al. 2003a; Figure 3).

Among mammals and other vertebrates, social group size appears to be an important predictor of parasite risk (Côté & Poulin 1995, Davies et al. 1991). At the level of single host populations, parasite prevalence, intensity, and occasionally diversity have been shown to increase with group size in a wide range of host taxa including prairie dogs (Hoogland 1979), mangabeys (Freeland 1979), cliff swallows (Brown & Brown 1986), bobwhites (Moore et al. 1988), and feral horses (Rubenstein & Hohmann 1989). A few studies have found similar patterns in cross-species comparative analyses including studies of birds (Poulin 1991) and fishes (Ranta 1992). However, the association between group size and infection risk is often confounded with other host traits (e.g., Rózsa 1997, Clayton & Walther 2001). In particular, because both host density and social group size have been shown to correlate positively with parasite prevalence and diversity, it is difficult to determine the relative importance of social contacts versus host density for parasite transmission (Arneberg 2002, Morand & Poulin 1998).

It is important to note that associations between parasite transmission and group size will depend on other host and parasite traits (Figure 1). For example, Moore et al. (1988) found consistent relationships between bobwhite covey size and helminth intensity only for directly transmitted parasites with relatively short life cycles. Furthermore, using a meta-analytical approach, Côté & Poulin (1995) found strong positive correlations between vertebrate group size and both the prevalence and intensity of directly transmitted helminths, whereas the intensity of

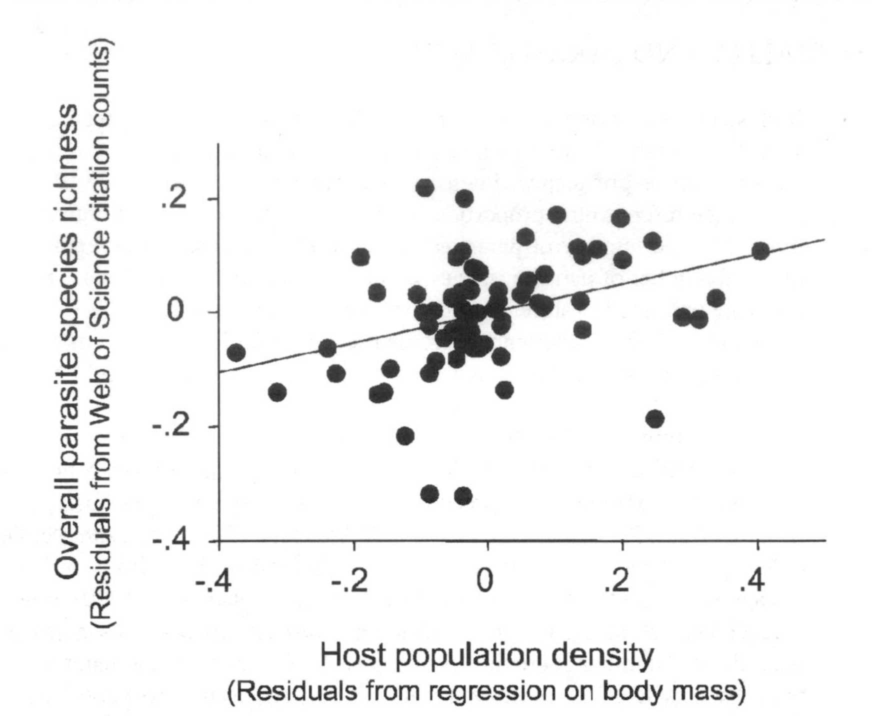

Figure 3 Effects of host density on overall parasite species richness in primates. Plot shows independent contrasts calculated using CAIC (Purvis & Rambaut 1995) and Purvis' (1995) composite estimate of primate phylogeny. Parasite species richness includes parasites in six functional classes (helminths, protozoa, viruses, bacteria, fungi, and arthropods), from free-ranging hosts only, collated from the published literature (Nunn et al. 2003a). Sampling effort was included as a covariate in statistical analyses using information on the number of citations for the different species from the Web of Science citation index, years 1975 to 2001. Results remain significant when using species values and controlling for body mass and geographic range size ($b_{popln\text{-}density} = 0.17$, $F_{1,75} = 5.49$, $P = 0.02$, two-tailed; Nunn et al. 2003a).

infection by indirectly transmitted parasites decreased with social group size. Host traits such as movement and territoriality may further confound the relationship between parasitism and host group size. For example, a field study comparing intestinal parasite loads among 11 species of African bovids (Ezenwa 2002) showed that the prevalence of coccidian parasites increased with group size only among host species with closed group structures (and not for species with high rates of intergroup exchange). The same study also found that nematode infections were more prevalent among territorial antelopes compared with nonterritorial species, possibly as a result of increased exposure to parasites resulting from infective stages accumulating in the environment (Ezenwa 2002). These results highlight the fact that many other features of host and parasite ecology are important to identifying the effects of group size.

A common modeling approach to investigating heterogeneity in patterns of social contacts in human populations is to group individuals into classes (e.g., social status, degree of sexual activity) and describe contacts among classes in terms of a "mixing matrix," where the entries in each of the cells describe the frequency distribution of contacts per unit time (Blower & McLean 1991). The most important prediction gained from these models is that the pattern of contacts between different activity classes has a major impact on parasite spread (Jacquez et al. 1988). Specifically, a high degree of mixing within an activity class results in a more rapid initial spread but a lower population-wide prevalence than a high degree of mixing among different activity classes. Despite their importance in human epidemiology, mixing matrices have not been applied to animal social and mating systems because detailed information for their construction (contact rates within and among social classes or mating groups) has generally not been available.

An important question related to parasite spread in socially structured populations involves identifying individuals that are at greatest infection risk. Parasitism is likely to correlate with dominance rank, age, sex, and mating status (Hausfater & Watson 1976, Muller-Graf et al. 1996) because these factors influence habitat use, the frequency of intraspecific contacts, and the effectiveness of immune defenses. For example, Halvorsen (1986) showed that dominant reindeer were more frequently exposed to nematode infections because they consumed more vegetation. Among African antelopes, territorial males were exposed to more parasites than bachelor males or females (Ezenwa 2002), possibly owing to the immunosuppresive effects of testosterone or the accumulation of parasites on resident male territories. Courchamp et al. (1998) showed that among feral cat populations, older males (with greater dispersal and a higher number of aggressive encounters) were more likely to be infected with feline immunodeficiency virus (FIV). Although processes that underlie individual differences in infection risk have been identified for some species, understanding their relative importance in a cross-species context is necessary for a broader understanding of factors that determine patterns of parasite occurrence.

Nearly all empirical research on infectious disease and sociality in wild populations has focused on opportunities for transmission in different mating and social systems (solid line in Figure 1). An important area for future research involves the effect of infectious diseases on mammalian sociality (dashed line in Figure 1). There is growing evidence that parasites represent a strong selective force for the evolution of mating systems and social interactions (e.g., Møller et al. 2001). Perhaps the best example from mammals is the observation that increased presence of ectoparasites (flies, ticks, and other arthropods) increases host tendency to form large groups, possibly as a way of avoiding high parasite loads through the dilution effect (Mooring & Hart 1992, Rubenstein & Hohmann 1989). These results suggest that parasitism can shape host sociality, but no studies have shown that species evolve behaviors that decrease group size following high parasite pressure.

SEXUALLY TRANSMITTED DISEASES

STDs are increasingly recognized as an important parasite group with potentially large impacts on host reproduction and evolution, in many cases increasing the chances of sterility (Lockhart et al. 1996, Smith & Dobson 1992). Despite apparent overlap between social and mating systems in mammals, and potentially similar mechanics of sexual and social contact, the characteristics and dynamics of STDs differ from many other infectious diseases. The pathogens causing such diseases have smaller host ranges, longer infectious periods, and are less likely to cause host mortality or induce protective host immunity (Lockhart et al. 1996, Oriel & Hayward 1974, Smith & Dobson 1992). Animals with promiscuous mating systems (or species in which females engage in frequent extrapair copulations) are predicted to experience a greater risk of acquiring parasites through sexual contact. However, empirical patterns illustrating potential links between host mating behavior and infectious disease risk have not been well documented among mammals or other vertebrates. The dynamics of most STDs cannot be understood without considering heterogeneity in sexual activity (Anderson & May 1992). For this reason, population models developed to predict HIV dynamics and control have focused on human sexual contact patterns (e.g., Anderson et al. 1988, 1989; Boily & Masse 1997), and this focus has extended to other human STDs such as gonorrhoea and syphilis (Garnett et al. 1997, Hethcote & Yorke 1984).

Characteristics of many STDs cause their dynamics to differ from other directly transmitted parasites. In particular, STDs tend to persist as endemic (rather than epidemic) infections with transmission relatively unaffected by increased host density or crowding. They have also been described as a unique class of pathogens well adapted to persisting in small, low density host populations (Smith & Dobson 1992), although their presence in large populations is certainly not theoretically precluded.

Mathematical models that incorporate heterogeneity in mating behavior show that STD transmission increases with increasing variance in partner exchange rates and that highly promiscuous individuals ("superspreaders") can facilitate STD persistence even when the mean number of sexual partners is low (Anderson & May 1992). Consistent with models that predict a higher risk of infection among more promiscuous subgroups, surveys of HIV and other STDs in human populations show that prevalence increases with increasing numbers of sexual partners per year (reviewed in Anderson & May 1992). One might expect this generalization to apply to wild mammals with polygynous mating systems with variance in male mating success at the population level being proportional to increased transmission of STDs. Using an individual-based simulation model of polygynous mating systems, Thrall et al. (2000) showed that variance in male mating success affects the spread of STDs only when the migration of females among mating groups is limited. Their model assumed that males varied in their attractiveness to females, that females had only one mate per breeding season, and that females could change groups between breeding seasons. Two mating system parameters were examined: variation in male mating success and variation in female fidelity to males. When females moved frequently among groups, their model demonstrated that variance in male mating success (meaning increasing skew in the number of females associated with any given male) had almost no effect on parasite spread. When intergroup movement was limited, parasites spread rapidly in groups where males monopolized a large number of females, but transmission was highly limited in smaller groups.

A second notable outcome of the model by Thrall et al. (2000) was that equilibrium STD prevalence was significantly greater in females than in males, with parasite prevalence in females increasing to an asymptote with increasing skew in male mating success (Figure 4*a*). Thus, when variance in male mating success was high, many males remained unmated, lowering the equilibrium prevalence among males relative to females. Using published data on two sexually transmitted retroviruses in wild primate populations, Nunn & Altizer (2003) tested the prediction that STD prevalence should be higher in females than in males among nonmonogamous species. Data from sexually mature adult primates showed that in a majority of the sample populations, seroprevalence was higher among females (Figure 4*b*), and prevalence differences for males and females were statistically significant and in the predicted direction when tested using a matched pairs t-test (Nunn & Altizer 2003). Although these analyses were consistent with the model predictions, alternative explanations (for example, sex-based differences in per contact transmission probabilities or disease susceptibility) are possible. Higher STD prevalence among females has also been reported among captive breeding primate colonies including sooty mangabeys and baboons (Fultz et al. 1990, Levin et al. 1988).

These differences in STD prevalence between males and females are more striking because theory predicts the opposite pattern for OIDs owing to a presumed sex-based difference in disease susceptibility. For example, prevalence is expected to be higher in males (Alexander & Stimson 1988, Bundy 1988) as a consequence of the energetic costs associated with competition for mates or the deleterious effects of testosterone on immunocompetence (Zuk 1990). Moore & Wilson (2002) showed that among mammal species where male-male competition is most extreme, male-biased mortality coincided with greater male susceptibility to parasitic diseases. This pattern may result from effects of testosterone on immunocompetence, but effects of body size on parasite infection can also explain male-biased parasitism. In this context, Moore & Wilson (2002) found that in general the larger sex suffered greater parasitism, regardless of gender. Further studies are needed to determine the consequences of sex-biased susceptibility for the evolution of mate choice, particularly with respect to traits that signal parasite infection.

STDs and the Evolution of Host Mating Strategies

An important epidemiological consequence of host sexual behavior is that "attractive" males are predicted to suffer a greater risk of STD infection (Graves & Duvall 1995, Thrall et al. 2000). Thus, parasites transmitted during mating may

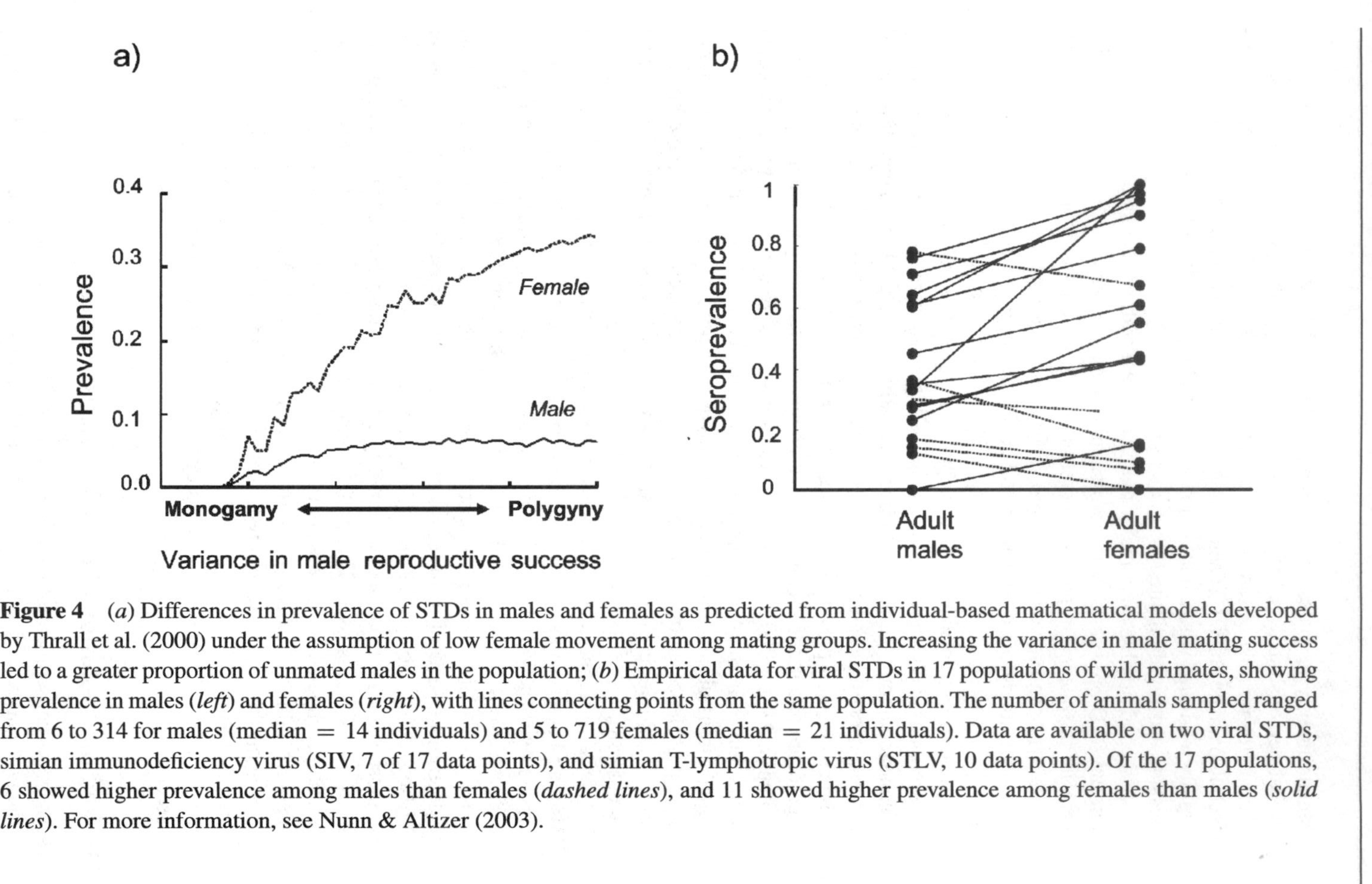

Figure 4 (*a*) Differences in prevalence of STDs in males and females as predicted from individual-based mathematical models developed by Thrall et al. (2000) under the assumption of low female movement among mating groups. Increasing the variance in male mating success led to a greater proportion of unmated males in the population; (*b*) Empirical data for viral STDs in 17 populations of wild primates, showing prevalence in males (*left*) and females (*right*), with lines connecting points from the same population. The number of animals sampled ranged from 6 to 314 for males (median = 14 individuals) and 5 to 719 females (median = 21 individuals). Data are available on two viral STDs, simian immunodeficiency virus (SIV, 7 of 17 data points), and simian T-lymphotropic virus (STLV, 10 data points). Of the 17 populations, 6 showed higher prevalence among males than females (*dashed lines*), and 11 showed higher prevalence among females than males (*solid lines*). For more information, see Nunn & Altizer (2003).

have debilitating effects on dominant males, hastening their replacement by subordinates. Studies of baboons, for example, revealed a strong positive correlation between social and reproductive status and parasite loads (Hausfater & Watson 1976). Interestingly, the theoretical studies of Thrall et al. (2000) suggested that STDs would not limit the evolution of male traits that increased polygyny unless females suffered reproductively by joining multi-female groups. Even in the presence of a sterilizing STD, more attractive males still had higher reproductive success than less attractive males (although the selective advantage for polygyny was much lower in the presence of an STD).

Should monogamy be the optimal mating strategy in the presence of a potentially sterilizing STD? Thrall et al. (1997) addressed this question directly by modeling mating events that were associated with both a per-contact transmission probability and a fertilization probability. They showed that optimal strategies for males and females could differ substantially in the presence of an STD, indicating that parasites alone have the potential to influence the evolution of sex-based differences in mating behavior. For example, when both transmission rates and STD prevalence were high, monogamy was always the optimal strategy for females, but the best strategy for males was to mate with as many females as possible. Overall these results confirmed that STDs spread more rapidly in promiscuous mating systems. However, even though monogamy always resulted in the lowest parasite levels, it was not always the favored strategy owing to reproductive benefits that arise from promiscuous mating.

Evolution of Sexual Transmission and Virulence

In a full coevolutionary model, host social and sexual behavior should interact with pathogen transmission and virulence. With regard to pathogen virulence, STDs range from those that are relatively benign to those that are highly virulent, either causing high mortality or extreme sterility (Lockhart et al. 1996). In general, STD virulence is expected to be higher when extrapair copulations are common than when monogamy predominates. This idea has been discussed with respect to human sexual behavior and the evolution of HIV (Ewald 1994).

Transmission modes themselves may evolve depending on host social and mating behavior. Using differential equation models, Thrall & Antonovics (1997) derived conditions under which an STD could invade a host population and displace a pathogen transmitted by nonsexual means (OID). Invasion by the STD was easier when the equilibrium host population size with an OID was relatively small. Conversely, an OID could invade more easily if the equilibrium population size with an STD was larger. Overall, these results reflect the general expectation that sexual transmission should be favored in low density populations, whereas nonsexual transmission should be favored at high densities (Anderson & May 1992, Smith & Dobson 1992). Based on more realistic assumptions related to the importance of host population density, Thrall et al. (1998) proposed the concept of a social-sexual crossover point (SSCP) associated with parasite transmission

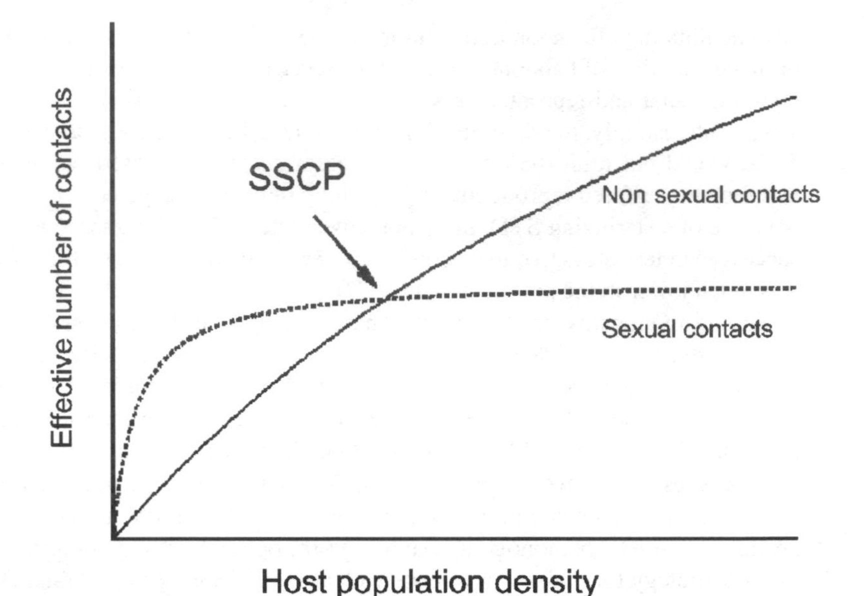

Figure 5 Relationship between effective contact number (the number of contacts per unit time that actually result in disease transmission) and host population density (see Thrall et al. 1998). The arrow indicates the social-sexual crossover point (SSCP) where the number of nonsexual contacts exceeds the number of sexual contacts. Because even at low population densities, individuals will still actively seek out sexual contacts for reproductive purposes, the number of sexual contacts is generally assumed to initially increase more rapidly with host density, but to reach an asymptote at lower numbers (owing to the greater handling time associated with sexual versus nonsexual contacts).

(Figure 5). These formulations assumed that (*a*) as population density increases, social and sexual contacts also increase; (*b*) the number of sexual contacts will initially increase more rapidly with density than the number of social contacts (at low population densities, individuals still seek mates); and (*c*) at higher densities, the number of sexual contacts will rapidly saturate (owing to longer durations associated with sexual contacts) but the number of social contacts will continue to increase. Thus, the SSCP represents a critical host population density at which the numbers of social and sexual contacts are equal (Figure 5). Clearly, the host density at which the SSCP occurs could vary considerably depending on the details of host social and mating structure, but empirical studies of this relationship remain a challenge for future research. As predicted, Thrall et al. (1998) found that increased sexual transmission was always favored if the equilibrium population size was less than the SSCP; otherwise, nonsexual transmission was favored.

HOST DEFENSES AND BEHAVIORAL AVOIDANCE

If the probability of infection increases with group size or promiscuity, then highly social or promiscuous hosts should experience more intense selection in favor of barriers (behavioral or immunological) to parasite transmission (Freeland 1976, Loehle 1995). For example, grooming, preening, and selective foraging have been suggested as parasite avoidance mechanisms in social vertebrates (Hart 1990, Loehle 1995, Moore 2002), although it is important to note that behaviors such as grooming may actually facilitate the transmission of some parasites while reducing the transmission of others. Ironically, if highly social hosts evolve more elaborate defenses as obstacles to parasite infection and impacts (Møller et al. 2001), this may eliminate expected relationships between parasitism and host sociality (the "ghost of parasitism past").

In vertebrates, the innate component of the immune system, including phagocytic cells such as monocytes and neutrophils, controls the immediate host response to general classes of pathogens. As such, white blood cell (WBC) counts may reflect a baseline defense against parasite invasion, particularly for those WBCs that target specific pathogen groups. The strength of innate immunity should therefore increase among taxa that experience high parasite pressure (Møller et al. 1998), especially if defenses are costly to maintain (Nordling et al. 1998, Sheldon & Verhulst 1996). However, it is important to note that other factors (including age and stress) will influence WBC counts, and other components of host immunity play a role in antiparasite defense.

Using data acquired from healthy zoo primates, Nunn et al. (2000) showed that mating promiscuity explained significant variation in leukocyte counts (including lymphocytes, neutrophils, monocytes, and eosinphils), whereas the effects of sociality, life history, and habitat use were nonsignificant. In a separate comparative study of carnivores, host promiscuity, sociality, and longevity explained significant variation in leukocyte counts across species (Nunn et al. 2003b). Moreover, a strong allometric relationship involving neutrophils was found in both primates and carnivores so that larger-bodied hosts harbored a greater neutrophil abundance, possibly indicating greater parasite exposure. Collectively, these results are consistent with experimental immunological research that demonstrates that innate immune defenses offer protection against pathogen invasions, and they draw attention to correlates between host life history, behavior, and immunity by showing that hosts more likely to encounter sexually transmitted pathogens had higher WBC counts.

A variety of behavioral traits may operate in conjunction with the immune system to limit exposure to parasites (Hart 1990, Loehle 1995). Strategies to avoid nonsexual parasites have been discussed extensively in primates, including alteration of ranging patterns (Di Bitetti et al. 2000, Hausfater & Meade 1982), ingestion of medicinal plants (Huffman 1997), and avoidance of recent immigrants that may harbor novel parasites (Freeland 1976). Behavioral avoidance of fecal-contaminated areas by selective foraging has been reported for domestic grazing ungulates, where the risk of ingestion of fecal-borne parasites is high (e.g., Hutchings et al. 1998, Moe et al. 1999). Interestingly, some wild territorial bovids also avoid dung while foraging (Ezenwa 2002), possibly owing to increased risk of exposure to nematode infections among resident animals. Grooming rates have been shown to correlate positively with the risk of parasitism in both ungulates and primates. In primates, allogrooming is concentrated in regions of the body inaccessible to self-grooming, further suggesting that this behavior plays an important antiparasite function (e.g., Barton 1985).

Parasites with transmission modes for which behavioral counterstrategies may be relatively ineffective are predicted to select more strongly for increased immune defenses. In the case of STDs, for example, behaviors that reduce the risk of transmission may result in lower reproductive success (Thrall et al. 1997, 2000). Several possible behavioral mechanisms of resistance to STDs have been proposed. Before copulation, individuals may inspect potential partners and avoid those with signs of infection, but many STDs involve carrier states with no visible signs (Holmes et al. 1994) consistent with theoretical models showing that both parasite and host have congruent interests in obscuring infection status (Knell 1999, Thrall et al. 1997). After copulation, behavioral mechanisms may be effective and less costly. Thus, postcopulatory oral-genital grooming has been shown to reduce STD transmission in male rats (Hart et al. 1987), and postcopulatory urination has been suggested as a mechanism to reduce STD transmission in humans (Donovan 2000, Hooper et al. 1978). In a comparative study of primates, however, postcopulatory genital grooming and urination showed no correlation with mating promiscuity (Nunn 2003, Nunn & Altizer 2003). Interestingly, the two primate radiations where genital grooming is more common are the small-bodied species (lemurs and callitrichids), suggesting that physical constraints may limit the evolution of this behavior in large-bodied species.

CONFOUNDING FACTORS

As noted throughout this review, a large number of ecological, life history, and behavioral traits of mammals should interact to influence parasite dynamics and diversity. For example, body mass is thought to result in increased parasite diversity because larger-bodied hosts represent larger "habitats" and provide more niches for colonization (e.g., Kuris et al. 1980, Poulin 1995). Many host characteristics predicted to influence parasite risk are themselves correlated across taxa, posing complications for comparative studies of multiple factors. As a case in point, large-bodied hosts of some mammalian orders, such as primates, tend to be terrestrial (Clutton-Brock & Harvey 1977, Nunn & Barton 2001) and most have "slow" life histories (e.g., increased longevity, delayed age at first reproduction). Disentangling the effects of body mass, life history, substrate use, and social organization therefore requires multivariate statistical models tested across multiple host and parasite groups.

Confounding variables are problematic in both field and comparative studies, but in comparative studies, two additional issues arise. First, closely related hosts may harbor similar numbers of parasites because of common ancestry rather than similar behavioral or ecological traits. This effect may result from specialist parasites that cospeciate with their hosts, and from geographical proximity among hosts that share generalist parasites. Methods for incorporating phylogenetic history are now well developed (Harvey & Pagel 1991, Martins & Hansen 1996), although debate on when correction for phylogenetic relatedness is overly conservative, or perhaps even misleading, continues (Harvey & Rambaut 2000, Westoby et al. 1995). Second, host species may differ in the size and diversity of their parasite communities because of uneven sampling effort, and many studies have shown that parasite species richness (the number of parasite species per host) is correlated positively with the degree to which host species have been examined for parasites. The most common approaches to control for uneven sampling effort are to use residuals from a linear regression of parasite species richness against host sample size or other measures of sampling effort (by a log-log transformation; Gregory 1990, Poulin 1998) or to include sampling effort directly as a predictor in multivariate models.

IMPLICATIONS FOR MAMMALIAN CONSERVATION

Management of parasites and infectious disease has increasingly become a focus in conservation biology (Cleaveland et al. 2002) because parasites can threaten already-reduced populations and because infectious diseases can trigger catastrophic declines in otherwise robust host populations. Severe negative impacts from introduced pathogens such as rabies (African wild dogs), canine distemper (African lions, black-footed ferrets), and phocine distemper (harbor seals) have occurred in recent decades, and reports of parasite outbreaks in wild populations are on the rise (Funk et al. 2001). Environmental factors such as habitat fragmentation, increased contact between wildlife and domesticated species, and climate change may further increase parasite prevalence and impacts (Daszak et al. 2000, Harvell et al. 2002). Nevertheless, global assessments of the causes and patterns of extinction risk often relegate the impact of parasites to "other causes" (MacPhee & Fleming 1999). Surprisingly, the *2002 IUCN Red List* (Hilton-Taylor 2002) does not include comprehensive records of parasites that threaten wild host species even though many listed species are known to have experienced recent declines or challenges from infectious diseases (Figure 6).

Increasing human population size and encroachment on native habitats will influence the impacts and emergence of infectious diseases in wildlife in several ways (Daszak et al. 2000, Dobson & Foufopoulos 2001). First, encroachment by humans alters animal foraging and social behavior, leading to increased stress and greater risk of acquiring infectious disease. Second, crowding animals onto wildlife reserves may further increase rates of parasite transfer among species. Finally,

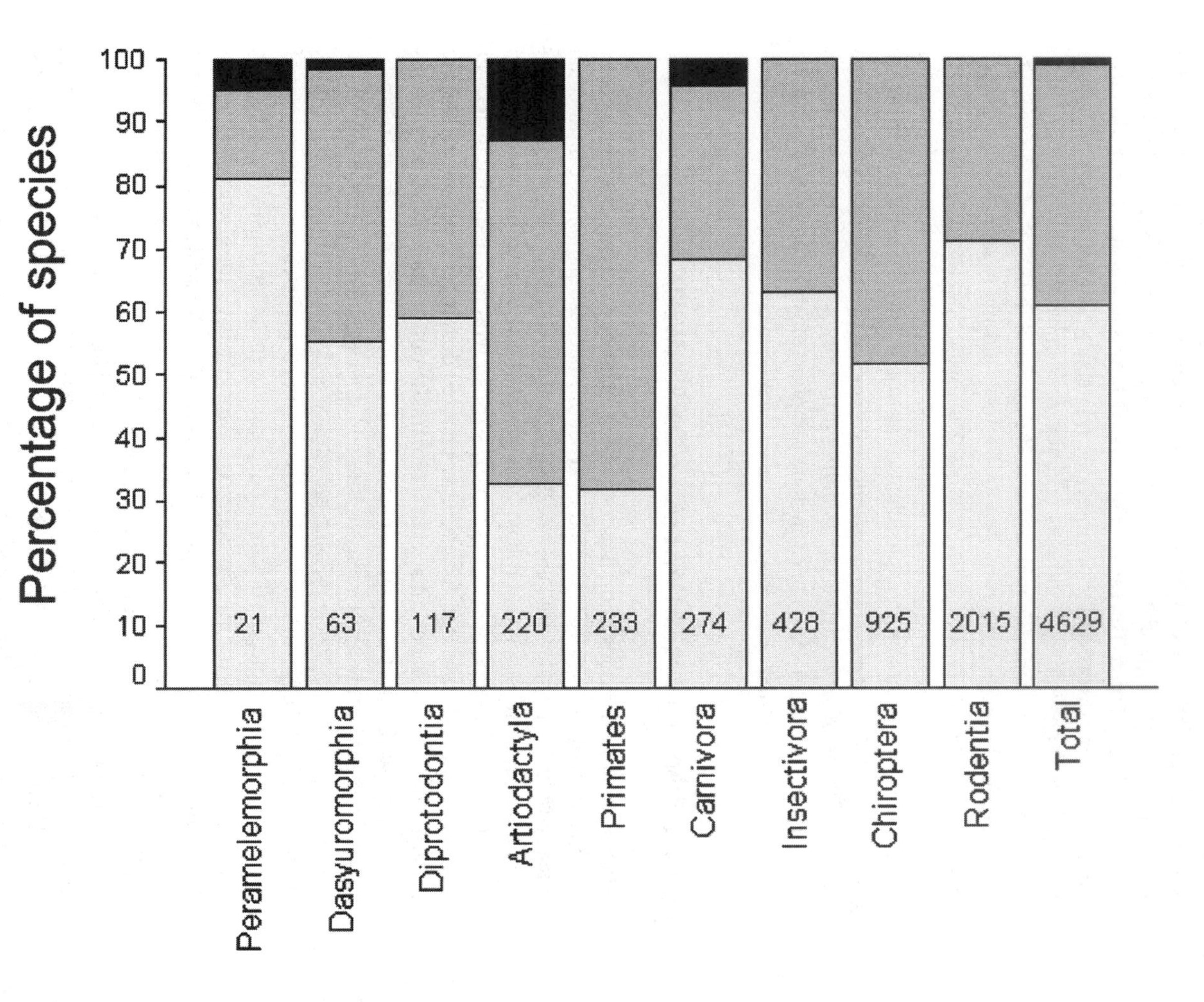
Percentage of species
100
90
80
70
60
50
40
30
20
10
0
21
63
117
220
233
274
428
925
2015
4629
Peramelemorphia
Dasyuromorphia
Diprotodontia
Artiodactyla
Primates
Carnivora
Insectivora
Chiroptera
Rodentia
Total

infectious disease may spread from domesticated species to their close relatives in the wild. In fact, many species listed as at risk from infectious disease in the *IUCN Red List* (Figure 6) have largely acquired their parasites from domesticated species.

Which Parasites Pose the Greatest Threats?

The potentially large number of parasites shared by humans, domesticated animals, and wildlife creates severe problems for conservation management. As populations decline or become fragmented, specialist parasites may be lost, and generalist parasites present in overlapping reservoir populations may pose a more significant threat to endangered species (Cleaveland et al. 2002). In carnivores, for example, most extinctions or near-extinctions are caused by generalist pathogens rather than specialists (Woodroffe 1999). During the past decade, rabies and canine distemper virus, both of which infect domestic dogs as reservoir hosts, have emerged as significant pathogens of wild carnivores in the Serengeti ecosystem. The persistence and impact of both of these viruses is expected to increase because domestic dog populations are growing rapidly in many African countries (Cleaveland 1998). Thus, an understanding of parasite specificity, with detailed understanding of host geographical and niche overlap, should play an important role in conservation efforts.

Transmission mode is also important for determining which parasites pose the greatest threats to declining host populations. For many directly transmitted parasites, persistence is unlikely at low densities without an alternative host or reservoir, and parasite-driven host extinction is questionable. However, the expectation is quite different for STDs, as these are characterized by frequency-dependent transmission and high levels of host sterility (Thrall et al. 1993, Lockhart et al. 1996). Parasites transmitted via mobile vectors or contaminated water may also pose unusually high risks to small or threatened populations, as these may spread rapidly to multiple host species given favorable environmental conditions (Woolhouse et al. 2001).

The Importance of Host Characteristics

For identifying host traits associated with both the risk of parasite infection and extinction risk, carnivores represent a well-studied group (Funk et al. 2001). Small

Figure 6 Relative risk of infectious disease as a conservation threat in selected mammalian orders. White represents the proportion of nonthreatened species in each clade, gray and black combined represent proportion of threatened (including species classified as extinct in the wild, critically endangered, endangered, vulnerable, lower risk–conservation dependent, and lower risk–near threatened), and black only represents the proportion of species where infectious disease is identified as a threatening process. Numbers at the bottom of each bar refer to the number of species in each clade. Clade definitions follow Wilson & Reeder (1993) and threat data are from Hilton-Taylor (2000).

and declining carnivore populations are especially prone to extinction risk by infectious diseases (Woodroffe 1999), and the interactive effects of small population size and parasite infections on extinction risk are observed in many carnivore species including African wild dogs and black-footed ferrets (Williams et al. 1992). One hypothesis is that mammalian orders with the largest numbers of domesticated species (or species that associate with humans; e.g., ungulates, carnivores, and rodents) may experience disproportionate risks of pathogen introduction (Figure 6). For example, even though canine distemper virus did not originally extend to felids other than in captive populations, in 1994, an outbreak in African lions that originated in domestic dogs reduced the lion population by about 30%. This virus also migrated from the Serengeti to another population in the Masai Mara, where eventually 55% of lions became seropositive (Roelke-Parker et al. 1996). Thus, certain mammal groups may be especially hard hit because domesticated animals and humans act as reservoirs of virulent parasites (Wallis & Lee 1999, Woodford et al. 2002).

Parasites and Biodiversity

Relative to other processes (such as habitat destruction, invasive species, overharvesting, and pollution), we know far too little about the effects of parasites on the processes of mammalian declines and extinction. Wilcove et al. (1997) assessed the influences of each of these sources of threat on approximately 2500 imperiled and federally listed species in the United States. Of information available on 1880 species, infectious disease represented the least important threat. It is doubtful that parasites play such a minor role, and it remains to be determined if these numbers reflect poor knowledge of infectious disease dynamics in natural populations rather than the actual impact of parasites in these groups.

The concern over parasite specificity and transmission highlights the importance of understanding broad patterns of parasite community diversity in wildlife. At present, only a small proportion of parasites infecting mammals have been identified and quantified in wild populations. For example, the African wild dog has received tremendous scientific study and popular attention (see Woodroffe et al. 1997) owing to observed catastrophic effects of parasites (such as rabies wiping out a pack of 21 individuals in less than two months; Kat et al. 1995), yet only a small number of parasites are known to potentially influence rates of population decline. Furthermore, the epidemic nature of certain pathogens makes recording parasite communities challenging. Some populations may not be affected for long intervals but then experience high fatality rates over a period of days, as was recently evidenced by outbreaks of phocine distemper in northern European seals (Jensen et al. 2002).

In considering the totality of biodiversity, it is also important to include the impacts of mammalian extinctions on the biodiversity represented by their parasites. Many parasites are specific to endangered or threatened mammals, and may themselves become extinct with their specific hosts (Gompper & Williams 1998). Moreover, hosts that lose their parasites during population bottlenecks or in captive breeding programs may also lose their ability to respond to future parasite threats following relaxed selection for immune defenses (Cunningham 1996). Lobbying efforts dedicated to preventing losses of mammalian diversity perhaps should take up the cause in defending pathogen biodiversity as an integral component of free-living host communities, particularly given the role parasites may play in shaping host phenotypic and genetic composition or their importance as potential sources of pharmaceuticals (Durden & Keirans 1996).

FUTURE DIRECTIONS

Interdisciplinary collaboration is crucial to advance our understanding of infectious disease ecology and evolution in mammalian populations. When dealing with complicated and poorly understood systems, it is always tempting to attribute causation to the most easily measured factor. In spite of recent advances in serology, PCR, and other detection methods, parasite infection—and its ecological and genetic correlates—are factors that are among the hardest to measure in natural populations. We are therefore left with a high probability of either over- or understating the importance of parasite infections based on superficial impressions. Rigorous studies are needed to understand the ecological and evolutionary roles that parasites play in natural populations. With this in mind, how can theory and data be more explicitly linked to study host and parasite features in mammalian systems? We outline several priorities for future research in this area, focusing particularly on increased correspondence between variables in theoretical models and empirical data from wild hosts and the development of new methods to assess the joint coevolution of host and parasite traits.

(1) **Expand empirical records of parasites from wild mammal populations.** Baseline data on the prevalence of parasites in wild mammals are critical to examining links between host behavior and parasite transmission. For most mammals, extensive data are rarely collected for nondomesticated animals outside of captive settings. For example, a recent survey of published records of nonexperimental parasite infections across 200 primate species (C. Nunn & S. Altizer, unpublished data) showed that the vast majority of records were obtained from captive animals (sampled in zoos, research centers, or semifree ranging populations). There is also a need for a centralized data repository for records of micro- and macroparasites in wild animals and monitoring programs that assess the prevalence, transmission, and population-level impacts of parasites infecting a wide range of mammalian species.

(2) **Improve the correspondence between behavioral data and model parameters.** At the present time, it is difficult to relate categorically defined mating systems (e.g., polygyny, serial monogamy) and social structures (e.g., solitary, fission-fusion communities) to the spread of parasites in wild populations. More precise measures of parameters suggested by theoretical

models are needed from wild mammal populations, including inter- and intragroup contact rates, dispersal rates, contact durations, and better measures of variance in mating success. Moreover, model parameters that define contacts leading to parasite transmission must reflect biologically realistic and estimable processes, a goal that can only be achieved by increasing interactions between behavioral ecologists and epidemiologists.

(3) **Include multiple factors in comparative analyses.** Increasing numbers of comparative studies are including measures of sampling effort and controlling for host phylogeny. However, to date most comparative studies of factors affecting parasite occurrence have examined only a small number of explanatory variables. Important variables to consider in the context of parasite diversity studies include not only host traits, such as life history and sociality, but also parasite traits involving transmission mode and host range. For example, the diversity of STDs should be influenced by factors that are different from those important to parasites transmitted by fecal-oral routes. Ignoring the confounding effect of transmission mode may therefore result in failure to detect traits important to patterns of parasite diversity.

(4) **Develop trait-based coevolutionary models for host-parasite interactions.** A major challenge involves developing modeling approaches to explore the joint coevolution of transmission mode, mating and social systems, and pathogen virulence. One approach is to acquire phylogenetic information on coevolving host and parasite lineages to examine the correlated evolution of host and parasite traits (Harvey & Keymer 1991, Morand et al. 2000). A complementary modeling approach would use individual-based rules for association and dissociation of males and females to generate a wide range of social and mating structures (e.g., Cohen 1969). A key issue is how to represent genetic variation for host behavior in a biologically realistic way that results in observed social and mating systems. Behavioral ecologists have extensively discussed verbal and optimality-based models, but overlook the fact that it is unrealistic to describe selection on genes that directly translate to specific mating systems. To our knowledge, no empirical studies have examined whether genetic variation in either host social/mating behavior or parasite effects on such behavior exists in wild mammal species.

(5) **Evaluate the role of spatial and metapopulation processes.** Spatial considerations are important in defining the geographic scales at which social and sexual interactions take place. There are numerous situations in wild animal populations where substructuring occurs at multiple scales. For example, hamadryas baboons (*Papio hamadryas*) have a polygynous mating system, where related males and their harems are themselves organized into larger clans; these clans are organized into yet larger groups, which may interact socially in various ways. A number of critical questions relating to the ways in which the relative transmission of STDs versus OIDs might

scale with the level of social hierarchy arise from considering these social structures. How do the overall population dynamics and persistence of parasites depend on interactions within and among these levels of organization? How would this affect group dynamics and the types of social/sexual systems that evolve? It has been suggested that interactions between primate social systems and the distribution of resources and predators will determine social organization (Dobson & Lyles 1989). Clearly, these ideas apply with equal force to pathogens and parasites.

(6) **Examine the links between host sociality, parasite infection, and extinction risk.** Infectious diseases have become an important threat to wildlife populations, and links between sociality and parasite transmission suggest that host behavior might play an indirect role in species extinctions. Despite the possible interaction among these factors, to our knowledge no studies have examined characteristics of parasites associated with extinction risk or whether host social organization influences the spread of novel pathogens. For example, are social species more likely to be threatened by novel parasites than nonsocial species? These questions can be answered by integrating information already available from veterinary field surveys with the results of empirical studies on host behavior and population dynamics. An understanding of the interactions between sociality, disease transmission, and extinction risk could provide important insights for wildlife conservation.

ACKNOWLEDGMENTS

This work was conducted as part of the "Understanding the Ecology and Evolution of Infectious Diseases in Mammalian Mating and Social Systems" working group supported by the National Center for Ecological Analysis and Synthesis, a Center funded by NSF (Grant #DEB-94-21535), the University of California at Santa Barbara, and the State of California. C.N. was supported by an NSF Postdoctoral Research Fellowship, J.A. was supported in part from NIH Grant #GM 60766-01, and P.H.T. gratefully acknowledges the support of a Queen Elizabeth II Fellowship from the Australian Research Council. C.N. and S.A. are thankful for support from NSF (Grant #DEB-0212096), and we thank Andrew Davis for editorial assistance.

The *Annual Review of Ecology, Evolution, and Systematics* is online at http://ecolsys.annualreviews.org

LITERATURE CITED

Alexander J, Stimson WH. 1988. Sex-hormones and the course of parasitic infection. *Parasitol. Today* 4:189–93

Anderson RM, Blythe SP, Gupta S, Konings E. 1989. The transmission dynamics of the human immunodeficiency virus type 1 in the male homosexual community in the United Kingdom: the influence of changes in sexual behaviour. *Philos. Trans. R. Soc. London Ser. B* 325:45–98

Anderson RM, May RM. 1979. Population biology of infectious diseases: Part 1. *Nature* 280:361–67

Anderson RM, May RM. 1992. *Infectious Diseases of Humans: Dynamics and Control*. New York: Oxford Univ. Press. 757 pp.

Anderson RM, May RM, McLean AR. 1988. Possible demographic consequences of AIDS in developing countries. *Nature* 332:228–34

Andersson M. 1994. *Sexual Selection*. Princeton, NJ: Princeton Univ. Press. 624 pp.

Antonovics J, Iwasa Y, Hassell MP. 1995. A generalized model of parasitoid, venereal, and vector based transmission processes. *Am. Nat.* 145:661–75

Arneberg P. 2002. Host population density and body mass as determinants of species richness in parasite communities: comparative analyses of directly transmitted nematodes of mammals. *Ecography* 25:88–94

Arneberg P, Skorping A, Grenfell B, Read AF. 1998. Host densities as determinants of abundance in parasite communities. *Proc. R. Soc. London Ser. B* 265:1283–9

Barton R. 1985. Grooming site preferences in primates and their functional implications. *Int. J. Primatol.* 6:519–32

Begon M, Hazel SM, Baxby D, Bown K, Cavanagh R, et al. 1999. Transmission dynamics of a zoonotic pathogen within and between wildlife host species. *Proc. R. Soc. London Ser. B* 266(1432):1939–45

Bininda-Emonds ORP, Gittleman JL, Purvis A. 1999. Building large trees by combining phylogenetic information: a complete phylogeny of the extant Carnivora (Mammalia). *Biol. Rev.* 74:143–75

Blower SM, McLean AR. 1991. Mixing ecology and epidemiology. *Proc. R. Soc. London Ser. B* 245:187–92

Boily MC, Masse B. 1997. Mathematical models of disease transmission: a precious tool for the study of sexually transmitted diseases. *Can. J. Publ. Health* 88:255–65

Brown CR, Brown MB. 1986. Ectoparasitism as a cost of coloniality in cliff swallows (*Hirundo pyrrhonota*). *Ecology* 67:1206–18

Bull JJ. 1994. Virulence. *Evolution* 48:1423–37

Bundy DAP. 1988. Sexual effects on parasite infection—gender-dependent patterns of infection and disease. *Parasitol. Today* 4:186–89

Carranza J, Garcia-Muñoz AJ, Vargas JD. 1995. Experimental shifting from harem defense to territoriality in rutting red deer. *Anim. Behav.* 49:551–54

Clayton DH, Walther BA. 2001. Influence of host ecology and morphology on the diversity of Neotropical bird lice. *Oikos* 94:455–67

Cleaveland S. 1998. The growing problem of rabies in Africa. *Trans. R. Soc. Trop. Med. Hyg.* 92:131–34

Cleaveland S, Hess GR, Dobson AP, Laurenson MK, McCallum HI, et al. 2002. The role of pathogens in biological conservation. See Hudson et al. 2002, pp. 139–150

Clutton-Brock TH. 1989. Mammalian mating systems. *Proc. R. Soc. London Ser. B* 236:339–72

Clutton-Brock TH, Harvey PH. 1977. Primate ecology and social organization. *J. Zool.* 183: 1–39

Clutton-Brock TH, Parker GA. 1992. Potential reproductive rates and the operation of sexual selection. *Q. Rev. Biol.* 67:437–56

Cohen JE. 1969. Natural primate troops and a stochastic population model. *Am. Nat.* 103: 455–77

Côté IM, Poulin R. 1995. Parasitism and group size in social animals: a meta-analysis. *Behav. Ecol.* 6:159–65

Courchamp F, Artois M, Yoccoz N, Pontier D. 1998. Epidemiology of feline immunodeficiency virus within a rural cat population. *Epidemiol. Infect.* 121:227–38

Cunningham AA. 1996. Disease risks of wildlife translocations. *Conserv. Biol.* 10: 349–53

Daszak P, Cunningham AA, Hyatt AD. 2000. Emerging infectious diseases of

wildlife—threats to biodiversity and human health. *Science* 287:443–49
Davies CR, Ayres JM, Dye C, Deane LM. 1991. Malaria infection rate of Amazonian primates increases with body weight and group size. *Funct. Ecol.* 5:655–62
Di Bitetti MS, Vidal EM, Baldovino MC, Benesovsky V. 2000. Sleeping site preference in tufted capuchin monkeys (*Cebus apella nigritus*). *Am. J. Primatol.* 50:257–74
Dobson AP, Foufopoulos J. 2001. Emerging infectious pathogens in wildlife. *Philos. Trans. R. Soc. London Ser. B* 356:1001–12
Dobson AP, Lyles AM. 1989. The population dynamics and conservation of primate populations. *Conserv. Biol.* 3:362–80
Dobson AP, Lyles A. 2000. Black-footed ferret recovery. *Science* 288:985–88
Dobson AP, Meagher M. 1996. The population dynamics of brucellosis in the Yellowstone National Park. *Ecology.* 77(4):1026–36
Donovan B. 2000. The repertoire of human efforts to avoid sexually transmissible diseases: past and present. Part 2: Strategies used during or after sex. *Sex. Transm. Infect.* 76:88–93
Dugatkin LA. 2001. *Model Systems in Behavioral Ecology*. Princeton, NJ: Princeton Univ. Press. 551 pp.
Dunbar RIM. 1988. *Primate Social Systems.* Ithaca, NY: Cornell Univ. Press. 373 pp.
Durden LA, Keirans JE. 1996. Host-parasite coextinction and the plight of tick conservation. *Am. Entomol.* 42:87–91
Eisenberg JF. 1981. *The Mammalian Radiations*. Chicago: Univ. Chicago Press. 610 pp.
Emlen ST, Oring LW. 1977. Ecology, sexual selection, and the evolution of mating systems. *Science* 197:215–23
Ewald PW. 1994. *Evolution of Infectious Disease*. Oxford, UK: Oxford Univ. Press. 298 pp.
Ezenwa VO. 2002. *Behavioral and nutritional ecology of gastrointestinal parasitism in African bovids*. PhD thesis. Princeton Univ., New Jersey. 161 pp.
Freeland WJ. 1976. Pathogens and the evolution of primate sociality. *Biotropica* 8:12–24
Freeland WJ. 1979. Primate social groups as biological islands. *Ecology* 60:719–28
Fultz PN, Gordon TP, Anderson DC, McClure HM. 1990. Prevalence of natural infection with simian immunodeficiency virus and simian T-cell leukemia virus type I in a breeding colony of sooty mangabey monkeys. *AIDS* 4:619–25
Funk SM, Fiorello CV, Cleaveland S, Gompper ME. 2001. The role of disease in carnivore ecology and conservation. In *Carnivore Conservation*, ed. JL Gittleman, S Funk, D Macdonald, RK Wayne, pp. 443–66. Cambridge, UK: Cambridge Univ. Press. 690 pp.
Garnett GP, Aral SO, Hoyle DV, Cates W, Anderson RM. 1997. The natural history of syphilis. Implications for the transmission dynamics and control of infection. *Sex. Transm. Dis.* 24:185–200
Geffen E, Gompper ME, Gittleman JL, Luh HK, Macdonald DW. 1996. Size, life-history traits, and social organization in the Canidae: a reevaluation. *Am. Nat.* 147:140–60
Getz WM, Pickering J. 1983. Epidemic models: Thresholds and population regulation. *Am. Nat.* 121:892–98
Gittleman JL. 1996. *Carnivore Behavior, Ecology and Evolution*, Vol. 2. Ithaca, NY: Cornell Univ. Press. 644 pp.
Gompper ME, Williams ES. 1998. Parasite conservation and the black-footed ferret recovery program. *Conserv. Biol.* 12:730–32
Graves BM, Duvall D. 1995. Effects of sexually transmitted diseases on heritable variation in sexually selected systems. *Anim. Behav.* 50:1129–31
Gregory RD. 1990. Parasites and host geographic range as illustrated by waterfowl. *Funct. Ecol.* 4:645–54
Halvorsen O. 1986. On the relationship between social status of host and risk of parasite infection. *Oikos* 47:71–74
Hart BJ, Korinek E, Brennan P. 1987. Postcopulatory genital grooming in male rats: prevention of sexually transmitted infections. *Physiol. Behav.* 41:321–25

Hart BL. 1990. Behavioral adaptations to pathogens and parasites: five strategies. *Neurosci. Biobehav. Rev.* 14:273–94

Harvell CD, Mitchell CE, Ward JR, Altizer S, Dobson A, Samuels MD. 2002. Climate warming and disease risks for terrestrial and marine biota. *Science* 296:2158–62

Harvey PH, Keymer AE. 1991. Comparing life histories using phylogenies. *Philos. Trans. R. Soc. London Ser. B* 332:31–39

Harvey PH, Pagel MD. 1991. *The Comparative Method in Evolutionary Biology*. New York: Oxford Univ. Press. 239 pp.

Harvey PH, Rambaut A. 2000. Comparative analyses for adaptive radiations. *Proc. R. Soc. London Ser. B* 355:1–7

Hausfater G, Meade BJ. 1982. Alternation of sleeping groves by yellow baboons (*Papio cynocephalus*) as a strategy for parasite avoidance. *Primates* 23:287–97

Hausfater G, Watson DF. 1976. Social and reproductive correlates of parasite ova emissions by baboons. *Nature* 262:688–89

Hethcote HW, Yorke JA. 1984. *Gonorrhea Transmission Dynamics and Control*. New York: Springer-Verlag. 105 pp.

Hilton-Taylor C. 2000. *2000 IUCN Red List of Threatened Species*. Morges, Swit.: IUCN

Hilton-Taylor C. 2002. *2002 IUCN Red List of Threatened Species*. Morges, Swit.: IUCN

Holmes KK, Sparling PF, Mardh PA, Lemon SM, Stamm WE, et al. 1994. *Sexually Transmitted Diseases*. New York: McGraw-Hill. 1079 pp.

Hooper RR, Reynolds GH, Jones OG, Zaidi A, Wiesner PJ. 1978. Cohort study of venereal disease. I: The risk of gonorrhea transmission from infected women to men. *Am. J. Epidemiol.* 108:136–44

Hoogland JL. 1979. Aggression, ectoparasitism, and other possible costs of prairie dog (Sciuridae, *Cynomys* spp.) coloniality. *Behaviour* 69:1–35

Hudson PJ, Rizzoli A, Grenfell BT, Heesterbeek H, Dobson AP, eds. *The Ecology of Wildlife Diseases*. New York: Oxford Univ. Press. 197 pp.

Huffman MA. 1997. Current evidence for self-medication in primates: A multidisciplinary perspective. *Year. Phys. Anthropol.* 40:171–200

Hutchings M, Kyriazakis I, Anderson D, Gordon I, Coop R. 1998. Behavioral strategies used by parasitized and non-parasitized sheep to avoid ingestion of gastro-intestinal nematodes associated with faeces. *J. Anim. Sci.* 67:97–106

Jacquez JA, Simon CP, Koopman J, Sattenspiel L, Perry T. 1988. Modeling and analyzing HIV transmission: the effect of contact patterns. *Math. Biosci.* 92:119–99

Jensen T, van de Bildt M, Dietz HH, Andersøn TH, Hammer AS. 2002. Another phocine distemper outbreak in Europe. *Science* 297:209

Jones KE, Purvis A, MacLarnon A, Bininda-Emonds ORP, Simmons N. 2002. A phylogenetic supertree of the bats (Mammalia: Chiroptera). *Biol. Rev.* 77:223–59

Kat PW, Alexander KA, Smith JS, Munson L. 1995. Rabies and African wild dogs in Kenya. *Proc. R. Soc. London Ser. B* 262:229–33

Knell RJ. 1999. Sexually transmitted disease and parasite-mediated sexual selection. *Evolution* 53:957–61

Knell RJ, Begon M, Thompson DJ. 1996. Transmission dynamics of *Bacillus thuringiensis* infecting *Plodia interpunctella*: a test of the mass action assumption with an insect pathogen. *Proc. R. Soc. London Ser. B* 263:75–81

Kuris AM, Blaustein AR, Alio JJ. 1980. Hosts as islands. *Am. Nat.* 16:570–86

Levin BR. 1996. The evolution and maintenance of virulence in microparasites. *Emerg. Infect. Dis.* 2:93–102

Levin J, Hilliard J, Lipper S, Butler T, Goodwin W. 1988. A naturally occurring epizootic of Simian Agent 8 in the baboon. *Lab. Anim. Sci.* 38(4):394–97

Liu FG, Miyamoto MM, Freire NP, Ong PQ, Tennant MR, et al. 2001. Molecular and morphological supertrees for eutherian (placental) mammals. *Science* 291(5509):1786–89

Lockhart AB, Thrall PH, Antonovics J. 1996. Sexually transmitted diseases in animals:

ecological and evolutionary implications. *Biol. Rev. Camb. Philos. Soc.* 71:415–71

Loehle C. 1995. Social barriers to pathogen transmission in wild animal populations. *Ecology* 76:326–35

MacPhee RDE, Fleming C. 1999. Requiem Aeternam: The last five hundred years of mammalian species extinctions. In *Extinctions in Near Time*, ed. RDE MacPhee, pp. 333–71. New York: Plenum. 394 pp.

Martins EP, Hansen TF. 1996. The statistical analysis of interspecific data: a review and evaluation of phylogenetic comparative methods. In *Phylogenies and the Comparative Method In Animal Behavior*, ed. EP Martins. pp. 22–75. New York: Oxford Univ. Press. 415 pp.

Messenger SL, Molineux IJ, Bull JJ. 1999. Virulence evolution in a virus obeys a trade-off. *Proc. R. Soc. London Ser. B* 266:397–404

Mills S, Benjarattanaporn P, Bennett A, Na Pattalung R, Sundhagul D, et al. 1997. HIV risk behavioral surveillance in Bangkok, Thailand: sexual behavior trends among eight population groups. *AIDS* 11:S43–51

Moe S, Holand O, Colman J, Reimers E. 1999. Reindeer (*Rangifer tarandus*) response to feces and urine from sheep (*Ovis aries*) and reindeer. *Rangifer* 19:55–60

Møller AP, Dufva R, Allander K. 1993. Parasites and the evolution of host social behavior. *Adv. Stud. Behav.* 22:65–102

Møller AP, Dufva R, Erritzoe J. 1993. Host immune function and sexual selection in birds. *J. Evol. Biol.* 11:703–19

Møller AP, Merino S, Brown CR, Robertson RJ. 2001. Immune defense and host sociality: a comparative study of swallows and martins. *Am. Nat.* 158:136–45

Moore J. 2002. *Parasites and the Behavior of Animals*. New York: Oxford Univ. Press

Moore J, Simberloff D, Freehling M. 1988. Relationships between bobwhite quail social-group size and intestinal helminth parasitism. *Am. Nat.* 131:22–32

Moore SL, Wilson K. 2002. Parasites as a viability cost of sexual selection in natural populations of mammals. *Science* 297:2015–18

Mooring MS, Hart BL. 1992. Animal grouping for protection from parasites: Selfish herd and encounter-dilution effects. *Behaviour* 123:173–93

Morand S, Poulin R. 1998. Density, body mass and parasite species richness of terrestrial mammals. *Evol. Ecol.* 12:717–27

Morand S, Hafner MS, Page RDM, Reed DL. 2000. Comparative body size relationships in pocket gophers and their chewing lice. *Biol. J. Linn. Soc.* 70:239–49

Muller-Graf CDM, Collins DA, Woolhouse MEJ. 1996. Intestinal parasite burden in five troops of olive baboons (*Papio cynocephalus anubis*) in Gombe Stream National Park, Tanzania. *Parasitology* 112(5):489–97

Nordling D, Andersson M, Zohari S, Gustafsson L. 1998. Reproductive effort reduces specific immune response and parasite resistance. *Proc. R. Soc. London Ser. B* 265:1291–98

Nunn CL. 2003. Behavioral defenses against sexually transmitted diseases in primates. *Anim. Behav.* In press

Nunn CL, Altizer S. 2003. Sexual selection, behavior and sexually transmitted diseases. In *Sexual Selection in Primates: New and Comparative Perspectives*, ed. PM Kappeler, CP van Schaik. In press

Nunn CL, Altizer SM, Jones KE, Sechrest W. 2003a. Comparative tests of parasite species richness in primates. *Am. Nat.* In press

Nunn CL, Barton RA. 2001. Comparative methods for studying primate adaptation and allometry. *Evol. Anthropol.* 10:81–98

Nunn CL, Gittleman JL, Antonovics J. 2000. Promiscuity and the primate immune system. *Science* 290:1168–70

Nunn CL, Gittleman JL, Antonovics J. 2003b. A comparative study of white blood cell counts and disease risk in carnivores. *Proc. R. Soc. London Ser. B* 270:347–356

O'Keefe KJ, Antonovics J. 2002. Playing by different rules: the evolution of virulence in sterilizing pathogens. *Am. Nat.* 159:597–605

Oriel JD, Hayward AHS. 1974. Sexually-transmitted diseases in animals. *Brit. J. Ven. Dis.* 50:412–20

Ostfeld RS. 1985. Limiting resources and territoriality in microtine rodents. *Am. Nat.* 126:1–15

Packer C, Altizer S, Appel M, Brown E, Martenson J. 1999. Viruses of the Serengeti: Patterns of infection and mortality in African lions. *J. Anim. Ecol.* 68(6):1161–78

Poulin R. 1991. Group-living and infestation by ectoparasites in passerines. *Condor* 93:418–23

Poulin R. 1995. Phylogeny, ecology, and the richness of parasite communities in vertebrates. *Ecol. Monogr.* 65:283–302

Poulin R. 1998. Comparison of three estimators of species richness in parasite component communities. *J. Parasitol.* 84:485–90

Purvis A. 1995. A composite estimate of primate phylogeny. *Philos. Trans. R. Soc. London Ser. B* 348:405–21

Purvis A, Rambaut A. 1995. Comparative analysis by independent contrasts (CAIC): an Apple Macintosh application for analysing comparative data. *Comp. Appl. Biosci.* 11: 247–51

Ranta E. 1992. Gregariousness vs. solitude: another look at parasite faunal richness in Canadian freshwater fishes. *Oecologia* 89:150–52

Roberts MG, Dobson AP, Arneberg P, de Leo GA, Krecek RC. 2002. Parasite community ecology and biodiversity. See Hudson et al. 2002, pp. 63–82

Roelke-Parker ME, Munson L, Packer C, Kock R, Cleaveland S. 1996. A canine distemper virus epidemic in Serengeti lions (*Panthera leo*). *Nature* 379:441–45

Rózsa L. 1997. Patterns in the abundance of avian lice (Phthiraptera: Ambylcera, Ischnocera). *J. Avian Biol.* 28(3):249–54

Rubenstein DI, Hohmann ME. 1989. Parasites and social behavior of island feral horses. *Oikos.* 55:312–20

Sheldon BC, Verhulst S. 1996. Ecological immunology: Costly parasite defences and trade-offs in evolutionary ecology. *Trends Ecol. Evol.* 11:317–21

Smith G, Dobson AP. 1992. Sexually transmitted diseases in animals. *Parasitol. Today* 8:159–66

Smuts BB, Cheney DL, Seyfarth RM, Wrangham RW, Struhsaker TT. 1987. *Primate Societies*. Chicago: Univ. Chicago Press. 578 pp.

Thrall PH, Antonovics J. 1997. Polymorphism in sexual versus non-sexual disease transmission. *Proc. R. Soc. London Ser. B* 264:581–87

Thrall PH, Antonovics J, Hall DW. 1993. Host and pathogen coexistence in vector-borne and venereal diseases characterized by frequency-dependent disease transmission. *Am. Nat.* 142:543–52

Thrall PH, Antonovics J, Bever JD. 1997. Sexual transmission of disease and host mating systems: within-season reproductive success. *Am. Nat.* 149:485–506

Thrall PH, Antonovics J, Wilson WG. 1998. Allocation to sexual versus nonsexual disease transmission. *Am. Nat.* 151:29–45

Thrall PH, Antonovics J, Dobson AP. 2000. Sexually transmitted diseases in polygynous mating systems: prevalence and impact on reproductive success. *Proc. R. Soc. London Ser. B* 267:1555–63

Trivers RL. 1972. Parental investment and sexual selection. In *Sexual Selection and the Descent of Man*, ed. B Campbell, pp. 136–79. Chicago: Aldine

van Schaik CP, Dunbar RIM. 1990. The evolution of monogamy in large primates: a new hypothesis and some crucial tests. *Behaviour* 115:30–62

van Schaik CP, van Noordwijk MA, Nunn CL. 1999. Sex and social evolution in primates. In *Comparative Primate Socioecology*, ed. PC Lee, pp. 204–40. Cambridge, UK: Cambridge Univ. Press. 424 pp.

Wallis J, Lee DR. 1999. Primate conservation: The prevention of disease transmission. *Int. J. Primatol.* 20:803–26

Westoby M, Leishman MR, Lord JM. 1995. On misinterpreting the phylogenetic correction. *J. Ecol.* 83:531–34

Wilcove DS, Rothstein D, Dubow J, Phillips A, Losos E. 1997. Quantifying threats to imperiled species in the United States. *Bio-Science* 48:607–15

Williams ES, Thorne ET, Appel MJG, Oakleaf B. 1992. Canine distemper in black footed ferrets (*Mustela nigripes*) from Wyoming. *J. Wildl. Dis.* 24:385–98

Wilson DE, Reeder DM, ed. 1993. *Mammal Species of the World.* Washington, DC: Smithson. Inst. Press. 1206 pp.

Woodford MH, Butynski TM, Karesh WB. 2002. Habituating the great apes: the disease risks. *Oryx* 36:153–60

Woodroffe R. 1999. Managing disease threats to wild mammals. *Anim. Conserv.* 2:185–93

Woodroffe R, Ginsberg JR, Macdonald D. 1997. *The African Wild Dog.* Morges, Swit.: IUCN. 166 pp.

Woolhouse MEJ, Taylor LH, Haydon DT. 2001. Population biology of multihost pathogens. *Science.* 292:1109–12

Zuk M. 1990. Reproductive strategies and disease susceptibility: an evolutionary viewpoint. *Parasitol. Today* 6:231–32

Annu. Rev. Ecol. Evol. Syst. 2003. 34:549–74
doi: 10.1146/annurev.ecolsys.34.011802.132400

First published online as a Review in Advance on July 30, 2003

The Community-Level Consequences of Seed Dispersal Patterns

Jonathan M. Levine[1] and David J. Murrell[2]
[1]*Department of Ecology, Evolution, and Marine Biology, University of California, Santa Barbara, California 93106; email: levine@lifesci.ucsb.edu*
[2]*Center for Population Biology, Imperial College at Silwood Park, Ascot, Berkshire SL5 7PY, United Kingdom; email: d.murrell@imperial.ac.uk*

Key Words abundance, aggregation, distribution, diversity

■ **Abstract** Because it lays the template from which communities develop, the pattern of dispersed seed is commonly believed to influence community structure. To test the validity of this notion, we evaluated theoretical and empirical work linking dispersal kernels to the relative abundance, distribution, dispersion, and coexistence of species. We found considerable theoretical evidence that seed dispersal affects species coexistence by slowing down exclusion through local dispersal and a competition-dispersal trade-off, yet empirical support was scant. Instead, most empirical investigations examined how dispersal affects species distribution and dispersion, subjects with little theory. This work also relied heavily on dispersal proxies and correlational analyses of community patterns, methods unable to exclude alternative hypotheses. Owing to the overall dichotomy between theory and empirical results, we argue that the importance of dispersal cannot be taken for granted. We conclude by advocating experiments that manipulate the seed dispersal pattern, and models that incorporate empirically documented dispersal kernels.

INTRODUCTION

Although ecologists have long appreciated the importance of dispersal for the spread and persistence of populations (Harper 1977, Howe & Smallwood 1982, Skellam 1951), the last decade has witnessed a surge of interest in how this phase of the life cycle influences community structure (Bullock et al. 2002, Cain et al. 2000, Clobert et al. 2001, Nathan 2003, Nathan & Muller-Landau 2000, Wang & Smith 2002). Theoretical models have been an important motivating force, with numerous studies emphasizing the importance of spatial structure in influencing species interactions. Particularly tantalizing are results suggesting that local seed dispersal or competition colonization trade-offs favor the coexistence of competitors (Murrell et al. 2001, Pacala 1997, Rees et al. 1996, Tilman 1994). Results such as these are mirrored in the tropical forest literature, where theories emphasizing the importance of limited dispersal have gained prominence (Hubbell

2001). Meanwhile, empirical ecology is increasingly implicating dispersal as an important control over species diversity. Numerous recent experimental and observational results suggest that local communities are seed limited, with diversity limited largely by the regional species pool (Cornell 1993, Srivastava 1999, Turnbull et al. 2000). Interest has also been heightened by dispersal-mediated processes in conservation biology, including habitat fragmentation (Hanski & Gilpin 1997), species ability to migrate with climate change (Clark 1998), and the spread of biological invasions (Drake et al. 1989).

Although the influence of dispersal on community structure is only beginning to be rigorously examined, ecologists seem to expect an important causal relationship. This is likely because dispersal is believed to set the template from which community patterns develop, and is well documented to influence population spread and persistence (Hanski & Gilpin 1997, Harper 1977, Skellam 1951). However, spread and persistence are very different response variables from most measures of community structure, including patterns of abundance, distribution, and coexistence. More importantly, the template laid by dispersal, often referred to as the seed rain, is influenced by other potentially more important factors, including the distribution, density, and fecundity of parent plants (Clark et al. 1998, Platt 1975), as well as landscape features that trap seeds (Schneider & Sharitz 1988) (Figure 1). Moreover, as noted by several authors (Nathan & Muller-Landau 2000, Schupp & Fuentes 1995, Wang & Smith 2002), a number of important life history processes occur between dispersal and the progression to adult plants. Species interactions and environmental factors strongly influence these transitions and can significantly change the template laid by dispersal (Figure 1). Thus, unless populations are substantially seed limited, the importance of the dispersal kernel for the abundance, distribution, and diversity of species should not be assumed a priori.

In this paper, we critically review theoretical and empirical work relating patterns of dispersal to spatial patterns in communities in an attempt to provide a critical framework for research in this area. More specifically, we ask how important the specific seed dispersal kernel or seed shadow is for explaining relative abundance, distributions, and coexistence in natural communities. By dispersal kernel, we mean the probability density function describing the probability of seed transport to various distances from the parent plant. Although both theoretical and empirical approaches address the importance of seed dispersal for community structure, a large gap exists in the coverage and motivation behind the different methods. As we demonstrate, the modeling work is focused largely on coexistence, motivated by interest in how incorporating spatial processes changes the predictions of earlier nonspatial models. Meanwhile, the empirical work focuses largely on dispersal's effect on species distributions, motivated by natural history observations related to dispersal kernels, the movement of animal vectors, seed trapping patterns, or the clumping of parent plants. Thus, of primary interest in our review is how empirical results match the predictions of theory.

To encourage a greater exchange of ideas and questions between empirical and theoretical approaches, we organize our review around the influence of dispersal kernels on four key features of community structure: patterns of relative

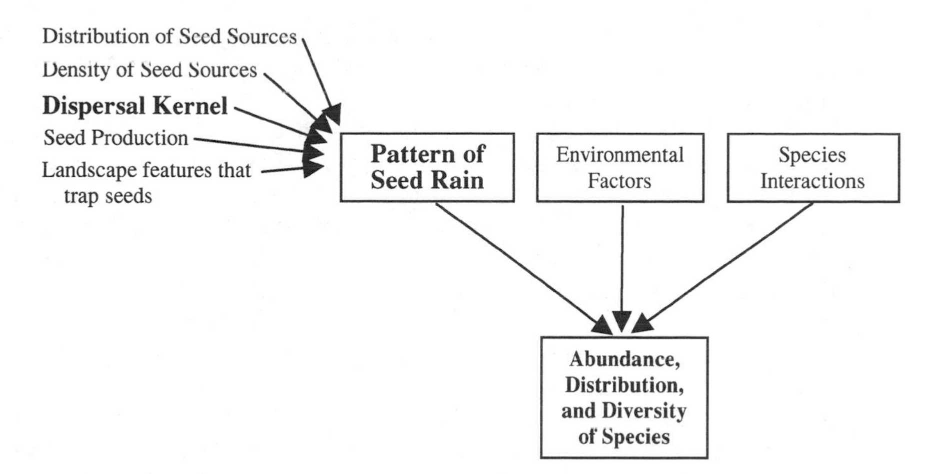

Figure 1 Pathway by which the shape of species dispersal kernels can influence the abundance, distribution, and diversity of species. Shown in bold is the link between dispersal kernals and community patterns that we examine in this review. We emphasize the multitude of factors other than dispersal kernels influencing the seed rain, and the multitude of factors other than seed rain influencing the abundance, distribution, and diversity of species.

abundance, distribution, dispersion, and coexistence. For each of these features, we first chart theoretical predictions on the importance of dispersal and then review evidence for these predictions in the empirical literature. We located our papers by electronically searching the Science Citation Index and examining references therein. Because of the large body of theoretical work on coexistence, we divide this literature into two sections, one describing the effects of local dispersal, the other focused on competition-dispersal trade-offs. For similar reasons, we review the work examining dispersal effects on dispersion separately from work on other types of distribution patterns. We do not review dispersal effects on persistence and spread (Hanski & Gilpin 1997, Howe & Smallwood 1982, Shigesada & Kawasaki 2002, Skellam 1951), the evolution of dispersal (Clobert et al. 2001), or the effects of colonization on community patterns achieved through dormancy or high fecundity. Still, it should be noted that both dormancy and fecundity combine with dispersal to influence colonization (Harper 1977).

RELATIVE ABUNDANCE

Theoretical Background

Over the past decade, it has become increasingly common to regard local populations as embedded within a larger metapopulation structure. In classic (Levins 1969) and contemporary (Hanski & Gilpin 1997) metapopulation models, species abundance at the metapopulation scale is a function of the colonization and extinction of individual subpopulations or patches. Thus, species ability to disperse may be an important determinant of relative abundance—which species are common and which are rare.

The expectation that dispersal is positively related to abundance also emerges in several other types of models (Turnbull et al. 2000). In spatial mass effect models (Shmida & Ellner 1984), species occupy marginally unfavorable habitats because of seed input from other, more favorable, locations. Greater input caused by increased dispersal can enhance abundance in these systems. Similarly, if community dynamics can be conceived as a competitive lottery (Chesson & Warner 1981), species can dominate systems by having high colonization rates, achievable through effective dispersal or high fecundity. Greater abundance of better-dispersing species is also predicted under the Janzen-Connell hypothesis (Connell 1971, Howe 1989, Janzen 1970, Schupp & Fuentes 1995), where dispersal away from the parent plant confers greatly reduced density-dependent mortality (Ellner 2001, Law et al. 2003, Pacala & Silander 1985). Although similar predictions emerge from these models collectively, it is the metapopulation predictions that have been most influential in motivating empirical tests of how abundance at the metapopulation scale relates to dispersal ability (Eriksson 1997, Eriksson & Jakobsson 1998).

These tests, however, may not have considered that long-range dispersal is sometimes disadvantageous in models. For example, if the landscape is variable in (abiotic) quality, depending on the spatial scale of favorable and unfavorable patches, short-range dispersal may lead to higher abundance than long-range dispersal (Bolker 2003, Dockery et al. 1998, Travis & Dytham 1999). In these cases the importance of remaining in a good patch more than outweighs the increased intraspecific competition often resulting from short-range dispersal.

Empirical Evidence for a Relationship Between Dispersal and Abundance

In addition to theoretical predictions, empirical work at the population level would seem to suggest that species ability to disperse is an important predictor of relative abundance, at least in some systems. A number of seed addition experiments (see Turnbull et al. 2000 for review) suggest that for early successional systems in particular, seed limitation is common, and thus greater dispersal should confer greater abundance. Nonetheless, in our review of empirical evidence, we found little support for the expectation that relative abundance was controlled by dispersal patterns. For example, Rabinowitz & Rapp (1981) found no relationship between abundance and dispersal under field conditions in a Missouri prairie. Similarly, Eriksson & Jakobsson (1998) found that dispersal mode was unrelated to abundance and geographic range size for over 80 grassland plant species in Sweden. Also in Sweden, the ability of 17 species to disperse on the fur of animals was unrelated to their occupancy rates in seminatural grasslands (Kiviniemi & Eriksson 1999). In the herbaceous flora of central England, seed dispersal as estimated through terminal velocity was very weakly related to species abundance as assessed by their UK range (Thompson et al. 1999). Thompson et al. (1999) and Eriksson & Jakobsson (1998) reviewed the literature relating dispersal mode to geographic range size more generally and found similarly ambiguous results.

Support for the notion that dispersal dictates commonness and rarity might seem to come from annual plant communities. It is often found that common species are small seeded, whereas large-seeded species are consistently rare (Guo et al. 2000, Levine & Rees 2002, Maranon & Grubb 1993, Rees 1995, Rees et al. 1996). However, although small-seeded species may be better colonizers, this is more likely the result of a relationship between seed size and fecundity rather than seed size and dispersal. Small-seeded species consistently produce more seeds (Jakobsson & Eriksson 2000, Leishman 2001, Rees 1995, Shipley & Dion 1992, Turnbull et al. 1999), yet seed size is very poorly related to dispersal (Carey & Watkinson 1993, Westoby et al. 1996).

How can we explain the absence of a relationship between abundance and seed dispersal when experiments often show seed-limited populations? One likely possibility is that the evidence for seed limitation is more a product of low seed production or a low density of adult plants than of poor dispersal. In addition, numerous other processes, such as competition, dormancy, and habitat variation, may simply be more important than dispersal in controlling relative abundance patterns in communities (Figure 1). Because dispersal can clearly influence the rate of species arrival, it may be more important in controlling temporal patterns of abundance after disturbance. In a simulation study, Hovestadt et al. (2000) showed that long-range dispersers were more abundant in the colonization phase immediately after disturbance, but depending on model details, short-range dispersers subsequently obtained a higher relative abundance. Indeed, on lake islands in Sweden, species lacking traits for water, wind, or animal dispersal are rare or absent early in succession (Rydin & Borgegard 1991), but become abundant years later. Despite these temporal patterns, we conclude that little empirical evidence supports the expectation that interspecific differences in dispersal control commonness and rarity at metapopulation scales.

SPATIAL DISTRIBUTION

Theoretical Background

The distance and direction a plant disperses its seeds should play an important role in the distribution of that species. However, as illustrated in Figure 1 (left), the influence of the dispersal kernel will depend strongly on the importance of other factors influencing the seed rain. For example, Levine (2003) modeled a streamside plant assemblage where species occur in discrete habitat patches linearly arrayed along the channel. Species dispersed 60% of their seeds to other patches and 95% of those seeds to patches downstream. In such a system, one naturally expects greater deposition downstream than upstream. However, Levine demonstrated that with a uniform distribution of parent plants throughout the system and a reasonable dispersal kernel, most seed input comes from the one or two patches upstream of the target patch. Thus, downstream of the first few patches in the system, seed input increases only marginally. Still, this relatively small downstream variation in seed deposition could drive downstream increases in population size and diversity, but only if the component populations are highly seed limited, as resulting from very low fecundity or recent disturbance. This result demonstrates two points introduced by Figure 1. First, the relationship between the seed-dispersal kernel and the seed rain can be complicated, deserving significantly more attention from both theoreticians and empiricists. Second, for patterns of seed rain to influence species distributions, the component populations must be constrained by seed availability and not species interactions or environmental factors.

The Importance of Seed-Trapping Agents

In contrast to theory, numerous empirical studies have related seed deposition to species distributions, yet in most of the work, spatial variation in deposition is not easily related to the dispersal kernels. Instead, it is driven largely by landscape elements that trap seeds or propagules. Differences in seed morphology still influence deposition patterns but do so largely through their interaction with the trapping agents. For example, Schneider & Sharitz (1988) tracked the dispersal of tree seeds in a swamp forest in the southeastern United States. Seeds were initially gravity dispersed, but then as water levels rose, they were redistributed by hydrochory to locations against emergent structures including twigs, trees, logs, branches, knees, and stumps. Some of these substrates, such as trees and tree knees, provided elevated stable microsites, and tree seedlings were disproportionately found in these locations. Importantly, seed morphology influenced deposition and seedling patterns across species. The small, angular cypress seeds were more frequently trapped by knees than the larger ellipsoid tupelo fruits. Consequently, 45% of the cypress seedlings occurred on knees in comparison to only 27% of tupelo seedlings.

Several studies conducted in northern Swedish rivers have experimentally released propagules or propagule mimics and correlated their deposition with species distributions and diversity. Although not a seed-dispersal study, Johansson & Nilsson (1993) tagged uniformly sized ramets of the vegetatively dispersing *Ranunculus lingua* and correlated their deposition with the presence of established plants. They found that ramet deposition was disproportionately high in curves and at obstacles along the river, also the locations where established stands tended to be found (see Andersson et al. 2000 for a similar approach focusing on diversity). Patterns such as these may relate to propagule characteristics. Nilsson et al. (1991) found that long-floating species tended to be more abundant in areas that trapped floating seed mimics, whereas the distribution of short-floating species was unrelated to trapping patterns. Danvind & Nilsson (1997) found no correlation between floating time and distribution patterns along an alpine Swedish river.

Seed deposition driven by trapping has also been correlated with species distributions in nonriverine studies. Rabinowitz (1978) hypothesized that the different sized seeds of the three mangrove dominants in Panama were trapped at different points along the tidal elevation gradient, matching the elevational distribution of adult trees. Support was provided by the experimental transport of mangrove propagules outside their tidal zone, which showed survivorship across the elevation gradient. Nonetheless, recent work (W. Sousa & B. Mitchell, manuscript in review) estimating the dispersal kernels in the same mangrove forests has found no evidence for tidal sorting of propagules. In a similar system, Rand (2000) found that the restricted distribution of several salt marsh species to particular tidal elevations was more a function of competition and physiological stresses than seed deposition patterns. In contrast, in the Negev desert of Israel, *Anastica hierochuntica* forms reticulate distribution patterns that relate to seed deposition in the cracks that form with the wetting and drying of the soil (Friedman & Stein 1980). Peart & Clifford (1987) attributed the segregation of grass species across different soil types to the interaction of the soil texture and cracking with the different awn types of the various species.

In these studies, patterns of seed deposition result from structures or topography that traps or sorts propagules. Seed morphology then influences not only how seeds move, but maybe more importantly, where seeds stop, a less appreciated and more difficult to predict component of the dispersal kernel. The realized distribution of seeds may therefore be very different than the potential distribution predicted by the dispersal kernel alone. In addition, the local community surrounding a dispersing individual may strongly influence the dispersal kernel by trapping seeds and influencing the size of the disperser through competition. Consequently, further work clarifying the relationship between dispersal kernels and patterns of deposition (Figure 1, left) is particularly important for clarifying its influence on distributions (Alcantara et al. 2000).

Distributions Driven by Dispersal Kernels

Relatively few studies related seed-dispersal kernels to distribution patterns. In one of the only studies relating dispersal to the seed rain, Dalling et al. (2002) demonstrated that spatial patterns of seed rain for many neotropical pioneer trees were driven largely by limited dispersal. The resulting seed rain patterns were an important predictor of seedling abundance in forest gaps. Similarly, Platt & Weiss (1977) examined field-derived dispersal kernels and competitive traits of fugitive plants living on badger mounds in an Iowa prairie. The species segregate across a habitat gradient in mound density such that the more poorly dispersing species dominate the portion of habitat with closely spaced mounds, whereas better-dispersing species dominate the area with distantly spaced mounds. Parallel gradients in soil moisture and interspecific differences in physiological tolerance maintain the species segregation.

Several riverine studies begin with the assumption that asymmetric downstream dispersal drives longitudinal variation in seed deposition in these systems (but see Levine 2003). They then attribute downstream increases in diversity to dispersal. Nilsson et al. (1994) found that tributaries of the Vindel river in Sweden contained only a subset of those species found in the main stem, a result they attributed to the downstream transport and accumulation of seeds. Honnay et al. (2001) analyzed spatial patterns of diversity in plant communities along dendritic networks of small forest streams and came to similar conclusions (also see Friedman & Stein 1980). These studies, however, are simply correlational analyses of community patterns. No measurements of seed dispersal, deposition, or limitation are made, and thus as Nilsson et al. (1994) acknowledge, other processes, particularly changing environmental conditions along the downstream gradient, could also explain the diversity pattern. A more manipulative approach was taken by Levine (2001), who documented downstream increases in diversity and individual plant abundance along the Eel river in California. Experimentally released propagule mimics successfully dispersed to suitable microsites downstream, and the streambed assemblages were seed limited. Most importantly, when seed supply was experimentally equalized across the downstream gradient, there was no greater colonization of the most downstream habitats as compared with those 5 km upstream. Nonetheless, even with all this evidence, whether downstream dispersal is sufficient to drive downstream increases in seed deposition remains untested in this and other work.

Considering the relative rarity of hydrochory in nature, our review documented a disproportionate number of studies where water dispersal influenced species distributions. In part, this can be attributed to the obvious directional movement of water in many systems. In addition, the plant communities that line rivers and lakeshores are readily disturbed, likely preventing processes such as competition from exerting their full impacts. Last, the flooding that occurs in these habitats can export most of the seed, favoring seed limitation (Levine 2001). Despite the apparent importance of dispersal for distributions in systems with hydrochory, few empirical studies in these or other habitats tested for seed limitation or examined the processes occurring between seed arrival and the progression to adult plants. Thus, regardless of whether the distribution patterns are argued to result from trapping agents or dispersal kernels, the empirical evidence as a whole is highly correlational, often unable to exclude alternative explanations for the putative dispersal-driven distribution patterns.

DISPERSION

The aggregation of species in communities can influence the rate of competitive displacement (Stoll & Prati 2001), patterns of resource availability (Pastor et al. 1999), and macroecological patterns including species-area curves and species-abundance distributions (Chave et al. 2002). Because local dispersal is an obvious source of clumping, numerous studies attribute adult aggregation to restricted seed transport. However, as is true with distribution patterns more generally, a number of conditions have to be met before dispersal controls dispersion patterns (Schupp & Fuentes 1995). Not only must the component populations be so sparsely arrayed that local dispersal drives clumping of dispersed seeds, but density-dependent mortality cannot be so severe as to eliminate the aggregation of seedlings.

Models Where Local Dispersal Generates Aggregation

Bleher et al. (2002) used simulation models to show that local dispersal and, to a lesser degree, low adult density strongly influenced the clumping of forest trees. Bleher et al. accomplished this by constraining fecundity such that all species effectively produce one offspring, but the relevance of such growth rates to natural systems is unclear. Presumably, with different fecundity across species, some populations would grow and their patches would coalesce, whereas others would decline and their clumps contract. Like Bleher et al., Chave et al. (2002) examined clumping patterns in neutral models (where birth rates just balance death rates), and showed greater clumping with increased local dispersal. Chave et al. also explored the influence of local dispersal in models where stable coexistence is achieved through a competition-fecundity trade-off and density dependence. They showed that when these other processes generate coexistence, local dispersal can strongly influence clumping, which causes the species-area curve to rise more gradually than when species are more diffusely spread through the habitat. Dispersal-generated clumping can also influence ecosystem processes. In a simulation model of boreal forest where species differ in decomposability, Pastor et al. (1999) showed how local dispersal generates species aggregation that alters the spatial pattern of nitrogen availability.

Further support for the influence of local dispersal on clumping is explicitly or implicitly provided by studies demonstrating its impact on coexistence (reviewed below). Still, it is important to remember that dispersal is just one of several spatial processes that shape the pattern of individuals across the landscape (Figure 1) (Schupp & Fuentes 1995). For example, because interactions between individuals also tend to be localized in space, offspring experience intense kin competition with local dispersal. Theory shows that local interactions can cause the realized pattern of adults to be random, overdispersed (spatially segregated or evenly spaced) or clumped (Ellner 2001, Law et al. 2003, Molofsky et al 2002, Pacala & Silander 1985). Generally, to achieve aggregation in homogeneous space with local dispersal, interactions must occur on a spatial scale similar to that of the dispersal kernel (Ellner 2001, Law et al. 2003, Pacala & Silander 1985). If dispersal occurs over much larger scales than neighborhood interactions then an even spacing of conspecific individuals is expected (Law et al. 2003). Alternatively, processes other than dispersal (Figure 1), such as spatial heterogeneity in the external environment, or reduced pollination of isolated individuals, may also generate aggregation. In sum, theory suggests that depending on the spatial scale of density dependence, local dispersal can drive a wide range of dispersion patterns, patterns that can also be explained by variation in environment.

Empirical Evidence

A large body of empirical work documents local seed dispersal and an aggregated distribution of adults, and suggests a causal link between the two (e.g., Bleher & Bohning-Gaese 2001, Fragaso 1997, Westelaken & Maun 1985). However, as argued by Schupp & Fuentes (1995) and put forth by theoretical work more generally, because of the primacy of processes occurring between seed arrival and maturity to adult plants, this correlative evidence is insufficient to demonstrate dispersal-driven patterns. Indeed, Overton (1996) documented local dispersal of mistletoe, but this had no effect on the spatial autocorrelation in the number of mistletoe per tree. More generally, numerous studies have found that the aggregated distribution of seeds and seedlings following local dispersal tends to disappear as the seedlings mature (Barot et al. 1999, Houle 1995, Rey & Alcantara 2000, Schupp & Fuentes 1995). We thus concentrate our review on several more quantitative and comparative approaches to assessing the importance of dispersal for aggregation patterns.

Because the major alternative to dispersal for explaining a clumped species distribution involves patchy environmental variables, one approach for testing the importance of local dispersal is to use multiple regression to ask whether species are more clumped than can be explained by spatial variation in the environment alone. Svenning (2001) found that aggregation of four palm species in an Andean rainforest could not be explained by variation in forest structure, topographic-edaphic conditions, altitude, or aspect. This, coupled with greater densities of seedlings near adult plants, caused the investigator to implicate dispersal as a control over the clumped distribution of these species. With a similar approach, Svenning & Skov (2002) found that in a managed forest in Denmark, 20 of 60 understorey plant species exhibited aggregation unexplainable by environmental parameters and thus attributable to dispersal. Moreover, animal- and wind-dispersed species showed less clumping than species with less-efficient dispersal modes. In contrast to these results, the predominance of environmental variables in influencing tree clumping patterns was suggested for the palm *Borassa aethiopum* in an African savanna (Barot et al. 1999) and for species occupying South Pacific island (Webb & Fa'aumu 1999) and Malaysian (Plotkin et al. 2002) forest.

Similar approaches have been used to attribute species turnover through space to dispersal. Tuomisto et al. (2003) found that decreasing floristic similarity with distance in Amazonian rainforests could not be entirely explained by changing environments with distance, pointing to the potential importance of local dispersal. Moreover, the importance of geographic distance alone (a proxy for local dispersal) versus environmental factors was comparatively less important in pteridophytes than in the more poorly dispersed Melastome shrubs, further supporting the role of dispersal. Similarly, Condit et al. (2002) used predictions of Hubbell's (2001) neutral model to argue that local dispersal could explain patterns of species turnover in tropical rainforests over the scale of 0.2 to 50 km.

An alternative approach for assessing how seed transport influences patterns of aggregation assumes that dispersal mode is a predictor of dispersal kernels (Willson 1993), and compares clumping across species differing in dispersal mode. Working in a Costa Rican dry forest, Hubbell (1979) found that the rate at which density declined with distance from adults was steepest for species dispersed by mammals, followed by wind, and then by birds and bats. Condit et al. (2000) found that wind- or explosively dispersed species were less clumped than animal-dispersed species in a Panamanian forest, but this difference was nonsignificant. Nonetheless, in a Malaysian forest, dipterocarps, with their poorly dispersed seeds were more aggregated than nondipterocarps (Condit et al. 2000). Also studying Malaysian rain forest trees, Plotkin et al. (2002) described how seed-dispersal mode influences the spatial distribution of large and small individuals within a cluster. In earlier work, Plotkin et al. (2000) showed clustering of individuals independent of topography, suggestive of dispersal driven aggregation.

Results for species other than tropical trees are more equivocal. Nieder et al. (2000) found that whether epiphytes were dispersed via wind or animals was unrelated to their degree of clumping. In contrast to the Hubbell (1979) results, but at a much smaller scale (2 m × 0.5 m plots), Myster & Picket (1992) found that bird-dispersed trees invading an old field were more clumped than those dispersed by wind or mammals. The perching behavior of birds presumably generates clumps of seed at this smaller spatial scale. A number of other studies have demonstrated that animal dispersal causes very local aggregation of seeds (Howe 1989).

Limitations of Current Approaches

One very important limitation of the work relating aggregation to dispersal is that nearly all of the evidence is correlational or indirect. These results are thus best regarded as patterns that generate hypotheses, rather than definitive support for the importance of dispersal. A number of key assumptions are implicit when dispersal is assessed as the residual clumping unexplained by environmental variables (as in Svenning 2001, Svenning & Skov 2002, Tuomisto et al. 2003). Most significantly, we must assume that all potentially important environmental variables, past or present, are accounted for in the statistical model. Any unmeasured variables must at least be correlated with those that are quantified. This is probably an unreasonable assumption because an historical process such as a forest gap four decades ago could easily influence clumping patterns today, yet unless it left a signature in currently measured environmental variables, such an effect would be incorrectly attributed to dispersal. In addition, these studies assume that positive density dependence, as might arise though pollination or shared soil mutualists, is not important.

Other concerns exist for studies using dispersal mode as a proxy for the dispersal kernel or mean dispersal distance. Dispersal mode is often shared across members of a family, such as wind dispersal in the Asteraceae, and is thus likely to be correlated with other factors that could influence clumping, such as seed production or competitive ability. Indeed, Horvitz & LeCorff (1993) found that within the tropical understorey herbs of the Marantaceae, ant- versus bird-dispersed plants did not differ in their degree of clumping. The potential power of alternative experimental approaches for assessing the influence of dispersal patterns on aggregation is emphasized by Turnbull et al. (1999). When annual species of limestone grassland were dispersed randomly by the investigators, the community that developed still showed strong aggregation, indicating that environmental variables exert important controls over dispersion patterns in the system.

A more general problem in this work is that aggregation is a scale-dependent measure, and thus species clumped at one scale may be randomly distributed at another. Thus, if the question is whether dispersal is important to the degree of aggregation, then the measurement of aggregation needs to be made over scales similar to the spatial scale of dispersal. Aggregations at scales larger than dispersal may indicate the importance of larger-scale processes such as environmental heterogeneity. For example, Peres & Baider's (1997) hypothesis that clumping in Brazilnut trees is dispersal driven may seem reasonable considering that the spatial scale of the seed-dispersal kernel generated by agoutis matches the scale of tree clumping. By contrast, it is not surprising that dispersal mode does not influence plant clumping at the scale of 10 × 10 km grid cells dividing the United Kingdom (Quinn et al. 1994).

COEXISTENCE THROUGH LOCAL DISPERSAL, AGGREGATION, AND SEGREGATION

Classic Lotka-Volterra competition models assume that dispersal and species interactions are spatially unrestricted. Much of the recent interest in the influence of species dispersal on community structure has been motivated by theoretical work relaxing this assumption. Results suggest that local dispersal often slows rates of competitive displacement (Bolker & Pacala 1999, Law et al. 2003, Murrell & Law 2003). This, along with coexistence achieved through a competition-colonization trade-off has generated a surge of interest in spatial processes in plant communities (Murrell et al. 2001, Murrell & Law 2003). However, as we explain below, the literature relating local dispersal to coexistence is plagued by ambiguity over the temporal scales over which local dispersal influences coexistence.

Models Showing Local Dispersal Effects on Coexistence

An increasingly popular perception among ecologists has been that local dispersal promotes coexistence by causing the spatial segregation of heterospecific individuals across a landscape (e.g., Green 1989, Murrell et al. 2001, Pacala 1997, Pacala & Levin 1997,Weiner & Conte 1981). In homogeneous environments (as assumed in most community theory), mathematical models for two-species competition show that local dispersal contributes more to the aggregation of conspecifics through parent-offspring proximity than it does to aggregation of heterospecifics (Murrell & Law 2003). This segregation of species reduces the relative frequency of inter-versus intraspecific interactions (Pacala 1997); more generally, weaker inter-versus intraspecific competition favors coexistence (Chesson 2000). Local interactions are also key in these models (Bolker & Pacala 1997, 1999; Dieckmann et al. 2000; Law et al. 2001, 2003; Murrell & Law 2003). If species interact with other individuals in a system over a sufficiently large spatial scale, local dispersal will have little effect on the frequency of heterospecific versus conspecific interactions.

However, the importance of such spatial segregation for long-term coexistence has recently been challenged on several counts. First, it is apparent that inequality in dispersal kernels between otherwise similar species will lead to the exclusion of the more aggregated species (the shortest disperser, all else being equal) (Bolker & Pacala 1999, Durrett & Levin 1998, Murrell & Law 2003, Pacala 1986). Second, even if dispersal is symmetric, segregation on its own is not enough to prevent exclusion where it is predicted in the nonspatial case (Chesson & Neuhauser 2002, Durrett & Levin 1998, Ghandi et al. 1998, Neuhauser & Pacala 1999, Takenaka et al. 1997). In these models, the dynamics may be thought of as having two phases: the initial phase where monospecific clusters build up owing to local dispersal, and the second phase where these clusters start to interact. It is the second phase that is most important for determining the long-term (equilibrium) outcome of competition, and in particular it is the interactions at the cluster boundaries that are of greatest importance (Chesson & Neuhauser 2002, Ghandi et al. 1998). Here, heterospecific interactions are relatively frequent and the stronger competitor wins. Thus, clusters of the stronger species will slowly but surely overwhelm the clusters of the weaker species. It is notable that the speed with which this occurs is dependent on the geometry of the cluster boundary; the larger the curvature of the cluster interface the faster the rate of exclusion (Ghandi et al. 1998). The curvature of the clusters is greatly dependent on the dispersal kernel, providing a mechanistic link between the dispersal distance and the rate of exclusion.

Thus, rather than a stabilizing force for coexistence, localized dispersal may be thought of as an equalizing process because it slows down the dynamics without changing the expected equilibrium (Chesson 2000, Ghandi et al. 1998). Nonetheless, by slowing exclusion to such long timescales, other processes such as immigration (Hubbell 2001) and selection (Aarssen 1984, Park & Lloyd 1955, Pimentel 1968) may act to maintain diversity. In neutral models explored by Hubbell (2001) and Chave et al. (2002), local dispersal slows the rate at which species drift to extinction. Because these models also incorporate external immigration or speciation, this slowing of displacement increases species richness by changing the balance between extinction and speciation (or colonization). In doing so, local dispersal also changes the species-abundance distribution (Chave et al. 2002). Similar results are achieved when species coexist through density dependence or fecundity-competition trade-offs.

In a comparatively rare example of theory examining the effects of local dispersal in spatially heterogeneous environments, Snyder & Chesson (2003) found that local dispersal increased the degree of coexistence as compared with that achieved under global dispersal. This result requires niche differentiation among species and dispersal occurring over smaller spatial scales than the environmental autocorrelation. However, in the model, environmental heterogeneity is fixed, and as the authors caution, longer-range dispersal may be more important in temporally heterogeneous environments.

A possibly more surprising result of segregation is that, at least in discrete-space models where interactions occur in very small neighborhoods (nearest neighbor on a grid), local dispersal can hinder coexistence (Bolker et al. 2003, Neuhauser & Pacala 1999). This occurs when species are competitively unequal but intra- is greater than interspecific competition, allowing coexistence in the nonspatial analog. Nonetheless, this effect only serves to reduce the amount of coexistence where it is marginal in the first place.

Much of the above theory is based on the well known Lotka-Volterra competition equations, which are extended to include local interactions and local dispersal. Clark & Ji (1995) used a more mechanistic, patch-based model to show that local dispersal between neighboring patches and a disturbance regime that reset patches periodically could aid coexistence as long as there was spatial variation in seed deposition (caused by both seed limitation and local dispersal), and a nonlinear relationship between fecundity and patch density. As in a spatial-storage effect (Chesson 2000), high seed production in relatively uncrowded patches more than outweighs the poor seed production in crowded patches, allowing stable coexistence. Incorporating a more mechanistic approach, such as that used by Clark & Ji, while still including local interactions and local dispersal, would be of much value in extending the theory on dispersal and coexistence.

Field Evidence that Local Dispersal Influences Coexistence

Motivated by the results of recent theory, a large number of empirical studies now speculate that local interactions favor coexistence (e.g., Hubbell et al. 1999, Tuomisto et al. 2003). However, largely owing to the only recent development of the models, the rigorous empirical tests required to justify such claims have yet to be completed. Rees et al. (1996) used models fit to monitoring data of sand dune annuals in Britain to show that species suffered intense intraspecific competition but little interspecific competition in the field. They attributed this result to the high degree of intraspecific aggregation found in the system. The degree to which this clumping resulted from local dispersal versus species specialization on different habitat types was unclear. Stoll & Prati (2001) examined competitive interactions among species experimentally planted in clumped versus random distributions. Consistent with aggregation enhancing the impacts of intra-versus interspecific interactions, competitively inferior species performed better, and competitively superior species performed worse in the aggregated treatment. These results are similar to those in a longer-term study by Schmidt (Rejmanek 2002, Schmidt 1981), which showed that over a period of three years, intraspecific aggregation allowed the exotic *Solidago canadensis* to coexist with the native *Urtica dioicia*, whereas without the aggregation, *S. canadensis* was quickly excluded. Both studies follow from a history of applied research examining how spatial patterning influences the impacts of weeds on crops (Garrett & Dixon 1998). We emphasize, however, that local dispersal need not always be important. Webb & Peart (2001) used models parameterized with field data to suggest that local dispersal was relatively unimportant for tree coexistence in a Bornean tropical forest.

We conclude from published work that local dispersal does not favor the long-term coexistence of species but instead simply slows the rate of displacement. Thus, if we want to attribute the coexistence of species in a natural community to the local nature of dispersal, we need to first clarify the timescales over which we are attempting to explain coexistence and then demonstrate that local dispersal is sufficient to explain coexistence over those timescales. This is very different than demonstrating the almost trivial result that local dispersal slows down displacement. Incorporating timescales will almost undoubtedly require the use of models parameterized with field data, as advocated in our conclusions.

COEXISTENCE THROUGH COMPETITION-DISPERSAL TRADE-OFFS

Ecologists have long known that a superior competitor and superior colonizer can coexist in homogeneous model systems (Hastings 1980, Holmes & Wilson 1998, Horn & MacArthur 1972, Hutchinson 1951, Levins & Culver 1971, Tilman 1994). However, the recent demonstration that this mechanism of coexistence extends to any number of species (Tilman 1994) has generated a surge of interest in this area. Coexistence occurs because the superior competitor lacks the colonization ability to fill all available habitats, leaving space for more poorly competing but better-colonizing species. Although colonization is a function of both fecundity and dispersal ability, most theory has assumed a competition-fecundity trade-off (Hastings 1980, Tilman 1994); only more recently has dispersal distance been considered explicitly (Bolker & Pacala 1999, Dytham 1994, Holmes & Wilson 1998, Murrell & Law 2003). Although our primary interest is coexistence through competition-dispersal trade-offs, we also review relevant theoretical results incorporating competition-fecundity trade-offs.

Assumptions Underlying Theoretical Results

The result that numerous species can coexist through a competition-colonization trade-off (Tilman 1994) depends on several questionable assumptions. First, there must be a strict hierarchy of competition inversely related to colonization. Second, the colonization of a stronger competitor must always eliminate any established weaker competitor from its location, and do so prior to its reproduction (Tilman 1994, Yu & Wilson 2001). This rapid displacement assumption seems particularly unrealistic for many plants (Levine & Rees 2002, Yu & Wilson 2001) because seedlings are unlikely to outcompete established individuals. Relaxing this assumption recovers so-called lottery models, where competition for empty sites occurs at the juvenile stage (Yu & Wilson 2001). In such lottery models, the competition-colonization trade-off is insufficient to produce coexistence, although coexistence can be achieved with environmental heterogeneity, fecundity-dispersal trade-offs, or stochastic variation in seed arrival (Chesson & Warner 1981, Comins & Noble 1985, Kisdi & Geritz 2003, Kohyama 1993, Yu & Wilson 2001).

Although the above assumptions pertain to competition-colonization trade-offs generally, the nature of competition is also key to coexistence produced specifically by a competition-dispersal trade-off. Species with longer-range dispersal can coexist with superior competitors (Bolker & Pacala 1999, Law & Dieckmann 2000, Murrell & Law 2003), but this requires a relatively large asymmetry in competition (Murrell & Law 2003). Without strongly asymmetric competition, the species possessing an optimal combination of competition and colonization dominates the system. A similar competitive hierarchy is required in competition-fecundity trade-off models (Adler & Mosquera 2000, Geritz et al. 1999). If this assumption is met, Holmes & Wilson (1998) show that species with greater dispersal can also coexist with those that are simultaneously better competitors and more fecund. Nonetheless, this result was only found for a small set of parameter combinations.

The ability of the competition-colonization trade-off to generate coexistence is also sensitive to the density and fecundity of the superior competitor. If the superior competitor has fecundity high enough that it leaves only a small fraction of unoccupied patches, the inferior competitor is unlikely to persist (Bolker et al. 2003). High density can also be achieved through high dispersal, which reduces parent-offspring competition, favoring a more even distribution of individuals across space (Bolker & Pacala 1997, Ellner 2001, Law et al. 2003). However, this more diffuse distribution of individuals may still leave temporary gaps that the weaker competitor can exploit as long as it more rapidly completes its life cycle. In fact, short-range-dispersing weaker competitors can coexist with a long-range-dispersing dominant by having a higher turnover rate (Bolker & Pacala 1999). In sum, the main conclusion from the theory is that while competition dispersal trade-offs favor coexistence, this requires several assumptions that may not apply to real systems.

Empirical Support

Like the theoretical work, most of the empirical studies examining competition-colonization trade-offs and their implications for coexistence examine fecundity-competition trade-offs (Levine & Rees 2002). Evidence for a dispersal-competition trade-off is sparse and comes largely from systems with regular disturbance. Brewer et al. (1998) found only weak evidence for a competition-dispersal trade-off among clonal grasses coexisting in a regularly disturbed salt marsh habitat. Instead they attributed coexistence to interspecific differences in physiological tolerance of gap conditions. Yeaton & Bond (1991) showed that ant dispersal gives a competitively inferior South African fynbos shrub an advantage in colonizing the open areas after disturbance. However, Markov chain models suggested that this dispersal advantage was not sufficient to explain long-term coexistence with the competitive dominant shrub in the system. In some of the best support for coexistence achieved through a competition dispersal trade-off, Platt (1975) showed that the species occupying mature Iowa prairie (the competitive dominants) tended to disperse more poorly than "fugitive" species living primarily on badger mound disturbances. However, dispersal capacity and propagule production were positively correlated; thus, the source of the colonization advantage is unclear.

The best support for coexistence achieved through a competition-dispersal trade-off comes from models parameterized with field data, although even here evidence is not definitive. Working in temperate forest with two codominant tree species, Nanami et al. (1999) documented clumping in the gravity-dispersed and dioecious-competitive dominant, which they hypothesize favored the persistence of the bird-dispersed competitive inferior. Although this is simply a pattern analysis, they followed the work with mathematical models testing the suitability of a competition-dispersal trade-off as a coexistence mechanism in the system (Nanami et al. 2000). Results indicated that because male trees of the competitive dominant do not drop seeds below the parent canopy, this creates gaps that are differentially colonized by the better-dispersing competitive inferior. Still, this result depends strongly on the dioecy, restricting its generality to other systems.

In Pacala et al.'s (1993) analysis of their SORTIE forest simulation model, which they parameterized with field-derived measures of vital rates, coexistence in Eastern U.S. forest trees is achieved in part through a documented trade-off between dispersal and the ability to cast and survive shade. Although other trade-offs are also important to coexistence in their simulation, Ribbens et al. (1994) show a tremendous impact of changing mean dispersal distance on coexistence and dominance. Still, although a number of studies implicitly or explicitly invoke competition-colonization trade-offs, relatively little definitive empirical evidence exists to support coexistence achieved through this mechanism.

CONCLUSIONS

Our review of the literature uncovered a dichotomy of support for the importance of seed dispersal for community patterns. Although we found an ever expanding body of theory suggesting that seed dispersal affects species coexistence through local dispersal and a competition-dispersal trade-off, empirical support was scant. Instead, most empirical investigations examined how dispersal and seed-trapping agents affect species distribution and dispersion, subjects with little theory. In addition, this work relied heavily on dispersal proxies and correlational analyses of community patterns, methods unable to exclude alternative hypotheses. Interestingly, this relatively tepid support for dispersal's influence on community patterns is in contrast to work at the population level, where dispersal is well appreciated to strongly influence fitness, colonization, spread, and persistence (Harper 1977, Howe & Smallwood 1982, Skellam 1951).

Thus, the main conclusion of our review is that it may be premature to expect that patterns of seed dispersal strongly influence community structure. Of course, species without any dispersal could not spread beyond a founding individual, but it remains unclear whether the specific pattern of dispersed seed is a strong determinant of relative abundance, distribution, dispersion, and coexistence in natural systems. Although this lack of clarity stems from the weak empirical support in the literature, this weak support could be interpreted in several ways. It may be that the shape of the dispersal kernel is important, but current methodological approaches are too correlational to definitively demonstrate its impacts on patterns in the field. Alternatively, dispersal may be less important than we commonly believe. Resolution of these alternatives requires more rigorous approaches to understanding the importance of dispersal.

Improving the Empirical Evidence

Because estimating and manipulating seed-dispersal kernels in the field is difficult, testing the importance of dispersal kernels for community structure can be challenging. In principle, the ideal empirical study would (*a*) document seed shadows and dispersion patterns, (*b*) correlate these with patterns of community structure, and then (*c*) demonstrate that experimentally manipulating the seed shadow changes population or community structure (Schupp & Fuentes 1995). Although examples of each of these steps can be found in the literature, we found no single study that performed all of them. Instead, most of the empirical work relating dispersal to community properties tended to be correlational, often involving proxies for dispersal kernels, such as dispersal mode. These studies thus relied heavily on several potentially precarious assumptions, including the fact that dispersal mode is a reasonable predictor of dispersal kernels and that aggregation unrelated to current environmental variables is attributable to dispersal. These assumptions may be valid in much of the work (Willson 1993), but without further support some conclusions may be incorrect. This is illustrated by the often-cited example where dispersal influences patterns of mangrove tree zonation through the tidal sorting of propagules (Rabinowitz 1978). Although Rabinowitz's hypothesis has intuitive appeal, recent work actually quantifying the seed-dispersal kernels of mangrove trees in the same forests suggests the tidal sorting of propagules is unlikely to generate the zonation (W. Sousa & B. Mitchell, unpublished data).

How can we more definitively demonstrate that dispersal influences relative abundance, distribution, dispersion, and coexistence? To this end, we strongly encourage experiments that directly manipulate the seed-dispersal kernel, as advocated in more specific cases by Bolker & Pacala (1999), Pacala & Rees (1998), and Schupp & Fuentes (1995). Such experiments are not substitutes for quantifying dispersal kernels and community patterns, but they are complementary and uniquely poised to definitively test the mechanisms suggested by the patterns. The basic methodology involves manipulating the seed-dispersal kernel in replicate plots and comparing the resulting community pattern (relative abundance, distribution, coexistence) to that in an unmanipulated control. The kernel is manipulated by first collecting all seed produced in a plot and then depending on the research question, dispersing it randomly or locally. Community patterns in an additional treatment, where seed is dispersed following the estimated natural kernels, can be compared with the unmanipulated control to test for any artifacts associated with seed handling. This basic approach has been successfully used to examine the population-level consequences of the dispersal kernel in a tropical forest tree (Augspurger & Kitajima 1992).

If dispersal controls commonness and rarity, then forcing all species' dispersal to follow the across-species mean or median kernel should cause the more common species to decline and the rarer species to increase relative to controls. If dispersal kernels control aggregation or other distribution patterns, then dispersing all seed globally should eliminate, or at least begin to homogenize, the distribution or aggregation patterns (e.g., Levine 2001, Turnbull et al. 1999). If species coexist through local dispersal, then dispersing seeds globally should enhance competitive displacement (Bolker & Pacala 1999). Lastly, if species coexist through a competition-dispersal trade-off, then equalizing the dispersal kernel across species should result in the displacement of species (Pacala & Rees 1998).

The experiments will be most tractable with communities of annual or short-lived perennial plants where dispersal occurs over relatively small spatial scales. Still, even with longer-lived organisms, trends in the predicted direction with early life stages (Augspurger & Kitajima 1992) can be used to bolster correlational evidence. In addition, seed addition experiments located at different locations relative to existing distributions could follow the rationale of the experiments above but be conducted over larger spatial scales. For example, in studies attributing adult clumping to local dispersal, experiments should show seed limitation away from the clump. In other words, if seeds did travel further, adults could establish. Last, we encourage experiments conducted in real ecological habitats where other potentially important factors are free to exert their impacts. Our review does not question whether dispersal has any effect on abundance, distribution, and diversity, but rather questions the importance of such an effect in comparison to other factors (Figure 1).

Future Directions for Theory

Theory has shown that without interspecific trade-offs, aggregation does not lead to stable coexistence (Bolker & Pacala 1999, Durrett & Levin 1998, Murrell & Law 2003, Neuhauser & Pacala 1999, Takenaka et al. 1997). Yet the models used to produce this result are based largely on the Lotka-Volterra competition equations, and most incorporate linear density dependence. Incorporating more complex processes such as nonlinear competition may change some of these predictions or at the very least lead to new strategies for coexistence (Bolker & Pacala 1999).

However, few models have taken this approach (see Pacala 1986 for a rare exception). Incorporating processes such as nonlinear effects of density may be at the expense of analytical tractability, but the techniques available to spatial theory are sophisticated enough to make it possible (Dieckmann et al. 2000, Tilman & Kareiva 1997).

Theory has also focused largely on the question of coexistence. Yet other community dynamics require closer attention from theoretical ecologists. For example, the work of Chave et al. (2002) showed that the scale of dispersal is the most important influence on many measures of community structure. Still, it is not known if these results are robust to inequality in dispersal ability across species. Although it is often assumed that longer-range dispersers are the more abundant, this prediction might change when the landscape is spatially heterogeneous, favoring short-range dispersal (Bolker 2003, Travis & Dytham 1999).

Last, empirical work should motivate interesting modeling questions. Most notably, we found a surprising number of empirical studies that documented the importance of trapping agents in controlling patterns of seed arrival. Interestingly, whereas clumping that results from local dispersal should enhance intraspecific aggregation and thus slow displacement, clumping that results from general trapping agents could increase heterospecific contact. Alternatively, trapping may segregate species if seed morphology influences trapping patterns, as in Schneider & Sharitz (1988). How trapping agents, their species specificity, and their spatial arrangement influence abundance, distribution, and coexistence are issues ripe for theoretical exploration.

Closing the Gap Between Theory and Empirical Work

We documented a clear dichotomy between the empirical and theoretical literature relating dispersal kernels to community patterns. Closing this gap is essential to clarifying the importance of dispersal. To this end, we strongly advocate the development of mathematical models parameterized with empirically documented seed-dispersal kernels and other realistic demographic parameters (as in Wu & Levin 1994). Such approaches can uniquely address whether the predictions of the mathematical models reasonably describe the dynamics of real ecological communities. Such approaches will also force investigators to ask the key long-term questions such as what maintains the dispersal-driven patterns, or what creates seed limitation. Models should also prove particularly informative for exploring the influence of seed-dispersal kernels, parent plant density, and seed production on seed deposition across the landscape. The relationship between dispersal kernels and seed rain is an underexplored, but critical, linkage between dispersal and community patterns.

The potential importance of seed dispersal for community dynamics has long been acknowledged, and recent theoretical results support this expectation. However, the results of our review suggest that the importance of dispersal cannot be taken for granted; empirical support for the theoretical predictions is largely lacking. A firmer understanding of the role of dispersal in community structure is achievable if we begin with more rigorous field approaches directly manipulating seed-dispersal kernels. These results must then by coupled with the predictions of models incorporating empirically derived dispersal kernels.

The *Annual Review of Ecology, Evolution, and Systematics* is online at http://ecolsys.annualreviews.org

LITERATURE CITED

Aarssen LW. 1984. Ecological combining ability and competitive combining ability in plants: toward a general evolutionary theory of coexistence in systems of competition. *Am. Nat.* 122:707–31

Adler FR, Mosquera J. 2000. Is space necessary? Interference competition and limits to biodiversity. *Ecology* 81:3226–323

Alcantara JM, Rey PJ, Valera F, Sanchez-Lafuente AM. 2000. Factors shaping the seedfall pattern of a bird-dispersed plant. *Ecology* 1937–50

Andersson E, Nilsson C, Johansson ME. 2000. Plant dispersal in boreal rivers and its relation to the diversity of riparian flora. *J. Biogeogr.* 27:1095–106

Augspurger CK, Kitajima K. 1992. Experimental studies of seedling recruitment from contrasting seed distributions. *Ecology* 73:1270–84

Barot S, Gignoux J, Menaut J. 1999. Demography of a savanna palm tree: predictions from comprehensive spatial pattern analysis. *Ecology* 80:1987–2005

Bleher B, Bohning-Gaese K. 2001. Consequences of frugivore diversity for seed dispersal, seedling establishment and the spatial pattern of seedlings and trees. *Oecologia* 129:385–94

Bleher B, Oberrath R, Bohning-Gaese K. 2002. Seed dispersal, breeding system, tree density and the spatial pattern of trees—a simulation approach. *Basic Appl. Ecol.* 3:115–23

Bolker BM. 2003. Combining endogenous and exogenous variability in analytical population models. *Theor. Popul. Biol.* In press

Bolker BM, Pacala SW. 1997. Using moment equations to understand stochastically driven spatial pattern formation in ecological systems. *Theor. Popul. Biol.* 52:179–97

Bolker BM, Pacala SW. 1999. Spatial moment equations for plant competition: understanding spatial strategies and the advantages of short dispersal. *Am. Nat.* 153:575–602

Bolker BM, Pacala SW, Neuhauser C. 2003. Spatial dynamics in model plant communities: What do we really know? *Am. Nat.* 162:135–48

Brewer SJ, Rand T, Levine JM, Bertness MD. 1998. Biomass allocation, clonal dispersal, and competitive success in three salt marsh plants. *Oikos* 82:347–53

Bullock JM, Kenward RE, Hails RS. 2002. *Dispersal Ecology*. Malden, MA: Blackwell. 458 pp.

Cain ML, Milligan BG, Strand AE. 2000. Long-distance seed dispersal in plant populations. *Am. J. Bot.* 87:1217–27

Carey PD, Watkinson AR. 1993. The dispersal and fates of seeds of the winter annual *Vulpia ciliata*. *J. Ecol.* 81:759–67

Chave J, Muller-Landau HC, Levin SA. 2002. Comparing classical community models: theoretical consequences for patterns of diversity. *Am. Nat.* 159:1–23

Chesson P. 2000. Mechanisms of maintenance of species diversity. *Annu. Rev. Ecol. Syst.* 31:343–66

Chesson P, Neuhauser C. 2002. Intraspecific aggregation and species coexistence. *Trends Ecol. Evol.* 17:210–11

Chesson P, Warner RR. 1981. Environmental variability promotes coexistence in lottery competitive systems. *Am. Nat.* 117:923–43

Clark JS. 1998. Why trees migrate so fast: confronting theory with dispersal biology and the paleorecord. *Am. Nat.* 152:204–24

Clark JS, Ji Y. 1995. Fecundity and dispersal in plant populations: implications for structures and diversity. *Am. Nat.* 146:72–111

Clark JS, Macklin E, Wood L. 1998. Stages and scales of recruitment limitation in southern Appalachian forests. *Ecol. Monogr.* 68:213–35

Clobert J, Danchin E, Dohndt AA, Nichols JD. 2001. *Dispersal*. Oxford: Oxford Univ. Press. 452 pp.

Comins HN, Noble IR. 1985. Dispersal, variability, and transient niches: species coexistence in a uniformly variable environment. *Am. Nat.* 126:706–23

Condit R, Ashton PS, Baker P, Bunyavejchewin S, Gunatilleke S, et al. 2000. Spatial patterns in the distribution of tropical tree species. *Science* 288:1414–18

Condit R, Pitman N, Leigh EG, Chave J, Terborgh J, et al. 2002. Beta-diversity in tropical forest trees. *Science* 295:666–69

Connell JH. 1971. On the role of natural enemies in preventing competitive exclusion in some marine animals and in rain forest trees. In *Dynamics of Populations*, ed. PJ den-Boer, GR Gradwell, pp. 298–312. Wageningen, Neth.: Cent. Agric. Publ. Doc.

Cornell HV. 1993. Unsaturated patterns in species assemblages: the role of regional processes in setting local species richness. In *Species Diversity in Ecological Communities*, ed. RE Ricklefs, D Schluter, pp. 243–52. Chicago/London: Univ. Chicago Press

Dalling J, Muller-Landau HC, Wright J, Hubbell SP. 2002. Role of dispersal in the recruitment limitation of neotropical pioneer species . *J. Ecol.* 90:714–27

Danvind M, Nilsson C. 1997. Seed floating ability and distribution of alpine plants along a northern Swedish river. *J. Veg. Sci.* 8:271–76

Dieckmann U, Law R, Metz JAJ. 2000. *The Geometry of Ecological Interactions: Simplifying Spatial Complexity*. Cambridge: Cambridge Univ. Press. 564 pp.

Dockery J, Hutson V, Mischaikow K, Pernarowski M. 1998. The evolution of slow dispersal rates: a reaction diffusion model. *J. Math. Biol.* 37:61–83

Drake JA, Mooney HA, DiCastri F, Groves RH, Kruger FJ, et al. 1989. *Biological Invasions: A Global Perspective*. Chichester, Engl.: John Wiley. 525 pp.

Durrett R, Levin S. 1998. Spatial aspects of interspecific competition. *Theor. Popul. Biol.* 53:30–43

Dytham C. 1994. Habitat destruction and competitive coexistence: a cellular model. *J. Anim. Ecol.* 63:490–91

Ellner SP. 2001. Pair approximation for lattice models with multiple interaction scales. *J. Theor. Biol.* 210:435–47

Eriksson O. 1997. Colonization dynamics and relative abundance of three plant species (*Antennaria dioca*, *Hieracium pilosella* and *Hypochoeris maculata*) in semi-natural grasslands. *Ecography* 20:559–68

Eriksson O, Jakobsson A. 1998. Abundance, distribution and life histories of grassland plants: a comparative study of 81 species. *J. Ecol.* 86:922–33

Fragaso JM. 1997. Tapir-generated seed shadows: scale dependent patchiness in the Amazon rain forest. *J. Ecol.* 85:519–29

Friedman J, Stein Z. 1980. The influence of seed-dispersal mechanisms on the dispersion of *Anastatica hierochuntica* (Cruciferae) in the Negev desert, Israel. *J. Ecol.* 68:43–50

Garrett A, Dixon PM. 1998. When does the spatial pattern of weeds matter? Predictions from neighborhood models. *Ecol. Appl.* 8:1250–59

Ghandi A, Levin S, Orszag S. 1998. Critical slowing down in time to extinction: an example of critical phenomena in ecology. *J. Theor. Biol.* 192:363–76

Geritz SA, van der Meijden E, Metz JAJ. 1999. Evolutionary dynamics of seed size and seedling competitive ability. *Theor. Popul. Biol.* 55:324–43

Green DF. 1989. Simulated effects of fire, dispersal and spatial pattern on competition within forest mosaics. *Vegetatio* 82:139–53

Guo Q, Brown JH, Valone TJ, Kachman SD. 2000. Constraints of seed size on plant distribution and abundance. *Ecology* 81:2149–55

Hanski IA, Gilpin ME. 1997. *Metapopulation Biology*. San Diego: Academic. 512 pp.

Harper JL. 1977. *Population Biology of Plants*. London: Academic. 892 pp.

Hastings A. 1980. Disturbance, coexistence, history and competition for space. *Theor. Popul. Biol.* 163:491–504

Holmes EE, Wilson HB. 1998. Running from trouble: long-distance dispersal and the competitive coexistence of inferior species. *Am. Nat* 151:578–86

Honnay O, Verhaeghe W, Hermy M. 2001. Plant community assembly along dendritic networks of small forest streams. *Ecology* 82:1691–702

Horn HS, MacArthur RH. 1972. Competition among fugitive species in a harlequin environment. *Ecology* 53:749–52

Horvitz CC, LeCorff J. 1993. Spatial scale and dispersion pattern of ant- and bird-dispersed herbs in two tropical lowland rain forests. *Vegetatio* 107/108:351–62

Houle G. 1995. Seed dispersal and seedling recruitment: the missing link(s). *Ecoscience* 2:238–44

Hovestadt T, Poethke HJ, Messner S. 2000. Variability in dispersal distances generates typical successional patterns: a simple simulation model. *Oikos* 90:612–19

Howe HF. 1989. Scatter- and clump-dispersal and seedling demography: hypotheses and implications. *Oecologia* 79:417–26

Howe HF, Smallwood J. 1982. Ecology of seed dispersal. *Annu. Rev. Ecol. Syst.* 13:201–28

Hubbell SP. 1979. Tree dispersion, abundance, and diversity in a tropical dry forest. *Science* 203:1299–309

Hubbell SP. 2001. *The Unified Neutral Theory of Biodiversity and Biogeography*. Princeton/Oxford: Princeton Univ. Press. 375 pp.

Hubbell SP, Foster RB, O'Brien ST, Harms KE, Condit R, et al. 1999. Light-gap disturbances, recruitment limitation, and tree diversity in a neotropical forest. *Science* 283:554–57

Hutchinson GE. 1951. Copepodology for the ornithologist. *Ecology* 32:571–74

Jakobsson A, Eriksson O. 2000. A comparative study of seed number, seed size, seedling size and recruitment in grassland plants. *Oikos* 88:494–502

Janzen DH. 1970. Herbivores and the number of tree species in tropical forests. *Am. Nat.* 104:501–28

Johansson ME, Nilsson C. 1993. Hydrochory, population dynamics and distribution of the clonal aquatic plant *Ranunculus lingua*. *J. Ecol.* 81:81–91

Kisdi E, Geritz SAH. 2003. On the coexistence of perrenial plants by the competition-colonization trade-off. *Am. Nat.* 161:350–54

Kiviniemi K, Eriksson O. 1999. Dispersal, recruitment and site occupancy of grassland plants in fragmented habitats. *Oikos* 86:241–53

Kohyama T. 1993. Size structured tree populations in gap forest—the forest architecture hypothesis for the stable coexistence of species . *J. Ecol.* 81:131–43

Law R, Dieckmann U. 2000. A dynamical system for neighborhoods in plant communities. *Ecology* 81:2137–48

Law R, Murrell DJ, Dieckmann U. 2003. On population growth in space and time: spatial logistic equations. *Ecology* 84:252–62

Law R, Purves DW, Murrell DJ, Dieckmann U. 2001. Dynamics of small-scale spatial structure in plant populations. In *Integrating Ecology and Evolution in a Spatial Context*, ed. J Silvertown, J Antonovics, pp. 21–44. Oxford: Blackwell Sci.

Leishman MR. 2001. Does the seed size/number trade-off model determine plant community structure? An assessment of the model mechanisms and their generality. *Oikos* 93:294–302

Levine JM. 2001. Local interactions, dispersal, and native and exotic plant diversity along a California stream. *Oikos* 95:397–408

Levine JM. 2003. A patch modeling approach to the community-level consequences of directional dispersal. *Ecology* 84:1215–24

Levine JM, Rees M. 2002. Coexistence and relative abundance in annual plant assemblages: the roles of competition and colonization. *Am. Nat.* 160:452–67

Levins R. 1969. Some demographic and genetic consequences of environmental heterogeneity for biological control. *Bull. Entomol. Soc. Am.* 15:237–40

Levins R, Culver D. 1971. Regional coexistence of species and competition between rare species. *Proc. Natl. Acad. Sci. USA* 68:1246–48

Maranon T, Grubb PJ. 1993. Physiological basis and ecological significance of the seed size and relative growth rate relationship in Mediterranean annuals. *Funct. Ecol.* 7:591–99

Molofsky J, Bever JD, Antonovics J, Newman JT. 2002. Negative frequency dependence and the importance of spatial scale. *Ecology* 83:21–27

Murrell DJ, Law R. 2003. Heteromyopia and the spatial coexistence of similar competitors. *Ecol. Lett.* 6:48–59

Murrell DJ, Purves DW, Law R. 2001. Uniting pattern and process in plant ecology. *Trends Ecol. Evol.* 16:529–30

Myster RW, Pickett STA. 1992. Effects of palatability and dispersal mode on spatial patterns of trees in old fields. *Bull. Torrey Bot. Club* 119:145–51

Nanami S, Kawaguchi H, Kubo T. 2000. Community dynamic models of two dioecious tree species. *Ecol. Res.* 15:159–64

Nanami S, Kawguchi H, Yamakura T. 1999. Dioecy-induced spatial patterns of two codominant tree species, *Podocarpus nagi* and *Neolitsea aciculata*. *J. Ecol.* 87:678–87

Nathan R. 2003. Seeking the secrets of dispersal. *Trends Ecol. Evol.* 18:275–76

Nathan R, Muller-Landau HC. 2000. Spatial patterns of seed dispersal, their determinants and consequences for recruitment. *Trends Ecol. Evol.* 15:278–85

Neuhauser C, Pacala SW. 1999. An explicitly spatial version of the Lotka-Volterra model with interspecific competition. *Ann. Appl. Probab.* 9:1226–59

Nieder J, Engwald S, Klawun M, Barthlott W. 2000. Spatial distribution of vascular epiphytes (including hemiepiphytes) in a lowland Amazonian rain forest (Surumoni Crane Plot) of southern Venezuela. *Biotropica* 32:385–96

Nilsson C, Ekblad A, Dynesius M, Backe S, Gardfjell M, et al. 1994. A comparison of species richness and traits of riparian plants between a main river and its tributaries. *J. Ecol.* 82:281–95

Nilsson C, Gardfjell M, Grelsson G. 1991. Importance of hydrochory in structuring plant communities along rivers. *Can. J. Bot.* 69:2631–33

Overton JMC. 1996. Spatial autocorrelation and dispersal in mistletoes: field and simulation results. *Vegetatio* 125:83–98

Pacala SW. 1986. Neighborhood models of plant population dynamics. 2. Multi-species models of annuals. *Theor. Popul. Biol.* 29:262–92

Pacala SW. 1997. Dynamics of plant communities. In *Plant Ecology*, ed. MJ Crawley, pp. 532–55. Oxford: Blackwell Sci.

Pacala SW, Levin SA. 1997. Biologically generated spatial pattern and the coexistence of competing species. In *Spatial Ecology: The Role of Space in Population Dynamics and Interspecific Interactions*, ed. D Tilman, P Kareiva, pp. 204–32. New York: Princeton Univ. Press. 368 pp.

Pacala SW, Rees M. 1998. Models suggesting field experiments to test two key hypotheses explaining successional diversity. *Am. Nat.* 152:729–37

Pacala SW, Silander JA Jr. 1985. Neighorhood models of plant population dynamics. I. Single-species models of annuals. *Am. Nat.* 125:385–411

Pacala SW, Canham CD, Silander JA. 1993. Forest models defined by field measurements: I. The design of a northeastern forest simulator. *Can. J. For. Res.* 23:1980–88

Park T, Lloyd M. 1955. Natural selection and the outcome of competition. *Am. Nat.* 89:235–40

Pastor J, Cohen Y, Moen R. 1999. Generation

of spatial patterns in boreal forest landscapes. *Ecosystems* 2:439–50

Peart MH, Clifford HT. 1987. The influence of diaspore morphology and soil surface properties on the distribution of grasses. *J. Ecol.* 75:569–76

Peres CA, Baider C. 1997. Seed dispersal, spatial distribution and population structure of Brazilnut trees (*Bertholletia excelsa*) in southeastern Amazonia. *J. Trop. Ecol.* 13:595–616

Pimentel D. 1968. Population regulation and genetic feedback. *Science* 159:1432–37

Platt WJ. 1975. The colonization and formation of equilibrium plant species associations on badger disturbances in a tall-grass prairie. *Ecol. Monogr.* 45:285–305

Platt WJ, Weis IM. 1977. Resource partitioning and competition within a guild of fugitive prairie plants. *Am. Nat.* 111:479–513

Plotkin JB, Chave J, Ashton PS. 2002. Cluster analysis of spatial patterns in Malaysian tree species. *Am. Nat.* 160:629–44

Plotkin JB, Potts MD, Leslie N, Manokaran N, LaFrankie J, Ashton PS. 2000. Species-area curves, spatial aggregation, and habitat specialization in tropical forests. *J. Theor. Biol.* 207:81–99

Quinn RM, Lawton JH, Eversham BC, Wood SN. 1994. The biogeography of scarce vascular plants in Britain with respect to habitat preference, dispersal ability and reproductive biology. *Biol. Conserv.* 70:149–57

Rabinowitz D. 1978. Early growth of mangrove seedlings in Panama, and an hypothesis concerning the relationship of dispersal and zonation. *J. Biogeogr.* 5:113–33

Rabinowitz D, Rapp JK. 1981. Dispersal abilities of seven sparse and common grasses from a Missouri prairie. *Am. J. Bot.* 68:616–24

Rand T. 2000. Seed dispersal, habitat suitability and the distribution of halophytes across a salt marsh tidal gradient. *J. Ecol.* 88:608–21

Rees M. 1995. Community structure in sand dune annuals: Is seed weight a key quantity? *J. Ecol.* 83:857–63

Rees M, Grubb PJ, Kelly D. 1996. Quantifying the impact of competition and spatial heterogeneity on the structure and dynamics of a four species guild of winter annuals. *Am. Nat.* 147:1–32

Rejmanek M. 2002. Intraspecific aggregation and species coexistence. *Trends Ecol. Evol.* 17:209–10

Rey PJ, Alcantara JM. 2000. Recruitment dynamics of fleshy-fruited plant (*Olea europaea*): connecting patterns of seed dispersal to seedling establishment. *J. Ecol.* 88:622–33

Ribbens E, Silander JA, Pacala SW. 1994. Seedling recruitment in forests: calibrating models to predict patterns of tree seedling dispersion. *Ecology* 75:1794–806

Rydin H, Borgegard S. 1991. Plant characteristics over a century of primary succession on islands: Lake Hjalmaren. *Ecology* 72:1089–101

Schmidt W. 1981. Über das Konkurrenzverhalten von *Solidago canadensis* und *Urtica dioicia*. *Verh. Ges. Ökol.* 9:173–88

Schneider RL, Sharitz RR. 1988. Hydrochory and regeneration in a Bald Cypress-Water Tupelo swamp forest. *Ecology* 69:1055–63

Schupp EW, Fuentes M. 1995. Spatial patterns of seed dispersal and the unification of plant population ecology. *Ecoscience* 2:267–75

Shigesada N, Kawasaki K. 2002. Invasion and the range expansion of species: effects of long-distance dispersal. In *Dispersal Ecology*, ed. JM Bullock, RE Kenward, RS Hails, pp. 350–73. Malden: Blackwell Sci. 458 pp.

Shipley B, Dion J. 1992. The allometry of seed production in herbaceous angiosperms. *Am. Nat.* 139:467–83

Shmida A, Ellner S. 1984. Coexistence of plant species with similar niches. *Vegetatio* 58:29–55

Skellam JG. 1951. Random dispersal in theoretical populations. *Biometrika* 38:196–218

Snyder RE, Chesson P. 2003. Local dispersal can facilitate coexistence in the presence of permanent spatial heterogeneity. *Ecol. Lett.* 6:301–9

Srivastava DS. 1999. Using local-regional richness plots to test for species saturation: pitfalls and potentials. *J. Anim. Ecol.* 68:1–16

Stoll P, Prati D. 2001. Intraspecific aggregation alters competitive interactions in experimental plant communities. *Ecology* 82:319–27

Svenning JC. 2001. Environmental heterogeneity, recruitment limitation and the mesoscale distribution of palms in a tropical montane rain forest (Maquipucuna, Ecuador). *J. Trop. Ecol.* 17:97–113

Svenning JC, Skov F. 2002. Mesoscale distribution of understorey plants in temperate forest (Kalo, Denmark): the importance of environment and dispersal. *Plant Ecol.* 160:169–85

Takenaka Y, Matsuda H, Iwasa Y. 1997. Competition and evolutionary stability of plants in a spatially structured habitat. *Res. Popul. Ecol.* 39:67–75

Thompson K, Gaston KJ, Band SR. 1999. Range size, dispersal and niche breadth in the herbaceous flora of central England. *J. Ecol.* 87:150–55

Tilman D. 1994. Competition and biodiversity in spatially structured habitats. *Ecology* 75:685–700

Tilman D, Kareiva P. 1997. *Spatial Ecology: The Role of Space in Population Dynamics and Interspecific Interactions.* New York: Princeton Univ. Press. 368 pp.

Travis JMJ, Dytham C. 1999. Habitat persistence, habitat availability and the evolution of dispersal. *Proc. R. Soc. London Ser. B* 266:723–28

Tuomisto H, Ruokolainen K, Yli-Halla M. 2003. Dispersal, environment, and floristic variation of western Amazonian forests. *Science* 299:241–44

Turnbull LA, Crawley MJ, Rees M. 2000. Are plant populations seed-limited? A review of seed sowing experiments. *Oikos* 88:225–38

Turnbull LA, Rees M, Crawley MJ. 1999. Seed mass and the competition/colonization tradeoff: a sowing experiment. *J. Ecol.* 87:899–912

Wang BC, Smith TB. 2002. Closing the seed dispersal loop. *Trends Ecol. Evol.* 17:379–85

Webb EL, Fa'aumu S. 1999. Diversity and structure of tropical rain forest of Tutuila, American Samoa: effects of site age and substrate. *Plant Ecol.* 144:257–74

Webb CO, Peart DR. 2001. High seed dispersal rates in faunally intact tropical rain forest: theoretical and conservation implications. *Ecol. Lett.* 4:491–99

Westelaken IL, Maun MA. 1985. Spatial pattern and seed dispersal of *Lithospermum caroliniense* on Lake Huron sand dunes. *Can. J. Bot.* 63:125–32

Westoby M, Leishman M, Lord J. 1996. Comparative ecology of seed size and dispersal. *Philos. Trans. R. Soc. London. Ser. B* 351:1309–18

Weiner J, Conte PT. 1981. Dispersal and neighborhood effects in an annual plant competition model. *Ecol. Model.* 13:131–47

Willson MF. 1993. Dispersal mode, seed shadows, and colonization patterns. *Vegetatio* 107/108:261–80

Wu J, Levin SA. 1994. A spatial patch dynamic modeling approach to patterns and process in an annual grassland. *Ecol. Monogr.* 64:447–64

Yeaton RI, Bond WJ. 1991. Competition between two shrub species: dispersal differences and fire promote coexistence. *Am. Nat.* 138:328–41

Yu DW, Wilson HB. 2001. The competition-colonization trade-off is dead; long live the competition-colonization trade-off. *Am. Nat.* 158:49–63

Annu. Rev. Ecol. Evol. Syst. 2003. 34:575–604
doi: 10.1146/annurev.ecolsys.34.011802.132428

First published online as a Review in Advance on August 6, 2003

THE ECOLOGY AND EVOLUTION OF SEED DISPERSAL: A Theoretical Perspective

Simon A. Levin
Department of Ecology and Evolutionary Biology, Princeton University, Princeton, New Jersey 08544; email: slevin@princeton.edu

*Helene C. Muller-Landau
National Center for Ecological Analysis and Synthesis, 735 State Street, Santa Barbara, California 93101; email: hmuller@nceas.ucsb.edu

*Ran Nathan
Department of Life Sciences, Ben-Gurion University of the Negev, Beer-Sheva 84105, Israel; email: rnathan@bgumail.bgu.ac.il

*Jérôme Chave
Evolution et Diversité Biologique, CNRS, Université Paul Sabatier, bâtiment IVR3, F-31062 Toulouse, France; email: chave@cict.fr

Key Words spatial ecology, long-distance dispersal, mechanistic models, invasion speed, population dynamics

■ **Abstract** Models of seed dispersal—a key process in plant spatial dynamics—have played a fundamental role in representing dispersal patterns, investigating dispersal processes, elucidating the consequences of dispersal for populations and communities, and explaining dispersal evolution. Mechanistic models of seed dispersal have explained seed dispersion patterns expected under different conditions, and illuminated the circumstances that lead to long-distance dispersal in particular. Phenomenological models have allowed us to describe dispersal pattern and can be incorporated into models of the implications of dispersal. Perhaps most notably, population and community models have shown that not only mean dispersal distances but also the entire distribution of dispersal distances are critical to range expansion rates, recruitment patterns, genetic structure, metapopulation dynamics, and ultimately community diversity at different scales. Here, we review these developments, and provide suggestions for further research.

*Order of the last three authors was determined by a random-number generator.

INTRODUCTION

Dispersal is defined as the unidirectional movement of an organism away from its place of birth. In sedentary organisms such as all plants and some animals, dispersal is mostly confined to a short early stage of the life cycle. In higher plants, individuals move in space mostly as seeds. Many plant species can also move through vegetative growth, but this kind of movement is not as common and typically induces relatively minor spatial change. Although nonvegetative dispersal units can be diverse and can be more appropriately described by more specific botanical terms (see van der Pijl 1982), here we use the term seed dispersal as a general expression for the dispersal of the reproductive unit of a plant.

To understand dispersal, we need to measure its spatial patterns, to explore the mechanisms that generate them, and to examine their consequences. We thus start with an overview of the empirical evidence for patterns and processes of dispersal and then discuss models describing these. The theoretical implications and explanations of dispersal are discussed in subsequent sections, starting with population spread, moving briefly through other aspects of population and community dynamics, to the evolution of dispersal. We end with a synthesis of main conclusions and directions for future research.

SEED DISPERSAL PATTERNS AND PROCESSES

Dispersal Mechanisms

The great variety of dispersal-aiding morphologies attracted the attention of naturalists as early as Aristotle (384–322 BC) and Theophrastus (371–286 BC) (Thanos 1994). For a long time, the study of seed dispersal was either anecdotal or speculative, with attempts to explain the selective value of each and every detail of a dispersal unit (Ridley 1930, van der Pijl 1982). The most commonly used classification system of dispersal syndromes is based on the agent or vector of dispersal, typically inferred from seed morphology. The principal agents of dispersal are either abiotic (wind and water) or biotic (animals and the plant itself), and the dispersal syndromes are termed, respectively, anemochory, hydrochory, zoochory, and autochory (van der Pijl 1982).

The vector-based method, and similarly any other classification of dispersal morphologies, can be refined to account for more subtle differences in the morphology of the dispersal unit, its potential dispersers, and adaptive features (van der Pijl 1982). However, as the level of detail in classification increases, the promise of theoretical generalization recedes. Furthermore, general classification methods tend to miss, and even misconstrue, important characteristics of the seed dispersal process. A key point is that dispersal is seldom mediated by a single dispersal agent and is not confined to the primary movement of seeds from the plant to the surface (Phase I dispersal) (Chambers & MacMahon 1994, Watkinson 1978). Rather, it also entails subsequent movements (Phase II dispersal) that can be mediated by other dispersal agents. The common practice of using the morphological dispersal syndrome to distinguish short- from long-distance dispersal is therefore questionable. In fact, the actual processes responsible for long-distance dispersal (LDD) are only loosely correlated with those interpreted from seed morphology (Higgins et al. 2003).

A second major limitation of traditional classification schemes stems from the ambiguous relationships between the morphological dispersal syndrome and the contribution of the dispersal process to plant fitness (disperser effectiveness sensu Schupp 1993). Dispersal agents, even within restrictively classified groups, differ markedly in their effectiveness both quantitatively (numbers and distances of dispersed seeds) and qualitatively (treatment and deposition of seeds) (Schupp 1993). An extreme example for the importance of the quality of the dispersal agents is directed dispersal, or the disproportionate arrival of seeds to favored establishment sites in which survival is relatively high (Howe & Smallwood 1982). Wenny (2001) provides many examples and a thorough discussion of this phenomenon, which seems to be more common than previously believed.

Spatial Patterns

Dispersal is encapsulated in the seed dispersion pattern (Nathan & Muller-Landau 2000), most commonly measured in a two-dimensional setting, though it can also be measured in one (e.g., Thébaud & Debussche 1991) or three (Nathan et al. 2002b, Tackenberg 2003) spatial dimensions.

Seed dispersion patterns depend on adult dispersion patterns, their geometry and fecundity, and on the variation in the direction and distance of dispersal events. As such, they can be very complex, hence difficult to quantify. The most common practice utilizes a ground network of seed traps (Greene & Calogeropoulos 2002); direct observations (e.g., Watkinson 1978) and genetic analyses (e.g., Godoy & Jordano 2001) are used much less frequently. Dispersal studies usually cover an area where most, but not all, dispersal occurs. Quantifying seed dispersion patterns at increasingly large scales is exceedingly more difficult because more uncertainty is associated not only with the fate of rare events but even with the identity of the mechanisms operating at these scales (Higgins et al. 2003). The seed trap method soon becomes unfeasible because of the extremely huge sampling area required to detect rare LDD events (Greene & Calogeropoulos 2002). Thus, the quantification of LDD is extremely challenging (Cain et al. 2000, Nathan et al. 2003). Unconventional methods that focus on individual movements and methods that couple modeling and empirical tools are the most promising ways to estimate LDD; see Greene & Calogeropoulos (2002), Nathan et al. (2003), Wang & Smith (2002) for recent reviews.

Seed dispersion patterns reflect the totality of all individual dispersal events in a population, whereas the dispersal curve summarizes the distribution of distances traveled by seeds. Dispersal curves can in principle form any kind of distribution; cases of directed dispersal discussed above, for example, may generate complex multimodal dispersal curves (Schupp et al. 2002). The majority of empirical seed dispersal data, however, fit a relatively simple, unimodal leptokurtic distribution, characterized by a peak at or close to the source, followed by a rapid decline and a long, relatively "fat" tail (Kot et al. 1996, Willson 1993). In the following section, we define relevant terms and discuss the mathematics of dispersal curves.

Many studies evaluate dispersal based on postdispersal (seedlings, young plants, or even adults) dispersion patterns. Although this kind of data can be collected in a cost-effective manner, it may not provide a reliable way to reconstruct dispersal. This is because data interpretation should address uncertainties involved not only with dispersal, but also with predispersal (e.g., pollination, seed production, and predispersal seed loss), and postdispersal (e.g., seed predation, germination, and seedling competition) processes (Nathan & Muller-Landau 2000, Schupp & Fuentes 1995). Such coupling is especially challenging for large-scale studies, with virtually no information available on establishment processes that follow LDD (Nathan 2001), despite their crucial importance for plant population dynamics.

THEORETICAL MODELS OF SEED DISPERSAL

Phenomenological Models

Ultimately, we argue that understanding of dispersal requires the development of mechanistic models that can explain observed patterns. We begin, however, with a characterization of those patterns through purely phenomenological models.

SEED DISPERSAL CURVES Data on dispersal can be represented either by the frequency distribution of dispersal distances or by the two-dimensional distance function of postdispersal seed densities. The two types have been coined, respectively, distance distribution and dispersal kernel (Nathan & Muller-Landau 2000), or one- and two-dimensional probability density functions (Cousens & Rawlinson 2001).

Mathematically, a dispersal kernel is expressed in Cartesian coordinates by $P(x, y)dxdy$, the probability that a seed released at point (0, 0) lands in a square of size $dxdy$ centered at the deposition site (x, y). Given the symmetries in this problem, it is easier to deal with polar coordinates r, θ, where $r = \sqrt{x^2 + y^2}$ is the distance between the release point and the deposition site, and θ is the radial angle. If dispersal is isotropic, the probability of landing in an annulus of width dr at a distance r from the point source is $2\pi rP(r)dr$. In certain cases, dispersal is directed along one preferential direction, and it can be suitably modeled as a one-dimensional process. Then, $P(x)dx$ is the probability that a seed starting at point 0 lands in the segment of length dx centered at x. It is assumed throughout this section that one-dimensional kernels are functions of x, while two-dimensional kernels are functions of r. The *seed shadow* $N(x)$ is the product of the dispersal kernel P and the total number of seeds dispersed Φ.

Dispersal curves can be estimated from seed dispersion patterns by taking into account the location, geometry, and fecundity of adults, and the directionality of the dispersal process, though the latter is usually ignored. A general problem with such estimation is the identification of the specific source location of dispersed seeds. This has led researchers to select isolated individuals so that the seed source location is not ambiguous (e.g., Lamont 1985). However, more generally, adult plants tend to form conspecific aggregations; hence, seed shadows of neighboring individuals typically overlap. Methods for resolving the overlapping seed shadow problem are discussed in Clark et al. (1998b, 1999), Nathan & Muller-Landau (2000), and Ribbens et al. (1994).

Traditionally, three functional forms for the distance distribution were commonly fitted to dispersal data: the Gaussian, the negative exponential, and the inverse power law. The Gaussian distribution is well defined but does not fit the leptokurtic distributions that are commonly observed. The negative exponential, probably the most commonly used functional form, has a fatter tail, but not fat enough to accommodate many LDD data. The inverse power law has a fat tail, but $P(r) = a/r^{\beta}$ goes to infinity as r goes to zero (if $\beta > 0$); hence only functions such as $P(r) = a/(1 + br)^{\beta}$ can be used as general forms for dispersal kernels. A general parametric formulation encompassing all these models is

$$P(r) = \frac{a}{r^{\beta}} \exp(-br^{\alpha}) = a \exp(-br^{\alpha} - \beta \ln(r)), \tag{1}$$

where a, b, α, and β are parameters (Turchin 1998, p. 200). Recent studies have proposed fat-tailed distributions such as the (two-dimensional Student *t*) *2Dt* (Clark et al. 1999)

$$P(r) = a(1 + br^2)^{-\alpha}. \tag{2}$$

For $\alpha = 1$, this is the Cauchy distribution (Shaw 1995). These dispersal kernels are strongly leptokurtic, with a disproportionately large fraction of the seeds dispersed far.

Short- and long-distance dispersal can be associated with different dispersal mechanisms (see Seed Dispersal Patterns and Processes above); hence the overall distribution of dispersal distances in a population may be best represented by stratified modeling, i.e., mixing several dispersal kernels such as two exponential distributions (Higgins & Cain 2002).

MOTIVATION OF MODEL FORMS Phenomenological models are chosen mainly—if not entirely—on their ability to fit the data. However, particular modeling forms can also be justified by general assumptions about the dispersal process. The simplest example rests on a homogeneous deposition model, which implies an exponential dispersal kernel. In one space dimension, assume that a fraction ρ of the dispersed seeds that have not yet settled are deposited between x and $x + dx$, while the remaining fraction are dispersed further. This yields a differential equation for $P(x)$, $dP(x)/dx = -\rho P(x)$, whose solution on the half-line $x \geq 0$ is the exponential

$$P(x) = \rho \exp(-\rho x). \tag{3}$$

with mean dispersal distance L equal to $1/\rho$.

A simple generalization of the decay model introduces a distance-dependent decay rate, such that (in two dimensions) at point r, a seed lands at rate $\rho(r) \sim r^{\alpha-2}$. In other words, the deposition rate decreases with distance from the release point if $\alpha < 1$, and it increases with the distance from the release point if $\alpha > 1$. This produces the distribution

$$P(r) = a \exp(-br^{\alpha}), \quad r > 0. \tag{4}$$

This corresponds to Equation 1 for $\beta = 0$ and has been suggested by many authors (Clark et al. 1998b, Ribbens et al. 1994, Taylor 1978, Turchin 1998) as a practical generalization of the exponential and Gaussian models. The particular form of the decay rate could be interpreted as an intrinsic property of the disperser: the smaller the parameter α, the fatter the tail of the distribution. This could correspond, for example, to changes in the behavior of an animal disperser, which deposits more seeds near parent plants than far from them (even corrected for the area effect).

Mechanistic Models

Beyond simply justifying particular forms for phenomenological models, truly mechanistic models of seed dispersal can predict exact seed distributions (including parameter values of dispersal curves) from characteristics of the dispersal process. Mechanistic models of seed dispersal by wind have a long history, building upon available theory on wind advection. Models of seed dispersal by animals are less well developed in large part because such models require quantification of detailed behavioral information.

MODELS OF SEED DISPERSAL BY WIND Modeling the movement of seeds dispersed by wind is analogous to modeling the movement of pollen, fungal spores, particulate pollution, etc.; thus the methodology from other fields, especially fluid dynamics, can be applied directly. One relevant set of models are plume models, specifically the tilted Gaussian plume model developed by Okubo & Levin (1989) and related work by Greene & Johnson (1989).

The tilted Gaussian plume model incorporates the joint influences of wind advection and gravity on seed movement (Okubo & Levin 1989). Advection is characterized by the horizontal velocity u (in one dimension), and seeds fall at a terminal velocity V_t that reflects the balance between gravity and friction (Greene & Johnson 1989). Seeds are released from a point of height H, disperse under the influences of diffusion and advection, and thus follow wind movement like a plume. If seed movement were entirely deterministic, then it would be determined entirely by wind speed, the height of release, and terminal velocity (Nathan et al. 2001 and references therein):

$$x = Hu/V_t. \tag{5}$$

The dispersal kernel thus peaks at x. This case ignores vertical, latitudinal, and longitudinal variation in wind speed. Of course, in reality, stochastic effects due to fluctuations in wind speed lead to much greater variation in dispersal distances and cause the dispersal kernel to widen. Pasquill & Smith (1983) incorporated some of this variation by considering not only advection by wind along direction x, but also diffusion along the y and z axes (respectively, cross-wind and vertical directions). They were interested in the final surface distribution of light particles—particles for which the influence of gravity could be ignored (see Okubo et al. 2001a for a recent review).

Okubo & Levin (1989) modified this example to include the influence of gravity, which becomes important for heavier particles and seeds. To take account of nonzero terminal velocity, V_t, they replaced the height term H by the expected height at distance x, that is $H - V_t x/u$. They derived a cross-wind integrated distribution that corresponds to the solution of the deterministic model mentioned above when the variance in vertical wind speed σ_w^2 goes to zero. Okubo & Levin (1989) further assumed that $t = x/u$ because the seeds are displaced at constant velocity u along the x axis. Thus, the tilted Gaussian plume model takes the form

$$P(x) = \frac{V_t}{\sqrt{2\pi} u \sigma_w} \exp\left(-\frac{(H - V_t x/u)^2}{2\sigma_w^2}\right). \tag{6}$$

This assumes that the parameters are constant in time, uniform in space, and do not vary from one seed to another. However, in reality, such variability is substantial and has important consequences for dispersal. For example, the shape of a wind-dispersal kernel will be very different if horizontal wind speed is assumed constant or to vary like Brownian motion with drift. In this example, if the wind speed varies around its mean much faster than the duration of the seed's flight, then standard tools from the theory of Brownian motion can be invoked (see below). Greene & Johnson (1989) suggested a simple generalization of the ballistic wind dispersal model (Equation 5) in which different seeds experience different wind speeds. Further, they suggested the lognormal distribution as an empirical fit to the distribution of wind speeds. From that, they deduced the one-dimensional dispersal kernel

$$P(x) = \frac{1}{x\sqrt{2\pi}\sigma_u} \exp\left(-\frac{(\ln(V_t x/u_g H))^2}{2\sigma_u^2}\right). \tag{7}$$

Here, u_g and σ_u^2 denote, respectively, the geometric mean and the variance of horizontal wind velocity u.

Spatial variation in wind speed and direction is another important issue. For example, horizontal wind speed is usually a function of height that increases roughly exponentially within a plant canopy and logarithmically above it. The influence of this vertical wind profile is discussed in Nathan et al. (2002a) in the context of the seed dispersal simulator WINDISPER (Nathan et al. 2001). WINDISPER assumes a lognormal distribution of horizontal wind speeds u, and Gaussian distributions of vertical wind speeds w (truncated to exclude net upward movements), height of release H, and seed terminal velocity V_t. For either a logarithmic (for trees in open landscapes) or an exponential (for trees within a dense forest) wind profile, there is an analytical solution for the distance traveled by an individual seed under any combination of parameter values (Nathan et al. 2002a).

The above models provide insights into dispersal processes in various ways. Increasing model complexity and relaxing critical structural assumptions expose different layers of dispersal processes. Okubo & Levin (1989) showed that the horizontal wind speed, seed terminal velocity, and release height are the key determinants of short-distance dispersal; and there is no need to increase model complexity if the goal is to predict the local dispersal of most seeds. Nathan et al. (2001) showed that, empirically, these factors vary considerably in their relative effects. The variation in wind speed components explained most (86%) of the variation in dispersal distances of *Pinus halepensis* seeds at their study site, whereas variation in the two biological parameters (V_t and H) together explained much less (9%). Wind dispersal models should therefore include accurate estimates of wind conditions. Moreover, seeds of isolated trees are expected to travel much farther than identical seeds released from identical trees within a dense forest. These differences have important implications for tree dynamics at small and large scales (Nathan et al. 2002a).

Whereas all the above models closely match the observed dispersal data at the local scale, their relevance to LDD has been questioned (Bullock & Clarke 2000, Greene & Johnson 1995, Nathan et al. 2002a). This is because LDD of tree seeds, for example, critically depends on the fine details of turbulence structure within the forest. Nathan et al. (2002b) applied a coupled Eulerian-Lagrangian approach to model the three-dimensional flight trajectories of wind-dispersed tree seeds within and above the forest canopy. Their approach resolves the effects of canopy turbulence and explicitly incorporates excursions whose timescales are on the order of seconds. This model was parameterized from high-frequency wind measurement data for a site at Duke Forest and tested against dispersal data collected along a 45-m tower. Predictions closely matched observed data for five wind-dispersed tree species. This study revealed the crucial role for LDD of seed uplifting, by turbulent wind updrafts that result in temporally and spatially autocorrelated deviations in vertical wind velocity. Tackenberg (2003), using a stochastic Lagrangian simulator of seed flight, fed with observed sequences of high-frequency wind measurements, demonstrated that uplifting was also critical for LDD of grass seeds and concluded that LDD in grasslands is promoted by thermal updrafts.

Variation in horizontal and vertical wind speeds can be incorporated along with most of the features of the above models into a unifying random walk model of wind dispersal. It has all the same properties of WINDISPER apart from accounting for variable vertical wind profile yet is also analytically tractable (Portnoy & Willson 1993, Tufto et al. 1997, Turchin 1998). To compute the dispersal kernel in the one-dimensional case, assume that the seeds perform a biased Brownian motion vertically starting from a release point at height H. As before, seeds are dispersed by wind. Assume a downwind dispersal velocity u with variance (per unit time) in distance traveled σ_u^2 and vertical mean wind speed w with variance σ_w^2. Using Brownian motion theory (Karlin & Taylor 1981), one can find the distribution of deposition times $P(t)$:

$$\hat{P}(t) = \sqrt{\frac{H^2}{2\pi\sigma_w^2 t^3}} \exp\left(-\frac{1}{2\sigma_w^2 t}(H - wt)^2\right). \tag{8}$$

The dispersal kernel is obtained from the distribution of deposition times by the equation

$$P(x) = \int_{t=0}^{\infty} \hat{P}(t) p(x, t)\, dt, \tag{9}$$

(see also Tufto et al. 1997, Equation 12), which yields

$$P(x) = \frac{H}{\pi\sigma_u\sigma_w} \exp\left(\frac{xu}{\sigma_u^2} + \frac{Hw}{\sigma_w^2}\right) \sqrt{\frac{u^2\sigma_w^2 + w^2\sigma_u^2}{x^2\sigma_w^2 + H^2\sigma_u^2}}$$

$$\times K_1\left(\left(\frac{x^2}{\sigma_u^2} + \frac{H^2}{\sigma_w^2}\right)^{1/2} \left(\frac{u^2}{\sigma_u^2} + \frac{w^2}{\sigma_w^2}\right)^{1/2}\right), \tag{10}$$

where $K_1(x)$ is the modified Bessel function of the second kind (Gradshteyn & Ryzhik 2000). Yamamura (2002) showed that if one assumes that the distribution of deposition times (Equation 8) is a gamma distribution, the dispersal kernel is also a Bessel function. This model has many different limiting behaviors as the parameters are varied. If σ_u^2 goes to zero, the deterministic formula is recovered. To analyze this model further, introduce the nondimensional variables: $\rho = uH/\sigma_u^2$, $\eta = w\sigma_u^2/u\sigma_w^2$, $\varphi = w/u$, $X = x/H$, and $X_c = \rho^{-1}(\sqrt{1+\eta\varphi} - 1)^{-1}$. For large distances, the dispersal kernel is $P(X) \sim X^{-3/2}\exp(-X/X_c)$. When the distance X_c gets very large, then $P(X) \sim X^{-3/2}$.

Portnoy & Willson (1993) developed the more difficult case of a two-dimensional, radially symmetric model with a horizontal drift velocity u and a horizontal noise term described by the variance σ_u^2. The two-dimensional model leads essentially to the same results as the one-dimensional model. With nondimensional parameterization defined by ρ, η, and φ as above, and $R = r/H$, their result is

$$P(R) = I_0(\rho R)\frac{\rho R}{\sqrt{R^2\eta + \varphi}}\left(\frac{1}{R^2\rho + \rho\eta\varphi^{-1}} + \sqrt{\frac{1+\eta\varphi}{R^2 + \eta\varphi^{-1}}}\right)$$

$$\times \exp\left(\rho\eta - \rho\sqrt{(1+\eta\varphi\eta)(R^2 + \eta\varphi^{-1})}\right), \tag{11}$$

where $I_0(a)$ is the modified Bessel function of the first kind, which behaves as $I_0(a) \approx (2\pi a)^{-1/2}\exp(a)$ (to within 5% for $a > 4$). We can also develop this expression for large values of R, and the dominant behavior is similar to that in the one-dimensional case.

MODELS OF SEED DISPERSAL BY ANIMALS Skellam (1951) demonstrated that wind dispersal was inadequate for explaining observed rates of advance of invading species, and that LDD in some cases must have involved animal vectors. Recent studies (Vellend et al. 2003) reinforce this observation. Typical models of seed dispersal by animals extend the approach described to combine a component describing animal movement and a component describing time until seed deposition. Such parameters can be estimated from statistical distributions of observed animal movements and gut retention times (or handling times, etc.) (Murray 1988). More mechanistic approaches seek to understand the rules that govern animal movements and feeding behaviors (Anderson 1982, Kareiva 1990, Neubert et al. 1995).

A very simple theoretical model of dispersal by animals makes the assumptions that (*a*) animals move randomly in space, and (*b*) deposit seeds at a constant rate during their movement. Again, the probability of displacement by a distance r of an animal t time units after it has picked up a seed follows a Gaussian distribution. The distribution of deposition times is given by an exponential function $P(t) = \rho \exp(-\rho t)$, ρ being the deposition rate as above. The resulting seed dispersal kernel, again, is obtained through Equation 9 (Broadbent & Kendall 1953, Turchin & Thoeny 1993):

$$P(r) = \frac{\rho}{2\pi D} K_0 \left(r \sqrt{\frac{\rho}{D}} \right), \tag{12}$$

where D is the diffusion rate for the dispersers (the animals). More complicated models could be constructed to include effects of landscape structure and variable deposition rates.

Model Assessment

A critical step in the development of models is the assessment of reliability of their assumptions and predictions. Model assessment should include evaluation of four components: structural assumptions, parameter estimates, and primary and secondary predictions (Bart 1995). In phenomenological models, structural assumptions involve, for example, the type of function and whether or not to mix two functions. In mechanistic models, structural assumptions involve, for example, whether seeds are dispersed under exponential or logarithmic wind profiles and whether seeds carried by animals are deposited at random or with some bias. Parameter estimation is crucial in both phenomenological and mechanistic approaches. Yet, good estimates for the parameters of a model are necessary but not sufficient for a good model: Ultimately, model predictions need to be compared with data. The primary predictions of dispersal models are the dispersal distances themselves, or more commonly, the predicted postdispersal seed densities or the proportions of seeds arriving at a seed trap location. Because parameters of mechanistic models are evaluated independently of the dispersal data, any dispersal data could serve to test primary predictions. Phenomenological models, however, necessitate dispersal data for their calibration; but their primary predictions can be tested against a random subset of the data not used for calibration (i.e., cross-validation).

Secondary predictions involve indirect features of the system, such as the shape of the dispersal kernel predicted by mechanistic models.

IMPLICATIONS OF DISPERSAL FOR RATES OF SPECIES ADVANCE

One of the primary motivations for studying dispersal is to understand the rates of spread of species (e.g., Okubo & Levin 2001, Turchin 1998). Models linking dispersal and rates of advance have a long history in ecology and evolutionary biology. In this section, we review classical models of species spread, recent extensions of these models, and empirical tests of the theory.

Classical Diffusion Models

One of the earliest applications of models of dispersal in ecology and evolutionary biology was to the spread of advantageous alleles. Fisher (1937) first considered the problem in genetics, writing the equation

$$\frac{\partial P}{\partial t} = D\frac{\partial^2 P}{\partial x^2} + rP(1 - P) \tag{13}$$

for the frequency P of the favored allele entering a new (linear) habitat. Here D is the diffusion coefficient, and r is the intrinsic rate of natural increase. The assumptions underlying this equation are that individuals move via random walk, in small steps, with no directional bias, and that population growth is logistic (Okubo & Levin 2001). In reality, one should write coupled equations for genotypes, since it is individuals rather than genes that move, and derive Equation 13 from these equations (Aronson & Weinberger 1978), but the reduced model (13) is a good approximation.

Based on Equation 13, Fisher reasoned that the eventual (asymptotic) rate of spread of the allele would be

$$2\sqrt{Dr}. \tag{14}$$

Kolmogorov et al. (1937) established this result formally, for general growth functions $f(P)$ (see Okubo et al. 2001b).

In an ecological setting, the seminal paper extending these results was that of Skellam (1951), who was interested in the rates of advance of invading species. Skellam embedded the problem in a broader framework that considered more complicated dispersal kernels in multiple spatial dimensions. He also derived the equivalent of Equation 13, for the population density P.

The nonlinear term can be ignored in the determination of asymptotic speeds, because densities are low enough at the front that growth is essentially exponential. For more general growth functions, this is also true, but nonlinearities can be important early in spread if there are multiple possible equilibria, resulting in the equation

$$\frac{\partial P}{\partial t} = D\frac{\partial^2 P}{\partial x^2} + rP. \tag{15}$$

No real front forms in this case because population size is growing (or declining) exponentially. Skellam addressed this by assuming that there is a cut-off level P* below which either the spreading population would not be detected, or population density would be too low to sustain spread, and defined the position of the front as the value(s) of x where population density equals that threshold. By symmetry, even in one dimension there will be two such points, and their locations will advance asymptotically at the speed given in Equation 14. Equation 13 assumes no bias in movement, such as might arise owing to advective forces like winds or water flow, or simply tactic movement in the case of animal or microbial dispersal. For constant advection, the speed is given by Equation 14 relative to a frame of reference moving with the advection.

In two or three spatial dimensions, the essential problem and conclusions are the same, at least in homogeneous habitats. The basic equation, in two dimensions, without advection, is given by

$$\frac{\partial P}{\partial t} = D_x\frac{\partial^2 P}{\partial x^2} + D_y\frac{\partial^2 P}{\partial y^2} + f(P). \tag{16}$$

If diffusion is isotropic, then a point release will eventually spread in circular fronts, at the speed given by Equation 14; nonisotropic diffusion will lead to elliptical patterns with different rates of spread in different directions. Without density dependence, arbitrary initial distributions can be treated as superpositions of point releases, yielding the same result; with density dependence, the picture becomes a bit more complicated, and the details again must depend on the specific form.

Empirical Tests of the Theory

Everything discussed so far relates to the general reaction-diffusion model (e.g., Equation 13) and its extensions. How good are the assumptions underlying this model, and how well do the predictions agree with observations? In general, the assumption of random walk is a convenience, representing ignorance of local stochastic factors that determine actual movements. The convenience can be justified, however, if in some statistical sense, the observed patterns do not differ from the predictions of the theory.

Skellam examined a number of empirical examples—in particular, the oaks in England following the recession of the glaciers, and the muskrats introduced into Bohemia by an incautious Czech prince. Although much of his focus was in interpreting patterns and mechanisms, Skellam did conclude that the spread of oaks was too rapid to be explained by simple diffusion, confirming the calculations of Clement Reid (Clark et al. 1998a, Reid 1899, Turchin 1998). Rates of Holocene postglacial spread of many temperate tree species were estimated by Davis (1976), Huntley & Birks (1983), Delcourt & Delcourt (1987), and MacDonald (1993), among others. The pollen record shows an individualistic response of tree species, with average rates of spread on the order of 200 m/yr. Reid's paradox (Reid 1899, Skellam 1951) highlights the discrepancy between these estimated high spread rates, and the observed dispersal distances of nearly all species, which typically average no more than a few tens of meters. We return to this in Tails and Rates of Advance of Populations below.

Nonlocal Transport, and Non-Gaussian Kernels

Equation 13 describes a diffusion approximation for a stochastic process, provided redistribution is the result of a large number of small steps. Alternatively, redistribution can be treated as a discrete event, with consequences that can be encapsulated in a single "kernel function." The approach is especially appropriate for situations where growth and dispersal occur sequentially rather than simultaneously, as for annual plants. Skellam (1951) developed the basic approach, in which a dispersal kernel represented the probability distribution for the terminal point of a seed released at a given position. Based on this, one can develop models of population redistribution while taking into account the details of influences on individuals. Mollison (1977) showed how consideration of redistribution kernels with nonlocal transport could change fundamentally the predictions about spread.

In recent years, a number of investigators have built on the framework developed by Skellam & Mollison. Notable among these have been Weinberger (1978), Aronson (1985), Fife & McLeod (1977), van den Bosch et al. (1988), van den Bosch (1990), Kot & Schaffer (1986), Kot et al. (1996), and Neubert et al. (1995, 2000). The simplest model follows annual populations that are sedentary during most of their lives, such as plants or intertidal sessile invertebrates, and that disperse their propagules at the end of a season. Thus the number of propagules n satisfies the equation

$$n_{t+1}(x) = \int_{-\infty}^{\infty} \phi(x, y) f(n_t(y))\, dy, \tag{17}$$

where the kernel $\phi(x, y)$ is the probability density function for the endpoint x of a dispersal event originating at y. Often, but not necessarily, $\phi = k(x - y)$, a function of $(x - y)$ alone. In this case, the probability of dispersing a given distance is independent of the point of origin (Neubert et al. 2000). The second term in the integral represents growth between dispersal events, or more precisely the number of propagules per adult.

The advantage of Equation 17 is that k need not take the Gaussian form, consistent with diffusive spread, and hence can capture LDD events and multimodal dispersal. Numerous examples in the plant literature illustrate the inadequacy of the Gaussian dispersal kernel, and the associated diffusion approximations (Cain et al. 1998; Davis 1976, 1987). Weinberger (1978) showed that, under reasonable conditions, the solutions to Equation 17 again converge to traveling waves, with speed

$$c = \min_{s \in S}\{(\ln(f'0)m(s))/S\}. \tag{18}$$

Here, $f'(0)$ is the growth rate of the population when it is rare, and

$$m(s) = \int_{-\infty}^{\infty} k(x)e^{sx}\,dx \tag{19}$$

is the moment-generating function for the kernel (Kot 1992; Kot et al. 1996; Neubert & Caswell 2000; Neubert et al. 2000; Weinberger 1978, 1982). S is the set of positive s for which the integral in Equation 19 converges (Neubert et al. 2000). Furthermore, to assure convergence to a wave front, one further assumes

$$0 \leq f(n) \leq nf'(0). \tag{20}$$

More generally one can consider delays owing to dispersal, overlapping generations, multiple dimensions, and other complications; but these are beyond the scope of this review. Equation 17 provides a natural bridge from explicit dispersal kernels to the problem of spread.

Tails and Rates of Advance of Populations

Whereas most seeds fall near the parent plant, and most larvae disperse short distances, it is the tail of the dispersal distribution that is of central importance in the spread of species (Mollison 1977, Skellam 1951, Turchin 1998). Both Skellam and Mollison emphasized the inadequacy of the diffusion approximation when long-distance transport is important. Reid's paradox (see Empirical Tests of the Theory, above) that rates of spread in the paleontological record far exceed that expected based on observed mean dispersal distances can be resolved if the tails of the distribution are sufficiently fat (Clark 1998, Clark et al. 1998a, Higgins & Richardson 1999).

Skellam (1951) and Mollison (1977) both emphasized the importance of focusing on the dispersal kernel, and all its moments, in the prediction of spread. The approach has been extended effectively by van den Bosch et al. (1988), van den Bosch (1990), and Kot et al. (1996). Kot et al. (1996) (see Equation 17) showed that one should distinguish three types of dispersal kernels, and that the diffusion approach works only for those with exponentially bounded tails.

IMPLICATIONS OF DISPERSAL FOR POPULATION DYNAMICS AND COMMUNITY ORGANIZATION

Implications of Dispersal for Population Dynamics

Dispersal plays an important role not only in range expansion (see Implications of Dispersal for Rates of Species Advance above) but also in determining the spatial and genetic structure of populations at local and landscape scales.

LOCAL POPULATION DYNAMICS Dispersal patterns directly affect the spatial structure of populations, equilibrium population densities, and rates of population dynamics. All other things being equal, shorter dispersal distances and more clumped seed deposition will tend to result in more clumped seedling and adult distributions (Hamill & Wright 1986), slower rates of exploitation of newly available sites, and lower equilibrium abundances (Bolker & Pacala 1999). In general, however, these effects are strongly modified by the spatial pattern of abiotic and biotic influences on establishment, growth, and survival (Schupp & Fuentes 1995). A species with long dispersal distances may nonetheless exhibit a clumped spatial distribution and low equilibrium abundance if the locations in which it can successfully establish are rare and clumped (Hamill & Wright 1986). Such spatial variation in the probability of seed success is ubiquitous in plant populations and must be considered in evaluating the impact of dispersal patterns (Schupp & Fuentes 1995). This is especially important when the probability of seed deposition itself depends upon habitat conditions (Nathan & Muller-Landau 2000). Disproportionately high seed deposition in favorable habitats (so-called "directed dispersal;" see Dispersal Mechanisms, above) has the potential to enhance greatly recruitment rates and thereby affect population dynamics (Wenny 2001).

Habitat favorability may itself not only affect but also be affected by the population spatial pattern. Proximity to parents or other conspecific adults may impact recruitment negatively if the activities of seed predators, pathogens, or other natural enemies are concentrated around parent trees (Connell 1971, Hammond & Brown 1998, Janzen 1970), or positively if parent trees provide a favorable microenvironment for recruitment (Tewksbury & Lloyd 2001) or are associated with higher local availability of mutualists such as mycorrhizae (Wilkinson 1997). Empirical studies have shown that many if not most plant species suffer increased mortality in areas of higher densities of conspecific seeds, seedlings, or adults (Harms et al. 2000, HilleRisLambers et al. 2002). Given a particular negative relationship between conspecific density and survival, the number and spatial pattern of successful recruits will depend upon seed dispersal patterns. In general, higher rates of seed dispersal will lead seeds to experience lower conspecific densities and thus higher survival. Depending on the exact dispersal and survival functions, seedling densities may decrease, increase (McCanny 1985), increase and then decrease (Janzen 1970), or remain unchanged with increasing distance from parents (McCanny 1985).

METAPOPULATION DYNAMICS Dispersal strategy has a very direct impact on a species's abundances and distribution among different subpopulations and its overall persistence in the whole metapopulation.

When patches are identical in suitability for the target species, dispersal rates among patches, within-patch population growth rates, and patch-carrying capacity alone will determine metapopulation dynamics since they determine rates of colonization and stochastic extinction. Most theoretical studies have used island models in which all patches (subpopulations) are equally connected through global dispersal. In these models, higher rates of dispersal alone lead to occupancy in greater numbers of subpopulations, increased mean subpopulation and overall metapopulation abundance, and longer-term persistence (Hanski 2001). Of course, these effects may be diminished or reversed if dispersal is associated with increased mortality (Hanski 2001). Spatially explicit metapopulation models in which dispersal rates among pairs of patches are distance-dependent provide additional insight into the relative importance of fecundity, short- and long-distance dispersal on colonization rates and thereby metapopulation dynamics. Higgins & Cain (2002) use such a model to demonstrate that fecundity, interpatch distances, and the rates and distances of LDD all had significant, interacting effects on colonization rates and overall dynamics, while short-distance dispersal was unimportant. They incorporated a stratified dispersal kernel (see Theoretical Models of Seed Dispersal, above), and their results imply that important gains in understanding can be made by relaxing the standard assumptions. Metapopulations with restricted dispersal and/or high degree of isolation can generate complex spatial patterns even in homogeneous environments (Hanski 2001).

When patches differ in such a way that competitive ability of a species varies among them, these differences will interact with dispersal to determine metapopulation structure and dynamics. Given stochastic and asynchronous variation in conditions among patches, dispersal can be an effective bet-hedging strategy, potentially allowing metapopulation persistence through dispersal from one transiently favorable site to another, even when the expected growth rate in all local populations is negative (Metz et al. 1983). Given fixed differences in conditions among patches, dispersal between source and sink patches can provide the means for a species to occur frequently outside the bounds of its fundamental niche (Pulliam 1988). In such source-sink metapopulations, dispersal is strongly asymmetrical from source to sink. Dispersal can provide a rescue effect (Brown & Kodric-Brown 1977) for small populations (in either sink or source habitats) facing high risk of extinction. Less frequently, dispersal may increase extinction risk of small isolated populations, especially in sink habitats, if the number of emigrants exceeds the number of immigrants. Matrix population models can provide a useful approach for approximating patch models in which colonization and survival rates differ among patches and among successive early recruitment stages (Horvitz & Schemske 1986).

GENETIC STRUCTURE Dispersal has profound effects on the genetic structure of populations (Malécot 1948; Wright 1943, 1969), and this can be used to infer the phylogeographic structure of plants (Petit & Grivet 2002). In plants, genes are dispersed either through haploid pollen or diploid seeds, and inheritance may be maternal (e.g., chloroplast DNA in angiosperms), paternal (e.g., chloroplast DNA in conifers), or biparental (e.g., nuclear DNA). Classical models of isolation by distance (Kimura 1953, Wright 1943) have been used to analyze the development of genetic structure of populations. In general, higher levels of long-distance gene flow act to reduce local genetic correlation, while higher levels of short-distance gene flow act to increase it (Wright 1969). These models have been generalized in various ways recently, for example to consider uniparentally inherited genes (Hu & Ennos 1999) or local density-dependence (Barton et al. 2002).

Implications for Communities

Just as seed dispersal is important to population structure and dynamics, it also is important to community structure and dynamics—specifically, to local species composition and its spatial and temporal turnover. We can divide the effects of seed dispersal on communities into two categories: those that result from general limits to dispersal or migration ability in all species, and those that result from differences among species in dispersal ability—differences associated with trade-offs between dispersal ability and other traits. We refer to these as effects of overall dispersal rates and dispersal trade-offs, respectively. These effects are clearly related, but have distinct theoretical implications.

OVERALL DISPERSAL RATES The effects on community structure and dynamics of overall dispersal or migration rates among local communities depend fundamentally on the nature of local competitive interactions. In the extreme (neutral) case in which all species are competitively equivalent, dispersal and speciation rates alone determine local and global diversity patterns. In the opposite extreme of such strict competitive hierarchies that only one species is suited for any given site, dispersal has little effect on diversity.

In neutral models, increasing dispersal increases the number of species competing for sites, increasing local community species richness and evenness (alpha diversity, sensu Whittaker 1972) while decreasing turnover among communities (beta diversity) and total metacommunity species richness (gamma diversity) because species can drift to extinction faster (Chave et al. 2002, Hubbell 2001). These effects are exactly analogous to those found in neutral models in population genetics discussed above. Building on those results (Kimura 1953, Malécot 1948), we can compute the expected species-area curves, relative abundance distributions, rates of turnover in space, and other measures of community structure directly from dispersal and speciation rates (Bramson et al. 1996, 1998; Chave & Leigh 2002).

As local community dynamics become less neutral, with increased local competitive differences among species, dispersal rates become relatively less important to diversity patterns but still can exert substantial effects. When species have competitive differences that vary among areas, the effects of dispersal are weaker than in the neutral model. Because continuing immigration can allow some locally inferior species to persist in areas where they would otherwise be eliminated, local diversity can be enhanced in a community-level "mass effect" (Shmida & Ellner 1984). Where local competitive ability is negatively frequency-dependent (e.g., owing to life-history niche differences or negative density-dependent recruitment), the effects of dispersal are stronger than in the neutral model because immigrants that represent locally new types are disproportionately advantaged (Chave et al. 2002). In contrast, in communities in which local competitive dynamics are positively frequency-dependent, increased dispersal is most likely to reduce local as well as total diversity. Extensive dispersal among communities homogenizes species composition, and eventually makes competitive ability dependent on global rather than local abundances, thus facilitating domination by the single most abundant species (Amarasekare 2000, Karlin & MacGregor 1972, Levin 1974).

The importance of dispersal in real communities can be tested empirically in a number of ways (Nathan & Muller-Landau 2000). The strongest test involves experimental manipulation of seed dispersal patterns, but except for seed addition experiments (which manipulate seed number as well as spatial pattern), such tests are rare. Where experiments are impractical, simulations using empirically parameterized population and community models are a good alternative (Ribbens et al. 1994). Comparison of empirical population and community patterns with those expected under models with varying influences of seed dispersal provides a final, albeit weaker, means of assessing the relative importance of dispersal (Condit et al. 2002, Schupp & Fuentes 1995).

DISPERSAL TRADE-OFFS The models discussed above all assume that dispersal rates are equivalent for all species; but in reality, seed dispersal varies widely among plant species together with other characters. Not all trait combinations are found, however—allocation and design limitations result in trade-offs that limit possible combinations, and selection eliminates others. Strategic trade-offs between the ability to disperse long distances and other traits can thus potentially enable niche differentiation that contributes to stable, equilibrium, species coexistence.

Trade-offs between the ability to colonize new sites (involving high dispersal rates and/or fecundity) and the ability to compete for sites upon arrival have long been hypothesized to contribute to stable species coexistence (Skellam 1951). Theoretical studies have demonstrated how such trade-offs potentially can contribute to ecologically (Hastings 1980, Tilman 1994) and/or evolutionarily (Geritz et al. 1999) stable coexistence of many species by allowing competitively inferior species to persist as fugitives in areas unoccupied by more dominant but less vagile species. Importantly, such coexistence requires either strong competitive asymmetry, as in the classic competition-colonization trade-off model in which sites held by weak competitors are instantaneously taken over by arriving strong competitors (Hastings 1980, Tilman 1994), or strong demographic stochasticity, insuring that the superior competitor's seeds do not reach all sites (Kisdi & Geritz 2003). Alternatively, trade-offs between dispersal and other traits such as fecundity, may enable equilibrium coexistence in spatially variable habitats (Yu & Wilson 2001).

There are a number of ways to test for the presence of competition-colonization trade-offs among species and to evaluate their importance to community dynamics and structure. The most direct and definitive test is experimentally to add seeds of all species: If a competition-colonization trade-off is present, this should result in an increase in the abundance of a subset of species that are competitively dominant and are ordinarily seen later in succession (Pacala & Rees 1998). Two studies involving addition of seeds of multiple co-occurring grassland plant species both found that seed addition increased the relative abundances of larger-seeded species within the community (Jakobsson & Eriksson 2000, Turnbull et al. 1999). This is in accordance with the idea that larger-seeded species are the better competitors and the poorer colonizers, and that a competition-colonization trade-off mediated by seed size drives successional patterns in this community (Pacala & Rees 1998). This idea is further reinforced by the finding (Moles & Westoby 2002) that seed addition is more likely to increase abundance in larger-seeded species. The presence of competition-colonization trade-offs also can be documented by analyzing correlations among relevant traits (Leishman et al. 2000). This is a weaker test, especially since many such correlations, for example fecundity-establishment trade-offs, can result in merely neutral dynamics (Yu & Wilson 2001).

EVOLUTION OF DISPERSAL

As we have seen, dispersal strategies have important implications for populations and communities, and thus not surprisingly, for fitness. As a result we expect strong selection on dispersal-related traits. Theoretical work on dispersal evolution has a long history (Hamilton & May 1977, Van Valen 1971), and recent years have seen an explosion of studies in this area (Gandon & Michalakis 2001) in part because of the increasing ease of simulating ever more complex scenarios under which dispersal can evolve. These theoretical studies have increasingly clarified how different factors can affect selective pressures on dispersal, although there remains a dearth of corresponding empirical work to test or parameterize theoretical models (Ronce et al. 2001).

The payoff of a dispersal strategy depends fundamentally on the strategies of other individuals in the population because these determine the number of competitors encountered by dispersing and nondispersing individuals and their relatedness. Thus the evolution of dispersal must be considered in a game-theoretic context. We search for strategies that are evolutionarily stable strategies (ESS)—strategies that cannot be invaded by any other strategy (Maynard Smith 1982). Further, to be evolutionarily accessible via mutation by small steps under the standard assumptions on quantitative genetic traits with continuous genetic variation, such a strategy must also be an evolutionary attractor, that is, be convergence stable (Eshel 1983) and a neighborhood invader strategy (NIS), making it an ESNIS (Levin & Muller-Landau 2000). Such strategies, and also (attracting) evolutionary branch points, can be sought using methods of adaptive dynamics (Geritz et al. 1997). Alternatively, genetic structure may be modeled explicitly, for example, in terms of selection at one or more genes (Vincent & Brown 1988).

The major forces selecting for dispersal are kin competition, inbreeding depression, and spatiotemporal variability in environmental conditions. Because pollen dispersal distances typically exceed seed dispersal distances and thus are much more important in determining the level of inbreeding (Ennos 1994), inbreeding depression is likely to be relatively unimportant in selection for dispersal among plants; thus we do not consider it further. Here we review work demonstrating how the remaining factors contribute to dispersal evolution—work that helps explain differences in dispersal strategies among species and the coexistence of multiple dispersal types within communities. It is important to note that the optimal dispersal strategy will also depend on how dispersal trades off with other traits, and on coevolution between dispersal and other characters.

Kin Competition

A universal advantage of a strategy incorporating dispersal over a strategy of no dispersal is that dispersers can win sites from nondispersers, but nondispersers can never win sites from dispersers (Hamilton & May 1977). Dispersal can allow some individuals to escape competition with kin for the home site and instead potentially compete with and take over sites held by unrelated individuals having a different dispersal propensity. Thus, even in a model in which dispersal is costly, in which there is no variation in environmental quality or crowding and no inbreeding depression, non-zero dispersal rates will be selected. This result was powerfully demonstrated by Hamilton & May (1977) using an island model of an asexually reproducing annual plant species in which each patch had one individual. If dispersing individuals incur a survival reduction c (that is, they survive at rate 1-c relative to nondispersing individuals), the ESNIS fraction of offspring to disperse outside the parent patch, D, is

$$D = \frac{1}{1+c}. \tag{21}$$

Thus, if there is no cost to dispersal, all the offspring should disperse, and even if dispersal is very costly (almost always lethal), half the offspring should disperse.

For more general island models having multiple individuals per site and different breeding systems, the influence of kin competition on dispersal rates can be assessed by considering the relatedness of individuals within and among patches (Frank 1986, Gandon & Michalakis 2001). In general, as patch size increases, relatedness within the home patch decreases, the benefits of dispersal for escaping kin competition decrease, and thus the optimal dispersal fraction decreases (Comins et al. 1980). Outbreeding reduces the optimal dispersal fraction because it reduce relatedness within patches; indeed, for high cost of dispersal, the optimal dispersal rate in outbred populations is zero (Hamilton & May 1977, Taylor 1988).

Qualitatively similar results hold for rates of dispersal under more realistic, spatially explicit dispersal models. This was demonstrated early in stepping-stone models—models in which dispersing offspring go only to neighboring sites (Comins et al. 1980). As the number of sites to which dispersal occurs decreases, relatedness between the home site and these sites increases, reducing selection for dispersal. Thus, the ESNIS dispersal fraction declines as the spatial scale of dispersal decreases (Comins et al. 1980, Gandon & Rousset 1999).

If dispersal to different distances is controlled independently, then the evolutionarily stable dispersal strategy is one in which the fitness gains of dispersal to all distances is equilibrated (Rousset & Gandon 2002). In particular, Rousset & Gandon (2002) show that dispersal is associated with two types of costs: the direct cost paid by the disperser (in increased mortality or equivalent), and the indirect cost due to competition with related individuals. At the ESNIS set of dispersal strategies, the product of direct and indirect benefits should be the same at different dispersal distances (Rousset & Gandon 2002).

An aspect of kin competition that has only recently received theoretical attention is parent-offspring competition, although it was discussed early on by Hamilton & May (1977). The potential for parent-offspring competition arises when there are overlapping generations. When parents senesce—so that the probability of parental mortality changes with age—there is selection for an increase in the dispersal fraction with increasing maternal age (Ronce et al. 1998).

Spatiotemporal Variability in the Environment

Spatiotemporal variability in the environment leads to selection for increased dispersal, which allows for bet-hedging over uncertainty (Gadgil 1971). This is true only, however, when the quality of the environment varies in both space and time. Fixed spatial variation in habitat quality selects against dispersal when dispersal is not habitat-dependent, because dispersal tends to move individuals from better habitats with higher abundances to poorer habitats with lower abundances (Hastings 1983). Fixed, synchronous temporal variation alone also has no effect on selection for dispersal in simple models (Ellner & Shmida 1981).

The effect of habitat variability in the absence of kin competition was first demonstrated by Comins et al. (1980). In their island model of annual plants with discrete, nonoverlapping generations, there are an effectively infinite number of individuals per patch, and thus, relatedness within a patch is zero, and there is no kin competition selection for dispersal. Each patch has a probability x of becoming extinct in any given generation. Then the ESNIS dispersal fraction is

$$D = \frac{x}{1 - (1 - x)(1 - c)}, \tag{22}$$

where c is as before the survival cost of dispersal. Note that if dispersal is very costly, then the optimal dispersal fraction approaches the probability of extinction of a patch (Van Valen 1971).

Levin et al. (1984) extended this work to more general types of uncorrelated environmental variation. They showed that for arbitrary stationary distributions of environmental variation, identical among sites, the dispersal fraction will depend on the normalized harmonic mean of the distribution of site quality. As long as there is some probability of patch extinction, the dispersal fraction will always be nonzero, no matter the cost of dispersal. However, if there is zero probability of patch extinction (harmonic mean greater than zero), then there will be some threshold dispersal cost beyond which the ESNIS is zero dispersal (Levin et al.

1984). Chaotic population dynamics, which produce a special kind of endogenous variation among patches, also favor dispersal (Holt & McPeek 1996).

While the above models support the intuitive conclusion that dispersal should increase as environmental stability decreases, this need not always be the case. Ronce et al. (2000) find conditions under which dispersal rates can decrease as the extinction rate of patches increases. They relax the assumption that carrying capacity is always reached in one generation, and thus allow for potentially multiple generations of growth within patches (see also Levin et al. 1984 for an earlier version of such a model). This increases the benefits of staying home, and thus decreases the ESNIS dispersal fraction. Further, as the extinction rate increases, more patches are below carrying capacity, and as a result, there is a region of parameter space where dispersal rates decrease with increasing extinction rate (Ronce et al. 2000).

Thus far, we have considered the effects of environmental variability only on the potential for evolution of strategies of fixed dispersal fractions in the island model—that is, the fraction dispersed does not change with local conditions. Yet there is extensive evidence that in many systems, dispersal is dependent on local density or habitat quality (Travis & French 2000). Such conditional dispersal strategies will generally be advantageous whenever crowding varies among patches (Levin et al. 1984, Metz & Gyllenberg 2001, Travis & Dytham 1999). Their relevance to plant populations deserves further consideration.

Even more importantly, we must extend results on dispersal fractions in island models to dispersal strategies in explicitly spatial habitats. In the past few years, a number of studies have made this leap using spatially explicit simulation, and most consider both habitat heterogeneity and kin competition together. Qualitative conclusions regarding the effects of spatiotemporal variation are the same as in island models (Heino & Hanski 2001). However, explicit spatial structure allows for consideration of the effects of realistic levels of spatial autocorrelation in habitat as well, and this has produced some novel results. In autocorrelated landscapes, "fat-tailed" dispersal kernels are favored, while uniform distributions are favored in random landscapes (Hovestadt et al. 2000). When spatial and temporal variability exhibits red rather than white noise (increasing rather than constant variance with time or distance), there is reduced selection for dispersal (Travis 2001). In models with relatively constant spatial variation in landscape quality, different dispersal rates evolve in narrow corridors and on the boundaries of good habitats than in the centers of habitats, with potentially important implications for conservation (Travis & Dytham 1999).

INTEGRATION AND FUTURE DIRECTIONS

The growing interest in dispersal is matched by increasing recognition of the value of models in describing, exploring, and predicting dispersal processes and patterns (Bullock et al. 2002, Clobert et al. 2001, Levey et al. 2002).

Phenomenological models fitted to data describe the pattern of dispersal distances, and the successes and failures of different models together have illuminated the fact that the distribution of dispersal distances is generally strongly leptokurtic. We advocate increased caution when inferring dispersal from patterns of establishment: Pre- and postdispersal processes should be separated from dispersal because they can significantly alter dispersal patterns. Yet, for the same reason, it is crucial to investigate better how these three basic recruitment processes are interrelated, e.g., how fecundity affects LDD and how LDD itself affects seed survival. Of particular interest are habitat-specific seed deposition rates (see Dispersal Mechanisms and Spatial Patterns, above), which appear to be very important for many species and have rarely been accommodated in any phenomenological models.

Mechanistic models have provided important insight into which aspects of the dispersal process are most important in determining dispersal distances, in general, and LDD events, in particular. In the case of seed dispersal by wind, mechanistic models have demonstrated the importance of not only average wind velocities but also the correlation structure of windspeeds, especially the incidence of updrafts. Further development of mechanistic models, especially of seed dispersal by animals, is an important direction for future research. This requires, first and foremost, a solid knowledge of the natural history of the relevant dispersal processes—knowledge that can be used to inform decisions about which details are crucial to include in the model and which are not. Model development also requires high-quality data on the operative factors for parameter estimation and on dispersal patterns at small and large scales for testing model predictions. Theoretical tools and pre-existing models from probability theory, fluid mechanics, animal behavior (e.g., optimal foraging theory), and other fields within and beyond ecology are likely to be useful in designing, simplifying, and solving (finding closed-form solutions to) mechanistic models.

Theoretical studies of the implications of dispersal for range expansion, population dynamics, and community structure, as well as of dispersal evolution, indicate that the shape of the dispersal kernel is fundamental. Nevertheless, most conclusions to date have relied upon very simplified, unrealistic models of dispersal, in particular global, nearest-neighbor, or Gaussian dispersal models. This work has provided qualitatively useful results, indicating that, in general, higher rates of dispersal are associated with faster range expansion, presence in more subpopulations, higher local diversity (genetic and species), and lower turnover in space. Yet, further advances that more precisely characterize the importance of dispersal depend critically upon models that incorporate realistic dispersal patterns, including habitat-specific seed deposition. In addition, we emphasize the importance of rigorous model assessment.

New technological advances promise to reduce the typically high uncertainty involved with quantifying dispersal processes, especially those operating at large scales, for a larger array of species and dispersal modes. Although the magnitude of inherent (nonreducible) uncertainty (Clark et al. 2001) is unknown, a potential for high inherent uncertainty does exist. This raises the question of how plants, facing high uncertainty in the behavior of the (external) dispersal agent, can still gain some control over the distance their seeds travel and eventually their fate. Such questions provide an important avenue for future theoretical studies, encompassing both ecological and evolutionary aspects of seed dispersal research.

ACKNOWLEDGMENTS

The authors gratefully acknowledge support from the National Science Foundation awards IBN-9981620 (SAL and RN) and DEB-0083566 (SAL); the Andrew W. Mellon Foundation (SAL and JC); the Israeli Science Foundation award ISF 474/02, and the German-Israeli Foundation award GIF 2006–1032.12/2000 (RN); and a postdoctoral fellowship at the National Center for Ecological Analysis and Synthesis, a Center funded by the National Science Foundation award DEB-0072909, the University of California, and the Santa Barbara campus (HCM). The authors are deeply grateful to Amy Bordvik for her patience, dedication, and supreme professionalism. Without question, this review could not have been written without her.

The *Annual Review of Ecology, Evolution, and Systematics* is online at http://ecolsys.annualreviews.org

LITERATURE CITED

Amarasekare P. 2000. The geometry of coexistence. *Biol. J. Linn. Soc.* 71:1–31

Anderson DJ. 1982. The home range: a new nonparametric estimation technique. *Ecology* 63:103–12

Aronson DG. 1985. The role of diffusion in mathematical population biology: Skellam revisited. In *Mathematics in Biology and Medicine*, ed. V Capasso, E Grosso, SL Paveri-Fontana, pp. 2–18. Berlin: Springer-Verlag

Aronson DG, Weinberger HF. 1978. Multidimensional nonlinear diffusion arising in population genetics. *Adv. Math.* 30:33–76

Bart J. 1995. Acceptance criteria for using individual-based models to make management decisions. *Ecol. Appl.* 5:411–20

Barton NH, Depaulis F, Etheridge AM. 2002. Neutral evolution in spatially continuous populations. *Theor. Popul. Biol.* 61:31–48

Bolker BM, Pacala SW. 1999. Spatial moment equations for plant competition: understanding spatial strategies and the advantages of short dispersal. *Am. Nat.* 153:575–602

Bramson M, Cox JT, Durrett R. 1996. Spatial models for species area curves. *Ann. Probab.* 24:1727–51

Bramson M, Cox JT, Durrett R. 1998. A spatial model for the abundance of species. *Ann. Probab.* 26:658–709

Broadbent SR, Kendall DG. 1953. The random walk of *Trichostrongylus retortaeformis*. *Biometrics* 9:460–66

Brown JH, Kodric-Brown A. 1977. Turnover rates in insular biogeography: effect of immigration on extinction. *Ecology* 58:445–49

Bullock JM, Clarke RT. 2000. Long distance seed dispersal by wind: measuring and modelling the tail of the curve. *Oecologia* 124:506–21

Bullock JM, Kenward RE, Hails R, eds. 2002. *Dispersal Ecology.* Oxford, UK: Blackwell Sci.

Cain ML, Damman H, Muir A. 1998. Seed

dispersal and the Holocene migration of woodland herbs. *Ecol. Monogr.* 68:325–47
Cain ML, Milligan BG, Strand AE. 2000. Long-distance seed dispersal in plant populations. *Am. J. Bot.* 87:1217–27
Chambers JC, MacMahon JA. 1994. A day in the life of a seed: movements and fates of seeds and their implications for natural and managed systems. *Annu. Rev. Ecol. Syst.* 25:263–92
Chave J, Leigh EG. 2002. A spatially explicit neutral model of beta-diversity in tropical forests. *Theor. Popul. Biol.* 62:153–68
Chave J, Muller-Landau HC, Levin SA. 2002. Comparing classical community models: Theoretical consequences for patterns of diversity. *Am. Nat.* 159:1–23
Clark JS. 1998. Why trees migrate so fast: confronting theory with dispersal biology and the paleorecord. *Am. Nat.* 152:204–24
Clark JS, Carpenter SR, Barber M, Collins S, Dobson A, et al. 2001. Ecological forecasts: An emerging imperative. *Science* 293:657–60
Clark JS, Fastie C, Hurtt G, Jackson ST, Johnson C, et al. 1998a. Reid's paradox of rapid plant migration: dispersal theory and interpretation of paleoecological records. *BioScience* 48:13–24
Clark JS, Macklin E, Wood L. 1998b. Stages and spatial scales of recruitment limitation in Southern Appalachian forests. *Ecol. Monogr.* 68:213–35
Clark JS, Silman M, Kern R, Macklin E, HilleRisLambers J. 1999. Seed dispersal near and far: patterns across temperate and tropical forests. *Ecology* 80:1475–94
Clobert J, Danchin E, Dhondt AA, Nichols JD, eds. 2001. *Dispersal.* Oxford, UK: Oxford Univ. Press. 474 pp.
Comins HN, Hamilton WD, May RM. 1980. Evolutionarily stable dispersal strategies. *J. Theor. Biol.* 82:205–30
Condit R, Pitman N, Leigh EG Jr, Chave J, Terborgh J, et al. 2002. Beta-diversity in tropical forest trees. *Science* 295:666–69
Connell JH. 1971. On the roles of natural enemies in preventing competitive exclusion in some marine animals and in rain forest trees. In *Dynamics of Populations, Proc. Adv. Stud. Inst. Dyn. Numb. Popul., Oosterbeek, 1970,* ed. PJ den Boer, GR Gradwell, pp. 298–312. Wageningen, Neth.: Centre Agr. Publ. Doc.
Cousens RD, Rawlinson AA. 2001. When will plant morphology affect the shape of a seed dispersal "kernel"? *J. Theor. Biol.* 211:229–38
Davis MB. 1976. Pleistocene biogeography of temperate deciduous forests. *Geosci. Man* 13:13–26
Davis MB. 1987. Invasion of forest communities during the Holocene: Beech and hemlock in the Great Lakes region. In *Colonization, Succession, and Stability,* ed. AJ Gray, MJ Crawley, PJ Edwards, pp. 373–93. Oxford, UK: Blackwell
Delcourt PA, Delcourt HR. 1987. *Long-Term Forest Dynamics of the Temperate Zone.* New York: Springer-Verlag
Ellner S, Shmida A. 1981. Why are adaptations for long-range seed dispersal rare in desert plants? *Oecologia* 51:133–44
Ennos RA. 1994. Estimating the Relative Rates of Pollen and Seed Migration among Plant-Populations. *Heredity* 72:250–59
Eshel I. 1983. Evolutionary and continuous stability. *J. Theor. Biol.* 103:99–111
Fife PC, McLeod JB. 1977. The approach of solutions of nonlinear diffusion equations to travelling wave solutions. *Arch. Ration. Mech. Anal.* 65:335–61
Fisher RA. 1937. The wave of advance of advantageous genes. *Ann. Eugenics* 7:355–69
Frank SA. 1986. Dispersal polymorphisms in subdivided populations. *J. Theor. Biol.* 122: 303–9
Gadgil M. 1971. Dispersal-Population Consequences and Evolution. *Ecology* 52:253
Gandon S, Michalakis Y. 2001. Multiple causes of the evolution of dispersal. See Clobert et al. 2001, pp. 155–67
Gandon S, Rousset F. 1999. Evolution of stepping-stone dispersal rates. *Proc. R. Soc. London Ser. B* 266:2507–13
Geritz SAH, Metz JAJ, Kisdi E, Meszéna G.

1997. Dynamics of adaptation and evolutionary branching. *Phys. Rev. Lett.* 78:2024–27
Geritz SAH, van der Meijden E, Metz JAJ. 1999. Evolutionary dynamics of seed size and seedling competitive ability. *Theor. Popul. Biol.* 55:324–43
Godoy JA, Jordano P. 2001. Seed dispersal by animals: exact identification of source trees with endocarp DNA microsatellites. *Mol. Ecol.* 10:2275–83
Gradshteyn IS, Ryzhik IM. 2000. *Table of Integrals, Series, and Products*. San Diego: Academic
Greene DF, Calogeropoulos C. 2002. Measuring and modelling seed dispersal of terrestrial plants. See Bullock et al. 2002, pp. 3–23
Greene DF, Johnson EA. 1995. Long-distance wind dispersal of tree seeds. *Can. J. Bot.* 73:1036–45
Greene DF, Johnson EA. 1989. A model of wind dispersal of winged or plumed seeds. *Ecology* 70:339–47
Hamill DN, Wright SJ. 1986. Testing the dispersion of juveniles relative to adults: a new analytic model. *Ecology* 67:952–57
Hamilton WD, May RM. 1977. Dispersal in stable habitats. *Nature* 269:578–81
Hammond DS, Brown VK. 1998. Disturbance, phenology and life-history characteristics: factors influencing distance/density-dependent attack on tropical seeds and seedlings. In *Dynamics of Tropical Communities*, ed. DM Newbery, HHT Prins, ND Brown, pp. 51–78. Oxford, UK: Blackwell Sci.
Hanski I. 2001. Population dynamic consequences of dispersal in local populations and in metapopulations. See Clobert et al. 2001, pp. 283–98
Harms KE, Wright SJ, Calderón O, Hernández A, Herre EA. 2000. Pervasive density-dependent recruitment enhances seedling diversity in a tropical forest. *Nature* 404:493–95
Hastings A. 1980. Disturbance, coexistence, history, and competition for space. *Theor. Popul. Biol.* 18:363–73
Hastings A. 1983. Can spatial selection alone lead to selection for dispersal? *Theor. Popul. Biol.* 24:244–51
Heino M, Hanski I. 2001. Evolution of migration rate in a spatially realistic metapopulation model. *Am. Nat.* 157:495–511
Higgins SI, Cain ML. 2002. Spatially realistic metapopulation models and the colonization-competition tradeoff. *J. Ecol.* 90:616–26
Higgins SI, Nathan R, Cain ML. 2003. Are long-distance dispersal events in plants usually caused by nonstandard means of dispersal? *Ecology.* In press
Higgins SI, Richardson DM. 1999. Predicting plant migration rates in a changing world: the role of long-distance dispersal. *Am. Nat.* 153:464–75
HilleRisLambers J, Clark JS, Beckage B. 2002. Density-dependent mortality and the latitudinal gradient in species diversity. *Nature* 417:732–35
Holt RD, McPeek MA. 1996. Chaotic population dynamics favors the evolution of dispersal. *Am. Nat.* 148:709–18
Horvitz CC, Schemske DW. 1986. Seed dispersal and environmental heterogeneity in a neotropical herb: a model of population and patch dynamics. In *Frugivores and Seed Dispersal*, ed. A Estrada, TH Felming, pp. 169–86. Dordrecht, Neth.: Junk
Hovestadt T, Messner S, Poethke HJ. 2000. Evolution of reduced dispersal mortality and 'fat-tailed' dispersal kernels in autocorrelated landscapes. *Proc. R. Soc. London Ser. B* 268:385–91
Howe HF, Smallwood J. 1982. Ecology of seed dispersal. *Annu. Rev. Ecol. Syst.* 13:201–28
Hu XS, Ennos RA. 1999. Impacts of seed and pollen flow on population genetic structure for plant genomes with three contrasting modes of inheritance. *Genetics* 152:441–50
Hubbell SP. 2001. *The Unified Neutral Theory of Biodiversity and Biogeography*. Princeton, NJ: Princeton Univ. Press. 375 pp.
Huntley B, Birks HJB. 1983. *An Atlas of Past and Present Pollen Maps for Europe: 0–13,000 Years Ago*. Cambridge, UK: Cambridge Univ. Press
Jakobsson A, Eriksson O. 2000. A comparative

study of seed number, seed size, seedling size and recruitment in grassland plants. *Oikos* 88:494–502

Janzen DH. 1970. Herbivores and the number of tree species in tropical forests. *Am. Nat.* 104:501–28

Kareiva P. 1990. Population dynamics in spatially complex environments: theory and data. *Philos. Trans. R. Soc. London Ser. B.* 330:175–90

Karlin S, MacGregor J. 1972. Polymorphisms for genetic and ecological systems with weak coupling. *Theor. Popul. Biol.* 3:186–209

Karlin S, Taylor ME. 1981. *A Second Course in Stochastic Processes*. San Diego: Academic. 542 pp.

Kimura M. 1953. 'Stepping stone' model of population. *Ann. Rep. Nat. Inst. Genet. Jpn.* 3:62–63

Kisdi E, Geritz SAH. 2003. On the coexistence of perennial plants by the competition-colonization trade-off. *Am. Nat.* 161:350–54

Kolmogorov A, Petrovsky I, Piscounoff N. 1937. Etude de l'equation de la diffusion avec croisssance de la quantité de matière et son application à un probleme biologique. *Bull. Univ. d'Etat Moscou Ser. Int. Sect. A* 1:1–25

Kot M. 1992. Discrete-time traveling waves: ecological examples. *J. Math. Biol.* 30:413–36

Kot M, Lewis MA, van den Driessche P. 1996. Dispersal data and the spread of invading organisms. *Ecology* 77:2027–42

Kot M, Schaffer WM. 1986. Discrete-time growth-dispersal models. *Math. Biosci.* 80: 109–36

Lamont BB. 1985. Dispersal of the winged fruits of *Nuytsia floribunda* (Loranthaceae) *Aust. J. Ecol.* 10:187–93.

Leishman MR, Wright IJ, Moles AT, Westoby M. 2000. The evolutionary ecology of seed size. In *Seeds: The Ecology of Regeneration in Plant Communities*, ed. M Fenner, pp. 31–57. Wallingford, UK: CABI

Levey DJ, Silva WR, Galetti M, eds. 2002. *Seed Dispersal and Frugivory: Ecology, Evolution and Conservation*. Wallingford, UK: CABI. 511 pp.

Levin SA. 1974. Dispersion and population interactions. *Am. Nat.* 108:207–28

Levin SA, Cohen D, Hastings A. 1984. Dispersal strategies in patchy environments. *Theor. Popul. Biol.* 26:165–91

Levin SA, Muller-Landau HC. 2000. The evolution of dispersal and seed size in plant communities. *Evol. Ecol. Res.* 2:409–35

MacDonald GM. 1993. Fossil pollen analysis and the reconstruction of plant invasions. *Adv. Ecol. Res.* 24:67–110

Malécot G. 1948. *Les Mathématiques de l'Hérédité*. Paris: Masson. 63 pp.

Maynard Smith J. 1982. *Evolution and the Theory of Games*. Cambridge, UK: Cambridge Univ. Press

McCanny SJ. 1985. Alternatives in parent-offspring relationships in plants. *Oikos* 45: 148–49

Metz JAJ, de Jong TJ, Klinkhamer PGL. 1983. What are the advantages of dispersing: a paper by Kuno explained and extended. *Oecologia* 57:166–69

Metz JAJ, Gyllenberg M. 2001. How should we define fitness in structured metapopulation models? Including an application to the calculation of evolutionarily stable dispersal strategies. *Proc. R. Soc. London Ser. B* 268:499–508

Moles AT, Westoby M. 2002. Seed addition experiments are more likely to increase recruitment in larger-seeded species. *Oikos* 99:241–48

Mollison D. 1977. Spatial contact models for ecological and epidemic spread. *J. R. Stat. Soc. Ser. B. Met.* 39:283–326

Murray KG. 1988. Avian seed dispersal of three neotropical gap-dependent plants. *Ecol. Monogr.* 58:271–98

Nathan R. 2001. The challenges of studying dispersal. *Trends Ecol. Evol.* 16:481–82

Nathan R, Horn HS, Chave J, Levin SA. 2002a. Mechanistic models for tree seed dispersal by wind in dense forests and open landscapes. See Levey et al. 2002, pp. 69–82

Nathan R, Katul GG, Horn HS, Thomas SM, Oren R, et al. 2002b. Mechanisms of long-distance dispersal of seeds by wind. *Nature* 418:409–13

Nathan R, Muller-Landau HC. 2000. Spatial patterns of seed dispersal, their determinants and consequences for recruitment. *Trends Ecol. Evol.* 15:278–85

Nathan R, Perry G, Cronin JT, Strand AE, Cain ML. 2003. Methods for estimating long-distance dispersal. *Oikos*. In press

Nathan R, Safriel UN, Noy-Meir I. 2001. Field validation and sensitivity analysis of a mechanistic model for tree seed dispersal by wind. *Ecology* 82:374–88

Neubert MG, Caswell H. 2000. Demography. and dispersal: Calculation and sensitivity analysis of invasion speed for structured populations. *Ecology* 81:1613–28

Neubert MG, Kot M, Lewis MA. 1995. Dispersal and pattern formation in a discrete-time predator-prey model. *Theor. Popul. Biol.* 48:7–43

Neubert MG, Kot M, Lewis MA. 2000. Invasion speeds in fluctuating environments. *Proc. R. Soc. London Ser. B* 267:1603–10

Okubo A, Ackerman JD, Swaney DP. 2001a. Passive diffusion in ecosystems. See Okubo & Levin 2001, pp. 31–106

Okubo A, Hastings A, Powell TM. 2001b. Population dynamics in temporal and spatial domains. See Okubo & Levin 2001, pp. 298–373

Okubo A, Levin SA. 1989. A theoretical framework for data analysis of wind dispersal of seeds and pollen. *Ecology* 70:329–38

Okubo A, Levin SA. 2001. *Diffusion and Ecological Problems: Modern Perspectives*. New York: Springer-Verlag

Pacala SW, Rees M. 1998. Models suggesting field experiments to test two hypotheses explaining successional diversity. *Am. Nat.* 152:729–37

Pasquill F, Smith FB. 1983. *Atmospheric Diffusion*. New York: Wiley

Petit RJ, Grivet D. 2002. Optimal randomization strategies when testing the existence of a phylogeographic structure. *Genetics* 161:469–71

Portnoy S, Willson MF. 1993. Seed dispersal curves: behavior of the tail of the distribution. *Evol. Ecol.* 7:25–44

Pulliam HR. 1988. Sources, sinks, and population regulation. *Am. Nat.* 132:652–61

Reid C. 1899. *The Origin of the British Flora*. London: Dulua

Ribbens E, Silander JA Jr, Pacala SW. 1994. Seedling recruitment in forests: calibrating models to predict patterns of tree seedling dispersion. *Ecology* 75:1794–806

Ridley HN. 1930. *The Dispersal of Plants Throughout the World*. Ashford, UK: L. Reeve & Co.

Ronce O, Clobert J, Massot M. 1998. Natal dispersal and senescence. *Proc. Natl. Acad. Sci. USA* 95:600–5

Ronce O, Olivieri I, Clobert J, Danchin E. 2001. Perspectives on the study of dispersal evolution. See Clobert et al. 2001, pp. 341–57

Ronce O, Perret F, Olivieri I. 2000. Evolutionarily stable dispersal rates do not always increase with local extinction rates. *Am. Nat.* 155:485–96

Rousset F, Gandon S. 2002. Evolution of the distribution of dispersal distance under distance-dependent cost of dispersal. *J. Evol. Biol.* 15:515–23

Schupp EW. 1993. Quantity, quality and the effectiveness of seed dispersal by animals. *Vegetatio* 107/108:15–29

Schupp EW, Fuentes M. 1995. Spatial patterns of seed dispersal and the unification of plant population ecology. *Ecoscience* 2:267–75

Schupp EW, Milleron T, Russo SE. 2002. Dissemination limitation and the origin and maintenance of species-rich tropical forests. See Levey et al. 2002, pp. 19–33

Shaw MW. 1995. Simulation of population expansion and spatial pattern when individual dispersal distributions do not decline exponentially with distance. *Proc. R. Soc. London Ser. B* 259:243–48

Shmida A, Ellner S. 1984. Coexistence of plant species with similar niches. *Vegetatio* 58:29–55

Skellam JG. 1951. Random dispersal in theoretical populations. *Biometrika* 38:196–218

Tackenberg O. 2003. Modeling long distance dispersal of plant diaspores by wind. *Ecol. Monogr.* 73:173–89

Taylor PD. 1988. An inclusive fitness model for dispersal of offspring. *J. Theor. Biol.* 130: 363–78

Taylor RAJ. 1978. The relationship between density and distance of dispersing insects. *Ecol. Entomol.* 3:63–70

Tewksbury JJ, Lloyd JD. 2001. Positive interactions under nurse-plants: spatial scale, stress gradients and benefactor size. *Oecologia* 127:425–34

Thanos CA. 1994. Aristotle and Theophrastus on plant-animal interactions. In *Plant-Animal Interactions in the Mediterranean-type Ecosystems*, ed. M Arianoutsou, RH Groves, pp. 3–11. Dordrecht, Neth.: Kluwer Acad.

Thebaud C, Debussche M. 1991. Rapid invasion of *Fraxinus ornus* L. along the Herault river system in southern France: the importance of seed dispersal by water. *J. Biogeogr.* 18:7–12

Tilman D. 1994. Competition and biodiversity in spatially structured habitats. *Ecology* 75:2–16

Travis JMJ. 2001. The color of noise and the evolution of dispersal. *Ecol. Res.* 16:157–63

Travis JMJ, Dytham C. 1999. Habitat persistence, habitat availability and the evolution of dispersal. *Proc. R. Soc. London Ser. B* 266: 723–28

Travis JMJ, French DR. 2000. Dispersal functions and spatial models: expanding our dispersal toolbox. *Ecol. Lett.* 3:163–65

Tufto J, Engen S, Hindar K. 1997. Stochastic dispersal processes in plant populations. *Theor. Popul. Biol.* 52:16–26

Turchin P. 1998. *Quantitative Analysis of Movement: Measuring and Modeling Population Redistribution in Animals and Plants*. Sunderland, MA: Sinauer

Turchin P, Thoeny WT. 1993. Quantifying dispersal of southern pine beetles with mark recapture experiments and a diffusion-model. *Ecol. Appl.* 3:187–98

Turnbull LA, Rees M, Crawley MJ. 1999. Seed mass and the competition/colonization tradeoff: a sowing experiment. *J. Ecol.* 87:899–912

van den Bosch F. 1990. *The velocity of spatial population expansion*. PhD thesis. Univ. Leiden, Neth.

van den Bosch F, Zadoks JC, Metz JAJ. 1988. Focus expansion in plant-disease. II: Realistic parameter-sparce models. *Phytopathology* 78:59–64

van der Pijl L. 1982. *Principles of Dispersal in Higher Plants*. New York: Springer-Verlag

Van Valen L. 1971. Group selection and the evolution of dispersal. *Evolution* 25:591–98

Vellend M, Meyers JA, Gardescu S, Marks PL. 2003. Dispersal of *Trillium* seeds by deer: Implications for long-distance migration of forest herbs. *Ecology* 84:1067–72

Vincent TL, Brown JS. 1988. The evolution of ESS theory. *Annu. Rev. Ecol. Syst.* 19:423–43

Wang BC, Smith TB. 2002. Closing the seed dispersal loop. *Trends Ecol. Evol.* 17:379–85

Watkinson AR. 1978. The demography of a sand dune annual: *Vulpia fasciculata*. III. The dispersal of seeds. *J. Ecol.* 66:483–98

Weinberger HF. 1978. Asymptotic behavior of a model in population genetics. In *Nonlinear Partial Differential Equations and Applications*, ed. JM Chadam, pp. 47–96. Berlin: Springer-Verlag

Weinberger HF. 1982. Long-time behavior of a class of biological models. *SIAM J. Math. Anal.* 13:353–96

Wenny DG. 2001. Advantages of seed dispersal: a re-evaluation of directed dispersal. *Evol. Ecol. Res.* 3:51–74

Whittaker RH. 1972. Evolution and measurement of species diversity. *Taxon* 21:213–51

Wilkinson DM. 1997. The role of seed dispersal in the evolution of mycorrhizae. *Oikos* 78:394–96

Willson MF. 1993. Dispersal mode, seed shadows, and colonization patterns. *Vegetatio* 107/108:261–80

Wright S. 1943. Isolation by distance. *Genetics* 28:114–38

Wright S. 1969. *Evolution and the Genetics of Populations. Vol. 2: The Theory of Gene Frequencies*. Chicago: Univ. Chicago Press

Yamamura K. 2002. Dispersal distance of heterogeneous populations. *Popul. Ecol.* 44:93–101

Yu DW, Wilson HB. 2001. The competition-colonization trade-off is dead; long live the competition-colonization trade-off. *Am. Nat.* 158:49–63

Annu. Rev. Ecol. Evol. Syst. 2003. 34:605–32
doi: 10.1146/annurev.ecolsys.34.011802.132407

First published online as a Review in Advance on September 29, 2003

ANALYSIS OF RATES OF MORPHOLOGIC EVOLUTION

Peter D. Roopnarine
Department of Invertebrate Zoology & Geology, California Academy of Sciences, Golden Gate Park, San Francisco, California 94118-4599;
email: proopnarine@calacademy.org

Key Words evolutionary rates, microevolution, darwin, random walks

■ **Abstract** Rates of morphological evolution exist at several hierarchical levels. The most fundamental rate, termed instantaneous or intrinsic, is a measure of evolutionary change between consecutive generations. Comparisons of these rates between different characters or taxa are best done by measuring change proportionally or logarithmically. The variation of instantaneous rates over geological time, or between taxa, is a reflection of differing intrinsic factors (e.g., mutation rates) or environmental conditions, and may explain rates at higher levels of the hierarchy such as apparent variable rates of evolution and diversification. Such long-term rates, termed here microevolutionary rates, have been measured variously as factors of exponential change over time (the "darwin"), or have been scaled for comparison according to sample standard deviations (the "haldane"). The relationship between these long-term measures and instantaneous rates, however, is not constant, nor measurable, unless evolution is monotonic and directional or static. Those modes produce an inverse relationship between microevolutionary rates and the interval of time over which they are measured. Any introduction of apparent randomness into the morphological evolutionary series of a taxon, however, likewise produces an inverse relationship between rate and timescale. This latter result is largely a mathematical artifact predictable on the basis of the behavior of random walks. Failure to reject a null hypothesis of random walk for an evolutionary series therefore precludes the interpretation of microevolutionary rates as products of real evolutionary processes. Methods developed to use random walk statistics and microevolutionary rates as sources of information about evolutionary mode and instantaneous rates are problematic because they (*a*) depend on a nonlinear relationship between microevolutionary rates and elapsed time, and (*b*) are prone to spurious correlation.

Rates of evolution may be measured above the level of species, for example as rates of origination within monophyletic clades. Taxonomic rates of evolution assume genealogical information that is generally not available for the groups under consideration. Likewise, many classic examples of rates of evolution comprise taxa of unknown or dubious genealogical relationship and should not be considered valid unless viewed in phylogenetic frameworks.

INTRODUCTION

Evolution, the descent of organisms with modification from ancestors, is a time-dependent process. Whether time is measured relative to the organism, say in generations, or in absolute chronological units, evolutionary change unfolds over time. Given a change in condition or state of any quantity over a period of time, it is natural to inquire about the rate of change. The concept of evolutionary rate, therefore, continues to garner significant attention from a broad range of workers. Presentation of all the issues surrounding this concept would be a monumental task, and therefore, this review will be restricted to a discussion of the analysis of rates of morphological evolution. This approach omits at least two other very important issues of rate, namely an examination of the mechanisms underlying rate variation, and rates of molecular evolution. However, the proximal and immediate mechanisms underlying rate variation are fairly well understood today: rates of mutation, genetic variability, heritability, and intensity of selection; rate variability at the molecular level was treated recently in the review by Mindell & Thacker (1996). Instead, the primary focus of this review is on the analysis of empirically derived rates of evolution, for example those associated most commonly with long-term selection experiments and studies (Grant & Grant 2002), artificial selection, and fossil data (Gingerich 1993).

This review takes the point of view that evolutionary rates are best understood if classified hierarchically, and much of the difficulties in analysis have stemmed from our as yet incomplete understanding of the relationship between rate and timescale (the interval over which rate is measured). A hierarchical arrangement of evolutionary rates proceeds from the population level, where rates are best understood, all the way to what I term here the megaevolutionary level, and include: (*a*) Instantaneous rates—rates as measured at the fundamental scale of evolutionary rate, the organismal generation [the "intrinsic rate" of Gingerich (1993)] (Sheets & Mitchell 2001). (*b*) Lineage rates—intra- and interspecific rates measured on the timescale of years to millenia, as is common in long-term evolutionary and ecological studies and high resolution paleontological studies. I refer to these rates more generally as microevolutionary rates because even when measured between higher taxa, the assumption is made that the taxa together form a lineage. (*c*) Macroevolutionary rates—rates of clade evolution, including speciation (origination), and the dynamics of higher taxonomic groups. (*d*) Megaevolutionary rates—the overall rate of organismal evolution throughout Earth's history.

Most of the thought and empirical work on rates of evolution have dealt with instantaneous rates and, to a much greater extent, microevolutionary rates. Whether and how rates at these lowest levels may be translated into phenomena at the macro- and even megaevolutionary levels remains to be examined.

WHY STUDY EVOLUTIONARY RATES?

The primary motivation for the study of evolutionary rates is the concept that rate is a direct result of mode, whether evolutionary mode is driven by Darwinian selection (Maynard Smith 1976), or is only one component in a larger macroevoltionary theory (Eldredge & Gould 1972). An interesting contrast may be drawn between (*a*) the frequency with which "evolutionary rate" is discussed qualitatively or quantitatively in the modern literature, and (*b*) our understanding of rates of evolution in the hierarchy outlined above. The attention paid to the quantification of rates at the population-level, and within lineages over long intervals of geological time, reflects a general feeling that rates at these levels are important to understanding rates of evolution both across taxa, and throughout the hierarchy (Gingerich 1983).

The intent of this review is to examine and discuss various quantitative approaches to the analysis of rates at the two lowest levels of the hierarchy, and to highlight two very important issues: First, the measurement and analysis of evolutionary rate is absolutely dependent upon the timescale and evolutionary unit under consideration (Gingerich 1983, Gould 1984). I will argue that at the level of the species, and at timescales above the fundamental (that of the generation), measured rates of evolution are often mathematical artifacts and simply do not exist (Bookstein 1987). Second, there is no reason to doubt the meaningful existence and variation of rates above the level of the species, where clades may exhibit variable rates of origination, dependent upon individual ecologies, emergent properties, and historical contingency (Stanley 1985). It is also highly likely that the overall global rate of organismal evolution has varied as organisms have diversified in cellular complexity and size (Knoll & Bambach 2000) and new means of modifying genetic variance have evolved. We do not yet, however, fully understand the relationship among evolutionary rate variation at these various levels, nor how and why rates at one level may or may not scale to phenomena at higher levels.

Although a single evolutionary rate by itself is relatively uninteresting, comparisons among rates may be informative; the fact that rates vary within and among taxa suggests that rates are sensitive to a range of intrinsic and extrinsic factors (Simpson 1944). For example, it has been claimed that Cenozoic mammals exhibit higher rates of evolution than do marine invertebrates (Van Valen 1974). Haldane (1949) devised the first metric, the darwin, intended to permit comparisons of rates among such disparate taxa by using proportional rather than absolute measures of character change. Kurtén's (1959) observation, however, that rates are often inversely related to the timescale of their measurement, leads to both the search for additional standardization according to the temporal scale of measurement (Gingerich 1983, 1993), and to conflicting explanations of this curious relationship (Bookstein 1987, Gingerich 1983, Gould 1984, Sheets & Mitchell 2001). The primary conclusion of the present review is that whereas Haldane's darwin, and related measures such as the haldane

(Gingerich 1993), may account for phylogenetic and sample-statistic disparities, the inverse relationship between rate and timescale is a real mathematical property of long-term evolutionary series. This property essentially prohibits the comparison of rates across disparate timescales, and the mode of the evolutionary series is known a priori to be monotonically nonrandom (that is, does not conform to a random walk).

The Significance of Microevolutionary Rates

The measurement of a single "instantaneous rate" is no more than a measure of the amount of evolution that occurred between two consecutive generations or populations of a lineage. Comparisons of such rates within a single lineage of consecutive generations, however, may be dependent upon the amount of genetic variance in the population and the nature and/or intensity of selection (Maynard Smith 1989). Comparisons among populations are likewise informative, and observation of such rates over time could indicate changes in the amount of genetic variability available in the population, as well as changing selection regimes (Haldane 1949, Simpson 1944). In measuring rates of microevolution within a presumed lineage of Cenozoic horses, Simpson (1944) assumed that an examination of rate variation and variability would be informative of the evolution of the lineage. Two very important corollaries to this line of reasoning are: (*a*) the measured microevolutionary rates can be equated somehow to the unobserved/unpreserved instantaneous rates, the variabilities of which are due largely to mutation and selection (Haldane 1949), and (*b*) evolution of species (or genera in the cases of Simpson & Haldane) within the clade or lineage result from the microevolutionary processes of mutation and selection. This is another statement of the classical view that macroevolutionary phenomena are the result of the cumulative effects of microevolutionary processes (Dobzhansky 1970). The consequences of this foundation to the examination of evolutionary rates cannot be overstated. Haldane's (1949) suggestion of proportional rates and Kurtén's (1959) observation have been basic to the analysis of rates (Gingerich 1983, 1993; Hallam 1975; Hendry & Kinnison 1999; Sheets & Mitchell 2001), and have led to a questioning of the validity of microevolutionary rate measurement (Bookstein 1987, 1988). Three central and consequential arguments will be explored in this review, namely that the inverse relationship noted by Kurtén (1959) (*a*) supports the interpretation of microevolutionary rates as indicators of the ubiquity of microevolutionary processes across broad genealogical and temporal scales (Gingerich 1983, 1993; Hendry & Kinnison 1999), (*b*) is indicative of a mathematical artifact that, in turn, reflects the lack of a smooth transition between microevolutionary processes and macroevolutionary patterns (Gould 1984, Sheets & Mitchell 2001), or (*c*) is often indicative of a mathematical artifact that renders the measured microevolutionary rates devoid of meaningful information (Bookstein 1987).

INSTANTANEOUS AND MICROEVOLUTIONARY RATES

The fundamental unit of evolutionary rate is the magnitude of change in gene frequencies between the consecutive generations of an interbreeding group. This unit will be referred to at the scale of a single generation as the "instantaneous rate" (denoted κ from here on) (Sheets & Mitchell 2001). [Gingerich (1993) equated his quantity "intrinsic rate" with this metric, but intrinsic factors need not dominate population-level evolution, nor are instantaneous rates expected to be constant.] Instantaneous rate is generally concerned with quantitative characters because at the level of the population most change may be expected to be continuous and quantifiable. The magnitudes of change of continuously varying characters, and hence their instantaneous rates, reflect the often complex interplay of genotype and environment that underlies their evolution.

An instantaneous rate is the net effect of the sources of evolutionary variation, namely mutation, immigration, and selection, or in the absence of selection, genetic drift. These factors may act in concert to increase the magnitude of an instantaneous rate, or they may have counteracting and dampening effects. Similarly, the phenotypic expressions of different mutations may counteract each other, agents of selection may counteract one another, and genetic and phenotypic linkages may constrain the magnitude of phenotypic change, and hence, an instantaneous rate (Bull 1987, Charlesworth 1984a, Maynard Smith 1976). Nevertheless, it is the expected variability of instantaneous rates that has lead to the expectation of evolutionary rate variation at scales of multiple generations.

The impetus for studying rates of evolution over longer timescales stems largely from the desire to document the action and effects of natural selection over time and across broad phylogenetic data sets (Simpson 1944), and the question of exactly how effective natural selection is at these scales (Charlesworth 1984b, Lande 1976). There are also questions of how the mode of short-term evolution, or intensity of selection, might affect the rate of evolution (Hendry & Kinnison 1999). More important perhaps is the question of the distribution of rates during the existence of a species (Bookstein 1987, Hendry & Kinnison 1999, MacLeod 1991). For example, if rate is tied to mode, then punctuated equilibrium theory predicts that rates will fluctuate between low magnitude during the predominant times of evolutionary stasis, and greater magnitude during the briefer episodes of cladogenesis (Gould 1984). At broader phylogenetic scales, one may wonder if variable rates of evolution play any role in clade diversification, the survivability of clades during times of ecological crisis (for example, mass extinctions), and the comparative abilities of clades to recover from mass extinctions. It is commonplace today to compare rates among phylogenetically divergent and disparate taxa (Gingerich 1983, Hendry & Kinnison 1999, Sheets & Mitchell 2001, Van Valen 1974).

Implicit to all these studies of course is the assumption that rates measured at one scale are somehow equivalent to rates measured at all the other scales, both temporal and phylogenetic. The question of comparing phylogenetic apples to oranges was first addressed meaningfully by Haldane (1949) (at least in the context of rate) who devised a comparative measure of proportional changes, the "darwin." Gingerich (1983, 1993) revisited both the darwin and the effects of timescale, revising the former and utilizing the latter in an attempt to estimate instantaneous rates from time series of evolutionary change. Bookstein (1987) later argued that rate is in fact dependent upon the mode of evolution and in many cases is an artifact and cannot be quantified. The general conclusions to be drawn from a review of this broad range of literature are: (*a*) measured rates are not equivalent among different types of evolutionary series; (*b*) measured rates are sometimes dependent upon the mode of evolution, but this may arise from a misconception of the relationships among mode, rate, and timescale, or it may be owing to the preservational nature of fossil series; and (*c*) measured rates are dependent upon the timescale of measurement, and whether this is an artifact or a real evolutionary phenomenon depends entirely upon the mode of evolution and the method of measurement. Therefore, under no circumstances should it be assumed that rates measured from fossil time series are necessarily equivalent to instantaneous rates, nor that rates measured among different times series are in any way comparable.

MEASUREMENTS OF CHANGE

Haldane (1949) was the first to formalize a quantitative approach to evolutionary rate. He noted that if there is a series of fossil populations believed to form a lineage (an assumption to which I return below), then a rate of change of the mean may be measured. Haldane also asserted that if comparisons of resulting rates are to be made among different characters ("organs" more specifically for Haldane) or taxa, then the changes should be based upon percentage change rather than absolute differences (Haldane 1949, Lerman 1965). This idea may be traced back to at least Galton (1879) who noted that proportional changes in measurements on organisms may be more important than the absolute measures themselves (Wright 1968), hence the common recommendation of a logarithmic transform. A further recommendation is that mean differences be standardized according to sample standard deviations (Gingerich 1993, Haldane 1949, Lerman 1965). The final and perhaps most well-known of Haldane's recommendations is the "darwin," a unit intended to be a fundamental and standard measure of evolutionary rate. The darwin represents a change of a character by a factor of the mathematical constant e per million years. Haldane clearly had the intent of applying the darwin to presumed fossil lineages spanning millions of years, although the unit has been applied to much shorter records (e.g., Gingerich 1993, Hendry & Kinnison 1999, Reznick et al. 1997). As explained below (Microevolutionary Rates), comparisons both within and between lineages on such disparate timescales are somewhat difficult, often leading to the question of why rates measured from the fossil record are so much lower than those measured in the Recent (Gingerich 1983, Gould 1984, Reznick et al. 1997). The only major modification of measurement to be adopted since Haldane's darwin has been Gingerich's "haldane," essentially a difference of means standardized by the pooled sample standard deviations (Gingerich 1993), although various suggestions have been made to account for multiple characters and multivariate situations (Kurtén 1958, Lerman 1965).

Analysis of Instantaneous Rates

The measurement of population-level evolution as an instantaneous rate (κ) only takes on significance when we observe its variation over time, or compare it among lineages. Several questions may be asked of κ both within and between lineages of a single species: (*a*) Does the direction of evolution/selection [that is, positive and negative, "upwards" and "downwards" (Falconer 1989)] matter? (*b*) Is κ or its distribution ($F(\kappa)$) related to the mode of evolution? (*c*) In the case of a change of selection intensity or the direction of selection, is κ dependent on the number of generations elasped since the change? When comparing separate evolutionary series within a single lineage, the following additional questions should be addressed: (*a*) What is the variance of κ, and if the mode of evolution or selection does differ between series, then (*b*) do rates differ between modes of evolution? (*c*) Finally, do signed (that is, nonabsolute) rates differ between modes of evolution, and does κ therefore depend on the mode of evolution?

Because κ is measured between consecutive generations, it is easily derived from numerous neontological studies of evolutionary change (e.g., Endler 1986, Hendry & Kinnison 1999). To address broad questions of rate however, such as Simpson and Haldane's questions of the relationship between natural selection and rate, long series presenting distributions of rates are desirable. Falconer has presented several such suitable series, one of the most well known being the two-way selection for body weight in mice over 11 generations (Falconer 1953) (Figure 1). Measuring instantaneous rate as

$$\kappa = |\ln(x_j) - \ln(x_i)|, \tag{1}$$

where x_i and x_j represent population mean values, we observe that positive and negative rates do not differ significantly in either the upwards or the downwards series (t-tests, $P = 0.362$ and 0.845, respectively), and there is no dependence of κ on time elapsed since initiation of the experiment (Figure 2). A comparison between the two series likewise shows that there is no significant difference of κ between the two directions of selection (t-test, $P = 0.554$). There are not enough observations to test whether signed rates differ between series, but this seems unlikely. There is essentially no relationship between the direction of selection and evolutionary rate.

Longer series are available, for example Dudley's (1977) experiment on two-way selection for oil content in maize seeds (Figure 3). The upward series is significantly trended (runs test, $P = 0.01$), indicating successful selection for increased oil content, but there is no significant difference between positive and

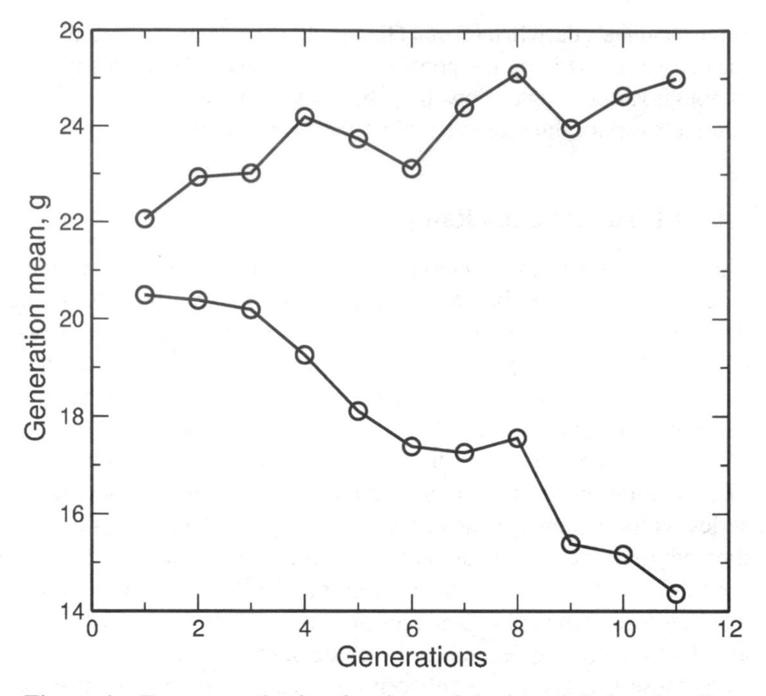

Figure 1 Two-way selection for six-week body weight in mice. Two lines represent selection for increased and decreased weight (Falconer 1953).

negative values of κ (t-test, $P = 0.362$). There is also no relationship between κ and time (generations) elapsed. The downward series, however, is not trended significantly (runs test, $P = 0.12$), that is, a null hypothesis of random walk cannot be rejected (Raup 1977). Once again there is no significant difference between positive and negative values of κ (t-test, $P = 0.846$). There is also no relationship between κ and generations elapsed, although there is an acceleration of change (increase in rate) toward the end of the series.

When comparing the two directions of selection in Dudley's experiment, however, we note that rates are significantly greater for the downward series ($\bar{\kappa} = 0.172 \pm 0.249$) than the upward series ($\bar{\kappa} = 0.052 \pm 0.054$) (Kolmogorov-Smirnov, $P < 0.0001$). Moreover, signed rate magnitudes also differ between the series, both for negative and positive rates (Kolmogorov-Smirnov, $P = 0.029$ and $P = 0.02$, respectively). There is no relationship between κ and the instantaneous direction of selection, so the results are best explained as a dependence of κ on the overall nature of selection for each series. Although the downward series was under strict artificial selection, the response of the population was nondirectional, most likely reflecting constraints set by a minimum oil content, and possibly also a rapid exhaustion of genetic variance. Neither the direction of the series nor the calculated instantaneous rates reflect the intensity of selection. The series of increasing oil

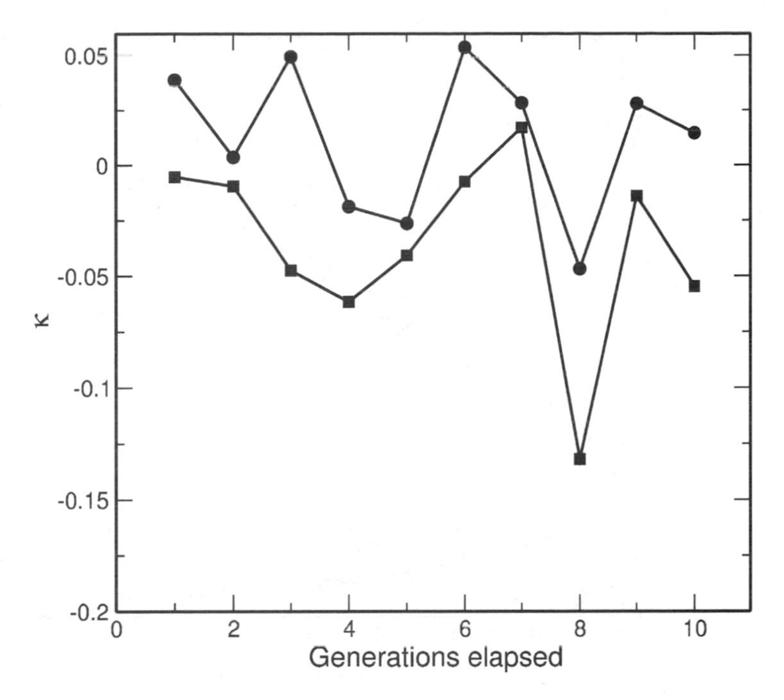

Figure 2 Instantaneous rates (κ) derived from selection lines in Figure 1. There is no dependence of κ on time (generations) elapsed.

content, however, responded significantly to the directional selection. The rates are significantly lower than those calculated for the nondirectional "downward" series, hence demonstrating a greater role for the direction of response, rather than the magnitude of response, at least in this experiment.

MICROEVOLUTIONARY RATES

Microevolutionary rates refer to measures of rate taken over extended periods of time within a single lineage. The questions concerning instantaneous rates apply here, but there are important analytical differences to be considered. First, series from which microevolutionary rates are calculated are generally incomplete, that is, measurements are based on populations or samples separated by more than a single generation. Most such series derived from neontological data tend to be short, based at most on a few years because the experiments are conducted in natural, not artificial settings, with all the logistic limitations of long-term ecological studies (Grant & Grant 2002, Hendry & Kinnison 1999). Longer series are generally derived from the fossil record, and in those cases gaps may represent time intervals of millenia to millions of years. Microevolutionary series of fossils also have the additional problem of comprising time-averaged samples of multiple populations.

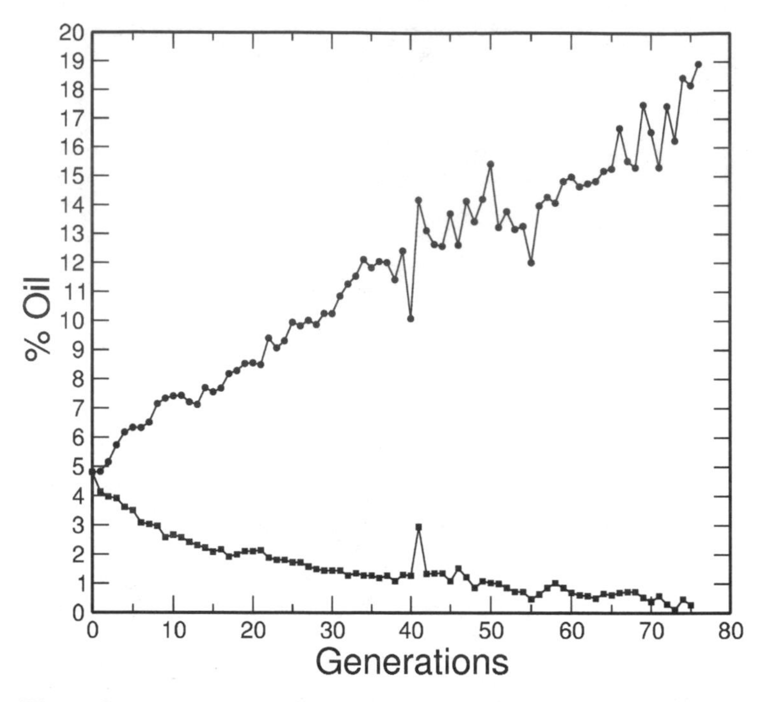

Figure 3 Two-way selection for increased and decreased oil content in maize seeds (Dudley 1977).

These factors complicate any straightforward interpretations and comparisons of microevolutionary rates to κ.

Units of Measure

The basic measure of microevolutionary rate is the amount of change accumulated per unit time, change sometimes being measured proportionally as the difference of logarithms:

$$r = \frac{x_j}{x_i} \frac{1}{\Delta t_{ij}}. \tag{2}$$

Extensions of r include the darwin as described earlier (Haldane 1949), where r is expressed relative to the mathematical constant e, mostly to facilitate comparisons among rates separated by vast intervals of geological time, or those derived from different taxa (Haldane 1949, Lerman 1965, Maynard Smith 1989). Haldane derived the darwin by considering that the mean value of the proportional rate of change for any given interval, $\frac{1}{x}\frac{dx}{dt}$, is

$$r_d = \frac{\ln(x_j) - \ln(x_i)}{\Delta t_{d,ij}}, \tag{3}$$

where r_d indicates rates measured in darwins. Time is measured in units of one million years ($\Delta t_{d,ij}$), making one darwin roughly equivalent to a proportional change of 1/1000 per 1000 years. Gingerich (1993) criticized the darwin for (*a*) being based on an arbitrary constant with no real biological meaning, (*b*) measuring rates in years (or millions of years) instead of the more fundamental generations, (*c*) being dimension-dependent, and (*d*) not taking the timescale of measurement into account. Gingerich instead pursued a suggestion by Haldane that a measure of rate be based on a metric of standard deviations, a unit that Gingerich termed appropriately the "haldane."

$$r_h = \frac{\ln(x_j) - \ln(x_i)}{s_{ij}\Delta t_{ij}}, \tag{4}$$

where s_{ij} is the pooled standard deviation of samples *i* and *j*. It is worth discussing Gingerich's criticisms of the darwin, as they generalize problems inherent in all quantitative measures of microevolutionary rate.

Problems with the Darwin

Basing the darwin on *e* is a natural consequence of considering proportional rather than absolute changes. The decision to use a factor of *e* per million years is a reflection of Haldane's original intention to apply this measure to the long-term evolution of putative fossil lineages. Haldane did recognize a disparity of magnitude between rates measured in the fossil record and instantaneous rates. For example, the greatest rate measured from Dudley's (1977) experiment, if generation time is converted to years (approximately three generations per year), is approximately 4.35 Md (megadarwins), whereas Haldane measured a maximum rate of 0.04 md (millidarwins) for Tertiary horses (see also Simpson 1944). Gingerich (1983) compiled rates ranging from 58.7 kd (kilodarwins) for selection experiments ($\Delta t_{d,ij}$ ranging from <1 to 4 years), down to 0.07 d for fossil invertebrates ($\Delta t_{d,ij}$ averaging almost 8 million years). Reznick et al. (1997) measured rates up to 45 kd in wild populations of Trinidadian guppies. The disparity among these rates can be traced to several causes. First, Gingerich is correct in pointing out the mathematical consequences of using years versus generations; but generation time is never known with certainty for fossil species, and it is also not always feasible to use "generations elapsed" when working outside of a laboratory setting (Hendry & Kinnison 1999). Moreover, if the results from Dudley's (1977) experiments were converted to a single generation per year, the differences would not alter the perception of disparity associated with differing temporal scales of measurement.

Second, Gingerich's point concerning the timescale of measurement is relevant whether time is measured in generations or years; Dudley's rates are instantaneous rates, κ, the rates of Reznick et al. (1997) are close to being κ rates (measured over 6.9 to 18.1 generations), while the rates measured for the fossil horses span geological intervals of 5 to 16 million years. Measures of evolutionary rate therefore appear to be dependent upon the timescale of measurement (Bookstein 1987, Gingerich 1983, Haldane 1949, Kurtén 1959, Roopnarine et al. 1999), a point to which we will return below.

Third, there is no a priori reason to expect equivalence of rates among different taxa. If the justification for comparing rates among taxa, or characters of taxa, is that the rates are somehow equivalent, that is, they are estimates of $F(\kappa)$, then account must be taken of the variation of the factors underlying κ. Even if measured on the timescale of a few generations, should we necessarily expect the linear dimensions of horses' teeth to change at rates approaching those for changing oil content in maize seeds? What are the comparable values of mutation rates, levels of pleiotropy, heritability, population size, intensity of selection, and so on? Heterogeneities of taxa and characters are not easily standardized by simple mathematical transformations, but uniformities of rates would have implications for unified interpretations of evolutionary processes across broad temporal and phylogenetic scales (Gingerich 1984).

The issue of dimensionality also presents a problem for the darwin. As Gingerich (1993) noted, rates measured in darwins are dimension-dependent, in that multidimensional features such as areas and volumes may be expected to evolve at rates that are powers (of two and three, respectively) of underlying linear dimensions and characters. Therefore one cannot make direct comparisons among rates that are measured on characters of different dimensionality. Once again, however, the question arises about the utility of making such comparisons among taxa when homologies are lacking. Comparing evolution between characters within a single taxon is perhaps approached more fruitfully by measuring the rate of evolution of their relationship, that is, the evolutionary rate of their allometric coefficient or relationship (however measured) (Haldane 1949). Kurtén (1958) attempted this with his measurement of evolutionary rates of taxonomic differentiation between species of hyaenids, but his use of a nonstandard "differentiation index," and expression in percentage changes per thousand years, does not permit easy comparison to rates measured in darwins.

TIMESCALE

Returning to the issue of timescale, Kurtén (1959) was the first to examine the relationship between evolutionary rate and the time interval over which the rate was measured. He noted a negative relationship between rates measured in darwins on Tertiary and Quaternary mammals and the timescale of measurement. Average rates declined from 12.6 d in Holocene mammals to 0.02 d in the Tertiary (increasing time intervals of measurement). Kurtén consequently suggested two explanations for the negative relationship, one biological and the other artifactual. First, the Pleistocene and Holocene were periods of relatively rapid and frequent climate changes, and the higher rates of evolution may be reflective of responses to subsequent intense selection. Slower rates derived from the older pre-Pleistocene mammals would instead be the result of more gradual and slower changes in the environment. Williams (1992), however, pointed out that the observed levels of stasis in many terrestrial species during and since the Pleistocene is difficult to reconcile with the reality of a rapidly changing environment, an observation that may be extended to Pleistocene marine invertebrates (Roopnarine 1995, Stanley & Yang 1987).

Conversely, Kurtén also hypothesized that the relationship between rate and timescale could be the result not of differing responses to monotonic selection, but rather responses to the fact that as the interval of time over which a rate is measured increases, so do the probabilities and frequencies of reversals and fluctuating directions of evolution. The accumulation of these reversals, although not measured or perhaps not even preserved/observed in the fossil lineage, nevertheless negates some of the net evolutionary change and results in lower measured evolutionary rates. This would of course require that the reversals be roughly as frequent as the direction of net change, calling into question the interpretation of "net" evolution from such a series, or that changes in the net direction be of greater magnitude than those associated with reversals. Neither of these issues has been addressed specifically to mammals, and the examples of instantaneous rates analyzed above in this review suggests a variety of responses of κ to selection. Nevertheless, the role of reversals has become generally accepted as one of the main features underlying the apparent relationship between rate and the relevant interval of time (Gingerich 1983, Gould 1984). Other mechanisms that undoubtedly play a role, but are much more difficult to quantify, include: (*a*) general constraints to change in a single character (Maynard Smith 1989), and (*b*) fossil taxa that may have experienced sustained high rates of evolution, say in the kilo- or megadarwin range, are probably not assumed to belong to the same genealogical lineage, and hence are not often the subjects of analyses of rates (Gingerich 1983). The mathematical roles of timescale and reversals are paramount to an explanation of Kurtén's observation, as will be shown below, whereas the issue of recognizing lineage continuity and monophyly is one that has rarely been addressed in studies of evolutionary rate (but see Hendry & Kinnison 1999). It is also explored below in this review.

Dependence on Time

Several alternative explanations have emerged since Kurtén's observation of the negative relationship between evolutionary rates and the interval over which they are measured. These explanations can generally be classified into two categories, those that suggest that the relationship is indicative of underlying evolutionary phenomena (Gingerich 1983, 1993; Gould 1984; Hendry & Kinnison 1999; Kurtén 1959; Sheets & Mitchell 2001), and others that claim varying degrees of mathematical artifact (Bookstein 1988, Gould 1984, Sheets & Mitchell 2001).

Gingerich (1983), in an analysis of over 500 published series measured over time intervals of 1.5 years to 350 million years, supported the latter of Kurtén's hypotheses. Gingerich concluded that the dependence of evolutionary rates on the interval over which they are measured results from: (*a*) a stable range of proportional differences and responses between genealogical samples, regardless of taxon or timescale, thereby leading to decreasing rates with increasing timescale, and (*b*) increasing frequencies of periods of stasis and reversals with increasing intervals of measurement leading to a dampening of evolutionary change, and hence, rate. The range of rates measured is also constrained by the fact that small rates may not be measurable or the resulting small character differences between taxa recognized, and very high rates will, over time, result in levels of divergence that obscure the genealogical relationship of the taxa. Gingerich (1983, 1984) therefore suggested that microevolutionary rates be standardized according to timescale prior to comparisons across phylogenetic and temporal scales.

Gould (1984) countered that rates measured in the Recent tend to be biased because the studies focus on examples of directional selection in the wild, or artificial selection, while ignoring the potentially numerous examples of populations that undergo very little generational change in the wild (that is, are essentially in stasis). This assertion cannot, of course, be verified until appropriate studies are carried out on numerous Recent populations, but the most recent compilation of neontological rates (Hendry & Kinnison 1999) exhibits the relationship between rate and interval of measurement (Figure 4). Furthermore, many series of fossil samples do exhibit incremental evolutionary changes of greater magnitude as stratigraphic resolutions increase without associated net evolutionary changes, for example, the "time of transition" in the *Globorotalia plesiotumida-tumida* transition published by Malmgren et al. (1983) and re-analyzed by Bookstein (1988) and Roopnarine (2001). The conclusion to be drawn from the Gingerich-Gould debate, therefore, is that while both agree that the dependence of rate on timescale is somewhat of an artifact, they ascribe the artifact to different evolutionary phenomena and processes. In fact, the conformity of Gingerich's large and phylogenetically heterogeneous data set to a single trend is more likely the combined and expected result of plotting a ratio of change to interval, versus interval (Gould 1984), along with the largely unconstrained errors inherent in any such heterogeneous study. It is not likely to be indicative of constant evolutionary process across vastly different timescales (Sheets & Mitchell 2001), nor can it address the question of any differences of process at the microevolutionary and macroevolutionary scales.

Bookstein (1987) later offered an alternative explanation of Kurtén's observations, but one based wholly on the notion of a mathematical artifact. The core of Bookstein's argument claims that no organic processes may be invoked as explanations of the evolutionary dynamics of fossil series unless a null hypothesis of a symmetric (unbiased) random walk can be discarded. Raup & Crick (1981) (see also Raup 1977) introduced the concept of using symmetric random walks as null models for testing the mode of evolution in such fossil series, and this idea has since been developed extensively (Bookstein 1987, 1988; Gingerich 1993; Roopnarine 2001; Roopnarine et al. 1999). In addition to its role in defining

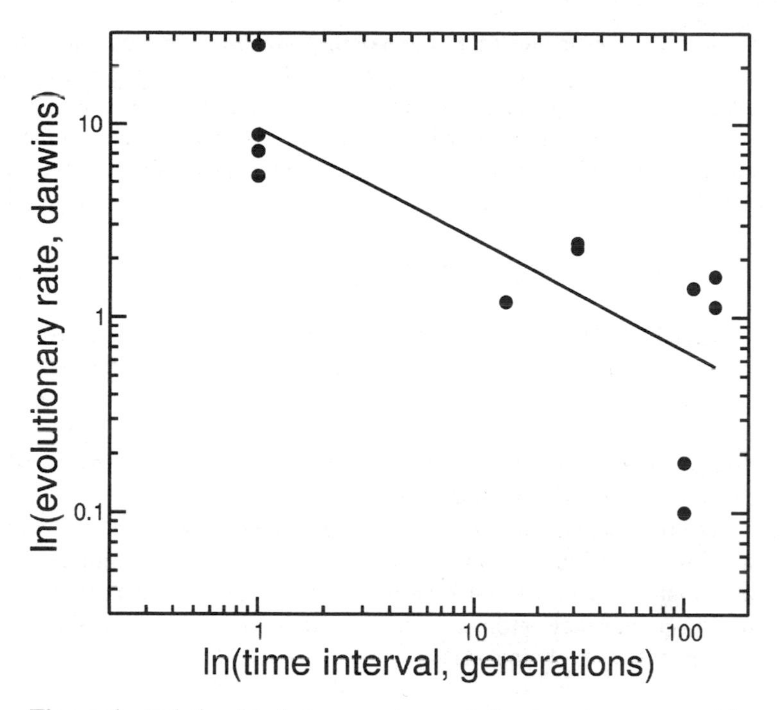

Figure 4 Relationship between microevolutionary rates, measured in darwins, and the interval of time over which they were measured (in generations). Data compiled from Hendry & Kinnison (1999).

evolutionary mode, the random walk hypothesis also plays a critical role in explaining the large-scale negative relationship between microevolutionary rates and the timescale over which they are measured. Bookstein (1987) pointed to the fact that unless the null hypothesis of symmetric random walk could be rejected for a particular fossil series, then no "rate" exists for that series. To understand this, consider the basic model of a symmetric random walk, where on average the frequency of reversals is equal to the frequency of changes in the direction of net evolution (a straightforward and simple illustration is provided by Maynard Smith 1989). The average expected displacement of a walk from its starting point, after a period of elapsed time τ, is a function of the square root of the time elapsed:

$$\langle x_\tau \rangle = x_0 \pm \kappa\sqrt{\tau}. \tag{5}$$

(Note that κ, the instantaneous rate of evolution, or its distribution $F(\kappa)$, serves here as step size.) The only true rate that may be derived from a series such as this is the instantaneous rate (or step size), which for biological series has a fixed lower limit, that being generation time (see also Gingerich 1993). Point estimates of rate made between steps that are separated by arbitrary intervals of time are more a function of the number of reversals not incorporated into the net measure than they are estimates of κ. Bookstein pointed out that the quantities claimed to be evolutionary rates derived from fossil series are, in fact, merely ratios of evolutionary change to elapsed time. The true rate of evolution is instead the limit of a series of such ratios as their denominators (elapsed time) become smaller. As an illustration, if we calculate a rate between the mean of the above estimate, $\langle x_\tau \rangle$ and the starting value of the series, then

$$r = \frac{|x_\tau - x_0|}{\tau} = \frac{\kappa}{\sqrt{\tau}}. \tag{6}$$

Gingerich (1993) termed these noninstantaneous rates "interval" or "net rates" to indicate that they are measured over intervals of time greater than the fundamental scale of the generation. Gingerich (1993) further suggested that the distribution of interval rates over time could serve as the basis for both distinguishing sustained directional evolution from (seemingly) random change, as well as estimation of the intrinsic rate, κ, for a lineage.

The two major issues, therefore, currently associated with the comparison and interpretation of microevolutionary rates are: (*a*) the proper scaling of interval rates to account for their negative relationship with timescale, and (*b*) the estimation of the truly comparable intrinsic, instantaneous, or in the case of monotonic evolution, average rate of evolution, κ.

ASSUMPTIONS AND ANALYSES OF EVOLUTIONARY RATES

Any fossil series that conforms statistically to the expectation of a random walk will yield interval rates that are a decreasing function of the square root of time. Conformation of an evolutionary series to random walk statistics is not meant to imply that the series is the outcome of random processes, but the often conflicting action of multiple evolutionary factors may frequently produce such series (Roopnarine et al. 1999). This idea underlies the notion of "no net change" or oscillating evolution (Eldredge 1989), and the appeal of a minimal number of assumptions of evolutionary processes forms the basis for using random walks as general models of character evolution (Bookstein 1988, Edwards & Cavalli Sforza 1964, Felsenstein 1988, Lande 1986). Deviation of walks from an unbiased symmetry leads to series that are either increasingly constrained or directional (biased) (Roopnarine et al. 1999, 2001), and such changes of evolutionary mode result in alterations of the relationship between interval rates, timescale, and κ.

Bookstein (1987, 1988) expected that it would be exceedingly difficult to reject the null hypothesis of symmetric random walk as an evolutionary mode for the majority of fossil series, with the consequence for microevolutionary rates that they are, in fact, uninformative and that their negative relationship with temporal scale of measurement is a mathematical artifact. The null hypothesis is indeed very difficult to reject, but the test constructed by Bookstein (1988) is generally too conservative for fossil data (Gingerich 1993, McKinney 1990, Sheldon 1993) and fails to account for the statistical consequences of incomplete preservation and time-averaging (Roopnarine et al. 1999) (see below). Nevertheless, tests of evolutionary mode with reasonable error rates may be constructed using the random walk approach, exclusive of evolutionary rate hypotheses (Roopnarine 2001).

Subsequent to Bookstein's arguments (1987, 1988), Gingerich (1993) constructed a test for mode based directly on his interpretation of the timescale effect (Gingerich 1983) and the properties of the random walk. The original paper on the effects of temporal scaling (Gingerich 1983) had demonstrated the negative linear relationship between the logarithms of evolutionary rate (measured in darwins) and time interval, which Gingerich (1993) assumed could be derived by referring to the previous equation:

$$\ln(r) = \ln(\kappa) - 0.5\ln(\tau). \quad (7)$$

The exponent of τ, however, can range mathematically between zero and one, suggesting evolutionary modes that range from strict stasis to sustained directionality,

$$\langle x_\tau \rangle = x_0 \pm \kappa\tau^0 = x_0 \pm \kappa, \quad \langle x_\tau \rangle = x_0 \pm \kappa\tau^1 \quad (8)$$

(Roopnarine et al. 1999). Consequently, Gingerich (1993) reasoned that these relationships could, in fact, be used to derive the mode of evolution of fossil series from the relationship between evolutionary rate (measured in haldanes) and the temporal interval of measurement. The prescribed procedure uses a robust maximum likelihood regression of ln(evolutionary rate) on ln(temporal interval), where the resulting slope is assumed to describe evolutionary mode ($-1 \rightarrow$ stasis, $-0.5 \rightarrow$ random, $0 \rightarrow$ directional), and the intercept is an estimate of instantaneous rate (κ).

The major recommendation that Gingerich drew from his results of analysis of series at multiple timescales (Gingerich 1983) is that the evolutionary rate reported depends on the time interval covered by the measurement, and that rates should therefore be adjusted appropriately for timescale. It is obvious, however, that only under circumstances where the mode of evolution is monotonically directional, are interval rates independent of the mode of evolution (that is, if the exponent of τ is denoted as α, then $\alpha = 1$) (Sheets & Mitchell 2001). In all other cases rate is a variably decreasing function of timescale ($\alpha < 1$). At the very least then, evolutionary mode should be determined prior to the calculation and analysis of evolutionary rates. There is currently only a single method for determining mode without explicit reference to logarithmic rates (Roopnarine 2001).

Subsequent to the determination of mode, Gingerich's results should suggest that the only sensible procedure with regard to rate would be the estimation of the instantaneous rate, κ. Nondirectional modes (stasis, random), however, yield interval rates that are mathematical artifacts because those net rates cannot account for reversals that must be common in nondirectional series. The interval rates of monotonically directional series, though, should, in fact, be estimates of κ. Therefore, if a series is determined to represent monotonic directional evolution, then interval rates may, in fact, form the basis for estimation of κ. Otherwise we recognize that any statistical deviation from monotonic directionality ($\alpha \neq 1$, $1 > \alpha \geq 0$) leads to mathematical uncertainty in the estimation of κ.

Effects of Fossilization

Series that deviate from the model of a symmetric random walk have a nonzero information content, and successive values may be correlated, or there may be biases in the directionality of the series. The two end-members of microevolutionary patterns, stasis and directionality, can both be modeled with biased random walks (Bookstein 1988, Roopnarine et al. 1999). Stasis is represented by series either with improbably high probabilities of reversal (Bookstein 1987, 1988), and/or reflecting boundaries (Roopnarine et al. 1999), with both mechanisms acting as forces of constraint (whether due to intrinsic factors such as developmental and functional constraints, or an external agent such as stabilizing selection) (Roopnarine 2001). Directional evolution may be modeled variously with biased random walks (Plotnick & Prestegaard 1995; Roopnarine et al. 1999, 2001) or linear functions with added noise (Roopnarine et al. 1999).

Fossil sequences are not, however, random walk simulations. Such sequences, even if they do form a genealogical succession, have been subjected to the fossilization process and almost invariably have been subjected to incomplete preservation and time-averaging (Gingerich 1985). Preservation is never perfect, regardless of preservation potential, and taphonomic biases aside, incompleteness presents a serious obstacle to the straightforward application of time series and random walk analyses to the calculation of microevolutionary rates for the following reason: As most time series are preserved at decreasing resolutions (increasing incompleteness), their statistics begin to approach those of symmetric random walks (Roopnarine et al. 1999, Schroeder 1991), regardless of the nature of the evolutionary mechanisms generating the series. This is a result of the loss of "information" from the series. Therefore, as the degree of preservational incompleteness increases, so does the probability that the preserved series will resemble a random walk. The effect of incompleteness on the estimation of rates depends on whether the modality of the series is determined more significantly by a bias in the direction of change (a biased probability), or whether $F(\kappa)$ is skewed in a particular direction. Directional evolutionary series are often modeled with a directional bias representing, for example, truncation selection. But as we saw with Dudley's results (Dudley 1977), $F(\kappa)$ can differ significantly between different selection regimes acting on the same lineage.

If κ is distributed uniformly or normally with respect to the direction of evolution, and only the direction of evolution is biased, then incompleteness will have a minimal effect on the estimation of κ from strictly static or directional series. As those types of series approach a symmetric random walk, however, which should be expected for fluctuating environments, selection, and so on, then the effect of timescale on interval rate will increase; the power of the relationship between r and τ will converge on 0.5, leading to a divergence between interval rates and $F(\kappa)$. The relationship between evolutionary change and the timescale of measurement can be measured independently of evolutionary rates, however, with the same expected range of α being $0 \rightarrow 1$. The relationship forms the basis of fractional random walks and grey noises (Mandelbrot 1999, Mandelbrot & Van Ness 1968, Mandelbrot & Wallis 1969, Schroeder 1991, Sokolov et al. 2002), and it seems possible that a randomization-based modification of a standard method for analyzing such series (Roopnarine 2001), rescaled range, or R/S analysis (Hurst 1951, Mandelbrot & Wallis 1969), could be used to transform distributions of interval rates appropriately to minimize their divergence from $F(\kappa)$, the distributions of true instantaneous rates. This idea has yet to be explored.

A skewed $F(\kappa)$ indicates a difference in the response to evolutionary change in different morphological directions resulting in directionality of the evolutionary series. The reasons for this may be multiple, including limited genetic variance, as well as functional or developmental constraints and limits to evolution in a particular direction. In the example of a directional series, the frequency of reversals may equal the frequency of change in the direction of net evolution, but reversals will account for less of the overall variance of the series. In this case the resulting distribution of interval rates will also be skewed, but it is unclear about how incompleteness of the series will affect the estimation of κ or $F(\kappa)$ from the relationship between rate and timescale.

These unknowns regarding estimation of $F(\kappa)$ from distributions of fossil interval rates could be addressed with appropriate simulations and analyses or with detailed long-term laboratory series. They cannot be resolved, however, by resorting to fossil series unless the details of preservation are known for the series, as is the mode of evolution during various intervals of the series (derived independently of rate calculations).

Finally, as is shown in the next section, the very concept of relating rates measured logarithmically as darwins or haldanes, to the interval over which they are measured, precludes any straightforward interpretation of the distribution of rates within the evolutionary series. The result is that neither evolutionary mode nor instantaneous rates can be derived from such analyses, and the conclusions drawn by Gingerich (1983) and Gould (1984) from the relationship of such rates to timescale may themselves be artifacts.

CAVEATS AND CORRECTIONS

Bookstein's null conclusions notwithstanding, there is a mathematical maneuver and a resulting correction that should be accounted for in subsequent analyses of microevolutionary rates. First, Haldane's suggestion of proportionality as the numerator in his measure of rate (x_2/x_1) was based on the argument that such ratio measurements are generally of greater precision than measures of absolute change ($|x_1 - x_2|$). This also permits equivalence and comparability of rates among diverse taxa (Gingerich 1985, Haldane 1949, Maynard Smith 1989). Gould (1984) pointed out correctly that comparison of darwins is valid regardless of timescale if the rate of change is exponential over time elapsed. It is curious to observe the subsequent development of the idea that the darwin predicts an exponential rate of evolution over time (Gould 1984, Hayami 1978, Hendry & Kinnison 1999), since Haldane's intent was to merely work with proportional rather than absolute changes (Gingerich 1985, Sheets & Mitchell 2001). There is no fundamental reason to believe that lineages exhibit exponential rather than absolute or additive changes over time, nor is there any fundamental need to work always with proportions (contrary to Sheets & Mitchell 2001). Haldane (1949), and later Bader (1955), both chose to work with percentages, but it was Kurtén (1959) who first explored use of the darwin extensively. The choice of a fundamental model of change is of great significance to the ability to detect evolutionary mode, however, as well as permitting the estimation of κ and the proper comparison of interval rates. This may be demonstrated most clearly by examining the consequences of both absolute and proportional change models.

Absolute and Proportional Changes

The model of absolute change assumes that the character of interest evolves as incremental additions (or subtractions) to the ancestral value (Roopnarine 2001). Given an average instantaneous rate of κ, and a monotonic mode of evolution, then the expected value of the character after τ_j time or generations have elapsed is

$$\langle x_j \rangle = x_0 \pm \kappa \tau_j^{\alpha}. \tag{9}$$

A simple measure of rate calculated as character change per unit time shows the predicted dependence of rate on timescale (according to random walk theory) and allows the recovery of κ and evolutionary mode (value of α) from a regression against time elapsed. If interval rates are calculated relative to the starting sample or point of the series only (Hendry & Kinnison 1999), and the series is first standardized by translating the origin to zero (that is, subtracting the starting value x_0 from all values), then the expected range of the series (R) at any given time j is

$$\langle R_j \rangle = x_j - x_0 = \kappa \tau_j^{\alpha}. \tag{10}$$

The corresponding interval rate is

$$\frac{R_j}{\tau_j} = \frac{\kappa \tau_j^{\alpha}}{\tau_j} \tag{11}$$

$$\Rightarrow \ln\left(\frac{R_j}{\tau_j}\right) = \ln \kappa + (\alpha - 1) \ln \tau_j. \tag{12}$$

Similarly, treating the changes within this series proportionally rather than linearly requires standardization of the series by scaling all values to the starting value (that is, dividing all values by the starting value x_0). The resulting rates are then reduced to

$$r_j = \frac{x_j}{x_0}\frac{1}{\tau_j} = \frac{\kappa \tau_j^{\alpha}}{\tau_j} \tag{13}$$

$$\Rightarrow \ln(r_j) = \ln\kappa + (\alpha - 1)\ln\tau_j = \ln\left(\frac{R_j}{\tau_j}\right), \tag{14}$$

which is the simple measure of absolute rate. The same result is obtained by logarithmically transforming all sample means and scaling to the origin by subtraction of $\ln x_0$.

Logarithmic Change

The haldane improves upon the above simple proportional rate by standardizing change according to sample standard deviations. Both the haldane and darwin, however, through the use of the difference of logarithms as a measure of proportionality, introduce a complication into the interpretation of the distribution of interval rates over time. When the difference of logarithms is substituted into the above equation (ignoring standardization to base e or standard deviations because these are effectively linear transformations), the rate is now expressed as

$$r_j = \frac{\ln(x_j) - \ln(x_0)}{\tau_j} = \frac{\ln\kappa + \alpha\ln\tau_j}{\tau_j}. \tag{15}$$

The attempt to linearize this relationship by subsequent log transformation (Gingerich 1983, 1993; Hallam 1975) yields

$$\ln r_j = \ln(\ln\kappa + \alpha\ln\tau_j) - \ln\tau_j. \tag{16}$$

This equation is equivalent to treating rates measured in darwins or haldanes as if those rates themselves are evolving according to the random walk model; that is, rate, and not character value, is changing according to a random walk model. The relationship between log-rates and log-timescale is therefore not equivalent to the relationship for rates derived from absolute and proportional changes, yet it is the one used commonly for the interpretation of microevolutionary rates (e.g., Hallam 1975, Gingerich 1983). The relationship here between rate and time interval is not a linear one, and there can be no straightforward recovery of evolutionary mode (α) or κ. Gingerich (1993) (see, e.g., Clyde & Gingerich 1994) has interpreted evolutionary modes from these plots, but Figure 5 shows that those interpretations are based on a similarity of scaling across values of α between plots of rate versus time, and log(rate) versus time. The interpretations must therefore carry substantial error, particularly at shorter timescales, due to both the nonlinearity of the functions, as well as the nonlinear convergence of the different evolutionary modes.

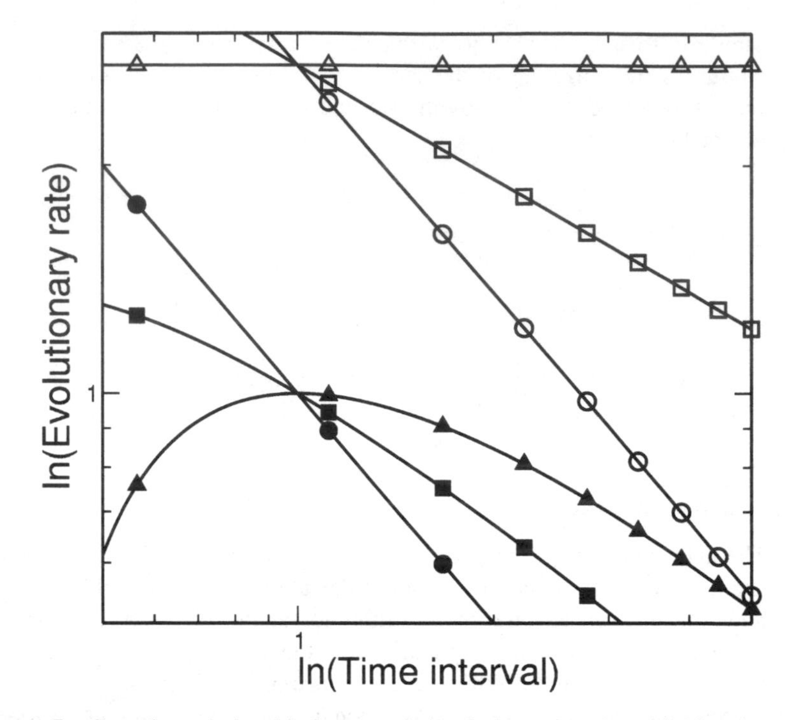

Figure 5 Complex relationship between ln(evolutionary rate) and ln(time interval). Upper curves (*open symbols*) represent three different modes of evolution, and were computed as the ratio logarithm of evolutionary change:time interval; (*circles*) monotonic stasis ($\alpha = 0$), (*squares*) random fluctuations ($\alpha = 0.5$), (*triangles*) monotonic directional selection ($\alpha = 1$). The lower curves (*closed symbols*) were computed according to the basis of the inverse relationship between rate and time, that is, as the logarithm of the above ratio. Closed symbols represent the identical modes of evolution as in the open-symbol curves.

Moreover, although κ may be estimated at the fundamental scale of one time unit from those plots, this may be accomplished only when the evolutionary mode is known without error or with very little error, and a proper curve fitted to the plot. Without this knowledge, plots of evolutionary rates (whether measured in darwins or haldanes) against time interval cannot be used to standardize comparisons of rate among different timescales, nor should they be used to estimate average or instantaneous rates of lineages.

Spurious Correlation

Sheets & Mitchell (2001) pointed to another serious concern with plots of log(rate) against time interval. The artifact of spurious correlation may appear in any relationship between a ratio and its denominator, and as Sheets & Mitchell (2001) demonstrated, spurious correlation might play a significant role in the perceived dependence of evolutionary rate on time. Their survey of several published data sets of long-term microevolutionary rates showed that the dependence on time often goes away when the spurious correlation between rate and time is accounted for; in other instances the dependence is noticeably weakened. The lack of a sustained mode of evolution was suggested as the most probable cause of the vanished dependence, and this is obvious if one expects the value of α to vary during the geological extent of a lineage. In fact, dynamic measurements of mode in single lineages suggest strongly that mode varies significantly both among characters in a single lineage (Roopnarine 2001, Roopnarine et al. 2003), as well as over time in a single character (Roopnarine 2001). There is also no reason to expect stability of $F(\kappa)$ during these intervals of time. Finally, one has to account also for the fact that Sheets & Mitchell (2001) solved the problem of spurious correlation for the logarithmic-type changes described above, where α and κ are not measurable directly from the relationship between rate and time. The authors do suggest the use of evolutionary change over time, which is equivalent to the equations suggested in the section above, Absolute and Proportional Changes, and is the approach used by Roopnarine (2001) and Roopnarine et al. (2003). It would be worthwhile to see the Sheets & Mitchell (2001) study repeated using proper measures of evolutionary change or proportional rate dependencies on time.

GENEALOGICAL ASSUMPTIONS

Implicit in all the calculations of evolutionary rates discussed in this review is the assumption that the time-ordered samples form a genealogical series, that is, a chain of ancestors and their descendants. This is obviously the case for modern samples where the lineages can be monitored carefully (e.g., Grant & Grant 2002). The situation is much less certain for historical and fossil samples. Although it is reasonable to assume that consecutive fossil samples of a species within a single stratigraphic section represent a single stratophenetic genealogy (Gingerich 1974), this is currently an untestable class of hypotheses (Roopnarine et al. 2002). The situation becomes even more tenuous when the supposed genealogy consists of species and higher taxa for which no genealogical or phylogenetic hypothesis exists. Hendry & Kinnison (1999) distinguished between rates derived from genealogical hypotheses, allochronic rates, and those derived from phylogenetic hypotheses involving sister-taxa, synchronic rates.

Allochronic rates are proper rates of evolution (dependent upon the accuracy of the genealogical hypotheses), whereas synchronic rates are really rates of divergence. Phylogenetic hypotheses are of course testable within a cladistic framework, but synchronic rates calculated between putative sister-taxa cannot be equated with allochronic rates derived from genealogical series. The major reason is that the mode of evolution of each taxon since divergence from their common ancestor is unknown, and calculations of microevolutionary rates are dependent upon knowledge of the mode of evolution. Several methods make use of random walk models for inferring rates of character change along phylogenetic branches (Felsenstein 1988), but these rates of divergence can only be equated with allochronic rates if within-taxon rate calculations are available. This is one area in which the presence of fossil taxa could be extremely useful.

Many of the evolutionary rates treated traditionally in the literature as allochronic rates must surely be synchronic rates. Considering one of the earliest and most significant examples, Simpson's Cenozoic horses (Haldane 1949, Simpson 1944), later analyses of some of those taxa (MacFadden 1985) show clearly that Simpson's data set does not comprise a uniform genealogy. Therefore, although Haldane's synchronic rates for the horses may be compared with other synchronic rates calculated across broad cladistic ranges, they should technically not be compared with allochronic rates, such as those calculated for single species or hypothesized ancestor-descendant pairs. Rates presented by Gingerich (1985), for example, are allochronic rates calculated on the basis of stratophenetic genealogical hypotheses and as such are rates of evolution, not divergence. The heterogeneity of large-scale comparisons of rates therefore stems from technical considerations, namely units of measurement and the magnitude of time intervals, as well as the phylogenetic nature of the rates themselves. This is a significant issue for measuring rates of evolution because at the very least, if we restrict ourselves to true allochronic rates, then we are forced to consider phylogenetically untestable genealogical hypotheses. However, if rates are calculated in a phylogenetic systematic framework, then those rates must be recognized as synchronic rates, suffering from the technical shortcomings outlined above of being interval rates based on unknown modes of evolution and varying timescales. Perhaps a robust approach could be developed combining phylogenetic analyses of taxa with documentable within-taxon records, thereby allowing the proper calibration of synchronic rates according to within-taxon allochronic rates.

Finally, it is common to see rates of taxonomic origination (Conklin 1920, Crick 1981, Kauffman 1978, Raup & Marshall 1980, Schaeffer 1952, Schopf et al. 1975, Stehli et al. 1969) or survivorship (Baumiller 1993, Kauffman 1978) described as rates of evolution. These measures may sometimes be correlated with underlying microevolutionary processes or species-level phenomena (Hecht 1965, Stanley 1973). Confirmation, however, can be derived only from the phylogenetic delineation of monophyletic clades (Novacek & Norell 1982) and an understanding of the history of those clades (Stanley 1973, 1985). Simpson (1944) was the first to suggest that rates of taxonomic turnover, or so-called rates of taxonomic evolution, may be substituted for rates of morphological evolution because taxonomic categories reflect summaries of morphological variability (Schaeffer 1952, Simpson 1944). Even if higher taxonomic groups do define discrete measures of within-group morphological variability, equating temporal change within, and among, such groups to evolution below the level of the species, or true macroevolution acting at and above the level of the species, requires phylogenetic support. This is a critical point because variable rates of evolution may be discussed meaningfully at macroevolutionary scales (Cracraft 1982, Stanley 1985) and above (Bowring et al.

1993, Knoll 1994), but the degree of independence among these levels (Eldredge 1982) cannot yet be established.

CONCLUSIONS

1. The rate at which a population-level character evolves is both a function of the mechanisms that generate variation within the population, as well as the nature of selection acting upon the character. The instantaneous rate of evolution is therefore a simple measure of character change between consecutive generations. Such character change may be measured as absolute additive differences or as proportional changes. The use of proportional changes (e.g. percentages, ratios) permits comparisons of rates among different lineages.
2. Units based on proportional changes, such as the darwin and haldane, must be applied with certain caveats. First, the darwin is sensitive to the dimensionality of the character being measured. Second, both measures exhibit an inverse relationship to the interval of time over which they are measured. If evolutionary series are modeled as random walks, then the inverse relationship is seen as a natural consequence of the rate at which such series change over time. Reversals, as well as biases in the direction and magnitude of instantaneous changes, control the rate of change. These factors, however, do not support previous explanations of the inverse relationship, namely a continuity of evolutionary processes across taxa and from the microevolutionary to the macroevolutionary scale, nor the opposing hypothesis which claims that the inverse relationship is a mathematical artifact indicative of stasis as a predominant mode. The artifactual nature of the relationship is specific to each lineage and its particular history of evolutionary mode.
3. Attempts have been made to compare microevolutionary rates across taxa by scaling according to time interval. Random walk-based techniques for addressing this problem have subsequently used interval rates to infer mode of evolution of a lineage, as well as its average or intrinsic rate of evolution. These attempts suffered from several flaws, including spurious correlation arising from the relationship between logarithmically transformed rates and time intervals, as well as the fact that evolutionary mode and average rate, as encompassed in random walk models of evolution, are not directly recoverable from the relationship between the log transformed quantities.
4. Distinctions have been made between within-lineage or allochronic rates of evolution, versus between-taxon or synchronic rates of divergence. Many past surveys of evolutionary rates have failed to make this important distinction. In cases where support for genealogical relationships among multiple taxa is lacking, allochronic rates should be abandoned in favor of phylogenetically supportable synchronic rates.

5. Rates of taxonomic turnover (origination, extinction) are not substitutes for true rates of evolution, unless the taxonomic categories involved are established as monophyletic clades. The relationships among rates of evolution at different levels of the phylogenetic hierarchy cannot be understood until they are measured and described at those levels for monophyletic clades.

The *Annual Review of Ecology, Evolution, and Systematics* is online at http://ecolsys.annualreviews.org

LITERATURE CITED

Bader RS. 1955. Variability and evolutionary rates in the Oreodonts. *Evolution* 9:19–140

Baumiller TK. 1993. Survivorship analysis of Paleozoic Crinoidea: effect of filter morphology on evolutionary rates. *Paleobiology* 19:304–21

Bookstein FL. 1987. Random walk and the existence of evolutionary rates. *Paleobiology* 13:446–64

Bookstein FL. 1988. Random walk and the biometrics of morphological characters. *Evol. Biol.* 9:369–98

Bowring SA, Grotzinger JP, Isachsen CE, Knoll AH, Pelechaty SM, Kolosov P. 1993. Calibrating rates of Early Cambrian evolution. *Science* 261:1293–98

Bull JJ. 1987. Evolution of phenotypic variance. *Evolution* 41:305–15

Charlesworth B. 1984a. The cost of phenotypic evolution. *Paleobiology* 10:319–27

Charlesworth B. 1984b. Some quantitative methods for studying evolutionary patterns in single characters. *Paleobiology* 10:308–18

Clyde WC, Gingerich PD. 1994. Rates of evolution in the dentition of Early Eocene *Cantius*: comparison of size and shape. *Paleobiology* 20:506–22

Conklin EG. 1920. The rate of evolution. *Sci. Mon.* 10:589–602

Cracraft J. 1982. A nonequilibrium theory for the rate-control of speciation and extinction and the origin of macroevolutionary patterns. *Syst. Zool.* 31:348–65

Crick RE. 1981. Diversity and evolutionary rates of Cambrio-Ordovician nautiloids. *Paleobiology* 7:216–29

Dobzhansky T. 1970. *Genetics of the Evolutionary Process*. New York: Columbia Univ. Press

Dudley JW. 1977. Seventy-six generations of selection for oil and protein percentage in maize. In *Proc. Int. Conf. Quant. Genet.*, ed. E Pollak, O Kempthorne, TB Bailey, pp. 459–73. Ames: Iowa State Univ.

Edwards AWF, Cavalli-Sforza LL. 1964. Reconstruction of evolutionary trees. In *Reconstruction of Evolutionary Trees*, ed. VH Heywood, J McNeill, pp. 67–76. London: System. Assoc.

Eldredge N. 1982. Phenomenological levels and evolutionary rates. *Syst. Zool.* 31:338–47

Eldredge N. 1989. *Macroevolutionary Dynamics. Species, Niches, and Adaptive Peaks*. New York: McGraw-Hill

Eldredge N, Gould SJ. 1972. Punctuated equilibria: an alternative to phyletic gradualism. In *Models in Paleobiology*, ed. TJM Schopf, pp. 82–115. San Francisco: Freeman, Cooper

Endler JA. 1986. *Natural Selection in the Wild. Monogr. Popul. Biol. 21*. Princeton, NJ: Princeton Univ. Press

Falconer DS. 1953. Selection for large and small size in mice. *J. Genet.* 51:470–501

Falconer DS. 1989. *Introduction to Quantitative Genetics*. New York: Wiley. 3rd ed.

Felsenstein J. 1988. Phylogenies and quantitative methods. *Annu. Rev. Ecol. Syst.* 19:445–71

Galton F. 1879. The geometric mean in vital and social statistics. *Proc. R. Soc.* 29:365

Gingerich PD. 1974. Stratigraphic record of

Early Eocene *Hyopsodus* and the geometry of mammalian phylogeny. *Nature* 248:107–9
Gingerich PD. 1983. Rates of evolution: Effects of time and temporal scaling. *Science* 222:159–61
Gingerich PD. 1984. Smooth curve of evolutionary rate: a psychological and mathematical artifact. *Science* 226:995
Gingerich PD. 1985. Species in the fossil record: concepts, trends, and transitions. *Paleobiology* 11:27–41
Gingerich PD. 1993. Quantification and comparison of evolutionary rates. *Am. J. Sci.* 293A:453–78
Gould SJ. 1984. Smooth curve of evolutionary rate: a psychological and mathematical artifact. *Science* 226:994–95
Grant PR, Grant BR. 2002. Unpredictable evolution in a 30-year study of Darwin's Finches. *Science* 296:707–11
Haldane JBS. 1949. Suggestions as to quantitative measurement of rates of evolution. *Evolution* 3:51–56
Hallam A. 1975. Evolutionary size increase and longevity in Jurassic bivalves and ammonites. *Nature* 258:493–96
Hayami I. 1978. Notes on the rates and patterns of size change in evolution. *Paleobiology* 4:252–60
Hecht MK. 1965. The role of natural selection and evolutionary rates in the origin of higher levels of organization. *Syst. Zool.* 14:301–17
Hendry AP, Kinnison MT. 1999. Perspective: the pace of modern life: measuring rates of contemporary microevolution. *Evolution* 53:1637–53
Hurst HE. 1951. Long-term storage capacity of reservoirs. *Trans. Am. Soc. Civil Eng.* 116:770–808
Kauffman EG. 1978. Evolutionary rates and patterns among Cretaceous Bivalvia. *Philos. Trans. R. Soc. London Ser. B* 284:277–304
Knoll AH. 1994. Proterozoic and Early Cambrian proteins: Evidence for accelerating evolutionary tempo. *Proc. Natl. Acad. Sci. USA* 91:6743–50
Knoll AH, Bambach RK. 2000. Directionality in the history of life: diffusion from the left wall or repeated scaling of the right? *Paleobiology* 26:1–14 (Suppl. S)
Kurtén B. 1958. A differentiation index, and a new measure of evolutionary rates. *Evolution* 12:146–57
Kurtén B. 1959. Rates of evolution in fossil mammals. *Cold Spring Harbor Symp. Quant. Biol.* 24:205–15
Lande R. 1976. Natural selection and random genetic drift in phenotypic evolution. *Evolution* 30:314–34
Lande R. 1986. The dynamics of peak shifts and the pattern of morphological evolution *Paleobiology* 12:343–54
Lerman A. 1965. On rates of evolution of unit characters and character complexes. *Evolution* 19:16–25
MacFadden BJ. 1985. Patterns of phylogeny and rates of evolution in fossil horses: Hipparions from the Miocene and Pliocene of North America. *Paleobiology* 11:245–57
MacLeod N. 1991. Punctuated anagenesis and the importance of stratigraphy to paleobiology. *Paleobiology* 17(2):167–88
Malmgren BA, Berggren W, Lohmann GP. 1983. Evidence for punctuated gradualism in the Late Neogene *Globorotalia tumida* lineage of planktonic foraminifera. *Paleobiology* 9:377–89
Mandelbrot B, Van Ness JW. 1968. Fractional brownian motions, fractional noises and applications. *SIAM Rev.* 10:422–37
Mandelbrot B, Wallis JR. 1969. Some long-run properties of geophysical records. *Water Resourc. Res.* 5:321–40
Mandelbrot BB. 1999. *Multifractals and 1/f Noise*. New York: Springer-Verlag. 442 pp.
Maynard Smith J. 1976. What determines the rate of evolution? *Am. Nat.* 110:331–38
Maynard Smith J. 1989. *Evolutionary Genetics*. New York: Oxford Univ. Press
McKinney ML. 1990. Classifying and analysing evolutionary trends. In *Evolutionary Trends*, ed. KJ McNamara, pp. 28–58. Tucson: Univ. Ariz. Press
Mindell DP, Thacker CE. 1996. Rates of molecular evolution: Phylogenetic issues and

applications. *Annu. Rev. Ecol. Syst.* 27:279–303

Novacek MJ, Norell MA. 1982. Fossils, phylogeny, and taxonomic rates of evolution. *Syst. Zool.* 31:366–75

Plotnick RE, Prestegaard KL. 1995. Time series analysis I. In *Nonlinear Dynamics and Fractals. New Numerical Techniques for Sedimentary Data*, ed. GV Middleton, RE Plotnick, DM Rubin, pp. 47–67. SEPM Short Course No. 36

Raup DM. 1977. Stochastic models in evolutionary paleontology. In *Patterns of Evolution*, ed. A Hallam, pp. 59–78. Amsterdam: Elsevier

Raup DM, Crick RE. 1981. Evolution of single characters in the Jurassic ammonite *Kosmoceras*. *Paleobiology* 7:200–15

Raup DM, Marshall LG. 1980. Variation between groups in evolutionary rate: a statistical test of significance. *Paleobiology* 6:9–23

Reznick DN, Shaw FH, Rodd FH, Shaw RG. 1997. Evaluation of the rate of evolution in natural populations of guppies *Poecilia reticulata*. *Science* 275:1934–37

Roopnarine PD. 1995. A re-evaluation of evolutionary stasis between the bivalve species *Chione erosa* and *Chione cancellata* (Bivalvia: Veneridae). *J. Paleontol.* 69(2):280–87

Roopnarine PD. 2001. The description and classification of evolutionary mode: a computational approach. *Paleobiology* 27:446–65

Roopnarine PD, Byars G, Fitzgerald P. 1999. Anagenetic evolution, stratophenetic patterns, and random walk models. *Paleobiology* 25:41–57

Roopnarine PD, Murphy MA, Buening N. 2002. Inferring the pattern and mode of species evolution in high resolution stratigraphic series. *8th Int. Conodont Symp. Eur., ECOS VIII, STRATA* 12

Roopnarine PD, Murphy MA, Buening N. 2003. Microevolutionary dynamics of the Early Devonian conodont *Wurmiella* from the Great Basin of Nevada. *Paleontol. Electr.* In review

Schaeffer B. 1952. Rates of evolution in the Coelacanth and Dipnoan fishes. *Evolution* 6:101–11

Schopf TJM, Raup DM, Gould SJ, Simberloff DS. 1975. Genomic versus morphologic rates of evolution: influence of morphologic complexity. *Paleobiology* 1:63–70

Schroeder M. 1991. *Fractals, Chaos, Power Laws*. New York: Freeman

Sheets HD, Mitchell CE. 2001. Uncorrelated change produces the apparent dependence of evolutionary rate on interval. *Paleobiology* 27:429–45

Sheldon PR. 1993. Making sense of microevolutionary patterns. In *Evolutionary Patterns and Processes*, ed. DR Lees, D Edwards, 14:19–31. London: Linnean Soc.

Simpson GG. 1944. *Tempo and Mode in Evolution*. London: Oxford Univ. Press

Sokolov IM, Klafter J, Blumen A. 2002. Fractional kinetics. *Phys. Today* 55:48

Stanley SM. 1973. Effects of competition on rates of evolution, with special reference to bivalve mollusks and mammals. *Syst. Zool.* 22:486–506

Stanley SM. 1985. Rates of evolution. *Paleobiology* 11:13–26

Stanley SM, Yang X. 1987. Approximate evolutionary stasis for bivalve morphology over millions of years. *Paleobiology* 13:113–39

Stehli FG, Douglas RG, Newell ND. 1969. Generation and maintenance of gradients in taxonomic diversity. *Science* 164:947–49

Van Valen L. 1974. Two modes of evolution. *Nature* 252:298–300

Williams GC. 1992. *Natural Selection. Domains, Levels, and Challenges*. New York: Oxford Univ. Press

Wright S. 1968. *Evolution and the Genetics of Populations: Genetic and Biometric Foundations*, Vol. 1. Chicago: Univ. Chicago Press

Annu. Rev. Ecol. Evol. Syst. 2003. 34:633–60
doi: 10.1146/annurev.ecolsys.34.011802.132425

First published online as a Review in Advance on August 6, 2003

DEVELOPMENT AND THE GENETICS OF EVOLUTIONARY CHANGE WITHIN INSECT SPECIES

Paul M. Brakefield,[1] Vernon French,[2] and Bas J. Zwaan[1]
[1]Institute of Biology, Leiden University, 2300 RA Leiden, The Netherlands; email: brakefield@rulsfb.leidenuniv.nl, zwaan@rulsfb.leidenuniv.nl
[2]Institute of Cell, Animal and Population Biology, University of Edinburgh, Edinburgh EH9 3JT, United Kingdom; email: vernon.french@ed.ac.uk

Key Words evo-devo, artificial selection, evolutionary genetics, morphology

■ **Abstract** Changes in genes and in developmental processes generate the phenotypic variation that is sorted by natural selection in adaptive evolution. We review several case studies in which artificial selection experiments in insects have led to divergent morphologies, and where further work has revealed information about the underlying changes at both the genetic and developmental levels. In addition, we examine several studies of phenotypic plasticity where multidisciplinary approaches are also beginning to reveal more about how developmental processes are modulated. Such integrated research will lead to a richer understanding of the changes in development that occur during evolutionary responses to natural selection, and it will also more rigorously examine how developmental processes can influence the tempo and direction of evolutionary change.

INTRODUCTION

Development is central to evolution because the processes of development translate genotypes into phenotypes; thus, developmental changes generate the variation on which natural selection can act—no variation, no evolution. Efforts to integrate evolutionary and developmental biology have gained much impetus in the past decade with the burgeoning understanding of developmental mechanisms. Much of this initiative has involved comparison of the development of major body features (e.g., segments, limbs, eyes) in a small number of model organisms that are widely disparate, both morphologically and taxonomically (Carroll et al. 2001, Davidson 2001, Raff 1996). These studies have discovered differences and also many surprising similarities in the developmental mechanisms employed by different phyla, but they do not address directly the developmental basis of variation in natural populations. Specific genes have been identified as central to the development of a trait through laboratory study of the consequences of major mutations, but does variation in these same genes also underpin variation in the trait arising within natural populations? This review seeks to demonstrate that understanding development of the phenotype is useful for those ecologists and evolutionary biologists interested in the process of adaptation because insights about development provide the potential for improved predictions about evolutionary change.

Some researchers have recently started to examine the development of subtle differences in morphology among more closely related organisms, including species of flies (Gormpel & Carroll 2003, Kopp & True 2002, Kopp et al. 2000, Simpson 2002, Skaer et al. 2002, Stern 1998, Sucena & Stern 2000, Sucena et al. 2003, True et al. 1999), nematodes (Felix et al. 2000), centipedes (Arthur & Kettle 2001), fish (Peichel et al. 2001), and salamanders (Parichy 1996). An important issue, however, is whether genes underlying the differences between even closely related species also contribute to the phenotypic variance for the same traits within a species. Few studies have attempted to understand directly how developmental mechanisms are modulated in an evolutionary response to selection within a particular species. Evolutionary geneticists study the genetic changes that underlie the differences in phenotype that result from such responses.

Relating genetic variation to modification of the developmental processes that generate the phenotype is becoming one major goal of evolutionary developmental biology, or "evo-devo" (Arthur 2002, Beldade & Brakefield 2002, Stern 2000), and the field will increasingly focus on the origins of ecologically and evolutionarily relevant phenotypic variation within species. The emphasis in current work is on morphological traits but this will grow to include the whole functional phenotype, which also encompasses life history and behavioral and physiological traits. Furthermore, there is likely to be an increasing interest in a multidisciplinary study of the evolution of phenotypic plasticity, which will help to understand the modulation of development (Frankino & Raff 2003, Pigliucci 2001, Schlichting & Pigliucci 1998, West-Eberhard 2003).

Artificial selection is a valuable tool of evolutionary geneticists. It can efficiently screen populations for the allelic variation and combinations of genes (genotypes) that underlie the generation of phenotypes of evolutionary relevance (Barton & Partridge 2000). These selected phenotypes can then be examined, both with respect to their fitness consequences and to the ways in which development has been modified. Scharloo (1983, 1987) was an early practitioner of the latter approach although he was unable to go beyond models of how the underlying developmental processes might have been modulated through genetic change to yield the selected phenotypes. The subsequent rapid expansion in knowledge of developmental mechanisms now allows explicit developmental analyses for at least certain examples of morphological evolution. One potential advantage of artificial selection from the evolutionary perspective is that it focuses attention on standing genetic variation within outbred stocks, which is probably more relevant to understanding microevolutionary processes in nature than is developmental change resulting from novel genetic variation induced by mutagenesis (Haag & True 2001).

The emphasis in this review is on the use of artificial selection, or comparisons between different locally adapted populations, in order to examine the genetic changes that underlie evolutionarily relevant phenotypic variation. We have limited the scope to work on insect species (which provide some of the clearest examples) and to studies that have at least started to explore how genetic change operates through the modulation of developmental processes. First, we introduce relevant aspects of the process of development and then we focus in some depth on particular case studies of microevolutionary change.

DEVELOPMENT: PRODUCING THE PHENOTYPE

In the development of a multicellular animal, typically from a fertilized egg, dramatic differences gradually appear between the cells. Specific cells will divide rapidly or slowly, equally or unequally, in random or highly predictable orientation. Some cells will enlarge, elongate, move, or die. The cells become spectacularly diverse in structure and function as they differentiate into epidermal, muscle, cartilage, nerve cells, and so on. These cellular differences appear appropriately in time and space, so that coherent and functional tissues and organs are formed, and they all result from selective gene expression (see Carroll et al. 2001, Davidson 2001). All cells of an individual contain all its genes (to a first approximation), but only a characteristic subset of them is active in each cell type, such that the cells will differ in the proteins produced and consequently in their properties. The central feature of development, then, is the coordinated control of gene expression, and this occurs primarily at the level of transcription, by the action of transcription factors (see below). A large number of gene products function as components of the array of developmental mechanisms, often called the toolkit (Carroll et al. 2001), that ensure that gene expression is appropriately controlled in the cells as development proceeds. Any genetic change that alters one of these developmental mechanisms may change the final organism—the phenotype.

In all embryos, differences in gene expression between the cells originate through some combination of two factors: initial heterogeneity within the fertilized egg and local interactions among the growing population of cells (Figure 1*A*, see color insert). Cell interaction in a developing embryo is mediated by the operation of signaling pathways (as illustrated in Figure 1*B* by the Notch pathway). Typically, a pathway consists of a transmembrane receptor that is activated by binding a specific extracellular ligand, initiating a cascade of protein modifications (e.g., phosphorylation, cleavage) that constitute a signal transduction pathway and end in the activation of a transcription factor. The ligand may be a transmembrane protein on an adjacent cell, or it may be secreted by cells that are nearby (as in the Hedgehog or Wingless pathways) or remote (as in the Insulin or Steroid Receptor pathways). In some cases, the ligand (e.g., a steroid) may directly penetrate the cell to interact with its receptor. The signal transduction pathways may be very complex and they may share components or interact in other ways to integrate the different signals that may be impinging on a cell.

Transcription of a eukaryotic gene is regulated by the binding of transcription factors to short sequences (binding sites) in the associated *cis*-regulatory DNA (Figure 1*B*). A transcription factor will typically have binding sites associated with many target genes and, reciprocally, a given gene will typically have binding sites for many different transcription factors, which will be both activators and repressors and will interact in characteristic ways upon binding to determine whether transcription occurs (Carroll et al. 2001, Davidson 2001). Hence, the *cis*-regulatory DNA acts to integrate the presence of different transcription factors within the cell. Furthermore, a gene's regulatory DNA may be very extensive and is typically organized into discrete and functionally independent modules, enhancers, each of which can initiate transcription in response to the appropriate activators (Davidson 2001). Hence, a gene, which may itself code for a transcription factor or a signaling pathway component, will be expressed in different contexts under different regulatory control, and will be performing different functions as development progresses. Clearly, a mutation in the coding region may alter the gene product, possibly affecting all functions, whereas a mutation in one of its *cis*-regulatory regions may change only one aspect of the expression pattern and affect only a single function (Stern 1998).

From an evolutionary perspective, it is significant that the signaling pathways used by animal cells are ancient and highly conserved, with the complete pathway being homologous across the metazoa, from arthropods to vertebrates (Carroll et al. 2001, Davidson 2001). Also, there are relatively few basic pathways (approximately 20—see Raff 2000) and they are used repeatedly, at different stages and in different parts of the developing embryo. Numerous genes code for transcription factors, but again, many of the major gene families are highly conserved and are recognizable across the animal kingdom. The same genes are, however, typically expressed in many different contexts during development (Carroll et al. 2001, Davidson 2001). Consideration of development suggests that developmentally important genes are likely to show both epistasis and pleiotropy, as the gene products interact in mediating cell interaction and controlling gene expression, repeatedly in the formation of different parts of the organism. Below we examine how changes in these genetic networks can produce evolutionary change in morphology, in particular, by describing three case studies.

EVOLUTION OF BRISTLES IN FLIES

An important issue for evolutionary developmental biology is to understand the origins of evolutionarily relevant variation in the phenotype: What is the genetic and developmental basis of the variation in a quantitative trait, and do the same genes also underlie trait differences between related species? Artificial selection on the phenotypic variation from natural populations can be used to generate large phenotypic differences between the selected lines, facilitating powerful genetic analysis to identify the relevant genes or gene regions. In *Drosophila*, development is understood in considerable detail and it can be determined whether the identified quantitative trait loci (QTL) do correspond to genes with known functions in development of that trait. A particularly well worked out example of this approach is the analysis of variation in bristle number in *D. melanogaster*.

The bristles of *D. melanogaster* are external sensory organs of the peripheral nervous system that have been studied for many years, from perspectives ranging from development to quantitative genetics. The number of sternopleural (ST) and abdominal (AB) bristles has been advocated as an ideal system for study of the nature of genetic variation (Mackay 1995). It is a classical quantitative genetic character for which ample genetic variation is described (Mackay 1995) and for which major mutations have been documented (Lindsley & Zimm 1992).

Development of Bristles

In *Drosophila*, many genes have known functions in development of the peripheral nervous system and these are plausible candidates for contributing to the standing genetic variation in bristle number. Bristle formation starts with the expression of genes of the *achaete-scute* complex (ASC) in a cluster of cells at a particular location in the epidermis. Detailed analysis has shown that specific enhancer sites in the ASC function to regulate expression for each of the cell clusters and some of the controlling transcription factors have been identified (Ghysen & Dambly-Chaudiere 1988, Gomez-Skarmeta et al. 1995). Expression of the ASC is required for neural development, and this expression gradually becomes restricted to one cell within the cluster through a lateral inhibition process mediated by the Notch signaling pathway (Figure 1). Briefly, binding of the ligand Delta to the Notch receptor results in the activation of transcription factors that upregulate *Notch*, but inhibits expression of the *Delta* and ASC genes. Only one cell of the initial cluster remains uninhibited, producing the ligand and expressing ASC genes, and this cell undergoes subsequent divisions to produce the cuticular bristle and associated sensory neuron. There are thus numerous potential places in the regulatory pathway that could be altered to generate phenotypic change in the bristle pattern.

Quantitative Genetics of Variation in Bristles

Several routes have been taken to finding and identifying the QTL relevant to variation in bristle number (Mackay 2001). For instance, after *P*-element mutagenesis, some of the insertions with large effect on bristle number were mapped to genes known to control either the ASC or the Notch pathway (Lyman et al. 1996). This result is suggestive but does not shed light on the issue of whether these genes are relevant to natural variation (but see Long et al. 2000).

Artificial selection for AB and ST bristle number has been conducted on strains of *D. melanogaster* recently established from the wild (Gurganus et al. 1999, Long et al. 1995). Crosses were then performed between the divergent selected populations, and marker genotypes were analyzed for their phenotype using QTL mapping techniques (Lynch & Walsh 1998). Many QTL were found, with effects that were variable but which, together, accounted for most of the difference in bristle number between the parental lines (Gurganus et al. 1999, Long et al. 1995). In both studies, most of the QTL mapped to the approximate positions of candidate genes that are components or controlling factors of either the ASC or the Notch pathway in the development of the peripheral nervous system (Gurganus et al. 1999, Long et al. 1995). The two sets of selected lines were further intercrossed to increase recombination in the QTL intervals, and thus improve the precision of mapping (Mackay 2001). In doing so, 26 QTL for bristle number were identified, 20 of which mapped to candidate gene positions (Nuzhdin et al. 1999; see also Dilda & Mackay 2002). In addition, by use of molecular markers and multiple backcrossing, putative QTL from the selected lines have been introgressed singly into homogeneous nonselected genetic backgrounds to attempt to confirm the existence of QTL mapping to specific candidate genes (Mackay 2001).

Another approach has been complementation testing using available mutants for the candidate genes. The rationale is that if the phenotypic effect of the mutant differs between crosses to the high and low selected lines, this failure to complement indicates that the candidate gene is indeed involved in causing the selection response (Mackay 2001). For both the AB- (Long et al. 1996) and the ST-selected lines (Gurganus et al. 1999), mutant alleles at some of the candidate loci failed to complement. Interpretation of these results is complex, however, as the noncomplementing locus may actually be allelic to the QTL in the selected lines, or it may be involved through a genetic interaction (i.e., epistasis). Indeed, mutants of some genes lying outside the QTL intervals did show significant interactions with the selected lines (Gurganus et al. 1999). Overall, the above results can be taken as a first indication of evolutionarily relevant variation in natural populations, which maps to genes with developmental functions that have previously been determined using mutations of large effects. However, even after high-resolution mapping, QTL map to the intervals containing many genes and the one truly affecting the trait may well not be the candidate gene (Mackay 2001).

Linkage disequilibrium mapping (Mackay 2001) was used to approach more closely to the quantitative trait nucleotide (QTN), in other words the polymorphism in the gene that is actually causing the phenotypic effect. Candidate gene regions of interest from wild-type flies were placed in a homozygous genetic background and polymorphisms within the candidate gene were associated with the bristle phenotype of carriers of these polymorphisms. This type of approach has proved successful for several genes [including ASC (Long et al. 2000, Mackay & Langley 1990), *Dl* (Long et al. 1998), *h* (Robin et al. 2002), and *sca* (Lyman et al. 1999)], with polymorphisms at these loci shown to be associated with divergence in bristle phenotypes. It is clear that very few systems allow such an elaborate genetic analysis as *D. melanogaster*. In the bristle example, the QTL mapping approach has drawn a direct connection between genes with a central role both in the development of the trait and in natural variation in that trait [including the genotype by environment interaction (see Gurganus et al. 1998)].

Considerable progress has been made in relating developmental mechanisms to naturally occurring variation in bristle patterns. However, an important disadvantage of bristle traits for evolutionary studies is that we have little understanding of the precise function of the bristles or of the relationship between the trait variation and fitness. In this respect, body size is a more promising trait.

EVOLUTION OF INSECT BODY SIZE

There is substantial (and rapidly increasing) information about the developmental control of growth and size, particularly in *Drosophila* and other insects (see below). Furthermore, body size is closely related to fitness (see Partridge & French 1996) and covaries with traits such as developmental time, growth rate, mating success, and progeny size.

That selection is acting on body size in natural populations is strongly indicated by the observation that genetic, latitudinal size clines have been described for several ectotherms from different taxa: Body size increases with increasing latitude (Endler 1977, Partridge & French 1996). The most convincing data come from studies on the cosmopolitan fruit fly *D. melanogaster*, in which latitudinal clines have been found on all major continents. The genetic and developmental basis of variation in body size has been much studied in *Drosophila* in the context of the geographical size clines and also of laboratory populations artificially selected for body size.

Developmental Control of Insect Body Size

Insects grow through a series of larval moult cycles that are controlled by changing titers of ecdysteroids and juvenile hormone (JH). The size of the adult depends on the dimensions of the cuticle secreted by the epidermis, and this is limited by the changes in hormone levels that occur through the last larval instar, to result in moulting—pupation in the case of holometabolous insects. When the larva reaches a critical size, the level of JH falls, allowing neurosecretory cells to release prothoracicotropic hormone (PTTH), which triggers ecdysteroid release, causing the larva to stop feeding and progress toward the moult (Nijhout 1994a).

The growth rate and final size of developing organs is controlled by organ-intrinsic as well as -extrinsic mechanisms (Bryant & Simpson 1984, Conlon & Raff 1999). In a holometabolous insect such as *D. melanogaster*, the adult epidermis grows as separate imaginal discs that fuse together and replace the larval epidermis during metamorphosis. Growth of the imaginal discs is intrinsically regulated (Bryant & Simpson 1984), as the discs will not grow beyond their normal size, even if pupation is delayed (Simpson et al. 1980). In addition, immature discs cultured in adult female hosts will terminate growth at a cell number close to that normally attained by the time of pupation (Bryant & Levinson 1985). The developing imaginal discs also communicate with the larval neurosecretory system because their damage and subsequent regenerative growth results in an extended larval period and a delay in the timing of the ecdysone release (Simpson et al. 1980, Berreur et al. 1979).

The intrinsic control of growth (cell number and cell size) is closely related to establishing patterns of cell fate within the developing imaginal disc. Several intercellular signaling pathways have been implicated in these processes. For example, the signals encoded by *decapentaplegic* and *wingless* are produced in specific narrow regions and specify cell fate across different axes of the wing blade. If these signals cannot be transduced the cells cannot grow or divide (Edgar & Lehner 1996), whereas their ectopic expression provokes local cell proliferation and pattern duplication (e.g., Zecca et al. 1996). *Wingless* negatively regulates expression of the *Drosophila* homologue of *myc* and a loss of function mutation in *dMyc* retards the growth and division of disc cells, whereas its overexpression promotes these processes (Johnston et al. 1999).

Apart from extrinsic control by ecdysteroids and intrinsic control by local cell interactions, it has become clear that imaginal disc growth is controlled through the highly conserved insulin/IGF pathway (Leevers 2001, Oldham et al. 2000). As discussed below, this pathway responds to external conditions, such as nutrition availability, to modulate both cell growth and division.

Various manipulations of signal transduction from the insulin receptor (Dinr) can increase or decrease wing size, altering both cell size and number (Leevers et al. 1996, Weinkove et al. 1999). Reduction of receptor activity and its overexpression cause decreases and increases in wing size, respectively, involving both cell size and number (Brogiolo et al. 2001). Members of a family of insulin-like peptides are expressed in various tissues (e.g., the brain and imaginal discs), and the overexpression of one of these putative ligands enlarges the adult, increasing both the number and size of its cells (Brogiolo et al. 2001). Thus, the insulin/IGF pathway regulates cell growth as well as cell division (Leevers 2001), and only one associated gene, ribosomal *S6 kinase*, has been shown to influence only cell size (see Oldham et al. 2000). The gene *chico*, which encodes the receptor substrate component of the insulin pathway, is implicated in fat storage in *Drosophila* (e.g., Bohni et al. 1999). This, together with work in the nematode, *Caenorhabditis elegans* (Guarente & Kenyon 2000), is strongly suggestive that the insulin/IGF signaling pathway plays a significant role in coordinating growth with the nutritional status of the developing larva (but see Oldham et al. 2000, Stern 2003).

Evolution of Body Size in *Manduca Sexta*

Much of our understanding of the hormonal control of insect moulting and metamorphosis comes from detailed studies on the large moth *Manduca sexta* (see Nijhout 1994a). As outlined above, adult size depends on the following variables during the last larval instar: the initial weight, the critical weight (influenced by the initial weight), the growth rate, the delay while JH level falls after attaining critical weight, and a further delay associated with photoperiodicity in PTTH release (which completes feeding and growth). These were all measured in the early

1970s for laboratory stocks recently taken from the wild. Reexamination of size and growth control of the cultures 30 years later showed a dramatic 50% increase in body size (D'Amico et al. 2001). This could be fully explained by increases in the critical weight, the growth rate, and the delay before PTTH could be released, whereas there had been no change in the weight at the start of the final instar or in the photoperiodicity of PTTH release (D'Amico et al. 2001). The genetic changes in body size had presumably occurred in response to changes in factors such as density and parasitism, and perhaps also to inadvertent selection for large individuals for breeding. It is fascinating (but perplexing) why only particular aspects of growth control were affected—increasing growth rate in the final instar, for example, but apparently not in earlier instars (D'Amico et al. 2001).

The *Manduca* study has defined changes in the developmental/physiological control of growth and size, but has not investigated their genetic basis: For this, we need to move to the extensive studies of the evolutionary genetics of body size in *Drosophila*.

Genetic Variation and the Evolution of Body Size in *Drosophila Melanogaster*

In *D. melanogaster*, latitudinal clines in body size have been found across the Middle East to Africa (Tantawy & Mallah 1961), Japan (Watada et al. 1986), North America (Coyne & Beecham 1987), Eastern Europe to Central Asia (Imasheva et al. 1994), Australia (James et al. 1995, James et al. 1997), and South America (van't Land et al. 1999). In all cases, populations from the higher latitudes give the bigger flies, even when all are reared in standard conditions. The similar evolutionary responses to life at higher latitudes and at lower laboratory temperatures implicates temperature as a selective agent on body size (Atkinson & Sibly 1997, Partridge & French 1996). Many climatic factors vary with latitude, but regression analysis for both the Australian (James et al. 1995) and South American (van't Land 1997) clines in *D. melanogaster* has shown the closest correlations to be between temperature, latitude, and body size. Moreover, in two independent studies, laboratory populations of *D. melanogaster* kept at different temperatures show rapid evolution—again toward genetically larger flies at the lower temperatures (Cavicchi et al. 1985, 1989; Partridge et al. 1994). There is also an intriguing parallel with the developmental response to temperature: Across a wide range of ectotherms, including *D. melanogaster*, there is an inverse relationship between rearing temperature and adult body size (see Atkinson 1994, Partridge & French 1996).

Body size in an adult insect depends on the dimensions of the cuticle and thus on the number and size of the underlying epidermal cells. These parameters are most conveniently analyzed on adult wings, where each epidermal cell secretes one hair (or trichome). Trichome counts showed that populations maintained at different temperatures had evolved their different body sizes through changes to cell size, with little or no effects on cell number. Thus, thermal evolution at the lower temperature produced flies with larger cells (Partridge et al. 1994). Intriguingly, the developmental response of size to rearing temperature was also shown to be mediated by changes in cell size, rather than cell number in the wings (Partridge et al. 1994), and also in the eyes and legs (Azevedo et al. 2002).

In contrast to the laboratory responses to temperature, size differences along the geographical clines involve both cellular parameters. In the Australian populations, the increase in size with latitude is mainly caused by an increase in the number of cells (James et al. 1995, Zwaan et al. 2000). In the South American cline, however, more than 40% of the size increase with latitude is caused by increased cell size, as compared with less than 20% in the Australian cline (Zwaan et al. 2000). Comparable differences have been found in the cellular basis of recently established clines for wing size in *D. subobscura* in North and South America (Calboli et al. 2003). These results indicate that increased body size (and wing size) is being selected for at lower temperatures, but the precise cellular basis is less important (Zwaan et al. 2000). Body size can be readily altered in the laboratory by artificial selection. Two independent selection experiments, targeting thorax length or wing area and using different base populations, both showed that an increase in fly size was achieved by increasing the number of cells, whereas the response to selection for a smaller adult was a reduction in cell size (Partridge et al. 1999). This interesting result suggests that, in establishing the Australian cline, selection from the founding population could have been mainly in the direction of an increase in body size, whereas in the expansion to form the South American cline, selection was for both smaller and larger body size (Zwaan et al. 2000) (Figure 2, see color insert).

Why do animals evolve to larger sizes at lower temperatures? Much of this topic is beyond the scope of this review (see Partridge & French 1996), but some points are directly relevant here. For instance, fat content and starvation resistance showed no correlation with latitude for the South American cline (Robinson et al. 2000), suggesting that, although overall size varies, body composition of the flies does not. Moreover, in both the thermal evolution laboratory lines and the populations from the Australian and South American clines, growth efficiency was higher for the populations that had adapted to lower temperatures (Robinson & Partridge 2001). This increased growth efficiency may result from differences in the acquisition and allocation of resources, producing the larger body size and shorter development time in a trade-off against other phenotypic characteristics. Such trade-offs need to be pinned down to specific developmental and physiological mechanisms (see Leroi 2001). Because the *Drosophila* insulin/IGF pathway is now known to regulate growth in relation to food availability (see above), genes associated with this pathway are plausible candidates for variation between lines differing in body size.

QTL mapping approaches have been taken to identify genetic variation associated with the size differences that have evolved among different geographical populations. In the comparison between populations at the two ends of the Australian cline, QTL were found on the second and third chromosomes (Gockel et al. 2002), including a continuous stretch of the third chromosome that contains genes from the insulin/IGF signaling pathway. Previously, five microsatellite loci were shown to vary with latitude (Gockel et al. 2001), but none of these loci was significantly associated with QTL [probably because of the low resolution (see also Merila & Crnokrak 2001)].

In analyzing the lines resulting from artificial selection on wing size, significant QTL were found on all major chromosomes and, again, some of these mapped to positions of genes in the insulin/IGF signaling pathway (B.J. Zwaan, B. Seifeid, G. Gedes & L. Partridge, unpublished results). The asymmetry in the cellular response in this selection experiment is highly relevant, as it might be anticipated that the large and small lines would identify genes with variation affecting cell number and cell size, respectively. Indeed, the QTL for large and small wing size do not map to the same positions. In selection experiments for fitness traits, asymmetrical responses are predicted (Roff 1997), and these may often be caused by variation in different genes contributing to the responses in different directions. It was also found that QTL from the lines selected for high or low bristle number did not correspond (Nuzhdin et al. 1999).

These studies of the evolution of body size in insects have thus explored genetical, developmental, and functional issues using variation among both natural and laboratory populations. They have detected examples of differences in the developmental mechanisms underlying examples of comparable changes in phenotype. Another variable morphological trait that has proved amenable to a multidisciplinary approach to understanding phenotypic diversity is the eyespot patterning on the wings of many species of butterfly.

EVOLUTION OF WING COLOR PATTERN IN BUTTERFLIES

The wings of butterflies and moths display striking, and often very intricate, patterns of colored scales that show great diversity across species and frequently also vary within a species (Nijhout 1991). The ultimate aim of studies on the evolutionary genetics and development of butterfly wing patterns is to understand the processes involved in generating the morphological diversity of color patterns observed in present-day species.

The lepidopteran wing originates as an internal imaginal disc within the larva, and as it grows by cell division and extension in the late larva and pupa, the cells acquire their different developmental fates with respect to subsequent scale formation and pigment synthesis (Nijhout 1991). Rows of specialized scale cells differentiate within the epidermal cell layer that forms each surface of the pupal wing, and each scale cell develops a large protrusion. A color pattern arises at late pupal stage, just before adult eclosion, as the cells at different locations on the wing surface synthesize and deposit different pigments in their scale cuticle. A long tradition of comparative analysis of wing color patterns, especially within the Nymphalid butterflies, has led to the concept of a groundplan (see Beldade & Brakefield 2002; Nijhout 1991, 2001) characterized by pattern elements such as transverse bands, chevrons, and eyespots. There has been much brilliant research on the evolutionary genetics of several types of wing pattern elements, especially those involved in the evolution of mimicry in species of *Papilio* and *Heliconius* (see Joron & Mallet 1998, McMillan et al. 2002, Nijhout 1994b). Unfortunately, current knowledge of development of pattern elements is largely limited to the eyespots (Beldade & Brakefield 2002, Brakefield & French 1999). Each eyespot consists of concentric rings of color, usually surrounding a central pupil located midway between wing veins. Eyespots can function in startling predators (Blest 1957), in deflecting predator attacks away from the vulnerable body (Lyytinen et al. 2003, Wourms & Wasserman 1985), and also in mate choice (Breuker & Brakefield 2002).

Evolutionary research on butterfly eyespots has built on earlier studies in ecological genetics on *Maniola jurtina* (Brakefield 1984, Brakefield & Shreeve 1992, Ford 1964). Information was gathered about patterns of phenotypic variation and its genetic basis, and about consequences for fitness in natural populations. However, as for many comparable studies, this system eventually proved frustrating because the mechanisms by which the phenotypes mapped onto genotypes remained elusive. Study of eyespot development began with Nijhout's (1980) demonstration of the signaling role of the center of a developing eyespot in *Precis* (= *Junonia*) *coenia* (see below). This work has been extended by use of new systems and by combining surgical manipulations with studies of gene expression patterns and the application of the tools of quantitative genetics (Beldade & Brakefield 2002).

Development of Eyespots

Surgical experiments, studies of gene expression, and analyses of wing pattern mutants in *Bicyclus anynana* have suggested that development of the butterfly eyespot proceeds by the initial specification of a central focus, followed by signaling to the surrounding cells and their subsequent synthesis of specific pigments (Brakefield et al. 1996, Brunetti et al. 2001).

Many genes are known to regulate wing development in *Drosophila* and study of their homologues has suggested that several of them have evolved additional functions in eyespot formation in butterflies (Carroll et al. 1994). For example, *Distal-less* (*Dll*) is expressed along the margin and in each subdivision of the wing disc in mid last larval instar (as in *Drosophila*), but then strong expression persists only in groups of cells that correspond to the centers of the future eyespot patterns (Figure 3, see color insert). Hence, focus formation correlates with *Dll* (and *engrailed*) expression and this appears to be established as a response to signals provided by *hedgehog* expression in flanking cells (Keys et al. 1999).

Ablation and transplantation of small regions of early pupal wing epidermis demonstrate that signals from the focus instruct the surrounding cells to form the eyespot pattern (Brakefield & French 1995, French & Brakefield 1995, Nijhout 1980). At this stage, several regulatory genes become expressed in nested rings around the focus, corresponding to the different fates of cells in forming the color pattern (Brunetti et al. 2001). The genes (e.g., *engrailed* and *spalt*) encode transcription factors that may control pigment synthesis, as strikingly illustrated by comparison of the gene expression and eyespot phenotypes of the wild-type and *Goldeneye* mutant of *B. anynana* (Figure 3). In several different butterfly species, the same transcription factors are expressed in the developing eyespot fields, but in different relative spatial domains and different relationships to the eyespot color scheme (Brunetti et al. 2001).

These observations show that genes (e.g., *engrailed*) involved in early establishment of the eyespot foci may also play later roles following focal signaling, and they also indicate a remarkable flexibility in the regulatory interactions downstream of focal signaling. This may have facilitated the diversification in the color composition of eyespots, perhaps following the evolutionary novelty (in some group of basal Lepidoptera) of forming a focus and a simple response, resulting in an undifferentiated spot pattern (Brunetti et al. 2001).

None of the genes known to be expressed in the focus encode an intercellular signal, so there is no direct information on the mechanism of focal signaling. The results of surgical experiments are broadly compatible with the simple gradient model: The focus produces a diffusible morphogen, the declining levels of which specify rings of future pigment synthesis over the surrounding wing epidermis (see Beldade & Brakefield 2002, Nijhout 1990, French & Brakefield 1995). Focal effects can extend for at least 100 cells across the early pupal wing epidermis, however, and this is farther by an order of magnitude than the demonstrated range of any intercellular patterning signal (e.g., in *Drosophila*). Thus, the mechanism by which the focus patterns the entire eyespot is likely to prove more complex than a single, long-range signal.

Linking Genetic Variation and Eyespot Development

Although the study of several spontaneous mutants of *B. anynana* has given information on eyespot development (Brakefield 1998, Brunetti et al. 2001, Monteiro et al. 2003), evolutionary genetical work has focused on the application of artificial selection.

The wild-type dorsal forewing of *B. anynana* has a small anterior and a large posterior eyespot, each with a central white pupil, a broad black disc, and an outer narrow gold ring (Figure 4*B*, see color insert). Directional selection has been applied in both upward and downward directions to several features of the large posterior eyespot. Eyespot size and color composition have both shown progressive responses to selection with heritabilities of approximately 50% (Monteiro et al. 1994, 1997a). The analysis of crosses between lines selected for ventral eyespot size suggest that 5–10 genes are involved in producing highly divergent phenotypes (Wijngaarden & Brakefield 2000). These observations all show that eyespot size and coloring behave as classic morphometric traits, but one other eyespot feature does behave differently in terms of genetic variation: Shape shows much lower heritability in lines selected for eyespots ellipsoidal in either of their axes

(Monteiro et al. 1997b,c). This result may indicate some developmental constraint on producing asymmetry in the focal signal or the epidermal response.

There is very little genetic correlation between eyespot size and color composition, as pairs of lines obtained by artificial selection had diverged only for the target feature of the eyespot. Although these features show closely similar genetic properties, reciprocal transplantations demonstrate a clear difference in their developmental basis (Beldade & Brakefield 2002; Monteiro et al. 1994, 1997a). When a focus is grafted ectopically between pupae from different selected lines, the size of the resulting eyespot is largely dependent on the identity of the donor (i.e., of the grafted focus), whereas its color composition depends on the identity of the responding host animal. Thus differences in eyespot size are attributable mostly to changes in properties of the focal signal, whereas those in color depend entirely on the sensitivity thresholds of the responding cells.

Linking Development to Specific Allelic Variation

Studies of gene expression patterns in developing butterfly wings have implicated some genes (e.g., *Dll*) and some developmental pathways (e.g., Hedgehog signaling) in eyespot formation, although direct evidence of their function awaits the use of methods of manipulating gene expression. Studies of gene expression and function cannot, however, identify those genes contributing to phenotypic variation that could be the basis of evolutionary change in eyespot pattern. The crucial issue in evolutionary developmental biology of identifying such genes has now also been examined in *B. anynana*.

Dll expression in late larval and pupal wings is associated with eyespot foci and the expression pattern changes in parallel with shifts in adult eyespot morphology (Brakefield et al. 1996) (Figure 3). In lines of *B. anynana* selected for the size of both dorsal forewing eyespots (see Figure 4*B*), *Dll* expression patterns have also diverged (Beldade et al. 2002a). Such correlations could occur through upstream genetic and developmental changes, but the study by Beldade et al. (2002a) tested whether variation in the *Dll* gene contributes directly to the responses in these selected lines. Informative molecular polymorphisms were identified in this gene and then F2 individuals from crosses between the selected lines were scored for both the parental origin of their *Dll* alleles and their eyespot size. In several crosses there was a clear association between the *Dll* genotype and eyespot phenotype, providing strong evidence that variation mapping to this gene contributes to phenotypic variation of potential relevance to evolutionary change within *B. anynana*. It is probable, but not yet demonstrated, that this variation lies in the *cis*-regulatory regions of the *Dll* gene.

Eyespot Development and the Flexibility of Morphological Change

In *B. anynana*, all developing eyespots have shown the same patterns of gene expression (Brunetti et al. 2001), and both forewing eyespots behave similarly in

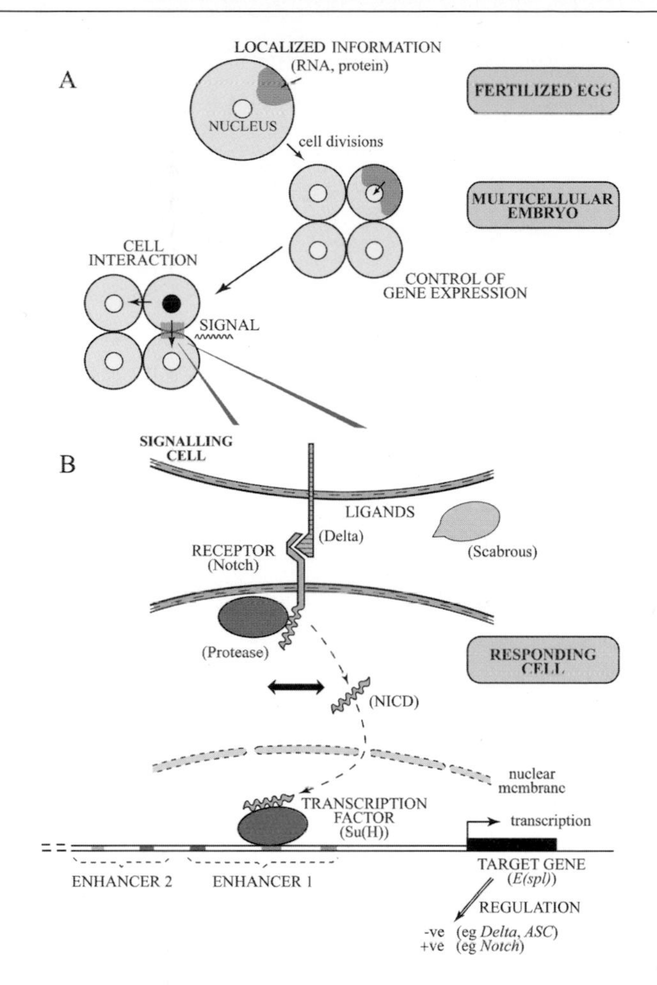

Figure 1 Developmental mechanisms. (*A*) Cartoon illustrating the two major ways in which cellular differences arise within an embryo. Typically, transcripts and/or proteins are localized in the egg cell as it forms in the ovary, and these control gene expression in embryonic cells derived from that portion of egg cytoplasm. Differences are then elaborated through cell signaling; signals may affect only immediate neighbors (as shown) or have different effects, depending on distance. (*B*) Cell signaling, illustrated by the conserved Notch pathway. The Notch receptor is unusual in being able to bind several ligands, including transmembrane Delta and secreted Scabrous, and in responding by being cleaved, generating a free intracellular fragment (Nintra or NICD) that enters the nucleus and activates the transcription factor [Su(H)]. This complex can then bind to specific sites (*color*) on enhancers in the *cis*-regulatory region of target genes [such as *E(spl)*], activating their transcription. In a more typical signaling pathway (e.g., Hedgehog) the bound receptor remains intact and initiates changes in a complex signal transduction pathway. Different signaling pathways interact extensively within the cell (*double-headed arrow*). Notch signaling inhibits neural development in the responding cell, as the transcription factor E(spl) then represses *Achete/Scute* (ASC) and *Delta* genes, while upregulating transcription of *Notch* itself.

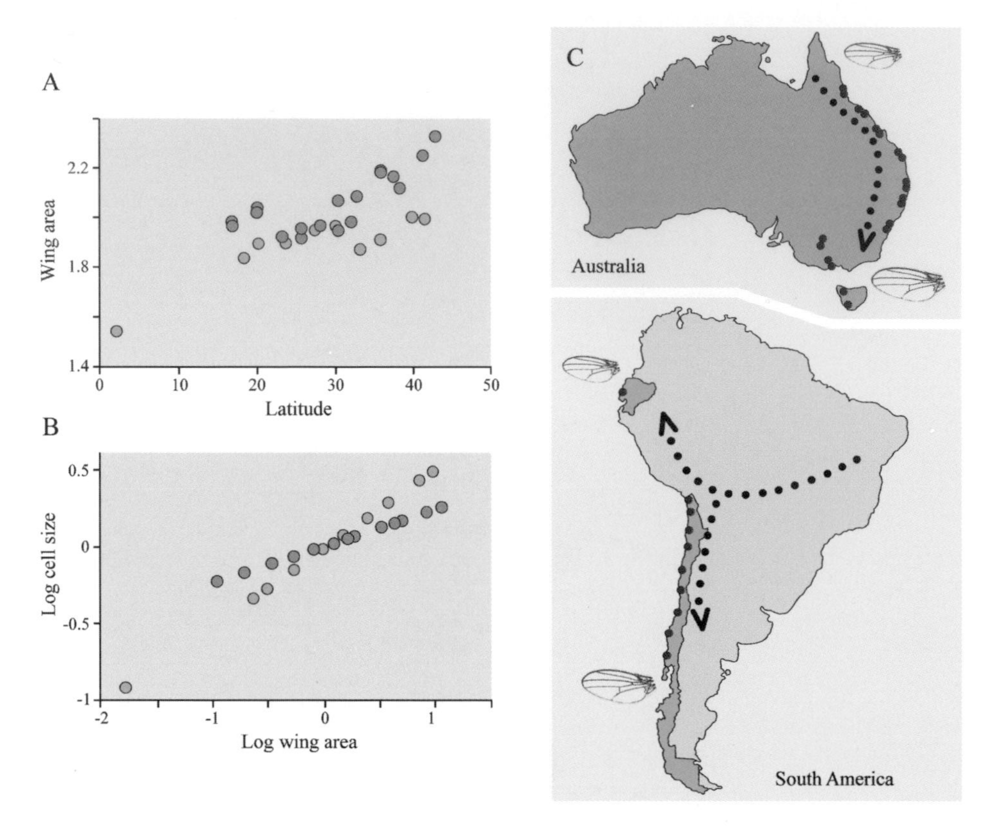

Figure 2 Size variation of *Drosophila melanogaster* along latitudinal clines. (*A*) Strong latitudinal clines for wing size are found on both the Australian (*orange*) and South American (*green*) continents. Flies from the different populations [red circles in (*C*)] were reared under standard temperature conditions. Rearing temperatures were different for the Australian (18°C) and the South American (25°C) populations, causing the general differences in size between continents in (*A*). Therefore, for (*B*), values for the traits were first standardized within continent. (*B*) The slope of the relationship between log(cell size) and log(wing area) is an estimate of the contribution of cell size to variation in wing area. Clearly, in South America (*green*), cell size contributes much more to the latitudinal variation in size than in Australia (*orange*). Log(wing area) and log(cell size) were first individually regressed on latitude and the predicted values saved. These predicted values are plotted in the graph to avoid confounding effects of interpopulation deviation from the clinal relationship. (*C*) The sample sites of *D. melanogaster* populations on each continent are shown as red circles, together with the inferred colonization routes by *D. melanogaster* in black (David & Capy 1988). Data are taken from James et al. (1995) and Zwaan et al. (2000); only those for females are shown, but closely similar patterns were observed for males.

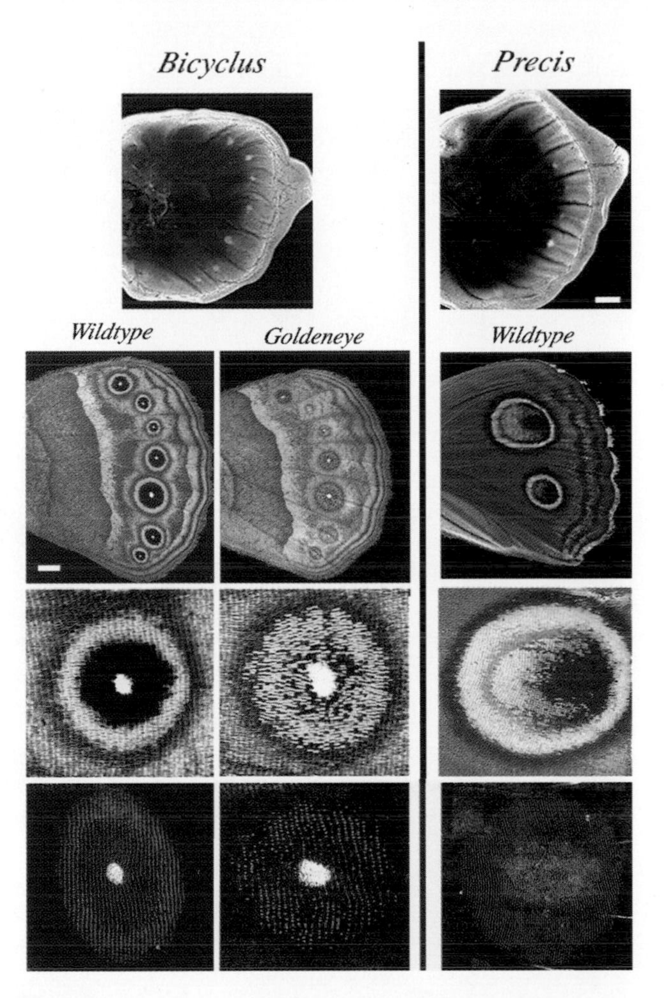

Figure 3 Butterfly eyespot formation and gene expression patterns. Left and right parts show the species *Bicyclus anynana* and *Precis* (= *Junonia*) *coenia*, respectively, with additional images for the *Goldeneye* mutant of *B. anynana*. The top row shows hindwing imaginal discs from wild-type final instar larvae, antibody stained to reveal *Distal-less* (*Dll*) expression (scale bar = 0.4 mm). Note the spots of strong *Dll* expression that correspond to the future signaling foci and the position of eyespots in the adult hindwing (second row, scale bar = 2 mm). The third row shows individual eyespots on the adult hindwing with, below, double labeling 16 h after pupation revealing rings of expression of *engrailed* (*en*)/*invected* (*green*) and *spalt* (*purple*). Both proteins are coexpressed in a central spot (the focus) in *B. anynana*, and the mutant also shows a change in expression corresponding to a near absence of black scales in the adult eyespot. The relationship between the expression of *en* and *spalt* and the scale pigmentation differs across species. Images from Brakefield et al. (1996) and Brunetti et al. (2001) courtesy of Craig Brunetti, Sean Carroll, Julie Gates, Steve Paddock, and Jayne Selegue.

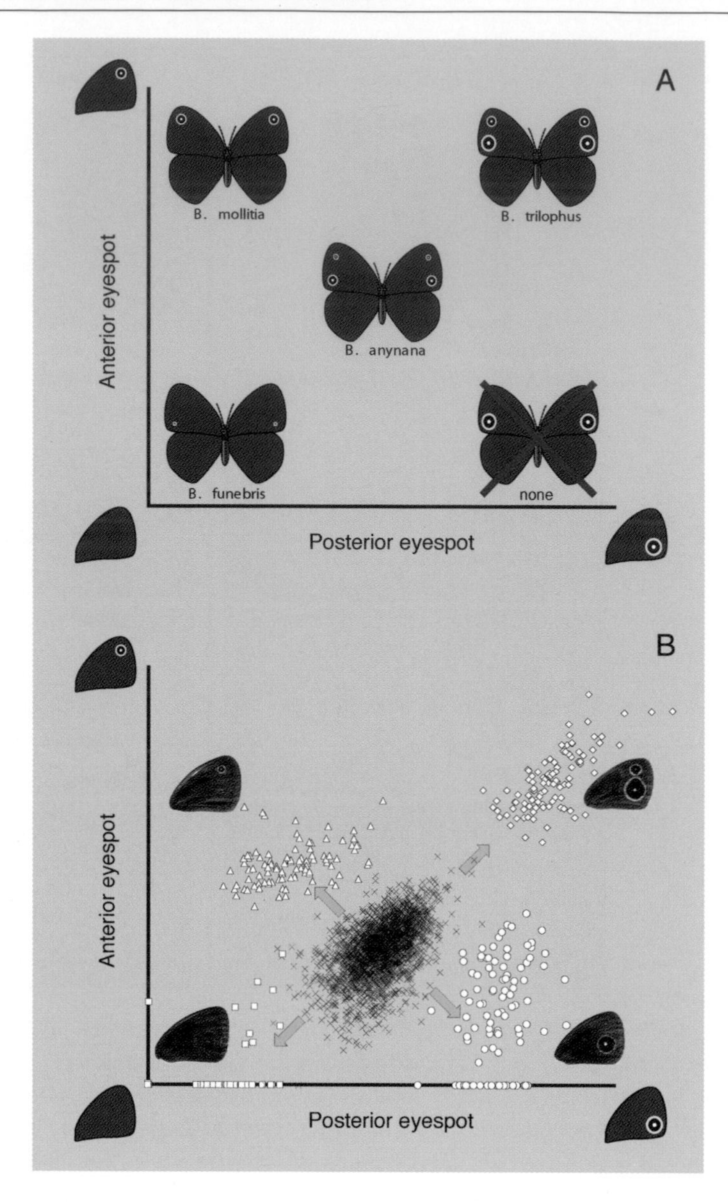

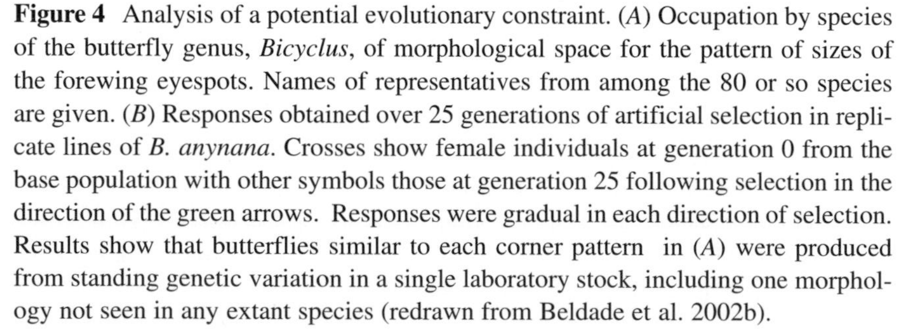

Figure 4 Analysis of a potential evolutionary constraint. (*A*) Occupation by species of the butterfly genus, *Bicyclus*, of morphological space for the pattern of sizes of the forewing eyespots. Names of representatives from among the 80 or so species are given. (*B*) Responses obtained over 25 generations of artificial selection in replicate lines of *B. anynana*. Crosses show female individuals at generation 0 from the base population with other symbols those at generation 25 following selection in the direction of the green arrows. Responses were gradual in each direction of selection. Results show that butterflies similar to each corner pattern in (*A*) were produced from standing genetic variation in a single laboratory stock, including one morphology not seen in any extant species (redrawn from Beldade et al. 2002b).

response to transplantation or ablation experiments (Brakefield & French 1995, French & Brakefield 1995). Furthermore, most single mutations affect all eyespots (Figure 3) (Brakefield 1998), and selection on one specific eyespot results in concerted changes in the other eyespots, especially those on the same wing surface (Beldade et al. 2002b, Beldade & Brakefield, 2003, Monteiro et al. 1994). The developmental coupling of the eyespots suggested that, although parallel changes could be readily accomplished by selection, it might be more difficult to uncouple eyespots or change them in different directions (Brakefield 1998). When this prediction was tested by selection on both the small anterior and large posterior forewing eyespots, however, both coupled and uncoupled changes were readily achieved (Beldade et al. 2002c). Final phenotypes after 25 generations were widely divergent in all directions of selection from the phenotypic range of the original base population (Figure 4*B*). This was especially noteworthy in one of the uncoupled directions (large anterior, small posterior eyespot) because none of the 80 or so extant species of *Bicyclus* show this phenotype (Figure 4*A*). Thus, although this phenotype might be suspected of being a "forbidden morphology," the experimental result indicates that its absence in nature is more likely to have resulted from lack of appropriate natural selection in any lineage.

Although eyespots share a common developmental mechanism, the pattern in relative size of the dorsal forewing eyespots can readily move through morphospace under appropriate selection regimes (Figure 4*B*). This result demonstrates variation in genes that differentially modify the causal mechanism that underlies eyespot formation, giving the eyespots individuality within the overall wing pattern. Analysis of eyespots on the other wing surfaces of the selected butterflies indicates a modular organization of the pattern (Beldade et al. 2002b, Beldade & Brakefield 2003). The ventral hindwing has a full series of seven eyespots, which show a conserved pattern of relative size in wild-type *B. anynana*. In response to the regimes of selection on the dorsal forewing, parallel changes occurred in the eyespot sizes on the ventral hindwing: The four most anterior changed in the direction of the anterior forewing eyespot, whereas the more posterior (especially the most anal two) shifted with the posterior forewing eyespot. Current work is exploring the genetic and molecular basis of this individuality and modularity in eyespot development (Monteiro et al. 2003). The flexibility apparent in the response to selection on eyespot size in *B. anynana* may indicate a long legacy of natural selection favoring the evolution of divergence among the eyespots, or subsets of eyespots, in butterflies (Beldade et al. 2002b).

Beyond the Eyespot

The examination of eyespot development has partly bridged the gap from genes to the eyespot phenotype and its functions in particular environments. We have restricted the discussion of lepidopteran wing patterns to studies concerned with the eyespot because, unfortunately, there is no comparable information about the development of other types of pattern element. From comparative analysis,

Nijhout has argued that signaling foci underlie formation of the colored bands and patches found in the much-studied, polymorphic and mimetic species of *Papilio* and *Heliconius (*Nijhout 1991, 1994b). However, these foci have yet to be demonstrated. In these species, allelic differences at a few major pattern loci give dramatic shifts in wing pattern (for references, see Nijhout 1994b, McMillan et al. 2002). When more is known about the mechanisms by which the *Papilio* and *Heliconius* patterns develop, and when the major loci become amenable to molecular study, we can hope for an exciting and fuller understanding of the adaptive evolution of mimicry. Furthermore, it will be fascinating to be able to compare the genetic and developmental mechanisms of evolutionary change in butterfly wing patterns with those of other morphologies, including bristle patterns in Diptera (Gompel & Carroll 2003, Simpson 2002, Stern 1998, Sucena & Stern 2000, Sucena et al. 2003) and the melanic wing patterns of species of *Drosophila* (True et al. 1999).

PHENOTYPIC PLASTICITY

Phenotypic plasticity is variation across environments in the phenotype developed from a given genotype. Its interest for ecologists and evolutionary biologists lies in being a means of adaptation to divergent environments, and it provides insight into understanding control of the stability of developmental mechanisms (Frankino & Raff 2003, Pigliucci 2001, Schlichting & Pigliucci 1998). From a genetic perspective, there is interest in the possibility of specific regulatory genes that mediate plasticity (genes for plasticity) and in how they might function. Few analyses of phenotypic plasticity in insects have included any examination of how developmental pathways are modulated. Here we discuss studies that illustrate the potential for a multidisciplinary approach, from genetic variation through hormonal regulation of development, to phenotype and function.

A particularly striking mode of phenotypic plasticity is polyphenism, where development can produce discrete, alternative phenotypes. In seasonal polyphenism, changing environmental cues lead to alternative adult phenotypes being produced by generations developing at different times of the year. There are several dramatic examples in butterflies (Shapiro 1976), including *B. anynana* (Brakefield 1997, Brakefield & French 1999, Beldade & Brakefield 2002).

Physiological Control of Polyphenic Development

In *B. anynana*, adults of the wet season form (WSF) have ventral wing surfaces with a pale band and conspicuous eyespots that may function to deflect bird attacks away from the vulnerable body (see Lyytinen et al. 2003). In contrast, dry season form (DSF) butterflies are uniformly brown in color, almost lacking ventral eyespots, and they rely on camouflage among brown leaves for survival. The results of field studies demonstrate that seasonal polyphenism in *Bicyclus* butterflies is adaptive (see Brakefield & French 1999). Field surveys (Brakefield & Mazzotta

1995, Brakefield & Reitsma 1991, Windig et al. 1994), together with controlled rearing experiments (Kooi & Brakefield 1999), reveal that temperature provides the predictable cue for the adult environment. Larvae experiencing high temperature develop as WSF, whereas those in cooler conditions form DSF adults.

Ecdysteroid hormones mediate the development of the seasonal forms in *B. anynana* (Koch et al. 1996), as well as those of some other species of butterfly (Koch 1992). The increase in ecdysteroid titer after pupation occurs at a later stage in pupae of the DSF of *B. anynana* than in those of the WSF. When animals are reared to produce the DSF and then microinjected as young pupae with ecdysteroid, the adult wing pattern is shifted toward the larger eyespots characteristic of the WSF (Koch et al. 1996). Understanding precisely how larval rearing temperature influences the secretion of ecdysteroids, and how the ecdysteroid titer in the early pupa then regulates eyespot development [and only on the ventral wing surface (see Brakefield et al. 1998)], are exciting challenges for the future.

The butterfly, *Precis* (=*Junonia*) *coenia*, also shows seasonal polyphenism, producing two forms differing in dorsal wing color. Rountree & Nijhout (1995a) found that response to photoperiod in the larval stage leads to differences in ecdysteroid titer in early pupae, and then to the divergent adult phenotypes. They also isolated a mutant that constitutively expresses only one phenotype (Rountree & Nijhout 1995b) and showed that the gene does not affect the endocrine system but alters the developmental response to the hormone in early pupae. Variation in the mode of control of such patterns of polyphenism is characteristic of perhaps the most intensively examined system, that involving variation in wing development in crickets (for general review, see Zera & Harshman 2001).

In the sand cricket, *Gryllus firmus*, there are two distinct phenotypes, a long-winged morph capable of flight (LWM) and a short-winged, flightless morph (SWM). The LWM has well-developed flight muscles, high lipid reserves, and underdeveloped ovaries, whereas in the SWM wing muscles are underdeveloped and there are low fat reserves but large ovaries. This polyphenism is thought to be a functional response to conflicting life history demands: reproduction or dispersal (Zera & Denno 1997). In different species and sometimes within species, the alternative phenotypes can be caused directly by genetic polymorphism, induced exclusively by environmental cues (such as crowding, temperature, photoperiod) or involve both mechanisms (Zera & Denno 1997). Although the morphs of *G. firmus* are discrete in phenotype, expression of wing morph is under polygenic control (Roff & Fairbairn 1999). Juvenile hormone (JH) is a key gonadotropic hormone in insects (Nijhout 1994a), and the differences in ovarian (and wing muscle) development between morphs may be caused by differences in JH titer (Zera & Cisper 2001, Zera & Huang 1999, Zera et al. 1998). In addition, ecdysteroid titers in the first week of adulthood are significantly higher in the SWM than in the LWM (Zera & Bottsford 2001). The dynamics and interaction between the hormones is complex, and more experimental work is needed to fully understand morph expression in this system (Fairbairn & Roff 1999, Zera 1999).

Within adults of both morphs, ovary and muscle size respond to feeding conditions (Roff & Gelinas 2003), and some LWM females change their phenotype through histolysis of their flight muscles and increase their ovarian growth and fecundity. Such histolyzed phenotypes can also be induced by topical application of a JH analogue (Zera & Cisper 2001), suggesting that there may be a common hormonal control of these suites of traits and that perhaps the same genes are responding both in juvenile development and in adult change.

Hormonal regulation of polyphenic variation has also been investigated in numerous other insects including species of aphid and hymenoptera (see Nijhout 1994a, 1999; Abouheif & Wray 2002). There have, however, been few studies in which geographical variation in polyphenism has also been examined. One such system is the development of horns in dung beetles, including the species *Onthophagus taurus*. Male beetles vary continuously in body size as a function of larval feeding conditions (Moczek 1998), and only males that exceed a critical size eventually form a pair of long, curved horns on their adult heads (Moczek & Emlen 1999). Smaller beetles, including females, remain almost hornless. Horns are used to prevent access to females, and the hornless males adopt different tactics to acquire matings (Emlen 1997, Moczek & Emlen 2000). Experiments involving the application of JH indicate that the horn dimorphism is controlled by threshold responses to JH titer at sensitive periods during the last larval instar (Emlen & Nijhout 1999, 2001).

O. taurus is endemic to the Mediterranean basin but was introduced in the 1960s to the eastern United States and, as part of a program to control cow dung, to Australia (Moczek & Nijhout 2002, Moczek et al. 2002). These introduced populations have now diverged in the critical body size above which horns are developed in the male beetles, probably in response to differences in ecology and levels of competition (Hunt & Simmons 1998). Evolution of this difference has been shown to involve both a change in sensitivity of the head tissue to JH and a change in the timing of the sensitive period during which the hormone titer is monitored (Moczek & Nijhout 2002, Moczek et al. 2002). It will indeed be exciting if the relevant genetic differences between the populations can be identified and to determine whether they are the same as those that underlie similar changes in critical size produced under artificial selection in a related beetle (Emlen 1996). Furthermore, there is an exciting challenge to be met in determining whether the hormone response can be linked directly to the developmental mechanisms of horn formation.

Genetics and Development of Polyphenism

Artificial selection has been used in *B. anynana* to survey genetic variation available for the evolution of phenotypic plasticity in wing pattern. Although field populations show classical seasonal polyphenism with rather discrete phenotypes (Windig et al. 1994), laboratory experiments demonstrate that the underlying reaction norms are continuous in form (Brakefield & Mazzotta 1995). Quantitative variation in ventral eyespot size at a single rearing temperature has provided the basis for artificial selection on plasticity. In general, the response is rapid and heritabilities are high, and there are positive genetic covariances between the target and the other ventral eyespots, and for the same eyespot across rearing temperatures (Holloway & Brakefield 1995, Holloway et al. 1993). In selection experiments that progressively increased (low line) or decreased (high line) rearing temperatures over the generations, the high line eventually developed the WSF phenotype across all temperatures, although plasticity remained (with high temperatures giving larger adult eyespots). In sharp contrast, the low line produced only butterflies lacking eyespots (DSF) at all temperatures (Brakefield et al. 1996).

Surveys of hormone titers indicate that these selected lines show a difference in the timing of the pupal ecdysteroid peak, similar to that seen between the unselected stock when reared at high and low temperature (Koch et al. 1996, Brakefield et al. 1998). The different seasonal phenotypes, whether genetically or environmentally determined, are associated with a divergent pattern of *Dll* gene expression in the wings of one-day-old pupae (Brakefield et al. 1996). This follows the time when surgical experiments indicate that the focus signals to specify the eyespot pattern.

Following selection on the ventral eyespot pattern in *B. anynana*, pairs of lines diverged (sometimes dramatically) in the elevation of their reaction norms. A response in the degree of phenotypic plasticity will, however, necessitate a change in reaction norm shape, which is only possible where there is genotype by environment interaction. Recent experiments targeting shape per se failed to yield either substantially steeper or shallower reaction norms, or ones of divergent shape (Wijngaarden & Brakefield 2001, Wijngaarden et al. 2002). Extreme changes in reaction norm elevation can evolve rapidly but, presumably owing to positive genetic covariances across environments, the same is not true for shape. It is not clear why this should be (see Wijngaarden et al. 2002), but interestingly, a recent genetic analysis of plasticity of wing and ovary size in relation to food supply in the cricket, *Gryllus assimilus*, has also indicated genetic variation mainly in the elevation and not in the slope of the relationship (Roff & Gelinas 2003, see also Emlen 1996).

For many of these examples of plasticity in insects, we need a deeper understanding of the developmental mechanisms—of the ways in which environmental stimuli and/or genetic variation influence the changing hormone concentrations and tissue sensitivities, and how these then regulate formation of various aspects of the phenotype (Brakefield et al. 1998, Nijhout 1999). It may then become clearer how particular instances of polyphenism and phenotypic plasticity have arisen and why they may be able to evolve more readily in some directions than in others.

DISCUSSION

Here, we discuss three areas of interest that follow on from the case studies we have reviewed and that represent future aims of, and challenges for, evolutionary developmental biology.

First, mapping of quantitative trait loci (QTL) has already provided a rich source of data on the genetics of complex traits (Mackay 1995, 2001). We review several of the studies of morphological traits that have begun to associate QTL variation with genes of known function in the development of the trait. Such an integration of data from gene mapping with known developmental pathways should gather pace as more systems become accessible to developmental investigation, gene expression studies, and genome-wide screens. This will give a more detailed mechanistic basis for properties of genes and genetic pathways, including epistasis and pleiotropy. Looking further ahead, a fusion of evolutionary developmental biology, quantitative genetics, and evolutionary functional genomics should provide a much deeper understanding of how functional phenotypes map onto genotypes in a variety of organisms. As this knowledge grows, mathematical modeling will become increasingly attractive as an approach to explore further the evolutionary dynamics of known gene networks and developmental processes (Rice 2002). Nijhout & Paulsen (1997) provided an early example based loosely on eyespot formation in butterflies, involving a one-dimensional diffusion gradient and threshold model and incorporating genetic variation for six developmental parameters. Although the biology was necessarily oversimplified, observations of emergent properties of the model illustrate the potential power of such approaches.

Second, complex morphological traits have numerous phenotypic dimensions. For such traits one can ask whether the generation of evolutionarily relevant variation in different features (e.g., size, pattern, or shape) of the trait is under the control of the same set of genes or whether different genes are involved. Similarly, are essentially independent developmental pathways involved, or is a single pathway regulated in a modular manner? Such ideas are implicit in thinking about developmental flexibility and its relationship to evolvability, or the capacity of a lineage to evolve (Kirschner & Gerhardt 1998, Leroi 2000, Wilkins 2002). Again, an increasingly detailed understanding of the genes and developmental processes involved in generating specific morphological and other phenotypic traits will enable increasingly robust predictions about the evolutionary dynamics of such modular structures (Wagner 1996, Wagner & Altenberg 1996). This goal is perhaps closest for certain morphological patterns comprising iterative repeats of a unit element such as the butterfly eyespot or *Drosophila* bristle. Two other examples from case studies we have covered in this review illustrate how developmental insights are likely to uncover important properties of evolvability. The two features of a butterfly eyespot, size and color, behave similarly with respect to the short-term responses to artificial selection, but their underlying development is very different (see Beldade & Brakefield 2002). Thus, although the estimates of genetic variances alone would lead to similar predictions about evolvability, the developmental changes may suggest different dynamics, at least for long-term responses to selection. Similarly, the asymmetry evident in the developmental mechanisms underlying selection responses for increased or decreased body size in *Drosophila* is also likely to have consequences for patterns of evolutionary change in this trait in natural populations.

Third, we have focused on case studies that reveal both genetic and developmental mechanisms of microevolutionary change, rather than on those underlying divergence across species or among phylogenetically distant lineages. Indeed, researchers who compare convergent patterns of morphological evolution (both loss and gain of structures) across related species or groups of species are also beginning to explore the predictability of the underlying changes in development (e.g., Abouheif & Wray 2002, Gompel & Carroll 2003, Kopp et al. 2000, Stern 1998, Sucerna et al. 2003, True et al. 1999). There is, however, an additional issue: Do the genetic changes that occur in such microevolutionary responses to selection involve essentially the same sets of genes and developmental pathways as those that underlie more macroevolutionary patterns of divergence? The examples we have discussed begin to suggest that some, perhaps most or even all, of the evolutionary change in morphology observed within species also maps to genes with central developmental functions, and perhaps to their regulatory elements (see also Stern 2000).

CONCLUDING REMARKS

Do ecologists and evolutionary biologists interested in adaptive evolution need to address developmental mechanisms at all, and if so, why? A more explicit way of formulating this question is to ask whether an understanding of the role of development in generating evolutionarily relevant variation in the phenotype can provide the potential for improving predictions of evolutionary change. We believe that further multidisciplinary analysis of examples of phenotypic variation of the type we have discussed will demonstrate that developmental, as well as genetic, insights are indeed important for predicting the paths of adaptive evolution and for understanding properties of evolvability. In particular, phenotypes map onto genotypes through the mechanisms of development, and study of the properties of development will yield additional information that is not available from estimates of genetic variances and covariances alone.

ACKNOWLEDGMENTS

We thank Tony Frankino for his helpful comments on the manuscript, and Martin Brittijn, Steve Paddock, and Steve Thurston for figures.

The *Annual Review of Ecology, Evolution, and Systematics* is online at http://ecolsys.annualreviews.org

LITERATURE CITED

Abouheif E, Wray GA. 2002. Evolution of the gene network underlying wing polyphenism in ants. *Science* 297:249–52

Arthur W. 2002. The emerging conceptual framework of evolutionary developmental biology. *Nature* 415:757–64

Arthur W, Kettle C. 2001. Geographic patterning of variation in segment number in geophilomorph centipedes: clines and speciation. *Evol. Dev.* 3:34–40

Atkinson D. 1994. Temperature and organism size—a biological law for ectotherms. *Adv. Ecol. Res.* 25:1–58

Atkinson D, Sibly RM. 1997. Why are organisms usually bigger in colder environments? Making sense of a life history puzzle. *Trends Ecol. Evol.* 12:235–39

Azevedo RBR, French V, Partridge L. 2002. Temperature modulates epidermal cell size in *Drosophila melanogaster*. *J. Insect Physiol.* 48:231–37

Barton N, Partridge L. 2000. Limits to natural selection. *BioEssays* 22:1075–84

Beldade P, Brakefield PM. 2002. The genetics and evo-devo of butterfly wing patterns. *Nat. Rev. Genet.* 3:442–52

Beldade P, Brakefield PM. 2003. Concerted evolution and developmental integration in modular butterfly wing patterns. *Evol. Dev.* 5:169–79

Beldade P, Brakefield PM, Long AD. 2002a. Contribution of *Distal-less* to quantitative variation in butterfly eyespots. *Nature* 415:315–18

Beldade P, Koops K, Brakefield PM. 2002b. Developmental constraints versus flexibility in morphological evolution. *Nature* 416:844–47

Beldade P, Koops K, Brakefield PM. 2002c. Modularity, individuality, and evo-devo in butterfly wings. *Proc. Natl. Acad. Sci. USA* 99:14262–67

Berreur P, Porcheron P, Berreur-Bonnenfant J, Simpson P. 1979. Ecdysteroid levels and pupariation in *Drosophila melanogaster*. *J. Exp. Zool.* 210:347–52

Blest AD. 1957. The function of eyespot patterns in the Lepidoptera. *Behaviour* 11:209–56

Bohni R, Riesgo Escovar J, Oldham S, Brogiolo W, Stocker H, et al. 1999. Autonomous control of cell and organ size by CHICO, a *Drosophila* homolog of vertebrate IRS1-4. *Cell* 97:865–75

Brakefield PM. 1984. The ecological genetics of quantitative characters of *Maniola jurtina* and other butterflies. In *The Biology of Butterflies*, pp. 167–90. London: Academic

Brakefield PM. 1997. Phenotypic plasticity and fluctuating asymmetry as responses to environmental stress in the butterfly *Bicyclus anynana*. In *Environmental Stress, Adaptation and Evolution*, ed. R Bijlsma, V Loeschcke, pp. 65–78. Basel: Birkhäuser Verlag

Brakefield PM. 1998. The evolution-development interface and advances with the eyespot patterns of *Bicyclus* butterflies. *Heredity* 80:265–72

Brakefield PM, French V. 1995. Eyespot development on butterfly wings: the epidermal response to damage. *Dev. Biol.* 168:98–111

Brakefield PM, French V. 1999. Butterfly wings: the evolution of development of colour patterns. *BioEssays* 21:391–401

Brakefield PM, Gates J, Keys D, Kesbeke F, Wijngaarden PJ, et al. 1996. Development, plasticity and evolution of butterfly wing patterns. *Nature* 384:236–42

Brakefield PM, Kesbeke F, Koch PB. 1998. The regulation of phenotypic plasticity of eyespots in the butterfly *Bicyclus anynana*. *Am. Nat.* 152:853–60

Brakefield PM, Mazzotta V. 1995. Matching field and laboratory environments: effects of neglecting daily temperature variation in insect reaction norms. *J. Evol. Biol.* 8:559–73

Brakefield PM, Reitsma N. 1991. Phenotypic plasticity, seasonal climate and the population biology of *Bicyclus* butterflies. *Ecol. Entomol.* 16:291–303

Brakefield PM, Shreeve T. 1992. Case studies in evolution. In *The Ecology of Butterflies in Britain*, ed. RLH Dennis, pp. 197–216. Oxford: Oxford Univ. Press

Breuker CJ, Brakefield PM. 2002. Female choice depends on size but not symmetry of dorsal eyespots in the butterfly *Bicyclus anynana*. *Proc. R. Soc. London Ser. B* 269:1233–39

Brogiolo W, Stocker H, Ikeya T, Rintelen F, Fernandez R, Hafen E. 2001. An evolutionarily conserved function of the *Drosophila* insulin receptor and insulin-like peptides in growth control. *Curr. Biol.* 11:213–21

Brunetti CR, Selegue JE, Monteiro A, French V, Brakefield PM, Carroll SB. 2001. The generation and diversification of butterfly eyespot color patterns. *Curr. Biol.* 11:1578–85

Bryant PJ, Levinson P. 1985. Intrinsic growth control in the imaginal primordia of *Drosophila*, and the autonomous action of a lethal mutation causing overgrowth. *Dev. Biol.* 107:355–63

Bryant PJ, Simpson P. 1984. Intrinsic and extrinsic control of growth in developing organs. *Q. Rev. Biol.* 59:387–415

Calboli FCF, Gilchrist GW, Partridge L. 2003. Different cell size and cell number contribution in two newly established and one ancient body size cline of *Drosophila subobscura*. *Evolution* 57:566–73

Carroll SB, Gates J, Keys DN, Paddock SW, Panganiban GEF, et al. 1994. Pattern formation and eyespot determination in butterfly wings. *Science* 265:109–14

Carroll SB, Grenier JK, Weatherbee SD. 2001. *From DNA to Diversity. Molecular Genetics and the Evolution of Animal Design*. Oxford: Blackwell

Cavicchi S, Guerra D, Giorgi G, Pezzoli C. 1985. Temperature divergence in experimental populations of *Drosophila melanogaster*. I. Genetic and developmental basis of wing size and shape variation. *Genetics* 109:665–89

Cavicchi S, Guerra D, Natali V, Pezzoli C, Giorgi G. 1989. Temperature-related divergence in experimental populations of *Drosophila melanogaster*. II. Correlation between fitness and body dimensions. *J. Evol. Biol.* 2:235–51

Conlon I, Raff M. 1999. Size control in animal development. *Cell* 96:235–44

Coyne JA, Beecham E. 1987. Heritability of two morphological characters within and among natural populations of *Drosophila melanogaster*. *Genetics* 117:727–37

D'Amico LJ, Davidowitz G, Nijhout HF. 2001. The developmental and physiological basis of body size evolution in an insect. *Proc. R. Soc. London Ser. B* 268:1589–93

David JR, Capy P. 1988. Genetic variation of *Drosophila melanogaster* natural populations. *Trends Genet.* 4:106–11

Davidson EH. 2001. *Genomic Regulatory Systems. Development and Evolution*. San Diego: Academic

Dilda CL, Mackay TFC. 2002. The genetic architecture of Drosophila sensory bristle number. *Genetics* 162:1655–74

Edgar BA, Lehner CF. 1996. Developmental control of cell cycle regulators: a fly's perspective. *Science* 274:1646–52

Emlen DJ. 1996. Artificial selection on horn length-body size allometry in the horned beetle *Onthophagus acuminatus* (Coleoptera: Scarabaeidae). *Evolution* 50:1219–30

Emlen DJ. 1997. Alternative reproductive tactics and male-dimorphism in the horned beetle *Onthophagus acuminatus* (Coleoptera: Scarabaeidae). *Behav. Ecol. Sociobiol.* 141:335–41

Emlen DJ, Nijhout HF. 1999. Hormonal control of male horn length dimorphism in the dung beetle *Onthophagus taurus* (Coleoptera: Scarabaeidae). *J. Insect Physiol.* 45:45–53

Emlen DJ, Nijhout HF. 2001. Hormonal control of male horn length dimorphism in *Onthophagus taurus* (Coleoptera: Scarabaeidae): a second critical period of sensitivity to juvenile hormone. *J. Insect Physiol.* 47:1045–54

Endler JA. 1977. *Geographic Variation, Speciation, and Clines*. Princeton, NJ: Princeton Univ. Press

Fairbairn DJ, Roff DA. 1999. The endocrine genetics of wing polymorphism in *Gryllus*. A response to Zera. *Evolution* 53:977–79

Felix MA, De Ley P, Sommer RJ, Frisse L, Nadler SA, et al. 2000. Evolution of vulva development in the Cephalobina (Nematoda). *Dev. Biol.* 221:68–86

Ford EB. 1964. *Ecological Genetics*. London: Chapman & Hall

Frankino WA, Raff RA. 2003. Evolutionary importance and patterns of phenotypic plasticity: insights gained from development. In *Phenotypic Plasticity, Functional and Conceptual Approaches*, ed. TJ DeWitt,

SM Scheiner. Oxford: Oxford Univ. Press. In press

French V, Brakefield PM. 1995. Eyespot development on butterfly wings: the focal signal. *Dev. Biol.* 168:112–23

Ghysen A, Dambly-Chaudiere C. 1988. From DNA to form—the achaete-scute complex. *Genes Dev.* 2:495–501

Gockel J, Kennington WJ, Hoffmann A, Goldstein DB, Partridge L. 2001. Nonclinality of molecular variation implicates selection in maintaining a morphological cline of *Drosophila melanogaster*. *Genetics* 158:319–23

Gockel J, Robinson SJW, Kennington WJ, Goldstein DB, Partridge L. 2002. Quantitative genetic analysis of natural variation in body size in *Drosophila melanogaster*. *Heredity* 89:145–53

Gomez-Skarmeta JL, Rodriguez I, Martinez C, Culi J, Ferresmarco D, et al. 1995. Cis-regulation of achaete and scute—shared enhancer-like elements drive their coexpression in proneural clusters of the imaginal discs. *Genes Dev.* 9:1869–82

Gompel N, Carroll SB. 2003. Genetic mechanisms and constraints governing the evolution of correlated traits in drosophilid flies. *Nature.* 424:931–35

Guarente L, Kenyon C. 2000. Genetic pathways that regulate ageing in model organisms. *Nature* 408:255–62

Gurganus MC, Fry JD, Nuzhdin SV, Pasyukova EG, Lyman RF, Mackay TFC. 1998. Genotype-environment interaction at quantitative trait loci affecting sensory bristle number in *Drosophila melanogaster*. *Genetics* 149:1883–98

Gurganus MC, Nuzhdin SV, Leips JW, Mackay TFC. 1999. High-resolution mapping of quantitative trait loci for sternopleural bristle number in *Drosophila melanogaster*. *Genetics* 152:1585–604

Haag ES, True JR. 2001. Perspective: From muntants to mechanisms? Assessing the candidate gene paradigm in evolutionary biology. *Evolution* 55:1077–84

Holloway G, Brakefield PM. 1995. Artificial selection of reaction norms of wing pattern elements in *Bicyclus anynana*. *Heredity* 74:91–99

Holloway GJ, Brakefield PM, Kofman S. 1993. The genetics of wing pattern elements in the polyphenic butterfly, *Bicyclus anynana*. *Heredity* 70:179–86

Hunt J, Simmons LW. 1998. Patterns of parental provisioning covary with male morphology in a horned beetle (*Onthophagus taurus*) (Coleoptera: Scarabaeidae). *Behav. Ecol. Sociobiol.* 42:447–51

Imasheva AG, Bubli OA, Lazebny OE. 1994. Variation in wing length in Eurasian natural populations of *Drosophila melanogaster*. *Heredity* 72:508–14

James AC, Azevedo RBR, Partridge L. 1995. Cellular basis and developmental timing in a size cline of *Drosophila melanogaster*. *Genetics* 140:659–66

James AC, Azevedo RBR, Partridge L. 1997. Genetic and environmental responses to temperature of *Drosophila melanogaster* from a latitudinal cline. *Genetics* 146:881–90

Johnston LA, Prober DA, Edgar BA, Eisenman RN, Gallant P. 1999. *Drosophila myc* regulates cellular growth during development. *Cell* 98:779–90

Joron M, Mallet JLB. 1998. Diversity in mimicry: paradox or paradigm? *Trends Ecol. Evol.* 13:461–66

Keys DN, Lewis DL, Selegue JE, Pearson BJ, Goddrich LV, et al. 1999. Recruitment of a *hedgehog* regulatory circuit in butterfly eyespot evolution. *Science* 283:532–34

Kirschner M, Gerhart J. 1998. Evolvability. *Proc. Natl. Acad. Sci. USA* 95:8420–27

Koch PB. 1992. Seasonal polyphenism in butterflies—a hormonally controlled phenomenon of pattern formation. *Zool. J. Physiol.* 96:227–40

Koch PB, Brakefield PM, Kesbeke F. 1996. Ecdysteroids control eyespot size and wing colour pattern in the polyphenic butterfly *Bicyclus anynana* (Lepidoptera: Satyridae). *J Insect Physiol.* 42:223–30

Kooi RE, Brakefield PM. 1999. The critical period for wing pattern induction in the polyphenic tropical butterfly *Bicyclus anynana* (Satyrinae). *J. Insect Physiol.* 45:201–12

Kopp A, Duncan I, Carroll, S. 2000. Genetic control and evolution of sexually dimorphic characters in *Drosophila*. *Nature* 408:553–59

Kopp A, True JR. 2002. Evolution of male sexual characters in the oriental *Drosophila melanogaster* species group. *Evol. Dev.* 4:278–91

Leevers SJ. 2001. Growth control: Invertebrate insulin surprises! *Curr. Biol.* 11:R209–12

Leevers SJ, Weinkove D, MacDougall LK, Hafen E, Waterfield MD. 1996. The *Drosophila* phosphoinositide 3-kinase Dp110 promotes cell growth. *EMBO J.* 15:6584–94

Leroi AM. 2000. The scale independence of evolution. *Evol. Dev.* 2:67–77

Leroi AM. 2001. Molecular signals versus the *loi de balancement*. *Trends Ecol. Evol.* 16:24–29

Lindsley DL, Zimm GG. 1992. *The genome of* Drosophila melanogaster. San Diego: Academic

Long AD, Lyman RF, Langley CH, Mackay TFC. 1998. Two sites in the *Delta* gene region contribute to natural occurring variation in bristle number in *Drosophila melanogaster*. *Genetics* 149:999–1017

Long AD, Lyman RF, Morgan AH, Langley CH, Mackay TFC. 2000. Both naturally occurring insertions of transposable elements and intermediate frequency polymorphisms at the *achaete-scute* complex are associated with variation in bristle number in *Drosophila melanogaster*. *Genetics* 154:1255–69

Long AD, Mullaney SL, Mackay TFC, Langley CH. 1996. Genetic interactions between naturally occurring alleles at quantitative trait loci and mutant alleles at candidate loci affecting bristle number in *Drosophila melanogaster*. *Genetics* 144:1497–510

Long AD, Mullaney SL, Reid LA, Fry JD, Langley CH, Mackay TFC. 1995. High resolution mapping of genetic factors affecting abdominal bristle number in *Drosophila melanogaster*. *Genetics* 139:1273–91

Lyman RF, Lai C, Mackay TFC. 1999. Linkage disequilibrium mapping of molecular polymorphisms at the *scabrous* locus associated with naturally occurring variation in bristle number in *Drosophila melanogaster*. *Genet. Res.* 74:303–11

Lyman RF, Lawrence F, Nuzhdin S, Mackay TFC. 1996. Effects of single *P*-element insertions on bristle number and viability in *Drosophila melanogaster*. *Genetics* 143:277–92

Lynch M, Walsh B. 1998. *Genetics and Analysis of Quantitative Traits*. Sunderland: MA: Sinauer

Lyytinen A, Brakefield PM, Mappes J. 2003. Significance of butterfly eyespots as an antipredator device in ground-based and aerial attacks. *Oikos* 700:373–79

Mackay TFC. 1995. The genetic basis of quantitative variation: number of sensory bristles of *Drosophila melanogaster* as a model system. *Trends Genet.* 11:464–70

Mackay TFC. 2001. Quantitative trait loci in *Drosophila*. *Nat. Rev. Genet.* 2:11–20

Mackay TFC, Langley CH. 1990. Molecular and phenotypic variation in the *achaete-scute* region of *Drosophila melanogaster*. *Nature* 348:64–66

McMillan WO, Monteiro A, Kapan DD. 2002. Development and evolution on the wing. *Trends Ecol. Evol.* 17:125–33

Merila J, Crnokrak P. 2001. Comparison of genetic differentiation at marker loci and quantitative traits. *J. Evol. Biol.* 14:892–903

Moczek AP. 1998. Horn polyphenism in the beetle *Onthophagous taurus*: larval diet quality and plasticity in parental investment determine adult body size and male horn morphology. *Behav. Ecol.* 9:636–41

Moczek AP, Emlen DJ. 1999. Proximate determination of male horn dimorphism in the beetle *Onthophagous taurus* (Coleoptera: Scarabaeidae). *J. Evol. Biol.* 12:27–37

Moczek AP, Emlen DJ. 2000. Male horn dimorphism in the scarab beetle, *Onthophagus taurus*: do alternative reproductive tactics favour alternative phenotypes? *Anim. Behav.* 59:459–66

Moczek AP, Hunt J, Emlen DJ, Simmons LW. 2002. Threshold evolution in exotic populations of a polyphenic beetle. *Evol. Ecol. Res.* 4:587–601

Moczek AP, Nijhout HF. 2002. Developmental mechanisms of threshold evolution in a polyphenic beetle. *Evol. Dev.* 4:252–64

Monteiro AF, Brakefield PM, French V. 1994. The evolutionary genetics and developmental basis of wing pattern variation in the butterfly *Bicyclus anynana*. *Evolution* 48:1147–57

Monteiro A, Brakefield PM, French V. 1997a. Butterfly eyespots: the genetics and development of the color rings. *Evolution* 51:1207–16

Monteiro A, Brakefield PM, French V. 1997b. The genetics and development of an eyespot pattern in the butterfly *Bicyclus anynana*: response to selection for eyespot shape. *Genetics* 146:287–94

Monteiro A, Brakefield PM, French V. 1997c. The relationship between eyespot shape and wing shape in the butterfly *Bicyclus anynana*: a genetic and morphometrical approach. *J. Evol. Biol.* 10:787–802

Monteiro A, Prijs J, Bax M, Hakkart T, Brakefield PM. 2003. Mutants highlight the modular control of butterfly eyespot patterns. *Evol. Dev.* 5:180–87

Nijhout HF. 1980. Pattern formation on lepidopteran wings: determination of an eyespot. *Dev. Biol.* 80:267–74

Nijhout HF. 1990. A comprehensive model for colour pattern formation in butterflies. *Proc. R. Soc. London Ser. B* 239:81–113

Nijhout HF. 1991. *The Development and Evolution of Butterfly Wing Patterns*. Washington, DC: Smithson. Inst. Press

Nijhout HF. 1994a. *Insect Hormones*. Princeton, NJ: Princeton Univ. Press

Nijhout HF. 1994b. Developmental perspectives on evolution of butterfly mimicry. *BioScience* 44:148–57

Nijhout HF. 1999. Control mechanisms of polyphenic development in insects. *BioScience* 49:181–92

Nijhout HF. 2001. Elements of butterfly wing patterns. *J. Exp. Zool.* 291:213–25

Nijhout HF, Paulsen SM. 1997. Developmental models and polygenic characters. *Am. Nat.* 149:394–405

Nuzhdin SV, Dilda CL, Mackay TFC. 1999. The genetic architecture of selection response: inferences from fine-scale mapping of bristle number quantitative trait loci in *Drosophila melanogaster*. *Genetics* 153:1317–31

Oldham S, Bohni R, Stocker H, Brogiolo W, Hafen E. 2000. Genetic control of size in *Drosophila*. *Proc. R. Soc. London Ser. B* 355:945–52

Parichy DM. 1996. Salamander pigment patterns: How can they be used to study developmental mechanisms and their evolutionary transformation? *Int. J. Dev. Biol.* 40:871–84

Partridge L, Barrie B, Fowler K, French V. 1994. Evolution and development of body size and cell size in *Drosophila melanogaster* in response to temperature. *Evolution* 48:1269–76

Partridge L, French V. 1996. Thermal evolution of ectotherm body size: Why get big in the cold? In *Animals and Temperature. Phenotypic and Evolutionary Adaptation*, ed. IA Johnston, AF Bennett, pp. 265–92. Cambridge, UK: Cambridge Univ. Press

Partridge L, Langelan R, Fowler K, Zwaan BJ, French V. 1999. Correlated responses to selection on body size in *Drosophila melanogaster*. *Genet. Res.* 74:43–54

Peichel CL, Nereng KS, Ohgi KA, Cole BLE, Colosimo PF, et al. 2001. The genetic architecture of divergence between threespine stickleback species. *Nature* 414:901–5

Pigliucci M. 2001. *Phenotypic Plasticity. Beyond Nature and Nurture.* Baltimore, MD: John Hopkins Univ. Press

Raff RA. 1996. *The Shape of Life*. Chicago: Univ. Chicago Press

Raff RA. 2000. Evo-devo: the evolution of a new discipline. *Nat. Rev. Genet.* 1:74–79

Rice SH. 2002. A general population genetic theory for the evolution of developmental interactions. *Proc. Natl. Acad. Sci. USA* 99:15518–23

Robin C, Lyman RF, Long AD, Langley CH, Mackay TFC. 2002. hairy: A quantitative trait locus for *Drosophila* sensory bristle number. *Genetics* 162:155–64

Robinson SJW, Partridge L. 2001. Temperature and clinal variation in larval growth efficiency in *Drosophila melanogaster. J. Evol. Biol.* 14:14–21

Robinson SJW, Zwaan B, Partridge L. 2000. Starvation resistance and adult body composition in a latitudinal cline of *Drosophila melanogaster. Evolution* 54:1819–24

Roff DA. 1997. *Evolutionary Quantitative Genetics*. New York: Chapman & Hall

Roff DA, Fairbairn DJ. 1999. Predicting correlated responses in natural populations: changes in JHE activity in the Bermuda population of the sand cricket, *Gryllus firmus. Heredity* 83:440–50

Roff DA, Gelinas MB. 2003. Phenotypic plasticity and the evolution of trade-offs: the quantitative genetics of resource allocation in the wing dimorphic cricket, *Gryllus firmus. J. Evol. Biol.* 16:55–63

Rountree DB, Nijhout HF. 1995a. Hormonal control of a seasonal polyphenism in *Precis coenia* (Lepidoptera, Nymphalidae). *J. Insect Physiol.* 41:987–92

Rountree DB, Nijhout HF. 1995b. Genetic control of a seasonal morph in *Precis coenia* (Lepidoptera, Nymphalidae). *J. Insect Physiol.* 41:1141–45

Scharloo W. 1983. The effect of developmental constraints on selection response. In *Organisational Constraints on the Dynamics of Evolution,* ed. G Vida, J Maynard-Smith. Manchester: Manchester Univ. Press

Scharloo W. 1987. Constraints in selective response. In *Genetic Constraints on Adaptive Evolution*, ed. V Loeschcke, pp. 125–49. Berlin: Springer-Verlag

Schlichting CD, Pigliucci M. 1998. *Phenotypic Evolution, a Reaction Norm Perspective*. Sunderland, MA: Sinauer

Shapiro AM. 1976. Seasonal polyphenism. *Evol. Biol.* 9:259–333

Simpson P. 2002. Evolution of development in closely related species of flies and worms. *Nat. Rev. Genet.* 3:907–17

Simpson P, Berreur P, Berreur-Bonnenfant J. 1980. The initiation of pupariation in *Drosophila*: dependence on growth of the imaginal discs. *J. Embryol. Exp. Morphol.* 57:155–65

Skaer N, Pistillo D, Gilbert J-M, Lio P, Wulbeck C, Simpson P. 2002. Gene duplication at the achaete-scute complex and morphological complexity of the peripheral nervous system in Diptera. *Trends Genet.* 18:399–405

Stern DL. 1998. A role of *Ultrabithorax* in morphological differences between *Drosophila* species. *Nature* 396:463–66

Stern DL. 2000. Perspective: evolutionary developmental biology and the problem of variation. *Evolution* 54:1079–91

Stern DL. 2003. Body size control: how an insect knows it has grown enough. *Curr. Biol.* 13:R267–69

Sucena E, Stern DL. 2000. Divergence of larval morphology between *Drosophila sechellia* and its sibling species caused by cis-regulatory evolution of *ovo/shaven-baby*. *Proc. Natl. Acad. Sci. USA* 97:4530–34

Sucena E, Delon I, Jones I, Payre F, Stern DL. 2003. Regulatory evolution of *shaven-baby/ovo* underlies multiple cases of morphological parallelism. *Nature* 424:935–38

Tantawy AO, Mallah GS. 1961. Studies on natural populations of *Drosophila*. I. Heat resistance and geographical variation in *Drosophila melanogaster* and *Drosophila simulans*. *Evolution* 15:1–14

True JR, Edwards KA, Yamomoto D, Carroll SB. 1999. *Drosophila* wing melanin patterns form by vein-dependent elaboration of enzymatic prepatterns. *Curr. Biol.* 9:1382–91

van't Land J. 1997. Latitudinal variation in *Drosophila melanogaster*. On the maintenance of the world-wide polymorphisms for *Adh, aGpdh* and *In(2L)t*. PhD thesis. Univ. Groningen. 153 pp.

van't Land J, van Putten P, Zwaan B, Kamping A, van Delden W. 1999. Latitudinal variation in wild populations of *Drosophila melanogaster* : heritabilities and reaction norms. *J. Evol. Biol.* 12:222–32

Wagner GP. 1996. Homologues, natural kinds and the evolution of modularity. *Am. Zool.* 36:36–43

Wagner GP, Altenberg L. 1996. Complex adaptations and the evolution of evolvability. *Evolution* 50:967–76

Watada M, Ohba S, Tobari YN. 1986. Genetic differentiation in Japanese populations of *Drosophila simulans* and *Drosophila melanogaster*. II. Morphological variation. *Jpn. J. Genet.* 61:469–80

Weinkove D, Neufeld TP, Twardzik T, Waterfield MD, Leevers SJ. 1999. Regulation of imaginal disc cell size, cell number and organ size by *Drosophila* class I_A phosphoinositide 3-kinase and its adaptor. *Curr. Biol.* 9:1019–29

West-Eberhard MJ. 2003. *Developmental Plasticity and Evolution*. New York: Oxford Univ. Press

Wijngaarden PJ, Brakefield PM. 2000. The genetic basis of eyespot size in the butterfly *Bicyclus anynana*: an analysis of line crosses. *Heredity* 85:471–79

Wijngaarden PJ, Brakefield PM. 2001. Lack of response to artificial selection on the slope of reaction norms for seasonal polyphenism in the butterfly *Bicyclus anynana*. *Heredity* 87:410–20

Wijngaarden PJ, Koch PB, Brakefield PM. 2002. Artificial selection on the shape of reaction norms for eyespot size in the butterfly *Bicyclus anynana*: direct and correlated responses. *J. Evol. Biol.* 15:290–300

Wilkins AS. 2002. *The Evolution of Developmental Pathways*. Sunderland, MA: Sinauer

Windig JJ, Brakefield PM, Reitsma N, Wilson JGM. 1994. Seasonal polyphenism in the wild: survey of wing patterns in five species of *Bicyclus* butterflies in Malawi. *Ecol. Entomol.* 19:285–98

Wourms MK, Wasserman FE. 1985. Butterfly wing markings are more advantageous during handling than during the initial strike of an avian predator. *Evolution* 39:845–51

Zecca M, Basler K, Struhl G. 1996. Direct and long-range action of a Wingless morphogen gradient. *Cell* 87:833–44

Zera AJ. 1999. The endocrine genetics of wing polymorphism in *Gryllus*: critique of recent studies and state of the art. *Evolution* 53:973–77

Zera AJ, Bottsford J. 2001. The endocrine-genetic basis of life-history variation: the relationship between the ecdysteroid titer and morph-specific reproduction in the wing-polymorphic cricket *Gryllus firmus*. *Evolution* 55:538–49

Zera AJ, Cisper G. 2001. Genetic and diurnal variation in the juvenile hormone titer in a wing-polymorphic cricket: implications for the evolution of life histories and dispersal. *Physiol. Biochem. Zool.* 74:293–306

Zera AJ, Denno RF. 1997. Physiology and ecology of dispersal polymorphism in insects. *Annu. Rev. Entomol.* 42:207–31

Zera AJ, Harshman LG. 2001. The physiology of life history trade-offs in animals. *Annu. Rev. Ecol. Syst.* 32:95–126

Zera AJ, Huang Y. 1999. Evolutionary endocrinology of juvenile hormone esterase: functional relationship with wing polymorphism in the cricket *Gryllus firmus*. *Evolution* 53:837–47

Zera AJ, Potts J, Kobus K. 1998. The physiology of life-history trade-offs: experimental analysis of a hormonally induced life-history trade-off in *Gryllus assimilis*. *Am. Nat.* 152:7–23

Zwaan BJ, Azevedo RBR, James AC, van't Land J, Partridge L. 2000. Cellular basis of wing size variation in *Drosophila melanogaster*: a comparison of latitudinal clines on two continents. *Heredity* 84:338–47

Annu. Rev. Ecol. Evol. Syst. 2003. 34:661–89
doi: 10.1146/annurev.ecolsys.34.011802.132417

First published online as a Review in Advance on August 6, 2003

FLEXIBILITY AND SPECIFICITY IN CORAL-ALGAL SYMBIOSIS: Diversity, Ecology, and Biogeography of *Symbiodinium*

Andrew C. Baker[1,2]
[1]*Wildlife Conservation Society, Marine Conservation Program, 2300 Southern Boulevard, Bronx, New York 10460*
[2]*Center for Environmental Research and Conservation, Columbia University, MC 5557, 1200 Amsterdam Avenue, New York, New York 10027; email: abaker@wcs.org*

Key Words bleaching, climate change, reef, scleractinia, zooxanthellae

"Whatever is flexible and flowing will tend to grow,
whatever is rigid and blocked will wither and die,"

—*Tao Te Ching*

■ **Abstract** Reef corals (and other marine invertebrates and protists) are hosts to a group of exceptionally diverse dinoflagellate symbionts in the genus *Symbiodinium*. These symbionts are critical components of coral reef ecosystems whose loss during stress-related "bleaching" events can lead to mass mortality of coral hosts and associated collapse of reef ecosystems. Molecular studies have shown these partnerships to be more flexible than previously thought, with different hosts and symbionts showing varying degrees of specificity in their associations. Further studies are beginning to reveal the systematic, ecological, and biogeographic underpinnings of this flexibility. Unusual symbionts normally found only in larval stages, marginal environments, uncommon host taxa, or at latitudinal extremes may prove critical in understanding the long-term resilience of coral reef ecosystems to environmental perturbation. The persistence of bleaching-resistant symbiont types in affected ecosystems, and the possibility of recombination among different partners following bleaching, may lead to significant shifts in symbiont community structure and elevations of future bleaching thresholds. Monitoring symbiont communities worldwide is essential to understanding the long-term response of reefs to global climate change because it will help resolve current controversy over the timescales over which symbiont change might occur. Symbiont diversity should be explicitly incorporated into the design of coral reef Marine Protected Areas (MPAs) where resistance or resilience to bleaching is a consideration.

INTRODUCTION

The unicellular algal symbionts found in reef corals and associated reef biota are critical to understanding the past evolution, present distribution, and future fate of coral reefs and the ecosystems they support (Cowen 1988, Hoegh-Guldberg 1999, Muscatine & Porter 1977). In fact, despite their fantastic abundance (a healthy coral reef might easily contain $>10^{10}$ algal symbionts per m^2), their relatively small overall biomass suggests these symbionts represent keystone species (Paine 1969, Power et al. 1996) on coral reefs—perhaps the only protists to play such a role.

Currently, eight genera in four (or five) classical orders of dinoflagellate are recognized as endosymbionts in marine invertebrates and protists (Banaszak et al. 1993, Trench 1997). *Symbiodinium* is the most studied genus in this paraphyletic group and is commonly found in shallow water tropical and subtropical cnidarians. These algae, commonly referred to as "zooxanthellae," are ubiquitous members of coral reef ecosystems (Rowan 1998, Taylor 1974, Trench 1993): Cnidarian species reported to contain *Symbiodinium* include many representatives from the class Anthozoa (including anemones, scleractinian corals, zoanthids, corallimorphs, blue corals, alcyonacean corals, and sea fans) and several representatives from the classes Scyphozoa (including rhizostome and coronate jellyfish) and Hydrozoa (including milleporine fire corals). *Symbiodinium* has also been identified from gastropod and bivalve mollusks [including tridacnid (giant) clams, heart cockles, and possibly, conch], large miliolid foraminifera (in the subfamily Soritinae), sponges, and a giant heterotrich ciliate (see Trench 1993 for review; Carlos et al. 1999; Lobban et al. 2002; Hill & Wilcox 1998; and references in Figure 1, see color insert). However, in many of these groups, symbionts have only been definitively identifid from a few representatives; only a few groups (scleractinian corals, soritid foraminifera, gorgonians, and tridacnid clams) have been surveyed using molecular techniques to an extent that might be considered marginally representative (Baker 1999; Baker & Rowan 1997; Goulet 1999; LaJeunesse 2002; Pawlowski et al. 2001; Pochon et al. 2001; Rowan & Powers 1991a,b; Santos et al. 2002a).

It is clear that additional diversity in *Symbiodinium* remains to be discovered (Baker 2003) and that most species are uncultured and undescribed (Rowan 1998, Santos et al. 2001). Many records of "zooxanthellae" present in invertebrates are based on only cursory observations or anecdotal reports. Moreover, in addition to the dominant populations of symbiotic dinoflagellates in these hosts, many unusual or novel variants may also occur as cryptic and unstable transients whose physiological or ecological importance is not yet clear (Goulet & Coffroth 1997, LaJeunesse 2001, Santos et al. 2001, Toller et al. 2001a).

The purpose of this review is to synthesize research of the past dozen years that has used molecular DNA techniques to quantify, classify, and study the distribution of diversity in *Symbiodinium*. In his original description of *Symbiodinium microadriaticum*, Freudenthal (1962) observed "a sound taxonomy of the zooxanthellae—one taking into account pure-culture studies and host-symbiont specificities—is needed for ecological work on coral and other zooxanthellae associations."

This statement remains a valid motivation for research on *Symbiodinium* today. However, with the rise of molecular methods for identification, certain avenues of research—notably those dealing with partner specificity, biogeography, and ecology—have advanced rapidly over the past few years in ways that Freudenthal could not have anticipated. This review emphasizes how molecular techniques have been applied to studying field-collected material from diverse hosts, environments, and locations—an approach that has revolutionized our understanding of how complex and flexible these associations can be. An implicit motivation behind much of this research has been to understand the role of symbiont diversity and/or flexibility in determining possible long- and short-term responses of coral reefs to environmental change and global warming. Consequently, this review focuses principally on scleractinian coral symbioses as the principal builders of contemporary coral reefs and the subjects of most of the research undertaken during this time.

DIVERSITY, PHYLOGENY, AND SYSTEMATICS OF *SYMBIODINIUM*

Diversity

There are currently eleven named species in the exceptionally diverse genus *Symbiodinium* (Figure 1). All four formally described species (*S. microadriaticum*, *S. pilosum*, *S. kawagutii*, and *S. goreaui*) have been distinguished using the morphological species concept (Freudenthal 1962, Trench 2000, Trench & Blank 1987), and a similar rationale has been used to name an additional six species ("*S. californium*," "*S. corculorum*," "*S. meandrinae*," "*S. pulchrorum*," "*S. bermudense*," and "*S. cariborum*") without formal description, although some of these names may be synonymous (Banaszak et al. 1993, Rowan 1998). A further species ("*S. muscatinei*") has been distinguished solely from molecular sequence data (LaJeunesse & Trench 2000). In addition, the described species *Gymnodinium linucheae* (Trench & Thinh 1995) is also recognized, on molecular grounds, to be a member of the genus *Symbiodinium* (LaJeunesse 2001, Wilcox 1998), and an additional *Symbiodinium* species (closely related to "*S. californium*") has been misidentified as *Gymnodinium varians* (LaJeunesse & Trench 2000).

Lack of observed sexual reproduction in this group precludes the use of the biological species concept to define species boundaries, although various genetic measures suggest these microalgae reproduce sexually (Baillie et al. 1998, 2000b; Belda-Baillie et al. 1999; Goulet & Coffroth 1997; LaJeunesse 2001; Santos et al. 2003b; Schoenberg & Trench 1980a). In documenting diversity in *Symbiodinium*, problems recognizing distinct species have been compounded by difficulties associated with the need to culture these microalgae for morphological description. However, increasing success in recognizing diversity using molecular methods has resulted in the de facto use of the phylogenetic species concept (Eldredge & Cracraft 1980) to distinguish taxa, particularly when they also have distinct ecological or host-specific distributions. Many of the molecular types thus identified are separated by genetic distances that are many times those found separating recognized species of free-living dinoflagellates. Consequently, a justifiable argument has been made that the genus *Symbiodinium* is speciose (Blank & Trench 1985a,b; Rowan 1998; Rowan & Powers 1992), consisting of several major clades or lineages ("subgenera") each containing multiple species. However, the use of sequence data to define species boundaries may well be confounded by the highly unusual nuclear characteristics of these dinoflagellates (e.g., Rizzo 1987) and their probably haploid nature (Santos & Coffroth 2003), making any informative conclusions premature at this stage.

Insufficient sampling also hinders our ability to determine exactly how many "species" exist within each of the principal *Symbiodinium* clades. The rate of discovery of novel molecular types continues to increase rapidly (Baker 2003) with current estimates reaching 100 or more (LaJeunesse 2001, 2002; LaJeunesse et al. 2003; T.C. LaJeunesse, unpublished data) each of which LaJeunesse (2001) considers distinct at the species level [but see S.R. Santos, T.L. Shearer, A.R. Hannes, M.A. Coffroth, submitted manuscript, for evidence of even finer subdivision, and Rodriguez-Lanetty (2003) for a quantitative estimate of lineage diversity]. At even finer taxonomic resolution, isozymes, randomly amplified polymorphic DNA (RAPDs), DNA fingerprinting, and microsatellites have reported extreme population-level variability in *Symbiodinium* (Baillie et al. 1998, 2000b; Goulet & Coffroth 1997, 2003a,b) and have demonstrated that hundreds of unique genotypes may exist for each of the taxa distinguished to date (Santos & Coffroth 2003).

Phylogeny

Molecular phylogenies of *Symbiodinium* have been dominated by the use of nuclear genes encoding ribosomal RNA (nrDNA). These studies have included the ribosomal small subunit (SSU) (Brown et al. 2002a, Carlos et al. 1999, Darius et al. 2000, Rowan & Powers 1992, Sadler et al. 1992), partial large subunit (LSU) (Baker 1999, Loh et al. 2001, Pawlowski et al. 2001, Pochon et al. 2001, Savage et al. 2002a, Toller 2001b, Van Oppen et al. 2001, Wilcox 1998), and internal transcribed spacers (ITS 1 and 2) and 5.8S regions (Brown et al. 2000, 2002; Hunter et al. 1997; LaJeunesse 2001, 2002; LaJeunesse et al. 2003; Savage et al. 2002a; Van Oppen et al. 2001). Recently, partial LSU chloroplast rDNA (cprDNA) sequences have been used to independently test the relationships inferred from nrDNA (Santos et al. 2002a,b, 2003a).

Together, these studies have recognized between four and ten distinct clades of *Symbiodinium*. Despite differences in the phylogenetic diversity of source material and semantic disagreement over which clades deserve their own name, a surprising degree of congruence between these phylogenies exists. In particular, the independent organellar marker (LSU cprDNA) used by Santos et al. (2002a) recovered a phylogeny that was not significantly different from established nrDNA phylogenies, although some uncertainty in the relative positions of clades *B*, *C*, and *F* was indicated (Baker 1999, Carlos et al. 1999, Darius et al. 2000,

LaJeunesse 2001, Loh et al. 2001, Pawlowski et al. 2001, Pochon et al. 2001, Rowan & Powers 1992, Savage et al. 2002a, Van Oppen et al. 2001, Wilcox 1998). Congruent datasets from independent nuclear and organellar sources strongly support the phylogeny presented in Figure 1, which recognizes seven distinct clades (*A* through *G*) and follows nomenclature established by Rowan & Powers (1991b), Baker (1999), Carlos et al. (1999), LaJeunesse & Trench (2000), Pochon et al. (2001), and Rodriguez-Lanetty (2003). *Symbiodinium F*, as the least well-resolved clade, would benefit from further analysis to determine the support for its highly divergent members (see Rodriguez-Lanetty 2003), including *S. kawagutii*, which earlier reports had placed in clade *C* (e.g., Banaszak et al. 2000, Carlos et al. 1999, Santos et al. 2001).

Systematics and Nomenclature

The usefulness of a name is reliably reflected by the extent to which it is employed by independent authors. In this context, the arbitrary *A*, *B*, and *C Symbiodinium* classification system introduced by Rowan & Powers (1991b) has proved remarkably useful: All molecular studies of *Symbiodinium* published to date have employed the same nomenclature to refer to the same lineages. However, recent disagreement over the expansion of Rowan & Powers' (1991b) original nomenclature to include *Symbiodinium* diversity not named by them has led to some confusion.

This confusion centers around the introduction of *Symbiodinium E* to refer to a group of symbionts found in Caribbean scleractinian corals (Toller et al. 2001a,b)—a practice followed by Goodson (2000), Brown et al. (2000, 2002a), Savage et al. (2002a), and Chen et al. (2003a,b) for corals containing similar symbionts from Thailand, St. Croix, Taiwan, and Hong Kong. Recognizing that *Symbiodinium D* had been used by Carlos et al. (1999) to refer to an unusual isolate from the interstitial water of a sponge, Toller et al. (2001a,b) distinguished newly discovered symbiont types as *Symbiodinium E*. However, sequence analysis of partial LSU sequences from both nuclear and chloroplast rDNA (Pochon et al. 2001, Santos et al. 2002a) reveal that the symbionts referred to as *Symbiodinium E* by Toller et al. (2001a,b) are distantly related members of the same clade as the *D*-type originally identified by Carlos et al. (1999), a finding supported by additional analysis (X. Pochon, unpublished data). Ironically, this lineage of symbionts was apparently documented from a Hawaiian scleractinian coral in Rowan & Powers' (1991b) original phylogeny, but it was not given a name and no sequence data were provided. Carlos et al. (1999) determined that the sponge isolate and the unusual coral symbiont of Rowan & Powers (1991b) shared a similar RFLP genotype, lending further support to the conclusion that they belong to the same clade.

While the naming of clades is wholly a question of semantics, and deciding which clades are worthy of names within an emerging phylogeny of *Symbiodinium* is a somewhat arbitrary process, names have value only if used consistently. The recommendation made here follows the practice of Baker (1999, 2001, 2003), Glynn et al. (2001), LaJeunesse (2001, 2002), Loh et al. (2001), Pawlowski et al. (2001), Pochon et al. (2001), Van Oppen et al. (2001), Santos et al. (2002a,b; 2003a), and LaJeunesse et al. (2003), Ulstrup & Van Oppen (2003), Van Oppen (2003), in using *Symbiodinium D* to refer to the single clade that includes both the unusual sponge isolate identified as *D* by Carlos et al. (1999) and the symbionts of certain scleractinian corals originally referred to as *D* by Baker (1999) and subsequently referred to as *E* by Toller (2001a,b). *Symbiodinium E* is used to refer to the clade that includes temperate symbionts isolated from Californian *Anthopleura* named "*S. californium*," and a free-living dinoflagellate misidentified as *Gymnodinium varians* cultured from a water sample taken from Wellington, New Zealand (41°S) (LaJeunesse & Trench 2000).

How diverse are the principal clades of *Symbiodinium*, and how are taxa within these clades related to one another? The most comprehensive *Symbiodinium* phylogenies to date (Pawlowski et al. 2001, Pochon et al. 2001) have been relatively successful in using LSU nrDNA to identify variation within all *Symbiodinium* clades except *B* and *E*. However, the phylogenetic structure of this variation is not well-resolved. In particular, a large cluster of closely related *B*- and *C*-types that have also been difficult to distinguish in other nuclear SSU and LSU rDNA datasets (Baker 1999, Darius et al. 2000, Rowan & Powers 1991b, Wilcox 1998) remains unresolved in these analyses. The extreme number of sequence variants in *Symbiodinium C* led Baker (1999) and Toller et al. (2001a) to conclude that although different *C*-types did exist within this clade, much of the variation represented artifacts of cloning (Speksnijder et al. 2001) and/or paralogous genes within the rDNA repeat (Rowan & Powers 1991a,b).

More recently, sequence analysis of more rapidly evolving nuclear ITS nrDNA sequences has resolved additional phylogenetic structure within all the principal clades ("subgenera") of *Symbiodinium* (LaJeunesse 2001, 2002; LaJeunesse et al. 2003; Rodriguez-Lanetty 2003; Savage et al. 2002a; Van Oppen et al. 2001). Although one subclade of *C* still contains a number of closely related sequence variants, LaJeunesse (2001) has argued this variation, rather than being artifactual, represents rapid diversification from a single ancestral taxon (Rodriguez-Lanetty 2003; T.C. LaJeunesse, submitted manuscript). It now appears clear that significant variation, not easily distinguished by cloning-based approaches, does indeed characterize this and other clades.

Figure 1 presents a comprehensive analysis of all 294 sequences of the D1-D2 region of *Symbiodinium* LSU nrDNA (~650 nt) available in Genbank as of May 2003. Certain clades not well-resolved in Figure 1 (particularly *B*, *C*, and *D*, for which many sequences are available) comprise many distinct taxa whose phylogenetic relationships are not unequivocally resolved in this analysis. Finer resolution of distinct types within these clades has been obtained using a variety of molecular screening techniques that include restriction fragment length polymorphism (RFLP) analysis of SSU and LSU nrDNA (Baker 1999;

Baker et al. 1997; Chen et al. 2003a,b; Glynn et al. 2001; Loh et al. 1998, 2001; Pochon et al. 2001; Rowan & Powers 1991a,b, 1992; Toller et al. 2001a,b; Wilcox 1998), temperature gradient gel electrophoresis (TGGE), and denaturing gradient gel electrophoresis (DGGE) of SSU nrDNA (Belda-Baillie et al. 2002, Carlos et al. 2000), DGGE analysis of ITS and 5.8S nrDNA (Baillie et al. 2000a; LaJeunesse 2001, 2002; LaJeunesse et al. 2003), single strand conformational polymorphism (SSCP) analysis of ITS nrDNA (Ulstrup & Van Oppen 2003, Van Oppen 2003, Van Oppen et al. 2001), length heteroplasmy of LSU cprDNA (Santos et al. 2001, 2002a,b, 2003a) and phylogenetic analysis of microsatellite flanking regions (S.R. Santos, T.L. Shearer, A.R. Hannes, M.A. Coffroth, submitted manuscript). An unfortunate consequence of the variety of different molecular markers and methods used in these *Symbiodinium* surveys is that direct comparisons of different datasets are not possible. A collaborative multiple-marker study of the diverse symbiont types thus far identified would be the only way to proceed in establishing a truly comprehensive phylogeny and a consensual nomenclature.

How do genetically different *Symbiodinium* vary functionally from one another? It is clear that closely related symbiont taxa can differ significantly in their physiological capacities (e.g., Chang et al. 1983; Iglesias-Prieto & Trench 1997; Warner et al. 1996, 1999) and host specialization (Diekmann et al. 2002, LaJeunesse 2002), making generalizations regarding the properties of particular clades (particularly the diverse clades *A*, *B*, *C*, and *F*) is premature at this stage (Goodson et al. 2001, Savage et al. 2002b; but see also Knowlton & Rohwer 2003, Rowan 1998). However, as molecular advances improve our ability to distinguish more closely related types, physiological patterns not apparent in earlier molecular studies (e.g., Savage et al. 2002b) may become apparent (e.g., Santos et al. 2002a). Functional diversity of *Symbiodinium* remains an important avenue of future research whose findings remain beyond the scope of a comprehensive review now.

SPECIFICITY OF HOSTS AND SYMBIONTS

Systematic Patterns of Symbiont Distribution

An overview of the distribution of the principal clades and subclades of *Symbiodinium* reported from different host taxa is shown in Figure 1. From this distributional dataset it is clear that associations between *Symbiodinium* and its various invertebrate and protist hosts exhibit specificity (sensu Dubos & Kessler 1963): patterns of association are clearly nonrandom and the ratio of observed combinations of hosts and symbionts compared with the range of possible combinations is very small (Rowan 1991; Trench 1988, 1992). However, an emerging picture of considerable flexibility on the part of both hosts and symbionts reveals that the idea of uniformly strict specificity—in which all hosts exclusively contain only one symbiont type—is incorrect (Trench 1988, 1992, 1993, 1996). One reason why the notion of strict specificity may have prevailed for so long is that *Symbiodinium* taxonomy from the 1950s to the early 1990s relied on the morphological study of cultured material. Culturing *Symbiodinium* reduces the diversity of heterogeneous isolates, favoring nonrepresentative members over previously dominant types (Santos et al. 2001). Early conclusions of strict symbiont specificity may therefore have been partly the result of a long-term culture that reduced the diversity of the original isolate to a single algal genotype that was mistakenly identified as "the" symbiont of a particular host species.

Scleractinian corals are among the most flexible hosts identified to date, containing symbionts from clades *A*, *B*, *C*, *D*, and *F*. However, this relative high diversity may be an artifact of sampling for these well-studied hosts. Benthic soritid foraminifera also show remarkable diversity, containing symbionts in clades *C*, *F*, and *G*. Anemones of various kinds show considerable symbiont diversity (members of clades *A*, *B*, *C*, *D*, and *E*), as do zoanthids (members of *A*, *B*, *C*, and *D*), milleporine fire corals (members of *A*, *B*, and *C*), and tridacnid clams (members of *A* and *C*). (See Figure 1.)

The relative ease with which diverse symbionts have been found in many different hosts suggests that we cannot as yet reject the null hypothesis that many (perhaps all) host taxa are able to associate with more than one type of *Symbiodinium*. Laboratory infection experiments suggest that hosts can become infected with a wide variety of different symbionts (Fitt 1984, 1985b; Schoenberg & Trench 1980c), and the fact that hosts in nature can also contain small numbers of (often unusual) minor symbionts within a generally homogeneous population of "normal" symbiont types suggests that what we perceive as specificity may be perhaps no more than the end result of competitive exclusion between symbionts (Goulet & Coffroth 1997, LaJeunesse 2002). Some of these minor types may be predominantly free-living *Symbiodinium* that are merely "moonlighting" as symbionts—they never dominate individual hosts under natural conditions but can occasionally emerge as the dominant species in culture (LaJeunesse 2002, Santos et al. 2001). We should be conservative in interpreting observed patterns of association as representing the full range of possible combinations. By implicitly recognizing specificity as a continuous variable we can distinguish which symbionts are probable (often dominant in a host) from which symbionts are only possible (rare in a host, but possibly dominant in different circumstances or different hosts). New molecular methods applied to field-collected material are paving the way for investigations of this kind (LaJeunesse 2001, 2002; LaJeunesse et al. 2003; Santos et al. 2003a; S.R. Santos, T.L. Shearer, A.R. Hannes, M.A. Coffroth, submitted manuscript). Additionally, unusual types arising in culture can fortuitously assist us in our task of recovering a comprehensive phylogeny of *Symbiodinium*. For example, Santos et al. (2003c) cultured an uncommon *Symbiodinium D* from an anemone isolate that may not be representative of its dominant populations in hospite.

Host Specificity Versus Symbiont Specificity

A new understanding arising largely from the application of molecular genetics suggests that some coral hosts are able to associate with a variety of distantly related symbiont types, while others are apparently restricted to a single symbiont type or subset of closely related types. Similarly, some algal symbionts are widely distributed and found in many hosts ("generalists"), while others appear endemic to particular locations and may be restricted to a particular host taxon ("specialists") (Rowan 1998, Baker 1999, Toller et al. 2001b, LaJeunesse 2002). We should therefore distinguish between host specificity (the specificity of hosts for a particular range of symbionts) and symbiont specificity (the specificity of symbionts for a particular range of hosts). In general, hosts appear more specific than symbionts: the mean number of combinations in which a given host is found tends to be less than the mean number of combinations in which a given symbiont is found (T.C. LaJeunesse, submitted manuscript).

A conceptual framework for coral-algal specificity is illustrated in Figure 2, in which examples (from the scleractinian corals) of all four combinations of

Host specificity
(specificity of hosts for particular symbionts)

Low ⟶ **High**

Symbiont specificity
(specificity of symbionts for particular hosts)

Low ↓ **High**

Symbiont specificity	Host specificity: Low	Host specificity: High
Low	*Montastraea faveolata* + *Symbiodinium B1* (LaJeunesse 2002)	*Meandrina meandrites* + *Symbiodinium B1* (LaJeunesse 2002)
High	*Colpophyllia natans* + *Symbiodinium* B_{cn} (Baker 1999)[1] *Acropora cervicornis* + *Symbiodinium* C_{ac} (Baker 1999)[2]	*Porites furcata* + *Symbiodinium* C_{pf} (Baker 1999) *Scolymia cubensis* + *Symbiodinium C11* (LaJeunesse 2002) *Madracis mirabiliis* + *Symbiodinium B13* (Diekmann et al. 2002).

Figure 2 Conceptual framework for symbiosis specificity. Host and symbiont specificity are represented as continuous variables ranging from strict specificity to relative flexibility (low specificity). Examples of each type of symbiosis are given for *Symbiodinium* in Caribbean scleractinian corals. [1]*Symbiodinium B6* of LaJeunesse (2002). [2]*Symbiodinium C12* of LaJeunesse (2002).

low and high host and symbiont specificity are represented. One finding which emerges from classifying symbioses in this way is that a single association can be both highly flexible and highly specific, depending on which partner is being discussed. For example, the Caribbean faviid coral *Colpophyllia natans* has been found hosting (at least) one *C*-type and two *B*-types, one of which appears highly specific for *Colpophyllia* (Baker 1999, Baker & Rowan 1997, LaJeunesse 2002). This association is thus composed of a generalist host with low symbiont specificity, together with two generalist symbionts and one host-specific symbiont. The wide range of interaction strategies (Goff 1982) found in *Symbiodinium* to date supports the notion that resultant symbiosis benefits from greater evolutionary and ecological potential through flexibility.

Intraspecific Symbiont Diversity

Examples of generalist host species that are able to contain more than one symbiont type represent particularly interesting special cases. Although initially documented in Pacific *Pocillopora* (Rowan & Powers 1991a), the significance of intraspecific symbiont diversity was first recognized in later studies of Caribbean *Montastraea* (Rowan & Knowlton 1995; Rowan et al. 1997; Toller et al. 2001a,b). These studies documented four different symbiont taxa (in four distinct clades of *Symbiodinium*) in three closely related host species, and they showed that these symbionts were distributed within single host species (and even within single coral colonies) in ecologically meaningful ways (see Ecology below). Examples of scleractinian species hosting multiple symbiont taxa are now common (Baker 1999, 2001; Baker et al. 1997; Glynn et al. 2001; LaJeunesse 2001, 2002; LaJeunesse et al. 2003; Pawlowski et al. 2001; Pochon et al. 2001; Santos et al. 2001; Van Oppen 2001). Baker (1999) conservatively reported that 38 of 107 species (36%) of scleractinian corals surveyed from the Caribbean, far eastern Pacific and Great Barrier Reef (GBR) contained multiple *Symbiodinium* types, a finding that has been supported in surveys of Caribbean reef invertebrates (LaJeunesse 2002). Intraspecific symbiont diversity is not confined to the scleractinian corals: similar flexibility within single species of host has been found in foraminiferans (Pawlowski et al. 2001, Pochon et al. 2001), anemones (Santos et al. 2003c), gorgonians (Coffroth et al. 2001; Santos et al. 2003b; S.R. Santos, T.L. Shearer; A.R. Hannes; M.A. Coffroth, submitted manuscript; but see also Goulet 1999, 2003a,b) and hydrocorals (Baker 1999, LaJeunesse 2002).

Exactly how deterministic is specificity? Do unusual symbionts occasionally appear in normally highly specific hosts, or do highly specific symbionts occasionally colonize unusual hosts? Strictly specific one-on-one partnerships are difficult to document with certainty because it is impossible to survey all hosts in all possible environments. Apparent strict specificity may merely reflect an established status quo in which a stable equilibrium has been reached. Disturbance, such as environmental change, bleaching, or disease might provide a window for opportunistic partnerships to become established (Baker 2001, 2002; Rowan 1998; Toller et al.

2001b) with significant implications for the biology of the recombinant symbiosis. In fact, because exhaustive range-wide sampling of individual species has not yet been (and probably never will be) undertaken, one might argue that we cannot as yet reject the hypothesis that all host species, at one time or another and in variously stable abundance, are able to maintain symbioses with various symbiont taxa to different degrees.

ECOLOGY OF *SYMBIODINIUM*

Studies of the biology and taxonomy of *Symbiodinium* have traditionally emphasized differences among host species rather than within them (Blank & Trench 1985a, 1986; Schoenberg & Trench 1980a,b,c; Trench 1988). A few early studies speculated on the potential significance of inter- and intraspecific symbiont diversity in explaining aspects of host biology (Dustan 1982, Gladfelter 1988, Kinzie 1970, Kinzie & Chee 1979, Jokiel & Coles 1977, Jokiel & York 1982, Sandeman 1988), but the requisite molecular tools to test these speculations were not then available. Because it has only recently been recognized that intraspecific symbiont diversity is relatively common in scleractinian corals, as well as in many other associations (see Specificity above), the consensus view has traditionally underplayed symbiont diversity.

Bathymetric Distribution of *Symbiodinium*

Initial investigations of intraspecific diversity focused on Caribbean scleractinian corals, which to date have been recorded in association with members of *Symbiodinium A*, *B*, *C*, and *D*. *Montastraea annularis* and *M. faveolata*, two principal builders of Caribbean reefs, contain members of all four of these clades in a predictable pattern of depth distribution: *Symbiodinium A*-, *B*-, and *D*-types are found in shallow water (0–6 m), while *Symbiodinium C*-types are found in deeper water (3–14 m) (Rowan & Knowlton 1995, Rowan et al. 1997, Toller et al. 2001b). *D*-types were also found in extremely deep (>35 m) colonies of *M. franksi* (Toller et al. 2001b). Further survey work documented a number of Caribbean scleractinian corals showing similar patterns (Baker 1999, 2001; Baker & Rowan 1997; Baker et al. 1997; Rowan 1998; but see Billinghurst et al. 1997, Diekmann et al. 2002). A community ecology approach to surveying the distribution of *Symbiodinium* in a suite of Caribbean invertebrate hosts also found depth zonation to be an emergent property of reef-wide patterns of symbiont distribution among a variety of invertebrate hosts (LaJeunesse 2002), drawing particular attention to the shallow-water occurrence of *Symbiodinium A* and its ability to produce UV-protecting mycosporine-like amino acids (MAAs) in culture (Banaszak et al. 2000).

The identification of "high-light" and "low-light" symbionts in scleractinian corals indicates that photoadaptation (Dustan 1982, Falkowski & Dubinsky 1981, Iglesias-Prieto & Trench 1994) is not the only way that corals can respond to changes in irradiance. Finding different symbionts in the same species of host from different habitats also provides an ecological explanation for the apparent lack of coevolution between hosts and symbionts and the failure of many coral species to pass symbionts directly to their offspring (Rowan & Knowlton 1995).

Why do corals contain multiple symbionts? Phenotypic characters of symbioses are presumably emergent properties of the genotypes of all partners involved. It therefore follows that the ability of invertebrate hosts to exist in symbiosis with several symbiont types creates a variety of combinations whose physiological abilities are likely to be much broader than those of a strict one-on-one association. Therefore, intraspecific symbiont diversity within single species of host provides invertebrate and protist hosts with the potential for dramatic phenotypic variability (Baker 1999, 2001; Baker et al. 1997; Buddemeier & Fautin 1993; Buddemeier & Smith 1999; Buddemeier et al. 1997; Douglas 2003; Kinzie 1999; Knowlton & Rohwer 2003; Rowan 1998; Rowan & Knowlton 1995; Rowan et al. 1997) challenging the conventional focus on the host as the fundamental unit of ecological diversity (Rowan & Knowlton 1995).

Baker (2001) tested the stability of depth-related patterns of zonation in the Caribbean by reciprocally transplanting corals between shallow and deep environments. He found that, after one year, patterns of symbiont distribution only reestablished themselves in upward transplants that had experienced bleaching as a result of acute exposure to high irradiance. These colonies survived well in comparison to colonies that were transplanted downwards, which did not bleach and did not change their symbionts. These data were interpreted as supporting the notion that bleaching promotes symbiont community change, which in turn can be beneficial to the coral host by allowing it to become repopulated with symbionts that are more suited to the new environment (see Climate Change below).

In general, high-light (shallow) environments are characterized by higher symbiont diversity than low-light (deep) environments, both in the Caribbean and the Pacific (Baker 1999, LaJeunesse 2002). This is most likely a consequence of environmental heterogeneity: shallow environments are more variable both in space and time than deeper environments. It remains to be seen whether additional, unusual symbionts can be found in extremely deep environments (up to 200 m, the deepest records for scleractinian corals, e.g., Zahl & McLaughlin 1959), but this may be unlikely given that extremely low-light environments can be found in shallow environments.

The relatively straightforward intercladal patterns of depth zonation found in some Caribbean scleractinian corals do not hold for their tropical Pacific counterparts (Baker & Rowan 1997; see also Rowan 1996). In tropical Pacific corals sampled to date, symbiont communities have been dominated by members of *Symbiodinium C* and *D*, with *Symbiodinium A*, *B*, and *F* being found only occasionally at higher subtropical (16–25°) or temperate (25–35°) latitudes (see Biogeography below) (Baker 1999; Baker & Rowan 1997; Darius et al. 1998, 2000; Loh et al. 1998, 2001; Rodriguez-Lanetty et al. 2001; Rowan & Powers 1991a; Van Oppen et al. 2001; A.C. Baker, submitted manuscript). Consequently, the relatively simplistic depth patterns of symbiont distribution involving *Symbiodinium A*, *B*, and *C* do not hold true in the tropical Pacific. Instead, several tropical Pacific scleractinians exhibit patterns of depth zonation involving different symbionts within clade *C* (Baker 1999, LaJeunesse et al. 2003, Van Oppen et al. 2001). For example, nine of 23 species (39%) of *Acropora* surveyed from the northern GBR contained two types of *Symbiodinium C* that showed consistent patterns of depth distribution, with one *C*-type in shallow colonies and a second *C*-type in deeper colonies (Baker 1999). Similar findings were reported by LaJeunesse et al. (2003) in their survey of scleractinian corals from the southern GBR, which found that nine of 25 species (36%) sampled from both shallow and deep habitats, but none of the *Acropora*, contained different symbionts.

Despite some well-documented cases of depth zonation in (particularly Caribbean) scleractinian corals, the fact remains that most of the intraspecific diversity documented to date has not been satisfactorily explained by any unifying deterministic mechanism. The stochastic and labile nature of many coral-algal symbioses may inhibit complete understanding of patterns of symbiont distribution especially considering that significant time lags may damp symbiont community response to environmental change (Toller et al. 2001a). The application of molecular techniques designed to resolve more closely related symbionts (LaJeunesse 2001, 2002; Santos et al. 2003a) has led to a rapid increase in the frequency and diversity of symbionts found within single species (Baker 2003), yet at the same time it has also been recognized that specificity, even for relatively flexible corals such as Caribbean *Montastraea*, exists. Understanding the incidence and distribution of much of this diversity represents the crucial next phase of this research.

Symbiont Diversity in Individual Hosts

In addition to within-species symbiont diversity, multiple types of *Symbiodinium* within single individuals and/or colonies have been documented in a number of hosts, including scleractinian corals (Baker 1999, 2001; Baker et al. 1997; Darius et al. 1998, 2000; LaJeunesse & Trench 2000; Rowan & Knowlton 1995; Rowan & Powers 1991b; Rowan et al. 1997; Van Oppen 2001), gorgonians (Coffroth et al. 2001; Santos et al. 2001, 2003b; but see also Goulet 1999, Goulet & Coffroth 2003a,b), tridacnid clams (Baillie et al. 2000a,b; Belda-Baillie et al. 1999; Carlos et al. 2000), and other hosts (Carlos et al. 1999).

In scleractinians, colony topography can create a landscape over which symbiont communities distribute themselves according to their photic optima (Rowan et al. 1997), producing patterns of zonation that are not unlike those found with depth (Rowan & Knowlton 1995). Rowan et al. (1997) tested the stability of these patterns by toppling vertical colonies of *Montastraea annularis*. Unlike Baker (2001), the bleaching response of these toppled colonies was not recorded, but the original within-colony symbiont distributions reestablished themselves within six months. Individual colonies can thus host dynamic communities of *Symbiodinium* that respond to changing environments. In a survey of scleractinian corals from the Caribbean, far eastern Pacific, and GBR, 25 of 38 species (66%) that contained multiple symbiont types also exhibited the same diversity at the colony level (Baker 1999). However, the fine-scale patterning of symbiont populations in Caribbean *Montastraea* in response to the external photic environment may represent an extreme example. Such flexibility and diversity at the colony level may be unusual; for example, Baker et al. (1997) found no evidence for within-colony diversity in *Acropora cervicornis*, despite clear patterns of depth zonation, and Ulstrup & Van Oppen (2003) found within-colony symbiont diversity in *A. tenuis* varied depending on which reef colonies were sampled.

Symbiodinium D in Scleractinian Corals

The distribution of (relatively low diversity) *Symbiodinium D* is unusual in many respects. Although it appears to be distributed throughout the tropics, it does not appear to be the dominant symbiont of any particular host species (Baker 1999, LaJeunesse 2002) and appears somewhat haphazard in its distribution. Baker (1999) documented *Symbiodinium D* at the transition depth between shallow *Symbiodinium A*-types and deep *C*-types in the scleractinian coral *Stephanocoenia intersepta*, and *D*-types have also been found to be dominant in extremely deep colonies of *Montastraea franksi* (Toller et al. 2001b), and in Taiwanese corals at the limit of their depth distribution (Chen et al. 2003a). This suggests that *Symbiodinium D* is favored in conditions where other symbionts are poorly suited. In a reciprocal transplant experiment, *Symbiodinium D* appeared as a novel symbiont taxon in two recovering colonies of scleractinian coral that had bleached as a result of transplantation from deep to shallow water (Baker 2001); and following a disease-related bleaching event, *Symbiodinium D* (as well as *Symbiodinium A*) was recorded in recovering colonies (Toller et al. 2001b). Glynn et al. (2001) reported that colonies of *Pocillopora* containing *Symbiodinium D* were unaffected by bleaching during the severe 1997–1998 El Niño event in the far eastern Pacific. Chen (2003a), Toller et al. (2001a), and Van Oppen et al. (2001) documented high abundance of *Symbiodinium D* in corals from inshore locations that were thought to be subject to significant terrestrial impacts. Chen (2003b) found *D*-types to be the only symbionts found in a coral characteristic of marginal habitats in Taiwan that were subject to high temperature variability. Taken together, these data indicate *Symbiodinium D* may be a weedy or opportunistic symbiont characteristic of recently stressed or marginal habitats, and/or bleached corals in the process of recovering their steady-state symbiont communities (Baker 1999, 2001; Rowan 1998; Toller 2001a; Van Oppen 2001). This may account for its unpredictable occurrence in surveys of scleractinians (and other hosts; e.g., Burnett 2002) and suggests that the abundance of *Symbiodinium D* might reflect coral community health by providing a useful signal of recent and/or recurrent stress events.

Free-Living *Symbiodinium*

Several studies have identified apparently free-living *Symbiodinium* in the waters or sediments surrounding potential invertebrate hosts (Carlos et al. 1999, Loeblich & Sherley 1979, Taylor 1983). However, to date no studies have systematically investigated the diversity of free-living *Symbiodinium* in the environment. The only relatively unambiguous records that are accompanied by reliable molecular identifications are members of *Symbiodinium E* misidentified as *Gymnodinium varians* (LaJeunesse 2001, Rowan & Powers 1992, Saldarriaga et al. 2001, Saunders et al. 1997, Wilcox 1998) and a *Symbiodinium* in clade *A* that was isolated and cultured from the interstitial waters of Hawaiian sands (Carlos et al. 1999). An unusual and highly divergent member of *Symbiodinium D* cultured from the interstitial water of a Palauan sponge may also represent a free-living type (Carlos et al. 1999). LaJeunesse (2002) sampled *Symbiodinium* from the digestive gland and the digestive tract of the queen conch *Strombus gigas* in Mexico and found members of both *Symbiodinium B* and *C* in the tract, but only the *C*-type in the gland, suggesting that the *B*-type represented a recent ingestion of a free-living symbiont.

Adult anemones have been shown to exchange symbionts with the environment (Kinzie et al. 2001), and tridacnid clams inoculated with cultured symbionts have also been successful in establishing new symbioses (Belda-Baillie et al. 1999, Fitt 1984). Because symbionts may be attracted to vacant symbiotic hosts (Fitt 1985a), our chances of detecting unusual free-living *Symbiodinium* as surface contaminants may be greater than might be otherwise expected; despite this, however, it is likely that an extraordinary diversity of free-living *Symbiodinium* remains to be identified.

BIOGEOGRAPHY OF *SYMBIODINIUM*

Latitudinal Variation in Coral-Algal Symbiosis

Symbiont distributions in scleractinian corals vary in different parts of the world (Baker & Rowan 1997, Rowan 1996). Some symbiont taxa are widely distributed, both among different hosts and across geographic regions (Burnett 2002, Loh et al. 2001, Rodriguez-Lanetty & Hoegh-Guldberg 2003), whereas other taxa show high host specificity or appear regionally endemic (Baillie et al. 2000b; Baker 1999; LaJeunesse 2001, 2002; LaJeunesse et al. 2003; Santos et al. 2003b,c).

Community-level surveys of *Symbiodinium* in scleractinian corals have shown that members of *Symbiodinium A*, *B*, and/or *F* are more common at higher latitudes worldwide, with *Symbiodinium C* more abundant in tropical latitudes (Baker 1999, Rodriguez-Lanetty et al. 2002, Savage et al. 2002a; see A.C. Baker, submitted manuscript) (Figure 3, see color insert). This pattern is true for both the Caribbean and the Pacific, despite the fact that *Symbiodinium A* and *B* are far more prevalent in tropical Caribbean corals than their Pacific counterparts (see Unusual Symbioses, below). Corals from the relatively isolated and high latitude Hawaiian archipelago have been documented to contain members of *Symbiodinium C* and *D* and possibly also *F* [*S. kawagutii* in *Montipora verrucosa* (now *M. capitata*), although this may be an artifact of culturing] (LaJeunesse 2001; Rowan & Powers 1991a; T.C. LaJeunesse, unpublished data). Based on the patterns presented here, future surveys of the Hawaiian archipelago (20°–29°N) seem likely to uncover further diversity in additional clades such as *Symbiodinium A*.

Latitudinal differences in the distribution of *Symbiodinium* have been harder to document at the level of individual host species than at the level of the reef community. The temperate anemone *Anthopleura elegantissima* hosts two species of *Symbiodinium* that vary their distribution along the Pacific coast of North America, with northern populations (43.5°–48.5°N) containing only *Symbiodinium* in clade *B* (sometimes in combination with a *Chlorella*-like green alga), and southern populations (33°–36°N) hosting mixtures of *Symbiodinium B* and *E* (LaJeunesse & Trench 2000). In the scleractinian corals, intraspecific surveys of *Plesiastrea versipora* (Baker 1999, Rodriguez-Lanetty et al. 2001), *Seriatopora hystrix*, and *Acropora longicyathus* (Loh et al. 2001) indicate that members of *Symbiodinium C* are common in tropical populations (with *S. hystrix* being dominated by *Symbiodinium D* at equatorial locations), but members of *Symbiodinium B* (*P. versipora*) or *A* (*A. longicyathus*) become more common at higher latitudes (23°–35°S). Some evidence of geographic variation within clade *C* was also reported (Loh et al. 2001). Taken together, these reports suggest that both temperature and light have important roles to play in determining *Symbiodinium* distribution (see review in A.C. Baker, submitted manuscript).

Unusual Symbioses in the Tropical Western Atlantic

Despite broad latitudinal patterns in the distribution of *Symbiodinium* worldwide, it is apparent that scleractinian corals in the tropical western Atlantic still host *Symbiodinium A* and *B* much more commonly than their Pacific counterparts at comparable latitudes. Why should coral-algal symbioses in the Caribbean more closely resemble those found at higher latitudes in the Pacific? Increased extinction rates and faunal turnover of Caribbean scleractinian corals that coincided with the Plio-Pleistocene onset of Northern Hemisphere glaciation (Budd 2000) may have selected for symbioses those that were suited not only to cooler temperatures (Stanley 1986), but more specifically to considerably higher seasonality (Jackson et al. 1993). Conditions in the tropical western Atlantic during this time may therefore have resembled those characterizing higher latitudes, resulting in a shift to symbiont communities involving clades *A* and *B* (Baker & Rowan 1997; LaJeunesse et al. 2003; A.C. Baker, C.J. Starger, T.R. McClanahan, P.W. Glynn, submitted manuscript). These symbionts have since diversified in the endemic Caribbean scleractinian coral fauna (T.C. LaJeunesse, submitted manuscript). Pulses of cooler water in tropical environments in the western Atlantic may thus have favored the evolution of an endemic Caribbean coral fauna whose symbioses were more characteristic of those found at higher latitudes (A.C. Baker, submitted manuscipt).

Large-scale symbiont community shifts, similar to those hypothesized for Caribbean corals, may also characterize contemporary reef corals experiencing rising sea surface temperatures and recurrent episodes of mass bleaching (A.C. Baker, C.J. Starger, T.R. McClanahan, P.W. Glynn, submitted manuscript.) The high incidence of *Symbiodinium D* in scleractinian corals from tropical locations (Baker 1999; LaJeunesse et al. 2003; Loh et al. 2001; Toller et al. 2001a,b; Van Oppen et al. 2001; A.C. Baker, submitted manuscript), recently bleached reefs (Glynn et al. 2001), and extreme high temperature environments (Baker et al. 2003), together with its absence from high latitude locations sampled to date (A.C. Baker, submitted manuscript; Baker 1999; Loh et al. 2001; Rodriguez-Lanetty et al. 2001; Savage et al. 2002) support the idea that symbiont community change may already be occurring in these affected ecosystems (see Climate Change below).

FLEXIBILITY OF INDIVIDUAL HOSTS

Controversy over the role of symbiont diversity in enabling hosts to mitigate environmental change revolves largely around how quickly symbiont change within individual hosts can occur. This has been investigated by experimental manipulation (e.g., Baker 2001, Kinzie et al. 2001, Rowan et al. 1997) and by field observations of individuals in response to bleaching (see Climate Change below), disease (Toller et al. 2001a), ontogeny, and seasonality.

Host Reproduction and Ontogeny

Scleractinian corals that do not supply their offspring with symbionts must perforce re-establish their symbioses with free-living *Symbiodinium*. The pool of symbionts upon which these "open" systems rely may be dependent on the diversity of potential symbionts found in other (non-scleractinian) hosts (e.g., Lee et al. 1995). The extent to which this is true may depend largely on the particular symbionts and hosts concerned, since it is clear that some symbiont types are found in a variety of taxonomically dissimilar hosts, but that others are highly specific (LaJeunesse 2002). For example, *Maristentor dinoferus*, a giant heterotrich ciliate recently discovered in Guam contains *Symbiodinium* that are indistinguishable (by lsrDNA analysis) from those found in many scleractinian corals (Lobban et al. 2002).

Corals that maternally transfer symbionts to their offspring might be expected to contain less diverse symbionts than corals whose larvae or juveniles are required to obtain them environmentally (Douglas 1998). Despite this logical expectation, little evidence for such patterns has been observed to date (Hidaka & Hirose 2000; Van Oppen 2003). This may reflect lack of reliable information on symbiont transmission for many coral species or may indicate that many corals are able to obtain symbionts from the environment throughout their life cycle (see Goulet & Coffroth 2003a,b).

Symbiont specificity during early ontogeny is a characteristic feature of invertebrate-algal symbiosis that has been well studied in anemones (Fitt 1984,

Kinzie & Chee 1979, Schoenberg & Trench 1980c), jellyfish (Colley & Trench 1983), gorgonians (Coffroth et al. 2001, Kinzie 1974), giant clams (Fitt 1985b), and scleractinians (Schwarz et al. 1999, Van Oppen 2001, Weis et al. 2001). Most of these studies investigated host specificity for cultured or freshly isolated symbionts from the same (homologous) or different (heterologous) host species, and only recently have field investigations using molecular methods been employed to survey the natural occurrence of symbionts in larval or juvenile hosts. In a field study of a Caribbean gorgonian, newly settled polyps initially acquired diverse symbionts whose identity appeared to depend on the settlement habitat; symbiont populations reverted to the original (maternal) populations after three to six months (Coffroth et al. 2001). Similar molecular comparisons of a second gorgonian found that newly settled polyps contained different symbiont communities than adult colonies (Santos et al. 2003a).

Studies of scleractinian corals have also demonstrated the capacity for diverse symbionts to be acquired during early ontogeny (Schwarz et al. 1999, Van Oppen 2001), but some symbionts may still be favored over others (Weis et al. 2001). These findings indicate that the early ontogeny of cnidarian hosts may allow more flexible associations than the later adult stage, but it is not yet clear to what extent this flexibility can be interpreted as an ecological strategy allowing juveniles to colonize new habitats, or whether it simply represents the early proliferation of unusual symbionts in vacant hosts before a steady symbiotic state is established. Such a phenomenon may also explain the novel symbionts that appear early in the recovery of severely bleached scleractinian colonies (Baker 2001, Toller et al. 2001a).

Adult hosts may also be able to exchange symbionts with the environment in varying degrees. This has been demonstrated in anemones (Kinzie et al. 2001), but to date there is little direct evidence for this in scleractinian corals, which are not easily rendered aposymbiotic (see review in Buddemeier et al. 2003).

Seasonal Changes in Symbiont Communities

Seasonal variation, especially at high latitudes, may also create environmental variation that favors different symbiont communities. Such changes may occur in benthic foraminifera (X. Pochon, unpublished data). Community surveys of one species of scleractinian coral in southern Taiwan documented a change in relative abundance of *Symbiodinium C* and *D* on a seasonal basis, being more dominated by *D*-types in the summer compared with winter (Yang et al. 2000). While predictable seasonal change has not been documented in individual hosts (e.g., Belda-Baillie et al. 2000), corals at high latitudes in variable environments may experience shifts between symbiont populations (Thornhill et al. 2003).

CORAL REEF BLEACHING AND CLIMATE CHANGE

Principal motivations for research into *Symbiodinium* diversity, biogeography, and ecology in coral reefs are the obligate nature of these symbionts for many of their hosts (including all tropical reef corals) and the central role they play in

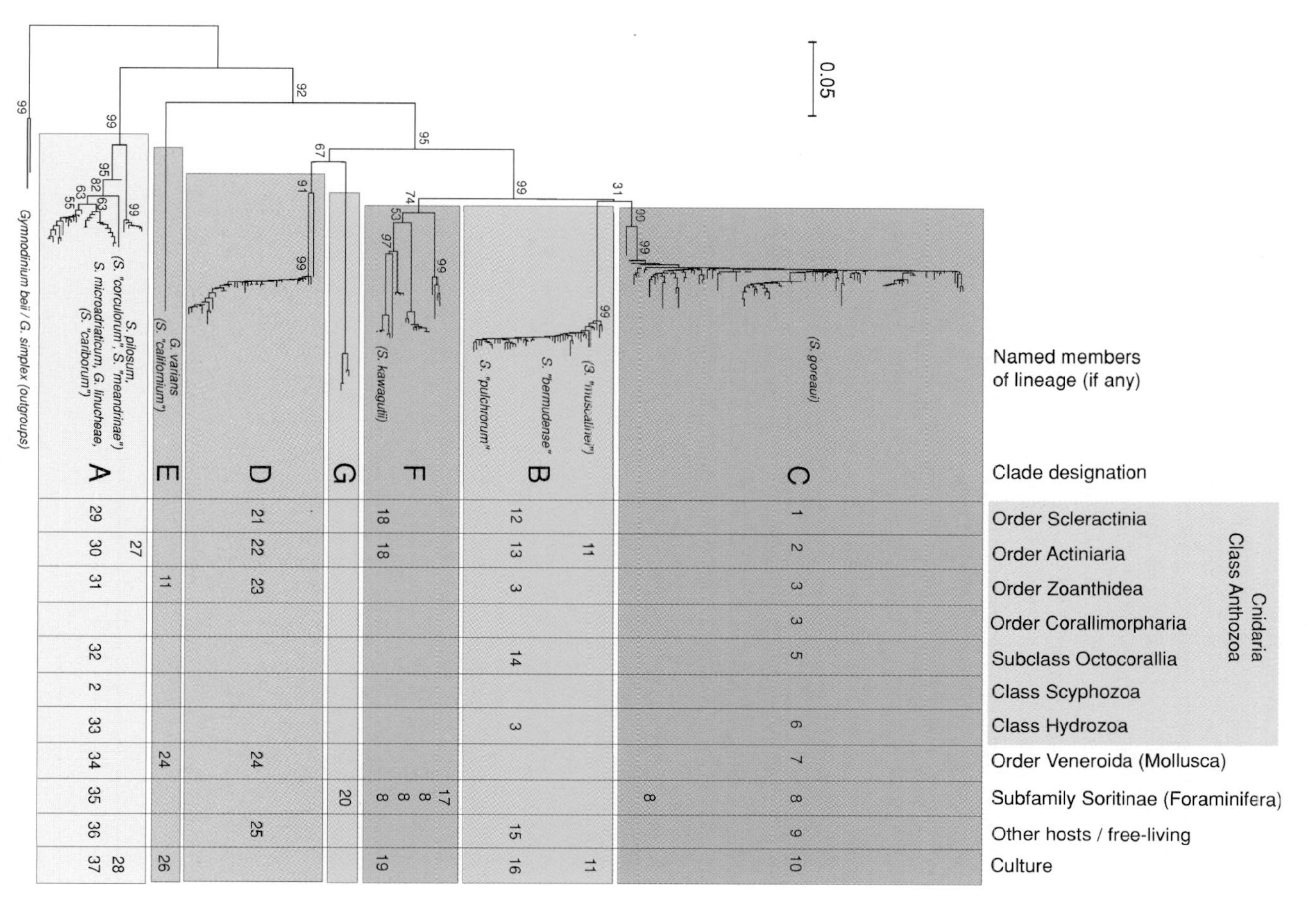

See legend on next page

Figure 1 Molecular phylogeny of *Symbiodinium* indicating seven principal clades *A–G* (*left*), the eleven named species (*center*), and the distribution of the principal clades (and subclades) of *Symbiodinium* among different host taxa (*right*). LSU nrDNA sequences (~600 nt) were aligned using Clustal X (Thompson et al. 1997) and phylogenetic analyses performed by neighbor-joining using PAUP 4.0 (Swofford 2002). Reliability of internal branches was assessed by 1000 bootstrap replicate analyses (numerical values shown at nodes). Scale bar represents number of substitutions per site. Position of named species in parentheses is included for illustrative purposes only; LSU rDNA sequence data not available for these taxa. Identifications based on cultured material are excluded for the purposes of host distribution. Details on phylogenetic reconstruction, Genbank Accession numbers, and literature references represented by numerals in right-hand columns can be found in our website supplementary materials (Follow the Supplemental Materials link from the Annual Reviews home page at http://www.annualreviews.org).

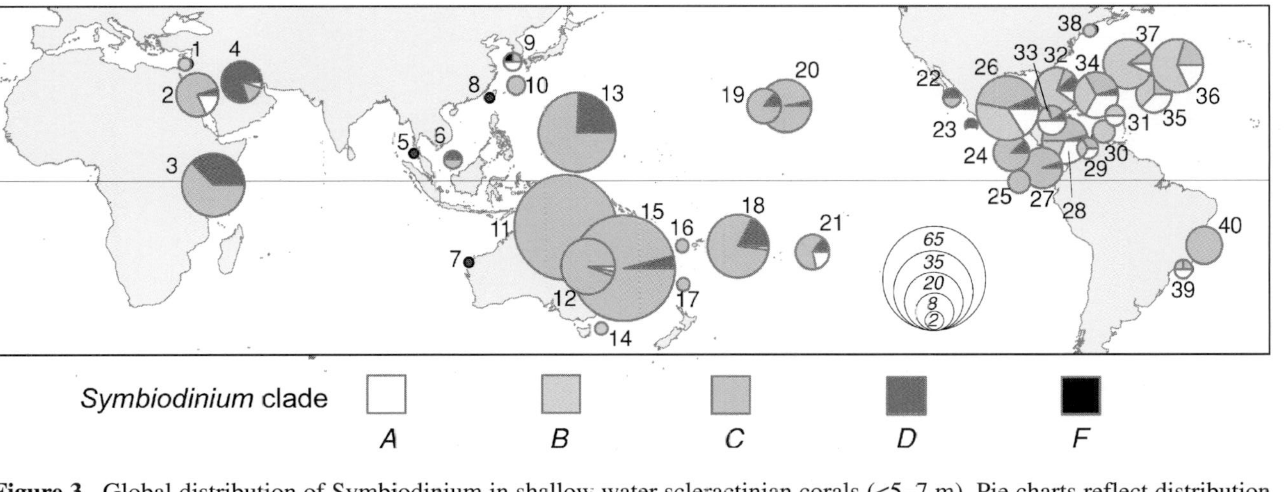

Figure 3 Global distribution of Symbiodinium in shallow water scleractinian corals (<5–7 m). Pie charts reflect distribution of different clades A, B, C, D, and F among species of coral host sampled, with the diameter of the pie chart approximately proportional to the square root of the number of species sampled (see inset scale). Numerals next to pie charts refer to data source listed in Table 2 (Follow the Supplemental Materials link from the Annual Reviews home page at http://www.annualreviews.org)). Modified from A.C. Baker (submitted manuscript).

understanding coral reef bleaching events. Increasingly frequent and severe episodes of mass coral bleaching and mortality over the past two decades as a result of warmer baseline temperatures and increasingly severe temperature anomalies such as El Niño (Brown 1997, Glynn 1993) suggest that reef ecosystems may be fast approaching a critical survival threshold (Hoegh-Guldberg 1999). To what degree does symbiont diversity explain the inherent spatial and systematic variability in coral bleaching, and to what degree will symbiont community change and/or recombination affect future bleaching scenarios?

Variability in Coral Bleaching

Recent reviews (Baker 2003, Buddemeier et al. 2003, Knowlton & Rohwer 2003) have emphasized the importance of coral bleaching in motivating contemporary research on the diversity of *Symbiodinium*. However, understanding how symbiont diversity affects the incidence and severity of bleaching has been difficult because we usually have little knowledge of how symbionts are distributed prior to bleaching events (but see Glynn et al. 2001; Rowan et al. 1997; A.C. Baker, C.J. Starger, T.R. McClanahan, P.W. Glynn, submitted manuscript). The possibility that symbiont taxa vary in their temperature sensitivities (and hence susceptibility to bleaching) was initially proposed in the absence of any supporting evidence (Buddemeier & Fautin 1993, Gladfelter 1988, Rowan & Powers 1991a, Sandeman 1988; but see Iglesias-Prieto et al. 1992). However, this hypothesis has since been confirmed in at least two studies of scleractinian coral: Rowan et al. (1997) demonstrated that loss of members of *Symbiodinium C* at the upper limits of their irradiance distribution explained otherwise enigmatic variability in bleaching of Caribbean *Montastraea* in 1995, whereas Glynn et al. (2001) showed that patchy bleaching in Pacific *Pocillopora* was explained by the preferential loss of *Symbiodinium C*-types and retention of *Symbiodinium D*-types. Bleaching severity may also correlate with symbiont diversity in affected hosts (Baker & Rowan 1997, LaJeunesse et al. 2003, Loh et al. 2000), but the resolution of these correlations will tend to be much lower than for direct comparisons in which specific symbiont identities are known. However, despite these uncertainties, differences in bleaching susceptibility among different symbiont taxa have clear implications for how symbiont distributions might be applied to predicting or understanding future bleaching incidence and severity (Baker 2003).

Changes in Symbiont Community Structure Following Bleaching

Changes in symbiont community structure following mass bleaching events can arise in three ways that are not mutually independent: (*a*) differential mortality of bleaching-susceptible combinations (true "Darwinian" adaptation, or natural selection); (*b*) quantitative change in the relative abundance of existing symbiont communities within colonies (symbiont "shuffling"); and (*c*) qualitative change by recombination with symbionts acquired from the environment (symbiont "switching"). The first process follows logically from mass mortality of bleached corals containing bleaching-susceptible symbionts (see Variability above). The latter two processes form part of the "adaptive bleaching hypothesis" (ABH) proposed by Buddemeier & Fautin (1993). Some confusion has arisen over the misunderstanding that the ABH applies exclusively to symbiont "switching" involving exogenous pools of prospective symbionts (Hoegh-Guldberg 1999, Hoegh-Guldberg et al. 2002); this is not the case (Baker 2002, Buddemeier et al. 2003).

Much of the controversy surrounding the ABH centers around how quickly different and/or novel symbiont communities can become established in individual hosts during a bleaching event and whether the "ecospecies" (Buddemeier et al. 2003) thus created is more likely to survive as a result (Douglas 2003, Hoegh-Guldberg 1999). Extensive mass mortality of reef corals following bleaching is used as de facto evidence against the hypothesis, and the lack of any evidence for partner recombination following natural episodes of bleaching (Hoegh-Guldberg et al. 2002) is also seen as evidence against the idea. However, few attempts to monitor symbiont communities during bleaching have been undertaken to date (Baker 2002), and the results of these studies have tended to support, rather than refute, the basic ideas of the ABH (Baker 2001, Glynn et al. 2001, Rowan et al. 1997, Toller et al. 2001a). The capacity of reef corals to become dominated by different symbionts following dramatic environmental change and bleaching has been documented in field experiments (Baker 2001, Rowan et al. 1997), and changes in symbionts can still occur (although these displacements may be only transitory) even when environmental conditions return to normal after bleaching (Toller et al. 2001a, see also Thornhill et al. 2003). However, none of these experiments have been able to distinguished quantitative change (in endogenous symbiont communities found within individual colonies) from qualitative change (involving exogenous populations of symbionts from the environment).

Recent studies indicate that bleaching does not represent the sudden breakdown of an otherwise stable relationship in which symbiont standing stocks are held constant. Instead, bleaching is an extreme version of a symbiont regulatory mechanism in which algal densities change in response to a changing environment (Fagoonee et al. 1999, Fitt et al. 2000, Kinzie 1999) over seasonal (and shorter) timescales. Symbiont community change by switching (an "open" symbiotic system) or shuffling (a "closed" symbiotic system) may occur without visible bleaching; severe bleaching may merely accelerate this process by promoting turnover of existing partnerships (Baker 2001, Buddemeier & Fautin 1993).

The Future of Coral Reefs

The extent to which symbiont diversity and flexibility will affect the long-term future of coral reefs in response to continued climate change is not yet clear. Although evidence exists indicating that diverse symbionts can significantly buffer the effects of climate change (by differential mortality of bleaching-susceptible combinations and/or symbiont switching or shuffling within hosts), the scales over which such mechanisms might occur are as yet unknown and many important questions remain. What fraction of existing symbioses on coral reefs might be considered bleaching susceptible, and to what degree does symbiont community structure on reefs remain constant over time? How long might it take for bleached reefs to recover from (the perhaps few) bleaching-resistant symbioses that survive, and might we nevertheless suffer irreversible ecological phase shifts as a result of significant coral loss (Done 1999, Hoegh-Guldberg 1999, Knowlton 1992)? Can we still expect ecological extinction of reefs in some areas, despite the longer-term persistence of coral reefs worldwide (Buddemeier & Smith 1999, Buddemeier et al. 1997, Smith & Buddemeier 1992)? How often do bleached individuals recover with quantitatively different (shuffled) or qualitatively different (switched) symbiont communities, and does bleaching accelerate either of these processes (Baker 2001, Buddemeier & Fautin 1993)? These are the underlying questions that must be addressed in order to establish the degree to which symbiont community change on reefs is possible over ecological (as opposed to evolutionary) timescales and whether or not individual colonies are able to modify their symbiont communities rapidly enough to adapt or acclimatize to a changing local climate.

The spatial resilience of coral reefs—the dynamic capacity of a reef to avoid thresholds at a regional scale (Nystrom & Folke 2001, Nystrom et al. 2000)—may be significantly increased by the diverse and flexible nature of symbioses involving *Symbiodinium*. Human efforts to increase the spatial resilience of coral reefs to bleaching through the creation and management of Marine Protected Area (MPA) networks (West & Salm 2003) should explicitly incorporate unusual and diverse habitats that maximize symbiont diversity into their design. When coral reef hosts are assessed over their full range of systematic, ontogenetic, ecological, and biogeographic gradients, the existence of unusual symbionts normally found only in uncommon host taxa, larval stages, marginal environments, or at latitudinal extremes may prove critical in understanding the long-term resilience of coral reef ecosystems to environmental perturbation.

ACKNOWLEDGMENTS

I would like to thank M. Coffroth, T. Goulet, R. Kinzie, N. Knowlton, T. LaJeunesse, X. Pochon, M. Rodriguez-Lanetty, S. Santos, and M. Van Oppen for their valuable comments on the manuscript, and C. Starger, J. Drew, and T. Baker for assistance in manuscript preparation. I acknowledge prior support from a University of Miami Maytag Fellowship, a Smithsonian Institution Predoctoral Fellowship, and a Lizard Island Doctoral Fellowship. I thank J. Fell, R. DeSalle, G. Amato, and D. Melnick for laboratory facilities, and P. Glynn, R. Rowan, and N. Knowlton for their teaching and advice. This paper was written with financial support from NSF (OCE-0099301 and INT-9908673), NOAA (NURC 2002–15B), and the Wildlife Conservation Society (Tiffany & Co. Foundation, McBean Family Foundation, Peretti Foundation, Gimbel Foundation, and H. Railton, contributors).

The *Annual Review of Ecology, Evolution, and Systematics* is online at http://ecolsys.annualreviews.org

LITERATURE CITED

Aisyah EN, Hoegh-Guldberg O, Hinde R, Loh W. 2000. The relationship between morphology and molecular variation of zooxanthellae from temperate Australian reefs. *Proc. Int. Coral Reef Symp., 9th, Bali*, p. 31 (Abstr.)

Baillie BK, Belda-Baillie CA, Maruyama T. 2000a. Conspecificity and Indo-Pacific distribution of *Symbiodinium* genotypes (Dinophyceae) from giant clams. *J. Phycol.* 36:1153–61

Baillie BK, Belda-Baillie CA, Silvestre V, Sison M, Gomez AV,et al. 2000b. Genetic variation in *Symbiodinium* isolates from giant clams based on random-amplified-polymorphic DNA (RAPD) patterns. *Mar. Biol.* 136:829–36

Baillie BK, Monje V, Silvestre V, Sison M, Belda-Baillie CA. 1998. Allozyme electrophoresis as a tool for distinguishing different zooxanthellae symbiotic with giant clams. *Proc. R. Soc. London Ser. B* 265: 1949–56

Baker AC. 1999. *The symbiosis ecology of reef-building corals.* PhD thesis. Univ. Miami. 120 pp.

Baker AC. 2001. Reef corals bleach to survive change. *Nature* 411:765–66

Baker AC. 2002. Is bleaching really adaptive?—reply to Hoegh-Guldberg et al. *Nature* 415:602

Baker AC. 2003. Symbiont diversity on coral reefs and its relationship to bleaching resistance and resilience. See Rosenberg & Loya 2003. In press

Baker AC, Jones SH, Lee TS. 2003. Symbiont diversity in Arabian corals and its relation to patterns of contemporary and historical environmental stress. *Fauna Saudi Arabia.* In press

Baker AC, Rowan R. 1997. Diversity of symbiotic dinoflagellates (zooxanthellae) in scleractinian corals of the Caribbean and eastern Pacific. *Proc. Int. Coral Reef Symp., 8th, Panama,* 2:1301–5

Baker AC, Rowan R, Knowlton N. 1997. Symbiosis ecology of two Caribbean acroporid corals. *Proc. Int. Coral Reef Symp., 8th, Panama,* 2:1295–300

Banaszak AT, Iglesias-Prieto R, Trench RK. 1993. *Scrippsiella velellae* sp. nov. (Peridiniales) and *Gloeodinium viscum* sp. nov. (Phytodiniales), dinoflagellate symbionts of two hydrozoans (Cnidaria). *J. Phycol.* 29:517–28

Banaszak AT, LaJeunesse TC, Trench RK. 2000. The synthesis of mycosporine-like amino acids (MAAs) by cultured, symbiotic dinoflagellates. *J. Exp. Mar. Biol. Ecol.* 249:219–33

Belda-Baillie CA, Baillie BK, Maruyama T. 2002. Specificity of a model cnidarian-dinoflagellate symbiosis. *Biol. Bull.* 202:74–85

Belda-Baillie CA, Baillie BK, Shimoike K, Maruyama T. 2000. Seasonal population dynamics of algal symbionts of acroporids and tridacnids in an Okinawa reef. *Proc. Int. Coral Reef Symp., 9th, Bali*, p. 31. (Abstr.)

Belda-Baillie CA, Sison M, Silvestre V, Villamor K, Monje V, et al. 1999. Evidence for changing symbiotic algae in juvenile tridacnids. *J. Exp. Mar. Biol. Ecol.* 241:207–21

Billinghurst Z, Douglas AE, Trapido-Rosenthal H. 1997. On the genetic diversity of the symbiosis between the coral *Montastraea cavernosa* and zooxanthellae in Bermuda. *Proc. Int. Coral Reef Symp., 8th, Panama,* 2:1291–94

Blank RJ, Trench RK. 1985a. Speciation and symbiotic dinoflagellates. *Science* 229:656–58

Blank RJ, Trench RK. 1985b. *Symbiodinium microadriaticum*: A single species? *Proc. Int. Coral Reef Congr., 5th, Tahiti,* 6:113–17

Blank RJ, Trench RK. 1986. Nomenclature of endosymbiotic dinoflagellates. *Taxon* 35: 286–94

Brown BE. 1997. Coral bleaching: causes and consequences. *Coral Reefs.* 16:S129–38

Brown BE, Dunne RP, Goodson MS, Douglas AE. 2000. Bleaching patterns in reef corals. *Nature* 404:142–43

Brown BE, Dunne RP, Goodson MS, Douglas AE. 2002. Experience shapes the susceptibility of a reef coral to bleaching. *Coral Reefs* 21:119–26

Budd AF. 2000. Diversity and extinction in the Cenozoic history of Caribbean reefs. *Coral Reefs* 19:25–35

Buddemeier RW, Baker AC, Fautin DG, Jacobs JR. 2003. The adaptive hypothesis of bleaching. See Rosenberg & Loya 2003. In press

Buddemeier RW, Fautin DG. 1993. Coral bleaching as an adaptive mechanism—a testable hypothesis. *BioScience* 43:320–26

Buddemeier RW, Fautin DG, Ware JR. 1997. Acclimation, adaptation and algal symbiosis in reef-building scleractinian corals. *Proc. Int. Conf. Coelenterate Biol., 6th, Netherlands,* 1:71–76

Buddemeier RW, Smith SV. 1999. Coral adaptation and acclimatization: a most ingenious paradox. *Am. Zool.* 39:1–9

Burnett WJ. 2002. Longitudinal variation in algal symbionts (zooxanthellae) from the Indian Ocean zoanthid Palythoa caesia. *Mar. Ecol.-Prog. Ser.* 234:105–9

Carlos AA, Baillie BK, Kawachi M, Maruyama T. 1999. Phylogenetic position of *Symbiodinium* (Dinophyceae) isolates from tridacnids (Bivalvia), cardiids (Bivalvia), a sponge (Porifera), a soft coral (Anthozoa), and a free-living strain. *J. Phycol.* 35:1054–62

Carlos AA, Baillie BK, Maruyama T. 2000. Diversity of dinoflagellate symbionts (zooxanthellae) in a host individual. *Mar. Ecol.-Prog. Ser.* 195:93–100

Chang SS, Prezelin BB, Trench RK. 1983. Mechanisms of photoadaptation in three strains of the symbiotic dinoflagellate *Symbiodinium microadriaticum. Mar. Biol.* 76: 219–29

Chen CA, Wei NV, Tsai WS, Fang LS. 2003a. Symbiont diversity in the scleractinian corals from tropical reefs and non-reefal communities in Taiwan. *Coral Reefs.* In press

Chen CA, Lam KK, Nakano Y, Tsai WS. 2003b. Stable association of a stress-tolerant zooxanthellae, *Symbiodinium* clade D, with the low-temperature tolerant coral *Oulastrea crispata,* (Scleractinia; Faviidae) in subtropical nonreefal coral communities. *Zool. Stud.* 42: In press

Coffroth MA, Santos SR, Goulet TL. 2001. Early ontogenetic expression of specificity in a cnidarian-algal symbiosis. *Mar. Ecol.-Prog. Ser.* 222:85–96

Colley NJ, Trench RK. 1983. Selectivity in phagocytosis and persistence of symbiotic algae by the scyphistoma stage of the jellyfish *Cassiopeia xamachana. Proc. R. Soc. London Ser. B* 219:61–82

Cowen R. 1988. The role of algal symbiosis in reefs through time. *Palaios* 3:221–27

Darius HT, Dauga C, Grimont PAD, Chungue E, Martin PMV. 1998. Diversity in symbiotic dinoflagellates (Pyrrhophyta) from seven scleractinian coral species: Restriction enzyme analysis of small subunit ribosomal RNA genes. *J. Eukaryot. Microbiol.* 45:619–27

Darius HT, Martin PMV, Grimont PAD, Dauga C. 2000. Small subunit rDNA sequence analysis of symbiotic dinoflagellates from seven scleractinian corals in a Tahitian lagoon. *J. Phycol.* 36:951–59

Diekmann OE, Bak RPM, Tonk L, Stam WT, Olsen JL. 2002. No habitat correlation of zooxanthellae in the coral genus *Madracis* on a Curaçao reef. *Mar. Ecol. Prog. Ser.* 227:221–32

Diekmann OE, Olsen JL, Stam WT, Bak RPM. 2003. Genetic variation within Symbiodinium clade B from the coral genus Madracis in the Caribbean (Netherlands Antilles). Coral Reefs 22:29–33

Done TJ. 1999. Coral community adaptability to environmental change at the scales of regions, reefs and reef zones. *Am. Zool.* 39:66–79

Douglas AE. 1998. Host benefit and the evolution of specialization in symbiosis. *Heredity* 81:599–603

Douglas AE. 2003. Coral bleaching—how and why? *Mar. Pollut. Bull.* 46:385–92

Dubos R, Kessler A. 1963. Integrative and disintegrative factors in symbiotic associations. *Symp. Soc. Gen. Microbiol.* 13:1–11

Dustan P. 1982. Depth dependent photoadaption by zooxanthellae of the reef coral *Montastrea annularis*. *Mar. Biol.* 68:253–64

Eldredge N, Cracraft J. 1980. *Phylogenetic Patterns and the Evolutionary Process: Method and Theory in Comparative Biology*. New York: Columbia Univ. Press. 349 pp.

Fagoonee I, Wilson HB, Hassell MP, Turner JR. 1999. The dynamics of zooxanthellae populations: A long-term study in the field. *Science* 283:843–45

Falkowski PG, Dubinsky Z. 1981. Light-shade adaptation of *Stylophora pistillata*, a hermatypic coral from the Gulf of Eilat. *Nature* 289:172–74

Fitt WK. 1984. The role of chemosensory behavior of *Symbiodinium microadriaticum*, intermediate hosts, and host behavior in the infection of coelenterates and mollusks with zooxanthellae. *Mar. Biol.* 81:9–17

Fitt WK. 1985a. Chemosensory responses of the symbiotic dinoflagellate *Symbiodinium microadriaticum* (Dinophyceae). *J. Phycol.* 21:62–67

Fitt WK. 1985b. Effect of different strains of the zooxanthella *Symbiodinium microadriaticum* on growth and survival of their coelenterate and molluscan hosts. *Proc. Int. Coral Reef Symp., 5th, Tahiti*, 6:131–36

Fitt WK, McFarland FK, Warner ME, Chilcoat GC. 2000. Seasonal patterns of tissue biomass and densities of symbiotic dinoflagellates in reef corals and relation to coral bleaching. *Limnol. Oceanogr.* 45:677–85

Freudenthal HD. 1962. *Symbiodinium* gen. nov. and *Symbiodinium microadriaticum* sp. nov., a zooxanthella: Taxonomy, life cycle, and morphology. *J. Protozool.* 9:45–52

Gladfelter EH. 1988. The physiological basis of coral bleaching. See Ogden & Wicklund 1988, pp. 15–18

Glynn PW. 1993. Coral reef bleaching-ecological perspectives. *Coral Reefs* 12:1–17

Glynn PW, Maté JL, Baker AC, Calderón MO. 2001. Coral bleaching and mortality in Panamá and Ecuador during the 1997–1998 El Niño-Southern Oscillation event: spatial/temporal patterns and comparisons with the 1982–1983 event. *Bull. Mar. Sci.* 69:79–109

Goff LJ. 1982. *Algal Symbiosis: A Continuum of Interaction Strategies*. New York: Cambridge Univ. Press. 216 pp.

Goodson MS. 2000. *Symbiotic algae: molecular diversity in marginal coral reef habitats.* Dr. Phil. thesis. Univ.York, UK. 155 pp.

Goodson MS, Whitehead LF, Douglas AE. 2001. Symbiotic dinoflagellates in marine Cnidaria: diversity and function. *Hydrobiologia* 461:79–82

Goulet TL. 1999. *Temporal and spatial stability of zooxanthellae in octocorals*. PhD thesis. State Univ. New York, Buffalo. 101 pp.

Goulet TL, Coffroth MA. 1997. A within-colony comparison of zooxanthellae in the Caribbean gorgonian *Plexaura kuna. Proc. Int. Coral Reef Symp., 8th, Panama*, 2:1331–34

Goulet TL, Coffroth MA. 2003a. Genetic composition of zooxanthellae between and within colonies of the octocoral Plexaura kuna, based on small subunit rDNA and multilocus DNA fingerprinting. *Mar. Biol.* 142:233–39

Goulet TL, Coffroth MA. 2003b. Stability of an octocoral-algal symbiosis over time and space. *Mar. Ecol. Prog. Ser.* 250:117–24

Hidaka M, Hirose M. 2000. A phylogenetic comparison of zooxanthellae from reef corals with different modes of symbiont acquisition. *Proc. Int. Coral Reef Symp., 9th, Bali*, p.33. (Abstr.)

Hill M, Wilcox T. 1998. Unusual mode of symbiont repopulation after bleaching in *Anthosigmella varians*: acquisition of different zooxanthellae strains. *Symbiosis* 25:279–89

Hoegh-Guldberg O. 1999. Climate change,

coral bleaching and the future of the world's coral reefs. *Mar. Freshw. Res.* 50:839–66

Hoegh-Guldberg O, Jones RJ, Ward S, Loh WK. 2002. Is bleaching really adaptive? *Nature* 415:601–2

Hunter CL, Morden CW, Smith CM. 1997. The utility of ITS sequences in assessing relationships among zooxanthellae and corals. *Proc. Int. Coral Reef Symp., 8th, Panama*, 2:1599–1602

Iglesias-Prieto R, Matta JL, Robins WA, Trench RK. 1992. Photosynthetic response to elevated temperature in the symbiotic dinoflagellate *Symbiodinium microadriaticum* in culture. *Proc. Natl. Acad. Sci. USA* 89:10302–5

Iglesias-Prieto R, Trench RK. 1994. Acclimation and adaptation to irradiance in symbiotic dinoflagellates. 1. Responses of the photosynthetic unit to changes in photon flux density. *Mar. Ecol. Prog. Ser.* 113:163–75

Iglesias-Prieto R, Trench RK. 1997. Acclimation and adaptation to irradiance in symbiotic dinoflagellates. II. Response of chlorophyll-protein complexes to different photon-flux densities. *Mar. Biol.* 130:23–33

Jackson JBC, Jung P, Coates AG, Collins LS. 1993. Diversity and extinction of tropical American mollusks and emergence of the isthmus of Panama. *Science* 260:1624–26

Jokiel PL, Coles SL. 1977. Effects of temperature on mortality and growth of Hawaiian reef corals. *Mar. Biol.* 43:201–8

Jokiel PL, York RH. 1982. Solar ultraviolet photobiology of the reef coral *Pocillopora damicornis* and symbiotic zooxanthellae. *Bull. Mar. Sci.* 32:301–15

Kinzie RA. 1974. Experimental infection of aposymbiotic gorgonian polyps with zooxanthellae. *J. Exp. Mar. Biol. Ecol.* 15:335–45

Kinzie RA. 1999. Sex, symbiosis and coral reef communities. *Am. Zool.* 39:80–91

Kinzie RA, Chee GS. 1979. The effect of different zooxanthellae on the growth of experimentally re-infected hosts. *Biol. Bull.* 156:315–27

Kinzie RA, Takayama M, Santos SR, Coffroth MA. 2001. The adaptive bleaching hypothesis: Experimental tests of critical assumptions. *Biol. Bull.* 200:51–58

Kinzie RAI. 1970. *The ecology of gorgonians (Cnidaria, Octocorallia) of Discovery Bay, Jamaica*. PhD thesis. Yale Univ. 107 pp.

Knowlton N. 1992. Thresholds and multiple stable states in coral reef community dynamics. *Am. Zool.* 32:674–82

Knowlton N, Rohwer F. 2003. Multi-species microbial mutualisms on coral reefs: The host as a habitat. *Am. Nat.* In press

LaJeunesse TC. 2001. Investigating the biodiversity, ecology and phylogeny of endosymbiotic dinoflagellates in the genus *Symbiodinium* using the ITS region: In search of a "species" level marker. *J. Phycol.* 37:866–80

LaJeunesse TC. 2002. Diversity and community structure of symbiotic dinoflagellates from Caribbean coral reefs. *Mar. Biol.* 141:387–400

LaJeunesse TC, Loh WKW, van Woesik R, Hoegh-Guldberg O, Schmidt GW, Fitt WK. 2003. Low symbiont diversity in southern Great Barrier Reef corals relative to those of the Caribbean. *Limnol. Oceanogr.* 48:2046–54

LaJeunesse TC, Trench RK. 2000. Biogeography of two species of *Symbiodinium* (Freudenthal) inhabiting the intertidal sea anemone *Anthopleura elegantissima* (Brandt). *Biol. Bull.* 199:126–34

Lee JJ, Wray CG, Lawrence C. 1995. Could foraminiferal zooxanthellae be derived from environmental pools contributed to by different coelenterate hosts? *Acta Protozool.* 34:75–85

Lobban CS, Schefter M, Simpson AGB, Pochon X, Pawlowski J, Foissner W. 2002. *Maristentor dinoferus* n. gen., n. sp., a giant heterotrich ciliate (Spirotrichea: Heterotrichida) with zooxanthellae, from coral reefs on Guam, Mariana Islands. *Mar. Biol.* 140:411–23

Loeblich AR, Sherley JL. 1979. Observations of the theca of the motile phase of free-living symbiotic isolates of *Zooxanthella microadriatica*. *J. Mar. Biol. Assoc. UK* 59:195–205

Loh W, Carter D, Hoegh-Guldberg O. 1998.

Diversity of zooxanthellae from scleractinian corals of One Tree Island (the Great Barrier Reef). *Proc. Aust. Coral Reef Soc. 75th Ann. Conf., Heron Island*, pp.141–51

Loh W, Sakai K, Hoegh-Guldberg, O. 2000. Coral zooanthellae diversity in bleached reefs. *Proc. Int. Coral Reef Symp., 9th, Bali*, p. 33. (Abstr.)

Loh WKW, Loi T, Carter D, Hoegh-Guldberg O. 2001. Genetic variability of the symbiotic dinoflagellates from the wide ranging coral species Seriatopora hystrix and Acropora longicyathus in the Indo-West Pacific. *Mar. Ecol. Prog. Ser.* 222:97–107

Muscatine L, Porter JW. 1977. Reef corals—mutualistic symbioses adapted to nutrient-poor environments. *BioScience* 27:454–60

Nystrom M, Folke C. 2001. Spatial resilience of coral reefs. *Ecosystems* 4:406–17

Nystrom M, Folke C, Moberg F. 2000. Coral reef disturbance and resilience in a human-dominated environment. *Trends Ecol. Evol.* 15:413–17

Ogden JC, Wicklund RI,eds. 1988. *Mass Bleaching of Coral Reefs in the Caribbean: A Research Strategy.* Natl. Undersea Res. Prog. Rep. 88–28

Paine RT. 1969. A note on trophic complexity and community stability. *Am. Nat.* 103:91–93

Pawlowski J, Holzmann M, Fahrni JF, Pochon X, Lee JJ. 2001. Molecular identification of algal endosymbionts in large miliolid foraminifera: 2. Dinoflagellates. *J. Eukaryot. Microbiol.* 48:368–73

Pochon X, Pawlowski J, Zaninetti L, Rowan R. 2001. High genetic diversity and relative specificity among *Symbiodinium*-like endosymbiotic dinoflagellates in soritid foraminiferans. *Mar. Biol.* 139:1069–78

Power ME, Tilman D, Estes JA, Menge BA, Bond WJ, et al. 1996. Challenges in the quest for keystones. *BioScience* 46:609–20

Rizzo PJ. 1987. Biochemistry of the dinoflagellate nucleus. In *The Biology of Dinoflagellates*, ed. FJR Taylor, 21:143–73. Oxford: Blackwell Sci.

Rodriguez-Lanetty M. 2003. Evolving lineages of *Symbiodinium*-like dinoflagellates based on ITS1 rDNA. *Mol. Phylogen. Evol.* In press

Rodriguez-Lanetty M, Cha HR, Song JI. 2002. Genetic diversity of symbiotic dinoflagellates associated with anthozoans from Korean waters. *Proc. Int. Coral Reef Symp., 9th, Bali*, 1:163–66

Rodriguez-Lanetty M, Hoegh-Guldberg O. 2003. Symbiont diversity within the widespread scleractinian coral *Plesiastrea versipora*, across the northwestern Pacific. *Mar. Biol.* In press

Rodriguez-Lanetty M, Loh W, Carter D, Hoegh-Guldberg O. 2001. Latitudinal variability in symbiont specificity within the widespread scleractinian coral Plesiastrea versipora. *Mar. Biol.* 138:1175–81

Rosenberg E, Loya Y, eds. 2003. *Coral Health and Disease*. Berlin: Springer-Verlag. In press

Rowan R. 1991. Molecular systematics of symbiotic algae. *J. Phycol.* 27:661–66

Rowan R. 1996. The distribution of zooxanthellae from different places. *Proc. Proc. Int. Coral Reef Symp., 7th, Guam*, p. 658. (Abstr.)

Rowan R. 1998. Diversity and ecology of zooxanthellae on coral reefs. *J. Phycol.* 34:407–17

Rowan R, Knowlton N. 1995. Intraspecific diversity and ecological zonation in coral-algal symbiosis. *Proc. Natl. Acad. Sci. USA* 92:2850–53

Rowan R, Knowlton N, Baker AC, Jara J. 1997. Landscape ecology of algal symbiont communities explains variation in episodes of coral bleaching. *Nature* 388:265–69

Rowan R, Powers DA. 1991a. Molecular genetic identification of symbiotic dinoflagellates (zooxanthellae). *Mar. Ecol. Prog. Ser.* 71:65–73

Rowan R, Powers DA. 1991b. A molecular genetic classification of zooxanthellae and the evolution of animal-algal symbiosis. *Science* 251:1348–51

Rowan R, Powers DA. 1992. Ribosomal RNA sequences and the diversity of symbiotic dinoflagellates (zooxanthellae). *Proc. Natl. Acad. Sci. USA* 89:3639–43

Sadler LA, McNally KL, Govind NS, Brunk CF, Trench RK. 1992. The Nucleotide sequence of the small subunit ribosomal RNA gene from *Symbiodinium pilosum*, a symbiotic dinoflagellate. *Curr. Genet.* 21:409–16

Saldarriaga JF, Taylor FJR, Keeling PJ, Cavalier-Smith T. 2001. Dinoflagellate nuclear SSU rRNA phylogeny suggests multiple plastid losses and replacements. *J. Mol. Evol.* 53:204–13

Sandeman IM. 1988. Coral bleaching at Discovery Bay, Jamaica: a possible mechanism for temperature-related bleaching. See Ogden & Wicklund 1988, pp. 46–48

Santos SR, Coffroth MA. 2003. Molecular genetic evidence that dinoflagellates belonging to the genus *Symbiodinium* Freudenthal are haploid. *Biol. Bull.* 204:10–20

Santos SR, Gutiérrez-Rodríguez C, Coffroth MA. 2003a. Phylogenetic identification of symbiotic dinoflagellates via length heteroplasmy in domain V of chloroplast large subunit (cp23S)-rDNA sequences. *Mar. Biotechnol.* 5:134–40

Santos SR, Gutiérrez-Rodríguez C, Lasker HR, Coffroth MA. 2003b. *Symbiodinium* sp. associations in the gorgonian *Pseudopterogorgia elisabethae* in the Bahamas: high levels of genetic variability and population structure in symbiotic dinoflagellates. *Mar. Biol.* 143:111–20

Santos SR, Kinzie RA, Sakai K, Coffroth MA. 2003c. Molecular characterization of nuclear small subunit (18S)-rDNA pseudogenes in a symbiotic dinoflagellate (*Symbiodinium*, Dinophyta). *J. Eukaryot. Microbiol.* In press

Santos SR, Taylor DJ, Coffroth MA. 2001. Genetic comparisons of freshly isolated versus cultured symbiotic dinoflagellates: Implications for extrapolating to the intact symbiosis. *J. Phycol.* 37:900–12

Santos SR, Taylor DJ, Kinzie RA, Hidaka M, Sakai K, Coffroth MA. 2002a. Molecular phylogeny of symbiotic dinoflagellates inferred from partial chloroplast large subunit (23S)-rDNA sequences. *Mol. Phylogenet. Evol.* 23:97–111

Santos SR, Taylor DJ, Kinzie RA, Sakai K, Coffroth MA. 2002b. Evolution of length variation and heteroplasmy in the chloroplast rDNA of symbiotic dinoflagellates (*Symbiodinium*, Dinophyta) and a novel insertion in the universal core region of the large subunit rDNA. *Phycologia* 41:311–18

Saunders GW, Hill DRA, Sexton JP, Andersen RA. 1997. Small subunit ribosomal RNA sequences from selected dinoflagellates: Testing classical evolutionary hypotheses with molecular systematic methods. *Plant Syst. Evol.* 11(Suppl.):237–59

Savage AM, Goodson MS, Visram S, Trapido-Rosenthal H, Wiedenmann J, Douglas AE. 2002a. Molecular diversity of symbiotic algae at the latitudinal margins of their distribution: dinoflagellates of the genus *Symbiodinium* in corals and sea anemones. *Mar. Ecol. Prog. Ser.* 244:17–26

Savage AM, Trapido-Rosenthal H, Douglas AE. 2002b. On the functional significance of molecular variation in *Symbiodinium*, the symbiotic algae of Cnidaria: photosynthetic response to irradiance. *Mar. Ecol. Prog. Ser.* 244:27–37

Schoenberg DA, Trench RK. 1980a. Genetic variation in *Symbiodinium* (=*Gymnodinium*) *microadriaticum* Freudenthal, and specificity in its symbiosis with marine invertebrates. I. Isoenzyme and soluble protein patterns of axenic cultures of *Symbiodinium microadriaticum*. *Proc. R. Soc. London Ser. B* 207:405–27

Schoenberg DA, Trench RK. 1980b. Genetic variation in *Symbiodinium* (=*Gymnodinium*) *microadriaticum* Freudenthal, and specificity in its symbiosis with marine invertebrates. II. Morphological variation in *Symbiodinium microadriaticum*. *Proc. R. Soc. London Ser. B* 207:429–44

Schoenberg DA, Trench RK. 1980c. Genetic variation in *Symbiodinium* (=*Gymnodinium*) *microadriaticum* Freudenthal, and specificity in its symbiosis with marine invertebrates. III. Specificity and infectivity of *Symbiodinium microadriaticum*. *Proc. R. Soc. London Ser. B* 207:445–60

Schwarz JA, Krupp DA, Weis VM. 1999. Late larval development and onset of symbiosis in the scleractinian coral *Fungia scutaria*. *Biol. Bull.* 196:70–79

Smith SV, Buddemeier RW. 1992. Global change and coral reef ecosystems. *Annu. Rev. Ecol. Syst.* 23:89–118

Speksnijder A, Kowalchuk GA, De Jong S, Kline E, Stephen JR, Laanbroek HJ. 2001. Microvariation artifacts introduced by PCR and cloning of closely related 16S rRNA gene sequences. *Appl. Environ. Microbiol.* 67:469–72

Stanley SM. 1986. Anatomy of a regional mass extinction: Plio-Pleistocene decimation of the western Atlantic bivalve fauna. *Palaios* 1:17–36

Swofford DL. 2001. PAUP*. Phylogenetic Analysis Using Parsimony (*and Other Methods). Version 4, Sunderland, MA: Sinauer

Taylor DL. 1974. Symbiotic marine algae: Taxonomy and biological fitness. In *Symbiosis and the Sea*, ed. CBW Vernberg, pp. 245–62. Columbia: Univ. S. C. Press

Thompson JD, Gibson TJ, Plewniak F, Jeanmougin F, Higgins DG. 1997. The Clustal X windows interface: flexible strategies for multiple sequence alignment aided by quality tools. *N. Acids Res.* 24:4876–82

Thornhill DJ, LaJeunesse TC, Schmidt GW, Fitt WK. 2003. Change and stability of symbiotic dinoflagellates in reef building corals. Int. Conf. Coelenterate Biol., 7th, Lawrence, 44 (Abstr.)

Toller WW, Rowan R, Knowlton N. 2001a. Repopulation of zooxanthellae in the Caribbean corals *Montastraea annularis* and *M. faveolata* following experimental and disease-associated bleaching. *Biol. Bull.* 201:360–73

Toller WW, Rowan R, Knowlton N. 2001b. Zooxanthellae of the *Montastraea annularis* species complex: Patterns of distribution of four taxa of *Symbiodinium* on different reefs and across depths. *Biol. Bull.* 201:348–59

Trench RK. 1988. Specificity in dinomastigote-marine invertebrate symbiosis: An evaluation of hypotheses of mechanisms involved in producing specificity. *NATO ASI Ser.* H 17:326–46

Trench RK. 1992. Microalgal-invertebrate symbiosis, current trends. In *Encyclopedia of Microbiology*, ed. J Lederberg, M Alexander, BR Bloom, 3:129–42. San Diego: Academic

Trench RK. 1993. Microalgal-invertebrate symbiosis—a review. *Endocytobiosis Cell Res.* 9:135–75

Trench RK. 1996. Specificity and dynamics of algal-invertebrate symbiosis. *Symbiosis* 96:16

Trench RK. 1997. Diversity of symbiotic dinoflagellates and the evolution of microalgal-invertebrate symbiosis. *Proc. Int. Coral Reef Symp., 8th, Panama*, 2:1275–86

Trench RK. 2000. Validation of some currently used invalid names of dinoflagellates. *J. Phycol.* 36:972

Trench RK, Blank RJ. 1987. *Symbiodinium microadriaticum* Freudenthal, *Symbiodinium goreauii* sp. nov., *Symbiodinium kawagutii* sp. nov. and *Symbiodinium pilosum* sp. nov.—gymnodinioid dinoflagellate symbionts of marine invertebrates. *J. Phycol.* 23: 469–81

Trench RK, Thinh LV. 1995. *Gymnodinium linucheae* sp. nov.—the dinoflagellate symbiont of the jellyfish *Linuche unguiculata*. *Eur. J. Phycol.* 30:149–54

Ulstrup KE, Van Oppen MJH. 2003. Geographic and habitat partitioning of genetically distinct zooxanthellae (*Symbiodinium*) in *Acropora* corals on the Great Barrier Reef. *Mol. Ecol.* In press

Van Oppen M. 2001. *In vitro* establishment of symbiosis in *Acropora millepora* planulae. *Coral Reefs* 20:200

Van Oppen MJH. 2003. Mode of zooxanthellae transmission does not affect zooxanthella diversity in acroporid corals. *Mar. Biol.* In press

Warner ME, Fitt WK, Schmidt GW. 1999. Damage to photosystem II in symbiotic dinoflagellates: A determinant of coral bleaching. *Proc. Natl. Acad. Sci. USA* 96: 8007 12

Van Oppen MJH, Palstra FP, Piquet AMT, Miller DJ. 2001. Patterns of coral-dinoflagellate associations in *Acropora*: significance of local availability and physiology of *Symbiodinium* strains and host-symbiont selectivity. *Proc. R. Soc. London Ser. B* 268:1759–67

Warner ME, Fitt WK, Schmidt GW. 1996. The effects of elevated temperature on the photosynthetic efficiency of zooxanthellae in hospite from four different species of reef coral: A novel approach. *Plant Cell Environ.* 19:291–99

Weis VM, Reynolds WS, deBoer MD, Krupp DA. 2001. Host-symbiont specificity during onset of symbiosis between the dinoflagellates *Symbiodinium* spp. and planula larvae of the scleractinian coral *Fungia scutaria. Coral Reefs* 20:301–8

West JM, Salm RV. 2003. Resistance and resilience to coral bleaching: implications for coral reef conservation and management. *Con. Biol.* 17:956–67

Wilcox TP. 1998. Large subunit ribosomal RNA systematics of symbiotic dinoflagellates: Morphology does not recapitulate phylogeny. *Mol. Phylogenet. Evol.* 10:436–48

Yang YA, Soong K, Chen CA. 2000. Seasonal variation in symbiont community composition within single colonies of *Acropora palifera. Proc. Int. Coral Reef Symp., 9th, Bali,* p. 36. (Abstr.)

Zahl PA, McLaughlin JJA. 1959. Studies in marine biology. IV. On the role of algal cells in the tissues of marine invertebrates. *J. Protozool.* 6:344–52

Subject Index

E

F

Q

R

CUMULATIVE INDEXES

CONTRIBUTING AUTHORS, VOLUMES 30–34

CHAPTER TITLES, VOLUMES 30–34

Volume 30 (1999)

Volume 31 (2000)

Volume 32 (2001)

Volume 33 (2002)

Volume 34 (2003)